Introduction to Satellite Remote Sensing

Introduction to Satellite Remote Sensing
Atmosphere, Ocean, Land and Cryosphere Applications

William Emery
*University of Colorado at Boulder,
Colorado, United States*

Adriano Camps
*Universitat Politècnica de Catalunya,
Barcelona, Spain*

ELSEVIER

Elsevier
Radarweg 29, PO Box 211, 1000 AE Amsterdam, Netherlands
The Boulevard, Langford Lane, Kidlington, Oxford OX5 1GB, United Kingdom
50 Hampshire Street, 5th Floor, Cambridge, MA 02139, United States

Notices
Knowledge and best practice in this field are constantly changing. As new research and experience broaden our
understanding, changes in research methods, professional practices, or medical treatment may become
necessary.

Practitioners and researchers must always rely on their own experience and knowledge in evaluating and using
any information, methods, compounds, or experiments described herein. In using such information or methods
they should be mindful of their own safety and the safety of others, including parties for whom they have a
professional responsibility.

To the fullest extent of the law, neither the Publisher nor the authors, contributors, or editors, assume any
liability for any injury and/or damage to persons or property as a matter of products liability, negligence or
otherwise, or from any use or operation of any methods, products, instructions, or ideas contained in the
material herein.

Library of Congress Cataloging-in-Publication Data
A catalog record for this book is available from the Library of Congress

British Library Cataloguing-in-Publication Data
A catalogue record for this book is available from the British Library

ISBN: 978-0-12-809254-5

For information on all Academic Press publications visit our
website at https://www.elsevier.com/books-and-journals

Working together
to grow libraries in
developing countries

www.elsevier.com • www.bookaid.org

Publisher: Candice G. Janco
Acquisition Editor: Louisa Hutchins
Editorial Project Manager: Emily Thomson
Production Project Manager: Mohana Priyan Rajendran
Designer: Christian J. Bilbow

Typeset by TNQ Books and Journals

Adriano Camps
Dedicated to my parents, siblings, wife, and daughters
William Emery
Dedicated to my wife and children

Contents

THE HISTORY OF SATELLITE REMOTE SENSING

1.1 THE DEFINITION OF REMOTE SENSING

Evelyn Pruitt coined the term "remote sensing" in the 1960s when she was working at the US Office of Naval Research. The term was intended to imply a measurement made by some indirect or "remote" means rather than by a contact sensor. In its application to satellite and aircraft instrumentation, remote sensing relied primarily upon either reflected or emitted electromagnetic radiation (optical and microwave) from the Earth to infer changes on the Earth's surface or in the overlying atmosphere. The fact that these inferences must be made from a by-product (either the reflected or the emitted radiation) of the surface or atmospheric process qualifies satellite data collection as "remote sensing."

Other applications such as the use of acoustic signals to map the internal character of the ocean and the solid Earth are often also considered as a remote sensing. In the past few decades, however, satellite and aircraft data analyses have become even more closely associated with the term remote sensing.

1.2 THE HISTORY OF SATELLITE REMOTE SENSING

1.2.1 THE NATURE OF LIGHT AND THE DEVELOPMENT OF AERIAL PHOTOGRAPHY

Aerial photography depends mainly on the use of reflected solar radiation to image the Earth's surface. Many developments in optics needed to take place before optical systems could be developed. Sir Isaac Newton conducted some of the earliest work on the nature of light during his time as the Lucasian Professor at the University of Cambridge.

He had reached the conclusion that white light is not a single entity. Every scientist since Aristotle had believed that white light was a basic single element, but the chromatic aberration in a telescope lens convinced Newton otherwise. When he passed a thin beam of sunlight through a glass prism, Newton noted the spectrum of colors that was formed. Newton argued that white light is really a mixture of many different types of rays, which are refracted at slightly different angles, and that each different type of ray produces a different spectral color. Newton was led by this reasoning to the erroneous conclusion that telescopes using refracting lenses would always suffer chromatic aberration. He therefore proposed and constructed a reflecting telescope (i.e., using mirrors).

In London, around 1862, Maxwell (Fig. 1.1) calculated that the speed of propagation of electromagnetic fields is that of the speed of light. He proposed that the phenomenon of light is therefore an electromagnetic phenomenon. Maxwell wrote the truly remarkable words:

Introduction to Satellite Remote Sensing. http://dx.doi.org/10.1016/B978-0-12-809254-5.00001-4

FIGURE 1.1

James Clerk Maxwell. Born: June 13, 1831 in Edinburgh, Scotland. Died: November 5, 1879 in Cambridge, Cambridgeshire, England.

We can scarcely avoid the conclusion that light consists in the transverse undulations of the same medium, which is the cause of electric and magnetic phenomena.

In the early part of the 19th century, Daguerre created the first photographic plate, which consisted of a thin film of polished silver on a copper base. Putting it into a container with iodine in it sensitized the surface of the silver; the iodine vapors reacted with the polished silver surface and formed a thin yellow layer of silver iodide. After a photograph was taken on the plate, it was developed by exposing the plate to magnesium vapor at 339 K. The vapor would only stick to the parts of the plate, which had been exposed to the light. The plate was then dipped in sodium thiosulfate to dissolve the unused silver iodide, and then rinsed in hot water to get rid of any remaining chemicals. Daguerreotypes, as these images came to be known, had the ability to capture fine detail, but due to their long exposure time they were constrained to motionless subjects.

On January 4, 1829, Niépce agreed to go into partnership with Louis Daguerre. Niépce died only 4 years later, but Daguerre continued to experiment. Soon he had discovered a way of developing photographic plates, a process, which greatly reduced the exposure time from 8 h down to half an hour. He also discovered that an image could be made permanent by immersing it in salt. Following a report on this invention by Paul Delaroche, a leading scholar of the day, the French government bought the rights to it in July 1839. Details of the process were made public on August 19, 1839, and Daguerre named it the daguerreotype (Fig. 1.2).

FIGURE 1.2

An early daguerreotype by John Plumbe of the east front elevation of the United States Capitol.

The application to topographic mapping was first suggested in 1849, and balloonist F. Tournachon undertook initial attempts in 1858 from a captive balloon a few hundred meters over Petit Bicetre in France using large silver plates as the camera (Fig. 1.3). Balloon photographs of Confederate positions during the American Civil War represent the first practical use of aerial photography. This is a good example of how war strongly motivates the rapid development of a new technology that could be used to gain an advantage over the enemy.

The invention of gelatin dry plates (film) by Maddox, in 1871, eliminated the need for transporting an entire darkroom on the balloon platform. Triboulet used dry plates in 1879 to photograph Paris from a free balloon. The size of the camera was also reduced which opened more opportunities for photography.

The English meteorologist E. Archibald took the first kite photographs in 1882. In 1889, R. Thiele, from Russia, mounted cameras on seven unmanned kites to produce a "panaramograph." In 1885, W. A. Eddy, an American meteorologist in New Jersey, reported the first kite photograph taken in the western hemisphere. He also developed a kite-camera system, which proved a useful supplement to balloon photography during the Spanish-American war. G. R. Lawrence, referred to as the "King of Kite Photography," used kite systems with cameras weighing up to 454 kg and negatives as large as 1.35 m × 2.4 m. He is particularly noted for his photograph of San Francisco just after the earthquake of 1906 (Fig. 1.4).

An innovative application of aerial photography was the attachment of cameras to carrier pigeons at the 1909 world's fair in Dresden, Germany (Fig. 1.5). These pigeons would fly over the fair and take

FIGURE 1.3

Felix Tournachon takes a picture of Petit Bicetre, France, in 1858.

FIGURE 1.4

Kite photograph of San Francisco after 1906 earthquake; the camera weighed 49 lbs and was held up by seven kites.

an exposure, which would then be developed and printed for sale to the attendees at the fair that can see themselves and the overall fairgrounds.

In 1908 a passenger appropriately collected the first aircraft still photographs with Wilbur Wright flying on a test flight in France (Fig. 1.6), while another passenger took the first aerial movies with Wilbur in the following year.

Similarly, Samuel Goddard collected the first rocket photos in 1926 during his experiments with rocketry.

FIGURE 1.5

Cameras on carrier pigeons took pictures at the 1909 Dresden World's Fair.

FIGURE 1.6

First aircraft photo by Wilbur Wright's passenger in France in 1908. In 1909, he made the first aerial movies.

1.2.2 THE BIRTH OF EARTH-ORBITING SATELLITES

In 1903, Konstantin Tsiolkovsky (1857—1935) published *Exploring Space Using Jet Propulsion Devices* (in Russian: Исследование мировыч Цространств реактивными Цриборами), which is the first academic treatise on the use of rocketry to launch spacecraft. He calculated the orbital speed required for a minimal orbit around the Earth at 8 km/s, and that a multistage rocket fueled by liquid propellants could be used to achieve this. He proposed the use of liquid hydrogen and liquid oxygen, though other combinations can be used.

In 1928, Slovenian Herman Potocnik (1892—1929) published his sole book, *The Problem of Space Travel—The Rocket Motor* (German: *Das Problem der Befahrung des Weltraums—der Raketen-Motor*), a plan for a breakthrough into space and a permanent human presence there. He conceived a space station in detail and calculated its geostationary orbit. He described the use of orbiting spacecraft for detailed peaceful and military observation of the ground and described how the special conditions of space could be useful for scientific experiments. The book described geostationary satellites (first put forward by Tsiolkovsky) and discussed communication between them and the ground using radio, but fell short of the idea of using satellites for mass broadcasting and as tele-communications relays.

In a 1945 *Wireless World* article, the English science fiction writer Arthur C. Clarke (1917—2008) described in detail the possible use of communications satellites for mass communications. Clarke examined the logistics of satellite launch, possible orbits, and other aspects of the creation of a network of world-circling satellites, pointing to the benefits of high-speed global communications. He also suggested that three geostationary satellites would provide coverage over the entire planet.

The world changed dramatically on October 4, 1957 with the successful launch and operation of the Russian Sputnik satellite, which was the first human created instrument to orbit the Earth. About the size of a basketball, it weighed only 183 pounds, and took about 98 min to orbit the Earth on its elliptical path. That launch ushered in new political, military, technological, and scientific developments. While the Sputnik launch was a single event, it marked the start of the space age and the US—USSR space race. This satellite carried no Earth-oriented sensors and only really sent out radio signals that were used to communicate with the satellite. It did demonstrate, however, that satellites could be launched from the Earth and operated on a continuous basis. The Sputnik launch changed everything. As a technical achievement, Sputnik caught the world's attention and the American public off-guard. Its size was more impressive than Vanguard's intended 3.5-pound payload. In addition, the public feared that the Soviets' ability to launch satellites also translated into the capability to launch ballistic missiles that could carry nuclear weapons from Europe to the United States. Then the Soviets struck again; on November 3, Sputnik II was launched, carrying a much heavier payload, including a dog named Laika. Table 1.1 lists all of the first satellites launched by 12 different countries starting with the Soviet Union launch of Sputnik-1 in 1957.

There were also a number of attempted first launches by many of these same countries before they were successful at launching a satellite and inserting it in to Earth orbit. Several other countries, including Brazil, Argentina, Pakistan, Romania, Taiwan, Indonesia, Australia, New Zealand, Malaysia, Turkey, Spain, Japan, India, Israel, France, Germany, and Switzerland (and others) are at various stages of development of their own small-scale launcher capabilities. This list grows a lot longer when you include nations and satellites that were launched by the capabilities of other nations.

Table 1.1 First Successful Satellite Launches by Country

Order	Country	Date	Rocket	Satellite
1	Soviet Union	October 4, 1957	Sputnik-PS	Sputnik-1
2	United States	February 1, 1958	Juno I	Explorer 1
3	France	November 26, 1965	Diamant-A	Astérix
4	Japan	February 11, 1970	Lambda-4S	Osumi
5	China	April 24, 1970	Long March 1	Dong Fang Hong I
6	United Kingdom	October 28, 1971	Black Arrow	Prospero
7	India	July 18, 1980	SLV	Rohini D1
8	Israel	September 19, 1988	Shavit	Ofeg 1
9	Russia	January 21, 1992	Soyuz-U	Kosmos 2175
10	Ukraine	July 13, 1992	Tsyklon-3	Strela
11	Iran	February 2, 2009	Safir-1	Omid
12	North Korea	December 12, 2012	Unha-3	Kwangmyongsong-3, unit 2

Today with the advent of small satellites such as "CubeSats" almost anyone can get a satellite payload into space. It is something the commercial remote sensing companies are taking a very close look at.

In the United States there had been studies going on as to how Earth-orbiting satellites could benefit the meteorological forecasting community in monitoring conditions on the Earth. These studies led to projects to create new satellites that would monitor the atmosphere. Some early launch failures in the United States delayed the launch of these new satellites, but eventually the first TIROS (Television and Infrared Observation Satellite) was launched and made operational in April of 1960. This satellite was spin stabilized which led to the fact that the Earth-oriented sensor (aligned with the spin access) could view only a limited portion of the Earth's latitude (Fig. 1.7).

This picture shows the location of the solar power panels on the outside of the satellite, which is typical of a spinning satellite.

Fig. 1.8 shows all of the different equipment on the first TIROS satellite and their roles in the operation of the satellite. The primary sensor was the wide-angle TV camera, which collected images of the Earth at approximately 750 km orbital altitude. A small infrared (IR) system was also used to collect some limited measurements through the narrow angle TV camera that also collected radiation in visible wavelengths. The receiving and transmitting antennas are shown, and all data collected were transmitted as analog signals down to the ground. A tape recorder on board was used to store these analog data so that they could be downlinked to the ground when the satellite was in view of a tracking ground station.

The solar cells and magnetic orientation coils were used to power and control the orientation of the spacecraft. Since it was spinning the spin axis would remain fixed in orientation, but the satellite was deployed in a moderate inclination orbit and it passed in and out of the Earth's magnetic poles and the magnetic coils were needed to unloaded torquing stresses that built up over time. The pointing of the main cameras was controlled by the spin stabilization and the camera pointed only to the latitudes from

FIGURE 1.7

TIROS I prototype on display at the Smithsonian National Air and Space Museum.

northern North America to South America. The rest of the time the camera pointed out into space providing no information on the rest of the Earth's surface. Thus, the TIROS satellites were incapable of observing the entire globe. This was a limitation of the spin stabilization at least as it was deployed in this fashion.

The TIROS camera was designed to depict the Earth's cloud cover as an indication of the weather systems over the Earth. Thus, these early satellite designs were driven primarily by meteorological considerations and the need for improved forecasting. An example of a TIROS image of an Atlantic storm is presented here in Fig. 1.9. Note the distortion of the latitude, longitude lines, which is caused by the curvature and rotation of the Earth as well as the pointing orientation of the satellite camera. Since all of the TIROS imagery were analog the correction of these geometric distortions was not possible using digital methods and mapping was done by overlaying "warped grids" that best matched the orientation of the global features. This type of mapping approach determined the lines on Fig. 1.9.

Viewing these early satellite, TV images became a method of discovery where images were located that depicted various important atmospheric processes. Another example is presented in Fig. 1.10, which shows a cloud streak that represents the jet stream over the land on the left as it passes over the ocean at the east coast of North America just south of Cape Blanco. The image shows the spreading of the jet stream over the central United States and thinning as it moves east. The cloud stream thickens once again as it moves eastward from the coast.

One of the big benefits of satellite imagery was the ability to view hurricanes from space. The thick cloud cover associated with a hurricane along with the characteristic eye of the storm made these

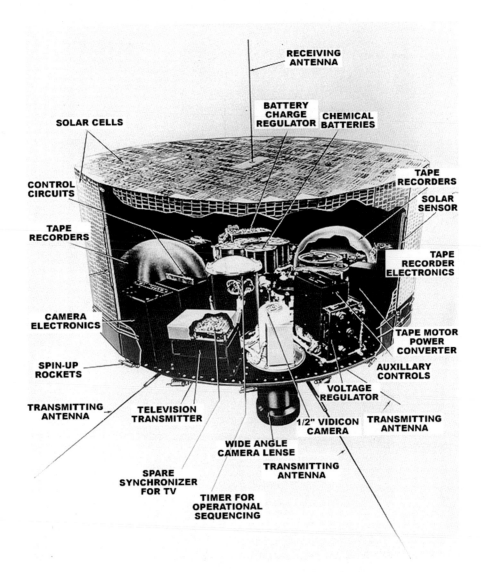

FIGURE 1.8

TIROS satellite equipment and instruments.

features easy to see as shown here in the distorted image of Fig. 1.11. Here the geometric distortion due to Earth's curvature and rotation are seen in the shape of the Earth's surface and the features. No latitude, longitude lines have been added to this image to further depict these distortions. As with Fig. 1.10, there are some dark lines showing boundaries of the central part of the image and a plus sign to show the center of the image. In this case, hurricane Betsy is in the southwestern corner of the

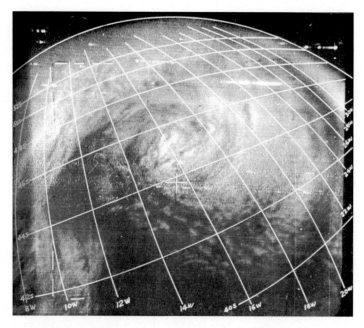

FIGURE 1.9

TIROS image of an Atlantic storm.

FIGURE 1.10

TIROS image of the jet stream south of Cape Blanco; the dark lines and cross in the middle of the image have been added to identify the jet stream. The data were collected on November 11, 1964, at 13:00 GMT.

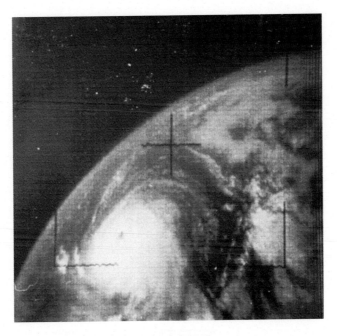

FIGURE 1.11

Hurricane Betsy north of the Bahamas from TIROS VII on September 4, 1965.

image. The slant view of the storm leads to an "apparent" closure of the eye of the storm due to the orientation.

Land surface features were also apparent in the early TIROS imagery when cloud cover was sufficiently low to make it possible to view the surface. As an example Fig. 1.12 presents an image of Lake Erie, which clearly shows some of the limitations to this type of imagery. Here the lake covers a number of satellite passes each of which has a slightly different exposures. This produces artificial striping in the image. Earth surface distortion continues to be a problem as shown by the elongated part of the lake in the southwest portion of the image. The presence of clouds in this same portion of the image also obscures the surface of the lake. Discontinuities in the cloud cover are introduced by the fact that the image is made up of sequential passes, which are not truly synoptic in coverage.

It is important to recognize that in this era of analog data relay all image processing was done using a collage of pictures printed with conventional means as shown in Fig. 1.13.

The initial TIROS satellites were relatively short-lived with satellites lasting only a few months each. By the end of the series, however, the satellites were lasting approximately a year and continuing to report data over this entire period.

To overcome the viewing limitations of the original TIROS series of satellites the next generation of spinning satellites was changed to have the camera pointing radially outward and the spin axis of the satellite turned 90 degrees relative to the original TIROS satellites. This new configuration was called the "wheel" satellite and a consequence of this change was the ability to collect a series of circular images that over the period of a day covered the entire surface of the Earth (Fig. 1.14). Operated by the

FIGURE 1.12

TIROS first satellite image of sea ice over the Gulf of St. Lawrence and St. Lawrence River, May 1960.

FIGURE 1.13

TIROS image data handling by photographic methods.

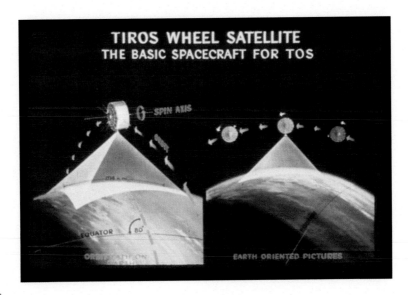

FIGURE 1.14

The scan operation of the Earth Satellite Science Administration wheel satellites (NOAA Photo Library).

newly formed Earth Satellite Science Administration (ESSA), these satellites were identified as ESSA satellites a practice that was followed when the agency evolved into the National Oceanic and Atmospheric Administration (NOAA).

The daily circular images of the Earth were pasted together to comprise a global image of the Earth and its atmosphere. While the resultant global image did not represent a synoptic picture of the Earth, it did represent a first look at daily images of the entire Earth (Fig. 1.15). Note how the edges of the

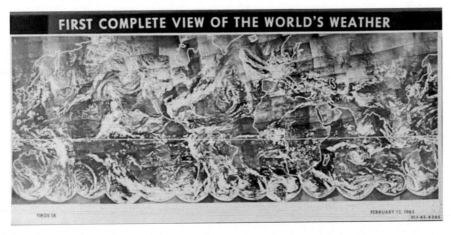

FIGURE 1.15

First global image from the Earth Satellite Science Administration wheel satellite (NOAA Photo Library), February 13, 1965.

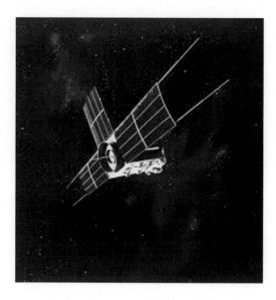

FIGURE 1.16

Three-axes stabilized ITOS satellite.

various component images are quite marked both by changes in brightness and in the expression of the individual features. Clouds in particular are not continuous, and major cloud features appear broken up due to the temporal discontinuities in the collection of the individual images.

In spite of these limitations, these global images became the standard to study atmospheric cloud conditions and to infer their connections to global weather patterns. These patterns could then be used to improve weather forecasting by being able to "see" the weather before it crossed the US coastline. Since the images were global similar forecasting improvements were possible for other parts of the world as well as the United States. This was a big improvement over the earlier TIROS program and there were a number of ESSA wheel satellites that operated between 1966 and 1971.

The next development in the evolution of operational weather satellites was the incorporation of spacecraft stability control. Developed as part of the ballistic missile program during the "cold war," three-axes stability systems were now available to control the pointing of the spacecraft without the need to spin the spacecraft. With this three-axes stabilization, it was possible to keep the Earth sensors always pointing at the Earth regardless of its position in the orbit. This made it possible to collect imagery over the entire Earth's surface from the same sensors at the same resolution Fig. 1.16.

Called the Improved TIROS Observing Satellite, or ITOS, this family of satellites brought in a new era of remote sensing. In addition to the three-axes stabilization, these satellites carried a new suite of optical radiometers which were scanning systems that collected reflected and emitted radiation from the Earth's surface line by line as the satellite moved along in its polar orbit. These first radiometers (Fig. 1.17) ushered in the new era of improved capabilities that became a standard approach to viewing the Earth.

To ensure data continuity, this satellite also carried cameras that could provide the same kind of data that had been previously available through its predecessor satellites. There were cameras

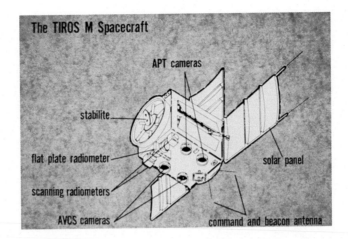

FIGURE 1.17

The ITOS satellite with its sensors and other equipment; note the introduction of solar panels. *APT*, applied picture transmission; *AVCS*, advanced vidicon camera system.

collecting data for two types of data relay: (1) the highest possible spatial resolution of the camera and (2) a lower resolution applied picture transmission (APT) that supplied lower resolution data to field stations using an omnidirectional antenna. The highest resolution required the use of a "tracking" antenna system that followed the spacecraft as it crossed over the receiving site. Such systems were at the time too complex and expensive for many field offices and ships at sea. The APT broadcast was set up to make it possible to broadcast a lower resolution version of the data using a VHF omnidirectional beacon that could easily be picked up by low-cost antenna systems. This tradition has been continued up to today and there are still APT transmissions providing lower resolution real-time imagery to these remote sites. The ITOS scanning radiometers were those that became the primary instruments for future satellites. Unlike cameras, these radiometers collected radiation one line at a time thus building up an image as the satellite continued forward in its orbit. These radiometers could also collect radiation in various spectral bands.

The first ITOS satellite demonstrated the utility of these new technologies and began a longer time series of polar-orbiting spacecraft, which were now called NOAA satellites after the name of the agency that operated them. A series of eight satellites with approximately the same suite of equipment filled in the years between 1970 and 1976. The practice was to designate the satellites as NOAA a, b, c, etc. when they were built, and then transition them to NOAA 1, 2, 3…etc. once they were operating on orbit. The fact that not all of the NOAA satellites achieved orbit or failed early on orbit led to the fact that alpha and numeric designations do not map one to one.

The details of the ITOS orbit are shown here in Fig. 1.18, which shows the satellite polar orbit relative to day/night and lists all of the sensor capabilities and gives the details of the orbit. Now the orbital altitude is about 1271 km and the orbit is Sun-synchronous with an 80-degree inclination in a retrograde orbit with a period of about 111 min.

A big change over this evolution of satellite capabilities was the size and weight of the spacecraft. The original TIROS satellites weighed about 150 kg, which increased to 250 kg with the change to the

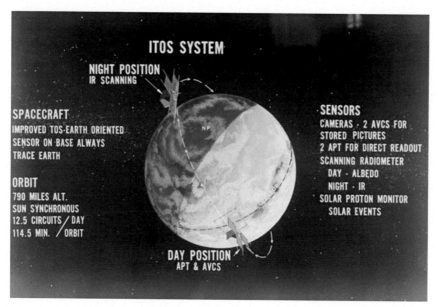

FIGURE 1.18

ITOS satellite operation. *APT*, applied picture transmission; *AVCS*, advanced vidicon camera system.

ESSA wheel satellites. The shift to three-axes stabilization increased the ITOS satellite up to 400 kg, which then increased by over a factor of three to the modern NOAA and Defense Meteorological Satellite Program (DMSP) satellites that weigh about 1500 kg.

The analog radiometer data from the NOAA satellites were digitized on the ground so that the images could be digitally processed and enhanced to geometrically correct the image geolocation and bring out various features in the atmosphere and on the ground. The geometric corrections for Earth curvature and rotation compensated for the distortions of satellite viewing. Additional corrections were also needed for satellite attitude and time, which influences the viewing angle.

Radiance enhancement was needed to bring out the weaker gradients in some of the radiometer channels such as the thermal IR patterns in the ocean. An example is shown here in Fig. 1.19, which is an image of the Gulf of Mexico and the east coast of Florida, which shows the warm water (dark gray shades) associated with the loop current in the Gulf of Mexico and the subsequent Gulf Stream off the east of Florida. The colder water closer to the shore off Florida represents the colder "shelf water" that flows southward inshore of the Gulf Stream. Colder waters also bound the dark pattern of the loop current in the center of the Gulf of Mexico.

This image has been remapped to correct for geometric distortion, which can be seen in the appearance of Florida at the edge of the image, which would be highly distorted if seen in satellite perspective. It is very difficult to quantitatively study features in satellite imagery without being able to "navigate" the imagery, which includes the geometric corrections for Earth curvature and rotation as well as corrections for spacecraft attitude and timing errors.

The ITOS and NOAA satellites carried two different radiometers. The primary instrument was the scanning radiometer (SR), which had an 8 km resolution and was limited to only three channels: (1) a

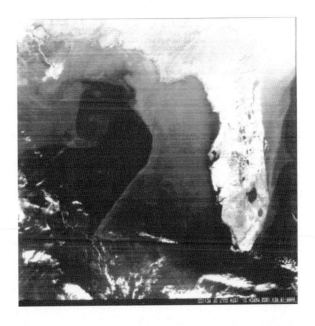

FIGURE 1.19

Very high-resolution radiometer thermal infrared image of the Gulf of Mexico, March 31, 1974 (NOAA Photo Library).

wide band visible, (2) a near-IR channel (0.7—1.1 μm), and (3) a thermal IR (11 μm) channel. The instrument was used to map clouds and later applied to the mapping of sea surface temperature (SST) using the 11 μm channel. A sophisticated processing system was developed that used a histogram method to filter out pixels dominated by clouds to produce SST over large 50 km boxes. This system was found to introduce a lot of errors by letting some cloudy pixels slip through and used an objective analysis (Cressman, 1959) routine that "filled" in erroneous data.

Another instrument flown on the NOAA satellites was the very high resolution radiometer (VHRR), which was the first instrument to demonstrate a real capability for being able to map SST. It had channels in the visible, the near-IR wavelengths, and the midrange IR and the thermal IR wavelengths (again 11 μm). Using the visible and near-IR channels for cloud clearing the VHRR data were then used to produce a 1 km resolution SST, which was the native resolution of the instrument. The image in Fig. 1.19 is from the VHRR sensor.

The biggest change in satellites and sensors came in the fall of 1978 with the advent of TIROS-N ("N" for new). An advanced version of this series of NOAA polar-orbiting satellites the last of which is still operating as this text is being written. These are the 1500 kg spacecraft referred to earlier where the added weight reflects greatly increased capabilities with these new spacecraft. They were fully digital systems that downlinked their data digitally. A new imager called the advanced very high resolution radiometer (AVHRR) became the workhorse radiometer on this spacecraft. With its basic 1 km footprint in four channels the AVHRR data have been used for a wide range of studies of ocean, land, and atmospheric processes. Over the subsequent three decades, this instrument has evolved from

Table 1.2 AVHRR-3 Channel Characteristics

	Ch. 1	Ch. 2	Ch. 3A	Ch. 3B	Ch. 4	Ch. 5
Spc Rg (μm)	0.58–0.68	0.7–1.0	1.58–1.64	3.55–3.93	10.3–11.3	11.5–12.5
Detector	Silicon	Silicon	InGaAs	InSb	HgCdTe	HgCdTe
Resolution (km)	1.09	1.09	1.09	1.09	1.09	1.09
S/N at 5%	9:1	9:1	20:1	–	–	–
NedT at 300 K	–	–	–	0.12 K	0.12 K	0.12 K
Temperature Range (K)	–	–	–	180–335	180–335	180–335

Ch., *channel; NedT*, noise equivalent delta temperature; *Spc Rg*, spectral range; *S/N*, signal-to-noise ratio. The S/N ratio is given at 5% reflectivity.

having only four channels to one that now has six different channels, is called AVHRR-3, and has the characteristics as described in Table 1.2.

The original four channels covered the visible (channel 1), the near-IR (channel 2), the mid-range IR (channel 3 only at 3.7 μm), and the 11 μm (channel 4) thermal IR. The first improvement in this sensor led to the AVHRR-2, which added the fifth channel at 12 μm. This channel was added to provide a "split-window" in the thermal IR to make it possible to correct for atmospheric water vapor attenuation of the thermal IR signal in computing SST. The nominal sensor spatial resolution of 1.09 km meant that all of the channels delivered images with essentially the same resolution.

Channel 3 is now broken into two parts. The approximately 3.7 μm channel (now called channel 3B) is continued at night, but during the day this channel shifts over to 1.6 μm to better resolve atmospheric aerosols and clouds. The visible and near-IR channels are widely used for mapping vegetation, snow cover, and atmospheric aerosols. These channels are also used for mapping snow and ice cover. The thermal IR channels are also used to compute land surface temperature in addition to SST.

The TIROS-N satellites also carried a variety of other instruments. The high-resolution IR sounder is the primary sensor in the TIROS operational sounder system that also includes data from the British stratospheric sounding unit , and the microwave sounding unit (MSU). Together these three instruments are used to retrieve atmospheric temperature and water vapor profiles for use in numerical model assimilation. Actually it was learned that it was better to directly assimilate satellite instrument radiances from this system into the numerical weather forecast models than it would be to retrieve temperature and water vapor profiles to be assimilated into the models.

Other instruments on TIROS-N (Fig. 1.20) are the search and rescue (SAR in Fig. 1.20B) and Argos data collection system (UHF data collection system antenna in Fig. 1.20A). Both of these systems collect data transmitted from the Earth's surface and use the Doppler shift of these signals to accurately locate these platforms. The Argos system also has the capability of collecting a limited amount (approximately 256 data words) of geophysical data collected on the platform.

This diagram shows how this spacecraft has considerable extra capacity and other sensors of opportunity have been flown on this satellite such as the Earth Radiation Budget Experiment instruments that flew only on NOAA-9. These TIROS-N satellites continued to carry the name of NOAA

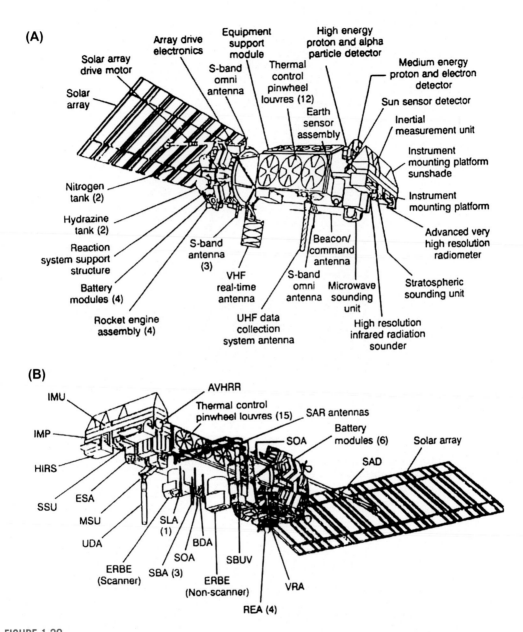

FIGURE 1.20

TIROS-N and ATIROS-N satellites schematic diagrams.

FIGURE 1.21

A TIROS-N satellite being worked on.

satellites. A picture of a TIROS-N satellite being worked on in storage is shown in Fig. 1.21, to give the reader a better appreciation of the size of these satellites.

A summary of the evolution of polar-orbiting environmental satellite (POES) weather satellites is given here in Fig. 1.22, which contains pictures of the important satellites and the relevant characteristics.

1.2.3 THE FUTURE OF POLAR-ORBITING SATELLITES

Prior to 2011 there were three separate US polar-orbiting satellite systems. The one that has been discussed most up to now is the one operated by NOAA. Originally these systems were developed by NASA, and then turned over to NOAA for operations and data analysis. More recently NOAA has gotten involved with the design and acquisition of these systems, and only uses NASA in a most formal basis to contract for these systems. There is also a parallel system operated by the US Department of Defense (DoD) known as the DMSP, which uses spacecraft very similar to the NOAA satellites, but flying a different suite of sensors. Finally, there is a set of NASA research polar-orbiting satellites that also provide data from similar Sun-synchronous orbits.

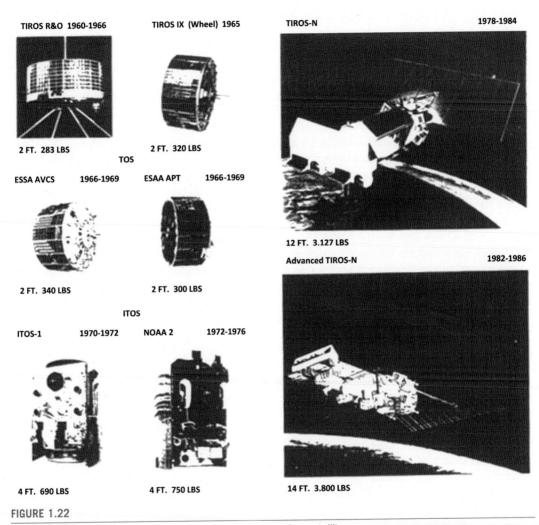

FIGURE 1.22

The evolution of polar-orbiting environmental satellite weather satellites.

In mid-1994 a US presidential directive dictated the convergence of these three systems into a single polar-orbiting satellite system that would fulfill the needs of the DoD, NOAA, and NASA. To be led by NOAA, an Integrated Project Office (IPO) was created that would be staffed by people seconded from each of the three component agencies. Each of these three agencies was given specific roles and responsibilities within this new National Polar-orbiting Operational Environmental Satellite System (NPOESS). These assignments are described below.

The Department of Commerce (NOAA) was designated to be the lead agency that will appoint the system program director who will be an NOAA employee who will act as the head of the executive committee (EXCOM) of the IPO/NPOESS program. NOAA will also have the lead responsibility for interfacing with national and international civil user communities, consistent with national security, and foreign policy requirements.

The DoD will have the lead agency responsibility to support the IPO in major system acquisitions necessary to the NPOESS. DoD will nominate the principal deputy system program director who will be approved by the NOAA system program director.

NASA will have the lead agency responsibility to support the IPO in facilitating the development and insertion of new cost-effective technologies that enhance the ability of the converged system to meet its operational requirements.

Originally all of the contracts were let for all of the NPOESS sensors before the contract was let for the spacecraft and ground systems to Northrop Grumman Corp. It was then decided to place all of the sensor contracts under Northrop Grumman Corp. as well. These sensors were all designed by the commercial contractors in response to a list of Environmental Data Records (EDRs) that were created by the NPOESS IPO. This EDR list contained a list of observables along with requirements such as resolution, precision, and accuracy as defined by the IPO.

The original schedule for the NPOESS program was seriously delayed and budgets overrun by exploding sensor budgets and inabilities to meet planned time schedules. The combined overruns led to a DoD mandated congressional review that led to a revision in scope agreed upon by the principals of NOAA, the DoD, and NASA. A dramatically scaled down and delayed program emerged with plans for an interim NASA sponsored NPOESS preparatory program satellite originally scheduled to be launched in 2006 was finally launched on October 28, 2011. Fortunately, most systems seem to be working properly.

In 2010 the program was completely reorganized by separating it into two parts; one funded via NOAA, but controlled by NASA employees at the Goddard Space Flight Center (GSFC). This segment is referred to as the Joint Polar Satellite System (JPSS) with a primary responsibility to the civilian part of the polar-orbiting satellite requirements. The other half was returned to the DoD and it is called the Defense Weather Satellite System. Each program has taken a very different path for the near future with the JPSS program using essentially a copy of the NPOESS Preparatory Project (NPP) satellite for its first satellite while the DoD has kept the larger satellite bus of the original NPOESS program. In 2011 the DoD decided to cancel its component of this program leaving only the JPSS program to fly a polar-orbiting weather satellite.

More recently the entire NPOESS program was canceled and reorganized as the JPSS. NOAA continued to lead the effort but contractual activists were carried out by NASA at the GSFC in Greenbelt, MD. The only satellite presently carrying some of the earlier NPOESS sensors was previously known as the NASA Preparatory Platform but now has been renamed in honor of Vern Suomi and is the Suomi satellite. JPSS-1 is scheduled to be launched sometime in 2017. It will be a copy of the Suomi satellite and carry the same suite of sensors [Visible/Infrared Radiometer Suite (VIIRS), Advanced Technology Microwave Sounder (ATMS), Cross-track Infrared Sounder (CrIS), Ozone Mapping and Profiler Suite (OMPS), and Cloud and Earth Radiant Energy System (CERES)].

The primary imager for NPOESS will continue on and is known as the VIIRS. It contributes to 23 EDRs and it is the primary instrument for 18 EDRs. VIIRS will combine a simple scan mechanism and radiometric fidelity of the earlier AVHRR with the high spatial resolution (0.65 km) of the Operational Line Scanner (OLS) operating on the DMSP satellites. VIIRS provides imagery of clouds in sunlit conditions in about a dozen visible channels as well as provide imagery in a number of IR channels for night and day cloud imaging products and a low-light sensing capability as was carried on DMSP.

VIIRS has multichannel imaging capabilities to support the acquisition of high-resolution atmospheric imagery and the generation of a variety of applied products including visible and IR imaging of

Table 1.3 Channel Characteristics for the Visible/Infrared Radiometer Suite

Band Name	Wavelength (nm)	Spatial Resolution (km)	Wavelength Type	Radiance Type	Focal Plane Type
M1	412	20	Visible	Reflective	VISNIR
M2	445	18	Visible	Reflective	VISNIR
M3	488	20	Visible	Reflective	VISNIR
M4	555	20	Visible	Reflective	VISNIR
M5	672	20	Visible	Reflective	VISNIR
M6	746	15	Near-infrared (IR)	Reflective	VISNIR
M7	865	39	Near-IR	Reflective	VISNIR
M8	1240	20	Shortwave IR	Reflective	SMWIR
M9	1378	15	Shortwave IR	Reflective	SMWIR
M10	1610	60	Shortwave IR	Reflective	SMWIR
M11	2250	50	Shortwave IR	Reflective	SMWIR
M12	3700	180	Midwave IR	Emissive	SMWIR
M13	4050	155	Midwave IR	Emissive	SMWIR
M14	8550	300	Longwave IR	Emissive	LWIR
M15	10,763	1000	Longwave IR	Emissive	LWIR
M16	12,013	950	Longwave IR	Emissive	LWIR
DNB	700	400	Visible	Reflective imaging	DNB
I1	640	80	Visible	Reflective imaging	VISNIR
I2	865	39	Near IR	Reflective imaging	VISNIR
I3	1610	60	Shortwave IR	Reflective imaging	SMWIR
I4	37,403	80	Midwave IR	Emissive imaging	SMWIR
I5	11,450	1900	Longwave IR	Emissive imaging	LWIR

DNB, *day night band;* LWIR, longwave infrared; SMWIR, *short- and mid-wavelength infrared;* VISNIR, visible/near infrared.

hurricanes, detection of wildfires, smoke, and atmospheric aerosols. VIIRS will have capabilities to produce higher resolution and more accurate measurements of SST then was possible from the heritage AVHRR instrument. VIIRS also has channels to measure ocean color products. The channels for VIIRS are shown here in Table 1.3.

Here we can see that VIIRS supplies a lot more data than the AVHRR and data processing systems must be increased in capacity and capability if they are going to be able to process and reduce all of these data. New algorithms had to be developed to take best advantage of the many new channels available with VIIRS that were not available with AVHRR. Fortunately, NASA's moderate resolution imaging spectrometer had most of these channels making it possible to gain a lot of experience with the use of these channels for computing various geophysical quantities.

The operation of the VIIRS radiometer is shown here as a flow chart in Fig. 1.23, which starts with the scanning telescope of the VIIRS instrument.

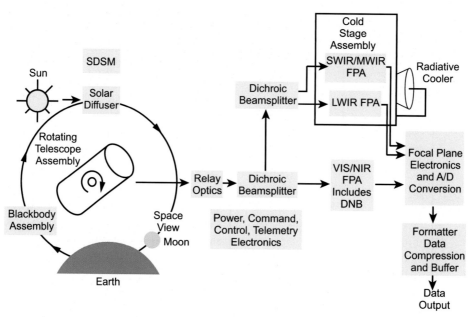

FIGURE 1.23

Flow chart of the Visible/Infrared Radiometer Suite.

The rotating telescope views the Earth's surface, the blackbody assembly is used for thermal calibration, the solar diffuser for visible/near-infrared (VISNIR) calibration, deep space for additional thermal calibration information, and finally the Moon for additional visible calibration. After collection by the telescope the data pass through some relay optics before being split by a dichroic beam splitter into the different wavelengths. The visible and near-IR energy goes to the VISNIR focal plane assembly (FPA), while the short and midrange IR go to one FPA and the longwave to another. Another beam splitter is employed to separate out the longwave infrared (LWIR) and the shortwave infrared (SWIR). Both of these two FPAs must be cooled with a radiative cooler to improve the performance of the FPA. The outputs from the FPAs are recorded electronically and converted to digital representations. These data are then further processed and reduced in volume.

1.2.3.1 The Cross-Track Infrared Sounder

Another instrument on the NPP satellite is the Cross-track Infrared Sounder (CrIS), which is an IR-based radiation sounder based on a Fourier transform interferometer. It has an 18.5 km resolution at nadir for the temperature profile, 15 km for the moisture profile, and 55 km for the atmospheric pressure profile, with a temperature accuracy of 1 K. CrIS is an interferometer that provides over 1000 spectral channels in the thermal IR with an improved spatial resolution, and it will be able to measure vertical temperature profiles with an accuracy approaching 1 K.

Along with the CrIS, the ATMS collects atmospheric information to measure temperature and moisture profiles at high (~daily) temporal resolution. ATMS is the only instrument not being developed by a contractor, but by NASA's GSFC. A conceptual picture of CrIS is shown here in Fig. 1.24.

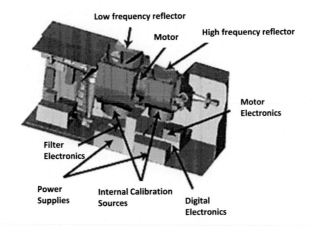

FIGURE 1.24

The Cross-track Infrared Sounder instrument conceptual diagram.

1.2.4 OTHER HISTORICAL SATELLITE PROGRAMS

1.2.4.1 The NIMBUS Program

One of the most important historical satellite programs was NASA's NIMBUS program that was a test platform for testing new instruments in space. Built as an early three-axes stabilized platform, the NIMBUS (raincloud) satellite, the first NIMBUS satellite (Fig. 1.25), was used to test the meteorological sensors that were later carried by the TIROS, ESSA wheel, and ITOS/NOAA satellites. Unlike

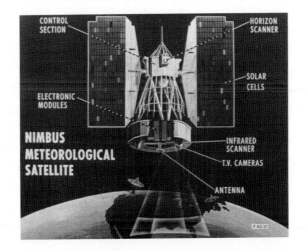

FIGURE 1.25

NIMBUS 1 satellite for testing meteorological sensors.

the ITOS satellites NIMBUS used a gravity gradient to maintain Earth orientation for all of the NIMBUS satellite instruments.

One of the most important NIMBUS purposes was the test of the multispectral scanner (MSS) that became the primary instrument of the Landsat satellites. In fact, the first three Landsat satellites used the NIMBUS bus to deploy these instruments.

The last of the NIMBUS series was NIMBUS 7 launched in late 1978 and was finally decommissioned in 1986. It carried the first ocean color imager the coastal zone color scanner (CZCS), and a quickly refurbished engineering model of the scanning multichannel microwave radiometer that flew earlier on the short-lived Seasat satellite. Unfortunately, NASA decided it would be too expensive to continue the NIMBUS program, which would need a new satellite platform, and this mechanism for routinely testing new satellite instrumentation was lost.

1.2.4.2 The Landsat Program

Landsat represents the world's longest continuously acquired collection of spaced-based, moderate resolution land remote sensing data. Four decades of Landsat imagery provides a unique resource for those who work in agriculture, geology, forestry, regional planning, education, mapping, and global change research. As mentioned above the land surface sensors were first launched and tested as part of the NIMBUS program. The Hughes Santa Barbara Research Center designed and built the first three MSSs. These first three Landsat satellites used the NIMBUS bus to simplify development and get the MSS imager into space as quickly as possible. At this stage this was called the Earth Resources Technology Satellites Program (1966), but this was changed to Landsat in 1975.

In 1979 by Presidential Directive 54, President Jimmy Carter transferred the Landsat operations from NASA to NOAA and recommended of a long-term operational system with four additional satellites beyond Landsat 3. At this time, it was also recommended that the Landsat program be transitioned to the private sector. This occurred in 1985 when the Earth Observation Satellite Company (EOSAT) was formed as a partnership between Hughes Aircraft and RCA. NOAA transferred the Landsat program to EOSAT with an agreement to operate the system for 10 years. EOSAT operated Landsat 4 and 5, had exclusive right to market Landsat data, and was to build Landsat 6 and 7.

A new and unique bus was developed for Landsat 4 and 5 (Fig. 1.26). Also a new instrument called the Thematic Mapper (TM) was developed for these two spacecraft. Thus, Landsat 4 and 5 carried both the MSS and TM imagers. The TM instrument added a thermal IR channel to the existing MSS channels.

They had completed the construction of Landsat 6, which had a new design and an improved sensor called the Extended Thematic Mapper (ETM). Landsat 6 was finally launched on October 5, 1993, but was lost in a launch failure. Processing of Landsat four and five data was resumed by EOSAT in 1994. It should be noted that in an effort to recover their costs EOSAT charges for the Landsat images was very high and basically cut out the majority of science users other than oil companies and the insurance industry.

Landsat 7 was a copy of Landsat 6, but by the time it was ready for launch the program had been transferred back to NASA who launched it on April 15, 1999. The value of the Landsat program was recognized by Congress in October, 1992 when it passed the Land Remote Sensing Policy Act (Public Law 102-555) authorizing the procurement of Landsat 7 and assuring the continued availability of Landsat digital data and images, at the lowest possible cost, to both traditional and new users of these data. This satellite continues to operate, but problems with the scan system led to the implementation of a scan line corrector in May of 2003.

FIGURE 1.26

Landsat 5 satellite.

Called the Landsat Data Continuity Mission (LDCM) Landsat 8 was launched February 11, 2013. Like the previous Landsat satellites, Landsat 8 represents a collaboration between NASA and the US Geological Survey (USGS). Landsat 8 represents a departure from previous Landsat satellites in that it carries two instruments that cover very different parts of the electromagnetic spectrum. Built by Ball Aerospace the Operational Land Imager (OLI) continues the long time series of land surface measurements, but using up-to-date technologies. The Thermal Infrared Sensor (TIRS) will provide unique thermal IR sensing capability for a Landsat satellite. The satellite bus was built by the Orbital Sciences Corp. and has a design life of 5 years, but carries enough fuel for 10 years of operations. The NASA contract for the LDCM satellite bus was originally awarded to the General Dynamics Advanced Information Systems (GDAIS) in April, 2008. Orbital Sciences subsequently acquired the spacecraft manufacturing division of GDAIS and therefore assumed the responsibility for the Landsat 8 satellite bus. The bus was built at Orbital's spacecraft manufacturing facility in Gilbert, Arizona.

The OLI was built by Ball Aerospace and Technologies Corp. and measures in the visible, near-IR, and shortwave IR portions of the electromagnetic spectrum. The images have a 15-m panchromatic and a 30-m multispectral spatial resolutions along a 185 km wide swath (Fig. 1.27).

OLI represents a real advancement in Landsat sensor technology and uses a technical approach demonstrated by the Advanced Land Imager Sensor flown on NASA experimental EO-1 satellite. Earlier Landsat instruments all used scan mirrors that stopped and started with each scan (see the optical instrument chapter). OLI instead uses long detector arrays with over 7000 detectors per spectral band, aligned across its focal plane to view across the swath. This "push-broom" scan approach results in a much more sensitive instrument providing improved land surface sensing with fewer moving parts. In addition, OLI has an improved signal-to-noise ratio as compared to previous Landsat instruments.

FIGURE 1.27

The data collection pattern of the Operational Land Imager (OLI) on Landsat 8.

The TIRS is designed to measure land surface temperature in two thermal bands using a new technology to detect heat. The motivation behind the TIRS instrument is a realization that state hydrology managers relied on the LDCM's predecessors, Landsat 5 (TM) and Landsat 7 (ETM), to track how land and water are being used in their state. Since nearly 80% of the freshwater in the western United States is used to irrigate crops, TIRS will become an invaluable tool for managing water consumption.

TIRS uses Quantum Well Infrared Photodetectors (QWIPs) to detect long wavelength radiation emitted by the Earth whose intensity depends on surface temperature. In these thermal IR portions of the spectrum, QWIPs are a new, lower-cost alternative to conventional longwave IR technologies. The QWIPs were developed at NASA's GSFC in Greenbelt, Maryland. The QWIPs are sensitive to two thermal IR wavelength bands, helping users separate the temperature of the Earth's surface from that of the atmosphere. Their design depends on quantum mechanics where gallium arsenide semiconductor chips trap electrons in an energy state "well" until the electrons are elevated to a higher state by thermal IR radiation of a certain wavelength is incident on it. The elevated electrons create an electrical signal that can be readout to form a digital image.

An overview of this sequence of satellites is best given by Table 1.4, which shows all of the Landsat satellites up to Landsat 8.

All of the Landsat satellites operate in a Sun-synchronous orbit with a 9:42 a.m. local equator crossing time. The swath width of the TM was 150 km, and the ground track separation at the equator was 2875 km. The OLI has increased the swath width to 185 km. Thus, any spot on the Earth is only covered approximately every 15 days. This low frequency of repeat coverage meant that cloud cover could dramatically reduce the amount of coverage of the Landsat instruments. The TM instrument added a number of channels to the MSS including a new thermal IR channel. In addition, the TM spatial resolution increased from 80 m with the MSS to about 30 m for the TM. Landsat 8 further improves on this spectral resolution by adding an independent TIRS instrument to cover the thermal IR channels while preserving the visible and near-IR channels with the OLI.

Table 1.4 Sequence of Landsat Satellites up to Landsat 8

Instrument	Picture	Launched	Terminated	Duration	Notes
Landsat 1		July 23, 1972	January 6, 1978	2 years, 11 months and 15 days	Originally named Earth resources technology satellite 1
Landsat 2		January 22, 1975	February 25, 1982	2 years, 10 months and 17 days	Nearly identical copy of Landsat 1
Landsat 3		March 5, 1978	March 31, 1983	5 years and 26 days	Nearly identical copy of Landsat 1 and Landsat 2
Landsat 4		July 16, 1982	December 14, 1993	11 years, 4 months 28 days	
Landsat 5		March 1, 1984	June 5, 2013	29 years, 3 months and 4 days	Nearly identical copy of Landsat 4. Longest Earth-observing satellite mission in history
Landsat 6		October 5, 1993	October 5, 1993	0 days	Failed to reach orbit
Landsat 7		April 15, 1999	Still active	16 years, 1 month and 8 days	Operating with scan line corrector disabled since May 2003
Landsat 8		February 11, 2013	Still active	2 years, 3 months and 12 days	Originally named Landsat data Continuity Mission from launch until May 30, 2013, when NASA operations were turned over to USGS

1.2.4.3 The Defense Meteorological Satellite Program

Smaller and lighter than the original TIROS, the 100-pound TIROS-derived RCA satellite was shaped like a 10-sided polyhedron, 0.58 m across, and 0.53 m high. A spinning motion, introduced on injection into orbit, was maintained on the early National Reconnaissance Office (NRO) weather satellites at about 12 rpm by small spin rockets. However, the spin axis was maintained perpendicular to the orbit plane by torquing the satellite against the Earth's magnetic field, the forces supplied through a direct-current loop around the satellite's perimeter. A ground command would cause the electric current to flow in the desired direction to generate the torque. Those few NASA officials who knew about it viewed the Air Force program as a no-risk test of a modified four-stage Scout with an "Earth-referenced" wheel-mode weather satellite. If it operated correctly, the RCA shuttered television camera (a photosensitive vidicon tube) would be pointed directly at the Earth once each time the satellite rotated. At the programmed interval, when IR horizon sensors indicated the lens was vertical to the Earth, the vidicon would take a picture of an 800-mile-square area of the surface below, with the image recorded on tape as an analog signal for later transmission to the ground. Launched into a Sun-synchronous 833 km height, circular polar orbit, the RCA television system would provide 100% daily coverage of the northern hemisphere at latitudes above 60 degrees, and 55% coverage at the equator. Readout of the tape-recorded pictures was planned to occur on each pass over the western hemisphere; at the ground stations, the video pictures of cloud cover over the Eurasian landmass would be relayed to the Air Weather Service's Air Force Global Weather Central collocated with Headquarters Strategic Air Command at Offutt AFB, near Omaha, Nebraska.

The second DMSP launch on August 23, 1962, resulted in success, although the Lockheed ground-control team failed at first to track the weather satellite. Each day at high noon the vehicle took pictures as it transited the Soviet Union. Weather pictures of the Caribbean returned by this vehicle 2 months later in October also proved crucial during the "Cuban Missile Crisis," permitting effective aerial reconnaissance missions and reducing the number of aerial weather reconnaissance sorties in the region.

The first DMSP weather satellite to be controlled at the ground stations manned by Air Force personnel was flight number three launched on February 19, 1963. At Vandenberg AFB, another Air Force team, the Systems Command 6595th Aerospace Test Wing, conducted launch operations. In this instance, the Scout booster upper stages again malfunctioned and placed the satellite in an orbit ill suited to strategic weather reconnaissance operations for more than a few months at best. In late April, the satellite's primary tape recorder control circuit failed and with it the storage of primary data for later commanded transmission, although direct vidicon readout continued for a few weeks more. A new experiment, however, continued to function nicely for many months: an IR radiometer that registered the Earth's background radiation and indicated the extent of nighttime cloud cover. At Global Weather Central, the 3rd Weather Wing used computer programs written by Air Weather Service personnel to produce crude operational maps of the cloud cover at night over the regions observed until January 1964. Indeed, the IR experiment proved so successful that it was mounted on all DMSP satellites through Block 4, eventually also providing measurements of cloud height and the Earth's heat balance.

Strategic weather reconnaissance might command the primary mission of the DMSP, but American military services wanted tactical weather data to meet a variety of operational needs. By 1963 it was plain that NASA's sophisticated, three-axes stabilized, low-altitude Nimbus-NOMSS (National Operational Meteorological Satellite System) satellite would be extensively delayed and, when finished, likely too complex and expensive to satisfy Defense Department meteorological requirements—tactical or strategic.

However, the political and bureaucratic climate in 1963 did not favor an all-military tactical weather satellite system. All of the military meteorological satellite requirements would continue to be furnished to NASA and the Department of Commerce. To assess and combine those requirements, in early 1964 the Defense Department established in the Air Staff a Joint Meteorological Satellite Program Office. After further agitation by the military services, however, the Defense Department approved a test of the defense meteorological satellite applied to tactical operations in the 1964 Strike Command Goldfire exercise at Fort Leonard Wood in southwest Missouri. Air Force Global Weather Central at Offutt AFB relayed weather reconnaissance pictures directly to the Army and Air Force users supporting ground and paratroop exercises at the fort, and for the deployment of fighter aircraft on a transatlantic flight. Later in the year, between November 24 and 26, Global Weather Central furnished tactical weather data over Central Africa to the Military Airlift Command, which proved crucial in the successful airlift of Belgian paratroopers from Europe to Stanleyville in the Congo, where hostages seized in an uprising were freed. The weather data proved to be of considerable value in these tactical operations, analysis revealed, but improvements were needed. Coverage had to be received daily at local ground stations before meteorologists could depend on a satellite as a primary source of data, and a resolution at the surface better than the three nautical miles provided by the DMSP Block-I satellites was judged "extremely desirable."

When US air strikes against North Vietnam commenced in February 1965, Det-14 personnel found themselves unable to meet the demand for weather information from the second Air Division and the Studies and Observation Group of the Military Assistance Command Vietnam, which conducted clandestine operations against North Vietnam. In response, the Air Force, with Defense Department approval, launched on March 18, 1965, a noontime military meteorological satellite that could be programmed to record and readout specific weather data in Southeast Asia to support tactical operations in the theater. In one of his last official acts in support of that effort, in January, US Army Colonel Haig planned and laid out the DMSP ground station at Tan Son Nhut Air Base, Saigon, in South Vietnam. The new station was erected and began operating in time to support the satellite launched in March. It furnished to military users, within 30 min of receipt, complete cloud cover data for North Vietnam, South Vietnam, and parts of Laos, China, and the Gulf of Tonkin.

These impressive results were enough to prompt action from Defense Department officials who now sought to break the NASA/Department of Commerce franchise and pursue openly a separate military weather satellite program for strategic and tactical applications. The DMSP program office in El Segundo, California would move from the Air Force Special Projects Office, to the Space Systems Division next door, in Air Force Systems Command, with Headquarters USAF and Systems Command assuming overall management responsibility for what was termed an "ongoing development/operational program." The Strategic Air Command would continue to launch the satellites and operate the DMSP control center and ground terminals in the continental United States; Air Weather Service would man the direct readout terminals overseas, while continuing to operate Air Force Global Weather Central and process DMSP strategic weather data at Offutt AFB. Perhaps anticipating an excess of public affairs enthusiasm on the Air Staff, he regretted to say that security restrictions precluded any public recognition of DMSP accomplishments.

Back in 1964, when tests of the meteorological satellite applied to tactical military operations at home and abroad began, the NRO approved modification of three additional satellites for direct readout. These 160-pound vehicles, identical in size and shape to their 45−54 kg Block 1 predecessors, also mounted improved IR radiometers and were known collectively as Block 2. Launched during 1965 and

FIGURE 1.28

Defense Meteorological Satellite Program Block 4 Satellite.

1966, two of them attained Earth orbit and provided tactical meteorological data for operations in Southeast Asia. A fourth satellite, the one equipped and launched expressly for tactical uses on May 20, 1965, came to be called Block 3. The reason for this curiosity, a "one-vehicle block," involved efforts to distinguish it from its Block 2 cousins that also supported the primary strategic cloud cover mission. Shortly before he stepped down as DMSP director and control of the DMSP passed to the Air Force Systems Command, in early 1965 Colonel Haig secured permission to begin the design of a more powerful military meteorological satellite that met more completely the demands of its customers.

The Block 4 satellite, slightly larger than those in Blocks 1 and 2, was 76 cm. in diameter, 74 in height, and weighed 79.4 kg (Fig. 1.28). Still spin stabilized, the satellite nonetheless provided improved weather coverage. Previously, the single 0.013 m focal length RCA vidicon television camera in Block 1 and 2 satellites furnished a nadir resolution of 5.5—7.4 km along a 2778 km swath. The resolution varied from 1.5 km at nadir to 5.5 km at the picture's edge. Besides a multisensor IR subsystem, Block 4 also incorporated a high-resolution radiometer that furnished cloud—height profiles. A tape recorder of increased capacity stored pictures of the entire northern hemisphere each day, while the satellite furnished real-time, direct local tactical weather coverage to small mobile ground or shipboard terminals.

The revolutionary Block 5 spacecraft (Fig. 1.29) took the form of an integrated system; it departed entirely from the TIROS-derived technology of its predecessors. The Block 5 design was based on the users' wish to receive a product in a form that approached as closely as possible the weather charts and maps that the meteorologists employed. Moreover, the product furnished the albedo of each scene, not its brightness, which varied enormously from full sunlight to partial moonlight. A survey of the industry and new technologies revealed line scanning sensors and advances in highly sensitive visible light and IR point (as opposed to array) detectors. Instead of using complicated electronics to scan the raster of a TV

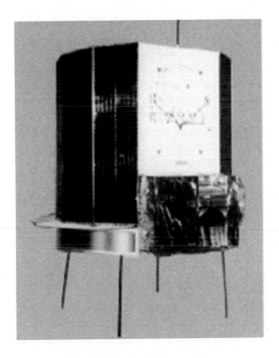

FIGURE 1.29

Defense Meteorological Satellite Program Block 5A Satellite.

camera, one now could let the motion of the satellite provide the scanning along the line of flight. That would require a spacecraft that always "looked down," rather than one that wheeled along its orbit. But a satellite stabilized on three axes would make possible acquiring a strip of imagery of indefinite length, imagery that could be rectified at will.

The Westinghouse "Operational Line Scanner" (OLS), as it came to be called, provided images of the Earth and its cloud cover in both the visual and IR spectral regions. With this system, nadir visual–imaging resolution at the Earth's surface improved to 0.56 km during daytime and 3.7 km at night through quarter-moonlight illumination levels. The higher resolution (less than 0.93 km) now satisfied the requirements of tactical users. The IR subsystem furnished 3.7 km resolution at the surface day and night, as well as cloud–height profile and identification of all clouds above or below a selected altitude, and heat-balance data. Complete global coverage was transmitted over encrypted S-Band digital data links. Block 5 simultaneously satisfied the meteorological needs of the military commander in the field for tactical support, while it met completely the "strategic" requirements of the NRO.

To achieve the pointing accuracy required for the Block 5 line scan sensor, the spacecraft employed a novel momentum-bias attitude-control system. It consisted of a momentum wheel and horizon scanner, and magnetic coils. The wheel and scanner controlled the pitch axis, while the magnetic coils controlled the roll and yaw axes, replacing the momentum dissipated by friction in the bearing between the momentum wheel and the main body of the spacecraft. The slab-sided, tube-shaped Block 5 satellite remained 0.76 m in diameter, but its height increased to 1.22 m and its weight rose to 104.3 kg. Three

Block 5A spacecrafts were built before military demands for greater tactical meteorological support dictated further change.

The DMSP program office, however, had introduced a requirement for an Earth-oriented pointing accuracy much greater than the one imposed on its predecessor Block 5C. In the design competition for the new spacecraft conducted between Boeing and RCA, only the latter firm was judged able to meet completely that requirement. A contract for five Block 5D satellites, signed with RCA in the fall of 1972, set a required launch date for the first of them in the fall of 1974. But the greater pointing accuracy and a complement of additional instruments also had increased the projected cost of these spacecraft as compared to their predecessors, and it introduced the risk of delays in development.

Whatever its pointing accuracy, and the numerical sleight-of-hand for "Block 5D" notwithstanding, in November 1972 the Office of Management and Budget (OMB) requested that the Departments of Commerce and Defense reexamine a consolidated civil and military polar-orbiting meteorological satellite program, and the possibility of using a single spacecraft to satisfy the demands of both. Either action could be expected to result in substantial dollar savings, and a steering group composed of representatives from NOAA, DoD, and NASA was formed once again to consider these questions. Since the technical capability of the existing Block 5C already exceeded the capability of a planned NOAA successor, TIROS-N, the group's report, issued in mid-1973, concluded that the greatest savings would be realized in a single national meteorological satellite system managed by the Air Force, using a standard DMSP Block 5D satellite. This military solution was quickly rejected by Henry Kissinger, President Nixon's National Security Advisor, who argued that it would violate the National Aeronautics and Space Act, which dictated a separation of military and civil spacefaring, and by officials made uneasy in the Department of State, who warned of adverse international repercussions. Subsequent interagency deliberations led by Air Force Under Secretary James W. Plummer, the director of the NRO, resulted in an agreement in July 1974 to achieve major cost savings by adopting a variant of the DMSP Block 5D military satellite for use in both the civil (replacing TIROS-N) and military polar-orbiting, low-altitude, meteorological space programs. The larger, joint-use version needed by the NOAA to support additional sensors, was identified as Block 5D-2. The five original Air Force-RCA spacecraft thus became DMSP Block 5D-1.

The Block 5D-1 design that had emerged back in the early 1970s resembled in appearance conventional Earth-oriented satellites of this period (Fig. 1.30). Sized to fit the space taken by the Burner IIA solid-propellant upper stage on the Thor, it was 1.5 m in diameter and 6 m long. The 5D satellite built by RCA consisted of three sections: a square precision-mounting platform on the forward end supported the sensors and other equipment required for precise alignment; in the center, a five-sided equipment-support module contained the bulk of the electronics and featured one or two pinwheel louvers on four sides for thermal control; and, at the aft end, a circular reaction and control-equipment support structure housed the spent third stage solid-propellant rocket motor and contained reaction-control equipment. A deployable, 1.83-by-4.88 m Sun-tracking solar array was also mounted aft, on this section. With its complement of additional sensors, the spacecraft weighed 522 kg, making it more than twice as massive as its Block 5C predecessors. To heft the additional weight into orbit, the program office contracted with Boeing for a new, larger, solid-propellant second stage. The original Burner IIA second stage, now adapted as a third stage and fixed to the satellite, was used during ascent to inject the vehicle into its circular, Sun-synchronous 724 km Earth orbit.

The electronic components of the follow-on satellites remained essentially the same as those in 5D-1, but the 5D-2 structure increased in length from 6 to 6.8 m. The extension increased the

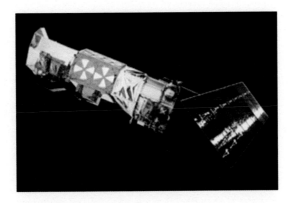

FIGURE 1.30

Defense Meteorological Satellite Program Block 5D-1 satellite.

downward-facing sensor-mounting area and lengthened the equipment-support module amidships. That module now contained a second 25.5 Ah battery and sported two or three pinwheel temperature control louvers on four of its five sides. The solar array mounted on the aft reaction control equipment-support structure also increased in size to 3 × 5 m, furnishing increased electrical power. Two important sensors were added to those in the 5D-1 complement: a topside ionospheric sounder provided detailed global measurements of the electron distribution in the Earth's ionosphere, and a microwave imager (flown on the last few 5D-2 satellites) defined the extent of sea ice and sea-state conditions (wave height and patterns) on the world's oceans. Withal, these changes increased the weight of the Block 5D-2 spacecraft to 813 kg a sum too great for the Thor/Burner booster combination.

Heated debates took place between officials in the program office and Aerospace Defense Command, the launch agency at that time, about adapting Thrust Augmented Thors to the task, just to keep a "blue suit" launch squadron. Ultimately, however, the launch vehicle selected for the 5D-2 meteorological satellite in 1980—after 16 months of vacillation—was the General Dynamics Atlas E, an improved version of the liquid-propellant intercontinental ballistic missile deployed briefly in the early 1960s.

One unique capability of the OLS was a "low-light" sensing capability. This requirement was to make it possible to view the Earth under moonlight lighting conditions. This capability has been used to study the lights of cities and other population centers as well as the progression of wildfires and anthropogenic burning of forests. One unique application was the capture of the Aurora Borealis in Fig. 1.31.

A variety of secondary sensors, some judged as "nice to have," appeared in different combinations on Block 5D-1 missions. Five of them frequently appeared on the spacecraft. An atmospheric density sensor measured the major atmospheric constituents (nitrogen, oxygen, and ozone) in the Earth's thermosphere on the daylight portion of each orbit. A precipitating electron spectrometer counted ambient electrons at various energies. A scanning IR radiometer furnished vertical temperature profiles, vertical water vapor profiles, and the total ozone concentration. A passive microwave-scanning radiometer profiled global atmospheric temperatures from the Earth's surface to altitudes above 30 km. Known as the Special Sensor Microwave/Imager (SSM/I) this instrument has been the source of a lot of science investigations.

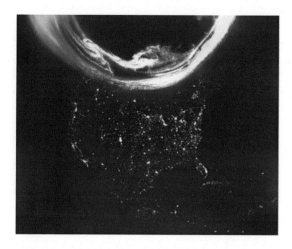

FIGURE 1.31

Defense Meteorological Satellite Program nighttime image of the Aurora Borealis taken by the first Block 5A satellite in 1971 (Note the lighted cities from Canada through Central America).

Early in its operation the DoD worked out with NOAA an agreement known as "shared processing" where the SSM/I data were decrypted and made freely available through NASA and NOAA data centers. Finally, a gamma-radiation sensor furnished by the Air Force Technical Applications Center detected nuclear detonations as part of the ongoing Integrated Operations NUDET Detection System.

Design studies of a still larger and heavier Block 5D-3 satellite began in the late 1970s, but funds for the military version were not appropriated until mid-1980. The 5D-3 satellites, though initially designed to be compatible for launch on NASA's Space Shuttle and be laser-hardened, ultimately would be launched on an unmanned expendable rocket. This satellite mounted an improved Westinghouse OLS and a larger combination of secondary sensors. The length of the satellite increased from 6.7 to 7.3 m, while the weight rose to 1033 kg. The RCA spacecraft consisted of the same basic components as its immediate predecessors, but included a larger solar array, three 50-Ah batteries, and a redesigned sunshade. The center section now sported four pinwheel temperature control louvers on four of its five sides. These and other design improvements combined to give the 5D-3 an anticipated mean mission lifetime on orbit of 5 years (60 months).

1.2.4.4 Geostationary Weather Satellites

In 1966 the first Advanced Technology Satellite (ATS1) was launched into a geostationary orbit and used a new instrument the "spin-scan camera" to sense radiative properties of a hemisphere under the satellite. As geostationary orbit dictated an increase in altitude from about 800 to 36,000 km the spin-scan camera had to be considerably more sensitive than the low Earth orbiter instruments. Also, since these early geostationary satellites were all spin stabilized the scanner had only a very limited time to collect the radiation from the Earth's surface.

An image from the spin-scan camera on ATS1 is presented here in Fig. 1.32, which clearly shows the cloud cover over this hemisphere. While the spin-scan camera was limited to visible imagery later versions of this instrument were extended to add IR images to provide night imagery of the Earth although at a lower spatial resolution than the polar orbiters.

FIGURE 1.32

Visible image from the spin-scan camera on ATS1.

In spite of the sampling limitations imposed by the geostationary orbit, it was demonstrated that the geostationary orbits were very useful for measuring the Earth due to the high frequency of temporal coverage. Since the satellite was stationary relative to the Earth a new scan mechanism was developed that stepped a mirror across the hemisphere making it possible to measure the Earth in a very rapid fashion. No longer were samples limited by the approximately 90 min repeat orbits of low Earth orbiting satellites, but scans could be collected every 30 min or hourly. This made it possible to examine a host of new atmospheric processes and geostationary satellites soon became the standard measurement method for meteorological studies and weather forecasting. All of the sides of this spinning satellite were covered with solar panels that would get equal amounts of solar illumination.

A typical spinning geostationary satellite is shown here in Fig. 1.33.

The first operational geostationary weather satellites were known as the Stationary Meteorological Satellites, which then evolved into the Geostationary Observing Environmental Satellites (GOES). The imager progressed from the limited channel spin-scan camera to a visible IR spin-scan radiometer, which collected hemispherical images in a variety of spectral bands. The final configuration of the GOES spinner satellites is shown here in Fig. 1.34, which differs from the earlier satellite in that the lower portion of the spacecraft was "despun" making communications with the ground a lot easier and allowing directive antennas and higher rates of data delivery to the ground.

A desire to improve the sensing capabilities of the geostationary satellites led to the development of a three-axes stabilized satellite that could "stare" at the Earth's surface collecting more radiation than would be possible during the limited time imposed by a spinning geostationary satellite. A despun satellite, however, introduced new problems. The thermal equilibrium provided by the spinning satellite was no longer available and the satellite had to be designed to withstand the thermal gradients imposed by a satellite being heated by the Sun on one side and cooled on the opposite side. This led to the satellite configuration seen here in Fig. 1.35, which is an artist's conception of GOES-8 the first of the operational US three-axes stabilized geostationary satellites.

FIGURE 1.33

Early Geostationary Observing Environmental Satellites spinner geostationary weather satellite.

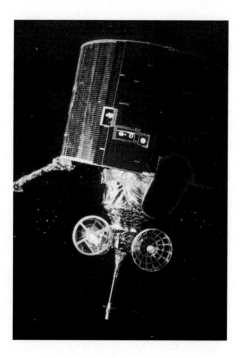

FIGURE 1.34

GOES satellite showing the despun communications antennas at the bottom.

FIGURE 1.35

Artist's impression of the GOES-8 satellite superimposed on a GOES west image.

Now the solar panels are now only on the sunlit side of the satellite while on the other side there is a "solar sail" to help with the stability of the platform. The new GOES imager was able to take advantage of the "starting" capability of this satellite and provided much improved radiative fidelity and accuracy.

As part of the weather forecast system the GOES satellites were used to collect the radiation from the Earth's surface and process it on the ground. The early spinner satellites relayed analog only data while the three-axes stabilized satellites provided digital data streams. The earlier data were digitized on the ground and an analyzed weather facsimile product was generated. This product was then uplinked to the same GOES satellite and broadcast as a low-resolution image to be used for forecast activities by remote forecasting locations. Known as WEFAX this product has continued to be available from geostationary satellites.

It is important to realize that the geostationary satellites were and are used to collect both imagery and data transmitted from platforms on the surface. In this way they acted as standard communications satellites providing the relay of data from the surface. They could handle a much larger data stream than the polar low Earth orbiters making it possible to collect both data and more recently accurate geolocation information from GPS receivers.

Today geostationary satellites are the primary observing platforms for all weather forecasting operations. In the United States, one GOES satellite orbits at 135°W off the United States. West coast to monitor storms and weather fronts approaching from their traditional position west of the United States, and the other satellite is located at 75°W off the eastern coast of the United States to monitor hurricane formation and their progression toward the US mainland. The European EUMETSAT organization operates METEOSAT at 0° as their primary weather satellite, while Japan operates the Geostationary

Meteorological Satellite at 145°E for severe weather in the Asian region. India operates INSAT at 83°E in the Indian Ocean, but the data are restricted to Indian weather operations. Japanese and European satellites are still spin stabilized, while the Indian and American satellites are three-axes stabilized. The European Space Agency claims that the spin stabilization provides (1) better geometric correction quality, (2) better spectral and geometric resolution, (3) better inversion availability, (4) moderate inversion availability, and (5) the same NEΔT as that of a slight larger aperture.

1.2.4.4.1 GOES-R

The next generation of geostationary satellites is already built and the first launch was November 19, 2016. GOES-R is a collaborative mission between NOAA and NASA and the satellite was built by Lockheed-Martin Corp. in Denver, CO. GOES-R will provide continuous imagery and atmospheric measurements of the Earth's western hemisphere and space weather measurements. It will carry six different instruments to measure both land and space parameters. The primary instrument is the Advanced Baseline Imager (ABI), which replaces the GOES imager and will image Earth's weather oceans and atmospheric environment (Fig. 1.36). ABI will view the Earth with 16 spectral bands (compared to the five on the GOES), including two visible channels, four near-IR channels, and 10 mid to thermal IR channels. It will deliver three times more spectral information, four times the spatial resolution and more than five times faster temporal coverage than the present system. ABI will deliver more than 65% of all the GOES-R mission data products.

ABI is designed to be used for a wide variety of applications that will utilize the 16 spectral bands and its higher space—time resolution. The instrument operates in two scan modes: one where it continuously takes an image of the entire full disk of the planet every 5 min, and a second one called the flex mode where it takes a full disk image every 15 min, an image of the continental United States every 5 min and smaller, more detailed images of areas where storm activity is present as often as every 30 s. This latter sampling is designed to see the formation of tornadoes as their cloud tops rise above the line of thunderstorm clouds.

FIGURE 1.36

The Advanced Baseline Imager.

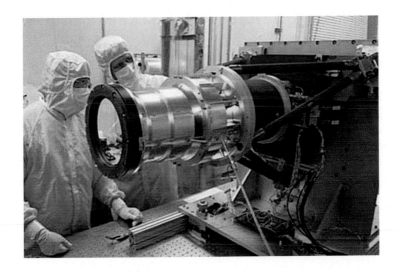

FIGURE 1.37

The Geostationary Lightning Mapper (http://www.goes-r.gov/spacesegment/glm-lightning-detect.html).

The ABI provides a dramatic improvement over present GOES imagery with an increase from 5 to 16 bands, visible spatial resolution going from 1 to 0.5 km and the thermal IR bands improving their spatial resolution from 4 to 2 km. All of these channels will have on-orbit calibration capabilities unlike the present GOES imager. These capabilities will be valuable in a number of nonweather applications as well. The rapid temporal sampling will be useful in mapping forest fires, volcanoes, floods, hurricanes, and storms that spawn tornadoes. It is estimated that benefits from the ABI may be as much as $4.6 billion over the lifetime of the series due to improved tropical cyclone forecasts, fewer weather related flight delays, airline incidences with volcanic plumes, improved production and distribution of electricity and natural gas, increased efficiency in irrigation, and higher protection rates for recreational boaters in the event of storms or hurricanes.

Another Earth-sensing instrument is the Geostationary Lightning Mapper (GLM), which is a single-channel, near-IR optical transient detector that can detect the rapid changes in an optical scene indicating the presence of lightning (Fig. 1.37). GLM measures total lightning activity continuously over the Americas and adjacent ocean regions with a near uniform spatial resolution of 10 km. GLM will provide early predictions of intensifying storms and severe weather events. GLM is unique in the collection of lightning measurement systems in that ground systems only provide cloud-to-ground lightning coverage, GLM includes both cloud-to-ground and cloud-to-cloud coverage of lightning events. In addition, ground-based systems are limited to land while GLM will operate over the ocean as well.

Another GOES-R instrument known as the extreme ultraviolet and X-ray Irradiance Sensors (EXIS) are designed to monitor solar irradiance in the upper atmosphere as the power and effect of the Sun's electromagnetic radiation per unit area of the atmosphere. EXIS will be able to detect solar flares that could disrupt communications and reduce satellite-based navigation accuracy by affecting satellites, high altitude aircrafts, and power grids on the Earth. EXIS is made up of two main sensors: the extreme ultraviolet sensor and the X-ray sensor. EXIS will be mounted on the Sun pointing platform which is on the yoke of the solar array. NOAA's Space Weather Prediction Center in Boulder, Colorado, will rely on products from EXIS to improve their warning of radio blackouts to improve communications and navigation systems.

The other three instruments on GOES-R are designed to measure the space rather than the Earth's environment. The magnetometer provides measurements of the space environment magnetic field that controls charged particle dynamics in the outer region of the magnetosphere. The geomagnetic field measurements provide important alerts and warnings to satellite operators and power utilities. The Solar Ultraviolet Imager (SUVI) is a telescope that monitors the Sun in the extreme ultraviolet portion of the electromagnetic spectrum. SUVI will provide full disk solar images around the clock. It replaces the current GOES Solar X-ray Imager (SXI) instrument and represents a change in both spectral coverage and spatial resolution over the SXI. Finally, the Space Environment In Situ Suite (SEISS) is comprised of four sensors that will monitor proton, electron, and heavy ion fluxes at geostationary orbit (35,786 km altitude). These four instruments are the Energetic Heavy Ion Sensor (EHIS), the Magnetospheric Particle Sensors-High and Low, and the Solar and Galactic Proton Sensor. The instrument suite also includes the data processing unit. Data from SEISS will drive the solar radiation storm portion of the NOAA's space weather forecast system.

1.3 STUDY QUESTIONS

1. What were the fundamental observational methods that needed to be developed before satellite remote sensing could become a reality?
2. Why were the first polar-orbiting weather satellites only able to view a small portion of the Earth's surface?
3. What change in the stabilization of the polar-orbiting weather satellites needed to be made to make it possible to observe the entire Earth?
4. What NASA satellite program was used to test new instruments designed for atmospheric, land, and ocean studies.
5. What satellite control system was needed to make it possible to continuously view the Earth's surface?
6. What is the difference between an early television image of the Earth and the image collected by a radiometer?
7. What is the difference between a radiometer and a spectrometer?
8. How do orbiting (polar and geostationary) satellites profile the atmospheric temperature and moisture?
9. How did the Landsat program begin?
10. What is the succession of instruments used in the Landsat program?

BASIC ELECTROMAGNETIC CONCEPTS AND APPLICATIONS TO OPTICAL SENSORS

2.1 MAXWELL'S EQUATIONS

James Clerk Maxwell was the first one to formally postulate the existence of electromagnetic (EM) waves. Maxwell combined ideas from other earlier researchers to formulate a set of partial differential equations that describe the properties of the electric and magnetic fields, and relate them to their sources. Maxwell derived a waveform of the electric and magnetic equations, revealing the wavelike nature of the electric and magnetic fields and demonstrating their symmetry. Since the speed of the EM waves predicted by the wave theory coincided with the measured speed of light, Maxwell concluded that light itself is an EM wave. The electric and magnetic waves are linked together, but one does not cause the other. They are linked together in the same way that time and space changes are linked in the theory of special relativity. Together, these fields form a propagating EM wave, which moves out into space. Individually the equations are known as Gauss's law, Gauss's law for magnetism, Faraday's law of induction, and Ampère's law with Maxwell's correction.

These equations together with the Lorentz force law constitute the complete set of classical electromagnetism. The Lorentz force law was actually derived by Maxwell under the name of "equation for electromotive force" and was one of an earlier set of eight Maxwell's equations.

In symbols these equations are given in Table 2.1.

In 1864 Maxwell derived the EM wave equation by linking the displacement current to the time-varying electric field associated with EM induction thus linking some earlier ideas. He commented that, "light and magnetism are affections of the same substance, and that light is an electromagnetic disturbance propagated through the field according to electromagnetic laws." Today this theory has been adapted to light propagation in a vacuum where one can have stable, self-perpetuating waves of oscillating electric and magnetic fields driving each other. The speed calculated for EM radiation exactly matches the speed of light since light is indeed one form of EM radiation bringing together the two fields of electromagnetism and optics.

2.2 THE BASICS OF ELECTROMAGNETIC RADIATION

EM radiation is a form of energy with the properties of a wave. The waves propagate through time and space in a manner rather like water waves, but (except for polarized light) oscillate in all directions perpendicular to their direction of travel (the wave front of a transverse wave).

Such a wave is characterized by two principal measures: wavelength and frequency (or period). The wavelength (λ) is the distance between two successive crests of the waves. The frequency (v or f)

Introduction to Satellite Remote Sensing. http://dx.doi.org/10.1016/B978-0-12-809254-5.00002-6

43

Table 2.1 Summary of Maxwell's Equations in Differential and Integral Forms

Name	Differential Form	Integral Form
Gauss's law	$\nabla \cdot \vec{E} = \frac{\rho}{\varepsilon_0}$ (2.1a)	$\oint_{\partial V} \vec{E} \cdot d\vec{A} = \frac{Q(V)}{\varepsilon_0}$ (2.1b)
Gauss's law for magnetism	$\nabla \cdot \vec{B} = 0$ (2.2a)	$\oint_{\partial V} \vec{B} \cdot d\vec{A} = 0$ (2.2b)
Maxwell–Faraday equation (Faraday's law of induction)	$\nabla \times \vec{E} = -\frac{\partial \vec{B}}{\partial t}$ (2.3a)	$\oint_{\partial S} \vec{E} \cdot d\vec{l} = -\frac{\partial Q_{B,S}}{\partial t}$ (2.3b)
Ampère's circuital law (with Maxwell's correction)	$\nabla \times \vec{B} = \left(\vec{J} + \frac{\partial \vec{E}}{\partial t} \right)$ (2.4a)	$\oint_{\partial S} \vec{B} \cdot d\vec{l} = \mu_0 I_S + \mu_0 \varepsilon_0 \frac{\partial \Phi_{E,S}}{\partial t}$ (2.4b)

where $\vec{E}$ is the electric field vector, $\vec{B}$ is the magnetic field vector, $\nabla \cdot$ is the divergence operator, $\nabla \times$ is the curl operator, $\partial/\partial t$ is the partial derivative with respect to time, $d\vec{A}$ is the differential vector element of a surface $\vec{A}$ with an infinitesimally small magnitude and direction normal to the surface S, $d\vec{l}$ is the differential vector element of path length tangential to the path, ε_0 is the dielectric permittivity of free space (8.854×10^{-12} F/m), μ_0 is the magnetic permeability of free space ($4\pi \times 10^{-6}$ H/m), ρ is the total charge density, and $\vec{J}$ is the total current density. Q(V) is the total charge contained in the volume ∂V, and I_s, $\Phi_{B,S}$, and $\Phi_{E,S}$ are the flux of the volumetric current density (J), magnetic field (B), and electric field (E) passing through a closed surface ∂S.

is the number of oscillations (cycles) completed per 1 s period. The period (T) of the wave is $T = 1/f$ with the frequency f in cycles/s.

In a vacuum, EM waves travel at a constant speed c, the speed of light ($c = 299,792,458$ m/s), which is the product of frequency and wavelength:

$$\lambda \cdot v \approx 300.000 \text{ km/s},\tag{2.5}$$

so the frequency and wavelength are inversely proportional to each other for an EM wave. These relationships are displayed here in Fig. 2.1.

Note in Fig. 2.1 that the electric and magnetic fields are in phase, and perpendicular to each other in orientation. Both are transverse waves with the same wave frequency and wavelength.

2.3 THE REMOTE SENSING PROCESS

The underlying basis for most remote sensing methods and systems is simply that of measuring the varying energy and/or frequency levels of a single entity, the fundamental unit in the EM force field known as the *photon*. As will be shown later, variations in photon energies are tied to the parameter *wavelength* or its inverse, *frequency*. Radiation from specific parts of the EM spectrum contains photons of different wavelengths whose energy levels fall within a discrete range of values. When any target material is excited by internal processes or by interaction with incoming EM radiation, it will emit photons of varying wavelengths whose radiometric quantities differ at different wavelengths in a way that can be used to diagnose the material. Photons may also be introduced by reflection and absorption. Photon energy received at detectors is commonly stated in power units such as Watts per square meter per wavelength unit (W/m^2 μm). The plot of variation of power with wavelength gives rise to a specific pattern or curve that is the *spectral signature* for the substance or feature being sensed.

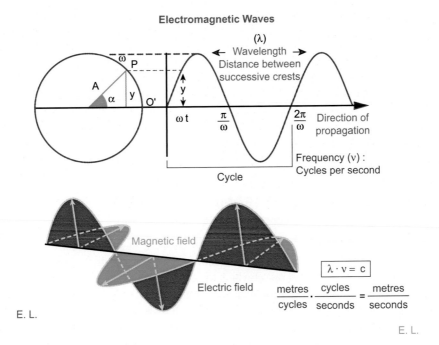

FIGURE 2.1

Electromagnetic wave showing both the magnetic and electric fields along with definitions of frequency and wavelength.

The photon is the physical form of a quantum, the basic particle of Quantum Mechanics, the part of Physics dealing with the very small, at atomic and subatomic levels. It is also described as the messenger particle for EM force or as the smallest bundle of light. This subatomic massless particle contains radiation *reflected*, *absorbed* or *emitted* either by matter when it is excited thermally, or by nuclear processes (fusion, fission), or by bombardment with other radiation. Photons move at the speed of light as waves and hence, have a "dual" nature. These waves follow a pattern that we described in terms of a sine (trigonometric) function, as shown in two dimensions in Fig. 2.1.

A photon is said to be quantized, in that any given one possesses a certain quantity of energy. Some other photons can have a different energy value. Photons as quanta thus show a wide range of discrete energies. The amount of energy characterizing a photon is determined using Planck's general equation:

$$E = h \cdot v, \tag{2.6}$$

where h is the Planck's constant ($h = 6.62607 \times 10^{-34}$ J·s) and v represents frequency. Photons traveling at higher frequencies are therefore more energetic. If an electron experiences a change in energy level from a higher-level E_2 to a lower-level E_1, a photon is emitted at a frequency given by:

$$\Delta E = E_2 - E_1 = h \cdot v, \tag{2.7}$$

Alternatively, the passage of an electron from E_1 to E_2 requires the absorption of a photon where v has some discrete value determined by $v_2 - v_1 = (E_2 - E_1)/h$. In other words, a particular energy change is characterized by emitting radiation (photons) at a specific frequency v and a corresponding wavelength at a value dependent on the magnitude of the change of energy.

The distribution of all photon energies over the range of observed frequencies is embodied in the term *spectrum* (a concept developed later). A photon with some specific energy level occupies a position somewhere within this range, i.e., lies at some specific point in the spectrum.

The remote sensing of color is a particular example of the remote sensing process since these wavelengths are restricted to the small portion of the EM spectrum that contains the visible bands (Fig. 2.2). All of this radiation is reflected radiation excited by incoming solar radiation as indicated in Fig. 2.3.

2.4 THE CHARACTER OF ELECTROMAGNETIC WAVES
2.4.1 DEFINITION OF RADIOMETRIC TERMS

Radiant energy (Q), transferred as photons, is said to emanate in short bursts (wave trains) from a source in an excited state. This stream of photons moves along lines of flow (also called rays) as a *flux* (ϕ), which is defined as the temporal rate at which the energy Q passes a spatial reference (dQ/dt). The flux concept is related to that of *power*, defined as the temporal rate of doing work or spending energy. The nature of the work that can be done is a combination of these: changes in motion of particles acted upon by force fields; heating; physical or chemical change of state. Depending on circumstances, the energy spreading from a point source may be limited to a specific direction (a beam) or can disperse in different directions.

Radiant flux density is the power per unit surface (W/m^2). The flux density is proportional to the squares of the amplitudes of the component waves. Flux density as applied to radiation coming from an external source to the surface of a body is referred to as irradiance (E); if the flux comes out of that

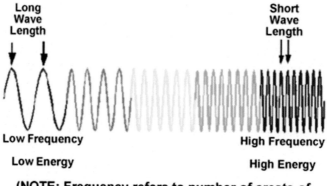

FIGURE 2.2

Heuristic diagram of visible wavelengths showing the breakdown of colors.

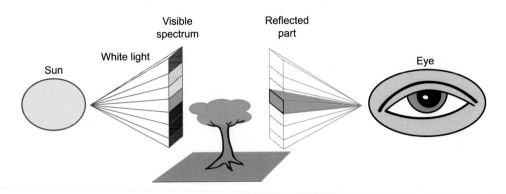

Reflection of colours

E. L.

FIGURE 2.3

Illustration of the reflection of colors by an Earth target.

body; its nomenclature is exitance (*M*) (see below for a further description). Mathematically the irradiance can be written as

$$E(x, y) = \frac{d\Phi}{dA} \; (\text{W}/\text{m}^2),$$ (2.8)

and depicted in Fig. 2.4.

The radiant exitance is the flux per unit area radiated, transmitted, or reflected by a surface and it can be written mathematically as

$$M(x, y) = \frac{d\Phi}{dA} \; (\text{W}/\text{m}^2),$$ (2.9)

and displayed in Fig. 2.5.

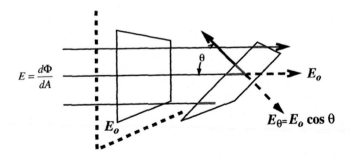

FIGURE 2.4

Definition of irradiance (Schott, 2007).

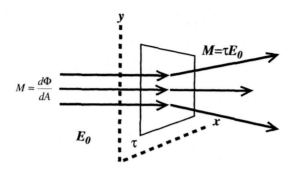

FIGURE 2.5

Definition of radiant exitance (Schott, 2007).

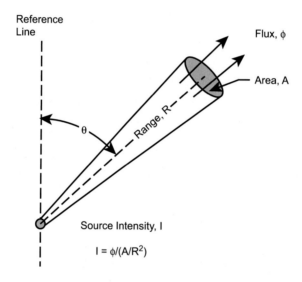

FIGURE 2.6

The definition of a solid angle for incoming radiation.

The notion of *radiant intensity* is given by the radiant flux per unit of solid angle (Fig. 2.6) in steradians (sr). Hence the radiant intensity is the power per unit solid angle and is given in (W/sr) and indicated by the symbol I. Thus, for a surface at a distance R from a point source, the radiant intensity I is the flux Φ [W/m^2] flowing through a cone of solid angle Ω on to the area A at a distance R, is given by $I = \Phi/(A/R^2)$. Note that the radiation is moving in some direction or pathway relative to a reference line as defined by the angle θ.

The radiant intensity can be written as

$$I(\theta, \varphi) = \frac{d\Phi}{dA} \ \left[\text{W/m}^2\right], \tag{2.10}$$

and shown graphically in Fig. 2.6.

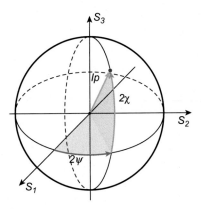

FIGURE 2.7

The Poincaré sphere is the parameterization of the last three Stokes' parameters in spherical coordinates.

From this, a fundamental EM radiation entity known as *radiance* (commonly noted as "*L*") can be derived. Radiance is defined as the radiant flux per unit solid angle leaving an extended source (of area *A*) in a given direction per unit projected surface area in that direction [W/(m² sr)].

$$L(x, y, \theta, \varphi) = \frac{d^2\Phi}{dA \cdot d\Omega \cdot cos\theta} = \frac{dE}{d\Omega \cdot cos\theta} = \frac{dM}{d\Omega \cdot cos\theta} \; \left[\mathrm{W}/\left(\mathrm{m}^2 \; \mathrm{sr}\right) \right]. \qquad (2.11)$$

Radiance is closely related to the concept of *brightness* as associated with luminous bodies. The magnitudes actually measured by remote sensing detectors are the radiances at different wavelengths leaving extended areas (which can "shrink" to point sources under certain conditions). Radiant fluxes that come out of sources (internal origin) are referred to as *radiant exitance* (*M*) or sometimes as "emittance" (now obsolete). Radiant fluxes that reach or "shine upon" any surface (external origin) are called *irradiance*. Thus, the Sun, a source, irradiates the Earth's atmosphere and surface.

The above radiometric quantities *Q*, Φ, *I*, *E*, *L*, and *M* apply to the entire EM spectrum. Most wave trains are polychromatic, meaning that they consist of numerous sinusoidal wave components of different frequencies. The bundle of varying frequencies (either continuous within the spectral range involved or a mix of discrete, but discontinuous monochromatic frequencies or wavelengths) constitutes a complex or composite wave. Any complex wave can be broken into its components by Fourier analysis, which extracts a series of simple harmonic sinusoidal waves each with a characteristic frequency, amplitude, and phase.

EM radiation can be *incoherent* or *coherent*. Waves whose amplitudes are irregular or randomly related are said to be incoherent; polychromatic light fits this state. If two waves of different wavelengths can be combined so as to develop a regular, systematic relationship between their amplitudes, they are said to be coherent; monochromatic light generated by lasers meets this condition. The above, rather abstract, set of ideas and terminology are important to the theorist. This synopsis is included mainly to familiarize the reader with these radiometric quantities in the event they are encountered in other reading.

If Eq. (2.4) is dot multiplied by $\vec{E}$ and Eq. (2.3) by $\vec{B}$, and subtracted, the following equation can be obtained:

$$\vec{E} \cdot \vec{B} - \vec{B} \cdot \vec{E} = \sigma |\vec{E}|^2 + \varepsilon_0 \frac{\partial \vec{E}^2}{\partial t} - \mu_0 \frac{\partial \vec{B}^2}{\partial t} = \sigma |\vec{E}|^2 + \frac{\partial}{\partial t}\left(\varepsilon_0 \vec{E}^2 + \mu_0 \vec{B}^2\right), \tag{2.12}$$

that can be reduced using vector calculus and the divergence theorem to

$$-\frac{\partial}{\partial t}\int_V \left(\varepsilon_0 \vec{E}^2 + \mu_0 \vec{B}^2\right) dV = \int_S (\vec{E} \times \vec{B}) d\vec{S} + \int_V \vec{j} \cdot \vec{E} dV, \tag{2.13}$$

This equation shows that the rate of decrease of energy in a volume equals the rate of energy crossing the surface of that volume plus the energy dissipated in the volume (which is a measure of work done by the field).

The Poynting vector $\vec{\Pi}$ can be written as

$$\vec{\Pi} = \vec{E} \times \vec{B}. \tag{2.14}$$

From Eq. (2.13), it can be seen that the integral of the Poynting vector over a closed surface has a physical significance. For a rapidly alternating field the Poynting vector does represent the flow across an isolated area.

One important consequence of Eq. (2.13) follows from the fact that the Poynting vector is parallel to $\vec{E} \times \vec{B}$; thus, the EM wave travels perpendicularly to the magnetic and electric field oscillations. While $\vec{E}$ and $\vec{B}$ are sinusoidal changing direction over time the Poynting vector is unidirectional, but varies from a maximum when $\vec{E}$ and $\vec{B}$ are at a maximum, to a value of zero when $\vec{E}$ or $\vec{B}$ are zero.

2.4.2 POLARIZATION AND THE STOKES VECTOR

Polarization describes the orientation of the oscillations in an EM wave. For typical transverse EM waves the polarization is the orientation of the oscillations perpendicular to the direction of wave travel. Most typical are linear polarizations where the oscillations are oriented in a single direction (either vertical or horizontal). Oscillations may also rotate as the wave travels (circular or elliptical polarizations[1]). These rotational polarizations can rotate either to the right (right-hand criterium) or left (left-hand criterium).[2]

The polarization of an EM wave is described by specifying the direction of the wave's electric field. According to Maxwell's equations, the direction of the magnetic field is uniquely determined for a specific electric field distribution and polarization. The Stokes parameters are a set of values that describe the polarization of EM radiation. George Gabriel Stokes first defined them in 1852 as a simpler approach to the description of incoherent or partially polarized radiation in terms

---

[1]Circular/elliptical polarization refers to the fact that the amplitude is constant or not as the wave propagates.
[2]Right/left-hand criteria refer to the sense of rotation as given by the fingers of the right/left hand, when the thumb points in the direction of propagation of the wave.

of its total intensity (I), (fractional) degree of polarization (p), and the shape parameters of the polarization ellipse.

The relationship of the Stokes parameters to intensity and polarization ellipse parameters is shown in the equations below and Fig. 2.7.

$$S_0 = I, \tag{2.15a}$$

$$S_1 = I \cdot p \cdot cos(2\psi) \cdot cos(2\chi), \tag{2.15b}$$

$$S_2 = I \cdot p \cdot sin(2\psi) \cdot cos(2\chi), \tag{2.15c}$$

$$S_3 = I \cdot p \cdot sin(2\chi), \tag{2.15d}$$

Here $I \cdot p$, 2ψ and 2χ are the spherical coordinates of the polarization state in the three-dimensional space of the last three Stokes parameters. The factor of two before ψ represents the fact that any polarization ellipse is indistinguishable from one rotated by 180 degree, while the factor of two before χ indicates that an ellipse is indistinguishable from one with the semiaxes lengths swapped accompanied by a 90 degree rotation. Alternatively, the following relationships can be derived from Eq. (2.15):

$$I = S_0, \tag{2.16a}$$

$$p = \frac{\sqrt{S_1^2 + S_2^2 + S_3^2}}{S_0}, \tag{2.16b}$$

$$2\psi = atan\left(\frac{S_3}{S_1}\right), \tag{2.16c}$$

$$2\chi = atan\left(\frac{S_3}{\sqrt{S_1^2 + S_2^2}}\right), \tag{2.16d}$$

and are often combined into a vector, known as the Stokes vector:

$$\vec{S} = \begin{bmatrix} S_0 \\ S_1 \\ S_2 \\ S_3 \end{bmatrix} = \begin{bmatrix} I \\ Q \\ U \\ V \end{bmatrix}. \tag{2.17}$$

The Stokes vector operates on intensities and spans the space of unpolarized, partially polarized, and fully polarized light. For comparison, consider the Jones vector (Yeh and Gu, 1999) that represents the relative amplitude and phase of the electric field in the x and y directions. It spans the space of fully polarized light, but is more useful for problems involving coherent light such as interference or diffraction. The four Stokes parameters do not form a preferred basis of the space, but rather were chosen because they can be easily measured or calculated.

The effect of an optical system on the polarization of light can be determined by constructing the Stokes vector for the input light and applying Mueller calculus, to obtain the Stokes vector of the light leaving the system.

$$\vec{S}_0 = \overline{\overline{M}} \cdot \vec{S}_i. \tag{2.18}$$

Similarly, when the light passes through a series of optical systems characterized by the Mueller matrices $\overline{\overline{M}}_1, \overline{\overline{M}}_2, \overline{\overline{M}}_3$, the Stokes vector of the light at the system output is given by:

$$\vec{S}_0 = \overline{\overline{M}}_3 \cdot \overline{\overline{M}}_2 \cdot \overline{\overline{M}}_1 \cdot \vec{S}_2. \tag{2.19}$$

Any Mueller matrix $\overline{\overline{M}}$ can be derived from the so-called Jones matrix $\overline{J}$ by the following relationship:

$$\overline{\overline{M}} = \overline{\overline{A}} \cdot (\overline{J} \otimes \overline{J}^*) \cdot \overline{\overline{A}}^{-1}, \tag{2.20}$$

where * is the complex conjugate, $\otimes$ is the Kronecker product, and

$$\overline{\overline{A}} = \begin{bmatrix} 1 & 0 & 0 & 1 \\ 1 & 0 & 0 & -1 \\ 0 & 1 & 1 & 0 \\ 0 & 1 & -1 & 0 \end{bmatrix}. \tag{2.21}$$

Below are shown some Stokes vectors for common states of polarization of light. Superscript "T" stands for transposed.

Polarization State	Stokes Vector	Jones Vector
Linearly polarized (horizontal)	$[\,1 \quad 1 \quad 0 \quad 0\,]^T$	$[\,1 \quad 0\,]^T$
Linearly polarized (vertical)	$[\,1 \quad -1 \quad 0 \quad 0\,]^T$	$[\,0 \quad 1\,]^T$
Right-hand circularly polarized	$[\,1 \quad 0 \quad 0 \quad -1\,]^T$	$\frac{1}{\sqrt{2}}[\,1 \quad -j\,]^T$
Left-hand circularly polarized	$[\,1 \quad 0 \quad 0 \quad 1\,]^T$	$\frac{1}{\sqrt{2}}[\,1 \quad j\,]^T$
Unpolarized	$[\,1 \quad 0 \quad 0 \quad 0\,]^T$	NA

2.4.3 REFLECTION AND REFRACTION AT THE INTERFACE OF TWO FLAT MEDIA

Whenever an EM wave passes from one medium to another one and the interface is flat there is both reflection and refraction as illustrated in Fig. 2.8.

2.4.4 BREWSTER'S ANGLE

Brewster's angle is also known as the polarization angle, and it is the angle of incidence at which an unpolarized EM wave (containing equal amounts of vertical and horizontal polarization, Fig. 2.9) separates into a vertically polarized EM wave that is transmitted through a surface, leaving the surface

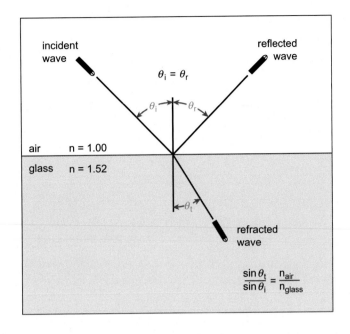

FIGURE 2.8

Reflection and refraction at a simple air/glass interface.

Adapted from http://www.mellesgriot.com/products/optics/oc_2_1.htm.

reflection with only the horizontal components of the incoming radiation. For an incoming vertically polarized EM wave there is no reflection. This angle is named after a Scottish physicist Sir David Brewster (1781−1868). At this angle the light with this particular polarization cannot be reflected as shown here in Fig. 2.9.

The condition for the Brewster's angle can be derived by forcing the reflection to be zero, which in terms of Snell's law can be written as

$$n_1 \cdot sin(\theta_i) = n_2 \cdot sin(\theta_t), \tag{2.22}$$

with $\theta_i = \theta_B$ and $\theta_i + \theta_t = 90°$.

$$n_1 \cdot sin(\theta_B) = n_2 \cdot sin(90° - \theta_B) = n_2 \cdot cos(\theta_B). \tag{2.23}$$

From this, the following expression can be derived:

$$\theta_B = arctan\left(\frac{n_2}{n_1}\right). \tag{2.24}$$

Thus, a dielectric (or a stack of dielectrics) placed at the Brewster's angle can be used as a polarizer.

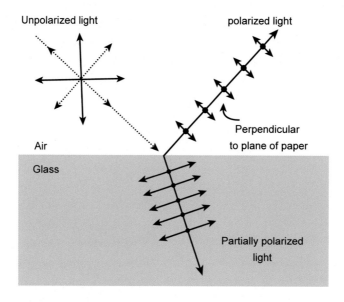

FIGURE 2.9

Illustration of the polarization of unpolarized light passing through and reflected off of glass.

For glass ($n_2 = 1.5$) in air ($n_1 = 1$), Brewster's angle for visible light is approximately 50 degree to the normal while for an air–water interface ($n_2 = 1.33$), it is approximately 53 degree.[3] Since refractive index for a given medium changes depending on the EM wavelength, as the Brewster's angle also varies with wavelength. If one desires to view the thermal emission from the sea surface the Brewster's angle is an optimal viewing angle for this application since a minimum of the longwave solar radiation will be reflected back into the viewing thermal infrared (TIR) radiometer.

2.4.5 CRITICAL ANGLE

If instead of going from a less dense to a more dense medium, our light ray travels from a medium with a greater density to one with lesser density, then the ray is bent away from the normal (Fig. 2.10).

For an EM wave at the critical incidence angle θ_c, the exit angle approaches 90 degree. This critical angle can be computed using Snell's law by setting the refraction exit angle to 90 degree and solving for θ_c, it can be obtained: $\theta_c = arcsin(n_2/n_1)$. Under this condition the EM wave does not exit the incident medium, thus it is called total internal reflection. For any angle of incidence less than the critical, part of the incident light will be transmitted and part will be reflected.

---

[3]This is a typical incidence angle used by many optical and passive microwave sensors, since it makes vertical polarization less sensitive to surface roughness effects.

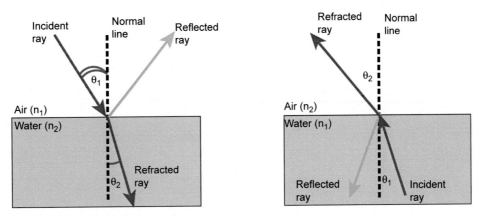

FIGURE 2.10

Condition of an electromagnetic wave traveling to a medium of lesser index of refraction (n).

2.4.6 ALBEDO VERSUS REFLECTANCE

It is important to discriminate between "albedo" and reflectance. The former is defined as the ratio of the amount of EM energy reflected by a surface to the amount of energy incident upon it. It differs from "spectral reflectance" since albedo is usually averaged over the visible range of the EM spectrum while spectral reflectance applies only to a specific region of the EM spectrum. Originally, the albedo (derived from the Latin albedo or whiteness) refers to reflected sunlight containing a full range of wavelengths. So, unless the albedo is restricted to some particular wavelength it refers to some average across the wavelength range of visible light.

Albedo also depends on the directional distribution of the incident radiation with the exception of Lambertian surfaces, which by their nature scatter radiation in all directions and therefore have an albedo that is independent of the incident angle of the irradiance. For other surfaces, it is most appropriate to specify a bidirectional reflectance distribution function (BDRF) to accurately characterize the scattering properties of a surface. It is common, however, to use albedo as a first approximation to the scattering.

Reflectance is the fraction of incident EM radiation that is reflected off a surface. This is in contrast to reflectivity, which is a property of the surface material itself. The spectral reflectance is a plot of the reflectance as a function of wavelength. Reflectivity is a directional property and most surfaces can be divided into those that give specular reflection and those that give diffuse reflection. For a specular surface such as a glass or a metal, the reflectivity will be nearly zero for all angles except for those normal to the surface. For diffuse surfaces such as flat white paint the reflectivity is uniform, and radiation is reflected at all angles equally or nearly equally. Such surfaces are said to be Lambertian. Few surfaces are solely Lambertian or specular and instead have a mixture of these properties (Fig. 2.11).

2.5 ELECTROMAGNETIC SPECTRUM: DISTRIBUTION OF RADIANT ENERGIES

As noted in the previous chapter, EM radiation (EMR) extends over a wide range of energies and wavelengths (frequencies). A narrow range of EMR extending from 0.4 to 0.7 μm, the interval

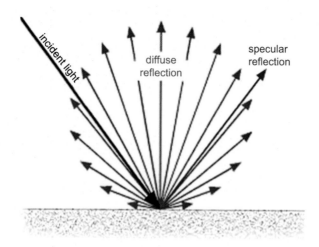

FIGURE 2.11

Lambertian scattering mixed with specular reflection.

detected by the human eye, is known as the *visible region* (also referred to as *visible light* but physicists often use that term to include radiation beyond the visible). White light contains a mix of all wavelengths in the visible region. It was Sir Isaac Newton who first in 1666 carried out an experiment that showed visible light to be a continuous sequence of wavelengths that represented the different colors the eye can see. He passed white light through a glass prism and found that as the radiation passes from one medium to another, it is bent according to what is known today as Snell's law. The index of refraction is dependent on the wavelength, so that the bending angle varies systematically from red (longer wavelength; lower frequency) to blue (shorter wavelength; higher frequency). The process of separating the constituent colors in white light is known as *dispersion* (Fig. 2.12). These phenomena apply to radiation of wavelengths outside the visible (e.g., a crystal's atomic lattice serves as a diffraction device that bends X-rays in different directions). The distribution of the continuum of all radiant energies can be plotted either as a function of wavelength or of frequency in a chart known as the electromagnetic spectrum (Fig. 2.13).

The EM spectral chart in Fig. 2.13 also indicates the atomic or molecular mechanisms for forming these different types of radiation; it also depicts the spectral ranges covered by many of the detector systems in common use today. This figure indicates that EMR, i.e., photon release, is produced in a variety of ways. Most involve actions within the electronic structure of atoms or in movements of atoms within molecular structures (as affected by the type of bonding). The fundamental physics involves changing the direction and/or magnitude of electric and magnetic fields in the source of the EMR in short time intervals.

One common mechanism is to excite an atom by heating or by electron bombardment, which causes electrons in specific orbital levels to momentarily move to higher energy levels; upon dropping back to the original level the energy gained is emitted as radiation at discrete frequencies given by

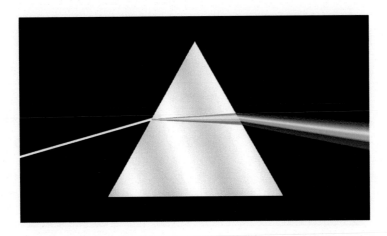

FIGURE 2.12

Prism showing the dispersion of white sunlight.

Eq. (2.9). EM waves are given off whenever such charged particles are accelerated, and these waves can interact with any other charged particles. Such EM waves carry energy, momentum, and angular momentum away from their source particle and can transfer these properties to matter with which they interact. The EM waves are "massless," but they are still affected by gravity. These EM waves continue to "radiate" without the continuing influence of their moving source particles that created them since they have achieved sufficient distance from their original source. This is often referred to as the "far field" as opposed to the "near field" which is still within the radius of influence of the source particle.

In the quantum theory of electromagnetism, EM radiation consists of photons: the elementary particles responsible for all EM interactions. Quantum processes provide additional sources of EM radiation, as mentioned earlier the transition of electrons to lower energy levels in an atom. The energy of an individual photon is quantized and is greater for photons of higher frequency. This relationship is given by Planck's law $E = h \cdot v$, where E is the energy per proton, v is the frequency of the photon, and h is Planck's constant. Thus, a gamma ray photon could carry as much as $\sim 200,000$ times the energy of a single photon of visible light. At high energies even the atom itself can be dissociated, releasing photons of short wavelengths. Photons themselves, in an irradiation mode, are capable of causing atomic or molecular responses in target materials that generate emitted photons (in the reflected light process, the incoming photons that produce the response are not necessarily the same photons that leave the target).

2.5.1 GAMMA, X-RAY, AND ULTRAVIOLET PORTIONS OF THE ELECTROMAGNETIC SPECTRUM

The shortest wavelengths are in the gamma, X-ray, and ultraviolet (UV) parts of the EM spectrum, while the near-infrared (NIR: 0.7−2.5 μm) covers the wavelengths just longer than the visible portion of the EM spectrum. As it will become apparent when explaining Planck's radiation law in

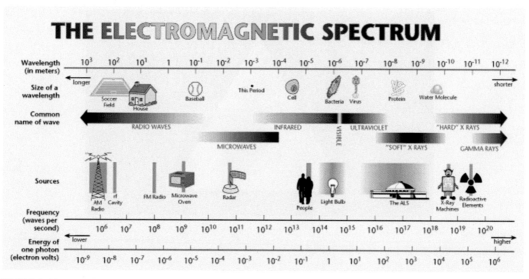

FIGURE 2.13

Electromagnetic spectrum.

Section 2.6.6, in the UV and in the NIR, the radiation received by the sensors is largely dominated by reflected solar radiation, which also dominates the visible wavelengths. They do not contain any thermal emissions, which distinguishes them from the mid- and longwave portions of the infrared (IR) region of the EM spectrum.

NIR radiation was discovered in 1800 by William Herschel, a musician and amateur astronomer who discovered the planet Uranus, because he wanted to know if any particular color was associated with heat from sunlight. He found that the heat maximum was beyond the red end of the spectrum. Herschel could not believe that light and his "radiant heat" were related, but here he was wrong. By 1835, Ampere had demonstrated that the only difference between light and what he named "infrared radiation" was their wavelength. This all came together with James Maxwell in 1864 who wrote.

"This velocity (of electromagnetic force) is so nearly that of light that it seems we have strong reason to conclude that light itself (including radiant heat and other types of radiation) is an electromagnetic disturbance in the form of waves propagated through the electromagnetic field according to electromagnetic laws."

Absorptions in the NIR are generated from fundamental vibrations by two processes: overtones and combinations. Overtones can be thought of as harmonics, so that every fundamental overtone will produce a series of absorptions at integer multiples of the wavelength. Combinations are a bit more complicated and NIR absorptions are at a higher state of excitement so that they require more energy than a fundamental absorption. These combinations arise from the sharing of NIR energy between two or more fundamental absorptions. While the number of possible overtones from a group of fundamental absorptions is limited to a few, a very large number of combinations will be observed.

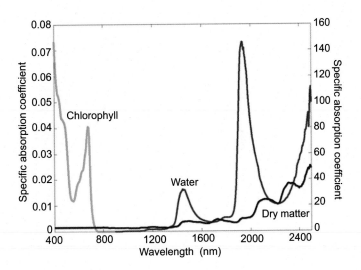

FIGURE 2.14

Specific absorption coefficient of chlorophyll a + b (cm^2/µg) on the left axis, water (cm^{-1}), and dry matter (cm^2/g) on the right axis (Jacquemoud et al., 2000).

One of the primary uses of NIR is to sense the health of the leaves in vegetation. While the visible channels respond to the green of the surface vegetation (response to chlorophyll), the NIR responds to the water carried in the mesophyll structure of the leaves. Thus, in the NIR wavelengths large peaks are observed where the NIR responds to the presence of water (Fig. 2.14). For comparison this plot shows the dry matter (Jacquemoud et al., 1999).

2.5.2 VISIBLE SPECTRUM

The visible portion of the spectrum (Fig. 2.15) includes an extremely limited number of wavelengths that are very important for remote sensing since they are the range of wavelengths viewed by human eyesight and hence the source of a lot of information for mankind. These wavelengths correspond to those of the maximum radiation coming from the Sun, and an atmospheric transmission window, so this radiation can reach the surface of the Earth, and be reflected back to outer space. Most of optical remote sensing is focused on the use of reflected radiation in these relatively few wavelengths. There is a lot of energy in these narrow bands, which are frequently divided into narrow channels to sense the different colors and their responses to radiation reflected by different features on the Earth's surface.

Visible remote sensing ranges from land surface geology and vegetation studies to ocean color imaging of ocean productivity. Each of these applications requires slightly different slices of the visible spectrum and each requires its own level of sensitivity. These various applications and their requirements will be discussed later.

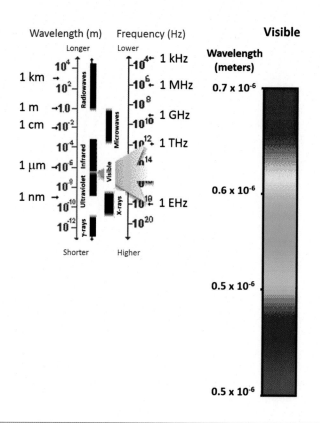

FIGURE 2.15

Visible spectrum.

2.5.3 THERMAL INFRARED SPECTRUM

The TIR portion of the EM spectrum covers wavelengths from 10^{-5} to 10^{-3} m and as shown in Fig. 2.16 includes both reflective and emissive channels. The shortest wavelengths have both reflected and emitted radiation while the longer wavelength channels represent only emitted radiation. One common practice to sort out the combination of reflected and emitted radiation in the shorter wavelength TIR channels is to use nighttime only data which then represents only emitted radiation for comparison with the longer wavelength thermally emitted only data (e.g., channels 3A and 3B in AVHRR3, Table 1.2, Section 1.2.1).

2.5.4 MICROWAVE SPECTRUM

The microwave portion of the EM spectrum refers to those wavelengths longer than those at the IR frequencies typically from 1 to 100 GHz although there are applications for frequencies beyond

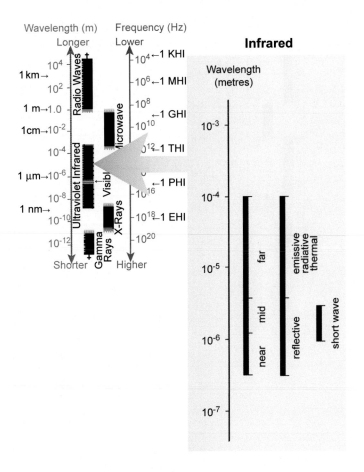

FIGURE 2.16

Infrared spectrum.

100 GHz (millimeter wave frequencies). As shown in Fig. 2.17 the microwave frequencies are separated into a strange assortment of unevenly spaced channels with nonsequential titles. This odd designation of naming and bandwidth size was selected during the Second World War to conceal the identity of this information. While one might think this strange designation would have been dropped after this conflict was over, this naming convention appears to have taken root and lives on in spite of the apparent wisdom to change it. Wavelengths range from 1 mm to 1.0 m. Another characteristic of microwave sensors is that the channels are typically dual-polarized, which corresponds to the first two Stokes elements thus taking advantage of differences seen in the polarization signatures of some land surface features. Other instruments may be fully polarimetric measuring the full Stokes vector. This is useful in discriminating vegetation, snow cover, surface roughness, ocean wave direction, or atmospheric hydrometer shape and orientation.

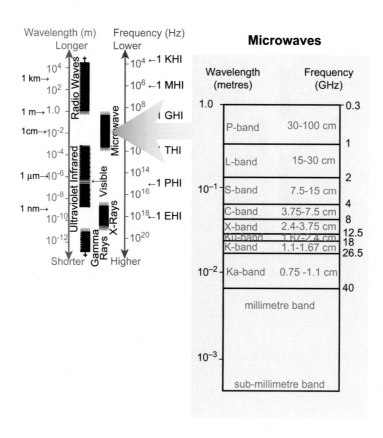

FIGURE 2.17

Microwave spectrum.

2.6 ATMOSPHERIC TRANSMISSION

Most remote sensing is conducted above the Earth either within or above the atmosphere. The gases in the atmosphere interact with solar irradiation, with radiation from the Earth's surface and with natural or artificial microwave radiation (often considered as radio frequency interference). The atmosphere itself is excited by EMR so as to become another source of released photons. Fig. 2.18 is a generic diagram showing relative atmospheric radiation transmission at different wavelengths.

Blue zones mark the minimal passage of incoming and/or outgoing radiation, whereas, white areas denote atmospheric transmission windows, in which the radiation does not interact much with air molecules and hence, is not absorbed or scattered by the atmosphere. Note the narrow "window" channels in the thermal and midrange IR. Also it should be pointed out that for most of the microwave channels the atmosphere is nearly transparent to the microwave radiation except at about 22 and 183 GHz where microwave absorption peaks due to water vapor and 55−60 GHz and 118 GHz where oxygen absorption bands exist.

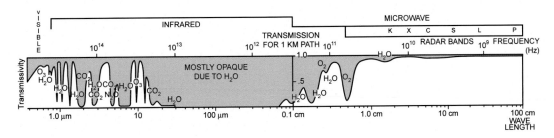

FIGURE 2.18

Atmospheric transmission spectrum.

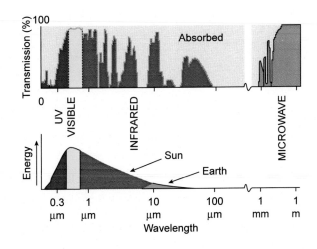

FIGURE 2.19

Atmospheric transmission windows and solar emission spectra.

It is also useful to view the atmospheric transmission windows relative to the solar emissions as presented here in Fig. 2.19.

As it can be seen energy in the UV, visible, and NIR is mostly transmitted through the atmosphere. However, in the midrange IR there are rather narrow windows in which the radiation passes through the atmosphere. As shown in Fig. 2.18, these narrow windows are due to the presence of different gases in the atmosphere. Most of the windows in the midrange and longer wavelength TIR portions of the spectrum are due to water vapor and carbon dioxide in the atmosphere.

In particular, there is a rather large window at about 10 μm, which passes the longer wavelength TIR energy through the atmosphere. This is the primary channel used to remotely sense both sea and land surface temperatures by the IR emissions from the Earth's surface. Due to the nature of IR radiation these emissions are necessarily from the skin of the land and ocean surfaces.

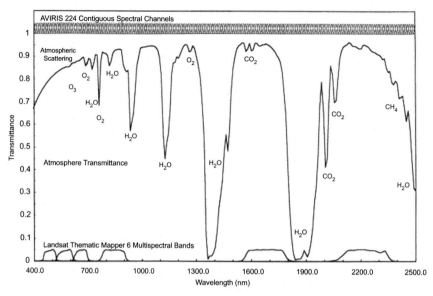

FIGURE 2.20

Atmospheric transmission spectrum for Airborne Visible/Infrared Imaging Spectrometer (blue); and the Landsat Thematic Mapper bands (green).

This next plot (Fig. 2.20), made with the Airborne Visible/Infrared Imaging Spectrometer hyperspectral spectrometer, gives a more detailed spectrum, made in the field looking up into the atmosphere, for the interval 0.4–2.5 μm (converted in the diagram to 400–2500 nm).

Most remote sensing instruments on airborne or space platforms operate in one or more of these windows by making their measurements with optics and detectors tuned to specific frequencies (wavelengths) that pass through the atmosphere. However, some sensors, especially those on meteorological satellites, directly measure absorption phenomena, such as those associated with carbon dioxide, CO_2, and other gaseous molecules. Note that the atmosphere is nearly opaque to EM radiation in part of the mid-IR and all of the far-IR regions. In the microwave region, by contrast, most of this radiation moves unimpeded through the atmosphere, so radar waves may reach the surface and EM emitted from the Earth's surface in the microwave portion of the spectrum can reach the top of the atmosphere and be sensed by satellite radiometers, except for the scattering and attenuation (caused by raindrops) that blocks the transport through the atmosphere and allows the rain to be detected by radars and microwave radiometers.

Backscattering refers to scattering of photons in all directions above the target in the hemisphere that lies on the source side. *Mie scattering* refers to directional scatting of radiation by atmospheric constituents (e.g., smoke aerosols, dust) whose dimensions are of the order of the EM wavelengths. *Rayleigh scattering* is quite isotropic and results from scattering by atmospheric constituents (e.g., molecular gases such as O_2, N_2 as well as other nitrogen compounds, and CO_2, together with water vapor) that are much smaller than the EM wavelengths. Rayleigh scattering increases with decreasing (shorter) wavelengths, causing the preferential scattering of blue light (blue sky effect); however, the red sky tones at sunset and sunrise result from significant absorption of shorter wavelength visible light owing to greater "depth" of the atmospheric path as the Sun is near the horizon. Particles much larger than the irradiation wavelengths give rise to *nonselective (wavelength-independent) scattering*. Atmospheric backscatter can, under certain conditions, account for 80%–90% of the radiant flux observed by a spacecraft sensor.

Remote sensing of the Earth traditionally has used reflected energy in the visible and NIR as well as emitted energy in the TIR and microwave regions to gather radiation that can be analyzed numerically or used to generate images whose variations represent different intensities of photons associated with a range of wavelengths that are received at the sensor. This gathering of a (continuous or discontinuous) range(s) of wavelengths is the essence of what is usually termed multispectral remote sensing. A hyperspectral imaging sensor has many more spectral channels (typically >10) than a multispectral sensor. The bands have narrower bandwidths, enabling the finer spectral characteristics of the targets to be captured by the sensor.

A hyperspectral imaging system is also known as an "imaging spectrometer." It acquires images in about 100 or more contiguous spectral bands. The precise spectral information contained in a hyperspectral image enables better characterization and identification of targets. Hyperspectral images have potential applications in fields such as precision agriculture (e.g. monitoring the types, health, moisture status and maturity of crops), coastal management (e.g. monitoring of phytoplankton, pollution, bathymetry changes).

In the reflected bands, it is also common practice to have a band that integrates radiation over all of the wavelengths, which is referred to as the panchromatic band. The advantage of this channel is that by integrating all of the wavelengths together it responds to a much stronger radiative signal and can have a greater spatial resolution due to the greater intensity of the radiometric signal.

Images made from the varying wavelength/intensity signals will show variations in gray tones in black and white versions or colors (in terms of hue, saturation, and intensity) in colored versions (also called "false color" as these colors are selected to emphasize the gradients and not reflect any true color on the Earth's surface). A pictorial (image) representation of target objects and features in different spectral regions and/or polarizations, usually using different sensors (commonly with band-pass filters) each tuned to accept and process the wave frequencies (wavelengths) that characterize each region, will normally show significant differences in the distribution (patterns) of color or gray tones.

2.6.1 SPECTRAL WINDOWS

The atmosphere has a strong influence on EM radiation, through the phenomena of absorption and scattering. The amount of radiation that *fails to* penetrate is due to the *opacity* of the atmosphere, and varies for radiation at different wavelengths. At certain wavelengths the atmosphere is transparent (or mostly transparent): these wavelengths define "windows" that can be used by space-borne imaging sensors (Fig. 2.21). Wavelengths at which the atmosphere is opaque (or mostly opaque) are used by space-borne sounding sensors to infer vertical properties in the atmosphere.

In practice, a range of atmospheric conditions such as weather (clouds), and dust particles affect the sensors "vision," particularly in tropical and polar regions. In recent years, new sensors have been developed, using microwaves with wavelengths between 1 cm and 1 m that can penetrate the atmosphere in almost all atmospheric conditions.

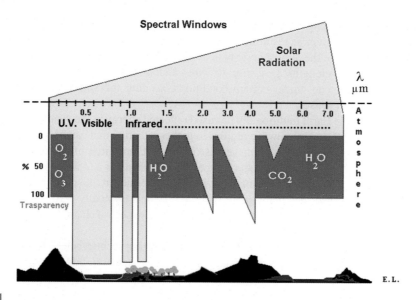

FIGURE 2.21

Atmospheric transmission windows (*marked in yellow*).

2.6.2 ATMOSPHERIC EFFECTS

The atmosphere exerts two principal effects on the radiation passing through it from either direction (down from space or up from the Earth's surface); absorption and scattering. Absorption is the process by which radiation is picked up by atmospheric molecules and atoms increasing the energy in the atmosphere. Scattering is the process by which radiative energy disperses off of atmospheric particles. These particles include gas molecules and atoms along with suspended aerosols and water droplets and ice particles in the atmosphere.

2.6.2.1 Beer–Lambert Absorption Law

Absorption attenuates the radiative signal depending on the constituent gases and aerosols components. The interaction between these atmospheric components and the radiation is expressed by:

$$\alpha = \sum_i m_i \cdot C_{\alpha,i} = \sum_i m_i \cdot C_{\text{geom},i} \cdot \xi_i, \tag{2.25}$$

where m_i is volumetric density, $C_{\alpha,i}$ and $C_{\text{geom},i}$ are the absorption coefficients (dependent on the ith species) and the geometric coefficient respectively, and ξ_i is the absorption efficiency, which can be larger than one.

The processes contributing to this absorption are atomic and molecular processes within the atmospheric gases and aerosol constituents. The atmospheric uptake of photons by shifts in electron energy states generally leads to a fairly rapid reemission of the equivalent energy, but at a different wavelength consistent with the temperature of the gas. Another energy absorption mechanism is the uptake of energy by crystalline processes at the molecular level. The amount of energy absorbed by the mechanism depends on the concentration and types of ions.

The atmospheric effects on the radiant energy transiting it include transmission, reflection, absorption (Fig. 2.21), and scattering. Transmission results only in refraction of the transmission lines, but absorption, as just explained, leads to the attenuation and eventual reemission of the energy.

2.6.2.2 Beer–Lambert Absorption Law: Opacity

Also known as the Beer–Lambert–Bouguer law, this is an empirical relationship that relates the absorption of light to the properties of the material through which the light is traveling. This law states that there is a logarithmic dependence between the transmission (or transmissivity), T, of light through a substance and the product of the absorption coefficient of the substance, and the distance the light travels through the material (i.e., the path length), l (Fig. 2.22).

For a nondispersive medium, the transmissivity can be written as:

$$T = \frac{I}{I_0} = 10^{-k \cdot l}, \tag{2.26}$$

where α is the absorption coefficient (Nep/m), and which can be expressed in terms of the absorptance, A, defined as

$$A = -\log_{10}\left(\frac{I}{I_0}\right). \tag{2.27}$$

Historically the Beer–Lambert law states that absorption is proportional to the light path length, and the concentration of the absorbing species in the medium.

The absorption coefficient A is one of the ways to express the absorption of EM waves. There are many other ways of writing the same relationship. For example, A can be expressed as the imaginary part of the refractive index, k, and the wavelength of the light in free space, λ_0 as;

$$A = \frac{4\pi k}{\lambda_0}, \tag{2.28}$$

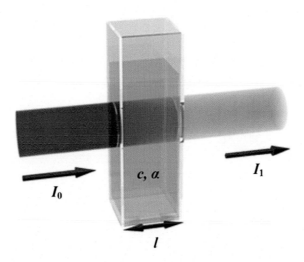

FIGURE 2.22

Diagram of Beer–Lambert absorption of light as it travels through a medium of width l.

There are five conditions that must be met for the Beer–Lambert law to be valid:

1. The absorbers must act independently of each other;
2. The absorbing medium must be homogeneously distributed in the interaction volume and must not scatter the radiation;
3. The incident radiation must consist of parallel rays, each traversing the same length in the absorbing medium;
4. The incident radiation should preferably be monochromatic, or have at least a width that is narrower than the absorbing transition; and
5. The incident flux must not influence the atoms or molecules; it should only act as a noninvasive probe of the species under study. In particular, this implies that the light should not cause optical saturation or optical pumping, since such effects will deplete the lower level and possibly give rise to stimulated emission.

2.6.2.3 Atmospheric Scattering

Fig. 2.23 summarizes the atmospheric effects on radiation. Scattering is represented by an isotropic distribution of the incoming radiation over the upper hemisphere. Depending on the particle size the three different scattering mechanisms mentioned previously must be considered.

Rayleigh scattering occurs when particles are very small as compared to the wavelength of the radiation. These could be particles such as small specks of dust or nitrogen and oxygen molecules. Rayleigh scattering causes shorter wavelengths of energy to be scattered much more than longer wavelengths. Rayleigh scattering is the dominant scattering mechanism in the upper atmosphere. The fact that the sky appears "blue" during the day is because of this phenomenon. As sunlight passes

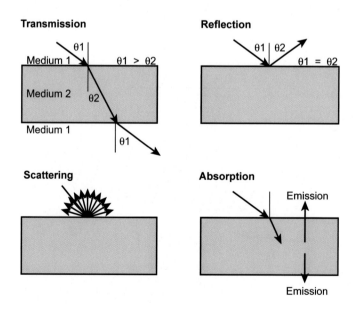

FIGURE 2.23

Summary of atmospheric effects on radiation.

through the atmosphere, the shorter wavelengths (i.e., blue) of the visible spectrum are scattered more than the other (longer) visible wavelengths. At sunrise and sunset, the light has to travel farther through the atmosphere than at midday and the scattering of the shorter wavelengths is more complete; this leaves a greater proportion of the longer wavelengths to penetrate the atmosphere and are less attenuated resulting in the yellow and red colors of sunrise and sunset.

Radiances associated with Rayleigh scattering are proportional to the inverse of the radiation wavelength to the fourth power and exhibit a wide scattering pattern as shown in Eq. (2.29):

$$\beta_r(\theta, \lambda) \propto \frac{1}{\lambda^4}\left(1 + \cos^2\theta\right). \tag{2.29}$$

Here β_r is the Rayleigh scattering coefficient, θ is the angle with respect to a reference line (Fig. 2.6), and λ is the wavelength of the radiation.

Mie scattering occurs when the particles are just about the same size as the wavelength of the radiation. Dust, pollen, smoke, and clouds are common causes of Mie scattering, which tends to affect longer wavelengths than those affected by Rayleigh scattering. Mie scattering occurs mostly in the lower portions of the atmosphere where larger particles are more abundant, and dominates when cloud conditions are overcast. Mie scattering is mostly forward scattering and the radiance associated with this scattering mechanism is highly variable as given here in Eq. (2.30):

$$\beta_r(\theta, \lambda) \propto \lambda^m f(\theta), \ m = 0.6... - 2. \tag{2.30}$$

The final form of scattering is referred to as nonselective scattering and it holds for particles that are much larger than the wavelength of the radiation. Water droplets, ice, and large dust particles can cause this type of scattering. Nonselective scattering gets its name from the fact that all wavelengths are scattered about equally. This type of scattering causes fog and clouds to appear white to our eyes because blue, green, and red light are all wavelengths scattered in approximately equal quantities (blue + green + red light = white light). This type of scattering is generally isotropic.

These atmospheric scattering processes are summarized in Fig. 2.24 as a function of EM wavelength and particle radius.

2.7 SENSORS TO MEASURE PARAMETERS OF THE EARTH'S SURFACE

The mechanisms of radiation leaving the Earth's surface are summarized here in Fig. 2.25 that covers both the radiation stimulated by incoming solar insolation (reflected radiation) and that in response to Earth's surface temperature (emitted radiation). These reflections and emissions are summarized by *colored arrows* that represent the various processes. Here the colors are only intended to discriminate the processes and do not refer to any color of visible radiation.

The incoming solar radiation transmitted through the atmosphere may reach the sensor after a direct reflection (scattering) from the Earth's surface such as on a clear day (A), or after being diffused in the atmosphere (B) such as in a foggy or cloudy sky. Under these conditions, solar radiation is also reflected (scattered) in the atmosphere (C), such as from the tops of the clouds, or after multiple reflections in the environment (G). On the other hand, the Earth's surface emits radiation according to the absolute surface temperature that can reach the sensor directly (D), or after several reflections (H).

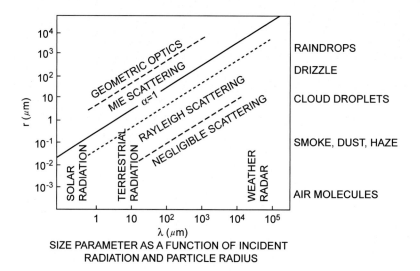

FIGURE 2.24

Summary of the atmospheric scattering mechanisms as a function of radiation wavelength and atmospheric size parameter. Also shown are the atmospheric constituents that contribute to these scattering processes.

Similarly, the atmosphere also emits, and this radiation may reach the sensor directly (F), or after a reflection on the surface of the Earth (E).

This set of mechanisms maps into remote sensing instrumentation in Eq. (2.31) where L indicates radiance and the subscripts refer to the mechanisms in Fig. 2.25.

$$\text{VIS/NIR} \qquad L \approx L_A \quad + L_B \quad + L_C \qquad\qquad\qquad\qquad + L_G \qquad\qquad (2.31a)$$

$$\text{LWIR} \qquad L \approx \qquad\qquad\qquad\qquad L_D \quad + L_E \ + L_F \qquad\quad + L_H \qquad (2.31b)$$

$$\text{MWIR} \qquad L \approx L_A \quad + L_B \ + L_C \ + L_D \qquad + L_E \quad + L_F \ + L_G \ + L_H \qquad (2.31c)$$

$$\text{Microwaves} \qquad L \approx L_A \qquad\qquad L_D \qquad\quad + L_E \ + L_F \qquad\qquad\qquad (2.31d)$$

As it will be shown later, under the Rayleigh—Jeans approximation, at microwave frequencies, L is proportional to the physical temperature T of the emitting body and therefore:

$$T \approx T_A + T_D + T_E + T_F, \qquad\qquad (2.32a)$$

$$T_{AP} \approx T_{Sun} + T_B + T_{SC} + T_{UP}. \qquad\qquad (2.32b)$$

Here VIS/NIR refers to visible and near-infrared radiation dominated by the direct, reflect, and scattered radiation from the Sun. MWIR is the midrange thermal infrared where both solar and Earth emissions are comparable; LWIR refers to the longwave thermal infrared channels in which the Earth's

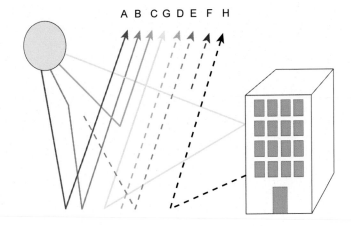

(A) Solar irradiance transmitted through the atmosphere and reflected by Earth's surface

(B) Solar radiation diffused by the atmosphere reflected by the Earth's surface

(C) Upwelling solar radiance diffused by the atmosphere

(D) Spontaneous thermal emission of the Earth's surface

(E) Downwelling atmospheric radiance (spontaneous emission) reflected at the Earth's surface

(F) Upwelling atmospheric radiation (spontaneous emission)

(G) Reflected solar irradiance reaching the sensor through multiple scattering

(H) Spontaneous thermal emission of bodies reaching the sensor through multiple scattering

FIGURE 2.25

Contributions to the radiance reaching a space sensor (Schott, 2007).

surface emission dominates, and the microwave region over which there a radiation components from both the Earth and the Sun. In Eq. (2.32) the radiances at microwave frequencies are related to temperatures (T) where the subscript AP refers now to "apparent" antenna temperature due to radiation reaching the radiometer antenna, Sun refers to the radiation coming from the Sun, B refers to the "brightness" temperature of the surface, SC refers to the downwelling (DN) radiation coming from the atmosphere and the Galactic and cosmic background that are "scattered" on the Earth's surface and attenuated by the atmosphere in the upwelling path, and UP refers to the upwelling atmospheric radiation.

2.8 INCOMING SOLAR RADIATION

The radiative energy coming from the Sun to the Earth is the ultimate source of energy for all natural processes that will be considered. The spectrum of the Sun's emissions is compared to the atmospheric windows we introduced earlier (Fig. 2.26) and we see that the highest energy windows are located at the shorter wavelengths while the wider windows, in terms of wavelength band, are located in the longer TIR wavelengths.

It is important to realize that even the very narrow windows in the shortwave part of the spectrum transmit a lot of solar radiation to the Earth's surface where about 50% of this shortwave radiation makes it to the Earth's surface. Here the radiation is converted into heat or biomass or some other form

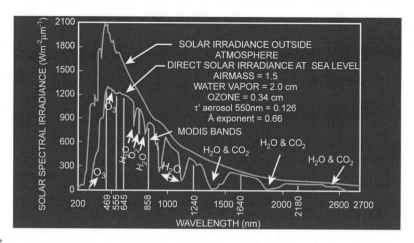

FIGURE 2.26

Top of the atmosphere solar irradiance spectrum outside the atmosphere (*blue line*) compared with atmospheric transmission windows (*yellow line*, https://objectivistindividualist.blogspot.com/2013/02/infrared-absorbing-gases-and-earths.html).

of energy equivalent. A much smaller amount is reradiated by the Earth (surface temperature of 300K) back into the atmosphere and up into space (Fig. 2.26).

2.9 INFRARED EMISSIONS

The radiation emitted by a surface is governed by Planck's law, which applies to theoretical black bodies which are ideal materials that absorb all the incident radiation at all frequencies and polarizations, and in thermodynamic equilibrium reemits it following Planck's law which is:

$$M_{\lambda,BB} = \frac{2 \cdot \pi \cdot h \cdot c^2}{\lambda^5} \cdot \frac{1}{e^{\frac{h \cdot c}{\lambda \cdot k_B \cdot T}} - 1} \quad \left[W/(m^2 \, \mu m) \right], \tag{2.33}$$

where k_B is the Boltzmann's constant ($1.3806 \times 10^{-23} \, m^2 \, kg \, s^{-2} \, K^{-1}$). The spectral curves for Planck's law for the Sun and Earth (considered as black bodies at 6000K and 288K, respectively) are presented here in Fig. 2.27.

In reality, no bodies are actually black bodies, and despite there are close approximations to reference black bodies, the reality is that all bodies are really "gray bodies" and do not emit all of the radiation consistently with their absolute temperature. The emissivity (e) of a body represents how well a real body approximates a blackbody. It is expressed as the ratio of the actual radiation at a given wavelength to that of a blackbody with the same absolute temperature:

$$0 \le e(\theta, \varphi, \lambda) = \frac{M_\lambda(\theta, \varphi, \lambda, T)}{M_{\lambda,BB}(\lambda, T)} \le 1, \tag{2.34}$$

where θ is the elevation angle and φ is the azimuth angle.

The wavelength of the maximum in Fig. 2.28 is a function of the absolute temperature and it expressed by Wien's Displacement law:

$$\lambda_{max} \cdot T = 2898 \, (\mu m \, K). \tag{2.35}$$

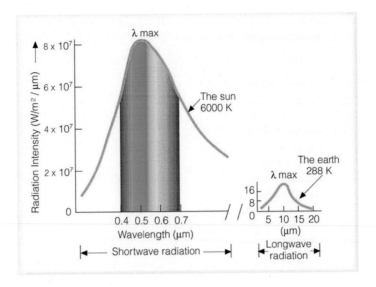

FIGURE 2.27

Planck emission curves for the Sun and Earth (*solid black*).

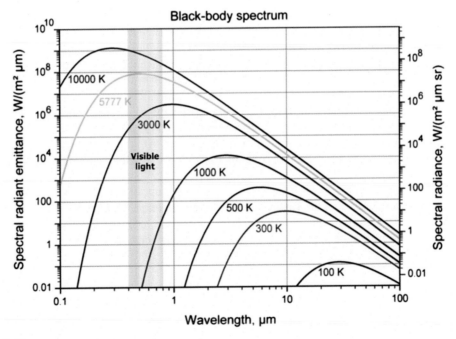

FIGURE 2.28

Emission curves illustrating Wien's displacement law.

Table 2.2 Emissivities for the 8–12 μm Portion of the Electromagnetic Spectrum

Material	Emissivity, e
Polished metal surface	0.006
Granite	0.815
Quartz sand, large grains	0.914
Dolomite, polished	0.929
Basalt, rough	0.934
Asphalt paving	0.959
Concrete walkway	0.966
A coat of flat black paint	0.970
Water, with a thin film of petroleum	0.972
Water, pure	0.993

This relationship is expressed graphically in Fig. 2.28 where the different curves represent different surface covers at various absolute temperatures ranging from Arctic ice at 220K to the Sun at 6000K. At the top of this diagram are the designations to those portions of the wavelength spectrum. Notice the large range of TIR wavelengths particularly as compared to the fairly narrow range of visible wavelengths.

Emissivity is a characteristic of the surface of the material. Emissivity is a measure of a material to both radiate and absorb energy. Recall that a blackbody is by definition a perfect emitter and a perfect absorber. High emissivities indicate materials that absorb and emit large proportions on incident radiation. Some emissivities are given here for various materials in Table 2.2 for the 8–12 μm region of EM spectrum, which covers the TIR portion of the spectrum.

Hence two surfaces with the same kinetic temperature, but different emissivities will have different radiant temperatures. Since remote sensing instrumentation can only observe the emitted temperature the emissivity of the surface material must be known to derive the kinetic temperature of the surface. When looking at the ocean's surface it is common to assume a very high emissivity, as suggested by Table 2.2.

Two other approximations to Planck's law that are important in remote sensing are the Stephan–Boltzmann's law and the Rayleigh–Jeans approximation. The latter applies to the microwave portion of the spectrum, which corresponds to the longer wavelengths shown in Figs. 2.21 and 2.22 as discussed earlier in this chapter (Eq. 2.32). This approximation is written as:

$$M_{\lambda,BB} \approx 2\pi \frac{c}{\lambda^4} \cdot k_B \cdot T \ \left[\mathrm{W}/(\mathrm{m}^2 \, \mu\mathrm{m}) \right], \tag{2.36}$$

where k_B is the Boltzmann's constant as given earlier. This equation is accurate to within 1% for frequencies $f \leq 112$ GHz and temperatures $T \leq 300$K. This equation makes it possible to compute microwave emissions from the knowledge of the absolute temperature (T). Similarly, for the TIR portion of the EM spectrum we can use the Stephan–Boltzmann's law:

$$M_{BB} = \frac{2\pi^5 \cdot k_B^4}{15 \cdot c^2 \cdot h^3} \cdot T^4 = \sigma \cdot T^4 \ \left(\mathrm{W}/\mathrm{m}^2 \right), \tag{2.37}$$

where $\sigma = 5.6698 \times 10^{-8}$ [W/(m² K⁴)] is Stefan–Boltzmann's constant.

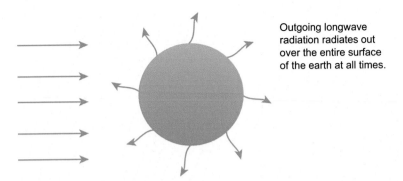

Outgoing longwave radiation radiates out over the entire surface of the earth at all times.

Incoming shortwave solar radiation shines on only hone-half of the earth at a time. This area if found as πR^2 where R is the radius of the earth. Thus, the heat flux received by the earth is 1.9 cal/cm²min •πR^2. The longwave radiation emitted from the earth is over the entire surface of the "sphere" with an area of $4\pi R^2$ the area of the sphere. Thus, the heat flux out is 0.49 cal/cm²min •$4\pi R^2$ which is approximately equal to the incoming heat flux and the temperature of the earth remains constant.

FIGURE 2.29

Radiative thermal equilibrium of the Earth's surface.

Over the annual cycle the incoming solar energy that makes it to the Earth's surface (about 50%) is balanced by the outgoing TIR energy emitted from the Earth's surface. While this emitted energy is at a much lower temperature than the incoming solar radiation it should be remembered that the incoming solar radiation only illuminates at a time (Fig. 2.29) while the Earth emits radiation over its whole surface. Thus, the thermal equilibrium is maintained between this incoming shorter wave solar radiation and the longer wavelength thermal emission from the Earth's surface.

The atmosphere transmits, absorbs (mainly by H_2O, dust, O_3), reflects (by clouds), and scatters radiation (by aerosols and gases); the Earth's surface reflects and absorbs radiation both short and long waves. The Earth reemits longwave radiation, which is selectively transmitted, reflected or absorbed by the atmosphere.

Examples of surface and atmospheric components are shown here in Fig. 2.30 which are composite images taken from geostationary satellite imagers. On the right is the "outgoing longwave radiation" which is the longwave radiation emitted by the Earth's surface while at the left is the higher peak representing the shorter wavelength incoming energy from the Sun.

The hemispherical image on the right has been separated in half with the right half showing the atmospheric water vapor content and the left half the outgoing longwave radiation. Land surface, clouds and water vapor have been individually labeled along with their average constituencies of the atmosphere on the right side. On the left clouds, surface and air (atmosphere) have been labeled to demonstrate the appearance of these features in reflected visible satellite imagery.

There are a few other radiative transfer terms that need to be defined in working with satellite data and atmospheric effects. The first is transmittance, which is the amount of radiant energy that is transmitted through the atmosphere

$$\tau(\lambda) = \frac{M_\tau}{E_i},$$

(2.38)

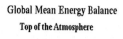

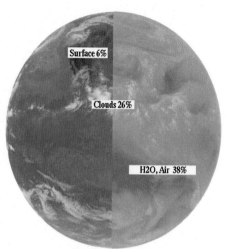

Global Mean Energy Balance
Top of the Atmosphere

Surface 6%

Surface 4%

Air 6%

Clouds 20%

Clouds 26%

H2O, Air 38%

Reflected Solar Radiation (30%)   Outgoing Infrared Radiation (70%)

NOAA/NESDIS/ORA/ARAD/ASPT

FIGURE 2.30

Reflected solar and outgoing longwave (infrared) radiation from the GOES satellite imager.

where M_τ is the radiant energy transferred by the atmosphere and the E_i represents the incoming radiation at the top of the atmosphere. Likewise, reflectivity, which represents the amount of radiant energy reflected, is expressed as

$$r(\lambda) = \frac{M_r}{E_i},$$
(2.39)

where the subscript "r" represents the reflected portion of the incoming radiation. Finally, the absorptivity is the amount of radiant energy absorbed by the atmosphere

$$\alpha(\lambda) = \frac{M_\lambda}{E_i},$$
(2.40)

Kirchhoff's law states that the absorptivity and emissivity are equal ($\alpha = e$), while in thermodynamic equilibrium the conservation of energy requires that

$$e + \tau + r = 1.$$
(2.41)

Real emission spectra show some variations from Planck's law that represent the different atmospheric constituent contributions. An example is presented here in Fig. 2.31 for a typical surface emission.

The CO_2 window is fairly wide revealing that it is rather well spread in the atmosphere as is water vapor. The ozone line is much narrower reflecting its variability in time and space.

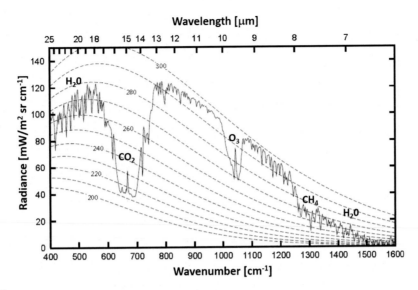

FIGURE 2.31

Earth surface emissions spectrum showing water vapor, ozone and carbon dioxide windows (http://acmg. seas.harvard.edu/people/faculty/djj/book/bookchap7.html).

2.10 SURFACE REFLECTANCE: LAND TARGETS

The primary remote sensing signal from the land surface takes the form of reflected visible energy. The amount and wavelength of reflectance depends on the character of the surface in terms of material and the amount of water content. Some examples of this reflectance as a function of wavelength are presented here in Fig. 2.32 where different curves represent different surface features displaying differing reflectances as a function of wavelength.

Reflectivity depends on absorption and scattering mechanisms that take place within the material. The absorption processes are electronic processes (absorption of a photon and reemission at a longer wavelength, i.e., heating), crystalline effects (atomic energy levels splitting the crystalline net, which also depends on the concentration of a given type of ion), and absorptions by charge transfer (inter-element transitions when a photon absorption causes the movement of an electron among ions). Scattering depends on the type of mixture between the reflecting materials, and also on the grain size where the larger the grain size, the larger the light absorption, and the lower the reflectance.

Notice that water bodies, either turbid or clear have relatively small reflectance in the shorter visible wavelengths and no reflectance at the longer wavelengths. As it is clear in Fig. 2.32 the amount of reflectance from a water surface is restricted to the shorter visible and NIR wavelengths. This reflectance is also a function of the amount of suspended matter in the water with clear water having less reflectance than turbid waters. The reflectance also depends on the phase of the water with a measurable increase in reflectance with the presence of ice and the actual crystalline form of the ice.

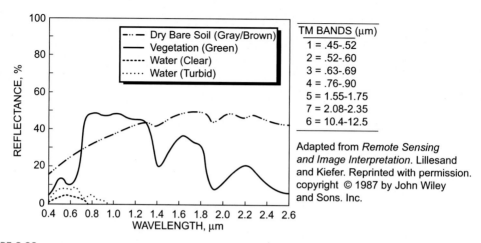

FIGURE 2.32

Reflectance of various surfaces as a function of wavelength.

From Hanel, R.A., Conrath, B.J., Kunde, V.G., Prabhakara, C., Revah, I., Salomonson, V.V., Wolford, G., 1972. The Nimbus 4 infrared spectroscopy experiment: 1. Calibrated thermal emission spectra. J. Geophys. Res. 77 (15), 2629–2641. http://dx.doi.org/doi:10.1029/JC077i015p02629.

Turbid waters lead to an increase in the visible reflectance which is the source of ocean color reflectance which has led to a whole new area of remote sensing of ocean color as an indicator of ocean primary productivity. The color largely reflects the presence of chlorophyll in the ocean as represented by the green reflectance from the upper ocean. Other biologically active substances have different signatures that complicate the ocean color signatures. Coastal waters with their sediment loads and frequent ice components further complicate the mapping of ocean color signatures in the coastal domain where the biological activity is generally most important.

This reflectance represents the characteristics of the upper few meters of the ocean rather than just the upper few microns as is reflected by the thermal emissions responsible for the remote sensing of IR sea surface temperatures. Another element that strongly alters the reflectance in the visible range is the amount of sediment content in the water. These features are only found in coastal regions usually associated with river discharges.

Dry bare soil has a relatively high reflectance in the visible, which increases going to the longer wavelengths. Green vegetation has a small peak in the green portions of the visible band, but shows the greatest reflectance in the NIR and midrange IR wavelengths.

The NIR reflectance can be used to discriminate between different types of vegetation such as coniferous and deciduous trees (Fig. 2.33). While both types of trees have very similar visible reflectances the coniferous trees do not attain as high values in the NIR beyond 0.7 μm as do the deciduous trees. This is because the coniferous trees simply maintain the same level of leaf health throughout the year while the deciduous trees go from a leaf-off stage in winter to a new and healthy leaf stage in summer. This increase in NIR reflectance dominates at the longer wavelengths near 0.9 μm.

Reflectance due to the presence of soil and rocks is moderate ($r \leq 0.3$) depending on the amount of vegetation covering it. Soil grain size and surface roughness also influence the amount of reflectance and shift the reflectance to the longer midrange IR channels. The relative constancy of the soil signature as compared to the two types of vegetation is seen in Fig. 2.34, which plots the spectral response of soil, dry and green vegetation reflectances as functions of wavelength.

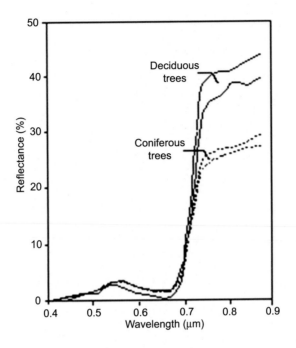

FIGURE 2.33

Canopy reflectance for coniferous and deciduous trees.

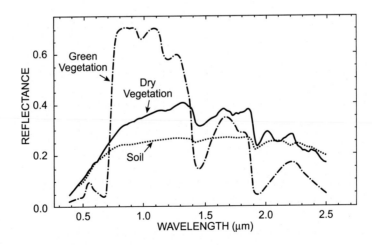

FIGURE 2.34

Vegetation and soil reflectance spectra.

After Clark, R.N., 1999. Spectroscopy of rocks and mineral, and principles of spectroscopy. In: Chapter 1: Manual of Remote Sensing. U.S. Geological Survey, MS 964, Fig. 1.18.

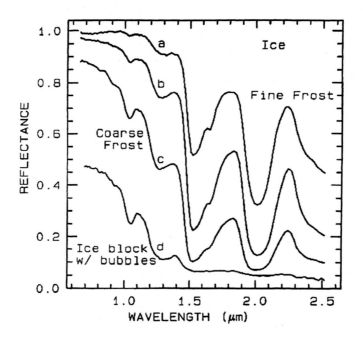

FIGURE 2.35

Reflectance of ice as a function of wavelength, ice type and electromagnetic wavelength (grain sizes: a = 50 μm, b = 200 μm, c = 400–2000 μm; http://speclab.cr.usgs.gov/PAPERS.refl-mrs/refl4.html).

Soil and dry vegetation show similar linear increases with wavelength at the shorter wavelengths up to about 0.65 μm. At this point the soil reflectance curve begins to flatten out at about 0.25, while the dry vegetation continues to increase to a maximum of about 0.4 at 1.3 μm before it begins a gradual decrease marked by a few minor subsequent maxima and minima, associated to the presence of water.

One of these at about 1.4 μm coincides with a major drop in the reflectance spectrum for green vegetation that has a maximum at about 0.8 μm. This peak follows a strong increase in reflectivity at about 0.7 μm. The peak in green vegetation reflectivity is rather broad covering the wavelength range from 0.7 to 1.2 μm. This represents a response to the health of the mesophyll structure in the leaves. Following the drop at 1.4 μm the green vegetation curve has a secondary maximum between 1.7 and 1.8 μm followed by a minimum at 1.9 μm. All three spectral curves exhibit a fall off towards the longer wavelengths.

Ice leads to a greater reflectance in the slightly longer wavelengths as shown here in Fig. 2.35. In this figure the different curves correspond to different wavelengths of the EM spectrum that would be used to sense the surface properties.

Ice types in Fig. 2.35 range from ice blocks to fine frost and the reflectance exhibits a rather large range. In curves a—c there is a sharp drop-off in reflectance at about 1.5 and 2.0 μm due to the water absorption.

2.10.1 LAND SURFACE MIXTURES

While it is nice to think of the surface as composite of various orthogonal surface types, that is again an abstraction. In reality the surface types are very mixed in space and time and any one scene needs to be

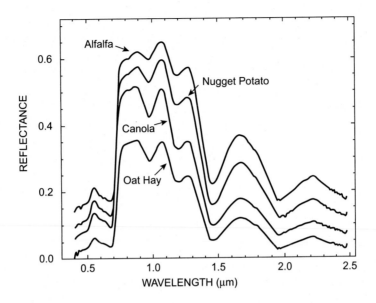

FIGURE 2.36

Reflectance spectra of four types of vegetation. Each curve is offset by 0.05 from the one below.

From Clark, R.N., 1999. Chapter 1: Spectroscopy of rocks and minerals, and principles of spectroscopy. In: Rencz, A.N. (Ed.), Manual of Remote Sensing, Volume 3, Remote Sensing for the Earth Sciences, John Wiley and Sons, New York, pp. 3–58.

thought of as a mixture of individual land cover types. In general, there are four types of mixtures. Linear mixtures where individual materials can be optically distinguished and there is no multiple scattering between components. Intimate mixtures where different materials are in close contact on a scattering surface. Depending on the optical properties of each component the resulting reflected signal is a highly nonlinear combination of the end-member spectra. Coating occurs when one material lies over another. Both coatings are scattering/transmitting depending on the optical properties of the material and the wavelength of radiation. Molecular mixtures occur on a molecular level such as a combination of two liquids or a liquid and a solid mixed together. Examples are water absorbed onto a mineral or gasoline spilled onto soil.

One of the most challenging set of spectral features to distinguish between is those of various plant species. Fig. 2.36 shows four plant spectra (offset for clarity) with shapes that are very similar. If the common background (often called the continuum) is removed, the chlorophyll absorption spectra for these and other plants can be seen in Fig. 2.37. Spectral shape matching algorithms such as those discussed in Clark et al. (1990) can discriminate between these spectra and compare them to spectral libraries.

2.11 STUDY QUESTIONS

1. A blackbody at 350K emits radiation.
   a. Determine the total radiance emitted by the object.
   b. Determine the radiance given off at a wavelength of 12 μm.
   c. Determine the wavelength and frequency of maximum emitted radiance.
   d. In which portion of the EM spectrum is this radiation?

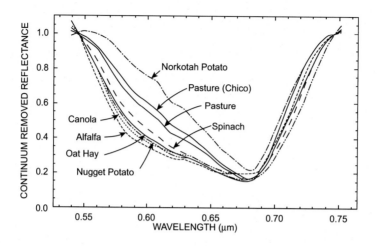

FIGURE 2.37

Continuum-removed chlorophyll absorption for eight vegetation types (including the four from Fig. 2.36) showing that the continuum-removed features can show subtle spectral differences.

From Clark, T.L., 1995. Marshall Space Flight Center Electromagnetic Compatibility Design & Interference Control (MEDIC) Handbook. CDDF Final Report, Project No. 93–15.

2. Another object at a temperature of 350K has an absorptance of 0.90. How much radiance is it giving off at a wavelength of 12 μm?

3. A disk of radius "a" is emitting radiation with intensity I_0. The disk is observed from a point at a distance "z" from the disk (the distance is measured from the center of the disk in a direction normal to the plane of the disk). Show that in the case of a large distance "z," the flux density to the disk at the observation point is given by

$$F = \pi \cdot I_0 \cdot \frac{a^2}{z^2}.$$

4. The atmosphere is found to have an absorption coefficient of 5×10^6 m^{-1} and a scattering coefficient equal to 1×10^6 m^{-1}. If the intensity of radiation entering a 10 km long section of the atmosphere is given by I_0, determine
   a. The optical depth of the section of the atmosphere.
   b. The transmissivity of the section of the atmosphere.
   c. The intensity of radiation leaving the section of the atmosphere.

5. A patch of ocean has a temperature of 25°C and an emissivity of 0.97 for radiation at a wavelength of 10 μm. Scattering may be neglected at this wavelength, but the atmosphere is considered to have an absorption coefficient of 5×10^6 m^{-1}. If the atmosphere is isothermal with a temperature of 20°C, determine the brightness temperature measured by a radiometer flying in an airplane at an altitude of 5 km above the surface. Consider absorption and emission in the atmosphere, assume the atmosphere emits as a blackbody, and assume the radiometer senses radiation only at a wavelength of 10 μm.

6. Which type of imager provides better sensitivity and why: a whisk-broom scanner or a push-broom? Are there other advantages of one type in front of the other? Please provide examples of missions using these types of scanners.

7. A lidar transmits laser pulses and measures the light radiation scattered back, and collected by a telescope. If the scattering is elastic (scattered radiation has the same wavelength as the transmitted one), and the wavelength of operation is green, to which type of atmospheric constituents it will respond? Why?

8. In a lidar, if the received signal is too weak, the signal-to-noise ratio can be increased by (mark all that apply):
   a. Increasing the transmitted power.
   b. Increasing the duration of the transmitted pulses.
   c. Increasing the aperture of the optical system (typically a telescope).
   d. Using a more sensitive photodetector.

9. The microwave part of the spectrum is nearly transparent, except for some resonances around 22 and 183 GHz, and 55−60 and 118 GHz, which are due to water vapor and oxygen. Will these bands be used for radar systems? Why?

10. In the abovementioned bands the increased atmospheric attenuation translates into an increased atmospheric emission. Which parameters can be inferred from?

OPTICAL IMAGING SYSTEMS

3.1 PHYSICAL MEASUREMENT PRINCIPLES

Optical remote sensing spans the range of electromagnetic wavelengths from the visible to the thermal infrared (IR) making images from both reflected and emitted radiation from the Earth's surface and overlying atmosphere. The various types of surface and atmospheric targets that both reflect and emit radiation are summarized here in Fig. 3.1. Note that the arrows going to the spacecraft represent both reflected and emitted radiation.

In many cases the characteristics of the surface targets are such that they can be discriminated by their optical spectral signatures. Land surface targets are generally imaged with reflected radiation from visible and near-IR wavelengths while the ocean's surface is usually imaged with the thermal IR since the interest is more in the ocean's surface temperature which is related to the thermally emitted radiation. In addition, the land does not retain heat the way the ocean does, so thermal imaging of the land surface is not as meaningful as that from the ocean's surface. Since the 1980s the measurement of ocean color features has made it possible to use the narrow reflected bands to estimate chlorophyll and other related substances in the ocean. It should be emphasized here that the ocean biological parameters are so

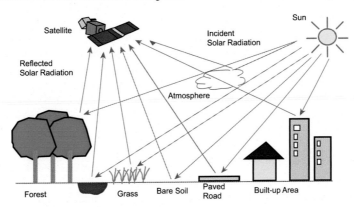

FIGURE 3.1

Earth surface targets for optical remote sensing (http://www.crisp.nus.edu.sg/~research/tutorial/optical.htm).

Liew, S.C. Principles of Remote Sensing, Centre for Remote Imaging, Sensing and Processing, National University of Singapore.

http://www.crisp.nus.edu.sg/~research/tutorial/rsmain.htm

Introduction to Satellite Remote Sensing. http://dx.doi.org/10.1016/B978-0-12-809254-5.00003-8

weak that a typical land sensor would "see" no gradients in the ocean. Also the ocean color signals are so weak that the measurements must be corrected for atmospheric attenuation, which is actually greater than the ocean color signal itself, and is even made worse for increased surface roughness.

Land targets range from bright surfaces such as roads, buildings, and parking areas to dark targets such as forests and crop fields ripe for harvest. In addition, land targets comprise bare soil, short grass, and shrubs. These surface subjects are again sensed by the reflected optical bands and require a high spatial resolution to discriminate important spatial changes in the nature of surface cover. Considerable effort has been spent in developing techniques to analyze images of the Earth's surface for different types of land cover and land use changes.

Visible channel sensing of the ocean's surface has been restricted to the study of biological productivity, which can only be sensed with narrowbands in the visible range of the EM spectrum. Even given these narrowbands the ocean primary productivity signal is so weak that it is masked by the atmospheric attenuation of these reflected signals from the ocean's surface. To compensate for this attenuation a global reference is used to correct for the atmospheric component. This works quite well in the low-productivity waters of the open ocean, but fails in the complex waters of the productive coastal regions known as case 2 waters. Here even the derived "ocean-color" algorithms fail to perform well against in situ calibration sources (Morel and Prieur, 1977).

One problem unique to ocean color measurements is "Sun glitter" which is caused by sunlight reflecting in a specular fashion from certain ocean waves to the sensor. This direct specular reflection is much stronger than the normally reflected signal and it easily blots out the ocean color signal of interest to the remote sensor. Since most of this Sun glitter is caused by the shortest ocean capillary waves caused by winds, it can be estimated from the knowledge of the local wind field.

Optical satellite sensors measure electromagnetic radiation in the visible through thermal IR portion of the spectrum as functions of location (x,y), time (t), and wavelength (λ). There are two types of radiation sensors: (1) imagers and (2) nonimagers. The first make two- and three-dimensional images of the target at a fixed wavelength or wavelength range. Nonimagers measure along the nadir, subsatellite track and stress wavelength and time resolution. This text focuses more on the imaging sensors and the images created.

These imaging sensors are generally referred to as "radiometers,"[1] which are instruments that collect radiant energy using a variety of techniques. These radiometers have some methods to separate wavelength contributions and quantify these various contributions with matched detectors. Some general dispersion methods used by radiometers to separate out the wavelengths are (1) prisms, (2) filter wheels, (3) grating spectrometers, and (4) interferometers. There are advantages and disadvantages associated with each approach, which will be discussed individually.

3.2 BASIC OPTICAL SYSTEMS

3.2.1 PRISMS

As introduced in Chapter 1, Sir Issac Newton was the first to use a prism to separate white light into the various colors of the visible spectrum. This dispersion effect is attributed to the wavelength-dependent refraction experienced by the light rays entering and exiting the prism (Fig. 3.2). The primary

---

[1]Optical radiometers, as opposed to microwave radiometers. Usually the distinction optical versus microwave is dropped, as it is implicit by the context.

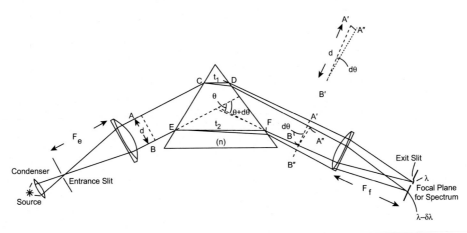

FIGURE 3.2

A prism as a radiative dispersion element (Slater, 1980).

advantages of the prism as a dispersing element are its relative simplicity, synopticity, and reliability. These advantages must be measured against the fact that prisms are heavy and cause a significant attenuation in radiative signal strength. Still for some applications the simplicity of operation along with the reliability of function lead to the choice of a prism as the means of radiation dispersion in the design of a remote sensing instrument. These applications are generally infrequent as the weight limitation of most spaceborne instruments precludes the use of prisms for wavelength dispersion.

3.2.2 FILTER-WHEEL RADIOMETERS

The next simplest and most widely used method of wavelength separation is the filter-wheel radiometer. In this type of radiometer, a "wheel" of radiative filters (Fig. 3.3) is cycled through the radiation path to separate out the different wavelengths. In this diagram, optical lenses have been used while in most remote sensing applications radiometers are constructed with mirrors to save space and weight.

Most of the other critical elements of filter-wheel radiometers are also shown in Fig. 3.3. This particular filter-wheel radiometer makes precise measurements of radiometric temperature in the mid- and longwave IR bands. The reference blackbodies are traceable back to National Institute of Standards and Technology (NIST) standards. First there is a collector for the radiation, which is usually known as the "scan mirror" which views the Earth target. Next comes the "chopper" which switches between the incoming radiation and the emitted radiation. The reference blackbody constitutes an internal calibration of the thermal IR channels. Next is a series of mirrors to focus the radiation and pass it through the filters for further optical processing before it reaches the detectors where it is converted into an electrical signal. This signal is amplified in a preamplifier before being transmitted further down the system. This diagram serves to demonstrate the positions and roles of most of the elements of a filter-wheel radiometer.

As a comparison a diagram of the scanning radiometer (SR), an early filter-wheel radiometer used on the US polar orbiting Earth satellites is presented in Fig. 3.4. Here mirrors are used for all of the optical processing steps. The optical design uses a Cassegrain mirror arrangement to focus the radiation on the dichroic filter that separates out the longwave radiation from the shorter visible wavelengths. This Cassegrain mirror arrangement requires the radiation from the secondary mirror to pass through a hole in the center of the primary mirror before going to the dichroic filter and detectors.

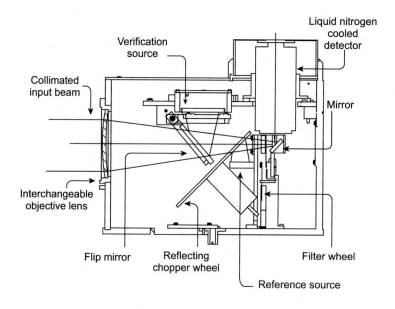

FIGURE 3.3

Filter-wheel radiometer (RAD-900 radiometer, http://www.optikos.com/products/rad-900/).

The large scan mirror rotates at a constant angular speed collecting radiation from the Earth's surface over a portion of this scan. Radiation from this scan mirror is focused by the primary mirror onto the secondary mirror, which then focuses the radiation through the dichroic on to the detector. A limit to this amount of radiation hitting the detector is the field stop, which restricts the amount of radiation that can hit the detector. This measured space allows only the wanted radiation in the center of the beam to reach the detectors cutting off the weaker "stray radiance" surrounding the stronger central core of the radiation.

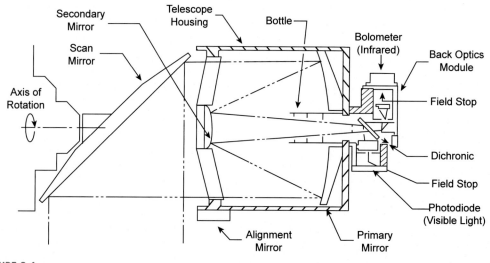

FIGURE 3.4

The scanning radiometer, an early filter-wheel radiometer.

In this case the filter wheel is the dichroic. This uses an optical glass substrate over which 20–50 thin (typically, 1 μm thick) layers of a special refractive index dielectric material (or materials in certain combinations) are deposited in a vacuum setup. This optical filter then selectively transmits a specific range or band of wavelengths, while reflecting the rest with nearly zero absorptivity. These dichroic filters can be either additive or subtractive color filters when operating in the visible range. Here a bolometer serves as the IR reference target instead of a pair of blackbodies usually used in higher precision IR radiometers. A bolometer measures IR radiation using the thermal electric effect whereby the incident thermal IR radiation increases the temperature which is then measured to estimate the value of the radiation. Again there is a field stop that limits the amount of the reference target that is viewed by the radiometer.

Note how the use of mirrors instead of lenses makes the optical "telescope" a lot more compact. Lenses require longer focal lengths, which can only be accommodated with longer optical bodies while mirrors can be located within close proximity to each other.

These two types of filter-wheel systems are shown here side-by-side in Fig. 3.5.

In addition to layer attenuation and chromatic aberration due to the wavelength dependence of the index of refraction, lenses also suffer from spherical aberration when parallel light rays focus at different points, depending on their distance to the optical axis of a spherical lens or mirror resulting in a blurry image. Positive/negative spherical aberration occurs when the outer light rays bend too much/little and focus at shorter/larger distances than the focal length. This aberration can be reduced by either reducing the optical system aperture and/or increasing the focal length (Fig. 3.6).

3.2.2.1 An Example: The Cloud Absorption Radiometer

Since filter-wheel radiometers are so widely used a more detailed example of one such system is given here. The cloud absorption radiometer (CAR, http://car.gsfc.nasa.gov/subpages/index.php?section¼Instrument&content¼Schematics) is designed to measure light scattered by clouds in 14 spectral bands between 0.34 and 2.3 μm. The scan mirror, rotating at 100 rpm, directs the light into a Dall–Kirkham telescope where the beam is split into nine paths. Eight light beams pass through beam splitters, dichroics, and lenses to individual detectors (0.34–1.27 μm), and finally are registered by eight data channels, which are sampled simultaneously, and continuously. The ninth beam passes through a spinning filter wheel to a Stirling cycle cooler. Signals registered by the ninth data channel are selected from among six spectral bands (1.55–2.30 μm). The filter wheel can either cycle through all six spectral bands at a prescribed interval (usually changing filters every fifth scan line), or lock onto any one of the six spectral bands and sample it continuously.

The CAR is a cross-track scanner with a scan mirror that rotates 360 degrees in a plane perpendicular to the direction of flight and the data are collected through a 190 degree field of view (FOV). In the normal mode of operation onboard the CV-580 aircraft, the CAR views 190 degrees of Earth-atmosphere scene around the starboard horizon. This configuration permits observations of both local zenith and nadir with as much as a 5 degree aircraft roll. In addition to the starboard viewing mode, the CAR instrument can now be rotated in-flight into three other viewing positions: downward-looking imaging mode, upward-looking imaging mode, and a dedicated bidirectional reflectance distribution function viewing mode.

The instrument incorporates several innovative features:

1. Since it is sometimes flown through clouds, there is the possibility that moisture may be deposited on optical surfaces, especially the scan mirror, producing large errors. The instrument is mounted outside the airplane and cannot be observed in flight. To check for water on the mirror, a unique detection system was devised. A thin beam of light is shone on the edge of the mirror, and the

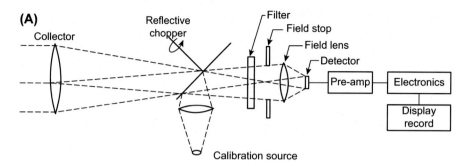

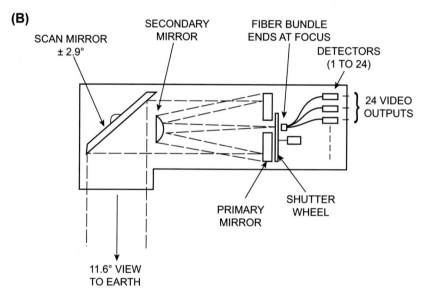

FIGURE 3.5

Example lens (A) and mirror (B) filter-wheel radiometer systems.

reflected beam is monitored by a photodiode. If any condensation appears on the mirror the reflected light scatters, reducing the photodiode's output and flagging data likely to be in error.

2. Another novel feature maintains low offset (ensuring that zero volts at the output always correspond to a zero-radiance input) by using the scan mirror as a type of radiation chopper. It works by forcing the electrical output to zero during each back scan while the detectors are all completely darkened by means of a scanner-synchronized moveable shutter. Long time constant coupling in the amplifier then ensures that data measured during the active part of the scan remains accurately related to this zero reference level.

A cutaway diagram of the CAR instrument in Fig. 3.7 shows the location of the filter-wheel housing relative to the scan mirror, the primary and secondary mirrors. This instrument uses the same Cassegrain optical design discussed earlier for the SR in Fig. 3.4. In this particular configuration, this design is referred to as a Dall–Kirkham telescope (King et al., 1986).

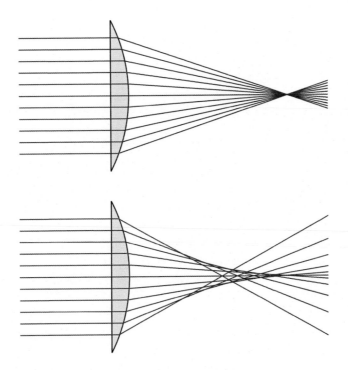

FIGURE 3.6

A perfect lens (top) focuses all incoming rays to a single point on the optic axis, but a real lens with spherical surfaces (bottom) focuses different rays to different locations on the optic axis. This is known as spherical aberration (https://commons.wikimedia.org/wiki/File:Spherical_aberration_2.svg).

A blow-up of the optical system in Fig. 3.8 clearly shows the complexity of the filter-wheel with its six positions and corresponding wavelengths.

The radiation having left the secondary mirror first passes through the defining field stop and then through a pair of specially designed dichroic beam splitters. The radiation reflected by the first dichroic (D1) enters branch I and is of wavelength range 0.34–0.87 μm. The radiation transmitted by D1 and reflected by D2 enters branch II where it is separated into three channels in a similar manner for filters in the 1.03–1.27 μm interval. The radiation transmitted by D2 enters branch III, where it is further defined spectrally by a filter wheel containing six narrowband interference filters in the wavelength range 1.55–2.30 μm.

3.2.2.2 Filters

There are essentially two types of filters: (1) absorption or (2) interference. For filter wheels only absorption lenses come into question. Filters can be designed to pass either long or short wavelengths, but most filters for satellite sensors are band-pass filters and pass radiation only in relatively narrow-bands. These filters are characterized by their central wavelength and the short and long wavelength cutoff limits. The shape of the filter is also important and it is generally described by the half-width of the transmission bandwidth that defines the spectral resolution of the filter and hence the radiometer. The

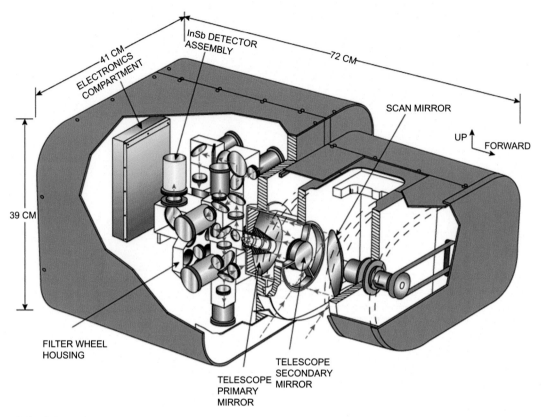

FIGURE 3.7

Cutaway diagram of the cloud absorption radiometer instrument.

Image modified by Gatebe and Gammage from King, M.D., Strange, M.G., Leone, P., Blaine, L.R., 1986. Multiwavelength scanning radiometer for airborne measurements of scattered radiation within clouds. J. Atmos. Oceanic Technol. 3, 513–522.

filter also attenuates the radiation passing through it, so some knowledge of its radiative efficiency is needed to evaluate the filter's characteristics. This is usually given as the signal-to-noise ratio (SNR) for the filter, which equals the band-pass energy/total energy outside the bandpass.

The colored absorption filters are either colored glass or dyed gelatin. The former has seen the widest application due to their advantages of (1) freedom from strictions, (2) they are optically flat, and (3) the ease of laying out parallel surfaces. In this design, all layers are deposited on one substrate. Typically, the absorption filters are broader in shape as opposed to the interference filters that can be designed to be narrower (Fig. 3.9).

These filters may be either broad or narrow *band-pass* filters. These filters may also be high bandpass (selectively removes longer wavelengths) or low bandpass (absorbs shorter wavelengths).

3.2.3 GRATING SPECTROMETER

While the term spectrometer accurately applies to any system that separates radiation by wavelength, it has been generally applied to the use of a grating to generate the interference pattern that separates the wavelengths (Fig. 3.10). Like the prism, this has the disadvantage that an array of detectors is needed

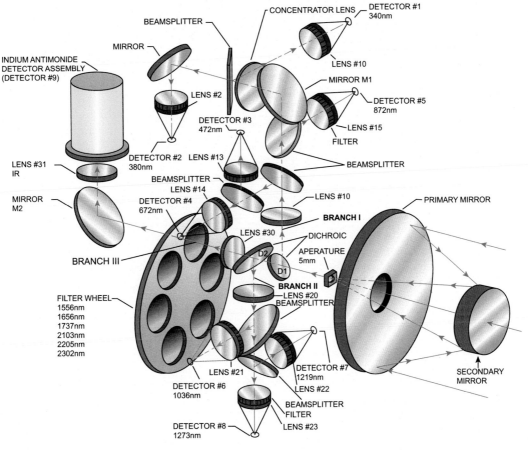

**SCHEMATIC ILLUSTRATION OF THE
CLOUD ABSORPTION RADIOMETER OPTICAL SYSTEM**

FIGURE 3.8

Optical system of the cloud absorption radiometer instrument.

Image modified by Gatebe and Gammage from King, M.D., Strange, M.G., Leone, P., Blaine, L.R., 1986. Multiwavelength scanning radiometer for airborne measurements of scattered radiation within clouds. J. Atmos. Oceanic Technol. 3, 513–522.

to measure the radiation from the separated rays. As a result, grating spectrometers are generally large and expensive. While a slit grating is depicted in Fig. 3.10 a spaceborne system Fig. 3.11 generally uses a serrated mirror to generate the spectral dispersion. The amount of dispersion depends on the spacing of these angular serrations. The vast majority of master gratings are formed in one of two ways: by mechanical ruling, in which a diamond is dragged across a metalized substrate to produce a series of parallel grooves, or by interference methods, in which the fringe pattern formed by two coincident laser beams exposed on a photosensitive blank, creating all grooves at once. The latter method produces gratings commonly termed "holographic." Since making one high-quality master grating can take weeks, without a replication process grating-based spectrometers would be commercially available.

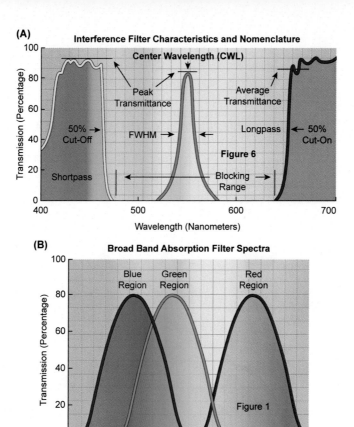

FIGURE 3.9

Spectral properties of interference (A) and absorption filters (B) (http://micro.magnet.fsu.edu/primer/java/filters/absorption/).

From OLYMPUS CORPORATION, http://www.olympus-lifescience.com/de/microscope-resource/primer/techniques/confocal/interferencefilters/.

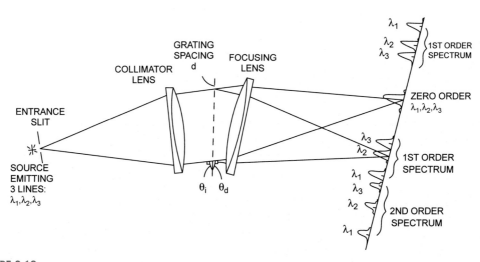

FIGURE 3.10

Grating spectrometer dispersion system (Slater, 1980).

(A)

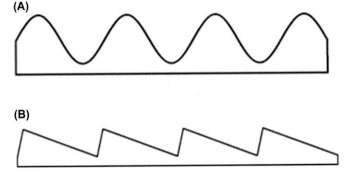

(B)

FIGURE 3.11

Two common groove patterns used for gratings.

The primary advantage of this type of a spectrometer is the truly synoptic separation of the radiation wavelengths also with a very precise definition of the wavelength separation set by the grating spacing (Fig. 3.11). The primary disadvantage of this system is the complex detector array required to sense this synoptic picture. This complex detector array also leads to the reality that a failure of any element of the detector array eliminates the radiation that belonged to this part of the spectrum. This fact together with the high-cost, complexity, and weight are often reasons that this option is not selected for satellite instrumentation applications.

3.2.4 INTERFEROMETER

A final technical approach to dispersion of the radiation wavelengths is in terms of an interferometer such as the Fabry–Perot system depicted here in Fig. 3.12 (Slater, 1980). In this instrument the dispersion mechanism is created by the spacing between the two "Fabry–Perot" mirrors. This spacing is stepped through a variety of intervals to separate out the desired wavelength signals.

The advantages of this type of wavelength dispersion system are that it has a very simple operation where only the spacing between the mirrors needs to be changed. This allows for a semi-infinite wavelength separation at the expense of a finite time between the wavelength samples. It also requires a simple detector system with a detector located only at the aperture point.

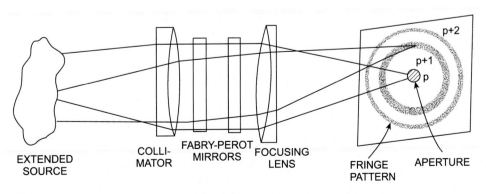

FIGURE 3.12

A Fabry–Perot interferometer (Slater, 1980).

The main disadvantage of this technique is the finite amount of time it takes to sample each wavelength, resulting in less than a truly synoptic sample of all of the wavelengths of interest, and the signal attenuation since it involves a number of mirrors and lenses. Additionally, the detector must be capable of responding to a wide range of wavelengths rather than being optimized for a particular wavelength as in the spectrometer. The relative simplicity of this system makes it attractive relative to the spectrometer, but it is much more complex than the filter-wheel radiometer.

3.3 SPECTRAL RESOLVING POWER; THE RAYLEIGH CRITERION

Spectral resolving power is given by $\lambda/\delta\lambda$, where λ is the wavelength and $\delta\lambda$ is the minimum wavelength separation. The "Rayleigh criterion" is satisfied when a minimum of the diffraction pattern of one image coincides with the maximum of another. Thus, the maximum spectral resolving power is proportional to the maximum optical path difference introduced in the beam or simply said the Rayleigh criterion specifies the minimum separation between two light sources for them to be resolved into two distinct objects. If the distance is greater, the two points are well resolved and if it is smaller they are regarded as not resolved (Fig. 3.13).

The central region of the profile, from the peak to the first minimum, is called the Airy disk and its angular radius is given by:

$$D \cdot sin\theta \approx 1.22 \cdot \lambda, \tag{3.1}$$

where $D = 2 \cdot a$ is the diameter of the aperture and λ is the electromagnetic wavelength. Using the small angle approximation that $sin\theta \approx \theta$ (where θ is measured in radians):

$$\theta \approx 1.22\lambda/D.$$

For the grating spectrometer the diffraction limit for a diffraction grating is given by:

$$R = N \cdot m, \tag{3.2}$$

where N is the number of ruled lines (or slits) being illuminated in a grating (or grooves in a space borne instrument), and m is the order of diffraction ($m = 0, 1, 2, 3 \ldots$). For the grating, the total width W is effectively limited to the width of the ruled region of the diffraction grating, since this is fully illuminated by the optical beam.

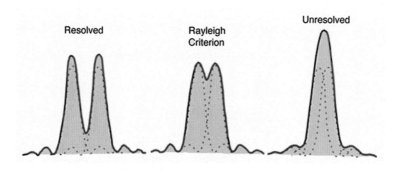

FIGURE 3.13

Resolved and unresolved wavelengths versus the Rayleigh criterion.

For a prism this diffraction limit is the base length times the refractive index, which can be written as:

$$R_{\text{prism}} = B \cdot \left| \frac{dm}{d\lambda} \right|, \tag{3.3}$$

where B is the length of the base of the prism (the side opposite the apex) and $dm/d\lambda$ the rate of change of the refractive index m with respect to wavelength λ. The refractive index of a prism at a given wavelength may be found by geometrical considerations from the angle of minimum deviation δ_{min} for that wavelength:

$$n = \frac{sin\left[\frac{1}{2}(A + \delta_{\text{min}})\right]}{sin\left(\frac{1}{2}A\right)} \tag{3.4}$$

where A is the apex angle of the prism. It is not possible to write a simple relationship between δ_{min} and λ as this depends on the precise material the prism is made of and how it is cut.

For the interferometer the distance between the reflecting surfaces gives this resolving power. found in modern satellite sensors with an increasing trend to achieve an increased spectral resolution using spectrometer and interferometers in spite of the increase in complexity of the optical systems. There is also a trend to require more channels than those that can easily be accomplished with the simple filter-wheel radiometer.

A nice summary of the role of the diffraction limit as applied to spatial resolution is given in Fig. 3.14, which is a log–log plot of aperture diameter versus angular resolution at various wavelengths. Also shown are various astronomical instruments at their operational wavelengths. It is interesting to note that the Hubble Space Telescope is almost diffraction limited at 0.1 arcsec as indicated by the blue star in the figure. The colored bands correspond to different wavelength bands used for remote sensing of various targets. The various apertures of different telescope systems are also included in this figure. Please note the change in units used to describe the different apertures since some earlier instruments were built to inch specifications rather than meters.

3.4 DETECTING THE SIGNAL

The next step is to get the spectrally separated radiation to appropriate detectors. This can be done through lenses or by detector positioning or, in the case of the multispectral scanner (MSS) and other sensors, by channeling radiation in specific ranges to fiber optics bundles that carry the focused radiation to an array of individual detectors. For the MSS, this involves six fiber optic leads for the six lines scanned simultaneously to six detectors for each of the four spectral bands, or a total of 24 detectors in all.

Once the radiation reaches the photodetector, assuming that all variables are constant within the spectral width ($\Delta\lambda$), the output voltage (V_0) is given by:

$$V_0 = \frac{L_i}{G_\#} \cdot A_{det} \cdot \Delta\lambda \cdot R, \tag{3.5}$$

where L_i is the incoming radiance reaching the sensor, A_{det} is the area of the photodetector which is related to the photodetectors' dimension ($A_{det} = l_{det}^2$), $\Delta\lambda$ is the spectral width, R is the photodetector's

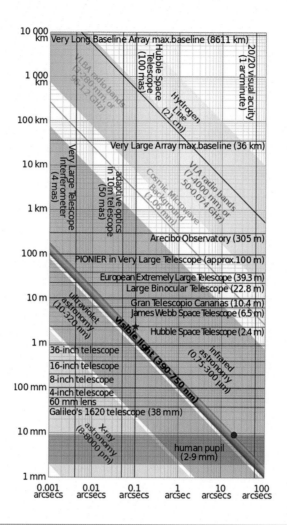

FIGURE 3.14

Log—log plot of aperture diameter versus angular resolution at the diffraction limit for various light wavelengths for various astronomical instruments. For example, the blue star shows that the Hubble Space Telescope is almost diffraction limited in the visible spectrum at 0.1 arcsec.

responsivity, that relates the output voltage and the input optical power (units [V/W]), $G_\#$ is the "G number" (units of [strad^{-1}]) which is related to the inverse of the FOV of the optical system,[2] and it is defined as:

$$G_\# \approx \frac{1 + 4 \cdot F_\#^2}{\tau_{opt} \cdot \pi},$$ (3.6)

---

[2]Not to be confused with the instantaneous field of view as seen by the photodetector that depends on its physical size (I_{det}), and it is approximately given by $IFOV \approx I_{det}/f$.

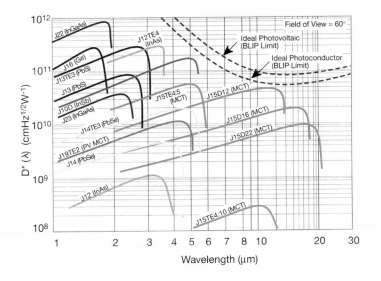

FIGURE 3.15

The sensitivity of various detector materials as functions of wavelength.

Image courtesy of Teledyne Judson technologies. http://www2.chem.uic.edu/tak/chem52413/notes8/notes8_13.pdf.

where τ_{opt} is the transmissivity of the optical system, and $F_\#$ is the "F number" defined as the ratio of the focal length (f) and the aperture (D) of the optical system:

$$F_\# = \frac{f}{D}. \tag{3.7}$$

Most detectors are made of solid-state semiconductors, metals, or alloys. A semiconductor has a conductivity intermediate between a metal and an insulator. Under certain conditions, such as the interaction with photons, electrons in the semiconductor are excited and moved from a filled energy level (in the electron orbital configuration around an atomic nucleus) to another level called the conduction band, which is deficient in electrons in the unexcited state. The resistance to flow varies inversely with the number of incident photons. The process is best understood by quantum theory. Different materials respond to different wavelengths and are thus spectrally selective.

In the visible light range, silicon and PbO are common detector materials. Photoconductor material in the near-IR includes PbS (lead sulfide) and InAs (indium-arsenic). In the mid-IR (3–6 μm), InSb (indium-antimony) is responsive. The most common detector material for the 8–14 μm range is Hg-Cd-Te (mercury-cadmium-tellurium). When the infrared sensors are operating it is necessary to cool the detectors to optimize the efficiency of electron release. Other detector materials are also used and perform under specific conditions. In Fig. 3.15 we show the photodetector specific detectivity (D^*) as a function of the wavelength, which is defined as:

$$D^*(\lambda) = \frac{\sqrt{A_{\text{photodetector}} \cdot B}}{NEP(\lambda)} \ \left(\text{cm} \sqrt{\text{Hz}}/\text{W}\right). \tag{3.8}$$

The specific detectivity is a figure of merit of the photodetectors' noise performance that allows us to compare them independently of their area ($A_{\text{photodetector}}$) and bandwidth (B). The *NEP* is the so-called noise equivalent power and it is defined as the ratio of the noise at the output, divided by the photodetector's responsivity R.

For the typically used photodiodes, the *NEP* can be computed as the ratio of the standard deviation of the dark current and the intrinsic responsivity:

$$NEP = \frac{\sigma_{darkness}}{R_i} = \frac{\sqrt{2q(I_{ds} + F \cdot M^2 \cdot I_{db}) \cdot B}}{\eta \cdot \dfrac{q \cdot \lambda}{c \cdot h} \cdot M}, \tag{3.9}$$

which depends on a number of physical parameters of the device. These are the dark current[3] that circulates through the surface (I_{ds}) or through the volume (I_{db}), η is the quantum efficiency (ratio of electrons reaching the output of the semiconductor divided by the number of electron–holes pairs generated by the incoming photons), and the multiplication factor (M), which is equal to $M = 1$ for a photodiode, and $M > 1$ for an avalanche photodiode. The term F is called the excess noise factor and it is equal to

$$F = k \cdot M + \left(2 - \frac{1}{M}\right) \cdot (1 - k), \tag{3.10}$$

with $k \approx 0.0033$, and accounts for the extra noise generated during the avalanche process that produces the amplification (M).

Other detector systems are less commonly used in remote sensing function in different ways. The list includes photoemissive, photodiode, photovoltaic, and thermal (absorption of radiation) detectors.

3.5 VIGNETTING

Vignetting is a "fact of life" for all optical systems and consists of a darkening of the corners of an image relative to the center of the image (Fig. 3.16). There are three types of vignetting that are characteristic of any optical systems, which are mechanical vignetting, optical vignetting, and natural vignetting. For digital systems there is an additional pixel vignetting.

Mechanical vignetting occurs when light beams emanating from object points located off-axis are partially blocked by external objects such as thick or stacked filters, secondary lenses, and improper lens hoods. The corner darkening can be gradual or abrupt, depending on the lens aperture (Fig. 3.17). Complete blackening is possible with mechanical vignetting.

Optical vignetting is caused by the physical dimensions of a multiple element lens. Rear elements are shaded by elements in front of them, which reduces the effective lens opening for off-axis incident light. The result is a gradual decrease of the light intensity towards the image periphery. Optical vignetting is sensitive to the aperture and can be completely cured by stopping down the lens. Two or three stops are usually sufficient.

Unlike the previous types, natural vignetting (also known as natural illumination falloff) is not due to the blocking of light rays. The falloff is approximated by the $\cos^4$ or "cosine fourth" law of illumination falloff. Here, the light falloff is proportional to the fourth power of the cosine of the angle

---

[3]Dark current refers to the current that circulates in the absence of incoming light to generate pairs of electron–holes.

FIGURE 3.16

An example of vignetting in a photograph.

FIGURE 3.17

An example of mechanical vignetting.

at which the light impinges on the film or sensor array. Wide-angle rangefinder designs and the lens designs used in compact cameras are particularly prone to natural vignetting. Telephoto lenses, retro focus wide-angle lenses used on SLR cameras, and telecentric designs in general are less troubled by natural vignetting. A gradual gray filter or postprocessing techniques may be used to compensate for natural vignetting, as stopping down the lens cannot cure it.

Pixel vignetting only affects digital systems and is caused by angle dependence of the digital sensors. Light incident on the sensor at a right angle produces a stronger signal than light hitting it at an oblique angle. Most digital cameras use built-in image processing to compensate for optical vignetting and pixel vignetting when converting raw sensor data to standard image formats such as JPEG or TIFF. The use of microlenses over the image sensor can also reduce the effect of pixel vignetting.

3.6 SCAN GEOMETRIES

Various kinds of scan geometries have been used to collect the radiation in these optical instruments. The simplest is a cross-track scanner, which consists of a rotating mirror (Fig. 3.18A) that rotates at a constant rate synchronized with the orbit of the satellite so that it increments the image each scan line.

Here the angular FOV represents the cross-track scan across the ground. As it will be discussed later, the rest of the mirror rotation is used to measure reference values such as a cold-space reference temperature as well as two blackbody targets one at ambient temperature and one at a controlled temperature for calibration purposes.

There are two types of mirror cross-track scanners: one that uses a mirror constantly rotating mirror (in the same sense) as depicted in Fig. 3.18A, while the other uses an optical system that oscillates back

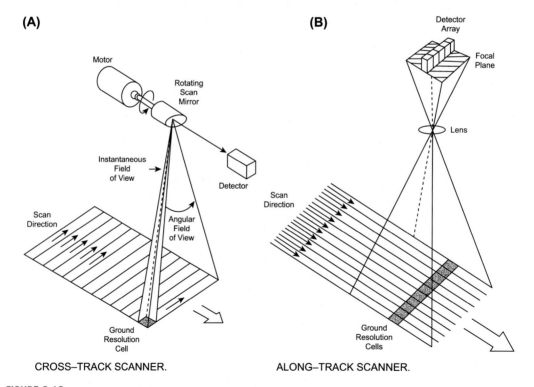

(A) CROSS–TRACK SCANNER.

(B) ALONG–TRACK SCANNER.

FIGURE 3.18

(A) Cross-track rotating mirror scanner. (B) Push-broom along-track scanner.

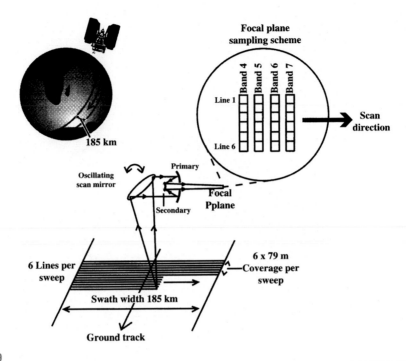

Focal plane
sampling scheme

Band 4
Band 5
Band 6
Band 7

Line 1

Scan
direction

Line 6

185 km

Primary

Oscillating
scan mirror

Focal
Pplane

Secondary

6 Lines per
sweep

6 x 79 m
Coverage per
sweep

Swath width 185 km

Ground track

FIGURE 3.19

Whisk-broom scanning on Landsat multispectral scanner (from Fig. 6.10 of [Schott, 2007]).

and forth across the scan line. The latter is called a whisk-broom scanner and these cross-track scanners all rely on the forward motion of the spacecraft to increment the image one line at a time.

Another type of scanner samples the entire line at the same time using a cross-track array, which is called a "push-broom" scanner (Fig. 3.18B). This configuration has the advantage that the samples are truly synoptic across the track. This has become a very popular method for optical sensor design. The type of cross-track scanner that oscillates back and forth across the track (Fig. 3.19) is the type of scanning mechanism using by the MSS, which flew on the Landsat series of satellites.

The more recent imager on Landsat is called the "Thematic Mapper (TM)" and produces a "bow-tie" effect on the scan as the satellite progresses along its track (Fig. 3.20). This effect compensates for the satellite movement as the line is scanned so that all lines are parallel, and perpendicular to the satellite ground track. While this effect requires additional image correction, it has the added benefit that this system increases the integration time since the mirrors do not have to rotate as fast as they did in the MSS.

Finally, there is a conical along-track scan configuration (Fig. 3.21).

This scan mechanism has the advantage of viewing the same spot on the Earth twice in rapid succession if the scan axis is not oriented strictly perpendicular to the ground track of the satellite. This is the scan method used in the along-track scanning radiometer–infrared radiometer on board the European Space Agency's (ESA) ERS-1/2 satellites (ATSR-IRR; Fig. 3.22). The principle of removing atmospheric effects in measurements by viewing the sea surface from two angles is the basis of the

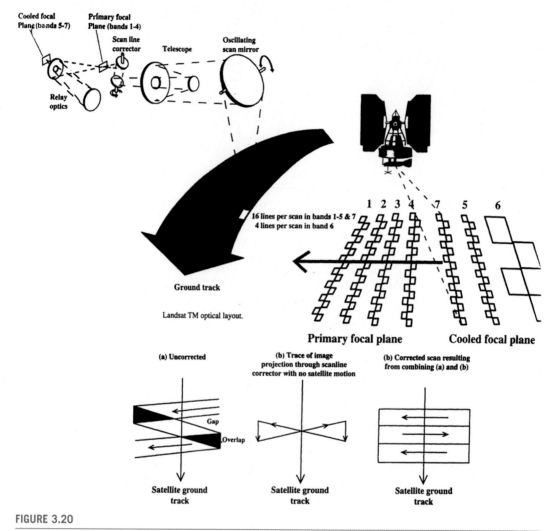

FIGURE 3.20

Bow-tie scanner Landsat Thematic Mapper (from Fig. 6.11, 6.12 and 6.13 of [Schott, 2007]).

family of (A)ATSR instruments. The sea surface temperature (SST) objectives are met through the use of thermal IR channels (centered on 1.6, 3.7, 10.85, and 12 μm), identical to those on ATSR 1 and 2. Atmospheric modeling for ERS-1 has shown that ATSR, with its thermal IR channels and two-angle viewing geometry (55 and 0 degree incidence angles in the fore and aft swaths), can achieve a global accuracy in SST better than 0.5K.

This same scan geometry is also being widely used in passive microwave imagers as it will be discussed later.

As an example of the Landsat series, we will look at the TM instrument as shown here in Fig. 3.23.

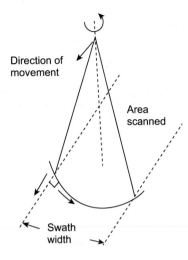

FIGURE 3.21

Conical scan.

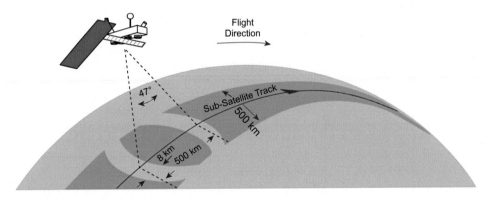

FIGURE 3.22

Scan geometry of the conical along-track scanning radiometer.

   This sensor is mounted in the Landsat spacecraft in a horizontal position with the sunshade pointing toward the Earth. Directly above the sunshade is the scan mirror surrounded by its drive mechanisms, control electronics, and scan mirror hardware. The TM IFOV is 42.5 μrad, which equates to a ground resolution of 30 m. The band ranges and their science applications are given here in Table 3.1.

   The temporal resolution of the TM, which has been flying since 1983, is approximately 15 days. As shown in Table 3.1 the TM has seven spectral bands including a rather broad thermal IR channel between 10.4 and 12.5 μm. The higher spatial resolution (~30 m) makes it possible to more accurately map surface features such as vegetation cover. Landsat TM has proven to be particularly useful in the study of surface expressions of geologic formations which do not change significantly within the 15-day repeat sample of the TM.

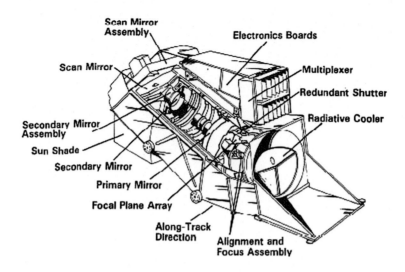

FIGURE 3.23

The Landsat Thematic Mapper sensor (https://eoportal.org/documents/163813/2855045/LS4-5_Auto2.jpeg).

Table 3.1 Landsat Thematic Mapper Bands and Their Applications

Band Number	Pixel Size	Spectral Range	NE$\Delta\rho$	Application
1	30 × 30 m	0.45–0.52 μm	0.8%	Shore water mapping
				Bare soil-vegetation mapping
				Deciduous-coniferous vegetation mapping
2	30 × 30 m	0.52–0.60 μm	0.5%	Vegetation health
3	30 × 30 m	0.63–0.69 μm	0.5%	Vegetation classification
4	30 × 30 m	0.76–0.90 μm	0.5%	Biomass studies
				Water contours delimitation
5	30 × 30 m	1.55–1.75 μm	1.0%	Clouds–snow differentiation
				Water content determination (vegetation, soil)
6	120 × 120 m	10.40–12.50 μm	0.5K	Temperature
				Vegetation health
7	30 × 30 m	2.08–2.35 μm	2.4%	Water temperature
				Mineral and oil geology

The latest in the Landsat series is Landsat 8 also known as the Landsat Data Continuity Mission (LDCM). Unlike its predecessors it carries the Operational Land Imager (OLI), a new instrument built by Ball Aerospace. The OLI is a push-broom scanner that has a focal plane with long arrays of photodetectors (Irons and Dwyer, 2010). A four-mirror anastigmatic telescope focuses incident

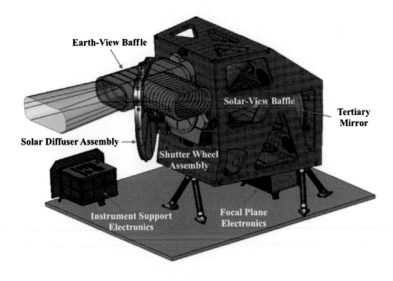

FIGURE 3.24

Drawing of the Operational Land Imager.

Courtesy of Ball Aerospace.

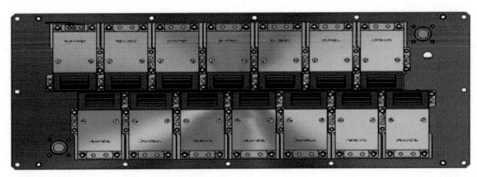

FIGURE 3.25

Drawing of the Operational Land Imager focal plane.

Courtesy of Ball Aerospace.

radiation onto the focal plane while providing a 15 degree FOV covering a 185 km across-track ground swath from the nominal LDCM observatory altitude (Fig. 3.24). Periodic sampling of the across-track detectors as the observatory flies forward along a ground track forms the multispectral digital images.

The detectors are divided into 14 modules arranged in an alternating pattern along the centerline of the focal plane (Fig. 3.25). Data are acquired from nearly 7000 across-track detectors for each spectral band with the exception of the 15 m panchromatic band that requires over 13,000 detectors. The spectral differentiation is achieved by interference filters arranged in a "butcher-block" pattern over the detector arrays in each module. Silicon PIN (SiPIN) detectors collect the data for the visible and near-IR spectral bands (Bands 1–4 and 8) while magnesium–cadmium–telluride (MgCdTe) detectors are used for the shortwave IR bands (bands 6, 7, and 9).

The OLI telescope views the Earth through a baffle extending beyond the aperture stop. A shutter wheel assembly sits between the baffle and the aperture stop. A hole in the shutter wheel allows light to enter the telescope during nominal observations and the wheel rotates when commanded to a closed position and acts as a shutter preventing light from entering the instrument. A second baffle, for solar views, intersects the Earth-view baffle at a 90-degree angle and a three-position diffuser wheel assembly dissects the angle. A hole in the diffuser wheel allows light to enter the telescope for nominal Earth observations. Each of the other two wheel positions introduces one of two solar diffuser panels to block the optical path through the Earth-view baffle. When the wheel is in either of these two positions, the solar-view baffle points at the Sun and a diffuser panel reflects solar illumination into the telescope. One position will hold a "working" panel that is exposed regularly to sunlight while the other position holds a "pristine" panel that is exposed infrequently and used to detect changes in the "working" panel spectral reflectance due to solar exposure. Additionally, two stimulation lamp assemblies are located just inside the telescope on the aperture stop. Each of the two assemblies holds six small lamps inside an integrating hemisphere and is capable of illuminating the full OLI focal plane through the telescope with the shutter closed. These assemblies, the shutter wheel, diffuser wheel, and stimulation lamp assemblies, constitute the OLI calibration subsystem.

3.7 FIELD OF VIEW

The FOV of an optical instrument is dictated by the wavelength of interest, and the angular resolution of the sensor optics. In terms of a digital camera, the FOV refers to the projection of the image on to the camera's detector array, which also depends on the camera lens' focal length. Hence the FOV (also called the field of view) is the angular extent of the observable world that is seen at any given moment. Humans have an almost 180 degree forward-facing FOV, while some birds have a complete or nearly complete 360 degree FOV. In addition, the vertical range of the FOV may vary. The range of visual abilities is not uniform across an FOV and varies from animal to animal. For example, binocular vision, which is important for depth perception, only covers 140 degrees of the field of vision in humans; the remaining peripheral 40 degrees have no binocular vision (because of the lack of overlap in the images from either eye for those parts of the FOV). The aforementioned birds would have a scant 10 or 20 degrees of binocular vision.

Angular FOV is typically specified in degrees, while linear FOV is a ratio of lengths. For example, binoculars with a 5.8 degree FOV might be advertised as having a (linear) FOV of 305 ft per 1000 yards or 102 mm per meter.

In terms of satellite remote sensing the instantaneous FOV (IFOV) angular resolution has a lower limit set by the diffraction limit which is λ/D, where λ is the wavelength and D is the diameter of the first target (lens or in the case of most satellite instruments the primary sensor mirror). Hence translated to ground resolution, assuming a flat Earth, the minimum $IFOV$ at normal incidence is:

$$IFOV_{min} \approx \lambda \cdot \frac{H}{D}, \qquad (3.11)$$

where H is the satellite height above the surface or the altitude. There is of course a relationship between the ground resolution, the satellite velocity, and the sampling time of the sensor. Thus, for a scanning sensor this limits the sampling time as a function of the satellite's orbit. For a push-broom configuration the scan line is collected simultaneously thus increasing the exposure time, but with the disadvantage that each sensor must be calibrated separately.

3.8 OPTICAL SENSOR CALIBRATION

There are two types of calibration that are important for optical and other types of satellite sensors. First there is an intensive prelaunch calibration and characterization intended to give the users of these data the best possible connection between the sensor radiation measurements and the FOV. Second, since all sensors degrade and drift with time and temperature, there is a need for a periodic calibration after the sensor has been launched in to orbit.

Here it is impossible to carry out the laboratory and controlled experiments used in the prelaunch configuration so a system referred to as "vicarious calibration" has been introduced. In this type of calibration, very uniform Earth targets are identified and field campaigns are conducted to measure and monitor the surface conditions that are coincident with the radiative measurements made by the satellite sensors. This assumes, of course, that the atmosphere is very clear and does not introduce any additional errors to this type of vicarious calibration.

Prelaunch calibration procedures and methods are first discussed, and then vicarious calibration approaches are introduced and discussed. The goal of all calibration efforts is to set and maintain over time the relationship between the satellite radiances and the in situ property being measured. Without a proper calibration, sensor radiance measurements are just numbers without any meaningful relationship to an in situ property or parameter.

3.8.1 VISIBLE WAVELENGTHS CALIBRATION

Short visible wavelengths are calibrated with an "integrating sphere" which is designed to create a hollow cavity with very high diffuse reflectivity, which generates a known uniform light condition that can be measured with the radiometer as part of its calibration (http://www.electro-optical.com/datashts/visible/ivs400.html). Also known as an Ulbricht sphere, the integrating sphere usually mounts a number of different light sources around its surface, which can be turned on or off to control the light intensity and frequency. The interior surface of the sphere is coated with white material that is nearly a Lambertian reflector creating a uniform diffuse reflectance of the interior lights that is then viewed by the radiometer through a view port through the sphere.

As an example the ISV400 is presented here, which is a compact, reliable, and easy to operate system specifically designed for test and calibration of detectors and sensor systems which respond in the visible light spectrum. It incorporates a 12″ diameter integrating sphere with a 4″ diameter exit port (Fig. 3.26). The interior is coated with barium sulfate to provide a uniform and Lambertian luminance within the sphere. Illumination is provided by four quartz halogen lamps whose variable output uses a closed loop control sensor and output attenuator to provide ease of use and stability. All four lamps are calibrated to a NIST traceable color temperature of $2950 \pm 25K$.

A silicon detector continuously monitors the output of the lamps and provides feedback to the attenuator, providing a very stable and repeatable performance. Control electronics and lamp power supplies are built in to a 5.25″ high, 19″ rack mount unit. Operation of the unit is from the front panel keypad or via the built-in IEEE-488 and RS-232 computer interfaces.

FIGURE 3.26

The ISV400 integrating sphere (http://www.electro-optical.com/html/datashts/visible/isv400.asp).

Integrating spheres have played fundamental roles in the prelaunch calibration of a number of sensors such as the advanced very high resolution radiometer (AVHRR), the ATSR, Moderate Resolution Imaging Spectroradiometer, Sea-viewing Wide Field-of-view Sensor, Landsat TM, and extended TM, and many others. As discussed in Gatebe et al. (2007), there are known problems inherent in the use of integrating spheres for instrument calibration. Quoting Hovis and Knoll (1983), the spectral irradiance from the sphere and the Sun peaks at different wavelengths (~ 805 and 550 nm, respectively). This causes a problem for instruments designed to look at the Earth in that calibration between these two wavelengths is done with different radiance shapes from the sphere and the Sun.

In the calibration of the CAR discussed earlier a series of integrating sphere sources (ISSs) were used at different light levels as determined by the number of lamps used (from 0 to 16). This test helped to establish the linearity of the instrument over its performance range and to give the full range of conversion from digital counts to radiance (Fig. 3.27).

A second test involved measuring the responsivity of the CAR (defined here as detector output per unit if incident power at a particular wavelength) at more than nine distances from the ISS aperture. This test ascertained the sensitivity of the calibration to the distance of separation between the CAR and the ISS. A third test involved determining the CAR responsivity across its angular scan range.

The ATSR (Fig. 3.28) calibration (Bachmann et al., 2014) was carried out with a Labsphere Uni-source 2000, 500 mm diameter, integrating sphere with a 200 mm exit aperture (Smith et al., 1999a).

This unit provided eight distinct output levels of uniform illuminance of the exit port. These were achieved with four lamps, two with 45 W, and two with 150 W equally spaced about the exit port. All lamps had a color temperature of 3000K and hence the spectral profile was independent of the lamp

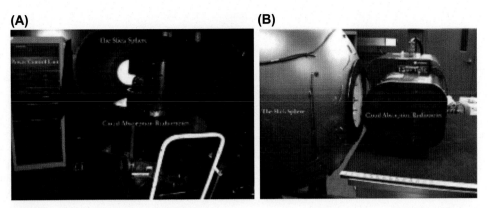

FIGURE 3.27

Setup for slick sphere and cloud absorption radiometer (CAR) for determining (A) responsivity across the scan range and (B) change in radiance with distance of separation, with the CAR at the closest distance from the slick sphere.

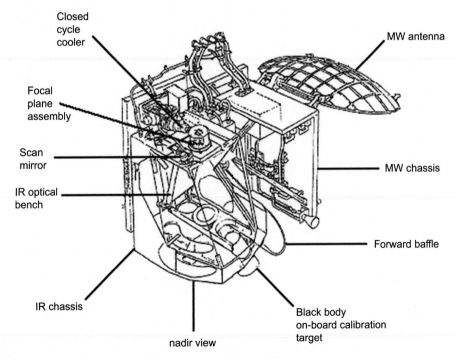

FIGURE 3.28

The along-track scanning radiometer sensor.

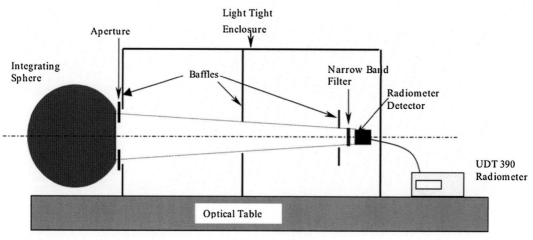

FIGURE 3.29

Experimental layout for the calibration with the Labsphere Unisource 2000 integrating sphere.

combination used in the test. Additional output levels were obtained by varying the area of the output aperture. This was done by inserting a blackened aluminum plate with a hole of the required diameter in front of the main aperture. A total of 27 apertures were used including the main aperture.

For the calibration, the radiometer detector was positioned on the sphere axis 1096 mm from the main aperture as shown in Fig. 3.29. Two detectors were used: a silicon detector with a 1 cm^2 active area for the 0.4−1.9 μm range, and a germanium detector with an active area of 0.2 cm^2 covering 0.8−1.8 μm. To eliminate any possible stray light from the calibration a tight enclosure was built and baffles inserted between the radiometer and the integrating sphere. The internal surface of the radiometer was blackened to minimize any stray light from the source. This arrangement reduced the background signal to the noise level of the radiometer.

In a similar calibration, narrowband filters were inserted in front of the detector to get the calibration at each ATSR wavelength. The filters were from the same batch as were used in the construction of the ATSR-2 and AATSR visible focal plane assemblies. For the 1.6 μm channel a commercial filter was used as no "witness" filter was available for this test. The spectral responses of these filters were measured at Bentham Instruments immediately after the sphere calibration was performed. The center wavelength, bandwidth, and peak transmissions of these filters are given here in Table 3.2. Here the "witness" version refers to the AATSR flight model filter as does the same connotation for ATSR-2.

The full aperture spectral radiances from the National Physical Lab (NPL) calibration and that carried out at Rutherford Appleton Lab (RAL) are compared in Table 3.3. The table shows very good agreement between RAL and NPL for 0.87 and 0.56 μm, respectively, but at 0.66 μm the RAL readings are consistently 6.5% lower than those from NPL which was later traced to a slight misalignment of the witness filter in the RAL setup. Better agreement (<2%) was achieved when the measurements were repeated at RAL taking care to make sure that the filters were optimally aligned.

At 1.6 μm, there is a significant difference between the RAL and NPL measurements. The differences and the lower accuracy of the NPL calibration highlight the difficulty in measuring radiances in the near-IR. The differences can be attributed to a combination of effects, errors in the filter transmission, area of the detector aperture and, more significantly, radiometric leaks and stray light.

Table 3.2 Along-Track Scanning Radiometer (ATSR) Channel Filter Characteristics

Channel (μm)	Filter Set	Mid Wavelength	Bandwidth (nm)	Transmission
0.56	AATSR flight	0.560	20.79	
	AATSR witness	0.560	20.09	0.4624
	ATSR-2 witness	0.558	21.22	0.8560
0.66	AATSR flight	0.660	20.13	
	AATSR witness	0.659	20.69	0.6771
	ATSR-2 witness	0.657	20.41	0.8275
0.87	AATSR flight	0.863	20.14	
	AATSR witness	0.863	20.25	0.7301
	ATSR-2 witness	0.863	21.65	0.8859
1.6	AATSR flight	1.594	62.88	
	Northern optics	1.610	166.79	0.6130

Table 3.3 Main Aperture Calibration Measurements by National Physical Lab and Rutherford Appleton Lab (in Parentheses)

| Power Level (W) | Integrating Sphere Radiance (mW/cm^2 sr) | | | |
	1.6 μm	0.87 μm	0.66 μm	0.56 μm
390	7.55 (9.94)	7.41 (7.46)	5.73 (5.26)	3.43 (3.42)
240	4.51 (6.01)	4.43 (4.45)	3.41 (3.16)	2.05 (2.05)
150	3.09 (4.04)	3.04 (3.03)	2.34 (2.13)	1.40 (1.38)
90	1.51 (1.95)	1.48 (1.48)	1.14 (1.06)	0.68 (0.69)
45	0.77 (0.98)	0.75 (0.74)	0.58 (0.53)	0.35 (0.35)

3.8.2 POLARIZATION FILTERS

Integrating spheres can also be used to test polarization filters that have been incorporated into a sensor. Such a test arrangement is shown here in Fig. 3.30, which was used for the ATSR 1, 2 and the AATSR. For this test a smaller 30 W integrating sphere was used. Two standard Ealing polarization filters were used, one for the visible range from 0.2 to 0.8 μm and the other to cover the near-IR range from 0.8 to 2.2 μm. The polarizers were mounted in rotatable mounts so that their variation in response could be measured as a function of polarization angle. A calcite polarizer was used to determine the orientation of the filters in the test setup (Fig. 3.30).

Before establishing the relative orientation of the polarizers, it was necessary to measure any polarization introduced by the integrating sphere. Radiometer readings were taken for all filter and detector combinations. The variation in measured signal, relative to the mean sphere output, was less than ±0.5% for all wavelengths (Fig. 3.31).

The polarization sensitivity of the radiometer was removed by performing the measurements with the detector set at orthogonal orientations. The variation at 1.6 μm in Fig. 3.31 is mainly due to sensor drift. The calcite polarizer was inserted into the light path and oriented so that the plane of the

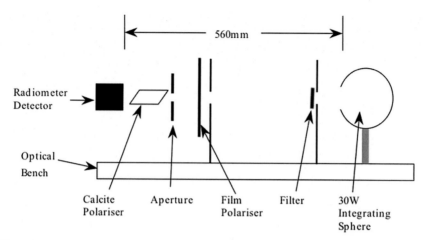

FIGURE 3.30

Test arrangement for calibrating the orientation of the thin film polarizing filters.

polarized light was perpendicular to the optical bench. Radiometer readings were again taken for each detector and filter combination. The results showed that the visible filter was well set up in the rotating mount and required no correction. The data for the IR filter showed that it was offset by −60 degrees in the mount and therefore a corresponding correction was required.

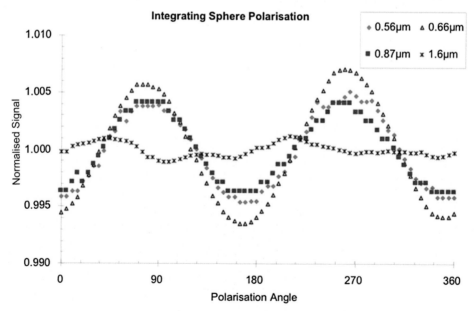

FIGURE 3.31

Measured variation of 30 W integrating sphere output with polarization angle.

3.9 LIGHT DETECTION AND RANGING

Light detection and ranging (LIDAR) is sometimes called "laser radar" in an abuse of language because its principles are similar to those of the radar systems that will be discussed later in this text. LIDAR is an optical remote sensing technique that measures the properties of scattered light to find the range or speed of distant targets, or the backscattering and attenuation of volume targets such as the atmosphere or the sea water. As in radar systems, the range to an object is determined by measuring the time delay between transmission of a (light) pulse and the detection of the reflected signal. LIDAR technology was first used in 1962 by Fiocco and Smultin reflecting a laser beam off the surface of the Moon and studying the turbidity in upper atmospheric layers. Later, in 1963 it was used by Ligda to perform the first cloud height and aerosols measurements. LIDAR has seen applications in archeology, geography, geology, geomorphology, seismology, remote sensing, and atmospheric physics.

3.9.1 PHYSICS OF THE MEASUREMENT

The main difference between LIDAR and radar is that LIDAR operates with much shorter wavelengths of the EM spectrum, in atmospheric transmission windows in the ultraviolet (UV), visible, and near-IR (e.g., $0.4-0.7, 0.7-1.5, 3-5$, and $9-13$ μm) (http://en.wikipedia.org/wiki/lidar). Thus, in general it is possible to detect a feature or object, which is about the size of the wavelength or larger. Thus, LIDAR is very sensitive to atmospheric aerosols and cloud particles and has many applications in atmosphere research and meteorology. However, an object needs to produce a dielectric discontinuity to reflect the transmitted wave. At radar (microwave or radio) frequencies, a metallic object produces a significant reflection. However nonmetallic objects, such as rain and rocks produce weaker reflections and some materials may produce nondetectable reflection at all, meaning some objects or features are effectively invisible at radar frequencies. This is especially true for very small objects (such as single molecules and aerosols).

Lasers provide one solution to these problems. The beam densities and coherency are excellent, and the wavelengths are much smaller. The basic atmospheric LIDAR equation (volumetric target) is given by Eq. (3.12):

$$P(\lambda, R) = P_0 \cdot \frac{c \cdot \tau}{2} \cdot \beta(\lambda, R) \cdot \frac{A_r}{R^2} \cdot exp\left\{ -2 \int_0^R \alpha(\lambda, R)dr \right\} \cdot \xi(\lambda) \cdot \xi(R), \qquad (3.12)$$

where $P(\lambda, R)$ is the received power at a wavelength λ from a range R, which is associated with a time delay $t = 2 \cdot R/c$, P_0 is the transmitted pulse power, τ is the pulse duration, $c \cdot \tau/2$ is the pulse "duration" in the range direction, $\beta(\lambda, R)$ is the backscattering coefficient, A_r is the effective area of the receiving system (typically a telescope), $\alpha(\lambda, R)$ is the absorption coefficient of the atmosphere, $\xi(\lambda)$ is the transmissivity of the receiver optics, and $\xi(R)$ is the so-called overlapping factor, which accounts for the volume intersection between the transmitted laser beam and the receiving cone, as illustrated in Fig. 3.32.

The detected output voltage of the LIDAR returns at each range gate is proportional to the received power (Eq. 3.12):

$$\begin{aligned}
\overline{V}_0 &= G_T \cdot R \cdot P_s + G_T(I_{ds} + M \cdot I_{db}) \\
&= G_T \cdot R\{P(R) + \xi(\lambda) \cdot P_{\text{back}}\} + G_T(I_{ds} + M \cdot I_{db}) \\
&= \underbrace{G_T \cdot R \cdot P(R)}_{s} + \underbrace{G_T \cdot R \cdot \xi(\lambda) \cdot P_{\text{back}} + G_T(I_{ds} + M \cdot I_{db})}_{V_{\text{off}}},
\end{aligned} \qquad (3.13)$$

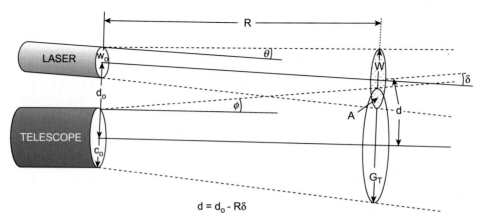

FIGURE 3.32

Graphical description of the overlapping factor as the intersection between the transmitted beam and receiver cone.

Credits: F. Rocadenbosch (UPC)

where G_T is the amplifier's voltage gain, P_{back} is the background power (power collected from the ambient light), S stands for the signal term, V_{off} stands for the offset term (to be compensated for), and all other terms have been previously defined.

The noise associated to the LIDAR measurements has three different contributions: the "shot" noise associated with the signal power (P), the "shot" noise associated to the background power, and the thermal noise $\left(\sigma_{thermal}^2\right)$[4]:

$$N = \sigma_V = \sqrt{2q \cdot G_T^2 \cdot F \cdot M^2 \cdot R \cdot P_S \cdot B + 2q \cdot G_T^2 (I_{ds} + F \cdot M^2 \cdot I_{db}) \cdot B + \sigma_{thermal}^2}, \qquad (3.14)$$

Since the SNR is typically very low, long incoherent averaging is required to increase it as in radar systems.

Different types of scattering are used for different LIDAR applications, most common are Rayleigh scattering, Mie scattering, and Raman scattering as well as fluorescence. The wavelengths are ideal for making measurements of smoke and other airborne particles (aerosols), clouds, and air molecules. A laser typically has a very narrow beam, which allows the mapping of physical features with very high resolution compared with radar. In addition, many chemical compounds interact more strongly at visible wavelengths than at microwaves, resulting in a stronger image of these materials. Suitable combinations of lasers can allow for remote mapping of atmospheric contents by looking for wavelength-dependent changes in the intensity of the returned signal.

LIDAR has been used extensively for atmospheric research and meteorology. With the deployment of the global positioning systems (GPS) in the 1980s precision positioning of aircraft became possible.

---

[4]Since the output voltage is proportional to the input power, the noise is proportional to the standard deviation of the output voltage, and not to the variance.

GPS-based surveying technology has made airborne surveying and mapping applications possible and practical. Many have been developed, using downward-looking LIDAR instruments mounted in aircraft or satellites.

3.9.2 OPTICAL AND TECHNOLOGICAL CONSIDERATIONS

There are in general two kinds of LIDARs: "incoherent" or direct detection of the power return, mainly an amplitude measurement, and "coherent" which uses the Doppler shift and must keep track of the phase information in each laser pulse. Coherent systems generally use optical heterodyne detection which is more sensitive than direct detection and allows operation at a much lower power levels, but at the expense of having more complex transceiver requirements.

In both coherent and incoherent LIDARs, there are two types of pulse models: micropulse and high-energy LIDARs. Micropulse systems have been developed as a result of the ever-increasing computer power available to process the sensor data combined with marked advances in laser technology. Micropulse systems typically operate at power levels that are "eye safe" meaning that they can operate without any additional safety precautions. High-power systems are common in atmospheric research where they are widely used for measuring atmospheric parameters such as cloud height, layering and densities of clouds, cloud particle properties, temperature, pressure, wind, humidity, trace gas concentrations (ozone, methane, nitrous oxide, etc.), aerosols.

The components of a typical LIDAR are as follows:

1. **Laser**. 600−1000 nm lasers are most common for nonscientific applications. They are inexpensive, but since they can be focused and easily absorbed by the eye the maximum power is limited by the need to make them eye-safe. Eye safety is often a requirement for most applications. A common alternative, the 1550 nm lasers are eye-safe at much higher power levels since this wavelength is not focused by the eye, but the detector technology is less advanced in this spectral region, so these wavelengths are generally used at longer ranges and lower accuracies. Airborne topographic mapping LIDARs generally use 1064 nm diode−pumped YAG lasers, while bathymetric systems generally use 532 nm frequency doubled diode−pumped YAG lasers because 532 nm penetrates water with much less attenuation than does 1064 nm. Variables in the individual systems include the ability to set the number of passes required through the gain (YAG, YLF, etc.) and Q-switch speed. Shorter pulses achieve better target resolution provided the LIDAR receiver detectors and electronics have sufficient bandwidth.
2. **Scanner and optics**. How fast images can be developed is also affected by the speed at which they can be scanned into the system. There are several different ways to scan the azimuth and elevation, including dual oscillating plane mirrors, a combination with a polygon mirror, a dual axes scanner. Optic choices affect the angular resolution and range that can be detected. A hole-mirror or beam splitter can be used to collect a laser return signal.
3. **Photodetector and receiver electronics**. Two different photodetector technologies are used in today's LIDARs: solid-state photodetectors, such as silicon avalanche photodiodes, or photomultipliers. The sensitivity of the receiver is another parameter that has to be balanced in a LIDAR design.
4. **Position and navigation systems**. LIDAR sensors mounted on mobile platforms such as airplanes or satellites require instrumentation to determine the absolute position and orientation (pointing angle) of the sensor. GPS and inertial measurement unit systems are primary types of systems used for this purpose.

3.9.3 APPLICATIONS OF LIDAR SYSTEMS

There are many different applications of LIDAR systems, but we will concentrate on those primarily in meteorology, which was one of the earliest applications of LIDAR remote sensing. The first LIDARs were used for studies of atmospheric composition, structure, clouds, and aerosols. Initially based on rube lasers, LIDARs for meteorology were constructed shortly after the invention of the laser.

Some modern LIDARs are as follows:

1. Elastic backscatter LIDAR is the simplest form of LIDAR and is typically used for studies of aerosols and clouds. The backscattered wavelength is identical to the transmitted wavelength, and the magnitude of the received signal at a given range depends on the backscatter coefficient of scatterers at that range and the extinction coefficients of the scatterers along the path to that range. The extinction coefficient is typically the quantity of interest.

2. Differential absorption LIDAR (DIAL) is used for range-resolved measurements of a particular gas in the atmosphere, such as ozone, carbon dioxide, or water vapor. The LIDAR transmits two wavelengths: an "on-line" wavelength that is absorbed by the gas of interest and an off-line wavelength that is not absorbed. The differential absorption between the two wavelengths is a measure of the concentration of the gas as a function of range. DIAL LIDARs are essentially dual-wavelength elastic backscatter LIDARs.

3. Raman LIDAR is also used for measuring the concentrations of atmospheric gases, but can also be used to retrieve aerosol parameters as well. Raman LIDAR exploits inelastic scattering to single out the gas of interest from all other atmospheric constituents. A small portion of the energy of the transmitted light is deposited in the gas during the scattering process, which shifts the scattered light to a longer wavelength by an amount that is unique to the species of interest. The higher the concentration of the gas, the stronger the magnitude of the backscattered signal.

4. Doppler LIDAR is used to measure wind speed along the beam by measuring the frequency shift of the backscattered signal. Scanning LIDARs have been used to measure atmospheric wind velocity in a large three-dimensional core. ESA's wind mission Atmospheric Dynamics Mission Aeolus (ADM-Aeolus) will be equipped with a Doppler LIDAR system to provide global measurements of vertical wind profiles. Doppler LIDAR systems are now beginning to be successfully applied in the renewable energy sector to acquire wind speed, turbulence, wind veer, and wind shear data. Both pulse and continuous wave systems are being used for these applications.

The number of spaceborne LIDARs has been very limited because of the reliability of lasers. So far, the only LIDAR-based mission is the NASA/CNES cloud-aerosol LIDAR and infrared pathfinder satellite observation (CALIPSO) mission, which belongs to the A-TRAIN constellation. CALIPSO combines an active LIDAR instrument with passive IR and visible imagers to probe the vertical structure and properties of thin clouds and aerosols over the globe. CALIPSO was launched on April 28, 2006, with the cloud profiling radar system on the CloudSat satellite. Fig. 3.33 shows an artist's view of the CALIPSO mission in the A-TRAIN constellation, a picture and a drawing of it, and sample LIDAR returns from the sea surface up to 30 km height. The dark blue regions underneath the high-reflectivity regions (in pink) correspond to regions where the SNR is too low because of the increased attenuation of the signal in previous regions. The main engineering parameters of the CALIPSO LIDAR are listed in Table 3.4.

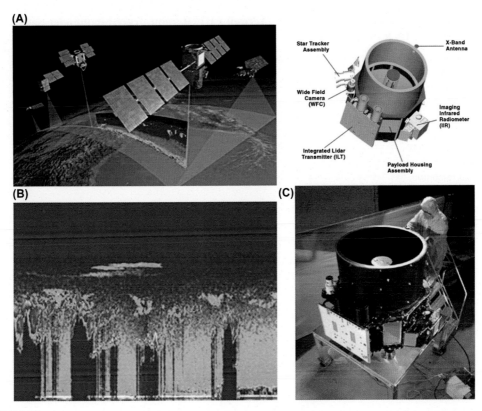

FIGURE 3.33

(A) Artist's view of the NASA cloud-aerosol LIDAR and infrared pathfinder satellite observation (CALIPSO) mission in the A-TRAIN constellation. (B) Example of data collected by CALIPSO's LIDAR in June 2006: data extend from sea level to 30 km. (C) CALIPSO satellite picture and drawing indicating its main components. Note the large telescope to collect the laser returns. (http://images.slideplayer.com/17/5360190/slides/slide_ 3.jpg).

3.9.4 WIND LIDAR

As just mentioned a wind LIDAR is a Doppler LIDAR that uses the frequency shift of the back-scattered signal to determine the wind velocity. This concept is depicted here in Fig. 3.34 adapted from Dobler et al., 2002.

This coherent Doppler wind LIDAR measures the frequency of the beat signal obtained by optically mixing the return signal with the local oscillator. As a consequence, both the local oscillator and the return signal must have narrow bandwidths to have sufficient coherent lengths. Thus, coherent LIDAR detection relies of the aerosol scattering with very narrow Doppler broadening meaning the LIDAR wind measurements apply only to those atmospheric regions with adequate aerosol loading. Since the Mie scattering due to aerosols is better suited to frequency analysis than is the molecular Rayleigh scattering the choice of LIDAR wavelength depends on the expected return signal and the expected ratio of

Table 3.4 Main Parameters of the Cloud-Aerosol LIDAR and Infrared Pathfinder Satellite Observation LIDAR

Parameter	Value/Description
Laser	Nd:YAG, diode-pumped, Q-switched, frequency-doubled
Wavelengths	532 nm, 1064 nm
Pulse energy	110 mJ/channel
Pulse repetition frequency	20.25 Hz
Receiver telescope	1.0 m diameter
Footprint/field of view	100 m/130 μrad
Vertical resolution	30–60 m
Horizontal resolution	333 m
Linear dynamic range	22 bits
Data rate	316 kbps

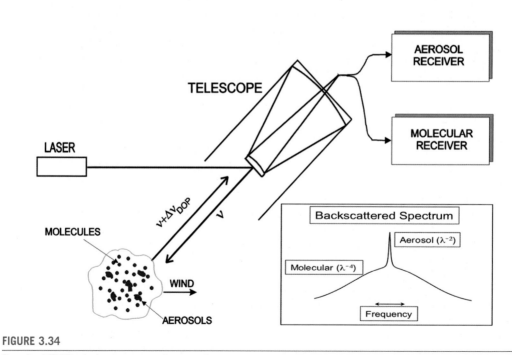

FIGURE 3.34

Doppler LIDAR wind measurement concept.

aerosol-to-molecular backscatter. The molecular scattering cross-section is proportional to λ^{-4}, and the aerosol signal is proportional to between λ^{-2} and λ^{+1}, depending on the wavelength and particle size and shape. Even if the aerosol returns decrease with increasing wavelength, the molecular background decreases much faster, so that the aerosol-to-molecular backscatter ratio becomes more favorable to the measurement. Therefore, longer wavelengths are desirable to minimize the influence of molecular (Rayleigh) scattering. Coherent Doppler LIDARs use laser wavelengths between 1 and 11 μm.

3.9.4.1 Vector Wind Velocity Determination

Vector wind measurements require radial velocity measurements from three independent "lines-of-sight" meaning you must have three LIDARs. If it can be assumed that there is no vertical velocity ($W = 0$), then only two LIDARs are needed. If horizontal homogeneity of the wind field can be assumed, then a LIDAR beam scanning technique can be used to determine the wind velocity.

The two main techniques are the velocity azimuth display (VAD) which is a conical scan LIDAR beam at a fixed elevation angle, and the Doppler beam swinging (DBS) which is a LIDAR pointing in the vertical which is tilted east and tilted north. These two methods are graphically displayed in Fig. 3.35.

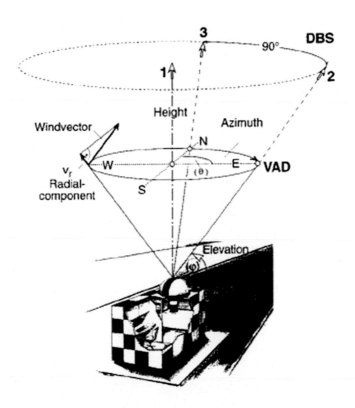

FIGURE 3.35

Schematic of the scan technique of a Doppler LIDAR. *DBS*, Doppler beam swinging; *VAD*, velocity azimuth display.

3.9.4.1.1 Velocity Azimuth Display LIDAR Vector Wind Method

The VAD scheme is a conical scan LIDAR beam at a fixed elevation angle (Fig. 3.35). For a ground-based LIDAR, positive u, v, w are defined as the wind blowing toward the East, North, and upward, and the positive radial wind $\vec{V}_R$ as the wind blowing away from the LIDAR. $\vec{V}_R$ consists of the following u, v, and w components:

$$\vec{V}_R = u \cdot sin\theta \cdot cos\varphi \cdot \hat{x} + v \cdot cos\theta \cdot cos\varphi \cdot \hat{y} + w \cdot sin\varphi \cdot \hat{z}, \tag{3.15}$$

where θ is the azimuth angle clockwise from the north, and φ is the elevation angle.

For each VAD scan the elevation angle φ is fixed and known, the azimuth angle θ is varied, but is also known. $\vec{V}_R$ is measured so the three unknowns u, v and w can be derived directly from fitting the data with the above Eq. (3.15).

3.9.4.1.2 Doppler Beam Swinging LIDAR Vector Wind Method

The DBS technique consists of pointing a LIDAR in the vertical up, and then tilting it east and north (Fig. 3.36).

If γ is the off-zenith angle then:

$$V_{RE} = u \cdot sin\gamma + w \cdot cos\gamma, \tag{3.16a}$$

$$V_{RN} = v \cdot sin\gamma + w \cdot cos\gamma, \tag{3.16b}$$

$$V_{RZ} = w, \tag{3.16c}$$

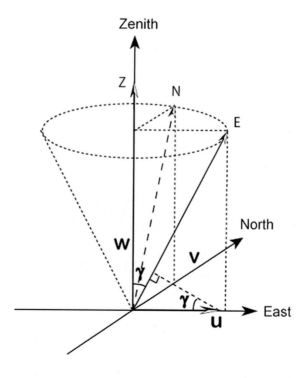

FIGURE 3.36

LIDAR orientation for the Doppler beam swinging wind vector.

then:

$$u = \frac{V_{RE} - V_{RZ} \cdot cos\gamma}{sin\gamma}, \tag{3.17a}$$

$$v = \frac{V_{RN} - V_{RZ} \cdot cos\gamma}{sin\gamma}, \tag{3.17b}$$

$$w = V_{RZ}, \tag{3.17c}$$

where V_{RZ}, V_{RE}, V_{RN} and are the vertical, tilted east, and tilted north radial velocities, respectively.

3.9.4.2 Direct Detection Doppler Wind LIDAR

Direct Detection Doppler (DDL) uses incoherent detection to measure the spectrum of returned signals. DDL can use aerosol/molecular scattering and/or resonance fluorescence to measure the wind from the ground to the upper atmosphere. There are several different ways to do the spectral analysis for the DDL method:

- Resonance fluorescence Doppler LIDAR uses the atmospheric atomic or molecular absorption lines as the frequency analyzer/discriminator.
- Direct detection Doppler LIDAR is based on molecular absorption edge filter, e.g., iodine (I_2) vapor filter, Na or K magnetooptic filter.
- Direct detection Doppler LIDAR is based on optical interferometer edge-filter, e.g., Fabry–Perot etalon transmission edge.
- Direct detection Doppler LIDAR is based on fringe patter imaging of an optical interferometer, e.g., FPI imaging.

3.9.4.3 LIDAR Wind Summary

Doppler wind techniques measure the wind velocity along the LIDAR beam requiring three independent radial velocity measurements from three independent lines of sight. Rather than point three different LIDARs in three different directions, we assume horizontal homogeneity of the wind field over the volume we are sensing and employ scanning LIDAR techniques to determine the vector wind. The two main scanning techniques are the VAD and the DBS methods. There are different wavelength requirements for coherent and incoherent detection LIDARs. Fig. 3.37 shows a commercial buoy carrying a VAD wind LIDAR (principles of operation explained below) to measure wind profiles up to 200 m height to optimize the selection of offshore aerogenerators.

An example of a spaceborne wind LIDAR is the upcoming ESA ADM-Aeolus that will be launched in 2017 and will provide for the first time global observation of wind profiles from space to further our knowledge of Earth's atmosphere and weather systems. Aeolus carries a single payload, the atmospheric laser Doppler instrument (ALADIN), a direct detection Doppler wind LIDAR operating at near UV wavelengths (355 nm). It comprises two main assemblies: (1) Transmitter: diode laser–pumped Nd:YAG laser, frequency tripled to 355 nm at 150 mJ pulse energy, 100 Hz pulse repetition and (2) Receiver: 1.5 m diameter SiC telescope, Mie channel (aerosol and water droplets) with Fizeau spectrometer, Rayleigh channel (molecular scattering). Fig. 3.38 shows an artist's view of the satellite ADM-Aeolus and its different subsystems.

FIGURE 3.37

Universitat Politècnica de Catalunya (UPC) designed EOLOS FLS200 buoy (manufactured by Eolos Floating Solutions) carrying a wind LIDAR to measure wind fields up to 200 m to optimize the location of offshore aerogenerators.

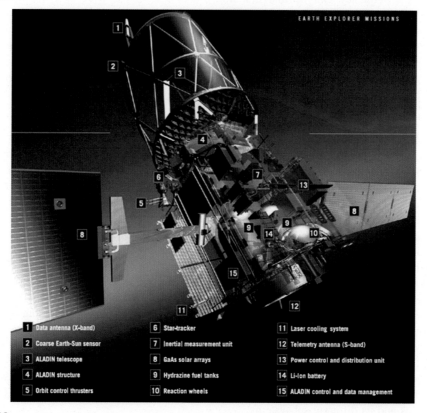

1 Data antenna (X-band)	**6** Star-tracker	**11** Laser cooling system
2 Coarse Earth-Sun sensor	**7** Inertial measurement unit	**12** Telemetry antenna (S-band)
3 ALADIN telescope	**8** GaAs solar arrays	**13** Power control and distribution unit
4 ALADIN structure	**9** Hydrazine fuel tanks	**14** Li-Ion battery
5 Orbit control thrusters	**10** Reaction wheels	**15** ALADIN control and data management

FIGURE 3.38

Atmospheric Dynamics Mission Aeolus payload components (http://www.esa.int/Our_Activities/Operations/ADM-Aeolus_operations). *ALADIN*, atmospheric laser doppler instrument.

From ESA. http://www.esa.int/spaceinimages/Images/2007/01/ADM-Aeolus_payload_components

3.10 STUDY QUESTIONS

1. Explain what the reflected radiance is in the visible part of the spectrum, its units, and how it can be computed from the surface's parameters, and the Sun irradiance.
2. At which wavelengths does the radiation of a blackbody at physical temperature equal to 300K and 6000K reach the maximum value?
3. The Landsat Thematic Mapper sensor covers the visible and near-IR parts of the spectrum in the following bands: TM-1 (0.45−0.52 µm), TM-2 (0.52−0.60 µm), TM-3 (0.63−0.69 µm), TM-4 (0.76−0.90 µm), TM-5 (1.55−1.75 µm), TM-6 (10.40−12.50 µm) and TM-7 (2.08-2.35 µm). Discuss which bands can be applied for the following applications and why:
   a. Vegetation health
   b. Forest fires detection
   c. Water bodies and coastline definition
   d. Crop classification
4. An optical instrument pointing at nadir from 500 km height has a CCD of 800×600 pixels in the focal plane of a convergent lens. Each photodetector has a size 10 µm $\times$ 10 µm, and the signal's bandwidth is 1 MHz. The lens diameter is 80 mm, and the focal distance 500 mm. Determine the pixel size over the Earth, the NEP if the specific detectivity is $D^*(\lambda) = 2 \times 10^{10} \frac{cm \sqrt{Hz}}{W}$, and the image size in kilometers.
5. A photodetector of 0.1 mm $\times$ 0.1 mm is mounted in a telescope with a focal length of 80 cm and an aperture of 10 cm diameter. Compute the pixel size when it observes the Earth's surface from a 780 km high satellite, and for an incidence angle of 45 degrees (major and minor axes) assuming a flat Earth and taking into account the Earth's curvature.
6. Compute the optical power collected by the photodetector of the previous question when it points to nadir, the surface is rough with a 30% reflectance, and the wavelength is $\lambda = 1.6$ µm. Note: assume that the Solar exoatmospheric radiation at this wavelength is 100 W/m^2 µm, the one way atmospheric transmissivity is 90%, the spectral width of the photodetector is 0.3 µm, and the transmissivity of the telescope optics is 80%.
7. Briefly explain the different types of LIDAR sensors to sense gas composition, and their principles of operation.
8. An optical push-broom sensor has a linear array of photodetectors in the focal plane of a convergent lens. If the photodetector's size is 0.75 mm $\times$ 0.75 mm, the system's bandwidth is 150 kHz, the diameter of the lens is 76.2 mm, and the focal length is 4 times the diameter of the lens, compute the NEP if the specific detectivity is $D^*(\lambda) = 3.31 \times 10^{10} \frac{cm \sqrt{Hz}}{W}$.
9. We want to design a multispectral airborne sensor (blue, green, red, and near-IR) for cartographic applications. The sensor will perform a push-broom scan and it has several parallel linear CCDs in the focal plane of the camera, one per band, as sketched in the following figure. The sensor has the following parameters:
   Linear CCD array with 12,000 pixels, photodetector size 6.5 µm $\times$ 6.5 µm, and a 6.5 µm spacing between them.
   Access time per line: 1 ms
   Focal length $f = 107$ mm and F# $= f/D = 4$, being D the aperture diameter of the optical system
   Transmissivity of the optics $\tau_{opt} = 0.68$
   Nominal flight altitude = 10 km
   Nominal flight speed = 140 m/s

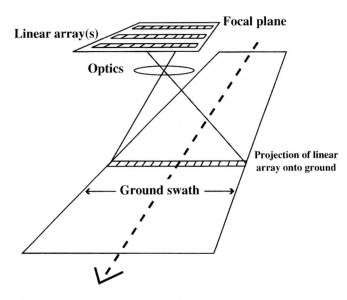

For the CCD located in the center of the focal plane (green) compute:

a. The swath width in the cross-track direction (ground swath in the figure), assuming a flat surface.

b. The angular extent of each individual photodetector of the CCD (IFOV) and the pixel size (GIFOV) for a pixel in the center of the swath.

c. To obtain images with an aspect ratio 1:1 in the swath center, obtain the frequency of acquisition of the lines of the CCD (in lines per second) at the nominal flight altitude and speed.

d. The optoelectrical conversion, access, and digitalization of the signals collected by the photodetectors must be performed in the period of acquisition of each line. If all these processes are performed consecutively without dead times, compute the integration time for the photodetectors.

e. Determine the bandwidth B associated with the inverse of the integration time of each photodetector.

f. If the specific detectivity is $D^* = 9.5 \times 10^{11}$ cm $(Hz)^{1/2}$/W, determine the NEP of each photodetector.

g. If the solar irradiance over the surface for the green band ($\lambda = 533{-}587$ nm) is 1400 W/m^2 μm under clear sky conditions, determine the minimum detectable reflectance change (NEΔρ) for a Lambertian surface. Note: the atmospheric transmissivity in the ascending path can be assumed to be one.

h. Taking into account that the maximum reflectance for rough surfaces is one (lossless Lambertian reflector), determine the number of bits required in the analog-to-digital conversion for the whole dynamic range so that the quantization step be smaller or equal to the NEΔρ of the sensor.

10. "CubeSats" are low-cost pico-satellites originally conceived by the California Polytechnic State University for teaching and training. However, they are now being considered also for small scientific and technology demonstrator missions. The basic CubeSat structure consists of a 10 cm side cube, with a maximum weight of 1.33 kg. All subsystems, including the onboard computer, electrical power supply, telecommunications, etc. have to fit inside, leaving only two boards for the payloads. We want to consider placing a CCD camera (photodiode length = 20 μm, spacing = 0.2 μm; VGA resolution: 640 × 480 pixels) in a CubeSat that will orbit the Earth in a circular orbit at 650 km height. If the maximum focal length is then 10 cm, and the lens diameter 1 cm, please compute:

a. The pixel's size at nadir (in m).
b. The image size (in km).
c. The image acquisition time (in s) for a maximum blurring of 1/10th of the pixel size.
d. With the acquisition time computed in (c) compute the maximum rotation speed for the same pixel blurring.
e. The bandwidth (Hz) if the readings have to be stabilized at 99.9% of their final value.
f. The NEP (in W).
g. The signal-to-noise ratio (in dB) when imaging the surface of the Earth (reflection coefficient = 0.4, Lambertian surface).
h. The number of bits required to sample each pixel of the image.
i. The amount of information in one image (without compression) and the time to download it to the ground station with a 9.6 kbps downlink.
j. If instead of a 1 unit CubeSat as in Fig. 3.39A, a three Unit one (3 times longer) is used so as to be able to increase the focal length by a factor of 3, how it will impact the pixel's size and the swath?

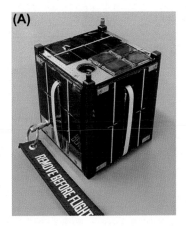

(A)

FIGURE 3.39A

Sample 1 Unit CubeSat.

Data:
Gravitational constant: $k = 3.986 \times 10^{14} \text{ m}^3/\text{s}^2$
Spectral range of the detectors: 0.4–0.7 μm

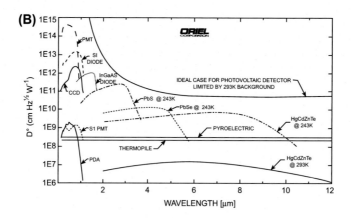

FIGURE 3.39B

Specific detectivity (cm $\sqrt{\text{Hz}}/\text{W}$) versus wavelength ($\mu$m) for different types of detectors.

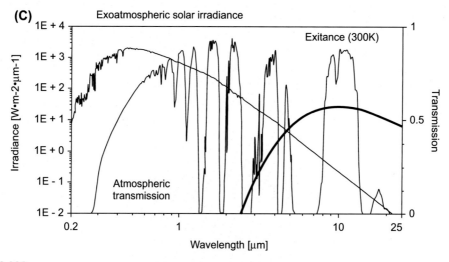

FIGURE 3.39C

Solar irradiance and atmospheric transmission versus wavelength (μm).

11. To measure the temperature of the Earth at global scale in the thermal IR band (10 μm, where there is an atmospheric transmission window), with a spatial resolution at nadir of 10 km, and a revisit time of 15 min, a constellation of spin stabilized geostationary satellites is required, taking advantage of the rotation to scan from East to West the images. The North to South scan is performed every 15 min completing the image, as sketched in Fig. 3.40. Please compute:
   a. The altitude of the satellite to be in geostationary orbit.

b. The number of satellites to have global coverage. How much overlap (at the Equator) will exist between the FOV of two adjacent satellites? What is the range of latitudes that can be measured?

c. The angular resolution of the system so that the spatial resolution at nadir is 10 km. What is the spatial resolution of a pixel over the same Meridian, but at a latitude of 45 degrees?

d. If the North–South scan only covers the Earth's disk, i.e., all lines have at least one pixel pointing towards the Earth, how many lines will the image have? If the image scan sweeps four lines simultaneously without overlapping, what is the integration time per pixel, and the minimum bandwidth required?

e. If the spectral radiance of a blackbody at 300K is $L_{300} = 9924$ W/(m^2 sr µm), and assuming that the average emissivity of the surfaces of interest at 300K is 90%, compute the optical power collected by the photodetector.

f. If the minimum detectable temperature change is 1K, and the spectral radiance of a blackbody at 301K is $L_{301} = 10,085$ W/(m^2 sr µm), compute the required NEP and the specific detectivity of the photodetector in cm $\sqrt{\text{Hz}}/\text{W}$.

g. How many bits will be required to quantize each pixel?

h. Compute the amount of information in kilobytes per second generated by each of the sensors in the different satellites.

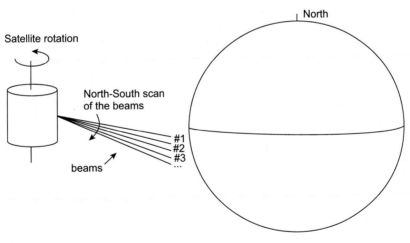

FIGURE 3.40

Geostationary spinning satellite.

Data:
- spin rate of each geostationary satellite: three revolutions per second
- spatial resolution at nadir: 10 km
- revisit time: 15 min (maximum interval between consecutive images, the time of acquisition of an image can be smaller than that)
- Longitude of the array of photodetectors: four

- Photodetector size: 0.2×0.2 mm
- Aperture diameter: 20 cm
- Wavelength: 10 μm
- Spectral width: 4 μm
- Atmospheric transmissivity: 85%
- Optical system transmissivity: 90%
- Dynamic range: $T_{min} = 150K$, $T_{max} = 320K$, $\Delta T = 1K$

MICROWAVE RADIOMETRY

4

Microwave radiometry is the science of the measurement of the noise emitted by bodies at a physical temperature higher than zero Kelvin. In this chapter, the physical principles will be presented, and the main technologies and techniques to perform these measurements.

4.1 BASIC CONCEPTS ON MICROWAVE RADIOMETRY
4.1.1 BLACKBODY RADIATION

As described in previous chapters, all bodies at a nonzero Kelvin temperature emit electromagnetic radiation. Gases radiate at discrete frequencies. According to quantum theory, each spectral line corresponds to the transition of an electron from an atomic energy level E_1 to another one E_2. The radiation is produced at a frequency given by Bohr's equation:

$$f_{1 \to 2} = \frac{E_1 - E_2}{h}. \tag{4.1}$$

where h is the Planck's constant: $h = 6.63 \times 10^{-34}$ J s. Alternatively, if a photon at a frequency $f_{1 \to 2}$ is absorbed, an electron is excited from energy level E_2 to energy level E_1.

When the interaction between gas atoms or molecules increases due to increased kinetic energy (higher temperature) or increased volumetric density, the energy levels split, and the emission is no longer produced at a single frequency $f_{1 \to 2}$, but at a band around that frequency $f_{1 \to 2}$. As an example, Fig. 4.1A and B show the water vapor and oxygen absorption bands around 22 GHz and from 55 to 65 GHz, respectively, for different atmospheric heights. It can be appreciated that at higher altitudes (lower atmospheric pressure) the bands are narrower than at sea level (higher atmospheric pressure). Their amplitude also decreases due to the shorter atmospheric path. In the case of O_2, except at very high altitudes, the bands even overlap among them and appear as a single wide band. This information will be used later to infer atmospheric temperature profiles.

For all materials at thermodynamic equilibrium, Kirchhoff's law (1860) states that all the absorbed power must then be reradiated, otherwise, if they radiate more power than they absorb, their physical temperature would decrease indefinitely, while if they absorb more power than they emit, their physical temperature would increase indefinitely.

A "black body" is an ideal body that acts as a perfect absorber, absorbing all the power impinging on it, at all frequencies, from all directions, and at all polarizations. Blackbodies do not exist in real life, but in the microwave part of the spectrum a close approximation to the blackbodies are the microwave absorbers used, for example, in the walls of anechoic chambers. In thermal equilibrium, the

Introduction to Satellite Remote Sensing. http://dx.doi.org/10.1016/B978-0-12-809254-5.00004-X

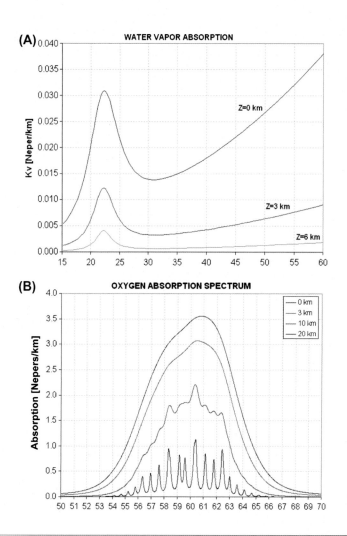

FIGURE 4.1

Atmospheric absorption due to oxygen at different atmospheric heights as a function of the frequency in GHz computed for a US standard atmosphere, temperatures and pressures (Gary and Hereford, 2013).

power absorbed by a blackbody is then reradiated isotropically, but with a frequency dependence which is given by Planck's law[1] (Planck, 1901):

$$B_f = \frac{2 \cdot h \cdot f^3}{c^2} \cdot \frac{1}{e^{\frac{h \cdot f}{k_B \cdot T_{ph}}} - 1} \quad (\text{W}/\text{m}^2 \text{ Hz strad}). \tag{4.2}$$

---

[1]Note that Eq. (4.2) is equivalent to Eq. (2.33), but expressed in terms of the frequency instead of the wavelength.

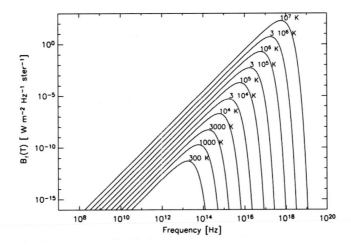

FIGURE 4.2

Planck's law for blackbodies at different physical temperatures.

(Ulaby et al., 1981).

In Eq. (4.2), B_f is called the spectral brightness density, whose physical meaning is the amount of power being radiated per unit of surface, per unit of spectral width, and per unit of solid angle. The frequency f is given in Hertz, k_B is the Boltzmann's constant $k_B = 1.38 \times 10^{-23}$J/K, T_{ph} is the absolute temperature in Kelvin, and c is the speed of light. A set of plots of the spectral brightness density versus frequency, for different temperatures is found here in Fig. 4.2.

As it can be seen, the higher the physical temperature, the larger the brightness is, and the higher the frequency where the brightness reaches its maximum. This maximum is given by Wien's displacement law (Eq. 2.35).

As it can also be noted from Fig. 4.2, this maximum typically occurs for frequencies above the thermal infrared; therefore, in the microwave part of the spectrum, where the so-called Rayleigh–Jeans law can be applied. Actually, the Rayleigh–Jeans law was obtained by Jeans based on classical mechanics, prior to Planck's quantum mechanics, but as for Eq. (2.36), it can be interpreted as a simplification of Planck's law for low frequencies:

$$B_f = \frac{2 \cdot h \cdot f^3}{c^2} \cdot \frac{1}{e^{\frac{h \cdot f}{k_B \cdot T_{ph}}} - 1} \approx \frac{2 \cdot h \cdot f^3}{c^2} \cdot \frac{k_B \cdot T_{ph}}{h \cdot f} = \frac{2 \cdot k_B \cdot f^2 \cdot T_{ph}}{c^2} = \frac{2 \cdot k_B \cdot T_{ph}}{\lambda^2}. \quad (4.3)$$

Note that there is a linear relationship between the spectral brightness density and the physical temperature of the body, and it is inversely proportional to the square of the wavelength.

In Fig. 4.3, we compare the Rayleigh–Jeans approximation and Planck's law. If λ and T meet the requirement that $\lambda \times T_{ph} \geq 0.77$ m K or $f/T_{ph} \leq 3.9 \times 10^8$ Hz/K, the error committed by the Rayleigh–Jeans approximation as compared to Planck's law is smaller than 1%. Thus, if the physical temperature is 300K, the frequency must be smaller than 117 GHz, which covers a large part of the microwave spectrum.

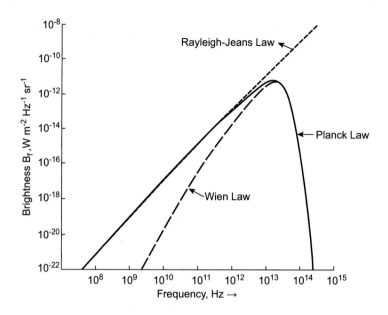

FIGURE 4.3

Planck's radiation law approximations: Rayleigh–Jeans law (low frequency limit, Eq. 4.3) and Wien's law (high frequency limit) (Ulaby et al., 1981).

Reproduced by permission from Ulaby, F., Moore, R.K., Fung, A.K., 1981. Microwave Remote Sensing. Active and Passive, Vol I: Microwave Remote Sensing Fundamentals and Radiometry. Artech House, Inc., Norwood, MA. © 1981 by Artech House, Inc.

4.1.2 GRAY-BODY RADIATION: BRIGHTNESS TEMPERATURE AND EMISSIVITY

A blackbody is an idealized body which is a perfect absorber and a perfect radiator. Real bodies, however, do not absorb all the incident power, a part is reflected, and part is transmitted into the body, which then is partially or totally absorbed. In thermodynamic equilibrium (temperature is constant), the power that is absorbed is then radiated. Therefore, a real body radiates less power than a blackbody at the same physical temperature: these bodies are called gray bodies. A semi-infinite material at a uniform temperature T is depicted in Fig. 4.4.

If the emitted brightness depends on the direction $B(\theta,\phi)$, a similar equation to that of the blackbody can be written:

$$B(\theta,\phi) = 2 \cdot \frac{k_B}{\lambda^2} \cdot T_B(\theta,\phi), \tag{4.4}$$

by replacing the physical temperature T_{ph} by the so-called "brightness temperature" $T_B(\theta,\phi)$. Since the brightness temperature of a gray body is smaller than that of a blackbody, the brightness temperature T_B is smaller than the physical temperature T_{ph}. The parameter relating both magnitudes is called the emissivity $e_f(\theta,\phi)$:

$$e_f(\theta,\phi) = \frac{B_f(\theta,\phi)}{B_{f,black\,body}} = \frac{T_B(f,\theta,\phi)}{T_{ph}}. \tag{4.5}$$

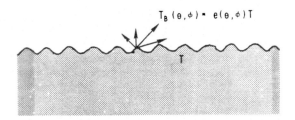

FIGURE 4.4

Brightness temperature of a semi-infinite medium at a uniform temperature (Ulaby et al., 1981).

Reproduced by permission from Ulaby, F., Moore, R.K., Fung, A.K., 1981. Microwave Remote Sensing. Active and Passive, Vol I:
Microwave Remote Sensing Fundamentals and Radiometry. Artech House, Inc., Norwood, MA. © 1981 by Artech House, Inc.

since $B_f(\theta,\phi) \le B_{f,\, black\, body}$, the emissivity is bounded by $0 \le e_f(\theta,\phi) \le 1$. The emissivity is zero for a perfect reflecting material, a lossless metal, and is one for a perfect absorber, the blackbody.

4.1.3 GENERAL EXPRESSIONS FOR THE EMISSIVITY

The scattering of a rough surface is characterized by its cross-section per unit area $\sigma^0_{p_0,p_s}(\theta_0, \phi_0, \theta_s, \phi_s)$. This parameter relates the scattered power in the (θ_s, ϕ_s) direction at p_s polarization, for an incident plane wave from the (θ_0, ϕ_0) direction at p_0 polarization. If $p_s = p_0 = h, v$, σ^0 is called the horizontal (h) or vertical (v) scattering coefficient. If $(\theta_s, \phi_s) = (\theta_0, \phi_0)$, σ^0 is the backscattering coefficient, while if $(\theta_s, \phi_s) = (\pi - \theta_0, \phi_0)$ σ^0 is the forward scattering coefficient. If $p_s \neq p_0$, σ^0 is the cross-polar scattering coefficient. By applying Kirchhoff's law to the rough surface case, Peake developed in 1959 the expressions for the horizontal and vertical emissivities $e_{h,v}(\theta_0, \phi_0)$ and the scattered downwelling temperature $T_{SC,h,v}(\theta_0, \phi_0)$:

$$e_{p_0}(\theta_0, \phi_0) = 1 - \frac{1}{4\pi \cos\theta_0} \iint \left\{ \sigma^0_{p_0,p_0}(\theta_0, \phi_0, \theta_s, \phi_s) + \sigma^0_{p_0,p_s}(\theta_0, \phi_0, \theta_s, \phi_s) \right\} d\Omega_s. \tag{4.6}$$

$$T_{SC,p_0}(\theta_0, \phi_0) = \frac{1}{4\pi \cos\theta_0} \iint \left\{ \sigma^0_{p_0,p_0}(\theta_0, \phi_0, \theta_s, \phi_s) + \sigma^0_{p_0,p_s}(\theta_0, \phi_0, \theta_s, \phi_s) \right\} T_{DN}(\theta_s, \phi_s) d\Omega_s \tag{4.7}$$

Eqs. (4.6) and (4.7) can be simplified in two cases of special interest: a specular surface and a completely rough surface.

4.1.3.1 Simple Emissivity Models: Emission From a Perfect Specular Surface

The scattering produced at a specular flat surface consists of the coherent reflection of the incident wave only. The cross-polar coefficients are zero, and the h and v scattering coefficients become delta functions:

$$\sigma^0_{p_0,p_s}(\theta_0, \phi_0, \theta_s, \phi_s) = 0, \tag{4.8}$$

$$\sigma^0_{p_0,p_0}(\theta_0, \phi_0, \theta_s, \phi_s) = 4\pi\Gamma(\theta_0, p_0)\frac{\cos\theta_0}{\sin\theta_{sp}}\delta(\theta_s - \theta_{sp})\delta(\phi_s - \phi_{sp}), \tag{4.9}$$

where (θ_{sp}, ϕ_{sp}) is the specular reflection direction.

Replacing Eqs. (4.8) and (4.9) in Eqs. (4.6) and (4.7) the following equations are readily obtained:

$$e_{p_0}(\theta_0, \phi_0) = 1 - \Gamma(\theta_0, p_0), \tag{4.10}$$

$$T_{SC,p_0}(\theta_0, \phi_0) = \Gamma(\theta_0, p_0) \cdot T_{DN}(\theta_0). \tag{4.11}$$

Fig. 4.5 illustrates this behavior. The horizontal polarization (solid line) and vertical polarization (dashed line) emissivity (left axis) and reflectivity (right axis) for flat surfaces with different dielectric constants are plotted versus the incidence angle. The following general trends can be observed:

- At nadir ($\theta_i = 0°$) the emissivity/reflectivity at horizontal and vertical polarizations are the same.
- For the horizontal polarization, the emissivity/reflectivity always decreases/increases monotonically down to zero/up to one at $\theta_i = 90°$.
- For the vertical polarization, the emissivity/reflectivity increases/decreases up to a given angle where it is a maximum/minimum, and then monotonically decreases/increases down to zero/up to one at $\theta_i = 90°$. The angle at which the emissivity is maximum/reflectivity is minimum (actually zero for a lossless medium) is called the Brewster angle. The Brewster angle increases for higher

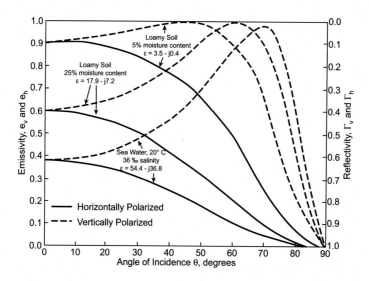

FIGURE 4.5

Horizontal (*solid line*) and vertical (*dashed line*) emissivity (left) and reflectivity (right) for flat surfaces with different dielectric constants.

(Ulaby et al., 1981).

dielectric constant values, although if there are losses, the emissivity/reflectivity does not reach the maximum/minimum value of one/zero.

As it will be seen later, the selection of the observation angle of the conical scanning microwave radiometers is very much linked to the Brewster angle over the ocean, to minimize the effect of the surface roughness (wind) on vertical polarization, and to be able to better perform atmospheric corrections, used in the horizontal polarization data resulting in more accurate surface wind speeds.

4.1.3.2 Simple Emissivity Models: Emission From a Lambertian Surface

The scattering coefficient from a perfectly rough surface, also called a Lambertian surface, depends only on the product of $\cos\theta_0 \cdot \cos\theta_s$:

$$\sigma^0_{p_0,p_0}(\theta_0, \phi_0, \theta_s, \phi_s) + \sigma^0_{p_0,p_s}(\theta_0, \phi_0, \theta_s, \phi_s) = \sigma^0_0 \cdot \cos\theta_0 \cdot \cos\theta_s, \qquad (4.12)$$

where σ^0_0 is a constant related to the dielectric properties of the scattering surface. The emissivity is then readily obtained by replacing Eq. (4.12) in Eq. (4.6):

$$e_{p_0}(\theta_0, \phi_0) = 1 - \frac{1}{4\pi \cos\theta_0} \iint \sigma^0_0 \cdot \cos\theta_0 \cdot \cos\theta_s \cdot d\Omega_s = 1 - \frac{\sigma^0_0}{4}, \qquad (4.13)$$

which is independent of the polarization and incidence angles.

Natural surfaces do not have either a perfect specular behavior, or a Lambertian behavior. They exhibit a mixed behavior, depending on the dielectric properties of the surface, and the surface roughness as compared to the wavelength.

4.1.4 POWER COLLECTED BY AN ANTENNA SURROUNDED BY A BLACKBODY

The power collected by an antenna from a direction (θ, ϕ) can be computed as:

$$dP(\theta, \phi) = A_{eff} \cdot B_f \cdot |F_n(\theta, \phi)|^2, \qquad (4.14)$$

where A_{eff} (m^2) is the effective area of the antenna, and $|F_n(\theta, \phi)|^2$ is the normalized copolar antenna radiation pattern. Therefore, the total power collected by the antenna is computed as:

$$P = \frac{1}{2} \int_{f_0-\frac{B}{2}}^{f_0+\frac{B}{2}} \iint_{4\cdot\pi} A_{eff} \cdot B_f \cdot |F_n(\theta, \phi)|^2 d\Omega \cdot df, \qquad (4.15)$$

where the integration in Eq. (4.15) takes place over the whole space (4π stereo-radians, Fig. 4.6) and over the receiver's noise bandwidth defined as $B = \int_{-\infty}^{+\infty} |H_n(f)|^2 df$, where $H_n(f)$ is the normalized receiver's (voltage) frequency response. The term ½ takes into account the fact that the antenna collects the power in one polarization only, while the thermal power emitted is randomly polarized.

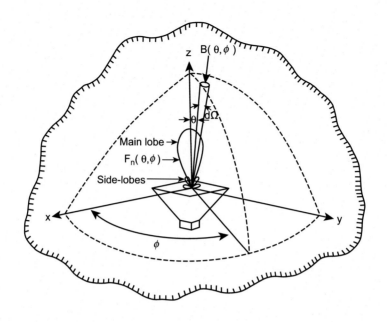

FIGURE 4.6

Geometry of the radiation incident over the antenna (Ulaby et al., 1981).

Reproduced by permission from Ulaby, F., Moore, R.K., Fung, A.K., 1981. Microwave Remote Sensing. Active and Passive, Vol I:
Microwave Remote Sensing Fundamentals and Radiometry. Artech House, Inc., Norwood, MA. © 1981 by Artech House, Inc.

Taking into account that $A_{eff} = \lambda^2/\Omega_{eq}$, where the antenna solid angle is defined as $\Omega_{eq} = \iint_{4 \cdot \pi} |F_n(\theta, \phi)|^2 d\Omega$, the total power collected by the antenna can be simply expressed as:

$$P = \frac{1}{2} \int_{f_0 - \frac{B}{2}}^{f_0 + \frac{B}{2}} \iint_{4 \cdot \pi} A_{eff} \cdot B_f \cdot |F_n(\theta, \phi)|^2 d\Omega \cdot df$$

$$= \frac{1}{2} \int_{f_0 - \frac{B}{2}}^{f_0 + \frac{B}{2}} \iint_{4 \cdot \pi} \frac{\lambda^2}{\Omega_{eq}} \cdot \frac{2 \cdot k_B \cdot T_{ph}}{\lambda^2} \cdot |F_n(\theta, \phi)|^2 d\Omega \cdot df = k_B \cdot T_{ph} \cdot B, \tag{4.16}$$

which is actually the same power as the one generated by a matched load at the same physical temperature, a result derived by Johnson and Nyquist in 1928. This means that for an ideal receiver of noise bandwidth B, the antenna delivers to the load the same power as a resistance at the same physical temperature. This is a very important result, and it will be emphasized when talking about the architecture of microwave radiometers and their internal calibration.

4.1.5 POWER COLLECTED BY AN ANTENNA SURROUNDED BY A GRAY BODY: APPARENT TEMPERATURE AND ANTENNA TEMPERATURE

The apparent temperature is an equivalent temperature related to the total brightness incident over the antenna, $B_i(\theta, \phi)$:

$$B_i(\theta, \phi) = 2 \cdot \frac{k_B}{\lambda^2} \cdot T_{AP}(\theta, \phi). \tag{4.17}$$

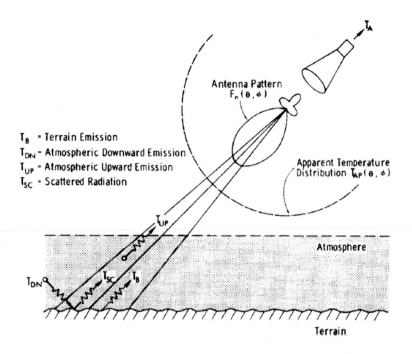

FIGURE 4.7

Relationship between the antenna temperature, the apparent temperature and the brightness temperature (Ulaby et al., 1981).

Reproduced by permission from Ulaby, F., Moore, R.K., Fung, A.K., 1981. Microwave Remote Sensing. Active and Passive, Vol I: Microwave Remote Sensing Fundamentals and Radiometry. Artech House, Inc., Norwood, MA. © 1981 by Artech House, Inc.

The apparent temperature depends on several terms related to the different sources irradiating the antenna. The relationship between them is shown in Fig. 4.7: the radiation emitted by the surface (land or sea) reaches the antenna attenuated by the atmosphere, the radiation emitted downwards by the atmosphere and reflected off the Earth's surface in the antenna direction, and the upward radiation emitted by the atmosphere itself. The sum of all these contributions weighted by the antenna pattern leads to:

$$P = \frac{1}{2} \int_{f_0 - \frac{B}{2}}^{f_0 + \frac{B}{2}} \iint_{4 \cdot \pi} A_{eff} \cdot B_i \cdot |F_n(\theta, \phi)|^2 d\Omega \cdot df$$

$$= \frac{1}{2} \int_{f_0 - \frac{B}{2}}^{f_0 + \frac{B}{2}} \iint_{4 \cdot \pi \Omega_{eq}} \frac{\lambda^2}{\lambda^2} \cdot \frac{2 \cdot k_B \cdot T_{AP}(\theta, \phi)}{\lambda^2} \cdot |F_n(\theta, \phi)|^2 d\Omega \cdot df \triangleq k_B \cdot T_A \cdot B,$$

(4.18)

which is similar to Eq. (4.16), but instead of the physical temperature T_{ph}, a thermodynamic variable called the "antenna temperature" (T_A, units of Kelvin) is defined.

In the frequency range from 1 to 10 GHz losses for a cloud-free atmosphere are very small and can be neglected. Consequently, the apparent brightness temperature T_{AP} can be approximated by the brightness temperature T_B.

4.2 THE RADIATIVE TRANSFER EQUATION

So far, only the radiation at two polarizations has been considered, that in Earth observations are usually the vertical (v) and the horizontal (h) ones. As discussed in previous chapters, the polarization state of an electromagnetic wave is completely described by four parameters, forming the so-called modified[2] Stokes emission vector $\bar{e}(\theta, \phi) = [e_v e_h e_U e_V]^T$, which is related to the electric fields incident in the antenna at vertical and horizontal polarizations [E_v and E_h (V/m)] by (Tsang, 1991, pp. 24−25):

$$\bar{T}_B = \begin{bmatrix} T_v(\theta, \phi) \\ T_h(\theta, \phi) \\ T_3(\theta, \phi) \\ T_4(\theta, \phi) \end{bmatrix} = T_s \begin{bmatrix} e_v(\theta, \phi) \\ e_h(\theta, \phi) \\ e_3(\theta, \phi) \\ e_4(\theta, \phi) \end{bmatrix} = \frac{\lambda^2}{\eta k_B B} \begin{bmatrix} \langle E_v E_v^* \rangle \\ \langle E_h E_h^* \rangle \\ 2\mathrm{Re}\langle E_v E_h^* \rangle \\ 2\mathrm{Im}\langle E_v E_h^* \rangle \end{bmatrix}, \tag{4.19}$$

where T_v and T_h are, as above, the brightness temperatures at vertical and horizontal polarizations, respectively; T_3 and T_4 are the so-called third and fourth modified Stokes elements, respectively; T_s is the surface's soil temperature (assumed to be constant with depth), λ is the electromagnetic wavelength, η is the wave impedance, k_B is Boltzmann's constant, B is the radiometer's noise bandwidth, and $<>$ stands for the expectation operator.

In Eq. (4.19) the third and fourth Stokes elements can also be computed not as complex cross-correlations, but as the difference between two absolute powers: $T_3 = T_{+45°} - T_{-45°}$, where $T_{\pm 45°}$ is the brightness temperatures at $\pm 45°$, and $T_4 = T_{LHCP} - T_{RHCP}$, where T_{LHCP} and T_{RHCP} are the brightness temperatures at left and right hand circular polarizations, respectively.

The problem of computing the modified Stokes emission vector on top of a generic layer, that may include absorption and scattering, is a very complex one. In the following paragraphs the complete formulation will be provided, and later the most common simplifications used will be presented.

4.2.1 THE COMPLETE POLARIMETRIC RADIATIVE TRANSFER EQUATION

The polarimetric emission can be computed by means of the radiative transfer equation (RTE) (Tsang, 1991) as:

$$\cos\theta \cdot \frac{d\bar{T}_B(\theta, \phi, z)}{dz} = -\bar{\bar{k}}_e(\theta, \phi)\bar{T}_B(\theta, \phi, z) + \bar{F}(\theta, \phi) \cdot T(z)$$
$$+ \int_0^{2\pi} \int_0^\pi \bar{\bar{P}}\left(\theta, \phi, \theta', \phi'\right)\bar{T}_B\left(\theta', \phi', z\right)d\Omega', \tag{4.20}$$

where $T(z)$ is the temperature distribution, and the Stokes vector $\bar{T}_B$ is related to the electric fields incident in the antenna at vertical and horizontal polarizations [E_v and E_h (V/m)], Eq. (4.20).

---

[2]The word "modified" is used because usually in optics the first Stokes parameter is the sum $I = T_v + T_h$, while the second one is the difference $Q = T_v - T_h$.

The extinction matrix $\overline{\overline{k_e}}$ is given by (Tsang, 1991):

$$\overline{\overline{k_e}} = \frac{2\pi n_0}{\kappa} \begin{bmatrix} 2\Im m\langle f_{vv}\rangle & 0 & \Im m\langle f_{vh}\rangle & -\Re e\langle f_{vh}\rangle \\ 0 & 2\Im m\langle f_{hh}\rangle & \Im m\langle f_{hv}\rangle & \Re e\langle f_{hv}\rangle \\ 2\Im m\langle f_{hv}\rangle & 2\Im m\langle f_{vh}\rangle & \Im m\langle f_{vv}+f_{hh}\rangle & -\Re e\langle f_{vv}-f_{hh}\rangle \\ 2\Re e\langle f_{hv}\rangle & -2\Re e\langle f_{vh}\rangle & \Re e\langle f_{hh}-f_w\rangle & \Im m\langle f_{vv}+f_{hh}\rangle \end{bmatrix} \tag{4.21}$$

where n_0 is the number of particles per unit of volume, $k = 2\pi/\lambda$ is the wave number, and f_{pq} are the forward scattering amplitude functions that provide the amplitude, phase, and polarization information of the scattered field at q-polarization $\overrightarrow{E_q} = \widehat{e}_q f_{pq}(\theta,\theta',\phi,\phi')e^{-jkr}/r$, when a plane wave at p-polarization $\overrightarrow{E_i} = \widehat{e}_p e^{-j\overrightarrow{k}_{inc}\cdot\overrightarrow{r}}$ is incident on each scatterer. The forward scattering amplitudes are computed as the average of the forward scattering amplitudes of each individual scatterer (branches, leaves, etc. in a canopy, or water drops, ice crystals, etc. in the atmosphere, for example) over the ensemble of sizes (a_0) according to their distribution [$N(a_0)$]:

$$<f_{pq}> = \int_0^{a_{max}} f_{pq}(a_0)N(a_0)da_0. \tag{4.22}$$

For example, in the case of water droplets $N(a_0)$ could be the Laws and Parsons (1943) distribution, among others, and a_0 is the radius of the water drops.

The emission vector $\overline{F}(\theta,\phi)$ in Eq. (4.20) is given by (Tsang, 1991, p. 282):

$$\overline{F}(\theta,\phi) = [k_{a1}(\pi-\theta,\pi+\phi) \quad k_{a2}(\pi-\theta,\pi+\phi) \quad -k_{a3}(\pi-\theta,\pi+\phi) \quad -k_{a4}(\pi-\theta,\pi+\phi)]^T,$$

$$k_{a1}(\theta,\phi) = k_{e11}(\theta,\phi) - \int \left[P_{11}\left(\theta,\phi,\theta',\phi'\right) + P_{21}\left(\theta,\phi,\theta',\phi'\right)\right]d\Omega',$$

$$k_{a2}(\theta,\phi) = k_{e22}(\theta,\phi) - \int \left[P_{12}\left(\theta,\phi,\theta',\phi'\right) + P_{22}\left(\theta,\phi,\theta',\phi'\right)\right]d\Omega',$$

$$k_{a3}(\theta,\phi) = 2k_{e13}(\theta,\phi) + 2k_{e23}(\theta,\phi) - 2\int \left[P_{13}\left(\theta,\phi,\theta',\phi'\right) + P_{23}\left(\theta,\phi,\theta',\phi'\right)\right]d\Omega',$$

$$k_{a4}(\theta,\phi) = -2k_{e14}(\theta,\phi) - 2k_{24}(\theta,\phi) + 2\int \left[P_{14}\left(\theta,\phi,\theta',\phi'\right) + P_{24}\left(\theta,\phi,\theta',\phi'\right)\right]d\Omega'.$$

$$\tag{4.23}$$

And the phase matrix $\overline{\overline{P}}\left(\theta,\theta',\phi,\phi'\right)$ in Eqs. (4.20) and (4.23) is given by (Tsang, 1991):

$$\overline{\overline{P}}\left(\theta,\theta',\phi,\phi'\right) = n_0 \begin{bmatrix} \langle|f_{vv}|^2\rangle & \langle|f_{vh}|^2\rangle & \Re e\langle f_{vv}f_{vh}^*\rangle & -\Im m\langle f_{vv}f_{vh}^*\rangle \\ \langle|f_{hv}|^2\rangle & \langle|f_{hh}|^2\rangle & \Re e\langle f_{hv}f_{hh}^*\rangle & -\Im m\langle f_{hv}f_{hh}^*\rangle \\ 2\Re e\langle f_{vv}f_{hv}^*\rangle & 2\Re e\langle f_{vh}f_{hh}^*\rangle & \Re e\langle f_{vv}f_{hh}^*+f_{vh}f_{hv}^*\rangle & -\Im m\langle f_{vv}f_{hh}^*+f_{vh}f_{hv}^*\rangle \\ 2\Im m\langle f_{vv}f_{hv}^*\rangle & 2\Im m\langle f_{vh}f_{hh}^*\rangle & \Im m\langle f_{vv}f_{hh}^*+f_{vh}f_{hv}^*\rangle & \Re e\langle f_{vv}f_{hh}^*+f_{vh}f_{hv}^*\rangle \end{bmatrix}.$$

$$\tag{4.24}$$

It should be noted that in the computation of the phase matrix (Eq. 4.24) the multiple scattering between the scatterers and between scatterers and the surface has to be accounted for. However,

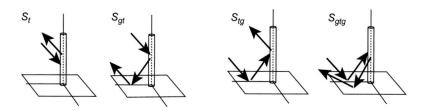

FIGURE 4.8

Four contributions to the scattered fields considered in the model. From left to right: trunk–trunk, trunk–ground, ground–trunk, and ground–trunk-ground.

usually, only scattering terms up to first-order scattering are included. For example, in the case of a vegetation structure over the ground only the following terms are included: (1) the direct scattered field, (2) the scattered field when the incident field is reflected off the soil surface, (3) the scattered field reflected off the soil surface when the incident field impinges directly on the element, and (4) the combination of (2) and (3) (Fig. 4.8). This assumption together with the far-field simplification helps to maintain the overall computation complexity at an affordable level.

Finally, the solution of Eq. (4.20) entails the application of the following boundary conditions: $\overline{T}_{up,p}(\theta,\phi,z=0)$ the upwelling surface's emission (height $= 0$), and $\overline{T}_{dn,p}(\theta,\phi,z=h)$ the down-welling emission at the top of the layer (height $= h$).

4.2.2 USUAL APPROXIMATIONS TO THE RADIATIVE TRANSFER EQUATION

The solution of Eq. (4.20) is very complex, and usually a number of approximations are made. These approximations assume that either the medium is a nonscattering one ($\overline{\overline{P}} = 0$) or that only the forward scattering is dominant. In this case if, in addition, the physical temperature of the medium is constant, the radiative transfer equation becomes

$$T_{ap,p} = T_{ph}|_{z=0} \cdot e_p(\theta,\phi) \cdot e^{-\tau_p/\cos\theta} + (1-\omega_p) \cdot T_{ph} \cdot \left(1 - e^{-\tau_p/\cos\theta}\right)$$
$$\cdot \left\{1 + \Gamma_p(\theta,\phi) \cdot e^{-\tau_p/\cos\theta}\right\}, \tag{4.25}$$

where the single scattering albedo ω_p is defined as:

$$\omega_v = \frac{\int\left[P_{11}\left(\theta,\phi,\theta',\phi'\right) + P_{21}\left(\theta,\phi,\theta',\phi'\right)\right]d\Omega'}{k_{e11}(\theta,\phi)}, \tag{4.26a}$$

$$\omega_h = \frac{\int\left[P_{12}\left(\theta,\phi,\theta',\phi'\right) + P_{22}\left(\theta,\phi,\theta',\phi'\right)\right]d\Omega'}{k_{e22}(\theta,\phi)}, \tag{4.26b}$$

and the opacity is defined as:

$$\tau_v = \int_0^{hly} k_{e11}(\theta,\phi)ds, \tag{4.27a}$$

$$\tau_h = \int_0^{hly} k_{e22}(\theta, \phi)ds, \tag{4.27b}$$

The physical interpretation of Eq. (4.25) is the following: the first term represents the surface's emission $\left(T_{B,p} = T_{ph}\big|_{z=0} \cdot e_p(\theta, \phi)\right)$, which is then attenuated on the way up by a factor $e^{-\tau_p/\cos\theta}$; the second term is called $T_{up,p}$ and represents the emission introduced by the losses in the medium $T_{ph} \cdot \left(1 - e^{-\tau_p/\cos\theta}\right)$, that gets scattered on the way up [$(1 - \omega_p)$ term], and finally the third term is equal to the second one, but is the downwelling radiation ($T_{dn,p} = T_{up,p}$), which includes other contributions that are reflected over the surface ($T_{SC,p} = \Gamma_p(\theta, \phi) \cdot T_{dn,p} = T_{SC,p}$) and are attenuated on the way up $\left(e^{-\tau_p/\cos\theta}\right)$. These other noise sources include the Sun and Moon radiation (if visible), the cosmic background, which is nearly homogeneous $T_{cos} = 2.7K$, and the galactic noise, very important below ~ 1 GHz, and strongly dependent on the antenna boresight direction, toward the galactic plane, or the galactic pole, which prevents useful observations below 1 GHz,

$$T_{dn} = T_{dn}^{atm} + (T_{gal} + T_{cos} + T_{Sun} + T_{Moon})e^{-\tau(0,\infty)\sec\theta}. \tag{4.28}$$

The largest solar flare ever recorded is depicted in Fig. 4.9 while the TRMM microwave imager (TMI) satellite's viewing angle matched the specular reflection angle from the Sun at 10.7 GHz (Remote Sensing Systems website, 2013). Fig. 4.10 shows the map of the galactic noise at 1.4 GHz this noise reflects over the Earth's surface and ends up reaching the sensor. Figures correspond to wind speed ranges 0–3, 3–6, 6–8, and 8–12 m/s. Note that for increasing wind speeds, the peak amplitude of the scattered galactic noise decreases, and the width of the band increases, due to the increased sea surface roughness.

We show the computed antenna temperature in Fig. 4.11 for a narrow beam antenna as a function of the elevation angle and frequency. Note the rapid increase below 1 GHz associated with the galactic noise contribution, and above ~ 10 GHz due to the increased atmospheric attenuation, which is especially relevant at around 22 and 60 GHz. Note also that plots for lower elevation angles indicate a larger antenna temperature due to the increased atmospheric path length, and attenuation.

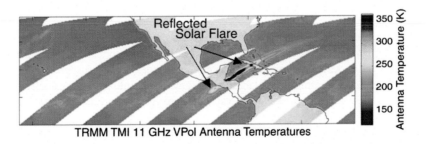

TRMM TMI 11 GHz VPol Antenna Temperatures

FIGURE 4.9

Largest solar flare event ever recorded erupted while the TRMM microwave imager (TMI) satellite's viewing angle matched the specular reflection angle (November 4, 2003, at $\sim$ 19:47 UTC) (Remote Sensing Systems website, 2013).

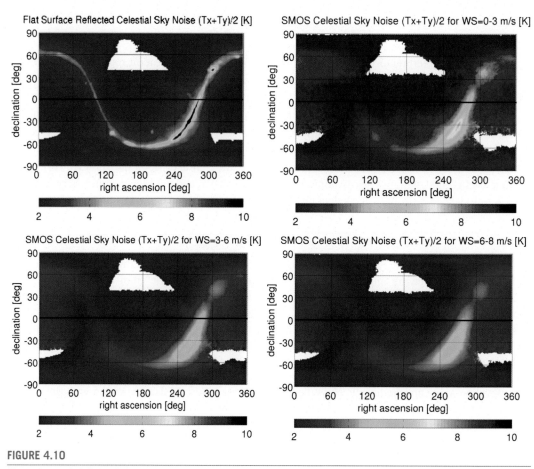

FIGURE 4.10

Sky noise at 1.4 GHz. The band corresponds to the galactic plane.

Solid curves correspond to the antenna temperature as a function of the elevation angle, for the mean galactic temperature. Upper dashed curve corresponds to the antenna temperature for the maximum galactic noise (antenna pointing to the galaxy center). Lower dashed curve corresponds to the antenna temperature for minimum galactic noise (antenna pointing to the galactic pole). Maxima around 22 and 60 GHz are due to the water vapor and oxygen absorption resonances, respectively (Blake, 1962).

Even though, strictly speaking, Eq. (4.25) is only valid in the low microwave frequency range, it is usually applied up to very high frequencies, provided the parameters involved are empirically adjusted. At microwave frequencies, for typical atmospheric conditions, scattering is minimal, except by liquid water drops varying in size from small drops in clouds and fog (1−10 μm in diameter) to the largest rain drops (up to ~5−10 mm). In these cases, the full RTE has to be solved, although often other assumptions, such as a horizontally stratified atmosphere, are applied for the sake of simplicity.

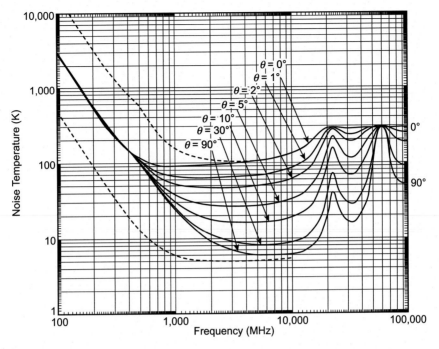

FIGURE 4.11

Antenna sky temperature for an ideal lossless antenna at the Earth's surface as a function of frequency and elevation angle.

(Ulaby et al., 1981).

In the following sections the behavior of the atmosphere, the soil, either bare or covered by vegetation or snow, or the sea, including the effects of wind, frozen or covered by oil slicks is explained.

4.3 EMISSION BEHAVIOR OF NATURAL SURFACES
4.3.1 THE ATMOSPHERE

The atmosphere acts as an attenuator and as a scatter. Up to about 300 GHz, the most significant gases affecting propagation are water vapor and oxygen. Depending on the frequency and, eventually, the presence of hydrometeors, scattering can be important.

To analyze the behavior of the atmosphere, let us first study the noise introduced by a homogeneous "slice" of the atmosphere using the model of a lossy transmission line of characteristic impedance Z_0 at constant temperature T_{ph} terminated by two matched loads at its ends (Fig. 4.12). Each matched load produces a noise power equal to $k_B \cdot T_{ph} \cdot B$, which at the other end appears attenuated by a factor $1/L$. In thermodynamic equilibrium, the transmission line must introduce an amount of noise equal to the difference between what the matched loads are giving ($k_B \cdot T_{ph} \cdot B$), and what they are receiving ($k_B \cdot T_{ph} \cdot B/L$): $k_B \cdot T_{ph} \cdot B \cdot (1 - 1/L)$.

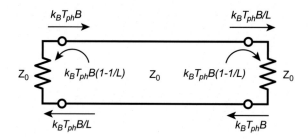

FIGURE 4.12

Lossy transmission line model to compute the noise introduced by a homogeneous "slice" of the atmosphere at constant temperature.

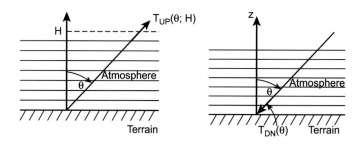

FIGURE 4.13

Horizontally stratified model of the atmosphere to compute T_{up}^{atm} and T_{dn}^{atm} (Ulaby et al., 1981).

Reproduced by permission from Ulaby, F., Moore, R.K., Fung, A.K., 1981. Microwave Remote Sensing. Active and Passive, Vol I: Microwave Remote Sensing Fundamentals and Radiometry. Artech House, Inc., Norwood, MA. © 1981 by Artech House, Inc.

Therefore, an absorbing atmosphere not only attenuates the radiation passing through it, but it also introduces an additional noise, which in the upwelling direction is called T_{up}^{atm}, and in the downwelling direction T_{dn}^{atm}.

In a more general, although simplified, case the atmosphere can be treated as a stratified series of attenuators each one at a given physical temperature, as illustrated in Fig. 4.13. The total upwelling temperature (Eq. 4.29) can be computed as the sum of the contributions from every single layer at height z and thickness $\sec\theta \cdot dz$ attenuated by all upper layers $e^{-\tau(z,H)\sec\theta}$, where $\tau(z',H) = \int_{z'}^{H} k_a(z)dz$ is called the atmospheric opacity. The brightness of each layer is the product of its physical temperature $T(z)$ times its emissivity, which by Kirchhoff's law is equal to the absorption coefficient $e(z) = k_a(z)$. Similarly, the downwelling temperature (Eq. 4.30) can be computed as the sum of the contributions from every single layer at height z and thickness $\sec\theta \cdot dz$ attenuated by all lower layers $e^{-\tau(0,z)\sec\theta}$, where $\tau(0,z) = \int_0^z k_a(z')dz'$.

$$T_{up}^{atm}(\theta, H) = \sec\theta \int_0^H k_a(z')T(z')e^{-\tau(z',H)\sec\theta}dz', \tag{4.29}$$

$$T_{dn}^{atm}(\theta, H) = \sec\theta \int_0^\infty k_a(z')T(z')e^{-\tau(0,z')\sec\theta}dz'. \tag{4.30}$$

Atmospheric attenuation can be due to gases, rain, clouds, and fog and in general hydrometeors in liquid phase. Due to the much lower dielectric constant of ice, and negligible losses, snow and ice do not introduce much attenuation, but they do introduce scattering and wave depolarization. In the following sections, the attenuation induced by different hydrometeors is detailed.

4.3.1.1 Attenuation by Atmospheric Gases

Up to 1 THz, the specific attenuation due to dry air and water vapor can be most accurately evaluated at any pressure, temperature, and humidity by summation of the individual resonance lines from oxygen and water vapor, together with small additional factors for the nonresonant Debye spectrum of oxygen below 10 GHz, pressure-induced nitrogen attenuation above 100 GHz, and a wet continuum to account for the excess water vapor absorption found experimentally.

The specific attenuation using the model is displayed in Fig. 4.14, calculated from 0 to 350 GHz, for a pressure of 1013 hPa, temperature of 15°C for the cases of a water vapor density of 7.5 g/m^3 (blue curve), and a dry atmosphere (red curve).

Near 60 GHz, many oxygen absorption lines merge together, at sea-level pressure, to form a single, broad absorption band, as shown in Fig. 4.14. At higher altitudes, the atmospheric pressure is lower, and the individual lines of the oxygen attenuation become resolved. Some additional molecular species (e.g., oxygen isotopic species, oxygen vibrationally excited species, ozone, ozone isotopic species, and ozone vibrationally excited species, and other minor species) are not included in the line-by-line prediction method. These additional lines are insignificant for typical atmospheres, but may be important for a dry atmosphere.

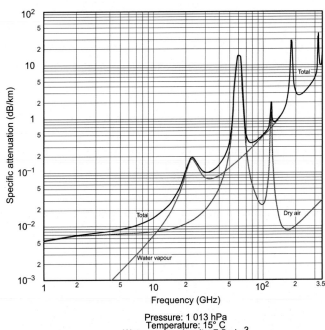

Pressure: 1 013 hPa
Temperature: 15° C
Water vapour density: 7.5 g/m^3

FIGURE 4.14

Specific attenuation due to atmospheric gases (Recommendation ITU-R P.676, 2012) for water vapor (*blue line*) and a dry atmosphere (*red line*).

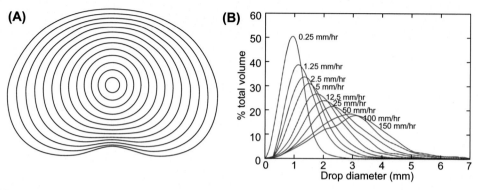

FIGURE 4.15

(A) Shape of a raindrop for equivalent radii: 0.25, 0.50, ..., 3.25 mm. (B) Laws and Parsons equivalent radii distribution versus rain rate (0.25, ..., 150 mm/h).

The specific gaseous attenuation is given by:

$$\gamma = \gamma_o + \gamma_w = 0.1820 \cdot f \cdot N''(f) \, (\text{dB/km}),\tag{4.31}$$

where γ_o and γ_w are the specific attenuations [in (dB/km)] due to dry air (oxygen, pressure-induced nitrogen, and nonresonant Debye attenuation) and water vapor, respectively, and where f is the frequency (GHz) and $N''f$ is the imaginary part of the frequency-dependent complex refractivity:

$$N''(f) = \sum_i S_i F_i + N''_D(f),\tag{4.32}$$

where S_i is the strength of the i-th line, F_i is the line shape factor, and the sum extends over all the lines,[3] and $N''_D(f)$ is the dry continuum due to pressure-induced nitrogen absorption and the Debye spectrum.

4.3.1.2 Attenuation by Rain

A detailed analysis of rain effects requires accounting for the scattering and attenuation introduced by each raindrop size, weighted by their relative distribution, as shown in Fig. 4.15 (Laws and Parsons, 1943) and calculated in Eq. (4.22) (see, for example, Duffo et al., 2009).

However, if we are interested in the simpler horizontally stratified nonscattering atmosphere model, the specific attenuation γ_R in (dB/km) as a function of the rain rate R (mm/h) can be computed using the power-law relationship:

$$\gamma_R = kR^\alpha,\tag{4.33}$$

where the coefficients $k_{h,v}$ and $\alpha_{h,v}$ for horizontal and vertical polarizations are shown in Figs. 4.16 and 4.17.

---

[3]For frequencies above 118.750343 GHz oxygen line, only the oxygen lines above 60 GHz complex should be included in the summation; the summation should begin at $i = 38$ rather than at $i = 1$.

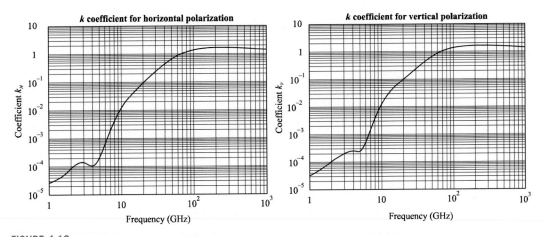

FIGURE 4.16

Coefficients for k_h (left) and k_v (right) (Recommendation ITU-R P.838-3).

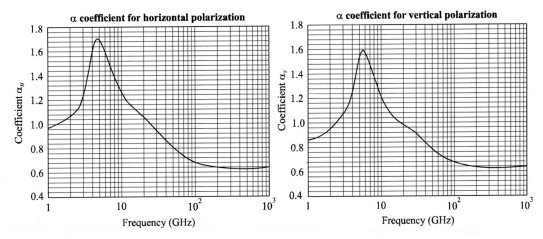

FIGURE 4.17

Coefficients for α_h (left) and α_v (right) (Recommendation ITU-R P.838-3).

4.3.1.3 Attenuation by Clouds and Fog

For clouds or fog consisting entirely of small droplets, generally less than 0.01 cm, the Rayleigh approximation is valid for frequencies below 200 GHz and it is possible to express the attenuation in terms of the total water content per unit volume. Thus the specific attenuation γ_c within a cloud or fog can be written as:

$$\gamma_c = K_l \cdot M \ (\text{dB/km}) \tag{4.34}$$

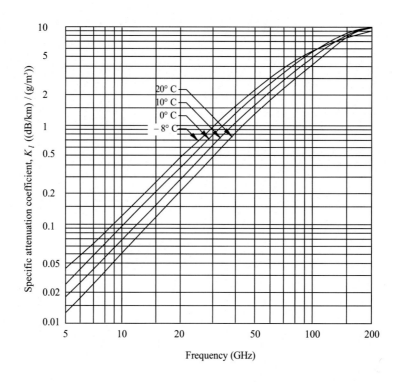

FIGURE 4.18

Specific attenuation by water droplets at various temperatures as function of frequency (Recommendation ITU-R P.840-5).

where K_l is the specific attenuation coefficient in (dB/km)/(g/m^3); and M is the liquid water density in the cloud or fog in (g/m^3).

At frequencies of the order of 100 GHz and above, attenuation due to fog may be significant. The liquid water density in fog is typically about 0.05 g/m^3 for medium fog (visibility on the order of 300 m), and 0.5 g/m^3 for thick fog (visibility on the order of 50 m).

The values of K_l at frequencies from 5 to 200 GHz and temperatures between $-8°C$ and 20°C are shown here in Fig. 4.18. For cloud attenuation, the curve corresponding to 0°C should be used.

To obtain the attenuation [in (dB)] due to clouds for a given probability, the statistics of the total columnar content of liquid water L (kg/m^2) or, equivalently, millimeters of precipitable water:

$$A = \frac{LK_l}{\sin \theta} \ (\text{dB}) \quad \text{for } 5° \leq \theta \leq 90° \tag{4.35}$$

where θ is the elevation angle and K_l is provided in Fig. 4.18.

4.3.2 THE IONOSPHERE

The ionosphere is a region in the uppermost portion of the Earth's atmosphere that extends from ~ 50 km altitude to more than ~ 600 km. It is composed by ionized particles and electrons formed by

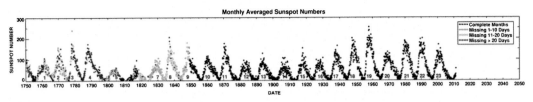

FIGURE 4.19

Wolf number since 1750.

the interaction of the solar wind with the very thin air particles that have escaped the Earth's gravity. Since the Sun is the source of energy to ionize these particles, solar activity plays a key role in the concentration of ions, which exhibit a diurnal and seasonal cycles, as well as a 11-year cycle linked to the solar activity (Fig. 4.19). The solar activity is measured in solar flux units (SFU) which is directly linked to the daily sunspot number (R):

$$SFU = 63.7 + 0.728 \cdot R + 0.00089 \cdot R^2, \tag{4.36}$$

being $1\,SFU \triangleq 10^{-22}\,W/m^2/Hz$.

The concentration of ionized particles is also very dependent on the altitude: the higher the altitude, the more energy the solar wind particles have, but the lower the atmospheric density, and the lower the altitude, the more attenuated the solar particles are, but the higher the atmospheric density. The combination of these two factors leads to a continuous vertical profile of the electron density [in (e^-/cm^3)], with local maxima and minima (Fig. 4.20), which leads to the definition of the layered structure sketched in Fig. 4.21, with its diurnal dependence.

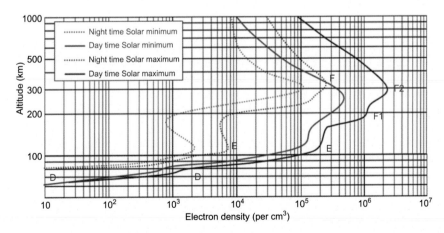

FIGURE 4.20

Diurnal variation of the ionospheric electron density profile as a function of the altitude (http://sidstation. loudet.org/ionosphere-en.xhtml).

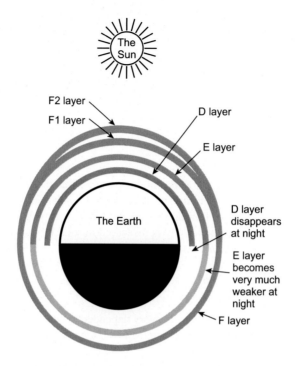

FIGURE 4.21

Simplified view of the layers in the ionosphere over the period of a day (http://www.radio-electronics.com/info/propagation/ionospheric/ionosphere.php).

Modified from http://www.radio-electronics.com/info/propagation/ionospheric/ionosphere.php

The ionosphere layers are summarized below:

- The **D-layer** is the lowest part of the ionosphere, and it appears from an altitude of 50–95 km. Due to the higher density it develops shortly after sunrise, but it also disappears shortly after sunset, reaching a maximum ionization when the Sun is at zenith.
- The **E-layer** extends from an altitude of 90–150 km. It also develops shortly after sunrise, but disappears a few hours after sunset. The maximum ionization of this layer is reached around midday. The ions in this layer are mainly O_2^+.
- The **sporadic E-layer** (Es) appears at an altitude that may vary anywhere between 80 and 120 km. It is still not known how this layer actually develops, but, it can appear at any time of the day, with a preference for the late morning and early evening, mostly during the summer months and briefly at midwinter, with the peak occurring in the early summer.
- The **F-layer** is the highest part of the ionosphere, extending between 250 and 500 km in altitude. Just before sunrise, the Sun illuminates the upper part of the atmosphere containing the F-layer, and due to an unclear physical mechanism, the sunlight causes the F-layer to split into two distinct layers called the F1-layer and F2-layer. The **F1-layer** is located between 150 and 200 km in altitude during daylight hours. The maximum ionization is reached at midday. The **F2-layer** is located between 250 and 450 km in altitude, extending occasionally beyond 600 km. At the

latitudes higher north or south of the equator, this layer is located at lower altitudes. The maximum ionization is usually reached in winter, starting 1 h after sunrise and remaining until shortly after sunset. However, it shows a great variability during the day, due to its high sensitivity to solar activity and major solar events. Ions in the lower part of the F-layer are mainly NO^+ and are predominantly O^+ in the upper part.

The ionosphere plays an important role in the propagation of radio waves (also for radar systems), especially at frequencies from 0.1 to 12 GHz. The main effects that take place on an Earth space path when the signal is passing through the ionosphere are (Recommendation ITU-R P.531-11, 2012) rotation of the polarization plane, also known as "Faraday rotation," due to the interaction of the electromagnetic wave with the ionized medium in the Earth's magnetic field along the path; absorption and emission due to ionospheric losses; group delay and phase advance of the signal due to the total electron content accumulated along the path; rapid variations of the amplitude and phase (scintillations) of the signal due to small-scale irregular structures in the ionosphere; a change in the apparent direction of arrival due to refraction; and Doppler effects due to nonlinear polarization rotations and time delays.

From all these effects, in microwave radiometry only the Faraday rotation, the absorption and the emission are of concern, mainly at low microwave frequencies. For active systems (radar, satellite navigation system, and other remote sensing systems using signals of opportunity) scintillation effects and group delay are also important, and for large bandwidth systems, dispersion effects (variation of the group delay with frequency) are also of concern. The effects that are important for microwave radiometry are analyzed below.

4.3.2.1 Faraday Rotation

The Faraday rotation is a rotation of the polarization plane due to the interaction between the electromagnetic wave, the ionized medium ("plasma") and the magnetic field of the Earth.

The magnitude of the rotation is given by (Le Vine and Abraham, 2002):

$$\psi_F = \frac{\pi}{cf^2} \int f_p^2(s)f_B(s)\cos(\Theta_B(s))ds, \tag{4.37}$$

where Θ_B is the angle between the direction of propagation and the Earth's magnetic field; $ds = \sec(\theta)$ dz, z is the normal to the surface, θ is the polar angle between the line of sight and the nadir; $f_p = \sqrt{\frac{N \cdot e^2}{4 \cdot \pi^2 \cdot \varepsilon_0 \cdot m}}$ is the plasma frequency; and $f_B = \frac{e \cdot B}{2 \cdot \pi \cdot m}$ is the electron gyro frequency, being the number of electrons per cubic meter, e and m the charge and mass of the electron, respectively, and B the Earth's magnetic field.

At L-band (1.413 MHz), for an altitude of 400 km, Eq. (4.37) becomes

$$\psi_F \approx 6950 \cdot B(h = 400 \text{ km}) \cdot \cos(\Theta_B) \cdot \sec(\theta) \cdot \text{VTEC}, \tag{4.38}$$

where B is in Tesla, ψ_F in degrees, and VTEC is the vertical total electron content (units of 10^{16} e/m^2) defined as the integral of the electron density (Fig. 4.20) in a vertical path:

$$\text{VTEC} = \int_0^h N_e(z) \cdot dz. \tag{4.39}$$

The rotation of the electric field emitted by the surface of the Earth implies that the electric fields measured in the antenna reference frame (x and y) do not correspond to the horizontal (h) and vertical (v) electric fields emitted by the Earth:

$$\begin{bmatrix} E_x \\ E_y \end{bmatrix} = \begin{bmatrix} A & B \\ -B & A \end{bmatrix} \begin{bmatrix} E_h \\ E_v \end{bmatrix}, \tag{4.40}$$

where $A = \cos(\psi_F)$ and $B = \sin(\psi_F)$.

This translates in a modification of the measured Stokes vector in the antenna reference frame:

$$\begin{bmatrix} T_{xx} \\ T_{xy} \\ T_{yx} \\ T_{yy} \end{bmatrix} = \begin{bmatrix} A^2 & AB & AB & B^2 \\ -AB & A^2 & -B^2 & AB \\ -AB & -B^2 & A^2 & AB \\ B^2 & -AB & -AB & A^2 \end{bmatrix} \begin{bmatrix} T_h \\ T_{hv} \\ T_{vh} \\ T_v \end{bmatrix}, \tag{4.41}$$

where $T_{hv} = C' < E_h E_v^* >$ and $T_{vh} = C' < E_v E_h^* >$ do not correspond to T_3 and T_4 as defined in Eq. (4.19). It should be noted that, while Eq. (4.41) can always be inverted, in general, if only the first two Stokes elements are measured:

$$\begin{bmatrix} T_{xx} \\ T_{yy} \end{bmatrix} = \begin{bmatrix} A^2 & B^2 \\ B^2 & A^2 \end{bmatrix} \begin{bmatrix} T_h \\ T_v \end{bmatrix}, \tag{4.42}$$

Eq. (4.42) cannot be inverted, since the matrix becomes singular for $\psi_F = 45°$.

A figure from Le Vine and Abraham (2002), Fig. 4.22, shows the global distribution of Faraday rotation for local time of 6 a.m. (left) and noon (right). The data are for high solar activity (June 1989),

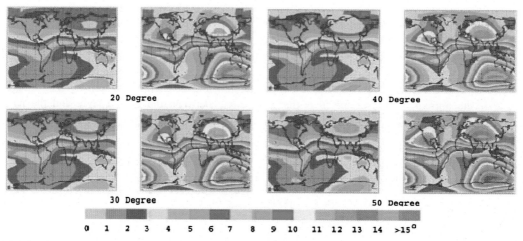

20 Degree   40 Degree

30 Degree   50 Degree

0 1 2 3 4 5 6 7 8 9 10 11 12 13 14 >15°

FIGURE 4.22

Global distribution of Faraday rotation for local time of 6 a.m. (left) and noon (right). The data are for high solar activity (June 1989), an altitude of 675 km, and looking perpendicular to the satellite heading to the right at incidence angles as indicated (20°, 30°, 40°, and 50°) (Le Vine and Abraham, 2002).

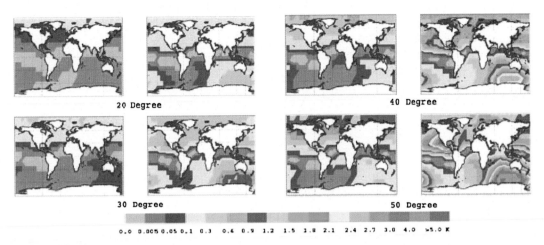

20 Degree   40 Degree

30 Degree   50 Degree

0.0 0.005 0.05 0.1 0.3   0.6 0.9 1.2 1.5 1.8 2.1 2.4 2.7 3.0 4.0 >5.0 K

FIGURE 4.23

Global distribution of the error in brightness temperature at L-band as a function of incidence angle due to neglecting Faraday rotation at 6 a.m. (left) and noon (right). The data are for high solar activity (June 1989; Rz = 158) and for a sensor at altitude of 675 km and looking perpendicular to the satellite heading to the right. The surface is ocean with $S = 35$ psu and $T = 20°C$ (Le Vine and Abraham, 2002).

an altitude of 675 km, and looking perpendicular to the satellite heading to the right at incidence angles as indicated (20°, 30°, 40°, and 50°). As can be clearly seen, the maximum Faraday rotation takes place at midlatitude, and when the local time is around noon (or slightly after), while it is minimum at 6 a.m. The effect is also more important for larger incidence angles.

Another figure (Fig. 4.23) from Le Vine and Abraham (2002) shows the impact of the Faraday rotation over the brightness temperature over the sea, for the same conditions as in Fig. 4.22. As it can be appreciated, the effect is relatively small, from a fraction of a Kelvin to a few Kelvin, but it is extremely large in terms of the magnitude of the parameter to be measured using L-band microwave radiometry, the sea surface salinity (SSS), whose sensitivity ranges from $\Delta T_B/\Delta SSS$ $\sim 0.25-0.5$ K/psu, and the required accuracy ~ 0.1 to 0.2 psu.

4.3.2.2 Ionospheric Losses: Absorption and Emission

Except for the amplitude scintillation, that may introduce rapid and large attenuations, especially at equatorial and polar regions (e.g., Fig. 4.24), in general, ionospheric absorption can be neglected for most applications. One of the applications in which these effects may be more serious is, again, SSS retrievals from L-band radiometric measurements. Due to the stringent accuracy requirements ($\sim 0.1-0.2$ psu), ionospheric losses may have to be taken into account, both for the attenuation of the brightness temperature, and for the extra noise added (see Fig. 4.10). Table 4.1 from Le Vine and Abraham (2002) shows the ionospheric attenuation and emission at 6 a.m. and 12 p.m., and their impact of the brightness temperature of the sea (SSS = 35 psu), sea surface temperature (SST = 20°C). As it can be easily recognized, the effect is always smaller than 12 mK, which assuming a worst case sensitivity of 0.25 K/psu, translates into a salinity error of <0.04 psu.

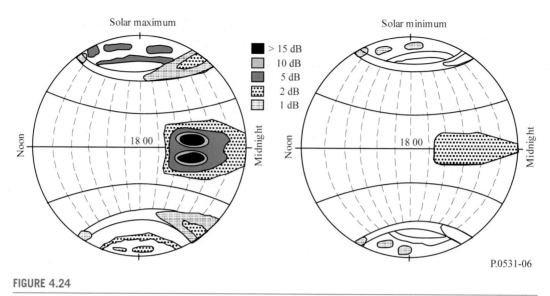

FIGURE 4.24

Depth of scintillation fading at 1.5 GHz during solar maximum and minimum years (Recommendation ITU-R P.531-11, 2012).

4.3.3 LAND EMISSION

In Monerris and Schmugge (2009) a summary of the land surface's emission is presented. The emissivity of land depends not only on the soil moisture, but also on the soil temperature (Choudhury et al., 1982; Wigneron et al., 2001), and the soil surface roughness (Mo and Schmugge, 1987; Escorihuela et al., 2007). It also depends on the vegetation canopy (Jackson et al., 1982; Ferrazzoli et al., 2002), the snow cover (Schwank et al., 2004), the relief (Mätlzer and Standley, 2000; Talone et al., 2007; Monerris et al., 2008), etc. In the following sections each of these effects are studied in more detail.

4.3.3.1 Soil Dielectric Constant Models

The dielectric constant determines the response of the soil to an incident electromagnetic wave. This response is composed of two parts, real and imaginary ($\varepsilon_r = \varepsilon' + j \cdot \varepsilon''$), which determine the wave velocity and the energy losses, respectively. In an inhomogeneous medium such as soils, the dielectric properties have a strong impact on its microwave emission. However, the relationship between the soil dielectric constant and the soil physical properties is not straightforward. The dependence of the dielectric constant for different soil types is clearly shown in Fig. 4.25. This figure clearly demonstrates that the larger the volumetric soil moisture, the larger the dielectric constant (both in real and imaginary parts).

A large number of studies have been performed during the last decades to discover this relationship since it plays an important part in the soil moisture retrieval algorithms from remote sensing data (Hipp, 1974: Wang and Schmugge, 2000; Topp et al., 1989; Hallikainen et al., 1985; Dobson et al., 1985; Roth et al., 1992; Mironov et al., 2004). Some of these models are simple empirical models in which data is fitted by a unique curve for all soils; others propose semiempirical approaches which take into account some soil physical properties. The dielectric constant of dry soils is almost

Table 4.1 Ionospheric Attenuation and Emission: SSS = 35 psu, SST = 20°C. June 6 a.m. (Left), 12 p.m. (Right). $T\uparrow = T_{up}$; $T\downarrow = T_{dn}$ **Scattered Over the Ocean**

Location			VTEC	τ	ΔT	$T\uparrow$	$T\downarrow$
Latitude	Longitude	Rz	(TECU)	(10^{-5} Np)	(mK)	(mK)	(mK)
Ionospheric Attenuation and Emission (6 a.m.)							
30 N	220 E	158.4	19.9	2.64	2.18	20.87	16.08
		8.5	6.4	1.45	1.20	4.75	3.51
0	220 E	158.4	20.6	2.47	2.04	23.01	18.22
		8.5	8.6	1.19	0.98	5.31	5.06
30 S	220 E	158.4	12.7	1.00	0.83	9.35	7.54
		8.5	3.6	0.30	0.25	1.23	0.98
30 N	330 E	158.4	24.9	3.51	2.89	31.42	23.98
		8.5	7.1	1.57	1.30	5.55	4.05
0	330 E	158.4	20.3	2.49	2.05	22.81	17.91
		8.5	8.0	1.11	0.91	4.79	4.39
30 S	330 E	158.4	9.5	0.65	0.53	5.38	4.60
		8.5	3.2	0.27	0.22	1.01	0.83
30 S	60 E	158.4	5.3	0.31	0.25	1.83	1.68
		8.5	2.2	0.21	0.17	0.65	0.48
75 N	160 E	158.4	18.7	3.32	2.74	17.70	22.86
		8.5	6.3	2.19	1.80	6.08	4.40
Ionospheric Attenuation and Emission (Noon)							
30 N	220 E	158.4	39.8	11.43	9.42	77.34	57.46
		8.5	14.2	5.00	4.12	18.49	13.52
0	220 E	158.4	45.0	13.42	11.06	99.21	78.06
		8.5	17.3	5.81	4.78	22.22	19.41
30 S	220 E	158.4	37.9	10.22	8.42	80.89	59.12
		8.5	9.0	3.26	2.69	10.32	7.59
30 N	330 E	158.4	43.5	12.37	10.20	89.77	67.04
		8.5	13.8	5.03	4.15	18.50	13.45
0	330 E	158.4	48.3	14.47	11.92	113.48	87.74
		8.5	17.7	5.93	4.88	23.24	19.51
30 S	330 E	158.4	35.7	9.98	8.22	75.06	54.73
		8.5	11.3	3.64	3.00	13.08	9.76
30 S	60 E	158.4	39.8	10.68	8.80	85.87	63.13
		8.5	9.8	3.46	2.85	11.82	8.57
75 N	160 E	158.4	20.1	6.11	5.034	24.57	28.29
		8.5	6.7	2.60	2.134	7.29	5.27

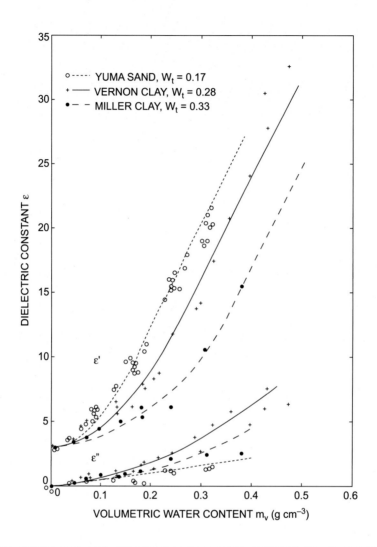

FIGURE 4.25

(A) Real and imaginary parts of the dielectric constant at 1.4 GHz of different soil types, as a function of the volumetric soil moisture content (Ulaby et al., 1986).

independent of temperature (Topp et al., 1980) and frequency. On the contrary, wet soils show a complex behavior depending on the interaction between soil, water, and air particles. Hallikainen et al. (1985) performed a series of dielectric constant measurements of five soils with different texture composition at frequencies between 1.4 and 18 GHz and found out that texture has a strong effect on the dielectric behavior, which is especially pronounced at frequencies below 5 GHz.

In the dielectric mixing model by Roth et al. (1992), differences in soil texture, and bound water and free water are ignored altogether. Other models use a semiempirical approach that contains a model of the complex dielectric constant and the volume fraction of each of the soil components. This kind of approach is used by two of the most widely used models presented in Wang and Schmugge (2000) and Dobson et al. (1985). The starting point of both of them is the dielectric mixing model by (Birchak et al., 1974):

$$\varepsilon^\alpha = V_s \varepsilon^\alpha + V_a \varepsilon_a^\alpha + V_{fw} \varepsilon_{fw}^\alpha + V_{bw} \varepsilon_{fw}^\alpha, \tag{4.43}$$

where $V_s(\varepsilon_s)$, $V_a(\varepsilon_a)$, $V_{fw}(\varepsilon_{fw})$, and $V_{bw}(\varepsilon_{bw})$ are the volume fraction (dielectric constant) of the solid phase, air, free water, and bound water in the soil, respectively. Eq. (4.43) can be rewritten as a function of the bulk density ρ_b, the particle density ρ_s, and the volumetric moisture m_v as:

$$\varepsilon^\alpha = 1 + \frac{\rho_b}{\rho_s} \left(\varepsilon_s^\alpha - 1 \right) + V_{fw} \varepsilon_{fw}^\alpha + V_{bw} \varepsilon_{fw}^\alpha - m_v, \tag{4.44}$$

The Wang and Schmugge (2000) model was proposed for 1.4 and 5 GHz frequencies and starts from Eq. (4.44) with $\alpha = 1$. This model provides separate dielectric constant equations for volumetric water content lower than or greater than the transition moisture w_t. The transition moisture is the moisture content at which the free water phase begins to dominate the soil hydraulics, and it is strongly dependent on the texture. The soil dielectric constant is then estimated as:

$$\varepsilon = \begin{cases} m_v \varepsilon_x + (P - m_v)\varepsilon_a + (1 - P)\varepsilon_r, & m_v < w_t, \\ w_t \varepsilon_x + (m_v - w_t)\varepsilon_{fw} + (P - m_v)\varepsilon_a + (1 - P)\varepsilon_r, & m_v > w_t, \end{cases} \tag{4.45}$$

with

$$\varepsilon_x = \begin{cases} \varepsilon_i + (\varepsilon_{fw} - \varepsilon_i)\dfrac{m_v}{w_t}\gamma, & m_v < w_t, \\ \varepsilon_i + (\varepsilon_{fw} - \varepsilon_i)\gamma, & m_v > w_t, \end{cases} \tag{4.46}$$

and $\gamma = -0.57 \cdot w_t + 0.481$, ε_i, ε_a, ε_{fw}, and ε_r the dielectric constants of the ice, air, free water, and rock, respectively, ε_x the dielectric constant of the initially adsorbed water, and P the soil porosity. The variable ε_{fw} is estimated using the Debye equation.

On the other hand, the Dobson et al. (1985) model starts from Eq. (4.43) and assumes that there is no distinction between bound and free water. Taking this into account, the new expression for the dielectric constant is

$$\varepsilon = \left(1 + \frac{\rho_b}{\rho_s} \left(\varepsilon_s^\alpha - 1 \right) + m_v^b \varepsilon_{fw}^\alpha - m_v \right)^{1/\alpha}. \tag{4.47}$$

The real and imaginary parts of the dielectric constant of soils are obtained separately using the percentage of sand S and clay C in the soil, and $\beta = (127.48 - 0.519 \cdot S - 0.152 \cdot C)/100$ for the real part, and $\beta = (1.33979 - 0.603 \cdot S - 0.166 \cdot C)/100$ for the imaginary part. For further information on dielectric constant models, please refer to Behari (2004), Chukhlantsev (2006), and in particular the new model from Mironov et al. (2004) seems to provide better soil moisture estimates from L-band radiometric data.

4.3.3.2 Bare Soil Emission

The emissivity of bare soils e_s is given by Eq. (4.11), reproduced here for convenience:

$$e_p(\theta) = 1 - \Gamma_{s,p}(\theta), \tag{4.48}$$

where $\Gamma_{s,p}$ is the total reflectivity at p-polarization in a chosen observation direction, which can be expressed in terms of the bistatic scattering coefficients σ_{pp} and σ_{pq} as follows (Fung, 1994):

$$\Gamma_{s,p}(\theta_i, \varphi_i) = \frac{1}{4\pi} \int_{2\pi} [\sigma_{pp}(\theta_i, \varphi_i, \theta_r, \varphi_r) + \sigma_{pq}(\theta_i, \varphi_i, \theta_r, \varphi_r)] d\Omega_r. \tag{4.49}$$

Subscripts "i" and "r" stand for the incident and scattered radiation, respectively, and θ and φ are the incidence and azimuth angles, respectively. The coefficient σ_{pp} takes into account the scattering in the same polarization, while σ_{pq} takes into account the scattering in the orthogonal polarization. The integration over the entire upper half space ($2 \cdot \pi$ strad in Eq. 4.49) and errors in the modeling of the bistatic scattering coefficients make the computation of Eq. (4.49) very prone to errors. For this reason, the half space model with a smooth surface is commonly adopted, usually assuming that it is uniform, and eventually including a dielectric constant and temperature profile. This model is very simple and thus suitable for soil moisture retrieval algorithms from remotely sensed data. In this case, the total reflectivity in Eq. (4.49) equals the reflection coefficient given by the Fresnel formulation.

Soil brightness temperature is related to soil emission through the soil effective temperature, T_{eff}, $T_{B,p} = e_p \cdot T_{eff}$. The theoretical effective temperature of a soil profile can be estimated as (Ulaby et al., 1986):

$$T_{eff} = \int_0^\infty T(z) \cdot \alpha(z) \cdot \exp\left\{ -\int_0^x \alpha\left(z'\right) dz' \right\} dz, \tag{4.50}$$

where $T(z)$ and $\alpha(z)$ are the thermodynamic temperature and the attenuation coefficient at a depth z. The attenuation is a function of the soil dielectric constant and of the microwave emission wavelength λ. Several simple formulations have been developed to estimate the soil effective temperature from soil properties, and soil moisture and temperature profiles. Choudhury et al. (1982) proposed a parameterization of T_{eff} based on the soil temperature at deep soil (T_∞) corresponding to a depth between 50 cm and 1 m, and on a surface temperature (T_{surf}) corresponding to a depth of 0—5 cm:

$$T_{eff} = T_\infty + C \cdot (T_{surf} - T_\infty), \tag{4.51}$$

The coefficient C was considered constant for a given frequency, and equal to 0.246 at L-band. On the other hand, Chanzy et al. (1997) presented a model for the soil effective temperature at L- and C-bands based on the air temperature, a deep soil temperature, and the brightness temperature measured at X-band and V-polarization. Wigneron et al. (1997) proposed a parameterization based on Eq. (4.51), but with a coefficient C dependent on the volumetric water content (w_s), and two semiempirical parameters. Another formulation using the soil dielectric constant instead of the volumetric water content was proposed by Holmes et al. (2006). The performance of some of these approaches is analyzed in Wigneron et al. (1997).

The effect of soil surface roughness on the brightness temperature has been an issue widely addressed in the literature. In Fung (1994) a theoretical physical model based on surface characteristics derived from the measured soil height profile is proposed. A simple empirical roughness model which takes into account only the coherent term of the scattering was reported in Choudhury et al. (1979):

$$\Gamma_{s,p} = \Gamma_{s,p}^* \exp\left[-(2k\sigma_s \cos \theta)^2 \right], \tag{4.52}$$

where $\Gamma_{s,p}^*$ is the Fresnel specular reflectivity of soil, k is the electromagnetic wave number, σ_s is the standard deviation of the surface height, and θ is the incidence angle. In Wang and Choudhury (1981) this model was reviewed, and another formulation was proposed:

$$\Gamma_{s,p} = \left[(1 - Q_s)\Gamma_{s,p}^* + Q_s\Gamma_{s,q}^*\right]\exp(-h_s\cos\theta^n). \tag{4.53}$$

In this case, two semiempirical parameters were included to model the effects of the polarization mixing Q_s, and surface roughness (h_s and n). The dependence of these parameters on surface properties such as correlation length or standard deviation of height is not yet clear. In Wigneron et al. (2001), Mo and Schmugge (1987), it was concluded that the $n = 2$ dependence in Eq. (4.53) is too strong at L-band. A value of $n = 0$ at both polarizations was found to be consistent with measurements in Wingeron et al. (2001), while in Escorihuela et al. (2007), different values of n for each polarization are found ($n = 1$ at horizontal, and $n = -1$ at vertical polarizations). Similarly, there are discrepancies on the value of the roughness parameter h_s. Some authors obtain h_s from experimental data by best-fit (Wingeron et al., 2001), while others propose expressions for h_s as a function of geophysical parameters. In Mo and Schmugge (1987), good results were obtained with two parameterizations of $h_s = A \cdot w_s^B \cdot (\sigma_s/l_c)^C$, with $A = 0.5761$, $B = -0.3475$, and $C = 0.4230$, or $h_s = A' \cdot (\sigma_s/l_c)^{c'}$, with $A' = 1.3972$ and $c' = 0.5879$. Finally, there is a general agreement on the value of the cross-polarization parameter Q_s, which has been found to be very small (0−0.12) at L-band.

Apart from these considerations, the effects of frequency and incidence angle on the roughness parameters have not been studied thoroughly. It was pointed out (Mo and Schmugge, 1987; Shi et al., 2002) that the roughness effects depend on both the frequency and the incidence angle. A parameterization of the surface reflectivity derived from data simulated for a wide range of soil water content and roughness properties using the integral equation model (Fung, 1994) has also been suggested. The surface reflectivity model in Shi et al. (2002) was tested in Schneeberger et al. (2004) and found not to be capable of explaining discrepancies between the ground truth and remotely sensed data. As a consequence, a new model was developed for describing the influence of the topsoil structure on the L-band emission as an impedance matching between the dielectric constants of soil and air (Mätzler, 2006). Fig. 4.26 shows the impact on the 1.4 GHz brightness temperature of different bare soils at nadir as a function of the volumetric soil moisture content, and the impact of the surface's roughness (Ulaby et al., 1986). As is clear, the larger the volumetric soil moisture content, the lower the brightness temperature. For the same volumetric soil moisture content, this effect is more noticeable for sandy soils, than for clays, due to the larger fraction of free water molecules in the interstices between soil grains. It can also be appreciated (Fig. 4.26B) that the surface's roughness masks the presence of the surface soil moisture, due to the decrease of the reflection coefficient (Eqs. 4.52 and 4.53).

Finally, it is worth noting that the emission being sensed is coming mostly from the soil's top layer due to the high attenuation that the electromagnetic waves being emitted undergo due to the presence of water. The soil penetration depth is given by $\delta_p \approx \frac{\lambda\sqrt{\varepsilon'}}{2\pi\varepsilon''}$ for materials with $\varepsilon''/\varepsilon' < 0.1$. The soil penetration depth as a function of the frequency and the volumetric soil moisture content is plotted in Fig. 4.27. At L-band ($\lambda \approx 20$ cm) the penetration depth is ~ 10 cm for 30% volumetric soil moisture.

4.3.3.3 Vegetated Soil Emission
If the remote sensor is placed above a canopy looking downward, the measured brightness temperature will contain not only information on the soil, but also on vegetation, since vegetation radiates its own

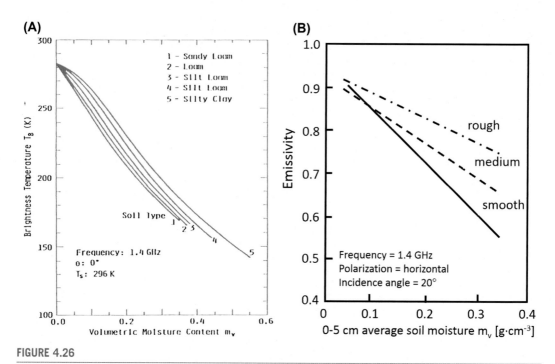

FIGURE 4.26

1.4 GHz brightness temperature at nadir for different soil types as a function of the volumetric soil moisture content (Ulaby et al., 1986).

energy and, moreover, attenuates and scatters the soil radiation. Chukhlantsev (2006) revised the theory and conducted experimental research over vegetated areas. Although the modeling of the land emission involves analytical solutions of the radiative transfer equation (Ferrazzoli et al., 2002), this approach is not easy to use with experimental data. Hence, the common practice is to use approximate formulas or semiempirical models in which the different components of the brightness temperature could be differentiated. The brightness temperature of a soil covered by vegetation is usually estimated as the contribution of three terms: the radiation from the soil that is attenuated by the overlying vegetation, the upward radiation from the vegetation, and the downward radiation from the vegetation, reflected by the soil, and attenuated by the canopy (Ulaby et al., 1986):

$$T_{B,p}^{\text{model}} = \left(1 + \frac{1 - e_{s,p}}{L_{veg}}\right)\left(1 - \frac{1}{L_{veg}}\right)(1 - \omega)T_{veg} + \frac{e_{s,p}}{L_{veg}}T_s, \qquad (4.54)$$

where T_{veg} and T_s are the physical temperatures of the vegetation and soil, respectively, $L_{veg} = \exp(\tau \cdot \sec\theta)$ is the attenuation due to the vegetation cover, $\tau = b \cdot \text{VWC}$ is the optical thickness, b is a factor (Van de Griend and Wigneron, 2004) that depends on the vegetation type and frequency, VWC is the vegetation water content, and ω is the single scattering albedo, which is a function of the polarization and the incidence angle. Actually, L_{veg} only follows the above expression at low incidence angles due to the larger anisotropy of the vegetation structure as the incidence angle increases, and the increased scattering effects. This formulation is known as the τ-ω model, and it is based on the single

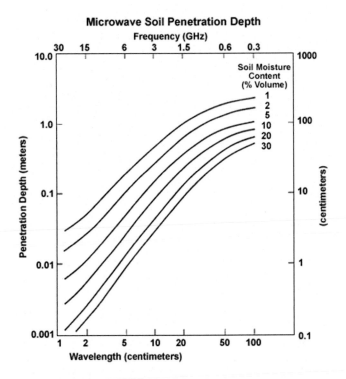

FIGURE 4.27

Soil penetration depth as a function of the frequency and the volumetric soil moisture content. (The source of this material is the COMET® Website at http://meted.ucar.edu/ of the University Corporation for Atmospheric Research (UCAR), sponsored in part through cooperative agreement(s) with the National Oceanic and Atmospheric Administration (NOAA), U.S. Department of Commerce (DOC). ©1997–2016 University Corporation for Atmospheric Research. All Rights Reserved.)

Adapted from Njoku, 1999

scattering approach proposed in Kirdiashev et al. (1979). The optical depth is related to the vegetation density and the frequency, while the single scattering albedo describes the scattering of the emitted radiation by the vegetation and is a function of plant geometry. This formula is actually only valid at low frequencies, and when scattering mechanisms are negligible or small, while at higher frequencies, the full polarimetric radiative transfer equation should be solved (Martinez-Vazquez et al., 2009).

Fig. 4.28 shows some simulated brightness temperatures at L-band for volumetric soil moisture 0% and 40%, and bare soil, and densely vegetated soil. As can be seen, for bare soil, the soil moisture change translates into a large brightness temperature change, while this is no longer the case when there is a very dense vegetation cover that masks the soil radiation. As can also be inferred, the presence of dense vegetation gets the vertical and horizontal brightness temperatures closer, while for bare soils, they are very much separated.

This polarimetric and multiangular information can be used to infer not only surface's soil moisture, but vegetation water content (vegetation optical depth) as well, as is routinely done now with the European Space Agency (ESA) SMOS mission.

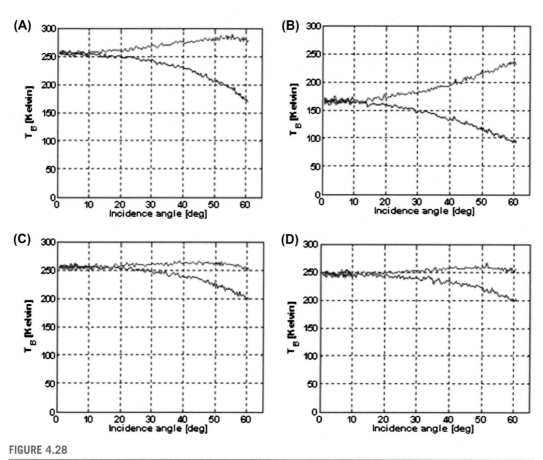

FIGURE 4.28

Simulated vertical (red) and horizontal (blue) brightness temperatures for (A, C) 0% and (B, D) 40% volumetric soil moisture, and (A, B) bare soil and (C, D) vegetated soil.

4.3.3.4 Snow-Covered Soil Emission

Water ice has a very different structure than liquid water, as do soils and rocks. For ordinary ice the real part ε' of the dielectric constant is independent of frequency from 10 MHz to about 300 GHz, and only slightly dependent on temperature (Mätzler and Wegmüller, 1987):

$$\varepsilon'_i = 3.1884 + 9.1 \cdot 10^{-4}(T - 273\text{K}); \quad 243\text{K} \le T \le 273\text{K}, \tag{4.55}$$

below 240K the temperature sensitivity decreases, and a constant value of $\varepsilon'_i \approx 3.10$ is found at $T < 100\text{K}$ (Gough, 1972). The imaginary part ε''_i is given by:

$$\varepsilon''_i = \frac{A}{f} + B \cdot f^C, \tag{4.56}$$

where f is the frequency in (GHz), and the empirical constants are $A = 3.5 \cdot 10^{-4}$, $B = 3.6 \cdot 10^{-5}$, and $C = 1.2$ at $T = -15°C$; and $A = 6 \cdot 10^{-4}$, $B = 6.5 \cdot 10^{-5}$, and $C = 1.07$ at $T = -5°C$.

Actually, ice water is one of the most transparent media at microwave frequencies with very low dielectric loss. In addition, new (and dry) snow has a large fraction of air ($\sim 90\%$), i.e., its density is quite low, which reduces the reflection coefficient in the air—snow interface, while old (and wet) snow is much more compact ($\sim 20\%$ air), is much denser and has a higher reflection coefficient [Table 4.2 from Gough (1972), Fig. 4.29].

Table 4.2 Typical Snow and Ice Densities (kg/m³)	
New snow (immediately after falling in calm)	50–70
Damp new snow	100–200
Settled snow	200–300
Depth hoar	100–300
Wind packed snow	350–400
Firn	400–830
Very wet snow and firn	700–800
Glacier ice	830–917

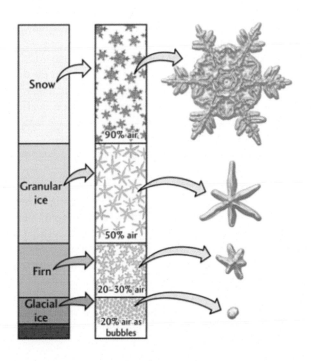

FIGURE 4.29

Snow density and typical grain size and shape (https://www.meted.ucar.edu/).

From NOAA

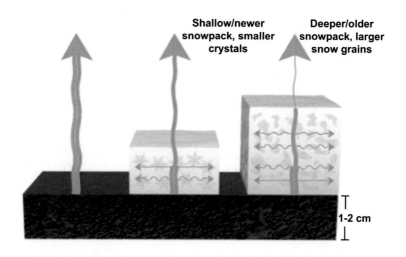

Shallow/newer snowpack, smaller crystals

Deeper/older snowpack, larger snow grains

1-2 cm

FIGURE 4.30

Snow emission and scattering of the soil radiation depends on the thickness, wetness, and ice crystal size (https://www.meted.ucar.edu/).

This means that the dominant mechanism in the radiative transfer equation is not the absorption, but the scattering in the ice crystals (dry snow). If there is liquid water in between the ice crystals (wet snow), absorption becomes more important.

The upwelling radiation emitted by the soil surface under the snow is altered significantly by the snowpack and it changes with time as illustrated in Fig. 4.30: the larger the thickness and/or the density of the snow layer, the larger the attenuation and scattering will be; the moister or older snowpack is, the larger the attenuation will be; and the amount of scattered radiation is affected by the grain size as compared to the microwave radiation: shallower/newer snowpack have smaller ice crystals and scatter less radiation than deeper/older snowpacks.

For dry snow a simple model using the snow single scattering albedo ω_{snow}, can be used:

$$T^p_{B,dry\ snow}(\theta) \approx (1 - \omega_{snow}) \cdot T_{soil}, \tag{4.57}$$

since $\Gamma^p_{air-snow}(\theta) \approx 0$. However, wet snow has a much larger emissivity than dry snow since the presence of water increases the absorption (losses), and therefore the emission, and the size of the snow grains is reduced, therefore, scattering is reduced accordingly. These two effects lead to an underestimation of the snow depth, unless properly accounted for.

Accurate snow state information can only be retrieved using multifrequency and polarimetric information, mainly the 19 and 37 GHz channels, and 85 GHz as well at high latitudes for thin snow layers. The microwave emissivity spectra for snow cover at 50° incidence angle is presented here as Fig. 4.31. As can be seen, wet snow has the highest emissivity due to the largest losses, and it is relatively constant with frequency, while in general, the emissivity decreases significantly

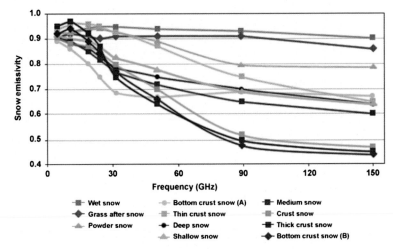

FIGURE 4.31

Microwave emissivity spectra for snow cover at 50° incidence angle (https://www.meted.ucar.edu/).

with increasing frequency since scattering effects become more significant as the wavelength decreases.

Finally, Fig. 4.32A and B show the surface's emissivity spectra at 53° vertical and horizontal polarizations, respectively, for different soil types:

- dry bare soil: which exhibits a very weak dependence on frequency, since the dielectric constant does not depend significantly on it,
- wet land: which exhibits a lower emissivity than bare soils, and a slightly higher dependence with frequency, due to the variation of the dielectric constant of water,
- desert: which at low frequencies behaves as dry bare soil, but the emissivity decreases with increasing frequency due to the larger scattering by the sand grains,
- canopy: which behaves as dry bare soil, but it decreases with increasing frequency due to the more important scattering in the vegetation structures (trunk, branches, and leaves) specially for vertical polarization.
- snow: which has the most significant decrease in the emissivity as frequency increases, due to scattering by the ice crystals, and
- ocean: whose emissivity actually increases with increasing frequency, mainly due to the decrease of the imaginary part of the dielectric constant (i.e., the water stops behaving as a bad conductor to start behaving more as a dielectric).

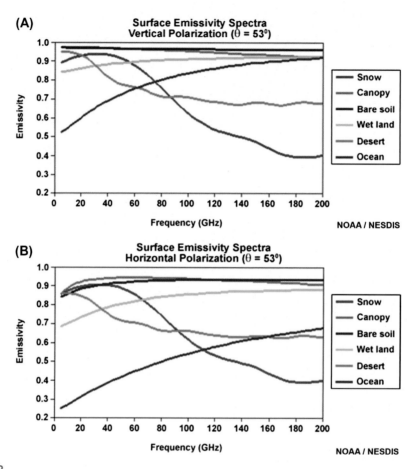

FIGURE 4.32

Surface's emissivity spectra at 53° (A) vertical and (B) horizontal polarizations, respectively, for different surfaces (https://www.meted.ucar.edu/).

4.3.3.5 Topography Effects

Topography (including Earth's curvature) refers to the large-scale surface roughness, as opposed to the small-scale surface roughness discussed in Eqs. (4.52) and (4.53). The effects can be summarized as follows:

- Assuming that the surface can be decomposed in facets, the **local incidence angle θ_l** differs from the global one θ (the one assuming a flat or ellipsoidal Earth):

$$\cos \theta_l = \widehat{n} \cdot \widehat{k} = \cos \theta \cdot \cos \alpha + \sin \theta \cdot \sin \alpha \cdot \cos \beta, \qquad (4.58)$$

where α and β are the tilt and azimuth angles of the facet. This affects the emission and reflection of the facet towards the sensor.

- In addition, the global reflectivities $[\Gamma_p(\theta)]$ and emissivities $[e_p(\theta)]$ are also affected by a rotation φ of the polarization reference frame:

$$\sin \varphi = \sin \beta \cdot \sin \theta_l. \tag{4.59}$$

As explained in Section 4.3.2, this effect translates into a polarization mixing effect described in Eq. (4.55), which is reproduced here for convenience ($A = \cos\varphi$, $B = \sin\varphi$). Please recall that in Eq. (4.41) subscripts h, v, hv, and vh refer to the local reference frame, while xx, yy, xy, and xy to the global one.

$$\begin{bmatrix} T_{xx} \\ T_{xy} \\ T_{yx} \\ T_{yy} \end{bmatrix} = \begin{bmatrix} A^2 & AB & AB & B^2 \\ -AB & A^2 & -B^2 & AB \\ -AB & -B^2 & A^2 & AB \\ B^2 & -AB & -AB & A^2 \end{bmatrix} \begin{bmatrix} T_h \\ T_{hv} \\ T_{vh} \\ T_v \end{bmatrix}. \tag{4.60}$$

It is usually assumed that $T_{hv} \approx 0$ and $T_{vh} \approx 0$, and therefore:

$$\begin{aligned} T_{xx} &= \cos^2 \varphi \cdot T_h + \sin^2 \varphi \cdot T_v, \\ T_{yy} &= \sin^2 \varphi \cdot T_h + \cos^2 \varphi \cdot T_v. \end{aligned} \tag{4.61}$$

However, it is worth noting that a nonzero third Stokes parameter appears, since:

$$T_{xy} = T_{yx} = \cos \varphi \cdot \sin \varphi (T_v - T_h). \tag{4.62}$$

- **Projection effects**: since the area of the facet that is projected towards the sensor depends on the local incidence angle as $\cos\theta_l$.
- **Shadowing** when a given facet is hidden from the direct sensor's view.
- **Multiple scattering** when either the radiation emitted by the surface or the atmospheric downwelling one undergo a number of "reflections" (scattering processes) in other facets of the surface before they propagate in the direction of the sensor.
- Finally, the **atmospheric downwelling and upwelling brightness temperatures** and the length of the **atmospheric path** and its associated **attenuation** are also affected and must be included accordingly in the radiative transfer equation (Section 4.2).

4.3.4 OCEAN EMISSION

The ocean surface emission depends on a number of factors. On one side, the sea water dielectric constant depends on the physical temperature, the salinity, and the frequency. On the other hand, the sea surface roughness, the presence of foam or oil slicks in the surface alters the interaction of the electromagnetic waves with the water surface. In this section, these different effects are revisited.

4.3.4.1 Water Dielectric Constant Behavior

Following Eq. (4.11), the emission of a calm ocean is given by the physical temperature times the emissivity:

$$e^p_{flat}(\text{SSS}, \text{SST}, f, \theta) = 1 - \Gamma^p_{spec,\text{H}_2\text{O}}(\varepsilon_r(\text{SSS}, \text{SST}, f), \theta). \tag{4.63}$$

It has been suggested that a bimodal relaxation time expression is the most appropriate description of the dielectric properties of water (Buchner et al., 1999):

$$\varepsilon_r = \varepsilon_\infty + \frac{\varepsilon_S - \varepsilon_2}{1 + j\omega\tau_D} + \frac{\varepsilon_2 - \varepsilon_\infty}{1 + j\omega\tau_2}, \tag{4.64}$$

where ε_r is the complex permittivity, ε_S is the relative permittivity at low frequencies (static region), ε_2 is the intermediate relative permittivity, and ε_∞ is the relative permittivity at high frequencies (optical permittivity). In addition, ω is the angular frequency in radians/second, and τ_D and τ_2 are relaxation times. Here, τ_D is relatively long (~ 18 ps at 0°C), due mainly to the rotational relaxation within a hydrogen-bonded cluster, but reduces considerably with temperature as hydrogen bonds are weakened and broken. τ_2 is small ($\sim 0.2{-}1$ ps) and less temperature dependent being determined mainly by the translational vibrations within the hydrogen-bonded cluster. In Fig. 4.33 we show plots of the dependence of these parameters on the physical temperature. Note that ε_∞, the dielectric permittivity at very high frequencies, does not change significantly with temperature.

The dielectric permittivity (real part of ε_r) and the dielectric loss (imaginary part of ε_r) of water between 0°C and 100°C, and from 300 MHz to 3 THz are shown in Fig. 4.34. As the temperature increases, the strength and extent of the hydrogen bonding decreases, which lowers the ε_S, allows the water molecule to oscillate at higher frequencies and reduces the rotation drag, thus reducing the friction and hence the dielectric loss. Most of the dielectric loss is within the microwave range of electromagnetic radiation (~ 1 to ~ 300 GHz), and the frequency for maximum dielectric loss is ~ 2.45 GHz, the frequency of operation of most microwave ovens.

The dilution of salt in water decreases the natural structure of the water so reducing the ε_S, in a similar manner to increased temperature. At the lower frequencies ions are able to respond and move with the changing potential so producing frictional heat and increasing the loss factor (imaginary part of ε_r). Thus, whereas water becomes a poorer microwave absorber with rising temperature, salty water becomes a better microwave absorber with rising temperature. The evolution of the dielectric permittivity (real part of ε_r) and the dielectric loss (imaginary part of ε_r) of salty water (10 psu) between 0°C and

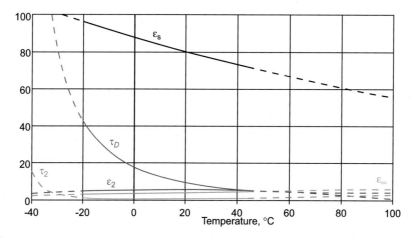

FIGURE 4.33

Dependence of the parameters of the water dielectric constant model with physical temperature.

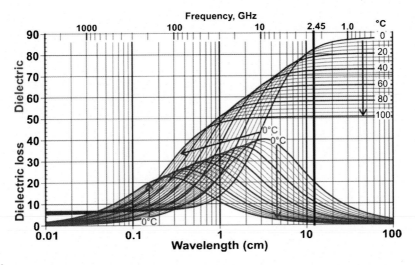

FIGURE 4.34

Dielectric permittivity (real part of ε_r) and the dielectric loss (imaginary part of ε_r) of freshwater between 0°C and 100°C, and from 300 MHz to 3 THz (http://www.lsbu.ac.uk/water/microwave.html).

100°C, and from 300 MHz to 3 THz are presented here in Fig. 4.35. As compared to Fig. 4.34, in the low microwave frequency range, the imaginary part of the dielectric constant does not tend to zero, but increases sharply. This property is actually used to sense SSS from radiometric measurements.

The curves in Fig. 4.36 show the evolution of the real and imaginary parts of the water dielectric constant at 1.4 GHz, for salinities ranging from 0 to 40 psu in 10 psu steps, for temperatures from −2°C to +40°C.

4.3.4.2 Calm Ocean Emission
4.3.4.2.1 Influence of the Salinity

The water dielectric constant computed in Section 4.3.4.1. can now be used to evaluate the emissivity and brightness temperature of a flat water surface as a function of the temperature, salinity, frequency, and incidence angle. The water emissivity and brightness temperatures, respectively, at 1.4 GHz from −2°C to 40°C, and from 0 to 40 psu are shown in Figs. 4.37 and 4.38. As can be easily seen, the behavior is highly nonlinear with the physical temperature. The water emissivity is approximately flat between 10 and 20 psu for physical temperatures up to about 25°C. However, the water nadir brightness temperature becomes insensitive to the physical temperature around 15°C and 35 psu. Indeed, this is good news for the SSS retrievals using microwave radiometry, since these conditions are close to the ones encountered in the open oceans, and the SST can be accurately estimated using thermal infrared sensors[4] or microwave radiometers at higher frequencies, the sea state correction will remain as the most important one.

---

[4]Although thermal infrared sensors measure the skin temperature, and the sea brightness temperature at L-band corresponds approximately to the emission from the top centimeter, the bulk and the skin temperatures are barely the same, unless when the water is in very calm conditions.

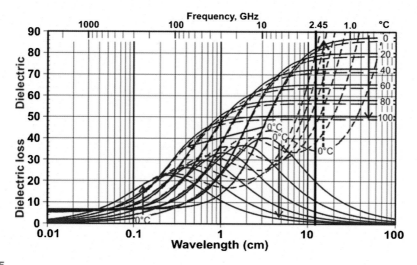

FIGURE 4.35

Dielectric permittivity (real part of ε_r) and the dielectric loss (imaginary part of ε_r) of 10 psu salty water between 0°C and 100°C, and from 300 MHz to 3 THz (http://www.lsbu.ac.uk/water/microwave.html).

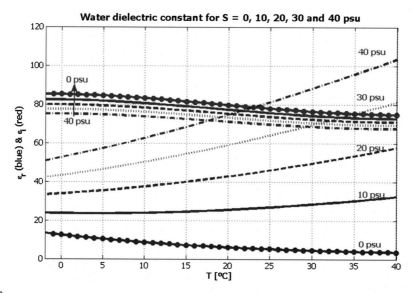

FIGURE 4.36

Dielectric permittivity (real part of ε_r) and the dielectric loss (imaginary part of ε_r) of salty water at 1.4 GHz from −2°C to 40°C, and from 0 to 40 psu.

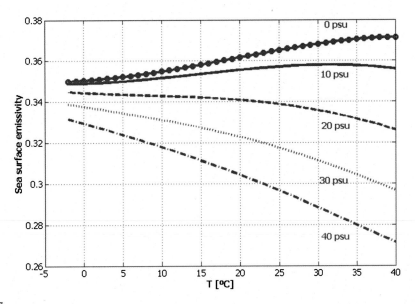

FIGURE 4.37

Water emissivity at 1.4 GHz from −2°C to 40°C, and from 0 to 40 psu.

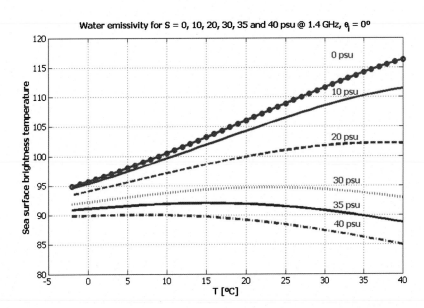

FIGURE 4.38

Water brightness temperature at $\theta_i = 0°$ and 1.4 GHz from −2°C to 40°C, and from 0 to 40 psu.

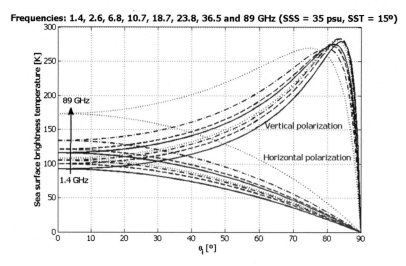

FIGURE 4.39

Brightness temperature evolution versus incidence angle at different frequencies from 1.4 to 89 GHz, 35 psu SSS, and 15°C.

4.3.4.2.2 Influence of Frequency

We use Fig. 4.39 to show the water brightness temperature at different frequencies from 1.4 to 89 GHz, 35 psu SSS, and 15°C versus incidence angle. The general behavior follows the trends of Fig. 4.5 with horizontal polarization monotonically decreasing with increasing incidence angle, and vertical polarization increasing with increasing incidence angle up to an angle where the brightness temperature reaches a maximum, and then decreases down to zero at 90°. Recall that this maximum is related to the Brewster angle, at which the reflectivity vanishes completely in the case of lossless materials, or is close to zero in the case of materials with losses. Note also that, for low incidence angles, as the frequency increases, the brightness temperature increases as well, due to the reduction of the dielectric constant (Fig. 4.35). That is, the sea water behaves less and less as a conductor (although a bad conductor) and starts behaving more as a dielectric.[5]

4.3.4.2.3 Influence of the Water Temperature

Here Fig. 4.40 shows the water brightness temperature at different frequencies from 1.4 to 89 GHz, for 35 psu SSS, and $\theta_i = 0°$ versus the SST. In this case, the behavior is more complex to understand intuitively than that in Fig. 4.39 because of the strong nonlinearity with temperature. Around 35 psu, at low microwave frequencies (e.g., 1.4 GHz), the sensitivity of the brightness temperature versus the physical temperature is zero. It becomes positive at 2.6, 6.8, and 10.7 GHz. At 18.7 GHz, the sensitivity is slightly positive or negative depending on the surface's temperature itself, and at 36.5 and 89 GHz, it is negative, that is, the emissivity decrease is not compensated by the physical temperature increase.

---

[5]Radio-waves attenuate much faster in the sea water than light waves.

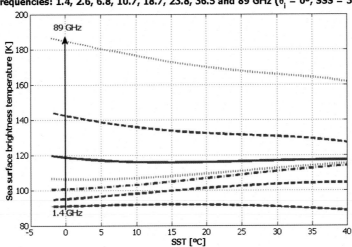

FIGURE 4.40

Brightness temperature evolution versus physical temperature at nadir for different frequencies from 1.4 to 89 GHz, 35 psu SSS, and 15°C physical temperature.

4.3.4.3 Influence of the Sea State

Over a perfectly calm sea, the wind has practically no friction but, as it slides over the water surface film, it makes it move. Above 0.23 m/s eddies and small ripples ("short" capillary waves using surface tension for a restoring force) are formed that make the water's surface rough, thus increasing the friction. The surface still looks glassy overall, but as the wind speed increases, gravity waves form and the waves become high enough to interact with the air flow and the surface starts to look rough. The wind then becomes turbulent just above the surface and starts transferring more energy to the waves. Due to the dispersive nature of the waves, as they propagate, energy is transferred from the shorter waves to the longer ones that keep propagating over long distances even when the wind has stopped blowing. This phenomenon is known as swell.

When the sea surface becomes rough (there are waves) the scattering of the electromagnetic waves over the surface becomes more diffuse and less specular, and therefore the emissivity increases. High-frequency microwave instruments (both active and passive) respond quickly to the "short" waves, showing a good correlation with the wind speed. However, at the lower end of the microwave spectrum (i.e., at L-band), due to the longer electromagnetic wavelengths involved, the interaction between the electromagnetic waves and the sea surface is also affected by the "long" sea waves. Therefore the radiometer response does not correlate so well with the wind speed, which ultimately limits the accuracy of the SSS retrievals from spaceborne microwave radiometers.[6]

In addition, when the sea waves' slope exceeds $\sim 15° - 20°$, waves break creating foam, which is a mixture of air and sea water, and therefore acts as a matching layer between the air and the sea water beneath, reducing the scattering coefficient and increasing the emissivity. This effect is particularly noticeable above 7 m/s when whitecaps start appearing in the sea surface. However, the whitecap

---

[6]Sea state can be compensated, at least partially, using synergetic radar (Chapter 5) or GNSS-R (Chapter 6) measurements.

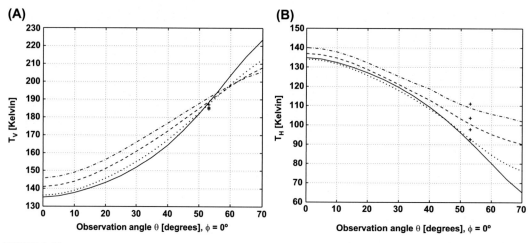

(A)

(B)

FIGURE 4.41

Modeled (A) vertical and (B) horizontal brightness temperatures at 37 GHz versus observation angle, for wind speeds of 3 m/s (*solid line*), 7 m/s (*dotted line*), 11 m/s (*dashed line*), and 15 m/s (*dash-dot line*). Wentz's (1992) geophysical model marked with crosses for comparison, with SST = 12°C and SSS = 33 psu.

coverage not only depends on the wind speed, but also on atmospheric stability $\Delta T \;\widehat{=}\; T_{water} - T_{atm}$ (by $\pm 9\%$ for $\Delta T = \pm 1°C$), water temperature T_{water} (water viscosity changes with T_{water}), wind duration, fetch, and water salinity (51% larger whitecap coverage on the sea than on freshwater).

The modeled brightness temperature of the sea surface at 37 GHz, for different wind speeds from 3 to 15 m/s as a function of the incidence angle, is compared to Meissner and Wentz (2004) geophysical model function at $\theta_i = 53°$ (Fig. 4.41). Close to nadir, from 7 to 11 m/s there is a sharp increase in the brightness temperature, which is due to the appearance of foam at the sea surface. In the range of incidence angle modeled, the behavior of the brightness temperatures is monotonic: increasing at vertical polarization and decreasing at horizontal polarization. The effect of wind speed is more noticeable for horizontal polarization, and therefore it will be used for wind speed retrievals. On the other hand, the lack of sensitivity at vertical polarization around 53°, which is related to the Brewster angle, is used as a reference to compensate for the atmospheric effects that affect in a similar way both polarizations.

4.3.4.3.1 Influence of the Look Angle

When the emitting surface is not completely random, but it has prreferred directions, as in the case of the sea surface driven by the wind, the solution of the RTE (Section 4.2) leads to a solution in which the Stokes elements do exhibit an angular dependence not only on the incidence angle, but on the azimuth angle as well defined as the angle between the look angle and the wind direction:

$$T_h(\theta, \phi) = T_{h0}(\theta) + T_{h1}(\theta)\cos\phi + T_{h2}(\theta)\cos(2\phi),$$
$$T_v(\theta, \phi) = T_{v0}(\theta) + T_{v1}(\theta)\cos\phi + T_{v2}(\theta)\cos(2\phi), \tag{4.65}$$
$$T_3(\theta, \phi) = T_{3,1}(\theta)\sin\phi + T_{3,2}(\theta)\sin(2\phi).$$

Intuitively, the $\cos\phi$ and $\sin\phi$ terms in Eq. (4.65) are induced by the different upwind ($\sigma_u^2 = 0.00316w$) and crosswind ($\sigma_c^2 = 0.003 + 0.00192w$) rms slopes, while the $\cos(2\phi)$ and $\sin(2\phi)$ terms are induced by the skewness ($c_{21} = 0.01 - 0.0086w$, $c_{03} = 0.04 - 0.033w$) and peakedness

($c_{40} = 0.40$, $c_{22} = 0.12$ and $c_{04} = 0.23$) terms in the *Gram−Charlier* probability density function of the sea surface slopes (Cox and Munk, 1954):

$$P = \frac{1}{\sqrt{2\pi}\sigma_u\sigma_c} \exp\left[-\frac{1}{2q_z^2}\left(\frac{q_x^2}{\sigma_u^2} + \frac{q_y^2}{\sigma_c^2}\right)\right]$$

$$\left\{ 1 - \frac{1}{2}c_{21}\left(\left(\frac{q_y}{\sigma_c}\right)^2 - 1\right)\frac{q_x}{\sigma_u} - \frac{1}{6}c_{03}\left(\left(\frac{q_x}{\sigma_u}\right)^3 - 3\frac{q_x}{\sigma_u}\right) + \frac{1}{24}c_{40}\left(\left(\frac{q_y}{\sigma_c}\right)^4 - 6\left(\frac{q_y}{\sigma_c}\right)^2 + 3\right)\right.$$

$$\left. + \frac{1}{4}c_{22}\left(\left(\frac{q_y}{\sigma_c}\right)^2 - 1\right)\left(\left(\frac{q_x}{\sigma_u}\right)^2 - 1\right) + \frac{1}{24}c_{04}\left(\left(\frac{q_x}{\sigma_u}\right)^4 - 6\left(\frac{q_x}{\sigma_u}\right)^2 + 3\right)\right\},$$

$$(4.66)$$

where w is the 10 m height wind speed, q_x/q_z and q_y/q_z are the slopes in the x and y-axes, respectively.

An example from Coriolis/WindSat (Anguelova and Gaiser, 2013, Fig. 4.42) shows the third Stokes parameter (T_3) plotted as a function of the relative azimuth angle for different wind speeds (color coded). Although having units of temperature (power), as we can infer, T_3 can be either positive or negative, depending on the relative azimuth angle, since it is not an absolute power measurement, but a

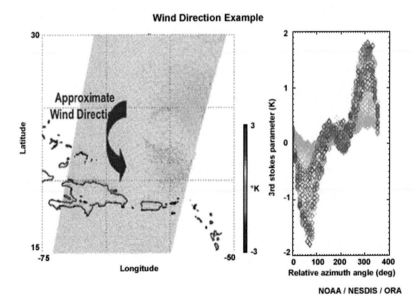

FIGURE 4.42

Hurricane Isabel on September 14, 2003, as seen by WindSat 18.7 GHz third Stokes parameter. Notice how the signal changes around the circulation pattern of the storm (Anguelova and Gaiser, 2013). The source of this material is the COMET® Website at http://meted.ucar.edu/ of the University Corporation for Atmospheric Research (UCAR), sponsored in part through cooperative agreement(s) with the National Oceanic and Atmospheric Administration (NOAA), U.S. Department of Commerce (DOC). ©1997−2016 University Corporation for Atmospheric Research. All Rights Reserved.

cross-correlation between the electric fields at horizontal and vertical polarizations. Since T_3 is less affected by atmospheric effects than T_v and T_h, it is more robust to infer the wind direction than T_v and T_h.

4.3.4.4 Emissivity of the Sea Surface Covered With Oil

Oil spills are of particular concern in environmental monitoring and microwave radiometry can help to monitor their extension and thickness. Once again, the change in the emissivity (or brightness temperature) provides the clue to measure it:

$$\Delta T_{B,sea}^p(\theta) = \left[e_{oil}^p(\theta) - e_{H_2O}^p(\theta) \right] \cdot T_{sea} = \left[\Gamma_{H_2O}^p(\theta) - \Gamma_{oil}^p(\theta) \right] \cdot T_{sea}, \tag{4.67}$$

This change is mainly due to two factors:

- the reduced water surface roughness as compared to that of the clean water, for the same wind speed (Cox and Munk, 1954). Since oil damps the short capillary waves, the surface becomes more specular, the reflectivity is increased, and the emissivity decreases;
- the reduced reflection coefficient due to the fact that the oil layer acts as an interface layer between the air and the water. The amount of matching depends on the oil characteristics and, most important, on the oil layer thickness. Partial reflection at the oil—water atmosphere interface leads to interference fringes that lead to minimum reflection coefficient values (maximum emissivity) for normal incidence when the oil thickness is equal to an odd number of a quarter of the electromagnetic wavelength of the oil: $(2 \cdot n + 1) \cdot \lambda/4$, and maximum reflection coefficient values (minimum emissivity) when it is equal to an even number of a quarter of the electromagnetic wavelength: $(2 \cdot n) \cdot \lambda/4$. This periodic increase is shown in Fig. 4.43 versus the oil film thickness in millimeters, for three different frequencies (5, 17, and 34 GHz) and it can be used to remove the oil film thickness ambiguity when trying to infer it from multifrequency microwave radiometry data. In Fig. 4.43, the incidence angle of the radiometer field of view (FOV) on the oil surface is 41° (5 GHz), 54° (17 GHz), and 50° (34 GHz).

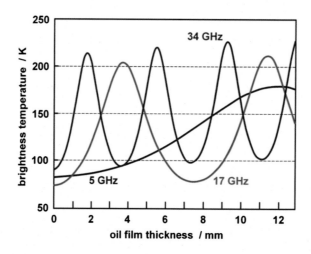

FIGURE 4.43

Oil spills cause a T_B increase periodic in frequency and oil spill thickness (http://www.seos-project.eu/modules/marinepollution/marinepollution-c02-ws02-p03-s.html).

Reproduced from SEOS tutorial. http://www.seos-project.eu/modules/marinepollution/marinepollution-c08-p02.html

FIGURE 4.44

Heavy fuel oil slick with a total volume of 17 m³, measured with a 32 GHz MWR Scanner. The angle of incidence of the radiometer field of view on the sea surface is 40°. The maximum brightness temperature variation is 17K.

Reproduced from SEOS tutorial. http://www.seos-project.eu/modules/marinepollution/marinepollution-c08-p02.html

A fuel oil spill as seen from a 32 GHz microwave radiometer is depicted in Fig. 4.44. Hot spots are clearly seen, indicating the presence of oil spills.

4.3.4.5 Emissivity of the Sea Ice Surface

When the sea water gets frozen the emissivity changes dramatically. While most materials have a dielectric constant between 1 and 4 (ice is 3.2), that of water is 80 approximately, which makes passive microwave sensing extremely sensitive to the presence of ice. Also, the detection of the melt onset can be easily detected, although there are problems afterwards when water and ice coexist.

The ice emissivity depends on a number of factors, including the frequency, the seasonal variation, that is whether the ice layer is of 1-year-old, several years old (Fig. 4.45A), or if it has melted and then refrozen (Fig. 4.45B). Additionally, the presence of snow layers complicates even more the measurement of the ice properties, since the snow scatters the ice's emitted radiance. While first-year ice emissivity is quite constant with frequency, multiyear ice's emissivity drops significantly with increasing frequency due to the increased scattering effects.

Despite the abovementioned effects, the ice—water contrast is so high, that the detection of the fraction of ice cover in the ocean is quite straightforward. It is based on the measurement of the distance between the brightness temperatures, or observables derived from them such as the polarization ratio (PR) and the gradient ratio (GR):

$$PR = \frac{T_{19,v} - T_{19,h}}{T_{19,v} + T_{19,h}}, \tag{4.68}$$

$$GR = \frac{T_{37,v} - T_{19,v}}{T_{37,v} + T_{19,v}}. \tag{4.69}$$

Fig. 4.46 shows the clustering of the different types of ice (first year and multiyear) and the ice-free surface, and how this scatter plot can be used to discriminate the type of surface, and also to infer the ice cover fraction for borderline pixels.

In Fig. 4.47 we show sample results of the application of this technique to infer ice cover maps in the Arctic and Antarctic oceans, while in Fig. 4.48 we show the ice cover map in the Antarctic Ocean during 1991.

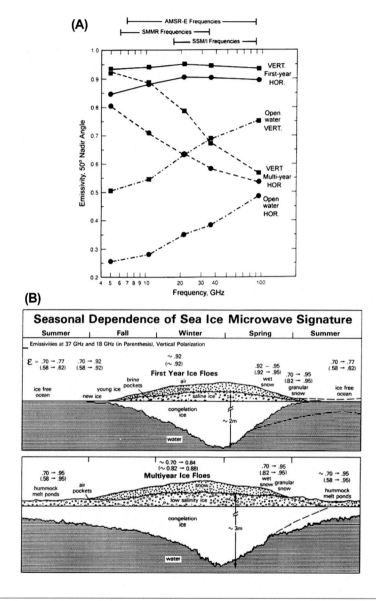

FIGURE 4.45

(A) Ice emissivity changes with frequency and type of ice layer (Svendsen et al., 1983), (B) the sea ice microwave signature also has a seasonal dependence (Comiso, 1986).

The new radiometers operating at the lowest frequency bands (e.g., SMOS at L-band) are capable of "seeing through" the ice layer, provided it is not too thick (<50 cm), and therefore are sensitivity to the sea surface emission below the ice layer as shown in Fig. 4.49. This allows us to determine the ice thickness, despite the cloud cover, which is quite persistent in the highest latitudes (Fig. 4.50).

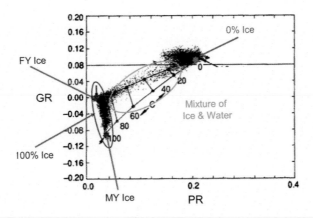

FIGURE 4.46

Sample plot of the gradient ratio (GR) versus polarization ratio (PR) with typical clustering of grid cell values (*small dots*) around the 0% ice (open water) point (*blue star*) and the 100% ice line (*circled in red*). Points with a mixture of ice and water (*circled in green*) fall between these two extremes.

Adapted from Fig. 10.2 of Steffen, K., Key, J., Cavalieri, D.J., Comiso, J., Gloersen, P., St. Germain, K., Rubinstein, I., 1992. The estimation of geophysical parameters using passive microwave algorithms. In: Carsey, F.D. (Ed.), "Microwave Remote Sensing of Sea Ice." American Geophysical Union Monograph 68, Washington, DC, pp. 201–231.

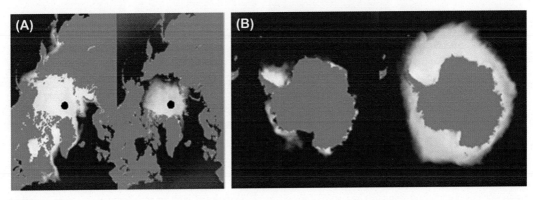

FIGURE 4.47

Sample seasonal variation of the ice cover in polar regions: (A) Arctic Ocean, (B) Antarctica detected by special sensor microwave imager. Note: the *black circle* in the Arctic corresponds to the area not measured by the sensor, despite being in a polar orbit.

4.4 UNDERSTANDING MICROWAVE RADIOMETRY IMAGERY

To understand how to interpret microwave radiometry imagery, in this section a few special sensor microwave imager (SSM/I) and advanced microwave sounding unit (MSU/AMSU) brightness temperature images are presented at different frequency bands. They have been selected to illustrate some of the effects described in Section 4.3 on the emission by natural surfaces and the atmosphere.

To begin with, Fig. 4.51 shows a Meteosat near-infrared image corresponding to September 29, 1997 at 18:00 h. This image has been selected, because it is the closest in time to the SSM/I afternoon

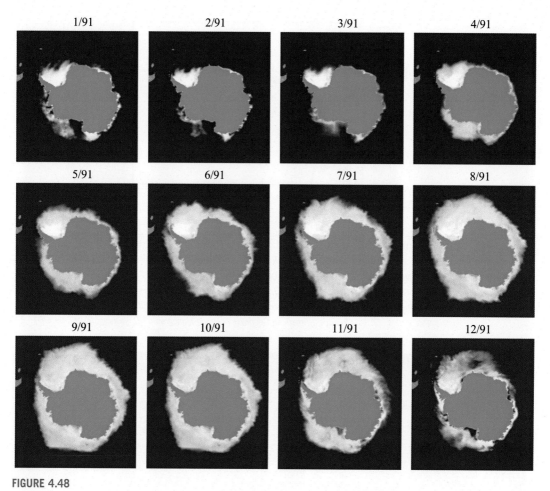

FIGURE 4.48

Seasonal evolution of the ice cover in the Antarctic Ocean during 1991.

overpass. In this image, clouds appear as white, since they are reflecting the solar radiation. Recall that optical and infrared sensors do not penetrate through clouds. Recall also, that while in the thermal infrared, the natural emission of bodies at a physical temperature different from zero is measured, and therefore sensors can operate during day and night, in the ultraviolet, visible, and near-infrared, they require the Sun as a source of illumination, and therefore they will only be sensing during the daytime.

The reader is requested to pay attention to the elongated cloud structure in the south east of Spain, marked in red, and to relate it with the microwave brightness temperatures, and to the geophysical parameters retrieved, notably rain. Please recall that microwaves, especially in the lower part of the spectrum, pass through the atmosphere and are able to sense the surface of the Earth.

The SSM/I is family of multifrequency (19, 22, 37, and 89 GHz) dual-polarization (horizontal and vertical, except at 22 GHz) radiometers that perform a conical scan to achieve a constant incidence

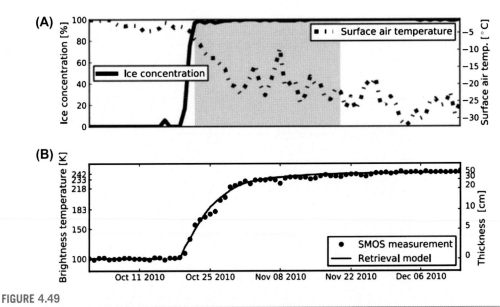

FIGURE 4.49

(A) Time series of advanced microwave scanning radiometer ice concentration and NCEP surface air temperature, and (B) Soil Moisture and Ocean Salinity (SMOS) observed and modeled brightness temperature with the corresponding ice thickness at 77.5°N, 137.5°E (indicated in Fig. 4.50). The particular ice growth period discussed in the text is indicated in gray. The modeled brightness temperature is based on an exponential model and the ice thickness from Lebedev's growth parameterization (Kaleschke et al., 2012).

angle over land of about 53°. Fig. 4.52 presents the different brightness temperatures as measured by the SSM/I (platform F14) at 19 GHz (horizontal and vertical polarizations), and 22, 37, and 89 GHz (vertical polarization only).

In Fig. 4.52, the first thing that becomes very apparent is the high contrast between land and water at all frequencies, being the water much "cooler" (in terms of the brightness temperature) than the land. As the frequency increases, this contrast reduces, which is mainly due to the increased atmospheric opacity, that masks both land and water bodies, except at very high latitudes—even for the highest frequency channels—where the water vapor density is very low, and so the atmospheric attenuation associated to it.

Over the ocean, strips of increased brightness temperature are quite visible around the equator and less towards midlatitudes. These strips are due to increased atmospheric opacity due to the water vapor, which is more important close to the equator, and in cloudy regions. This happens at all frequencies, but it is more evident at 22 GHz, where there is a water vapor absorption peak in the passive microwave spectrum. Sea ice is also clearly visible at higher latitudes north and south. Due to the sharp decrease of the sea water dielectric constant when water gets frozen, a large increase in the brightness temperature is observed.

Over land, bare soil regions such as the Sahara Desert clearly exhibit a much higher brightness temperature at vertical polarization than at horizontal polarization. Actually, warm deserts are identified when $T_{19v} - T_{19h} > 20K$ at the SSM/I 53° incidence angle. Over densely vegetated areas

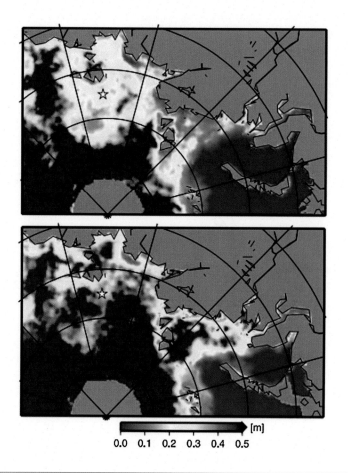

FIGURE 4.50

Sea ice thickness in the Russian Arctic on (top) November 1 and (bottom) November 15, 2010. The *white star* indicates the grid cell position for the time series analysis (Fig. 4.49) (Kaleschke et al., 2012).

such as the tropical rainforests over Africa and America, the high attenuation of the vegetation layer, and the polarization mixing introduced by the leaves and branches, increases the brightness temperatures, while at the same time makes both polarizations more similar. Snow is identified by the brightness temperature depression produced by the scattering in the snow layer of the radiation emitted by the land (e.g., the Andes). The so-called spectral polarization difference or $SPD = (T_{19v} - T_{37v}) + (T_{19v} - T_{19h})$ uses the 37 GHz channel at vertical polarization and both channels at 19 GHz to estimate the presence of snow. Since scattering increases with increasing frequency, when snow is present $T_{19v} > T_{37v}$, and also the difference $T_{19v} - T_{19h}$ increases. Actually, it has been empirically demonstrated that the snow depth can be directly related to the SPD by $SD_{[cm]} = 0.68 \cdot SPD - 0.67$, for all data, and by $SD_{[cm]} = 0.72 \cdot SPD - 1.24$, when $T_{MAX\ air} < 0$.

Finally, rain exhibits a particular behavior. For light to moderate rains, and low frequencies, absorption mechanisms dominate over scattering, and the brightness temperature increases. Scattering

FIGURE 4.51

METEOSAT near-infrared image of Europe and North Africa corresponding to September 29, 1997 at 18:00 h. Note the cyclonic structure over the Iberian Peninsula.

effects start to dominate when the frequency increases and/or the rain intensity increases (raindrop size increases as compared to the wavelength). This effect starts being observed first over land (e.g., Fig. 4.52C in the northern part of the Gulf of Guinea), than over the sea, since the radiation emitted by land is larger than that emitted by the ocean. In Fig. 4.52D and E additional "blue" spots can be appreciated also in South America, in equatorial region of the Atlantic Ocean, and in the south of the Iberian Peninsula, where a cold front event was taking place.

Sample brightness temperature images measured by the MSU/AMSU are presented here in Fig. 4.53. This sensor operates mainly in the frequency bands from 51.3 to 57.95 GHz, and 183 GHz, with additional channels at 23.8, 31.4, 89, and 159 GHz. The first band covers the atmospheric oxygen absorption band. Since the oxygen concentration is quite homogeneous all over the world, the attenuation due to the oxygen is rather constant, and the brightness temperatures measured at different channels are sensitive to the physical temperatures at different heights in the atmosphere.

Fig. 4.54 from http://www.ssmi.com/msu/msu_data_monthly.html?channel=2&type=absolute shows the so-called "weighting functions," which represent the relative contribution of each atmospheric layer to the antenna temperature measured from the satellite. As in the SSM/I, the 23.8 GHz channel is sensitive to water vapor, while the 31.4, the 89, and the 159 GHz channels are in transmission bands for calibration purposes.

The MSUs make measurements of microwave radiance in four different frequencies (or channels) ranging from 51.3 to 57.95 GHz. Of these four channels, MSU channels 2, 3, and 4 were used to construct the following data sets:

- The temperature of the lower stratosphere (**TLS**) is constructed from MSU channel 4, and it corresponds to a layer of the atmosphere from about 12 km to about 25 km above Earth's surface.
- The temperature of the troposphere stratosphere (**TTS**) is constructed from MSU channel 3 and corresponds to a layer of the atmosphere from about 3 km to about 20 km above the surface, with most of the weight coming from about 10 km.

- The temperature of the middle troposphere (**TMT**) is constructed using data from MSU channel 2 and corresponds to a layer of the atmosphere from the surface to about 15 km, with the peak weight at about 4 km.
- The temperature of the lower troposphere (**TLT**) is constructed from MSU channel 2 by subtracting measurements made at different angles from each other.

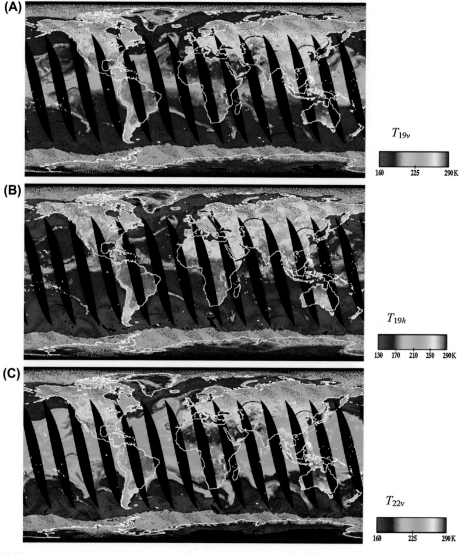

FIGURE 4.52

Special sensor microwave imager brightness temperatures at 19 GHz (*h*- and *v*-polarizations), and 22, 37, and 89 GHz (*v*-polarization only).

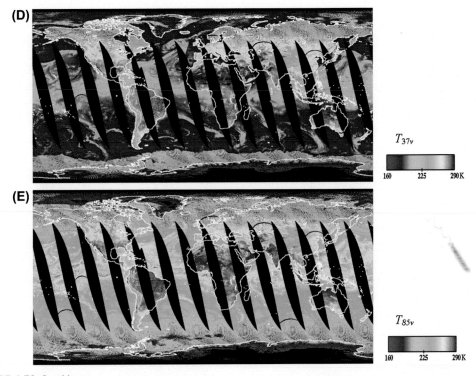

FIGURE 4.52 Cont'd

Beginning in 1998, a series of follow-on instruments to the MSUs, the advanced AMSUs, began operation. The AMSU instruments are similar to the MSUs, but they make measurements using a larger number of channels, thus sampling the atmosphere in a larger number of layers, including a number of layers higher in the stratosphere. By using the AMSU channels that most closely match the channels in the MSU instruments, the data set can be extended beyond 2005. For example, TLS is extended using AMSU channel 9, TTS is extended using AMSU channel 7, and TMT and TLS are extended using AMSU channel 5. With the AMSU series of instruments, temperatures can be monitored higher in the stratosphere using data channels 10 through 14.

4.5 APPLICATIONS OF MICROWAVE RADIOMETRY

In the previous sections the basic concepts of microwave radiometry have been reviewed, including the RTE, and the emission behavior of the atmosphere, and natural surfaces (land and ocean), and the perturbing effects of the ionosphere. The emission behavior has been linked to a number of phenomena that can be sensed by means of microwave radiometry. We summarize the dependence of the emissivity on different surfaces and that of rain in Fig. 4.55 (https://www.meted.ucar.edu/).

The separation of these bands is made on the basis of maximizing the sensitivity to the different parameters to be sensed. The relative sensitivity of the brightness temperatures to different geophysical parameters can be found in Fig. 4.56.

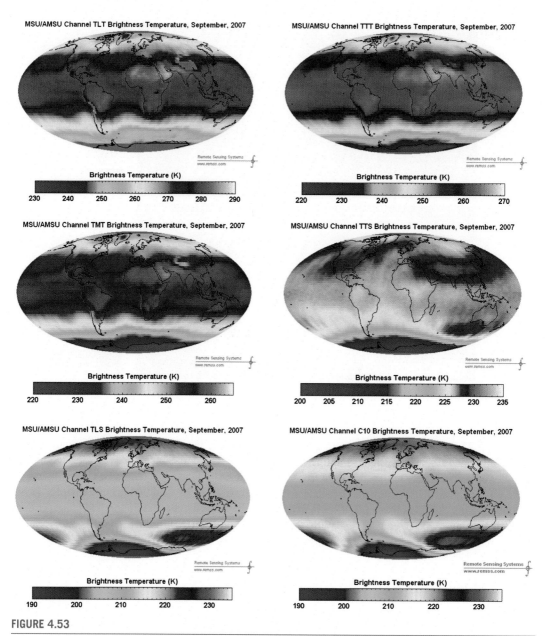

FIGURE 4.53

Brightness temperatures for different advanced microwave sounding unit (MSU/AMSU) channels (http://www.ssmi.com/msu/msu_data_monthly.html?channel=2&type=absolute).

The following figures show some sample geophysical products that are routinely derived from passive microwave observations. These examples are provided as samples of the different land, ocean, and atmosphere applications that will be described in more depth in the corresponding chapters of this book, using the different types of remote sensors. The first set of examples (Figs. 4.57–4.61)

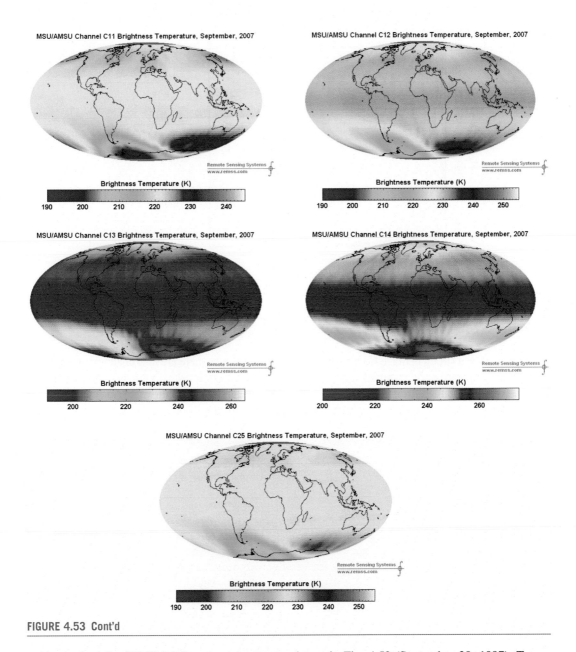

FIGURE 4.53 Cont'd

correspond to the SSM/I brightness temperatures shown in Fig. 4.52 (September 29, 1997). To understand these data, it is important to recall that the retrieved values correspond to average values over the antenna footprint, which does not mean that the value is constant over the whole footprint, and point measurements can be very different from the average values.

The second set of examples (Figs. 4.62–4.67) corresponds to WindSat brightness temperatures from October 29, 2012, when Hurricane Sandy hit New York City. Most of the geophysical parameter retrieval

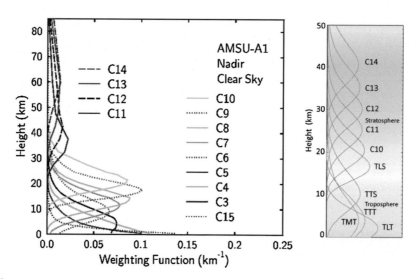

FIGURE 4.54

Weighting functions of the AMSU-A1 channels (http://www.ssmi.com/msu/msu_data_monthly.html?channel=2&type=absolute, http://amsu.ssec.wisc.edu/explanation.html).

algorithms have a strong heritage from previous missions, but since there are new channels and WindSat is the first full-polarimetric instrument, new information can be extracted. Among this new information, there are mainly the SST, thanks to the new 6.8 GHz channel, and the wind direction (in addition to wind speed) over the ocean thanks to the polarimetric capabilities of the WindSat radiometer.

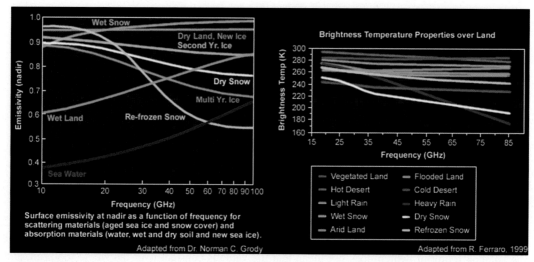

FIGURE 4.55

Summary of the dependence of the emissivity of different surfaces and that of rain (https://www.meted.ucar.edu/).

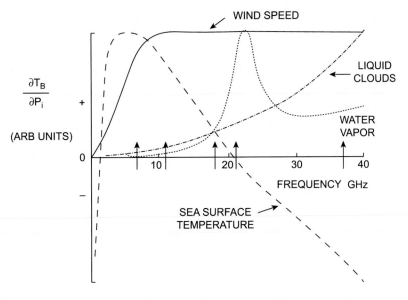

FIGURE 4.56

Normalized sensitivity of the brightness temperatures to different geophysical parameters.

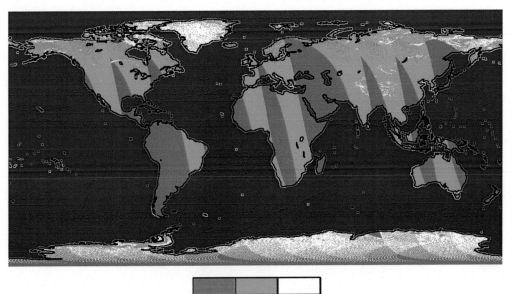

Missing Data | Land No Snow | Snow

FIGURE 4.57

Snow coverage (100% of footprint) as derived from satellite F-14 corresponding to September 29, 1997. Recall that the presence of snow is mainly determined by the scattering signature present in the 19 and 37 GHz channels.

Courtesy of National Oceanic and Atmospheric Administration.

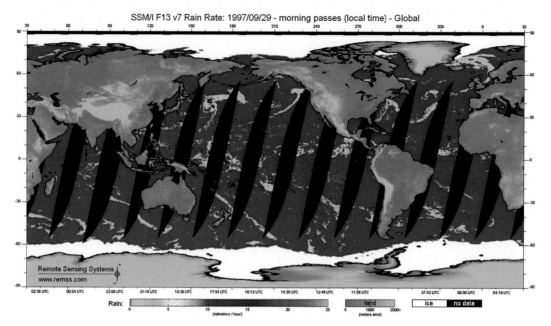

FIGURE 4.58

Rain intensity as derived from satellite F-13 corresponding to September 29, 1997. The rain intensity is mainly determined by the increased absorption at the lower frequency bands with increased rain rate, while at higher frequencies and/or rain rates scattering produces a depletion of the measured brightness temperatures.

Courtesy of Remote Sensing Systems.

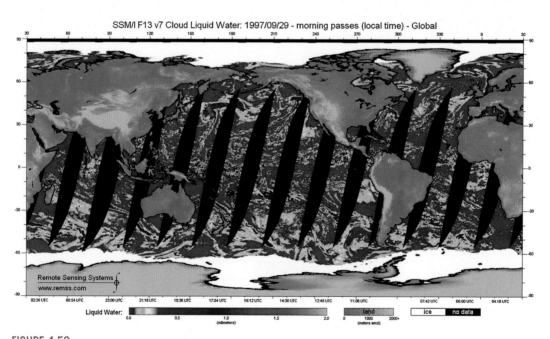

FIGURE 4.59

Cloud liquid water as derived from satellite F-13 corresponding to September 29, 1997. The cloud liquid water is mainly determined using the 22 and 37 GHz bands.

Courtesy of Remote Sensing Systems.

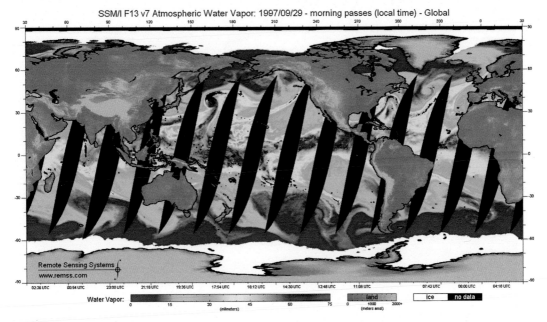

FIGURE 4.60

Atmospheric water vapor as derived from satellite F-13 corresponding to September 29, 1997. The atmospheric water vapor is mainly determined using the brightness temperature increase at the 22 GHz band.

Courtesy of Remote Sensing Systems.

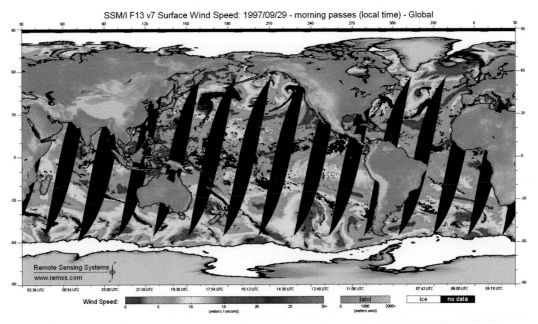

FIGURE 4.61

Surface wind speed as derived from satellite F-13 corresponding to September 29, 1997. The surface wind speed is mainly determined based on the brightness temperature increase at horizontal polarization at 37 GHz band with corrections for ocean surface temperature and atmospheric effects using the brightness temperatures at the 19, 22, and 37 GHz bands and vertical polarization, which are almost insensitive to surface roughness effects.

Courtesy of Remote Sensing Systems.

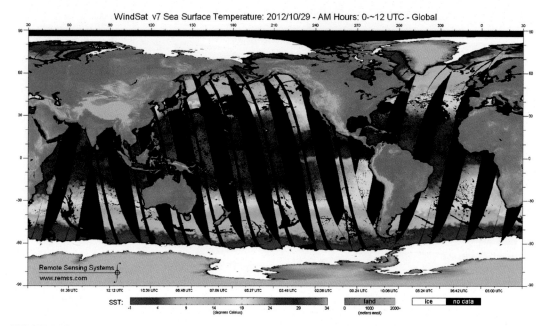

FIGURE 4.62

Sea Surface Temperature (SST) as derived from WindSat corresponding to October 29, 2012. The SST is mainly determined from the brightness temperature at vertical polarization at 6.8 GHz band. Note the absence of parameter returns in those regions where the presence of rain has been detected (Fig. 4.65).

Courtesy of Remote Sensing Systems.

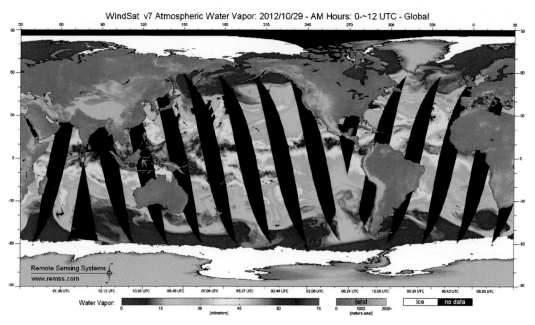

FIGURE 4.63

Atmospheric water vapor as derived from WindSat corresponding to October 29, 2012. Note the higher concentration of atmospheric water vapor in the equatorial regions, where evaporation is more intense, and in regions where either cloud liquid water (Fig. 4.64) or rain (Fig. 4.65) are detected later, and in particular in the North Atlantic Coast of the United States.

Courtesy of Remote Sensing Systems.

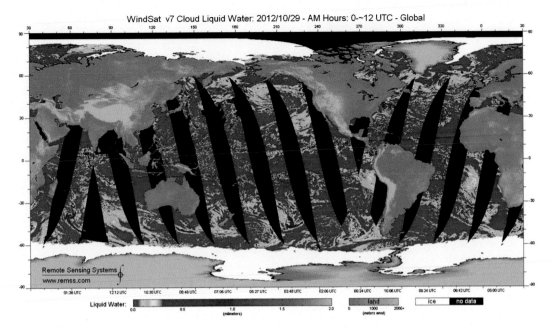

FIGURE 4.64

Cloud liquid water as derived from WindSat corresponding to October 29, 2012. Note the higher concentration of atmospheric water vapor in the equatorial regions, where atmospheric water vapor (Fig. 4.64) is present.

Courtesy of Remote Sensing Systems.

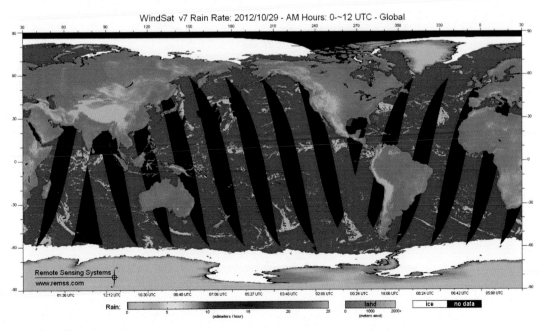

FIGURE 4.65

Rain rate as derived from WindSat corresponding to October 29, 2012. Note the higher concentration of atmospheric water vapor in the equatorial regions, where atmospheric water vapor (Fig. 4.64) is present.

Courtesy of Remote Sensing Systems.

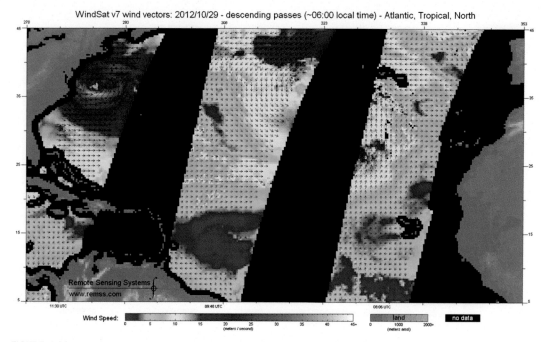

FIGURE 4.66

Zoom of Fig. 4.67 corresponding to the North Atlantic region. Hurricane Sandy can be clearly seen as well as the decreased wind speed (∼10 m/s) in the hurricane's eye.

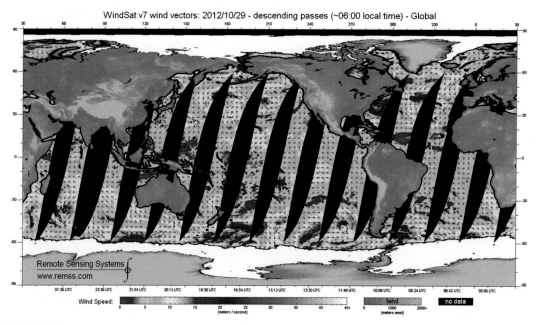

FIGURE 4.67

Wind vector maps derived from WindSat corresponding to October 29, 2012. Color indicates the wind intensity, and arrows indicate the wind direction. Wind direction is inferred thanks to the third Stokes parameter and thanks to two-look observations of the same pixel. Note the extreme high winds in the North East coast of the United States, overpassing 45 m/s.

Courtesy of Remote Sensing Systems.

4.6 SENSORS

4.6.1 HISTORICAL REVIEW OF MICROWAVE RADIOMETERS AND FREQUENCY BANDS USED

Table 4.3 is a nonexhaustive table (from Recommendation ITU-R RS.515-5) summarizing the main frequency bands and applications up to about 1 THz. The bands listed are not necessarily reserved and/ or protected for passive observations, either in part or completely. For example, the bandwidth from 1.400 to 1.427 GHz is nominally reserved exclusively for passive observations, but it does not necessarily mean that it is free of radio frequency interference (RFI). The band from 1.370 to 1.400 GHz is shared with other services, so it is likely to encounter other signals. Similarly, the band from 10.68 to 10.70 GHz is nominally protected for passive observations, but not the one from 10.60 to 10.68 GHz. The use of bandwidths larger than the strictly protected one is a trade-off between improved radiometric resolution and the probability to suffer from RFI. For ocean sensors, where RFI is usually less frequent, and better radiometric resolution is often required, it is not uncommon to use bandwidths larger than the ones reserved exclusively. The topic of RFI detection and mitigation is becoming a very important one due to wider use of wireless communications over larger bandwidths, and at higher frequencies. This topic will be discussed at the end of this chapter, since it is likely that most future microwave radiometers will include some sort of digital back end to detect, and eventually mitigate, RFI.

Table 4.4 presents a nonexhaustive list of microwave radiometers and missions, together with a short description of their applications and main performance parameters. This table is provided to illustrate the use of the different frequency bands and bandwidths for different applications, to be compared with Table 4.3. Later in this section, this table will be revisited again to look in more detail at the imaging and scanning configurations for the different applications and their relative occurrence.

Table 4.5 presents a noncomprehensive summary historical overview of spaceborne microwave radiometers. As can be noticed, there is a general trend towards the use of an increasing number of channels (and polarizations) and higher frequency bands, so as to achieve a better spatial resolution. However, this trend has been recently reversed with three missions ESA's SMOS Soil Moisture and Ocean Salinity mission (launch November 2009), NASA/CONAE Aquarius/SAC-D Ocean Salinity mission (launch June 2010), and NASA's SMAP (Soil Moisture Active/Passive) mission (launch January 2015), all three using a single polarimetric radiometer at L-band, and in the case of the Aquarius/SAC-D and SMAP missions an L-band radar.

4.6.2 MICROWAVE RADIOMETERS: BASIC PERFORMANCE

In Section 4.2 the four Stokes elements were derived from the electric fields measured in a single position. In this section, before the different types of microwave radiometers are discussed (total power or correlation radiometers, real or synthetic aperture), a general framework is presented to give a unified understanding of them. The definition of the Stokes elements given in Eq. (4.19) can be extended to the case where the electric fields are measured at different positions (i.e., by different antennas). For the sake of simplicity, it will be assumed that the surface under measurement is an extended source of random electromagnetic emission (thermal noise) spatially uncorrelated, that the system is narrowband, and the source is in the far-field of the pair of antennas collecting the electric fields.

In the general case, a correlation radiometer is formed by two antennas located at (x_m, y_m, z_m) and (x_n, y_n, z_n), and measures the cross-correlation between the signals received by each antenna. This

Table 4.3 Main Frequency Bands and Applications Up to 1 THz for Passive Observations

Frequency Band(s) (GHz)	Total Bandwidth Required (MHz)	Spectral Line(s) or Center Frequency (GHz)	Measurement (Meteorology-Climatology, Chemistry)	Typical Scan Mode N, C, L[a]
1.37–1.427	57	1.4	Soil moisture, ocean salinity, sea surface temperature, vegetation index	N, C
2.64–2.7	60	2.67	Ocean salinity, soil moisture, vegetation index	N
4.2–4.4	200	4.3	Sea surface temperature	N, C
6.425–7.25	350[d]	6.85	Sea surface temperature	N, C
10.6–10.7	100	10.65	Rain rate, snow water content, ice morphology, sea state, ocean wind speed	N, C
15.2–15.4	200	15.3	Water vapor, rain rate	N, C
18.6–18.8	200	18.7	Rain rates, sea state, sea ice, water vapor, ocean wind speed, soil emissivity and humidity	N, C
21.2–21.4	200	21.3	Water vapor, liquid water	N
22.21–22.5	290	22.235	Water vapor, liquid water	N
23.6–24	400	23.8	Water vapor, liquid water, associated channel for atmospheric sounding	N, C
31.3–31.8	500	31.4	Sea ice, water vapor, oil spills, clouds, liquid water, surface temperature, reference window for 50–60 GHz range	N, C
36–37	1,000	36.5	Rain rates, snow, sea ice, clouds	N, C
50.2–50.4	200	50.3	Reference window for atmospheric temperature profiling (surface temperature)	N, C
52.6–59.3	6,700[b]	Several between 52.6 and 59.3	Atmospheric temperature profiling (O_2 absorption lines)	N, C
86–92	6,000	89	Clouds, oil spills, ice, snow, rain, reference window for temperature soundings near 118 GHz	N, C
100–102	2,000	100.49	N_2O, NO	L
109.5–111.8	2,300	110.8	O_3	L
114.25–116	1,750	115.27	CO	L
115.25–122.25	7,000[b]	118.75	Atmospheric temperature profiling (O_2 absorption line)	N, L
148.5–151.5	3,000	150.74	N_2O, Earth surface temperature, cloud parameters, reference window for temperature soundings	N, L
155.5–158.5[c]	3,000	157	Earth and cloud parameters	N, C

164–167	3,000[b]	164.38, 167.2	N₂O, cloud water and ice, rain, CO, ClO	N, C, L
174.8–191.8	17,000[b]	175.86, 177.26, 183.31, 184.75	N₂O, water vapor profiling, O₃	N, C, L
200–209	9,000[b]	200.98, 203.4, 204.35, 206.13, 208.64	N₂O, ClO, water vapor, O₃	L
226–231.5	5,500	226.09, 230.54, 231.28	Clouds, humidity, N₂O (226.09 GHz), CO (230.54 GHz), O₃ (231.28 GHz), reference window	N, L
235–238	3,000	235.71, 237.15	O₃	L
250–252	2,000	251.21	N₂O	L
275–285.4	10,400	276.33 (N₂O), 278.6 (ClO)	N₂O, ClO	L
296–306	10,000	Window for 325.1, 298.5, (HNO₃), 300.22 (HOCl), 301.44 (N₂O), 303.57 (O₃), 304.5 (O¹⁷O), 305.2 (HNO₃)	Wing channel for temperature sounding Oxygen, HNO₃, HOCl, N₂O, O₃, O¹⁷O	N, L
313.5–355.6	42,100	313.8 (HDO), {315.8, 346.9, 344.5, 352.9} (ClO), {318.8, 345.8, 344.5} (HNO₃), {321.15, 325.15} (H₂O), {321, 345.5, 352.3, 352.6, 352.8} (O₃), {322.8, 343.4} (HOCl), {345.0, 345.4} (CH₃Cl), 345.0 (O¹⁸O), 345.8 (CO), 346 (BrO), 349.4 (CH₃CN), 351.67 (N₂O), 354.5 (HCN)	Water vapor profiling, cloud, wing channel for temperature sounding HDO, ClO, HNO₃, H₂O, O₃, HOCl, CH₃Cl, O¹⁸O, CO, BrO, CH₃CN, N₂O, HCN	N, C, L
361.2–365	3,800	364.32 (O₃)	Wing channel for water vapor profiling O₃	N, L
369.2–391.2	22,000	380.2 (H₂O)	Water vapor profiling	N, L
397.2–399.2	2,000		Water vapor profiling H₂O	N, L
409–411	2,000		Temperature sounding	L
416–433.46	17,460	424.7 (O₂)	Oxygen, temperature profiling O₂	N, L
439.1–466.3	27,200	442 (HNO₃), {443.1, 448} (H₂O), 443.2 (O₃), 452.09 (N₂O), 461.04 (CO)	Water vapor profiling, cloud HNO₃, H₂O, O₃, N₂O, CO	N, L, C
477.75–496.75	19,000	487.25 (O₂)	Oxygen, temperature profiling O₂	L
497–502	5,000	{497.6, 497.9} (BrO), 497.9 (N₂¹⁸O), 498.6 (O₃)	Wing channel for water vapor profiling BrO, N₂¹⁸O, O₃	L, N
523–527	4,000	Window for 556.9	Wing channel for water vapor profiling	N
538–581	43,000	{541.26, 542.35, 550.90, 556.98} (HNO₃), {544.99, 566.29, 571.0} (O₃), 556.93 (H₂O), 575.4 (ClO)	Water vapor profiling HNO₃, O₃, H₂O, ClO	N, L
611.7–629.7	18,000	620.7 (H₂O), 624.27 (ClO₂), {624.34, 624.89, 625.84, 626.17} (SO₂), {624.48, 624.78} (HNO₃), 624.77 (⁸¹BrO), 624.8 (CH₃CN), 624.98 (H³⁷Cl), 625.04 (H₂O₂), {625.07, 628.46} (HOCl), 625.37 (O₃), 625.66 (HO₂), 625.92 (H³⁵Cl), 627.18 (CH₃Cl), 627.77 (O¹⁸O)	Water vapor profiling, oxygen Oxygen, H₂O, ClO₂, SO₂, HNO₃, BrO, CH₃CN, (H³⁷Cl), H₂O₂, HOCl, O₃, HO₂, H³⁵Cl, CH₃Cl, O¹⁸O	L

Continued

Table 4.3 Main Frequency Bands and Applications Up to 1 THz for Passive Observations—cont'd

Frequency Band(s) (GHz)	Total Bandwidth Required (MHz)	Spectral Line(s) or Center Frequency (GHz)	Measurement (Meteorology-Climatology, Chemistry)	Typical Scan Mode N, C, L[a]
634–654	20,000	635.87 (HOCl), 647.1 ($H_2^{18}O$), 649.24 (SO_2), 649.45 (ClO), 649.7 (HO_2), 650.18 (^{81}BrO), 650.28 (HNO_3), 650.73 (O_3), 651.77 (NO), 652.83 (N_2O)	Wing channel for water vapor profiling HOCl, $H_2^{18}O$, SO_2, ClO, HO_2, BrO, HNO_3, O_3, NO, N_2O	L, N
656.9–692	35,100	658 (H_2O), 660.49 (HO_2), 687.7 (ClO), 688.5 (CH_3Cl), 691.47 (CO)	Water vapor profiling, cloud H_2O, HO_2, ClO, CH_3Cl, CO	L, N, C
713.4–717.4	4,000	715.4 (O_2)	Oxygen O_2	L
729–733	4,000	731 (HNO_3), 731.18 ($O^{18}O$)	Oxygen HNO_3, $O^{18}O$	L
750–754	4,000	752 (H_2O)	Water H_2O	L
771.8–775.8	4,000	773.8 (O_2)	Oxygen O_2	L
823.15–845.15	22,000	834.15 (O_2)	Oxygen O_2	N, C, L
850–854	4,000	852 (NO)	NO	L
857.9–861.9	4,000	859.9 (H_2O)	Water H_2O	L
866–882	16,000		Cloud, window	N, C
905.17–927.17	22,000	916.17 (H_2O)	Water H_2O	N, L
951–956	5,000	952 (NO), 955 ($O^{18}O$)	Oxygen NO, $O^{18}O$	L
968.31–972.31	4,000	970.3 (H_2O)	Water H_2O	L
985.9–989.9	4,000	987.9 (H_2O)	Water H_2O	L

[a]N, nadir; nadir scan modes concentrate on sounding or viewing the Earth's surface at angles of nearly perpendicular incidence. The scan terminates at the surface or at various levels in the atmosphere according to the weighting functions. L, limb; limb scan modes view the atmosphere "on edge" and terminate in space rather than at the surface, and accordingly are weighted zero at the surface and maximum at the tangent point height. C, conical; conical scan modes view the Earth's surface by rotating the antenna at an offset angle from the nadir direction.
[b]This bandwidth is occupied by multiple channels.
[c]This band is needed until 2018 to accommodate existing and planned sensors.
[d]This bandwidth is the required sensor bandwidth within the frequency range given in Column 1.
Extracted from Recommendation ITU-R RS.515-5.

Table 4.4 Nonexhaustive List of Microwave Radiometers and Missions, With Short Description of Their Applications and Main Performance

Instrument	Mission(s)	Launch Date/EOL Date	Applications	Observation Requirements and Techniques	Products Description
ATSR-MWS (along-track scanning radiometer-microwave sounder)	ERS-1 ERS-2 ENVISAT SENTINEL-3	17/07/1991–10/03/2000 21/04/1995–05/09/2011 01/03/2002–08/04/2012 16/02/2016–present	• Atmosphere (primary) • Land • Ocean • Snow • Ocean	**Geometry:** near-nadir looking (+35 km, −25 km from nadir) **Spatial resolution:** 22.4 km (fore), 21.2 km (aft) **Swath size:** ~20 km (1 pixel only) **Band:** 23.8 and 36.5 GHz **Antenna type:** offset parabola, two feeds **Bandwidth:** 200 MHz **Acquisition modes:** continuous	• WVC (water vapor column) • CLW (cloud liquid water), as correction terms for the radar altimeter signal. • MWR data also useful for surface's emissivity and soil moisture, sea ice, and snow.
MIRAS (microwave imaging radiometer by aperture synthesis)	SMOS (Soil Moisture and Ocean salinity) mission	2/11/2009–present	• Land • Ocean • Cryosphere	**Geometry:** snap-shot imaging >1000 × 1000 km² image **Spatial resolution:** ~40 km (varies with position in field of view) **Swath size:** ~ 1000 km **Band:** L (1.4 GHz) **Antenna type:** dual-polarization microstrip patches (69 antenna array) **Bandwidth:** 27 MHz **Acquisition modes:** Dual-polarization and full-polarimetric (after commissioning phase full polarimetric mode only)	• Soil moisture • Ocean surface salinity maps **By-products:** • RFI maps • High winds mapping (hurricane mapping) • Ice thickness maps (thin ice)
SSM/I (special sensor microwave imager)	DMSP-F8 DMSP-F10 DMSP-F11 DMSP-F12 DMSP-F13 DMSP-F14 DMSP-F15	20/06/1987–01/10/2006 17/12/1990–04/11/1997 28/11/1991–16/05/2000 29/08/1994–10/08/2007 24/03/1995–18/11/2009 04/04/1997–23/08/2008 12/12/1999–present	• Atmosphere • Land • Ocean • Snow	**Geometry:** conical scan, 53.1° incidence angle **Spatial resolution:** 13–50 km **swath size:** ~1400 km **Band:** 19.35, 22.235, 37.0 and 85.5 GHz (4 frequencies, 7-channels MW radiometer) **Antenna type:** offset parabola **Bandwidth:** varies with channel **Acquisition modes:** continuous global mapping	• Ocean surface wind speed • Ice cover/edge/age • Precipitation over land and water, • CLW & integrated WVC • Soil moisture • Land/sea surface temperature, • SWE (snow water equivalent)
SSM/T (Or SSM/T-1) (special sensor microwave temperature sounder)	DMSP-F5 DMSP-F7 DMSP-F10 DMSP-F11 DMSP-F12	06/06/1979–14/07/1980 18/11/1983–16/05/1988 01/12/1990–04/11/1997 28/11/1991–16/05/2000 29/08/1994–10/08/2007	• Atmosphere	**Geometry:** Cross-track scan ±36° from nadir (7 spots) **Swath:** 1500 km **Spatial resolution:** 174 km at nadir, up to 213 x 304 km **Bands:** 50.5, 53.2, 54.35, 54.9, 58.4, 58.825, and	• Atmospheric (0–30 km) vertical temperature profiles

Continued

Table 4.4 Nonexhaustive List of Microwave Radiometers and Missions, With Short Description of Their Applications and Main Performance—cont'd

Instrument	Mission(s)	Launch Date/EOL Date	Applications	Observation Requirements and Techniques	Products Description
	DMSP-F13 DMSP-F14 DMSP-F15	24/03/1995–18/11/2009 04/04/1997–23/08/2008 12/12/1999–present		59.4 GHz; **Bandwidth:** 400, 400, 400, 400, 350, 300, 250 MHz; **Antenna type:** reflector; **Acquisition modes:** continuous global mapping; **Polarization:** horizontal	
SSM/T-2 (special sensor microwave water vapor profiler)	DMSP-F12 DMSP-F13 DMSP-F14 DMSP-F15	29/08/1994–10/08/2007 24/03/1995–03/02/2015 04/04/1997–23/08/2008 12/12/1999–present	• Atmosphere	**Geometry:** cross-track scan $\pm40.5°$ from nadir (28 spots); **Swath:** 1500 km; **Spatial resolution:** 48 km; **Bands:** 91.655 ± 1.250, 150.0 ± 1.250, 183.3 ± 7, 183.31 ± 3.0, 183.31 ± 1.0 GHz; **Bandwidth:** 3000, 1500, 500, 1000, 1500 MHz; **Antenna type:** offset reflector; **Acquisition modes:** continuous global mapping	• Concentration of water vapor in the atmosphere
SSMI/S (special sensor microwave imager/sounder)	DMSP-F16 DMSP-F17 DMSP-F18 DMSP-F19 DMSP-F20	18/10/2003–present 04/11/2006–present 18/10/2009–present 03/04/2014–11/02/2016 Canceled	• Atmosphere • Land • Ocean • Snow	**Geometry:** conical scan, 53.1° incidence angle; **Spatial resolution:** (see below); **Swath size:** ~1700 km; **Bands/bandwidths/spatial resolution/polarization:** **1.** 50.3 GHz/400 MHz/38 × 38 km/V; **2.** 52.8 GHz/400 MHz/38 × 38 km/V; **3.** 53.596 GHz/400 MHz/38 × 38 km/V; **4.** 54.4 GHz/400 MHz/38 × 38 km/V; **5.** 55.5 GHz/400 MHz/38 × 38 km/V **6.** 57.29/350 MHz/38 × 38 km/RHCP; **7.** 59.4 GHz/250 MHz/38 × 38 km/RHCP; **8.** 150 GHz/1500 MHz/14 × 13 km/H; **9.** 183.31 ± 6.6 GHz/1500 MHz/14 × 13 km/H; **10.** 183.31 ± 3 GHz/1500 MHz/14 × 13 km/H; **11.** 183.31 ± 1 GHz/1500 MHz/14 × 13 km/H; **12.** 19.35 GHz/400 MHz/73 × 47 km/H; **13.** 19.35 GHz/400 MHz/73 × 47 km/V; **14.** 22.235 GHz/400 MHz/73 × 47 km/V **15.** 37 GHz/1500 MHz/41 × 31 km/H;	Replacing and merging SSMI, SSM/T and SSM/T-2 flown on DMSP up to F15 Multipurpose MW conical-scan imager with temperature/humidity sounding channels for improved precipitation

| TMI (TRMM microwave imager) | TRMM | 27/11/1997—present | • Atmosphere | 16. 37 GHz/1500 MHz/41 × 31 km/V;
17. 91.655 GHz/3000 MHz/75 × 75 km/RHCP;
18. 91.655 GHz/3000 MHz/75 × 75 km/RHCP;
19. 63.283248 ± 0.285271 GHz/3 GHz/75 × 75 km/RHCP;
20. 60.792668 ± 0.357892 GHz/3 GHz/75 × 75 km/RHCP;
21. 60.792668 ± 0.357892 ± 0.002 GHz/6 MHz/75 × 75 km/RHCP;
22. 60.792668 ± 0.357892 ± 0.006 GHz/12 MHz/75 × 75 km/RHCP;
23. 60.792668 ± 0.357892 ± 0.016 GHz/32 MHz/75 × 75 km/RHCP;
24. 60.792668 ± 0.357892 ± 0.050 GHz/120 MHz/75 × 75 km/RHCP
Antenna type: offset parabola
Acquisition modes: V, H, and/or RHCP polarizations
Same as SSMI/S in DMSP-16, but 1st 5 channels are H- instead of V-polarization

Geometry: conical scan (49°)
Spatial resolution: 4.6 × 7.2 km to 36.8 × 63.2 km
Swath size: 760 km
Bands/Bandwidths/Polarization/Footprint
1. 10.65 GHz/100 MHz/V/35.7 × 59.0 km
2. 10.65 GHz/100 MHz/H/36.4 × 60.1 km
3. 19.35 GHz/500 MHz/V/18.4 × 30.5 km
4. 19.35 GHz/500 MHz/H/18.2 × 30.1 km
5. 21.3 GHz/200 MHz/V/16.5 × 27.2 km
6. 37.0 GHz/2000 MHz/V/9.7 × 16.0 km
7. 37.0 GHz/2000 MHz/H/9.7 × 16.0 km
8. 85.5 GHz/3000 MHz/V/4.1 × 6.7 km
9. 85.5 GHz/3000 MHz/H/4.2 × 6.9 km
Antenna type: offset reflector
Acquisition modes: continuous | • Rain:
 • Very heavy/oceans (10 GHz)
 • Heavy/ocean (19.35 GHz)
 • Light rain (37 GHz)
 • Very light rain (85.5 GHz)
• Water vapor content (21.3 GHz). |
| WindSat | CORIOLIS | 06/01/2003—present | • Atmosphere
• Ocean | **Geometry:** conical scan (~50°–53° incidence angle, see below)
Spatial resolution: (see below) | • Sea surface temperature
• atmospheric water Vapor
• Surface wind speed vector |

Continued

Table 4.4 Nonexhaustive List of Microwave Radiometers and Missions, With Short Description of Their Applications and Main Performance—cont'd

Instrument	Mission(s)	Launch Date/EOL Date	Applications	Observation Requirements and Techniques	Products Description
				Swath size: 1025 km	• Cloud liquid water
				Bands/bandwidths/polarizations/incidence angle/footprint:	• Rain rate
				1. 6.8 GHz/125 MHz/V, H/53.5°/40 × 60 km	
				2. 10.7 GHz/300 MHz/V, H, ±45°, LHCP, RHCP/49.9°/25 × 38 km	
				3. 18.7 GHz/750 MHz/V, H, ±45°, LHCP, RHCP/55.3°/16 × 27 km	
				4. 23.8 GHz/500 MHz/V, H/53.0°/12 × 20 km	
				5. 37 GHz/2000 MHz/V, H, ±45°, LHCP, RHCP/53°/8 × 13 km	
				Antenna type: offset reflector	
				Acquisition modes: continuous	
AMSU-A	NOAA-15	13/05/1998–present (AMSU-A not nominal)		**Geometry:** cross-track scan, ±48.3° from nadir (30 IFOVs per line)	• Temperature (0–45 km)
	NOAA-16	21/09/2000–09/06/2014		**Spatial resolution:** 48 km (nadir)	• Water vapor profiles
	NOAA-17	24/06/2002–10/04/2013		**Swath size:** 2250 km	• Snow and ice
	NOAA-18	20/05/2005–present		**Bands/Bandwidths/Polarization:**	• Cloud liquid water
	NOAA-19	06/02/2009–present		**1.** 23800 MHz/270 MHz/V;	• Rain rate
	AQUA	04/05/2002–present		**2.** 31400 MHz/180 MHz/V;	
				3. 50300 MHz/180 MHz/V;	
				4. 52800 MHz/400 MHz/V;	
				5. 53596 ± 115 MHz/170 MHz/H;	
				6. 54400 MHz/400 MHz/H;	
				7. 54940 MHz/400 MHz/V;	
				8. 55500 MHz/330 MHz/H;	
				9. 57290.344 MHz/330 MHz/H;	
				10. 57290.344 ± 217 MHz/78 MHz/H;	
				11. 57290.344 ± 322.2 ± 48 MHz 36 MHz/H;	
				12. 57290.344 ± 322.2 ± 22 MHz 16 MHz/H;	
				13. 57290.344 ± 322.2 ± 10 MHz 8 MHz/H;	

Instrument	Satellites	Dates	Domain	Specifications	Applications
AMSU-B	NOAA-15 NOAA-16 NOAA-17 NOAA-18	13/05/1998–present (AMSU-B off sine 3/2011) 21/09/2000–09/06/2014 24/06/2002–10/04/2013 20/05/2005–present	• Atmosphere	**14.** 57290.344 ± 322.2 ± 4.5 MHz/3 MHz/H; **15.** 89000 MHz/6000 MHz/V **Acquisition modes: V and/or H polarizations** **Geometry:** cross-track scan, ±48.95° from nadir (90 IFOVs per line) **Spatial resolution:** 16 km (nadir) **Swath size:** 2250 km **Bands/bandwidths/polarization:** **16.** 89.0 GHz/1000 MHz/V; **17.** 150.0 GHz/1000 MHz/V; **18.** 183.31 ± 7.00 GHz/2000 MHz/V; **19.** 183.31 ± 3.00 GHz/1000 MHz/V; **20.** 183.31 ± 1.00 GHz/500 MHz/V; **Antenna type:** offset parabola	• Atmospheric moisture sounding
MHS	NOAA-18 NOAA-19 MetOp-A Metop-B	20/05/2005–present 06/02/2009–present 19/10/2006–present 17/09/2012	• Atmosphere	**Geometry:** cross-track scan, ±49.44° from nadir (90 IFOVs per line) **Spatial resolution:** 17 km (nadir) **Swath size:** 1500 km **Bands/bandwidths/polarization:** **1.** 89.0 GHz/2800 MHz/V; **2.** 157.0 GHz/2800 MHz/V; **3.** 183.311 ± 1.00 GHz/1000 MHz/H; **4.** 183.311 ± 3.00 GHz/2000 MHz/H; **5.** 190.311 GHz/2200 MHz/V **Antenna type:** offset parabola	• Atmospheric moisture sounding
AMR (advanced microwave radiometer)	JASON-2	20/06/2008–present	• Atmosphere	**Geometry:** nadir pointing only **Spatial resolution:** 25 km **Swath size:** 25 km **Band:** 18.7, 23.8, and 34.0 GHz **Antenna type:** parabolic reflector **Acquisition modes:** continuous Internal calibration only: 3 noise diodes	• Water vapor content • Cloud liquid water
Altika	SARAL	25/02/2013–present	• Atmosphere	**Geometry:** nadir pointing only **Spatial resolution:** 8 km **Swath size:** 8 km **Bands:** 23.8 and 37 GHz	• Water vapor content • Cloud liquid water

Continued

Table 4.4 Nonexhaustive List of Microwave Radiometers and Missions, With Short Description of Their Applications and Main Performance—cont'd

Instrument	Mission(s)	Launch Date/EOL Date	Applications	Observation Requirements and Techniques	Products Description
MADRAS	MEGHA-TROPIQUES	12/10/2011—present	• Atmosphere	**Bandwidths:** 400 MHz and 1 GHz **Antenna type:** parabolic reflector **Acquisition modes:** continuous **Geometry:** conical scan 56° **Spatial resolution:** (see below) **Swath size:** 1700 km **Bands/bandwidths/polarization/footprint:** 1. 18.7 GHz/100 MHz/V, H/40 km; 2. 23.8 GHz/200 MHz/V/40 km; 3. 36.5 GHz/500 MHz/V, H/40 km; 4. 89 GHz/1350 MHz/V, H/10 km; 5. 157 GHz/1350 MHz/V, H/6 km **Antenna type:** offset parabola **Acquisition modes:** continuous	• Rain over oceans • Integrated water vapor • Liquid water in clouds, • Convective rain over land and sea • Cloud top ice
Saphir	MEGHA-TROPIQUES	12/10/2011—present	• Atmosphere	**Geometry:** cross-track scan, up to 50° incidence angle **Spatial resolution:** 10 km (nadir) **Swath size:** 1700 km **Bands:** 183.31 ± 0.2, 183.31 ± 1.1, 183.31 ± 2.7, 183.31 ± 4.2, 183.31 ± 6.6, 183.31 ± 11 GHz **Antenna type:** offset parabola **Bandwidth:** 200, 350, 500, 700, 1200, 2000 MHz **Polarization:** horizontal **Acquisition modes:** continuous	• Atmospheric humidity sounder
SMR (submillimeter radiometer)	Odin	20/02/2001—present	• Atmosphere	**Geometry:** limb sounder **Spatial resolution:** 1.5–3 km (vertical from 5 to 100 km) × 300 km (horizontal) **Band:** 118.25—119.25, 486.1—503.9, and 541.0—580.4 GHz **Antenna type:** gregorian reflector **Bandwidth:** – **Acquisition modes:** continuous	• Atmospheric studies: Clorine monoxide, nitrous oxide, nitrogen dioxide, hydrogen peroxide, nitrous oxide, nitric acid, etc.
SMAP-radiometer	SMAP	31/01/2015—present	• Land		• Soil moisture

MLS	EOS-AURA	15/07/2004–present	• Atmosphere	**Geometry:** conical scan 40° **Spatial resolution:** 40 km **Swath size:** ~1000 km **Bands:** 1.4 GHz **Bandwidths:** 22 MHz **Polarization:** Th, Tv, T3, and T4 (full polarimetric) **Antenna type:** offset parabola **Acquisition modes:** continuous **Geometry:** limb sounder **Spatial resolution:** 1.5 km (vertical from 5 to 120 km) × 300 km (horizontal) **Bands:** 5 bands/36 subbands: 118 GHz (9 subbands), 190 GHz (6 subbands), 240 GHz (7 subband), 640 GHz (9 subbands) and 2.5 THz (5 subbands). **Bandwidths:** – **Polarization:** – **Antenna type:** scanning offset antenna (≤640 GHz) and telescope and scan mirror (2.25 THz) **Acquisition modes:** continuous	Temperature and pressure profiles (118 GHz) • Water vapor and nitric profiles (190 GHz) • Ozone and CO concentrations (240 GHz) • Nitrous acid, HCl, ClO, BrO, and sulfur dioxide (640 GHz) • OH concentration (2.25 THz)

Information compiled from https://www.wmo-sat.info/oscar/satellites, August 2016.

Table 4.5 Historical Evolution of Microwave Radiometers

Year	Platform/Instrument	1.4 GHz	6 GHz	10 GHz	18 GHz	21 GHz	37 GHz	50–60 GHz	90 GHz	160 GHz	183 GHz	Spatial Resolution (km)
1962	Mariner	X			X	X						1.300
1968	**Cosmos 243**			X	X	X	X					**37**
1970	Cosmos 384											13
1972	Nimbus-5 ESMR				X							25
	NEMS					X	X	X(3)				180
1973	**Skylab** S-193			X								16
	S-194	**X**										**115**
1974	Meteor						X					
1975	Nimbus-6 ESMR						X					20 × 43
	SCAMS					X	X	X(3)				150
1978	DMSP SSM/T							X(7)				175
1978	Tiros-N MSU							X(4)				110
1978	Nimbus-7 SMMR		X	X	X	X	X					18 × 27
	Seasat SMMR		X	X	X	X	X					22 × 35
1982	DMSP SSM/I				X	X	X		X			16 × 14
1986	NOAA AMSU-A					X	X	X(12)	X			50
	AMSU-B								X	X	X(3)	15
1992	DMSP SSM/T-2								X	X	X	50
2002	Aqua AMSR-E		X	X	X	X	X		X			
	AMSU-A					X	X	X(24)	X			
	HSB									X(1)	X(3)	
2002	Envisat MWR					X	X					
2003	Coriolis		X	X	X	X	X					
2009	**SMOS**	**X**										**30—100**
2010	**Aquarius**	**X**										**~70**
2014	**SMAP**	**X**										**10**

SMMR, scanning multichannel microwave radiometer.

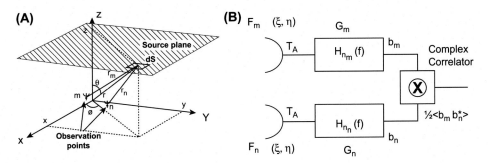

FIGURE 4.68

(A) Two antennas of a correlation radiometer. (B) Diagram of a correlation radiometer: two receiving chains and a complex correlator (Camps and Swift, 2002).

measurement is a sample of the visibility function (V) measured at a spatial frequency $(u_{mn}, v_{mn}, w_{mn}) = (x_n - x_m, y_n - y_m, z_n - z_m)/\lambda_0$ (Fig. 4.68A), being $\lambda_0 = c/f_0$ the wavelength at the central frequency of operation (f_0) and can be derived from the Van Cittert−Zernike theorem (Goodman, 1968, 1985):

$$V_{mn}^{pq} \triangleq V^{pq}(u_{mn}, v_{mn}, w_{mn}) \triangleq \frac{1}{k_B \sqrt{B_m B_n} \sqrt{G_m G_n}} \cdot \frac{1}{2} \left\langle b_m^p(t) b_n^{q*}(t) \right\rangle$$

$$= \frac{1}{\sqrt{\Omega_m \Omega_n}} \iint_{\xi^2 + \eta^2 \leq 1} \frac{T^{pq}(\xi, \eta) - T_{ph} \cdot \delta_{mn}}{\sqrt{1 - \xi^2 - \eta^2}} F_m(\xi, \eta) F_n^*(\xi, \eta)$$

$$\cdot \bar{r}_{mn} \left(-\frac{u_{mn}\xi + v_{mn}\eta + w_{mn}\sqrt{1 - \xi^2 - \eta^2}}{f_0} \right)$$

$$\cdot \exp\left(-j2\pi \left(u_{mn}\xi + v_{mn}\eta + w_{mn}\sqrt{1 - \xi^2 - \eta^2} \right) \right) d\xi d\eta,$$

(4.70)

where $b_{m,n}^{p,q}(t)$ are the analytic signals collected by the antennas m and n, at p- and q-polarizations, respectively, $(\xi, \eta) = (\sin\theta\cos\phi, \sin\theta\sin\phi)$ are the director cosines with respect to the X- and Y-axes respectively, $B_{m,n}$ and $G_{m,n}$ are the noise bandwidth and the power gain of the receiving chains (Fig. 4.68B), $\Omega_{m,n}$ and $F_{m,n}(\xi, \eta)$ are the equivalent solid angle and the normalized radiation *voltage* patterns of antennas m and n, and $T^{pq}(\xi, \eta)$ (Kelvin) are defined in Eq. (4.19) and are proportional to $T^{pq}(\xi, \eta) \propto \left\langle E_m^{ant, p}(\xi, \eta) E_n^{ant, q*}(\xi, \eta) \right\rangle$, where $E_{m,n}^{ant\ p,q}$ are the electric fields collected by antennas m and n at p- and q-polarizations, that are related to the electric fields at a given pixel (ξ, η) by geometrical relationships and the co- and cross-polar antenna patterns as defined by Ludwig's third definition (Ludwig, 1973; Claassen and Fung, 1974; Martin-Neira, 2001). The term $T_{ph} \cdot \delta_{mn}$ accounts for the thermal noise emitted by one receiver as coupled to the other ones (Corbella et al., 2004). This term appears in Eq. (4.70) if $m \neq n$.

The function $\tilde{r}_{mn}(t)$ is the so-called fringe-washing function that accounts for spatial decorrelation effects $\left(t = -\left(u_{mn}\xi + v_{mn}\eta + w_{mn}\sqrt{1 - \xi^2 - \eta^2}\right)/f_0\right)$ and depends on the normalized frequency response $H_{n_{m,n}}(f)$ of each channel (Thompson et al., 1986; Camps, 1996; Camps et al., 1999), and may include several amplifiers, filters, and frequency conversions:

$$\tilde{r}_{mn}(t) = \frac{e^{-j2\pi f_0 t}}{\sqrt{B_m B_n}} \int_0^\infty Hn_m(f)H_{n_n}^*(f)e^{j2\pi ft}df; \quad B_{m,n} = \int_0^\infty \left|H_{n_{m,n}}(f)\right|^2 df. \tag{4.71}$$

In an ideal system $\tilde{r}_{mn}(\tau) = 1$, $H_{n_m}(f) = H_{n_n}(f)$ and $F_m(\xi,\eta) = F_n(\xi,\eta)$, Eq. (4.70) reduces to a continuous Fourier transform between the so-called modified brightness temperature $(T^{pq}(\xi,\eta) - T_{ph}\cdot\delta_{mn})\cdot|F(\xi,\eta)|^2 \Big/ \left(\Omega\cdot\sqrt{1-\xi^2-\eta^2}\right)$ and $V^{pq}(u,v)$.

At this stage, the following particular cases can be considered:

- **Total power radiometer (TPR):** If both antennas have the same polarization (p) and are located in the same physical position $(u_{mn},v_{mn},w_{mn}) = (0,0,0)$ (single-polarization antenna), then the visibility sample $V^{pp}(0,0,0)$ is equal to the antenna temperature at p-polarization.
- **Polarimetric radiometer:** If both antennas have two orthogonal linear polarizations p- and q- and are located in the same physical position $(u_{mn},v_{mn},w_{mn}) = (0,0,0)$ (dual-polarization antenna), then the visibility sample $V^{pq}(0,0,0) = 1/2(T_3 + j\cdot T_4)$ provides a measurement of the third and fourth Stokes parameters.
- **Synthetic aperture radiometer[7]:** If both antennas have the same polarization, but are located at different positions then the visibility sample $V^{pp}(u_{mn},v_{mn},w_{mn})$ is equal to a sample of a Fourier transform[8] of the radiation intensity of the source at a spatial frequency (u_{mn},v_{mn},w_{mn}), which is the result known as the Van Cittert–Zernike theorem (Goodman, 1985). The first Stokes parameter $I = T_v + T_h$ can be recovered everywhere in the FOV.
- **Polarimetric synthetic aperture radiometer:** If both antennas have orthogonal linear polarizations, but are located at different positions, and the visibility sample $V_{mn}(u_{mn},v_{mn},w_{mn})$ is measured in any polarization combination, the full Stokes vector can be recovered for any pixel in the FOV.

4.6.2.1 Spatial Resolution
4.6.2.1.1 Real Aperture Radiometers
Imaging with real aperture radiometers (polarimetric or not) is achieved by scanning the antenna beam along the different "pixels" of the image. The radiometer's output provides the measurement at the particular pixel being pointed by the antenna beam. Along the scan direction, the radiometer's output must be sampled at least at the Nyquist rate that is twice per pixel size (antenna footprint), which will ultimately limit the maximum integration time and the achievable radiometric precision.

---

[7]This name has been subject to some polemics among some authors because it may be misleading in the sense that the synthetic aperture may be formed as in a synthetic aperture radar. However, since the term is widely used in the radio-astronomy and in the microwave radiometry communities, it is kept in this book.

[8]Usually the antennas are either aligned along the X-axis or on the XY plane, and therefore $(u_{mn},v_{mn},w_{mn}) = (u_{mn},0,0)$ (e.g., Le Vine et al. (1992) or $(u_{mn},v_{mn},w_{mn}) = (u_{mn},v_{mn},0)$ by design (e.g., Kerr et al., 2010; Font et al., 2010). Therefore, the relationship becomes either a one- or two-dimensional Fourier transform.

The scanning techniques most widely used in microwave radiometry are (Fig. 4.69) as follows:

1. **nadir-looking:** the antenna beam is fixed and points to the nadir direction. This configuration is typically used in the radiometers that sense the atmospheric water vapor for radar altimeters.
2. **limb sounders:** the antenna beam is fixed and points to the Earth's limb. This configuration is used in atmospheric limb sounders for gas concentration tracing.
3. **cross-track scan:** a zig-zag movement with varying incidence angle and pixel size, typically used in atmospheric sounders.
4. **conical scan** of the antenna beam with a constant incidence angle—typically between 50° and 55°, where the vertical polarization emissivity exhibits a minimum dependence with surface's roughness—and constant pixel size. This configuration is typically used in Earth imagers since it eliminates the angular dependence of the emissivity.
5. **push-broom antenna:** the antenna has several contiguous fixed, or fast switching, beams and the image is formed as the beams "sweep" the scene by the movement of the platform.

From all of them, the most widely used in imagers is the radiometer with a conical scan, a configuration successfully used in the SSM/I series and the cross-track in atmospheric sounders.

The antenna footprint determines the spatial resolution (pixel size), which is given by the antenna half-power beam width ($\Delta\theta_{-3 \text{ dB}}$), the distance from the radiometer to the target (r), and the incidence angle (θ_i), defined in Fig. 4.69:

$$\Delta x = r \cdot \Delta\theta_{-3 \text{ dB}}, \tag{4.72a}$$

$$\Delta y = r \cdot \frac{\Delta\theta_{-3 \text{ dB}}}{\cos \theta_i}. \tag{4.72b}$$

The antenna half-power beam width is usually approximated by $\Delta\theta_{-3 \text{ dB}} \approx \lambda_0/\text{Ø}$, where $\lambda_0 \triangleq c/f_0$ is the electromagnetic wavelength, and Ø is the maximum antenna size. We illustrate the antenna footprint from a low Earth orbit (LEO) and from a geostationary Earth orbit satellites at 400 and 35.760 km height in Fig. 4.70. This figure explains why microwave radiometers have a quite poor spatial resolution as compared to other sensors, even at very high frequencies.

The ideal antenna pattern of a microwave radiometer should have a conical shape with constant gain over the antenna beam width, and zero elsewhere. However, actual antenna patterns exhibit a main beam, whose width is usually defined at half-power with respect to the peak directivity, and side lobes that collect radiation from directions outside the main beam. The amplitude of the side lobes can be reduced by design of the distribution of the amplitude of the electric field over the antenna aperture ("aperture illumination"), so it is tapered so that it gradually decreases towards the edge of the aperture. The price to pay is a larger main beam, with the corresponding loss of spatial resolution (Eq. 4.72).

Instead of the relative amplitude of the side lobes, in microwave radiometry the figure of merit used is the so-called "main beam efficiency" (η_{ML}), defined as the fraction of the total power collected by the main beam (defined up to a given angle from the antenna boresight) with respect to the total power collected from all directions:

$$\eta_{ML} = \frac{1}{\Omega_A} \cdot \underset{\text{Main beam}}{\iint} t(\theta, \varphi) d\Omega, \tag{4.73}$$

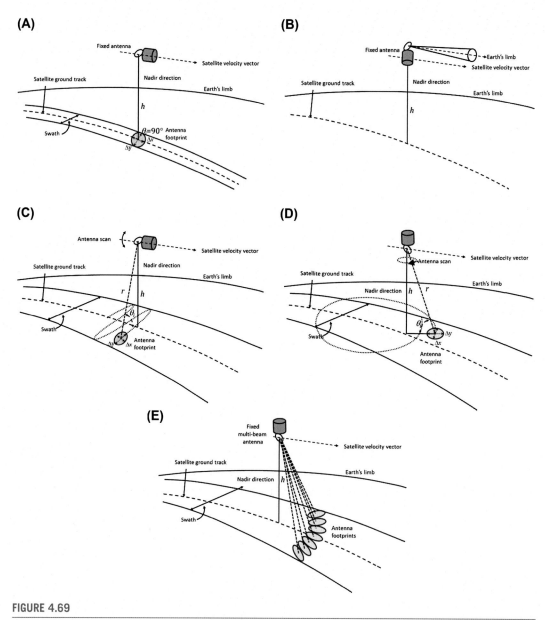

FIGURE 4.69

Typical microwave radiometer scanning configurations: (A) nadir looking, (B) limb sounders, (C) cross-track scanner, (D) conical scanner, and (E) push-broom.

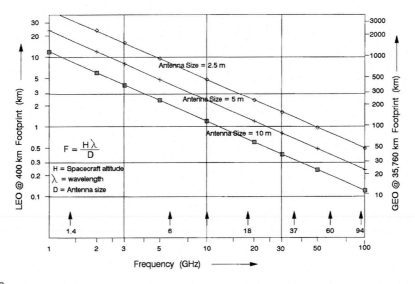

FIGURE 4.70

Antenna footprint at nadir from a low Earth orbit (LEO) and from a geostationary Earth orbit (GEO) satellites as a function of frequency for three different antenna sizes.

where Ω_A is the so-called antenna solid angle:

$$\Omega_A = \iint_{4\pi} t(\theta, \varphi)d\Omega, \tag{4.74}$$

and $t(\theta, \varphi)$ is the antenna radiation *power* pattern.

The evolution of the antenna main beam efficiency is a function of the off-boresight angle for different antenna aperture illuminations (Fig. 4.71). As it can be appreciated, the uniform illuminations achieve the highest main beam efficiencies if the beam is defined close to the boresight, since the uniform illumination lead to the narrowest beams. However, since it exhibits the highest side lobes (side lobe level: SLL $= -13.2$ dB), the main beam efficiency stops increasing quickly. On the other hand, highly tapered illuminations (smooth transition from the amplitude of the electric field in the center of the aperture toward the edge) provide very low side lobes, and ultimately achieve the highest main beam efficiencies. Typical well-behaved microwave radiometer antennas exhibit a main beam efficiency on the order of $\eta_{ML} \approx 0.96$ or even higher.

The resulting antenna temperature image can then be deconvolved to remove errors from radiation picked through the secondary lobes and by antenna finite cross-polarization or to improve the spatial resolution.

4.6.2.1.2 Synthetic Aperture Radiometers

In synthetic aperture radiometers, the brightness temperature images are obtained from the visibility function (Eq. 4.70) using Fourier synthesis techniques, and in the ideal case, when antenna patterns are

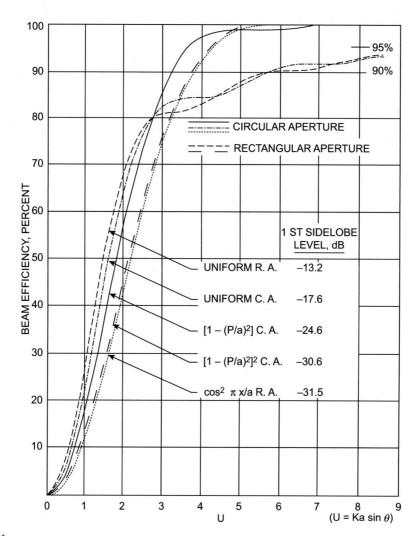

FIGURE 4.71

Main beam efficiency as a function of the antenna aperture illumination and the off-boresight angle parametrized in terms of $U = K_0 \cdot \sin\theta$. R.A. stands for rectangular aperture, and C.A. stands for circular aperture.

Adapted from Johnson, R.C., Jasik, H., 1993. Antenna Engineering Handbook. McGraw-Hill Professional.

identical, and spatial decorrelation is negligible, the relationship between the modified brightness temperature $\left(\breve{T}^{pq}(\xi, \eta)\right)$ and the visibility samples is a Fourier transform:

$$V^{pq} = \mathcal{F}\left\{\breve{T}^{pq}(\xi, \eta)\right\} = \mathcal{F}\left\{\frac{(T^{pq}(\xi, \eta) - T_{ph}) \cdot |F(\xi, \eta)|^2}{\Omega\sqrt{1 - \xi^2 - \eta^2}}\right\}. \qquad (4.75)$$

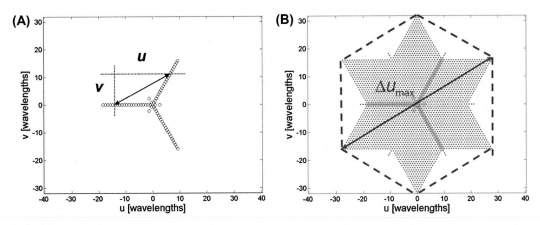

FIGURE 4.72

Microwave imaging radiometer by aperture synthesis (A) Y-shaped array topology and (B) associated (u,v) coverage.

However, the visibility function can only be sampled at some spatial frequencies (u_{mn}, v_{mn}) determined by the antenna positions (assumed from now on that lie on the XY plane). That is, the reconstructed brightness temperature image will suffer from the artifacts associated with discrete Fourier transforms (aliasing and ringing—or Gibbs phenomenon—at brightness temperature steps). Therefore, applying a window $W(u_{mn}, v_{mn})$ to the visibility samples $V(u_{mn}, v_{mn})$ prior to the image reconstruction will reduce the side lobes and will improve the main beam efficiency, at the expense of a widening of the synthetic beam, in a very similar way as the tapering of the antenna aperture impacts the antenna beam in a real aperture radiometer.

Actually, the behavior of a synthetic aperture radiometer can be interpreted as a phased array with an antenna element located at each (u,v) point, although the final response is not squared, since the brightness temperature is already a measure of the received power. Consequently, the impulse response of a synthetic aperture is named the "equivalent array factor," which can be computed analytically as:

$$AF_{eq}(\xi, \xi_0, \eta, \eta_0) = \Delta S \sum_{m,n} W(u_{mn}, v_{mn}) \cdot \tilde{r}\left(-\frac{u_{mn} \cdot \xi + v_{mn} \cdot \eta}{f_0}\right)$$
$$\cdot e^{j2\pi[u_{mn} \cdot (\xi - \xi_0) + v_{mn} \cdot (\eta - \eta_0)]},$$

(4.76)

where ΔS is the area of the (u_{mn}, v_{mn}) cells (Fig. 4.72B) that tessellate the (u,v) plane. As previously discussed, depending on the array configuration, synthetic aperture radiometers can be either one- or two-dimensional imagers:

- The **Electronically Steered Thinned Array Radiometer (ESTAR)** was initially proposed by Le Vine et al. in the early 1980s (LeVine and Good, 1983) and validated by an airborne system developed by the University of Massachusetts at Amherst (LeVine et al., 1992). It achieves **spatial resolution** in the along-track direction by the narrow beam of the stick antennas, and **in the cross-track direction by aperture synthesis.** Aliasing is avoided by setting the antenna spacing to half the electromagnetic wavelength ($\lambda_0/2$). The maximum achievable angular resolution in the case that a rectangular window is used is:

$$\Delta\theta_{-3\text{ dB}} \approx \frac{\lambda_0}{2L}, \tag{4.77}$$

being L the maximum array length. It is worth noting that—as compared to real aperture radiometers—since $V_{mn}^{pq} = V_{nm}^{pq*}$, the maximum spatial frequency coverage is twice the maximum array length, then the angular resolution is halved.

- The **microwave imaging radiometer by aperture synthesis (MIRAS)** aboard the ESA SMOS Earth Explorer Opportunity Mission achieves **two-dimensional spatial resolution** by sampling the visibility function in two dimensions with a Y-shaped array (Fig. 4.72A). To avoid aliasing the antenna elements spacing must be $d = 1/\sqrt{3}$ wavelengths (Camps, 1996). However, since in SMOS $d = 0.875 \geq 1/\sqrt{3}$ wavelengths, the Nyquist criterion is not satisfied and there is aliasing, which is the limiting factor of the instrument's FOV, in a similar way the diffraction lobes (grating lobes) limit the maximum angular scan of a real aperture antenna array. The angular resolution of the instrument is determined by the maximum spacing of the (u,v) spatial frequency points (Fig. 4.72B):

$$\Delta u_{\max} = 2 \cdot \sqrt{3} \cdot N_{EL} \cdot d, \tag{4.78}$$

being N_{EL} the number of antenna elements in each arm of the Y-array, spaced d wavelengths. For large arrays, since $\xi = \sin\theta \approx \theta$ for small θ, if the visibility samples are not tapered (i.e., rectangular window is used), the angular resolution is given by:

$$\Delta\theta_{-3\text{ dB}} \approx \Delta\xi_{-3\text{ dB}} \approx \frac{\pi}{2 \cdot \Delta u_{\max}}. \tag{4.79}$$

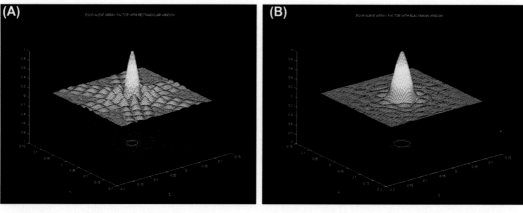

FIGURE 4.73

Microwave imaging radiometer by aperture synthesis normalized equivalent array factor or "impulse response" to a point target at nadir for (A) rectangular and (B) Blackman windowing.

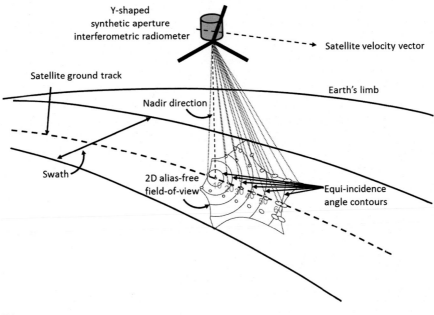

FIGURE 4.74

Scanning configuration of a Y-shaped two-dimensional synthetic aperture radiometer such as microwave imaging radiometer by aperture synthesis.

Out of boresight (ξ_0, η_0), due to the spatial decorrelation of the received electric field being cross-correlated (Bará et al., 1998), the angular resolution is slightly degraded in the radial direction by a factor $1 + \Delta\rho$, where $\Delta\rho$ is defined as:

$$\Delta\rho = \frac{B}{f_0} \cdot \sqrt{\xi_0^2 + \eta_0^2}. \tag{4.80}$$

Similarly to a real aperture radiometer, if, for example, a Blackman window with rotational symmetry and a null at the furthermost visibility samples is used to taper the visibility samples, the amplitude of the side lobes is decreased, the main beam efficiency is increased from 0.43 to 0.90, and the main beam is widened by a factor ~ 1.48, as shown in the equivalent array factors of Fig. 4.73.

To illustrate the imaging principle of a Y-shaped two-dimensional synthetic aperture radiometer as MIRAS in SMOS, we use Fig. 4.74. Each pixel is seen up to 78 times since it enters in the alias-free FOV (determined by the array topology and the antenna spacing) until it leaves it, under a wide range of incidence angles. This multi-look capability can be used to infer simultaneously several geophysical parameters even with a single frequency instrument (see Fig. 4.27).

4.6.2.2 Radiometric Resolution

In this section the radiometric resolution is assessed for ideal total power, and aperture synthesis radiometers, i.e., gain fluctuations are negligible. The radiometric precision is defined as the standard deviation of the random error due to the thermal noise. Assuming identical receivers with a rectangular frequency response, the standard deviation of the output of a correlation radiometer (Eq. 4.81) is given by (Thompson et al., 1986; Camps, 1996; Ruf et al., 1988; Camps et al., 1998a,b):

$$\sigma_{V_{r,i}}^2 = \frac{1}{2 \cdot B \cdot \tau_{eff}} \left\{ T_{sys}^2 \cdot \left[1 + \Lambda\left(\frac{2\Delta f}{B}\right) \right] + V_{r,i}^2 \cdot \left[1 + \Lambda\left(\frac{2\Delta f}{B}\right) \right] - V_{i,r}^2 \cdot \left[1 - \Lambda\left(\frac{2\Delta f}{B}\right) \right] \right\},$$

(4.81)

where $T_{sys} = T_A + T_R$ is the system's temperature, T_A is the antenna temperature, T_R is the receiver's noise temperature. The effective integration time $\tau_{eff} = \tau/Q$ depends on the integration time τ and the correlator type: $Q = 1$ for analog or multibit correlators, $Q = 1.29$ for two-bit by two-bit correlators, $Q = 2.46$ for 1 bit/2 level digital correlators, etc. when samples taken at Nyquist rate (Hagen and Farley, 1973. Other parameters are $\Delta f = f_0 - f_{LO}$, where f_0 is the center frequency and f_{LO} is the local oscillator frequency, f_0 and $\Lambda(x) = 1 - |x|$ for $|x| \le 1$ and zero elsewhere.

4.6.2.2.1 Real Aperture Radiometers

In the case of a single side-band receiver $\Delta f \ge B/2$, the standard deviation in each correlation reduces to:

$$\sigma_V = \sqrt{\sigma_{V_r}^2 + \sigma_{V_i}^2} = \frac{T_{sys}}{\sqrt{B \cdot \tau_{eff}}},$$

(4.82)

which is the equation of the radiometric precision of **an ideal TPR** (Ulaby et al., 1981). The radiometric precision of other types of real aperture radiometers will be studied in more detail in the following sections.

Since the third and fourth Stokes parameters can be computed as $T_3 = 2 \cdot \Re\{V_{mn}^{pq}\}$ and $T_4 = 2 \cdot \Im\{V_{mn}^{pq}\}$ (Eq. 4.19) and $T_{sys} \gg |V_{mn}^{pq}|$, the radiometric precision of **a real aperture polarimetric radiometer** is $\sqrt{2} \cdot \sigma_{V_{r,i}}$:

$$\sigma_{T_{3,4}} = \frac{\sqrt{2} \cdot T_{sys}}{\sqrt{B \cdot \tau_{eff}}}.$$

(4.83)

Actually, this is the same value as if the third and fourth Stokes parameters were computed as $T_3 = T_{+45°} - T_{-45°}$, and $T_4 = T_{LHCP} - T_{RHCP}$, being $T_{\pm 45°}$ and $T_{LHCP, RHCP}$ are the brightness temperatures at $\pm 45°$, and left/right hand circular polarizations.

4.6.2.2.2 Synthetic Aperture Radiometers

For an **aperture synthesis interferometric radiometer**, the radiometric precision is degraded because the aperture being synthesized is much larger than the actual collecting area. Alternatively, it can be understood as the consequence of the combination of the errors in all visibility samples when combined through the same Fourier synthesis process to form an image pixel. The radiometric precision of the modified brightness temperature (Eq. 4.75) can be expressed as (Camps et al., 1998a):

$$\Delta \breve{T} = \Omega_A \cdot \frac{\sqrt{3} \cdot d^2}{2} \cdot \frac{T_{sys}}{\sqrt{B \cdot \tau_{eff}}} \cdot \sqrt{N_V} \cdot \frac{\alpha_W \cdot \alpha_{LO}}{\alpha_F},$$

(4.84)

where N_V is the total number of different (u,v) points, α_W is a parameter that depends on the window used to taper the visibility samples and ranges between 0.45 and 1 (nonredundant case), α_{LO} is a parameter that depends on the type of demodulation used $\alpha_{LO}^{DSB} = \sqrt{2}(\Delta f = 0)$, $\alpha_{LO}^{SSB} = 1(\Delta f \ge B/2)$, and $1 \leq \alpha_F \leq \sqrt[4]{2}$ is a parameter that depends on filters' shape (from rectangular to Gaussian), and the correlation between errors (Bara et al., 2000) has been neglected. The radiometric precision of the brightness temperature degrades off-boresight with the obliquity factor and the inverse of the antenna radiation power pattern: $\sqrt{1 - \xi^2 - \eta^2} \big/ |F(\xi, \eta)|^2$.

4.6.2.3 Trade-off Between Spatial Resolution and Radiometric Precision

A last comment regarding the basic performance of microwave radiometers is the relationship between the achievable spatial resolution and the radiometric precision, which cannot be improved arbitrarily. For the sake of simplicity, in this text the derivation will be performed for a cross-track scanner with zero return time (e.g., an electronically steered array) and an ideal TPR, but similar developments can be applied for any other imaging configurations.

Let us assume that the platform's height is h, the ground speed is v, that the angular extent of the whole swath as seen from the sensor is the FOV, and the sensor's angular resolution is $\Delta\theta$. In this case, the total number of pixels in the swath is:

$$n = \frac{\text{FOV}}{\Delta\theta}. \tag{4.85}$$

Since the minimum spatial resolution is $\Delta x = h \cdot \Delta\theta$, the maximum time in which the whole swath must be scanned is:

$$T_{swath} = \frac{\Delta x}{v}, \tag{4.86}$$

and the maximum integration time per pixel is:

$$\tau = \frac{T_{swath}}{n} = \frac{h}{v} \cdot \frac{\Delta\theta^2}{\text{FOV}}, \tag{4.87}$$

which ultimately limits the radiometric precision:

$$\Delta T = \frac{T_{sys}}{\sqrt{B \cdot \tau}} = \frac{T_{sys}}{\sqrt{B}} \cdot \sqrt{\frac{v \cdot \text{FOV}}{h}} \cdot \frac{1}{\Delta\theta}. \tag{4.88}$$

That is, the better the angular resolution ($\Delta\theta$), the worse the radiometric precision (ΔT), and the product $\Delta x \cdot \Delta T$ turns out to be a constant that depends only on platform and instrument parameters:

$$\Delta x \cdot \Delta T = \frac{T_{sys}}{\sqrt{B}} \cdot \sqrt{h \cdot v \cdot \text{FOV}}. \tag{4.89}$$

Eq. (4.89) is known as the uncertainty principle of a microwave radiometer and determines the best achievable performance. Since $T_{sys} \geq T_A$ (the equality holds for an ideal noise-free receiver), and B is limited by the allocations of the radio-electric spectrum (see Section 4.6.1), and for a spaceborne instrument $v(h)$, to improve Δx and ΔT simultaneously, the only design parameter that remains free is the FOV, which determines the swath width. However, it should be noted that reducing the swath width translates into an increase of the revisit time, which could only be kept constant if more platforms and sensors are deployed, that is, a constellation of microwave radiometers is formed.

For a conical scanner, a similar development can be performed, since halving the angular resolution, translates into twice as many pixels in one antenna rotation, which has to rotate twice as fast to avoid gaps on the ground between consecutive rotations. The integration time being reduced by a factor of four, translates into an increase of the radiometric precision by a factor of two.

Similar developments can be derived for synthetic aperture radiometers as well, although in this case, the angular resolution can be modified during the processing by just adapting the window used to weight the visibility samples (Camps, 1996; Camps et al., 1998a,b):

$$\Delta\xi_{-3\text{ dB}} \cdot \Delta T \approx \Omega_A \cdot \frac{T_{sys}}{\sqrt{B \cdot \tau_{eff}}} \cdot \frac{\alpha_{LO}}{\alpha_F} \cdot d. \qquad (4.90)$$

So far, a unified perspective of the operation and basic instrument performance of all radiometer types has been provided. In the following sections, a more detailed explanation of the particulars of each radiometer type will be provided.

4.6.3 REAL APERTURE RADIOMETERS

In this section more advanced concepts of the sensor technology and calibration strategies are discussed, and it can be skipped by those readers not interested in the technological aspects.

4.6.3.1 Instrument Considerations
4.6.3.1.1 Antenna Considerations

In Section 4.6.2.2.1, antenna considerations related to the spatial resolution and scanning configuration were addressed. In this section, other aspects influencing the specification/design of the antenna (or the impact of a given antenna in the measurements) are described.

- Antenna losses (L_A) either generated by the finite conductivity of the metals forming them or by the connectors or transitions used to connect its output attenuate the noise power collected from the surroundings and introduce additional noise. This effect was studied in detail in Section 4.3.1 when analyzing the effect of atmospheric losses.

$$T_A' = \eta_\Omega \cdot T_A + (1 - \eta_\Omega) \cdot T_{ph,\,A}, \qquad (4.91)$$

where $\eta_\Omega = 1/L_A$ is the so-called antenna ohmic efficiency and equals one when the antenna is lossless ($L_A = 1$) and equals zero when losses are infinite.

For example, if $L_A = 0.5$ dB, then $\eta_\Omega = 0.90$, that is, 90% of the received radiation is passed to the receiver and 10% is added by the antenna. This is a very important effect since a 1°C variation of the physical temperature of the antenna will translate into a 0.1K drift of the radiometer's output. The only way to keep these variations under control is by proper thermal stabilization of the antenna itself, if feasible, or the parts of the antenna in which losses are generated, that is, in regions where the amplitude of the electric field is higher, in the connectors or cables connecting the antenna output and the receiver front-end. Fig. 4.75 shows the impact of the antenna losses in the radiometer's output as a function of the difference between the antenna temperature and the physical temperature of the antenna. As a consequence, errors are larger when this difference is larger, which usually happens when the radiometer is performing a cold sky calibration.

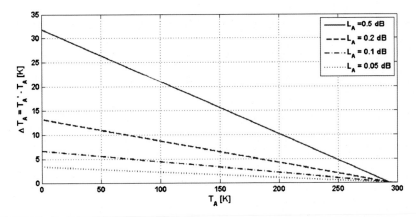

FIGURE 4.75

Impact of antenna losses in the radiometer's output as a function of the difference between the antenna temperature and the physical temperature.

- Polarization mixing occurs when the XY antenna reference frame is not properly aligned with the HV Earth's reference frame. This effect was also studied in Section 4.3.2.1 when analyzing the impact of the Faraday rotation. The solution consists of the inversion of Eq. (4.42), which is always feasible provided the rotation angle is different 45°. Otherwise, the matrix becomes singular in which case only Eq. (4.41) can be used.
- Finite antenna cross-polarization refers to the capability of the antenna to separate to incoming orthogonal polarizations. This effect will be analyzed in more detail in the next bullets.
- Finite antenna pattern beam width effects refer to the mix of the Stokes elements as measured by the sensor due to the spatial averaging over the antenna beam of the different contributions to the Stokes elements each with a different rotation angle (polarization mixing) and cross-polarization value. The measured Stokes vector $[T_v, T_h, T_3, T_4]^{\text{meas}}$ is then a linear combination of the nominal one $[T_v, T_h, T_3, T_4]^{\text{nominal}}$ averaged over the antenna beam. The electric fields measured by the antenna in the XY antenna reference frame can be related to the h and v electric fields by:

$$
\begin{bmatrix} \widehat{E}_x \\ \widehat{E}_y \end{bmatrix} = \begin{bmatrix} R_x \cos\alpha - C_x \sin\alpha & R_x \sin\alpha + C_x \cos\alpha \\ -C_y \cos\alpha - R_y \sin\alpha & -C_y \sin\alpha + R_y \cos\alpha \end{bmatrix} \begin{bmatrix} E_h \\ E_v \end{bmatrix},
\tag{4.92}
$$

where $R_{x,y}$ and $C_{x,y}$ are the co- and cross-polar antenna patterns at X- and Y-polarizations.

And the Stokes elements can then be computed as:

$$
\widehat{T}_{xx} = \frac{1}{\Omega_A} \iint_{4\pi} \left\{ |R_x|^2 T_{xx} + 2\text{Re}\{R_x C_x^*\} T_{yx} + |C_x|^2 T_{yy} \right\} d\Omega,
\tag{4.93a}
$$

$$
\widehat{T}_{yx} = \frac{1}{\Omega_A} \iint_{4\pi} \left\{ -C_y R_x^* T_{xx} + (R_y R_x^* - C_y C_x^*) T_{yx} + R_y C_x^* T_{yy} \right\} d\Omega,
\tag{4.93b}
$$

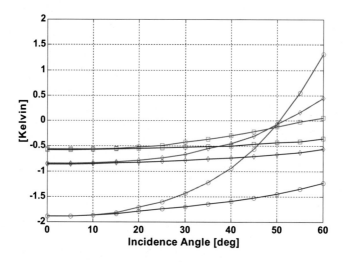

FIGURE 4.76

Impact of antenna array beam width: 20° (circles), 15° (diamonds), and 10° (squares) at horizontal (blue) and vertical (red) polarizations, for a radiometer observing the sea surface at 1.4 GHz, with SST = 15°C, SSS = 36 psu, $U_{10} = 0$ m/s.

$$\widehat{T}_{yy} = \frac{1}{\Omega_A} \iint\limits_{4\pi} \left\{ |C_y|^2 T_{xx} + 2\mathrm{Re}\left\{R_y C_y^*\right\} T_{yx} + |R_y|^2 T_{yy} \right\} d\Omega, \qquad (4.93c)$$

This effect for an antenna array of 4×4, 6×6, and 8×8 microstrip patches with triangular illumination, and a 20°, 15°, and 10° beam width, assuming the ocean surface at 1.4 GHz is illustrated in Fig. 4.76. At nadir, the bias between the brightness temperature at boresight and the antenna temperature computed from Eq. (4.93) is the same at both polarizations, but, as expected, decreases with decreasing antenna beam. As the incidence angle increases, the bias remains quite constant at horizontal polarization, but quickly increases at vertical polarization, being approximately zero for all beam widths around 50°. However, despite this bias, the sensitivities of the vertical and horizontal antenna temperatures versus salinity, SSS, temperature, and wind speed remained nearly constant.

As an example, the above considerations led to the final antenna design of the L-band AUtomatic RAdiometer (LAURA) developed by UPC in 2000, a 4×4 patch dual-polarization array to be used in a number of field experiments over the ocean and over land in preparation for the ESA SMOS mission. The antenna array consisted of an array of 4×4 microstrip patches with dielectric air to minimize losses ($L_A = 0.4$ dB, $\eta_\Omega = 0.91$) and was all thermally stabilized to have constant added thermal noise. Extreme care was taken during the design and manufacturing to optimize the antenna cross-polar pattern. The measured co-polar and cross-polar antenna patterns and the main antenna parameters are displayed in Fig. 4.77A and B.

4.6.3.1.2 Receiver Considerations

The basic topology of a microwave radiometer is that of a superheterodyne receiver, such as those used in communications. Since the input signal is very weak (the thermal noise):

$$N = k_B \cdot T_{sys} \cdot B, \qquad (4.94)$$

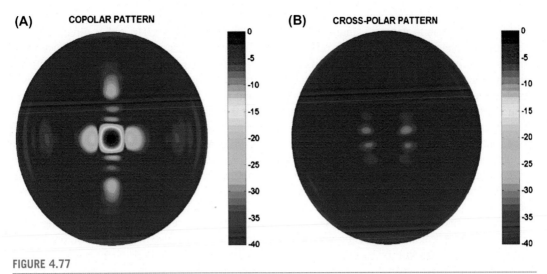

FIGURE 4.77

LAURA's measured (A) copolar and (B) cross-polar antenna pattern. Main parameters: $\Delta\theta_{-3\,dB} = 20°$ in both planes, side lobes: 26 and -19 dB (E- and H-planes), cross-polar <-40 dB in main beam, <-35 dB peak value, $\eta_{ML} = 0.965$, and losses $= 0.4$ dB.

Courtesy of Prof. Sebastián Blanch, UPC.

and the signal power at the detector's input ($P_{det,\ in}$) is typically between -20 and -30 dBm, the amount of gain must be very large. Values from 60 to 80 dB, or even larger, are not uncommon.

If the signal is sampled for digital processing, the power at the analog-to-digital converter (ADC) input must be even higher (and so the gain), typically around 0 dBm, which corresponds to $\sigma = 0.224 V_{eff}$ over 50 Ω. The signal power, the number of bits of the ADC, the ADC window, and the sampling frequency must be carefully selected to minimize the error.

Depending on the frequency, different technologies can be used: coaxial components are widely available below ~ 50 GHz, allowing an easy reconfigurability of the hardware, and "low cost"; microstrip circuits allow for small circuit layouts at low cost and are used at most at ~ 100 GHz due to the increased radiation losses; waveguide circuits exhibit very low losses, at a moderate cost, but circuits are difficult to reconfigure, and the selection of components is limited to ~ 300 GHz; and finally quasi-optical techniques become necessary at higher frequencies (>200 GHz) due to the lower losses at millimeter and submillimeter wavelengths, but circuits are designed for unique applications, and the cost is higher.

As in any receiver, there are passive and active components that perform different functions. Passive components include directional couplers to inject calibration and test signals, power splitters to split or combine signals, attenuators to set the signal power levels, matched loads used as terminations or temperature reference ($N = k_B \cdot T_{ph} \cdot B$), isolators to improve the matching, isolate or switch signals, filters to set the radiometer's bandwidth and filter out unwanted signals, mixers to downconvert signals for processing at a lower frequencies, and detectors to convert the input signal power into a DC voltage. Active components include radio frequency (RF) amplifiers, handle the low-incoming signal levels, have a low noise figure and operate typically from 100 MHz to 200 GHz; intermediate-frequency (IF) amplifiers handle low to high power signals that have been previously down-converted from RF to an IF frequency typically in the range 1 MHz to 1−2 GHz; video amplifiers handle the detected signals, have low noise and are highly stable, and operate typically from DC to 10 MHz; oscillators to generate sinusoidal signals, typically up to 200 GHz, either directly or as an harmonic of a lower frequency, that is mixed (multiplied) with the incoming signal to downconvert it into an IF (depending on the frequency and stability requirements, oscillators can be crystal oscillators, dielectric resonator oscillators, Gunn oscillators, klystron, or frequency synthesizers); noise diodes are used for calibration purposes generating high equivalent noise temperatures up to frequencies larger than 100 GHz; ADCs to convert the analog signal, either the detected one or the predetected one, into a digital word for storage or later digital signal processing[9]; sample-and-hold (S&H) circuits may be used to multiplex different outputs before they are digitized, voltage-to-frequency converters convert the detected analog signal into a frequency modulated signal to avoid transmission problems with different grounds, interferences, etc. and are used for integration times larger than 10 ms; and finally (diode) detectors convert the input analog signal into a DC signal that is proportional to the input power.

Detectors need to operate in the square law region, which requires operating at low power levels, typically from -20 to -30 dBm, and be stable in temperature. They also need to have a low linearity error, that is the function $V_0(P_i)$ must be a line with deviations less than $0.1\%-0.5\%$. Tunnel diodes are the best choice, with detector's constants[10] on the order of $C_d \sim 1$ mV/μW, but Schottky diodes are usually used as well because of the larger sensitivity, although they have a poorer temperature stability. In this case, differential configurations are used to improve temperature stability. Fig. 4.78 shows the typical transfer characteristic of a Schottky diode, showing the square and linear law regions.

The nonideal properties of the microwave components introduce errors in the observables that will require proper mitigation by design and calibration: losses attenuate the signal and add extra noise, since noise is correlated at multiport outputs, it may also induce correlation offsets in correlation radiometers; impedance mismatches cause signal reflections that add unwanted noise and add some extra losses; and finite directivity of directional couplers also injects unwanted signals. These effects introduce calibration errors, since the estimated slope and bias of the calibration line will be (slightly) different from the true ones during the measurements. This is an important topic, but requires advanced concepts of microwave engineering. The interested reader is referred to for a comprehensive analysis using the noise waves (Camps, 2010).

4.6.3.1.3 Sampling Considerations

Digitization effects can be separated into quantization and sampling effects. Mainly, the effects related with the quantization of the input signal (thermal noise) imply the loss of its statistical properties due to the nonlinear quantization process. Consequently, it is not possible to apply the well-known Gaussian statistical relationships to the quantified signal. This effect has a large impact when limited quantization levels are considered, and it can be mitigated by increasing the number of quantization levels. Nonlinear effect studies on Gaussian signals started with the early analysis of the spectrum of clipped noise by Van Vleck and Middleton (1966). In the late 1950s, Price (1958) published a work focusing on the relationship between the ideal correlation between two random signals with Gaussian

---

[9]If the detected signal is sampled, the number of bits must be high, typically between 12 and 16 bits per sample, so that the quantization noise is negligible in front of the radiometric resolution. If the predetected signal is sampled, often a much lower number of bits can be used, although the final processed value must be stored with much longer digital words. In the limit, in correlation radiometers, samples at just 1 bit are required to compute the complex cross-correlations.

[10]The detector constant plays a similar role as the responsibility (A/W) in photodetectors.

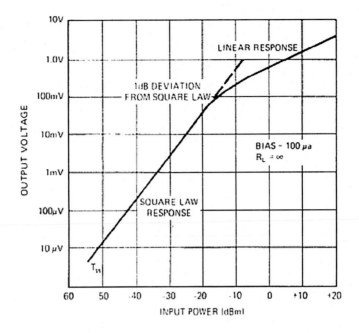

FIGURE 4.78

Typical transfer characteristic of a Schottky diode.

probability density function (PDF) and the correlation measured after a nonlinear manipulation of these random signals. This relationship can be used to study the effects of arbitrary quantization schemes on the correlation of two signals. Recall that a power measurement is just the cross-correlation at the origin, so the following considerations may be of general interest for all types of (digital) radiometers.

The discontinuities of the ADC transfer function impact the correlation spectrum by enlarging and distorting the spectrum of the input signals. At the same time, sampling also has an impact on the correlation of the two input signals by creating replicas of the previously distorted spectra. Therefore, additional noise can be added to the mean correlation value due to spectra replication, but the exact amount depends on the ratio between the sampling frequency and the signal's bandwidth.

$$\sigma_{R_{xy}}^2 = \frac{\sigma_p^2}{N_q} + \frac{2}{N_q^2} \sum_{n=1}^{N_q} (N_q - n) R_{xx}(nT_s) R_{yy}(nT_s), \tag{4.95a}$$

where, $\sigma_p^2 = \sigma_{g_1(x)}^2 \sigma_{g_2(y)}^2$ is the variance for each sample of the correlation ($N_q = 1$), and $R_{xx/yy}$ are the autocorrelation functions of the two signals being correlated. To mitigate this effect, the sampling frequency must be increased above the Nyquist by, at least, the factor given in Hagen and Farley (1973).

Moreover, the sampling period and its inaccuracies (skew and jitter in the sampling periods) of the ADC can affect and distort the sampled signal and so the cross-correlation value. For $V_{ADC} = 5\sigma_{x,y}$ and at least 15 quantization levels (4 bits) the quantization scheme neither spreads nor distorts the spectrum ($<1\%$ rejected band increase).

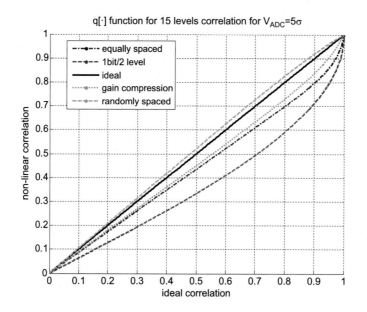

FIGURE 4.79

Relationship between the nonlinear and the ideal correlation for different digitization schemes (Bosch-Lluis et al., 2011).

To compare different sampling and quantization schemes an equivalent integration time is defined as that required to obtain the same resolution as in the ideal (analog) correlation. Hagen and Farley (1973) conducted important work on this topic, where the effective time was defined and some easy-to-calculate digitization configurations were analyzed in depth. In the case of quantifying with two levels (Price, 1958):

$$R_{xy}\left(\rho_{xy}(\tau)\right) = \frac{2}{\pi}\arcsin\left(\rho_{xy}(\tau)\right). \tag{4.95b}$$

When the quantization scheme is more complex, it can only be analyzed numerically; Fig. 4.79 shows some results when using 15 levels ($\log_2 15 = 3.9$ bits), $V_{ADC} = 5\sigma_{x,y}$, and different quantization schemes.

The root mean square error (RMSE) coefficient between $R_{xy}(\tau)$ and $\rho_{xy}(\tau)$ for different quantization levels and ratios of the ADC conversion window (V_{ADC}) to the standard deviation of the input signal ($\sigma_{x,y}$) is shown in Fig. 4.80. The minimum of each curve corresponds to the optimum configuration of V_{ADC} with respect to $\sigma_{x,y}$. As the number of quantization levels increases, the minimum gets closer to 0, and the $V_{ADC}/\sigma_{x,y}$ range where the curves remain close to 0 increases as well, since the function $q[\cdot]$ is linear over a wider range of input powers. For $V_{ADC} < 2\sigma_{x,y}$, clipping of the input signal(s) has a dominant impact on the non-linear correlation, and it is more important for 1 bit. When $V_{ADC}/\sigma_{x,y} \rightarrow \infty$, the RMSE asymptotically increases again up 14.2% (as for 1 bit/2 levels sampling) for all the quantization schemes. This effect is explained as a decrease of the effective number of bits, as fewer bits are being used as the $V_{ADC}/\sigma_{x,y}$ ratio increases. In the limit, only two levels are effectively used for quantization.

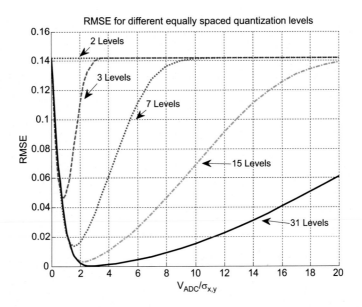

FIGURE 4.80

Root mean square error (RMSE) and analog-to-digital converter span window relationship for different equally spaced quantification levels (Bosch-Lluis et al., 2011).

4.6.3.2 Types of Real Aperture Radiometers

The simplest concept of a microwave radiometer is sketched in Fig. 4.81, and it is called the **total power radiometer** (TPR). An antenna collects the radiation (thermal noise) in a given band, which is amplified and filtered, and then detected and averaged (low-pass filtered). The detected output

$$V_0 = k_B \cdot \left(T_A' + T_R \right) \cdot B \cdot G \cdot C_d + Z, \tag{4.96}$$

is proportional to the input system temperature, times the receiver's noise bandwidth (predetection bandwidth) $B \triangleq \left[\int_0^\infty G(f)df \right]^2 / \left[\int_0^\infty G^2(f)df \right]$, times the chain's power gain, times the detector's constant C_d, plus an offset term Z originated by impedance mismatches or by the detector itself.

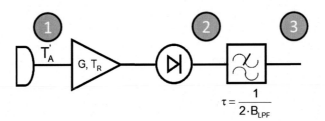

FIGURE 4.81

Basic block diagram of a total power radiometer. Note: due to practical implementation issues, this topology is seldom used, since it becomes unfeasible to have enough gain in front of the detector, without amplifier instability problems.

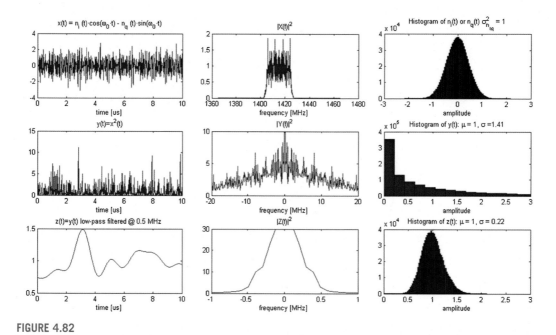

FIGURE 4.82

Evolution of the signals in a basic total power radiometer in first row: radio frequency input, second row: instantaneous detected power, third row: low-pass filtered detected power. Signal representation in first column: time domain, second column: frequency domain, third column: histogram.

To understand its operation, it is worth spending some time discussing Fig. 4.82, which shows a numerical simulation of the signals in Fig. 4.81 for a radiometer with center frequency 1.413.50 MHz, RF bandwidth 20 MHz, and a low-pass filter of 0.5 MHz. The RF signal (first row, number 1 in Fig. 4.82) is a 20 MHz band-pass signal with Gaussian in-phase and quadrature components of unity variance (third column). The RF spectrum (second column) is nearly flat from 1405 to 1425 MHz (oscillations are due to the finite averaging of the different realizations performed). The detected signal (second row, number 2 in Fig. 4.82) is the instantaneous power of the RF signal, obtained by squaring it sample by sample. Its spectrum consists of a delta function (average power) and an approximately triangular function of width ~ 40 MHz, which, due to the squaring process, is the result of the convolution of the quasirectangular RF frequency response with itself. Finally, the output of the integrator (third row, number 3 in Fig. 4.82), in this case a low-pass bandwidth of 0.5 MHz (low-pass filters of real instruments have much smaller bandwidths), is a smoothed version of the rapid fluctuations of the instantaneous power (second row), or equivalently most of the AC components have been removed due to the lower bandwidth. Therefore, the ratio of the standard deviations is equal to the ratio of the bandwidths $\sqrt{20\,\text{MHz}/0.5\,\text{MHz}} = 1.41 \big/ 0.22 = 6.4$.

Therefore, the radiometric precision is given by:

$$\frac{\sigma_T}{\overline{T}} = \sqrt{2} \cdot \sqrt{\frac{B_{LPF}}{B_{RF}}} = \frac{1}{\sqrt{B_{RF} \cdot \tau_{LPF}}}, \tag{4.97}$$

or

$$\sigma_T = \frac{\overline{T}}{\sqrt{B_{RF} \cdot \tau_{LPF}}} = \frac{T'_A + T_R}{\sqrt{B_{RF} \cdot \tau_{LPF}}}. \tag{4.98}$$

However, this expression is ideal, in the sense that it does not account for any receiver imperfections except for the thermal noise. The main limitations of the TPRs come from the gain fluctuations of the receiving chain (Tiuri, 1964), which can be long-term ones, such as thermal drifts, aging, etc. and can be corrected for by periodic calibration, and short-term ones ($\sim$ Hz), which are the ones that ultimately limit the radiometric precision, since the gain variation is attributed to a variation of the antenna temperature: $V_0 = k_B \cdot (T'_A + T_R) \cdot B \cdot (G + \Delta G) \cdot C_d + Z$.

Since the random gain fluctuations produce a variation of the measured system's temperature, that is uncorrelated with the random fluctuation of the magnitude being measured, the global radiometric precision of TPRs includes these two terms:

$$\sigma_T^2 = \frac{(T'_A + T_R)^2}{B \cdot \tau} + (T'_A + T_R)^2 \cdot \left(\frac{\Delta G}{G}\right)^2, \tag{4.99}$$

but it is usually governed by gain fluctuations $(\Delta G/G)^2 \cong \int_0^{+\infty} S_G(f) \cdot df$, where $S_G(f)$ is the spectrum of the gain fluctuations of $G(f)$, the radiometer's power transfer function.

Fig. 4.83 shows a typical spectrum of the gain fluctuations, which can be fitted by

$$S_G(f) = a_g + \frac{b_g}{f}, \tag{4.100}$$

where a_g is set by the white noise term, which depends on $B \cdot \tau$ and, b_g is fit to system measurements.

Gain fluctuations can be minimized by controlling the stability of the output voltage of the power supply and the physical temperature of the radiometer's environment. However, at high frequencies, it is difficult to build highly stable receivers with small gain fluctuations. A possibility is to eliminate the RF amplifier and minimize gain fluctuations using a mixer as a first stage (e.g., Fig. 8 in Hersman and Poe, 1981), but in this case the receiver's noise temperature becomes very large.

The **Dicke radiometer (DR)** implements a simple way to minimize the gain fluctuations (Dicke, 1946). The input is modulated using an input switch (the "Dicke switch") to commute between the

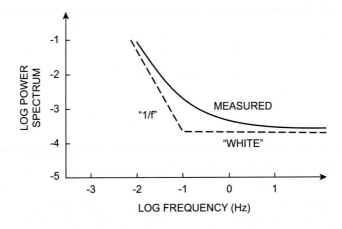

FIGURE 4.83

Typical gain fluctuations spectrum. Fluctuation power <1–10 Hz, requires calibration every <0.1–1 s.

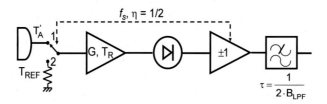

FIGURE 4.84

Basic block diagram of a Dicke radiometer.

antenna and a matched load at a known and stable physical temperature (T_{REF}) (Fig. 4.84). At the same time, the detected output is DC blocked and demodulated by multiplying it synchronously by ±1 before the low-pass filter, that acts as an integrator. With this synchronous demodulation technique, the output is no longer proportional to the T_{sys}, but to the difference between the antenna temperature (including ohmic losses) and the reference temperature as illustrated in Fig. 4.85:

$$V_0 = \frac{1}{2} \cdot k_B \cdot \left(T'_A - T_{REF} \right) \cdot B \cdot G \cdot C_d, \tag{4.101}$$

and provided the switching rate (f_s) is much higher than the bandwidth of the gain fluctuations' spectrum, so that $\Delta G/G|_{f = m \cdot fs} << \Delta G/G|_{f = 0}$, the impact of gain fluctuations in the radiometric precision is minimized (Thomsen, 1984):

$$\sigma_T = \sqrt{\frac{2 \cdot \left(T'_A + T_R \right)^2}{B \cdot \tau} + \frac{2 \cdot \left(T_{REF} + T_R \right)^2}{B \cdot \tau} + \left(T'_A - T_{REF} \right)^2 \cdot \left(\frac{\Delta G}{G} \right)^2}, \tag{4.102}$$

since they are no longer proportional to $\left(T'_A + T_R \right)$, but to $\left(T'_A - T_{REF} \right)$, which is much smaller. In Eq. (4.102), the "2" factors come from the fact that the antenna and the reference load are being measured during only half the integration time τ.

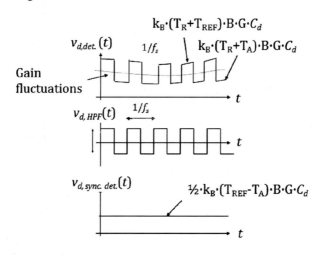

FIGURE 4.85

Synchronous demodulation minimizes gain fluctuation effects: signal at detector's output (upper panel), after the DC block (central panel) most gain fluctuations have been removed, and, after the synchronous demodulation an output proportional to $^1/_2$ of the difference between the antenna temperature and the reference temperature is obtained.

Eventually, the impact of gain fluctuations can be completely mitigated if $T'_A = T_{REF}$. When this condition is met, the radiometer is said to be "balanced." In this case, the radiometric precision is twice that of an ideal TPR (Eq. 4.98 or Eq. 4.102 with $\Delta G/G = 0$) due to the shorter integration time.

For the DR to perform well there are a number of conditions that have to be met: the switching frequency and the cut-off frequency of the DC block must be much higher than the maximum frequency of the gain fluctuations spectrum ("knee" in Fig. 4.83), and the video amplifier's bandwidth must be large enough so that the rise time (τ_{sw}) is much smaller that the switching period (e.g., $\tau_{sw} < 0.5 \cdot 10^{-2}/f_s$. In addition, usually T_{REF} is set to be higher than the highest T_A to be measured, so that the sign of the voltage after the synchronous demodulator does not change. This is achieved by thermal stabilization of the whole instrument at a physical temperature typically around 40–45°C.

If $T_A \neq T_{REF}$, the gain fluctuations are not totally compensated, and other "balancing" techniques must be devised, such as the **duty-cycle modulation** (Orhaug and Waltman, 1962), **automatic-gain control** (AGC) methods and **gain-modulation** methods (Thomsen, 1984), **two-reference temperature radiometer** (Hach, 1968), or noise injection techniques can be devised so that the balancing condition ($T'_A + T_I = T_{REF}$) is always met such as in the **noise-injection radiometer** (Goggins, 1967), using the **reference-channel radiometer** (Machin et al., 1952), or using other more sophisticated techniques taking advantage of digital processing techniques such as in the "**ultra-stable radiometer**" (Wilson et al., 2003, 2005). **Pseudo-correlation radiometers** such as the Planck LFI instrument (Mennella et al., 2003) or in the PAU hybrid radiometer/GNSS-reflectometer (Camps et al., 2007) also achieve some of the balancing benefits, but the noise is injected in a subtler way. Table 4.6 from Camps and Tarongi (2010) summarizes the main properties of the main types of radiometers.

Here Fig. 4.86 is taken from Camps and Tarongi (2010) and illustrates the relative performance of the different types of radiometers for $\Delta G/G = 0.01$, $T_R = 400K$, $T_{REF} = 318K$, $T_{ON} = 913K$, $T_{OFF} = 30K$, $B = 20$ MHz, $\tau = 1$ s, and $\tau_{AGC} = 1$ s. As it can be seen, for $\Delta G/G = 0.01$ the TPR performs the worst, and the DR with the reference channel the best. The performance of other types of radiometers is quite similar and quite constant over a wide range of T_A's. However, these relative performances are strongly dependent on hardware parameters ($\Delta G/G$, T_R, etc.) and the thermal stabilization of the whole instrument including, as much as possible, the antenna where a significant part of the ohmic losses originate.

Radiometric stability refers to the mid- or long-term variations of the radiometer's output for a constant input, not to the radiometric precision discussed before. In general, for well-designed instruments, DRs are more stable than total power ones, and noise injection radiometers (NIRs) are more stable than DRs. A fair intercomparison between a Ku-band (18.6–18.8 GHz) radiometer designed to operate either as total power (TPR), DR, or NIR is given here in Fig. 4.87A. It is said that the comparison is fair because the data presented is acquired exactly under the same conditions, same hardware, and same temperature (and temperature drifts), after the same initial calibration was performed (Camps and Tarongi, 2010). The first thing that becomes apparent is the much larger drifts of the TPR, as compared to the other two. The second thing is that the drifts for the DR and NIR, follow the same trend, with a larger temperature sensitivity for the DR than for the noise injection one. The measured temperature sensitivity is $\Delta T_{NIR}/\Delta T_{ph} \sim 4K/°C$, which demonstrates the exquisite care that must be paid to the thermal control. The figure of merit of the radiometer's stability is the so-called Allan's variance (Allan, 1987), originally conceived to assess the stability of clocks. It is defined as the variance of the difference of two consecutive averages y_{i+1} and y_i obtained over a time $\tau = M \cdot T_s$:

$$\sigma_y^2(\tau) = \frac{1}{2(N-1)} \sum_{i=1}^{N-1} (y_{i+1} - y_i)^2, \tag{4.103}$$

Table 4.6 Main Types of Microwave Radiometers and Basic Performance (Camps and Tarongi, 2010)

1. Total power radiometer

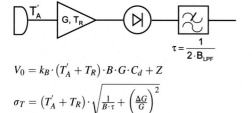

$$V_0 = k_B \cdot (T'_A + T_R) \cdot B \cdot G \cdot C_d + Z$$

$$\sigma_T = (T'_A + T_R) \cdot \sqrt{\frac{1}{B \cdot \tau} + \left(\frac{\Delta G}{G}\right)^2}$$

2. Dicke radiometer (unbalanced) (Dicke, 1946)

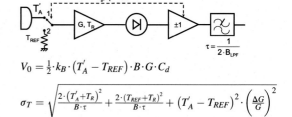

$$V_0 = \frac{1}{2} \cdot k_B \cdot (T'_A - T_{REF}) \cdot B \cdot G \cdot C_d$$

$$\sigma_T = \sqrt{\frac{2 \cdot (T'_A + T_R)^2}{B \cdot \tau} + \frac{2 \cdot (T_{REF} + T_R)^2}{B \cdot \tau} + (T'_A - T_{REF})^2 \cdot \left(\frac{\Delta G}{G}\right)^2}$$

3. Balanced Dicke radiometer by duty-cycle modulation (Orhaug and Waltman, 1962)

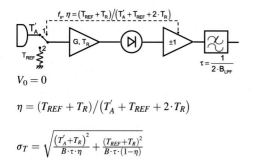

$$V_0 = 0$$

$$\eta = (T_{REF} + T_R)/(T'_A + T_{REF} + 2 \cdot T_R)$$

$$\sigma_T = \sqrt{\frac{(T'_A + T_R)^2}{B \cdot \tau \cdot \eta} + \frac{(T_{REF} + T_R)^2}{B \cdot \tau \cdot (1-\eta)}}$$

4. Balanced Dicke radiometer by gain modulation (Thomsen, 1984)

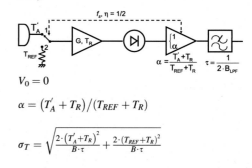

$$V_0 = 0$$

$$\alpha = (T'_A + T_R)/(T_{REF} + T_R)$$

$$\sigma_T = \sqrt{\frac{2 \cdot (T'_A + T_R)^2}{B \cdot \tau} + \frac{2 \cdot (T_{REF} + T_R)^2}{B \cdot \tau}}$$

5. Balanced Dicke radiometer by reference channel (Machin et al., 1952)

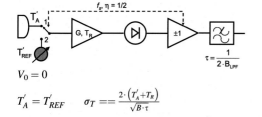

$$V_0 = 0$$

$$T'_A = T'_{REF} \qquad \sigma_T = = \frac{2 \cdot (T'_A + T_R)}{\sqrt{B \cdot \tau}}$$

6. Noise injection radiometer—variable noise (Goggins, 1967)

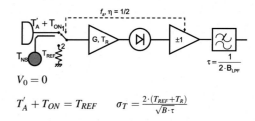

$$V_0 = 0$$

$$T'_A + T_{ON} = T_{REF} \qquad \sigma_T = \frac{2 \cdot (T_{REF} + T_R)}{\sqrt{B \cdot \tau}}$$

7. Noise injection radiometer—variable duty cycle (Hardy, 1973)

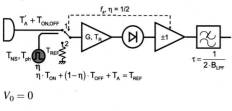

$$V_0 = 0$$

$$\eta \cdot T_{ON} + (1-\eta) \cdot T_{OFF} + T_A = T_{REF}$$

$$\sigma_T = \frac{2 \cdot (T_{REF} + T_R)}{\sqrt{B \cdot \tau}}$$

8. Hach radiometer (Hach, 1968)

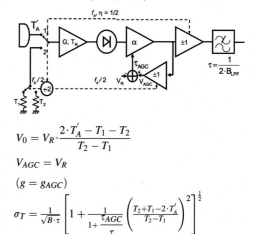

$$V_0 = V_R \cdot \frac{2 \cdot T'_A - T_1 - T_2}{T_2 - T_1}$$

$$V_{AGC} = V_R$$

$$(g = g_{AGC})$$

$$\sigma_T = \frac{1}{\sqrt{B \cdot \tau}} \left[1 + \frac{1}{1 + \frac{\tau_{AGC}}{\tau}} \left(\frac{T_2 + T_1 - 2 \cdot T'_A}{T_2 - T_1}\right)^2 \right]^{\frac{1}{2}}$$

$$\cdot \left[(T_2 + T_R)^2 + (T_1 + T_R)^2 + 2 \left(T'_A + T_R\right)^2 \right]^{\frac{1}{2}}$$

Table 4.6 Main Types of Microwave Radiometers and Basic Performance (Camps and Tarongi, 2010)—cont'd

9. Ultra-stable radiometer (Wilson et al., 2003, 2005)[a]

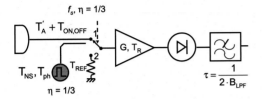

$$V_A = k_B \cdot (T'_A + T_{OFF} + T_R) \cdot B \cdot G \cdot C_d + Z$$
$$V_{A+N} = k_B \cdot (T'_A + T_{ON} + T_R) \cdot B \cdot G \cdot C_d + Z$$
$$V_{REF} = k_B \cdot (T_{REF} + T_R) \cdot B \cdot G \cdot C_d + Z$$

$$R^{NIR} \cong \frac{V_{REF} - V_A}{V_{A+N} - V_A} = \frac{T_{REF} - T'_A - T_{OFF}}{T_{ON} - T_{OFF}}$$

$$\sigma_T = \sqrt{\frac{3}{B \cdot \tau}} \cdot \left\{ (T_{REF} - T_{OFF} + T_R)^2 + \right.$$

$$(T'_A + T_R)^2 \cdot \left[1 - \frac{(T_{REF} - T_{OFF} - T'_A)}{T_{ON} - T_{OFF}} \right]^2$$

$$\left. + (T'_A + T_R + T_{ON} - T_{OFF})^2 \cdot \left[\frac{T_{REF} - T_{OFF} - T'_A}{T_{ON} - T_{OFF}} \right]^2 \right\}^{\frac{1}{2}}$$

11. Pseudo-correlation radiometer using Wilkinson power splitter (Camps et al., 2007)

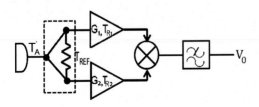

$$V_0 = k_B \cdot (T'_A - T_{REF}) \cdot \sqrt{B_1 \cdot B_2} \cdot \sqrt{G_1 \cdot G_2}$$

$$\sigma_T = \frac{T_A + T_R + 2 \cdot T_{ph}}{\sqrt{B \cdot \tau}}$$

10. Advanced pseudo-correlation radiometers (Planck LFI instrument) (Mennella et al., 2003).

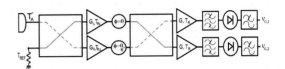

$$V_0 = V_{01} - V_{02}$$
$$= k_B \cdot \left[(T'_A + T_R) - r \cdot (T_{REF} + T_R) \right] \cdot B \cdot G \cdot C_d = 0$$

Gain modulation factor : $\quad r = \dfrac{T'_A + T_R}{T_{REF} + T_R}$

$$\sigma_T = \sqrt{2} \cdot \frac{T_A + T_{REF}}{\sqrt{B \cdot \tau}}$$

[a]*The above result can be optimized by adjusting the integration times as described (Camps and Tarongi, 2010).*

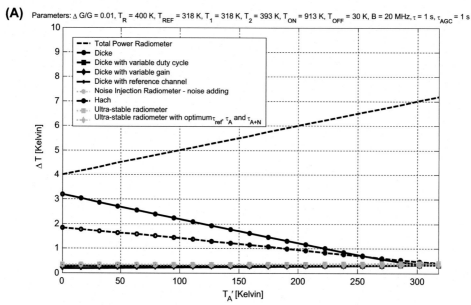

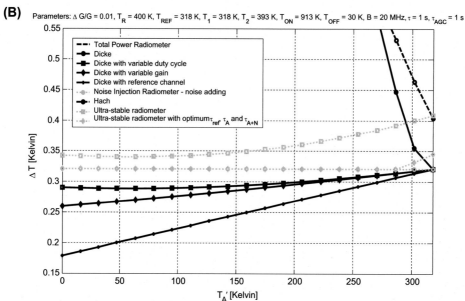

FIGURE 4.86

(A) Radiometric precision of different radiometers as a function of the antenna temperature (including antenna ohmic losses), and (B) zoom of Fig. 4.86A. For $\Delta G/G = 0.01$ the total power radiometer perform the worst, and the DR with the reference channel the best. The performance of other types of radiometers is quite similar and quite constant over a wide range of $T_A's$.

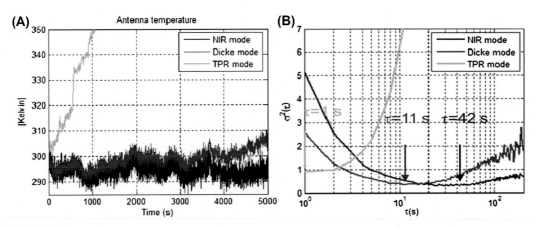

FIGURE 4.87

(A) Temporal evolution of radiometers' output reveals larger drifts for the total power radiometer (green), than for the Dicke radiometer (DR) (red), and greater DR drift than for the noise injection radiometer (black) (Camps and Tarongi, 2010). (B) Allan's variance computed for the time series in Fig. 4.87A.

where:

$$y_i = \frac{1}{M} \sum_{j=1}^{M} x((j + i \cdot M) \cdot T_s). \qquad (4.104)$$

For the example in Fig. 4.87A, the computed Allan's variance is shown in Fig. 4.87B. The minimum Allan's variances for the three cases occur at 1, 11, and 42 s for the TPR, the DR, and the NIR, respectively. If the radiometric precision at 1 s integration time are compared, the following results are obtained: $\sigma_{T,\,TPR} = 0.96$K, $\sigma_{T,\,DR} = 1.58$K, and $\sigma_{T,\,NIR} = 2.30$K, that is $\sigma_{T,\,DR} = 1.58 \cdot \sigma_{T,\,TPR}$ and $\sigma_{T,\,NIR} = 2.30 \cdot \sigma_{T,\,TPR}$, that compare well, although they are slightly worse, to the expected ratios for the ideal cases: $\sigma_{T,DR} = \sqrt{2} \cdot \sigma_{T,TPR}$ and $\sigma_{T,\,NIR} = 2 \cdot \sigma_{T,\,TPR}$.

The selection of the radiometer topology depends on one side on the radiometric resolution required for a particular application (e.g., ideally a TPR has a better resolution than a NIR) and on the stability, which determines the intercalibration period. Most scanning radiometers, either conical or cross-track scanners, are TPRs, because a two-point calibration (hot and cold external targets) can be performed in every scan. Other radiometers that cannot perform such a frequent calibration, e.g., the NASA/CONAE Aquarius 3 beam push-broom radiometer, the radar-altimetry companion radiometers for the "wet delay" correction, or the ESA SMOS radiometers to measure the "zero baseline," are NIRs, even though from time to time a satellite maneuver is performed to assess the instrument's stability.

4.6.3.3 Radiometer Calibration

Radiometer calibration refers to the process of finding the relationship between the radiometer's output (V_d) and the input antenna temperature (T_A), by means of well-known internal or external targets (Fig. 4.88A).

$$V_d = \frac{1}{a}(T_A - b), \qquad (4.105a)$$

$$T_A = a \cdot V_d + b, \qquad (4.105b)$$

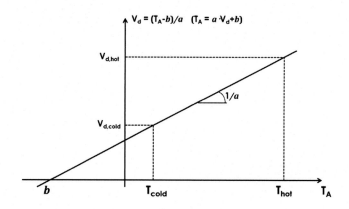

FIGURE 4.88

The radiometer calibration procedure consists of measuring the instrument's response to two known input targets: the "hot load" and "cold load" measurements.

This definition does include the correction for the antenna ohmic losses, but formally does not include the process of compensating for antenna pattern errors that will require some sort of deconvolution of the antenna beam. However, in the case of external calibration, some antenna pattern corrections must be required to ensure that the radiation is the one coming from the main beam:

$$T_A = \eta_{ML} \cdot \overline{T}_{ML} + (1 - \eta_{ML}) \cdot \overline{T}_{SL}, \tag{4.106}$$

where η_{ML} is the main beam efficiency defined in Eq. (4.73), and $\overline{T}_{ML}$ and $\overline{T}_{SL}$ are the average brightness temperatures in the main and side lobes, respectively.

Assuming that the radiometer's response is linear, the calibration procedure reduces to the determination of two points of a straight line (Fig. 4.88). One of the points is at a "high" input noise temperature such as a microwave absorber at ambient temperature (Fig. 4.89), and the other one at a "low" noise temperature, such as the sky or a microwave absorber embedded in liquid nitrogen, as described in Section 4.6.3.3.1. These points are called the "hot" and "cold" loads, respectively. From these two measurements:

$$T_{hot} = a \cdot V_{d,hot} + b, \tag{4.107a}$$

$$T_{cold} = a \cdot V_{d,cold} + b, \tag{4.107b}$$

the slope and the ordinate at the origin are determined:

$$a = \frac{T_{hot} - T_{cold}}{V_{d,hot} - V_{d,cold}}, \tag{4.108a}$$

$$b = T_{hot} - a \cdot V_{d,hot}, \quad \text{or } b = T_{cold} - a \cdot V_{d,cold}. \tag{4.108b}$$

If the radiometer's response is not linear, then additional input temperatures are required so as to determine the nonlinear transfer function.

Internal calibration targets, such as thermal loads (either ambient, hot, or cold loads), or noise sources, have the advantage of allowing for much faster and frequent calibrations, independent of the weather conditions, and potential external RFI. On the other hand external calibration targets, such as cold space (2.7K, except at some particular frequency bands, and in the direction of the galaxy plane, the Sun, or other "hot spots" in the sky, see Section 4.2.2), Earth targets used for vicarious calibration, such as some regions of the ocean, the Amazon rain forest, Dome C in Antarctica, or the atmosphere

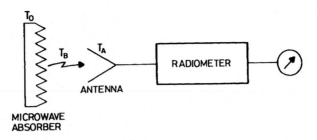

FIGURE 4.89

A microwave absorber at ambient temperature can be used as the "hot load" calibration target (Skou, 1989).

Reproduced by permission from Skou, N., 1989. Microwave Radiometer Systems: Design and Analysis, First Edition, Artech House, Inc., Norwood, MA. © 1989 by Artech House, Inc.

itself using "tip curves," allow for a true end-to-end calibration from the antenna to the ADC output, including changes in losses, physical temperature, etc. and can be used to periodically vicariously *calibrate* the internal calibration targets themselves. When it is not possible to perform an external calibration, or the atmosphere is so opaque that tip curves are not an option either, radiometers must rely solely in internal calibration targets.

Finally, a note again on the radiometer's thermal stability that must be controlled to a fraction of the expected calibration accuracy requirement, usually a small fraction of a 1°C.

4.6.3.3.1 External Calibration

4.6.3.3.1.1 Using Hot and Cold Targets. External calibration targets are designed to have an emissivity very close to one and a uniform distribution of the physical temperature, stabilized using heating and thermoelectric cooling, or accurately measured at ambient temperature. At microwave and millimeter-wave frequencies, the materials that exhibit a very high emissivity (reflectivity very close to zero) are the microwave absorbers either at ambient temperature (as "hot" loads) or cooled down (as "cold" loads) (Figs. 4.90–4.94).

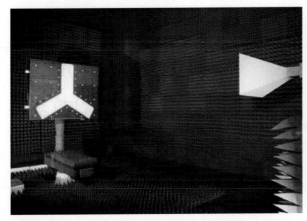
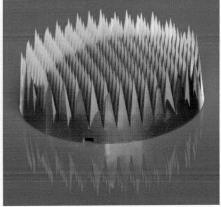

FIGURE 4.90

(Left) Sample microwave absorbers used in an anechoic chamber for antenna pattern characterization of the SMOSillo or miniature microwave imaging radiometer by aperture synthesis instrument to test the performance of the Soil Moisture and Ocean Salinity mission or for microwave radiometer calibration (right).

Left: courtesy of UPC Antenna Lab; right: courtesy of the Science and Technology Facilities Council UK—RAL Space.

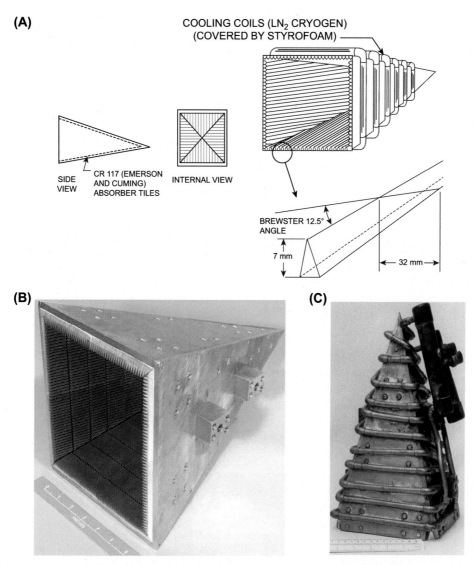

FIGURE 4.91

(A) Schematics of the microwave calibration target inverted pyramid used in scanning multichannel microwave radiometer (SMMR) (Skou, 1989). (B) SMMR target without cooling tubes. (C) Cooling tubes on the SMMR target.

(A) Reproduced by permission, from N. Skou, Microwave Radiometer Systems: Design and Analysis, First Edition, Norwood, MA: Artech House, Inc., 1989. © 1989 by Artech House, Inc.

A microwave absorber at a physical temperature T_0 emits a brightness temperature T_B very close to its physical temperature, and therefore, if the whole antenna pattern is surrounded by a microwave absorber at a physical temperature T_0, the antenna temperature will be $T_A = T_0$ as well. If T_0 is the ambient temperature, this provides the "hot load" calibration target.

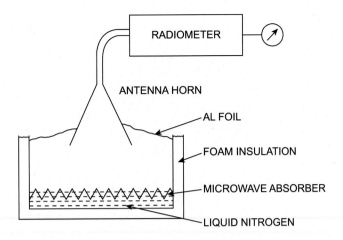

FIGURE 4.92

A cooled microwave absorber can be used as the "cold load" calibration target (Skou, 1989).

Reproduced by permission, from N. Skou, Microwave Radiometer Systems: Design and Analysis, First Edition, Norwood, MA: Artech House, Inc., 1989. © 1989 by Artech House, Inc.

The reflectivity of the microwave absorbers in the normal direction is usually quite small: values of −30, −40 dB are common. This is achieved by several means. First, the shape of the absorber is pyramidal, providing a gradual transition between the medium air and the medium "absorbing material." Second, the material is usually a foamy material that provides a gradual transition of the

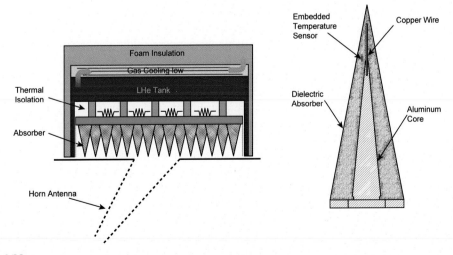

FIGURE 4.93

Mitigating "cold load" commonly induced errors (Kogut et al., 2006): array of conical microwave absorbers, consisting of a metal core covered by microwave absorber to improve thermal homogeneity weakly coupled to a superfluid liquid helium reservoir. A single absorbing cone consists of a metal core covered by microwave absorber. Thermometers embedded within the absorber monitor the temperature of the calibrator.

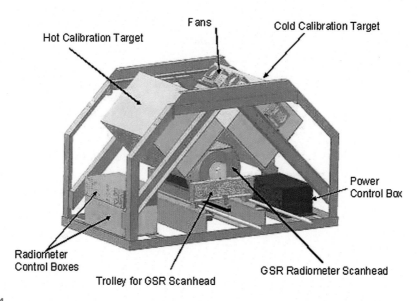

Hot Calibration Target

Fans

Cold Calibration Target

Power
Control Box

Radiometer
Control Boxes

Trolley for GSR Scanhead

GSR Radiometer Scanhead

FIGURE 4.94

Ground-based scanning radiometer (GSR) from National Oceanic and Atmospheric Administration Environmental Technology Lab, Microwave Systems Development Division http://cet.colorado.edu/instruments/gsr/.

dielectric constant. And third, the material is a lossy one (using graphite, ferrites, etc.) that attenuating the wave as it travels through it to the base of the pyramid. However, especially at low frequencies, having a low reflectivity requires very large pyramids, which are infeasible, or at least impractical in many systems. A way to reduce the absorbers' reflectivity is by using incidence at the Brewster angle, at which for vertical polarization, the reflection coefficient exhibits a minimum and eventually vanishes. An example of this technique is the variable temperature calibration target that was manufactured by JPL in the 1970s for the calibration of the scanning multichannel microwave radiometer (SMMR) (Fig. 4.91). The tiles of epoxy absorbing material inside the pyramid are at the Brewster angle, which is nearly frequency independent. The metallic enclosure provides good thermal conductivity to ensure a homogeneous temperature distribution, and the outside serpentine is a pipe in which liquid nitrogen circulates to cool down the absorbers so as to achieve the "cold target," or heated nitrogen gas up to 400K for the "hot target."

The "cold load" used as a second target of the calibration is usually a microwave absorber embedded in liquid nitrogen (LN2), which has a boiling temperature of 77K and an emissivity very close to one, and therefore does not affect the absorber's reflectivity. The concept is illustrated in Fig. 4.92.

However, there are a few known problems with this method: the surface of the boiling LN2 presents a complicated nonspecular surface to the radiometer antenna; microwave emission from the antenna, and the radiometer enclosure reflected over the LN2 surface back to the antenna increases the apparent antenna temperature; if absorber tips are exposed to the air, water condensing on the absorber increases the antenna temperature; and nitrogen vapor may cool down the antenna, and therefore the term

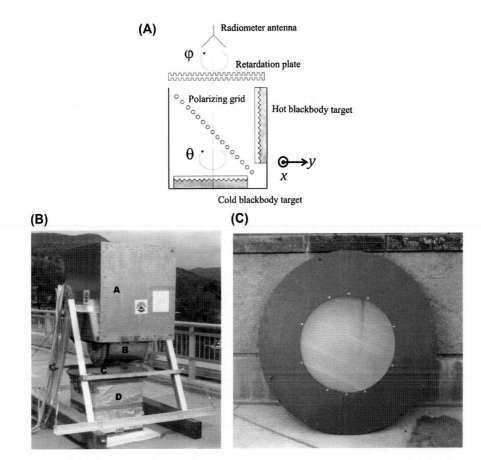

FIGURE 4.95

(A) Fully polarimetric calibration target developed at the National Oceanic and Atmospheric Administration (NOAA) Environmental Technology Lab (ETL), (B) NOAA fully polarimetric calibration setup: (a) polarimetric scanning radiometer (PSR) housing, (b) PSR scan head, (c) microwave retardation plate, and (d) linearly polarized standard, (C) Retardation plate of the NOAA/ETL fully polarimetric calibration standard. Plate is mounted in a wooden disk (Lahtinen et al., 2003).

associated to the noise generated by the antenna ohmic losses ($T_{ph} \cdot (1 - \eta_\Omega)$) decreases. The solution to some of these problems consists of tilting the antenna so that the antenna boresight is not perpendicular to the absorber's surface, and inserting a radiometer to divert the cold air and nitrogen vapor from reaching the antenna, or by cooling the absorber, with the antenna looking up, as illustrated in Fig. 4.93.

Finally, Fig. 4.94 shows the calibration system of the ground-based scanning radiometer (GSR) from NOAA, in which "hot" and "cold" targets are enclosed in thermal boxes for calibration purposes. The radiometer scan head points to them during the calibration, and then exits the box to perform the sky measurements at different elevation angles.

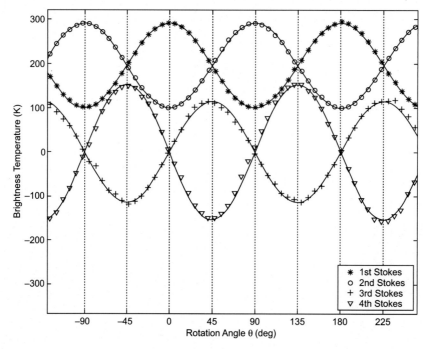

FIGURE 4.96

Sample fully polarimetric calibration as a function of the rotation angle (Lahtinen et al., 2003).

4.6.3.3.1.2 Fully Polarimetric Radiometer Calibration Using External Targets.

The output of a fully polarimetric radiometer can be expressed as the vector $\bar{r}$ shown in Eq. (4.109):

$$
\bar{r} = \begin{bmatrix} r_v \\ r_h \\ r_3 \\ r_4 \end{bmatrix} = \begin{bmatrix} g_{vv} & g_{vh} & g_{v3} & g_{v4} \\ g_{hv} & g_{hh} & g_{h3} & g_{h4} \\ g_{3v} & g_{3h} & g_{33} & g_{34} \\ g_{4v} & g_{4h} & g_{43} & g_{44} \end{bmatrix} \cdot \begin{bmatrix} T_v \\ T_h \\ T_3 \\ T_4 \end{bmatrix} + \begin{bmatrix} o_v \\ o_h \\ o_3 \\ o_4 \end{bmatrix} + \bar{n} = \bar{\bar{G}} \cdot \bar{T}_B + \bar{o} + \bar{n}, \qquad (4.109)
$$

The first term in the sum represents the weighting and cross-talk among the true four Stokes parameters. The sources of cross-talk are the antenna limited polarization isolation, cross-talk in the radiometer receiver, cross-talk in the correlator, or an unbalance in the signals before detection used to measure the third and fourth Stokes parameters in the polarization-combining method. The $\bar{o}$ and $\bar{n}$ terms represent the calibration offsets and instrument noise referred to the outputs.

A possible implementation of the calibrator using external targets is shown in Fig. 4.95 from Lahtinen et al. (2003). When the polarizing grid is along the X-direction (perpendicular to the plane of the figure, as shown $\theta = 0°$), it reflects toward the radiometer the hot blackbody target at X-polarization, while the cold blackbody target at X-polarization is reflected back to the blackbody, and the radiation at Y-polarization passes through. When the polarizing grid is along the Y-direction (parallel to the plane of the figure, perpendicular to the plot as drawn $\theta = 90°$), it reflects toward the radiometer the hot blackbody target at Y-polarization, while the cold blackbody target at Y-polarization is reflected back to the blackbody, and the radiation at X-polarization passes through. When the polarization grid is rotated an intermediate angle θ, a fraction of each polarization passes and the first three Stokes elements vary as:

$$(T_v \equiv)T_{xx} = T_{hot} \cdot \cos^2 \theta + T_{cold} \cdot \sin^2 \theta \tag{4.110a}$$

$$(T_h \equiv)T_{yy} = T_{hot} \cdot \sin^2 \theta + T_{cold} \cdot \cos^2 \theta \tag{4.110b}$$

$$T_3 = (T_{hot} - T_{cold}) \cdot \sin 2\theta \tag{4.110c}$$

The fourth Stokes parameter is only nonzero when a phase shift δ is induced among the two orthogonal polarizations:

$$T_4 = -(T_{hot} - T_{cold}) \cdot \sin 2\theta \cdot \sin \delta \tag{4.110d}$$

Experimental results for the polarimetric scanning radiometer are shown in Fig. 4.96.

4.6.3.3.1.3 Tip Curves. Tip curves are believed to be the most reliable technique for external calibration (Han and Westwater, 2000). They use the cosmic background temperature as an ultrastable, known cold source, which is $T_{cos} = 2.7K$ nearly constant above 2 GHz. The principle of the technique lies on the assumption of an effective "atmospheric height" h, so that when the radiometer is pointed to the zenith the downwelling brightness temperature is:

$$T_{dn} = T_{cos} + T_{atm} \quad (\text{depth} = h), \tag{4.111a}$$

and when it is pointed to a zenith angle θ, the downwelling brightness temperature is:

$$T_{dn} = T_{cos} + T_{atm}/\cos\theta \quad (\text{depth} = h/\cos\theta). \tag{4.111b}$$

The concept is sketched in Fig. 4.97.

If the atmosphere is measured at a number of zenith angles and the antenna temperature is plotted with respect to the "number of atmospheres," i.e., $\sec(\theta)$, a straight line will result. The extrapolated value at zero atmospheres corresponds to the cosmic background.

Tip curves rely on assumption of a "vertically stratified atmosphere," i.e., horizontally homogeneous. This assumption can be checked using a two-sided tipping curve, scanning on both sides of zenith, preferably from east to west, because it is the direction in which weather systems typically move. In addition, care must be taken with the lower zenith angles to avoid radiation being picked up from the ground by antenna side lobes. In some very high absorption peaks (e.g., O_2 around 55 GHz) the atmosphere is so opaque that it is not feasible to perform "tip curves." These radiometers must be calibrated by internal calibration targets only.

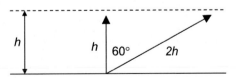

FIGURE 4.97

Calibration concept using a "tip curve." Number of "effective" atmospheres as a function zenith angle.

4.6.3.3.1.4 Earth Targets: Vicarious Calibration. Earth targets such as the Amazon rain forest, some parts of the ocean, or Dome C in Antarctica have been used for calibration of microwave radiometers because they appear as very stable targets over time. Among the class of methods that use Earth targets, the so-called "vicarious calibration" method was devised in 2000 (Ruf, 2000) to calibrate satellite microwave radiometers. It was used to identify long-term drifts in the TOPEX/POSEIDON radiometers used to correct altimeter measurements for the effects of water vapor, and it relies on the fact that Earth brightness temperatures measured from space have a well-known, repeatable, and predictable lower bound due to the specular emissivity of the sea surface in low humidity conditions. This minimum occurs at different SSTs, depending on the radiometer's frequency. Histogram analysis is used to find the coldest point in a set of satellite radiometer data, and this value is then compared to the theoretical model and then adjusted accordingly.

A word of caution with regard to the method is the constraint on the accuracy of a dielectric model for the sea surface, since recent models disagree by up to 2K in the 18−37 GHz range. This implies that the method does not provide an absolute calibration, but it allows us to track small changes and to perform precise intersatellite cross-calibration. The brightness temperature models at 37 GHz for horizontal and vertical polarizations, and 53.1° incidence angles for a flat sea, without clouds, and low-integrated water vapor are presented in Fig. 4.98A and B. As it can be easily seen, a minimum occurs, and this minimum is at a different SST. Fig. 4.98C shows the histograms of the offset between measured brightness temperature and the modeled one at 18, 21, and 37 GHz.

The evolution of the position of the minimum brightness temperature indicates the offset drift with respect to the model that can be tracked over long time series. This is exemplified in Fig. 4.99 by the evolution of the minimum brightness temperatures for the three channels of TOPEX/POSEIDON. The drift (almost linear) of the 18 GHz channel is clearly evident and traceable, which allows for an off-line compensation.

4.6.3.3.2 Internal Calibration

Internal calibration targets, such as thermal loads (either ambient, hot, or cold loads) or noise sources, have the advantage of allowing for much faster and frequent calibrations. In this section a brief overview is provided, using examples of commercial devices whenever possible, with their basic performance.

- Matched loads at ambient temperature (Fig. 4.100A). In this case the noise temperature T_N is equal to the physical temperature T_0. As for an external calibration target (microwave absorber) the key parameter is the reflection coefficient, which must be as low as possible, or the voltage standing wave ratio as close to one as possible.
- Solid-state hot noise sources consist of avalanche diodes (Fig. 4.100B) that produce an equivalent noise temperature T_N much higher than the ambient temperature:

$$T_N = \frac{N}{k_B \cdot B} \gg T_0, \qquad (4.112)$$

The amount of noise generated is usually expressed in decibels in terms of the excess noise ratio (ENR), defined as:

$$\text{ENR} = 10 \cdot \log\left(\frac{T_N}{T_0}\right), \qquad (4.113)$$

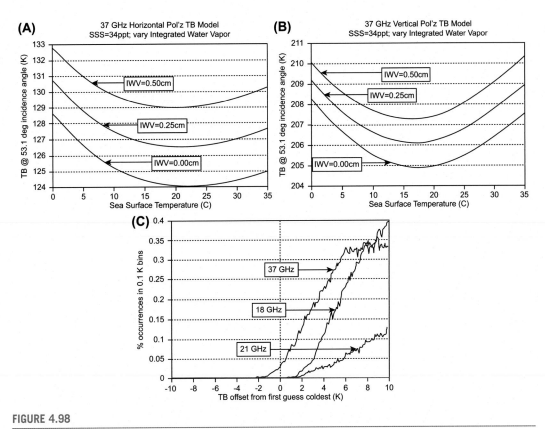

FIGURE 4.98

(A and B) Brightness temperature models at 37 GHz for horizontal and vertical polarizations, and 53.1° incidence angles for a flat sea, without clouds, and low-integrated water vapor. (C) Histograms of coldest brightness temperatures measured by the radiometer allow us to track small bias drifts (Ruf, 2000).

typical values of ENR are on the order of 12−15 dB. This amount of noise is generated by reverse polarizing an avalanche diode, with a typical voltage of 28 V. Since the typical ENR variations with temperature and voltage are 0.01 dB/°C and 0.1 dB/V, respectively, both the temperature of the noise source and the voltage of the power supply must be very stable. For example, a 0.1°C temperature variation or a 1 mV power supply voltage variation translate into a 0.001 dB ENR variation which, for ENR = 15 dB noise source corresponds to a T_N variation of 2.1K, which is unacceptable for most applications.

- Precision hot loads are ovenized matched loads (Fig. 4.100C) that provide a very accurate and stable noise temperature as determined by the physical temperature at which they are heated. Calibration tests performed at JPL with a precision hot load as these ones showed a repeatability within 0.1K rms over 2 months.
- Cryogenic loads (Fig. 4.100D) are precision matched loads cooled with liquid nitrogen at an operating temperature of 77.36K that provide a very accurate and stable noise temperature. As for

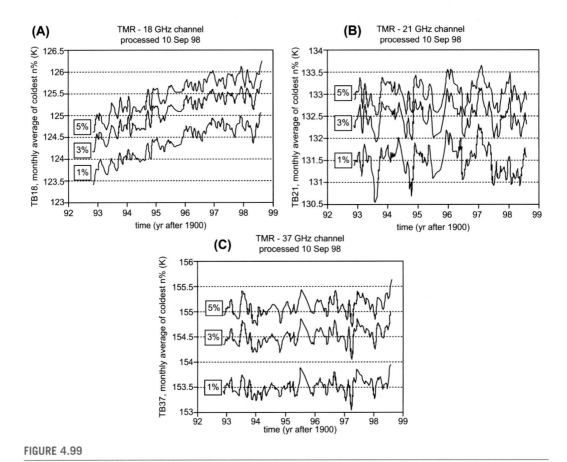

FIGURE 4.99

Evolution of the minimum brightness temperatures for the 18, 21, and 37 GHz channels (Ruf, 2000).

---

the precision hot loads their matching is much better than for the avalanche diodes, although the output noise temperature is fixed and cannot be controlled. The system performs well until the liquid nitrogen has evaporated, which will require to be refilled.

- Noise calibration systems (Fig. 4.100E) are the combination of a precision hot load warmed with an accuracy of ±0.2K and a cryogenic load cooled with liquid nitrogen, together with a precision switch into a single unit. Although the performance is much better than the avalanche diodes, but it is worse than the cryogenic loads or precision loads that have been isolated, because of the uncertainties introduced by the switch between them (Fig. 4.101).

- Solid-state cold noise sources or active cold loads (ACLs) were first described in 1981 (Frater and Williams, 1981). ACLs are active devices that generate an equivalent noise temperature lower than the physical temperature. The physical mechanism for this apparently counterintuitive behavior lies in the fact that active devices, when operated normally amplify the signal (and the noise) present at their input, while introducing additional noise themselves. However, in the

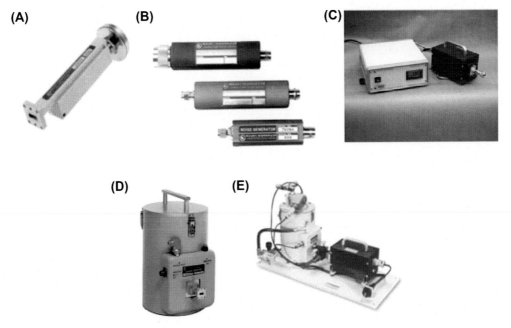

FIGURE 4.100

(A) Matched load at ambient temperature in waveguide, (B) solid-state hot noise sources, (C) ovenized hot load, (D) cryogenic cold load, and (E) noise calibration system.

From https://www.maurymw.com/Precision/Noise_Calibration_Systems_Accessories.php.

reverse direction, the signal (and noise) are usually strongly attenuated, and for most of them, the extra noise added is fairly small, so that the total noise emitted at their input ($T_{out,1}$) is lower than the physical temperature.

An amplifier is characterized by its two-port S-parameters (S_{ij}), and its four noise parameters: R_n the noise resistance, Γ_{opt} the optimum input complex reflection coefficient, and $F_{\min}$ the minimum noise figure (Fukui, 1966). The output of the amplifier is terminated with a matched load ($\Gamma_2 = 0$) at a temperature T_2. The available output noise temperature seen at the input of the amplifier (port 1) can be written as:

$$T_{out,\,1} = T_{rev} + A_{12} \cdot T_2,$$ (4.114a)

where the first term in Eq. (4.114a), T_{rev}, is the ideal lowest temperature when the load temperature T_2 is zero. T_{rev} can be expressed only in terms of the conventional noise parameters as (Weatherspoon and Dunleavy, 2006):

$$T_{rev} = \frac{T_k - T_{e,\min} \cdot \left(1 - \left|\Gamma'_{opt}\right|^2\right)}{1 - \left|\Gamma'_{opt}\right|^2},$$ (4.114b)

with:

$$T_{e,min} = T_0 \cdot \left(10^{F_{min}/10} - 1\right), \tag{4.114c}$$

$$\Gamma'_{opt} = \frac{S_{11}^* - \Gamma_{opt}}{\Gamma_{opt}S_{11} - 1}, \tag{4.114d}$$

$$T_k = 4T_0 R_n G_{opt}, \tag{4.114e}$$

$$G_{opt} = \frac{1}{Z_0} \Re\left\{\frac{1 - \Gamma_{opt}}{1 + \Gamma_{opt}}\right\}, \tag{4.114f}$$

and the reverse available gain of the amplifier:

$$A_{12} = \frac{|S_{12}|^2}{1 - |S_{22}|^2} \cdot \frac{1 - |\Gamma'_2|^2}{1 - |\Gamma_{1i}|^2}, \tag{4.114g}$$

$$\Gamma'_2 = \frac{\Gamma_2 - S_{22}^*}{1 - S_{22}\Gamma_2}. \tag{4.114h}$$

Since the input reflection coefficient of the ACL may be too high for radiometer calibration, an isolator may be used, at the expense of the extra thermal noise generated by the internal matched load in the isolator that is reflected at the ACL input and sent back to the input.

ACLs and programmable arbitrary wave generators have been used in prototypes of programmable partially correlated noise sources for the calibration of polarimetric radiometers (Ruf and Li, 2003).

4.6.3.3.3 Radiometer Linearity

So far, the radiometer's transfer function $V_0(T_A)$ has been assumed to be a linear function. However, systems may have a gain expansion behavior, and gain compression at the higher levels, and the best linearization results have to be found with an nth order polynomial. Usually a second-order polynomial suffices, but still software corrections are required to compensate for these "nonlinearities."

To measure the radiometer's linearity a variable and known antenna temperature must be present at the main antenna input that will be compared to the calibrated antenna temperature (radiometer's output). The nonconstant term of the difference between the known input and the calibrated output is the nonlinearity error. The constant term is just the average radiometric accuracy (systematic error). Two sample ways to achieve a variable noise temperature at the input are discussed below:

- Starting from a cold load (e.g., $\sim$98K, cold reference load in a cryostat), extra noise generated by a noise source with a large ENR followed by an adjustable attenuator L, is added through a directional coupler (C):

$$T_{IN} \approx T_{COLD} + \left\{T_{HOT} \cdot \frac{1}{L} + T_{ph} \cdot \left(1 - \frac{1}{L}\right)\right\} \frac{1}{C}. \tag{4.115}$$

The block diagram is shown in Fig. 4.101A. For example, assuming that ENR = 25 dB, that $T_{cold} = 98K$ (typical value from Maury cryogenic loads) and the step attenuator ranges from 15

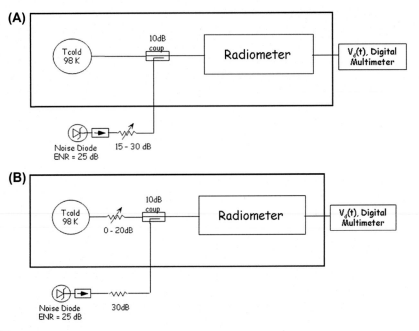

FIGURE 4.101

Sample block diagram for radiometer linearity tests (A) #1, and (B) #2.

to 30 dB in steps of 3 dB, the values of the noise temperature at the input of the radiometer are: 423.9, 276.2, 202.2, 165.1, 146.5, and 137.2K, that correspond to noise temperature steps of 147.7, 74.0, 37.1, 18.6, and 9.3K. That is, each step is very approximately half the previous one, from which the linearity of the instrument can be inferred with great accuracy. The accuracy is driven by the repeatability of the step attenuator divided by the coupling factor.

- Another method to perform the linearity correction is given in Colliander et al. (2007). It consists of inserting a variable, but extremely repetitive attenuator in series with the cold load in such a way that the output noise be $T_{IN} = T_{cold}/L + T_{ph} (1 - 1/L)$. When $L = 1$ (0 dB): $T_{IN} = T_{cold}$, while when L tends to infinite T_N tends to T_{ph} ($L = 20$ dB, $T_{ph} = 298K$, $T_{IN} = 296K$), being T_{IN} the noise temperature of the ensemble cryogenic load + variable attenuator. The attenuation L is adjusted in known steps and its physical temperature must be accurately measured. It determines the "antenna temperature" where the nonlinearity will be measured. On top of this noise, an additional noise level is added. This amount of noise is constant for all T_N values, and therefore this is an extremely stable method to assess the linearity of the system through the whole dynamic range of the radiometer. A schematic diagram of the measurement setup used in this technique is shown in Fig. 4.101B. The idea is to add a constant amount of noise on top of a noise level that can be varied. The benefit is the fact that the only critical parameter for the accuracy of the measurement is the stability of the added noise, which is relatively easily to achieve.

4.6.3.4 Radio Frequency Interference Detection and Mitigation

RFI includes spurious signals and harmonics from lower frequency bands, spread-spectrum signals overlapping the "protected" band of operation, or out-of-band emissions not properly rejected by the predetection filters due to its finite rejection. The presence of RFI increases the detected power and the estimated antenna temperature, leading to a degradation in the accuracy of the retrieved geophysical parameters, if they can be retrieved at all. RFI is specially a serious problem in populated land areas (Njoku et al., 2005; Ellingson and Johnson, 2006; Younis et al., 2007), although due to different spectrum regulations not all bands are affected the same way in all regions.

Current RFI detection and, eventually, mitigation methods in microwave radiometry include the following ones:

- **Time-domain** techniques blank ("eliminate") signal samples with power peaks larger than a predefined factor of the expected variance (Güner et al., 2007). However, the detected power is an average of the instantaneous power, therefore RFI peaks shorter than the integration time may pass undetected.
- **Frequency-domain** techniques blank subbands with power larger than a predefined factor of the expected variance (Güner et al., 2007). Similar to the time-domain techniques, the detected power is an average over a given bandwidth, therefore RFI power peaks narrower than the resolution bandwidth is "blurred" and may pass undetected.
- **Spectrogram** techniques (simultaneous time- and frequency-domain signal decomposition, e.g., Tarongi and Camps, 2011) use very long sequences of data to achieve fine resolution simultaneously in time and frequency. Image processing techniques such as edge detection of clusters of high power time—frequency bins are then used to detect anomalously high power time—frequency bins, which are then eliminated.

In these three cases, the signal power is estimated from the remaining signal samples, subbands, or time—frequency bins, properly scaled.

- **Statistical** techniques are based on the fact that the RFI-free radiometric signal should be a zero-mean random Gaussian variable. Therefore, the PDF and the statistical parameters are perfectly known. If the normality test is not passed, the whole sequence is eliminated. In the remote sensing literature, the Kurtosis method (ratio of the fourth moment and the square of the second moment, Misra et al., 2009) is the most widely test used, and there are time- and frequency-domain versions of it. To prevent a blind spot of the Kurtosis method, other methods have been studied such as the sixth-order moment (De Roo and Misra, 2008), the Shapiro—Wilk (SW) test (Güner et al., 2008), or in Tarongi and Camps (2010) several other normality tests were also carefully reviewed.
- **Polarimetric** techniques look for anomalous signatures in the third and fourth Stokes parameters, and, if found, the corresponding first and second Stokes parameters are discarded. These techniques have been applied to both real (Ellingson and Johnson, 2006) and synthetic aperture radiometers (Camps et al., 2011a; Kristensen et al., 2012).

In these two cases, all data are lost when RFI is detected.

- Finally, **wavelet** techniques can also be used to estimate the RFI signal without any priori knowledge of it and then cancel it (Camps and Tarongi, 2009). In this case, there is no signal loss, although a residual RFI may be present. Its performance depends mainly on the ratio of sampling frequency to signal bandwidth, the interference-to-noise ratio (INR), and the wavelet family and decomposition level.

Actually, it has to be acknowledged that many of these methods were first used in radio-astronomy, where the RFI problems where first experienced.

The performance of the different methods is characterized by two parameters:

- The **probability of a false alarm** (P_{fa}), or the "detection" of RFI in the absence of RFI (*Type I error* or rejection of a true hypothesis), which leads to the blanking of correct data, and
- The **probability of a missed detection** (P_{miss}), or the "no detection" of an RFI when there is RFI present (*Type II error* or acceptance of a false hypothesis), which leads to an erroneous measurement. In our context the term probability of detection (P_{det}) is usually used and it is defined as $P_{det} = 1 - P_{miss}$.

The objective is to obtain a low P_{fa} and a high P_{det}, but since both are strongly correlated, setting the value of one determines the value of the other one. To achieve a low P_{fa}, the threshold used to take the decision of "presence of RFI" in methods 1 to 5 must be high enough, but then many RFI are not detected and P_{miss} increases. The combined performance of a technique in terms of P_{det} and P_{fa} depends on the PDF of the signal values, and it is given by the receiver operating characteristic (or ROC) curves $P_{det} (P_{fa}(threshold))$. The ROC curves have been previously derived for many of the methods discussed in this book and will not be repeated here. The interested reader is referred, for example, to Misra et al. (2009) or Tarongi and Camps (2010).

The performance of the statistical tests is summarized in the following plots. In Fig. 4.102 we present the INR value required to obtain an ROC curve with a $P_{fa} = 0.1$ for $P_{det} = 0.9$ for different normality tests. This value is given for a pulsed sinusoidal signal with variable duty cycle. In general, the Kurtosis (K) test outperforms, i.e., it requires a lower INR for a given P_{fa} and P_{det}, except around the 0.5 duty cycle, where it has a blind spot. The JB (Jarque-Bera) and the K2 (D'Agostino K-square) do have a blind spot as well, since they are based on the K test. JB and K2 test results cannot be used with 1024 samples.

Around a duty cycle of 0.5, it is the SW test the one that has the best performance, followed by the Anderson−Darling (AD) test for 1024 samples, while AD tests present the best performance for 16,384 samples, since SW test performance is degraded due to averaging.

However, the presence of a blind spot is not restricted to pulsed sinusoidal RFI signals only. It does affect other pulsed signals that have a Kurtosis larger than three for a duty cycle of 100%. Results are shown in Fig. 4.103 for other signals, including pseudorandom-noise and telegraphic signals.

Finally, the determination of the presence or not of RFI, based on a finite number of signal samples, is prone to uncertainties that decrease as the number of samples increases. For example, for a confidence level of 0.05, the Kurtosis thresholds are $\sim 3 \pm 0.2$, which requires ~ 2000 samples, but for thresholds $\sim 3 \pm 0.1$, the number of samples required is ~ 6000.

In summary, the Kurtosis seems to be the best RFI detection algorithm for almost all kinds of interfering signals, although it has a blind spot for sinusoidal (chirp) signals of 0.5 of a duty cycle. The AD test is a complementary normality test that covers this blind spot, and has a very good performance for all the studied sample sizes. The combination of the Kurtosis and the AD tests seems capable to detect most types of RFI. The performance of the detection tests improves with the sample size and depends on the duty cycle of the pulsed RFI.

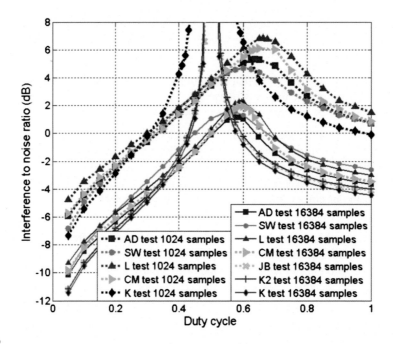

FIGURE 4.102

Normality test performance in the detection of a pulsed sinusoidal interference of 1024 samples (*dotted line*) and 16,384 samples (*solid line*) as a function of signal's duty cycle. Graphs represent the interference-to-noise ratio value required to obtain a receiver operating characteristic curve with a $P_{fa} = 0.1$ for $P_{det} = 0.9$. Results obtained from a Monte-Carlo set of 2^{15} simulations (Tarongi and Camps, 2010).

Spectrogram-based algorithms treat the radiometric signal's spectrogram (time−frequency representation of the signal) as an image and apply image processing techniques to detect the presence of anomalous features. The spectrogram is a powerful tool that has also been previously used in RFI detection in radio-astronomy (Offringa et al., 2010; Winkel et al., 2007). The presence of RFI can be detected by identifying power peaks, usually clustered in the time−frequency domain, larger than a given threshold above the variance (power) measured for the thermal noise in regions free of RFI, similarly as it is done for time- or frequency-domain blanking (Güner et al., 2007; Misra et al., 2009; Tarongi and Camps, 2011). An advantage of spectrogram methods—eventually in combination with normality tests either globally or at time−frequency bin level—is that pulsed RFI, which is wide in frequency, but narrow in time, or CW-like RFI, which is long in time, but narrow in frequency (see RFI present in Fig. 4.104) can be detected and eliminated, leaving a lot more of "clean" bandwidth to be processed. Spectrogram methods have also demonstrated their power in detecting RFI, that otherwise would have passed over inadvertently (Forte et al., 2013; Tarongi, 2013). More recently, improved performance has been achieved using the multiresolution Fourier transform (Querol et al., 2015, 2016).

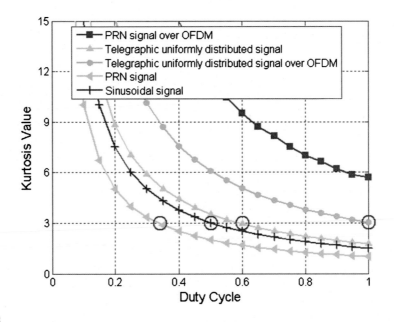

FIGURE 4.103

Other signals used in communications have a Kurtosis value of three for a determined duty cycle. *PRN*, pseudorandom noise; *OFDM*, orthogonal frequency division multiplexing (Forte et al., 2013).

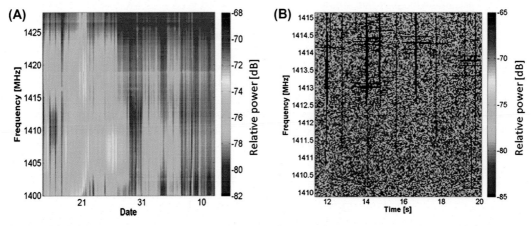

FIGURE 4.104

(A) Uncalibrated power spectrogram measured in Barcelona city during the months of May and June 2012. (B) Zoom of Fig. 4.104A starting on May 15, 2012, at $t_0 = 16{:}56{:}55$ h spanning from 1410 to 1415 MHz and from $t_1 = t_0 + 11$ s to $t_2 = t_0 + 20$ s. Detected radio frequency interference in the spectrogram using the MERITXELL instrument (Tarongi, 2013) are marked in black. Note: the power in each time–frequency bin is uncalibrated and it is expressed in relative units (dB).

In summary, there is not a single method that performs satisfactorily for all sorts of RFI signals, but combinations of them can do a decent job. The best candidates are the Kurtosis test, for its simplicity and good performance in most cases, in combination with the AD test to mitigate the blind spot. These methods can also be applied at the time−frequency bin level in the spectrogram before RFI mitigation algorithms are applied. The NASA SMAP mission radiometer, for example, implemented a combination of time and frequency diversity and Kurtosis to mitigate RFI (Entekhabi et al., 2010).

4.6.3.5 Example: Special Sensor Microwave Imager Radiometric and Geometric Corrections

From all the sensors listed in Table 4.4, the SSM/I has been the one that resulted in the greatest change in the way radiometers are used today. This sensor flew since 1987 continuously aboard the platforms DMSP F8 to F15 of the Department of Defense of the United States, and since then, merged with the SSM/T and SSM/T-2 and named as the SSMI/S, up to the DMSP F19. These platforms orbit the Earth in a Sun-synchronous polar orbit at $h = 890$ km, with an inclination of $98.8°$, and a period of 102 min. The radiometer is a seven channel instrument, with dual-polarization capabilities, except at 22 GHz, where only the vertical polarization is implemented. Table 4.7 summarizes the main instrument parameters: frequency bands, polarizations, antenna footprint, and integration time (Hollinger et al., 1987). Note that the antenna footprint size is inversely proportional to the instrument band. As compared to its predecessor the SMMR, the SSM/I does not have the 6.6 and 10.7 GHz channels, but it does include the 85.5 GHz one (see Table 4.4).

The imaging process is performed using a conical scan of the 79 cm diameter reflector. The rotation every 1.9 s and the $42°$ off-nadir angle lead to a swath width of 1400 km, and a constant incidence angle over the Earth of approximately $53.1°$, for which the emissivity at vertical polarization exhibits a minimum sensitivity with respect to surface roughness, which allows a simpler classification and geophysical parameter retrieval. Fig. 4.105A shows an artist's view of the SSM/I instrument, and Fig. 4.105B the geometry of observation, the antenna footprints and the sampling in scans A and B, while Fig. 4.106 presents the sampling strategy for the A and B scans. The sampling is produced every 12.5 km for scans A and B for the 85 GHz channel only (128 pixels per swath), while at 25 km for scans A only, for the rest of the channels (64 pixels per swath).

During the conical scan, in the farthermost points from the ground-track direction, the calibration is performed. At one swath edge the so-called "sky mirror" diverts the radiation from the sky into the antenna feed, providing the "cold load" calibration. At the other swath edge, a microwave absorber blocks the antenna feeders for the "hot load" calibration. The data acquired in between calibrations are

Table 4.7 Main Parameters of the Special Sensor Microwave Imager Instruments				
Frequency	19.35 GHz	22.235 GHz	37.0 GHz	85.5 GHz
Polarization	*h/v*	*v*	*h/v*	*h/v*
Integration time	7.95 ms	7.95 ms	7.95 ms	3.89 ms
Footprint (along-track)	69 km	50 km	37 km	15 km
Footprint (cross-track)	43 km	40 km	28 km	13 km

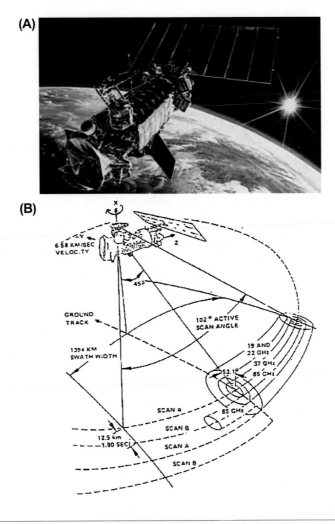

FIGURE 4.105

(A) Artist's view of special sensor microwave imager instrument (http://podaac.jpl.nasa.gov/SSMI), and (B) geometry of observation: antenna footprints and scans A and B (Hollinger et al., 1987).

corrected using the prior/later calibration data. However, if the calibration is wrong, then the entire line is discarded (see black arcs in Fig. 4.52).

Due to antenna/platform effects, the Earth incidence angle (EIA) can also vary along the swath position (Fig. 4.107A), which will require a proper compensation. Near the swath edges the feed collects partial power from the sky mirror, and the antenna temperature drops, which requires proper correction ($\sim$1K, see Fig. 4.107B).

Then, the brightness temperatures (T_B) are computed from the antenna temperatures (T_A) as the weighted sum of the central pixel and the adjacent pixels (including adjacent scans) at the same frequency and polarization. The information of the central pixel at the orthogonal polarization (h/v) is included as well in the algorithm to model the polarization mixing due to antenna misalignment and cross-polarization errors.

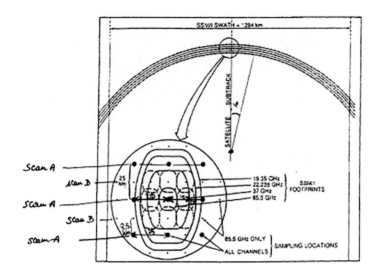

FIGURE 4.106

Sampling strategy for the A and B scans. See text for description (Hollinger et al., 1987).

The brightness temperatures recovered are stored in 2 byte integers, multiplied by 10, leading to a 0.1K resolution. All brightness temperatures must be between 50 and 350K. If the temperatures are outside this range or are erroneous, they are set to zero and not processed. If the time tag is wrong, it translates into a position error. If the calibration is wrong, the whole scan calibration is wrong.

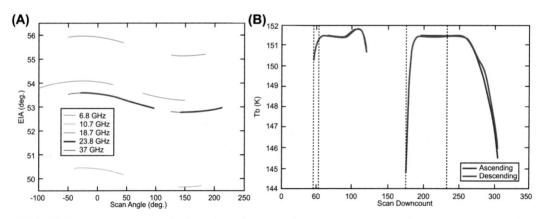

FIGURE 4.107

(A) Variation of the Earth's incidence angle (EIA) along the swath position. (B) Variation of the antenna temperature near the swath edges due to the collection of sky radiation scattered in the sky mirror (http://www.jcsda.noaa.gov/documents/meetings/WARSO2007/BettenhausenWindSat08061300.pdf).

Both the electronically scanning microwave radiometer (EMSR) and the SMMR used the "*drop-in-the-bucket*" technique to assign a radiometer value to a given pixel. This technique consisted of assigning to the central pixel the average value, which was averaged later with all the values in a single day. After analyzing more sophisticated techniques, that took into account the distance to the central pixel and performed interpolations with adjacent pixels, it was found that the improvement was not significant, while the computational load increased significantly. Finally, the "drop-in-the-bucket" technique was also used in the SSM/I. Geolocation errors can be important, even higher than 20 or 30 km (about half pixel, though!). Sources of errors are data processing software errors, inaccurate satellite ephemeris due to satellite tracking errors and, orbit prediction (specially under intense solar activity periods), and alignment error between SSM/I's boresight and instrument's boresight. The correction is performed by computing the average and filters the satellite ephemeris in a 7-day period centered in the orbit to be processed. The final accuracy achieved is $\sim$5 km.

4.6.4 SYNTHETIC APERTURE RADIOMETERS

This section deals with the novel imaging radiometers in which the image is formed by the cross-correlation of the signals collected by a number of antennas, as opposed to the pseudo-correlation radiometers in which the signals being correlated were collected by the same antenna.

In synthetic aperture radiometers the antennas can have either a wide beam or a narrow one, either in one or in two directions, and in some cases, antenna movement is required to achieve angular resolution by some sort of matched filtering. At present, there is only one satellite mission carrying a synthetic aperture radiometer, it is the ESA's SMOS mission and the MIRAS payload. This new technique is now being pursued in several other missions such as the NASA/JPL PATH and the Chinese Academy of Sciences GIMS.

As a reminder, Eq. (4.70) provides the expression of the cross-correlation (V_{mn}^{pq}) of the signals $b_{m,n}^{p,q}(t)$ collected by a pair of antennas m and n at positions $(x_{m,n}, y_{m,n}, z_{m,n})$, at p- and q-polarizations, respectively, and pointing towards the $+z$ direction:

$$
V_{mn}^{pq} \cong V^{pq}(u_{mn}, v_{mn}, w_{mn}) \cong \frac{1}{k_B \sqrt{B_m B_n} \sqrt{G_m G_n}} \cdot \frac{1}{2} \left\langle b_m^p(t) b_m^q * (t) \right\rangle
$$

$$
= \frac{1}{\sqrt{\Omega_m \Omega_n}} \iint_{\xi^2 + \eta^2 \le 1} \frac{T^{pq}(\xi, \eta) - T_p h \delta_{mn}}{\sqrt{1 - \xi^2 - \eta^2}} F_m(\xi, \eta) F_n^*(\xi, \eta)
$$

$$
\cdot \widetilde{r}_{mn} \left(-\frac{u_{mn}\xi + v_{mn}\eta + w_{mn}\sqrt{1 - \xi^2 - \eta^2}}{f_0} \right)
$$

$$
\cdot \exp\left(-j2\pi \left(u_{mn}\xi + v_{mn}\eta + w_{mn}\sqrt{1 - \xi^2 - \eta^2} \right) \right) d\xi d\eta. \tag{4.70}
$$

The terms in Eq. (4.70) have already been previously defined and are not be repeated here. Eq. (4.70) provides the basic observable of the different types of synthetic aperture radiometers that are presented in the next Sections.

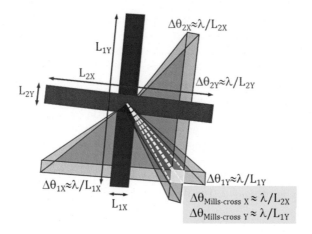

FIGURE 4.108

Principle of operation of the Mills cross radiometer.

4.6.4.1 Types of Synthetic Aperture Radiometers
4.6.4.1.1 Mills Cross

The Mills Cross is one of the simplest and oldest sparse-aperture microwave radiometers (Christiansen and Hogbom, 1985). It consists of two antennas forming a cross, as sketched in Fig. 4.108. Each antenna is very large in one dimension, thus generating a very narrow beam[11] ($\Delta\theta_{1Y} \approx \lambda/L_{1Y}$ and $\Delta\theta_{2X} \approx \lambda/L_{2X}$), while they are narrow in the other dimension, thus generating a very wide beam (($\Delta\theta_{1X} \approx \lambda/L_{1X}$ and $\Delta\theta_{2Y} \approx \lambda/L_{2Y}$). These beams that are very wide in one direction and narrow in the other one are called "fan beams." The "pencil beam" (narrow in both dimensions: $\Delta\theta_{1Y} \approx \lambda/L_{1Y}$ and $\Delta\theta_{2X} \approx \lambda/L_{2X}$) is formed from the cross-correlation of the signals collected by both antennas. Since the thermal noise signal is spatially uncorrelated, the only net contribution to the cross-correlation is coming from the thermal noise being collected by both antennas simultaneously, that is from the noise coming from the intersection of the two fan beams. Contributions coming from other directions do contribute to the detected output noise, but not to the cross-correlation. This is equivalent to a reduction of the measured antenna temperature (absolute value of the cross-correlation) by a factor:

$$\frac{\Omega_{Mills-cross}}{\sqrt{\Omega_1 \cdot \Omega_2}} = \frac{\dfrac{\lambda^2}{L_{1Y} \cdot L_{2X}}}{\sqrt{\dfrac{\lambda^2}{L_{1X} \cdot L_{1Y}} \cdot \dfrac{\lambda^2}{L_{2X} \cdot L_{2Y}}}} = \sqrt{\frac{L_{1Y}}{L_{1X}} \cdot \frac{L_{2X}}{L_{2Y}}}, \tag{4.116}$$

and an increase in the radiometric precision $\sigma_V = \sqrt{\sigma_{V_r}^2 + \sigma_{V_i}^2}$ (Eq. 4.82) by the inverse of that factor.

This is again, the radiometer's uncertainty principle. The amount gained in angular resolution is exactly the same factor the radiometric precision is degraded.

---

[11]Recall from Section 4.6.2.1 that the angular resolution is proportional to the electromagnetic wavelength and inversely proportional to the antenna size.

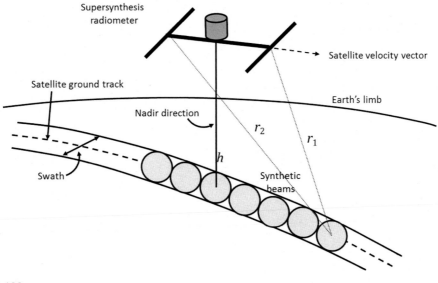

FIGURE 4.109

Concept of a supersynthesis radiometer or RadSAR. Long stick antennas are used to spatially filter a dwell line. Swath can be increased with a multibeam antenna as in a push-broom configuration.

4.6.4.1.2 Synthetic Aperture Radiometers using Matched Filtering

In the early 1970s the concept of space−time matched filtering was conceived in the former USSR (Melnik, 1972a,b), but it was not until the first half of the 1990s that the so-called supersynthesis radiometer (Komiyama et al., 1994, 1997; Komiyama and Kato, 1995, 1996; Komiyama, 1998) and the RadSAR (Edelson, 1994; Edelson et al., 1998) concepts were developed in Japan and in the United States. Both concepts are quite similar, in the sense that both require the movement of the antenna to form a brightness temperature image after matched filtering of the measured visibility function (as a function of time) with the impulse response of a point source in each single position of FOV to achieve spatial resolution.

If the antenna array is moving, this is equivalent to having a brightness temperature image $T^{pq}(\xi,\eta,t)$ moving in the opposite direction. For simplicity, it will be assumed that the array is a linear array at a height $z_{m,n} = h$, moving along the X-axis at constant speed v, and that $T^{pq}(\xi,\eta)$ is a point source at $(x_0,0,0)$, as illustrated in Fig. 4.109. It will be assumed as well that antenna patterns are identical, and that fringe-washing effects are negligible, which means that the maximum antenna spacing is much smaller than the coherence distance c/B.

Under these conditions, Eq. (4.70) can be rewritten in Cartesian coordinates as:

$$V^{pg}(u_{mn}, v_{mn}, t) = \frac{1}{\sqrt{\Omega_m \Omega_n}} \iint\limits_{\xi^2+\eta^2\leq 1} \frac{T^{pq}(\xi, \eta, t) - T_{ph}\delta_{mn}}{\sqrt{1 - \xi^2 - \eta^2}} F_m(\xi, \eta) F_n^*(\xi, \eta)$$

$$\cdot \tilde{r}_{mn}\left(-\frac{u_{mn}\xi + v_{mn}, \eta + w_{mn}\sqrt{1 - \xi^2 - \eta^2}}{f_0}\right)$$

$$\cdot \exp\left(-j2\pi\left(u_{mn}\xi + v_{mn}\eta + w_{mn}\sqrt{1 - \xi^2 - \eta^2}\right)\right) d\xi d\eta$$

$$= \frac{1}{\sqrt{\Omega_m \Omega_n}} \int_{-\infty}^{+\infty} \int_{-\infty}^{+\infty} \{T^{pq}(x, y, t) - T_{ph}\delta_{mn}\}$$

$$\cdot F_m(x, y) F_n^*(x, y) \exp\left(-j\frac{2\pi}{\lambda}(r_2(t)) - (r_1(t))\right) dx dy$$

$$= \frac{1}{\sqrt{\Omega_m \Omega_n}} T^{pq}(x_0, y_0) |F(x_0, y_0)|^2 \exp\left(-j\frac{2\pi}{\lambda}\left(\sqrt{(x_2 + vt - x_0)^2 + h^2}\right)\right.$$

$$\left. -\sqrt{(x_1 + vt - x_0)^2 + h^2}\right) dx dy \tag{4.117}$$

where $r_{1,2}(t)$ are the distances from the pixel being focused on the antennas. In Eq. (4.117) the phase of the "instantaneous visibility function" is:

$$\varphi(t) = -\frac{2\pi}{\lambda}\left(\sqrt{(x_2 + v\cdot t - x_0)^2 + h^2} - \sqrt{(x_1 + v\cdot t - x_0)^2 + h^2}\right), \tag{4.118}$$

from which the instantaneous frequency can be defined as:

$$f_{inst}(t) = -\frac{1}{2\pi}\cdot\frac{\partial\varphi(t)}{\partial t}, \tag{4.119}$$

although it should be estimated in practice from discrete differences from consecutive measurements $t_n = n\cdot\tau$:

$$f_{inst}(t_n) = -\frac{1}{2\pi}\cdot\frac{\varphi_V(t_n) - \varphi_V(t_{n-1})}{\tau}. \tag{4.120}$$

As in a synthetic aperture radar, the image focusing consists of applying a matched filter so as to compensate the phase modulation history of each target in the scene, while the phase of other targets is not, and their contribution vanishes during the integration. We present an example of the phase history of two targets at 0 and 25 km in Fig. 4.110, as would be observed by a supersynthesis radiometer at 800 km height, at 7.5 km/s speed, with a baseline of 4 m at a frequency of 1.413 GHz.

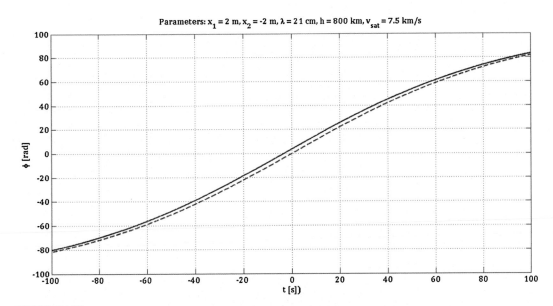

Parameters: $x_1 = 2$ m, $x_2 = -2$ m, $\lambda = 21$ cm, $h = 800$ km, $v_{sat} = 7.5$ km/s

FIGURE 4.110

Unwrapped phase history of the cross-correlation corresponding to two targets at positions $x_0 = 0$ km (red) and $x_0 = +25$ km (blue), which allows us to focus it and isolate it from the rest.

As in a synthetic aperture radar, the wider the elementary antenna beam, the longer the integration time, and the better the achievable angular resolution, which is given by Komiyama et al. (1994, 1997), Komiyama and Kato (1995, 1996), Komiyama (1998):

$$\varphi_s = 0.886 \frac{1}{d \cdot \varphi_A}, \tag{4.121}$$

where d is the baseline (antenna spacing), and ϕ_A is the antenna beam width in the along-track direction (the direction in which the synthetic aperture is formed). The factor 0.886 is due to the uniform weighting (windowing) of the visibility samples during the focusing, but may be larger if they are tapered to reduce the amplitude of the side lobes of the synthetic pattern.

One consideration that has to be taken into account is that the maximum baseline ($d = x_1 - x_2$) is smaller than the coherence distance (c/B) so that the correlation is not lost. One way to increase the maximum baseline is to reduce the bandwidth, but this reduces the radiometric precision. Alternatively, the integration (focusing) can be performed over a shorter period of time so that the coherence is preserved:

$$\left| t \right| \leq \frac{h \cdot \sin \theta_{i,\max}}{v}, \tag{4.122}$$

where:

$$\sin \theta_{i,\max} = \frac{c}{d \cdot B},\qquad(4.123)$$

although this translates into a poorer angular resolution. Another possibility consists of applying subbanding, i.e., dividing the original band into smaller ones so that the above condition is met.

Although computationally far more complex, an alternative solution consists of the use a variable true-time delay for each pixel to be focused so that the maximum baseline is no longer limited by the coherence distance. This concept is the "**Doppler-delay radiometer**" (Camps and Swift, 2001) which uses just three antennas, which was later extended to a larger number of antennas motion induced synthetic aperture radiometer (MISAR) to recover part of the radiometric resolution lost (Park and Kim, 2009a,b).

4.6.4.1.3 Synthetic Aperture Radiometers using Fourier Synthesis

As real aperture microwave radiometers, synthetic aperture microwave radiometers were first devised in radio-astronomy as a means to achieve high angular resolution, with "moderate size" antennas (Thompson et al., 1986). The very large array (VLA), a turning point in radio-astronomy using synthetic aperture microwave radiometers is depicted in Fig. 4.111.

FIGURE 4.111

(A) Aerial view of the very large array (VLA), New Mexico, Socorro (United States), and (B) view of the "elementary" antennas forming the VLA array.

Courtesy of NRAO.

(A)

(B)

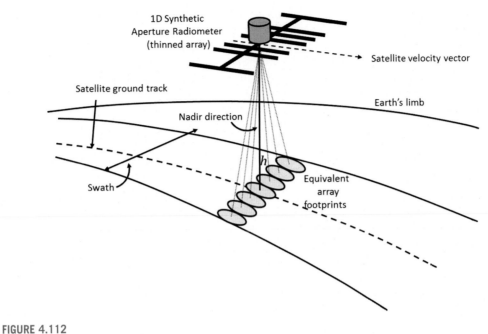

FIGURE 4.112

1D aperture synthesis radiometer scanning configuration is similar to a push-broom.

Its application to Earth observation was first proposed by LeVine and Good in 1983. This led to the concept of the ESTAR instrument (illustrated in Fig. 4.112) that performed aperture synthesis in the cross-track dimension and achieved angular resolution in the along-track dimension by means of long antennas. The concept was demonstrated in the late 1980s with the ESTAR instrument prototype designed and built at the University of Massachusetts—Amherst (Tanner, 1990; Tanner and Swift, 1993; Ruf et al., 1988).

Despite the fact that the main concept is similar, the differences between radio-astronomy and Earth observation are important:

- In radio-astronomy the antenna spacing is very large and the antennas are very directive, this decreases the antenna coupling, which allows the use of simpler image reconstruction approaches. Antenna coupling distorts the antenna voltage radiation pattern, which will have to be very well characterized and corrected for during the image reconstruction process. Differences in the antenna patterns have proven to be one of the main limitations in the accuracy error budget of synthetic aperture radiometers.
- In radio-astronomy, the FOV is very narrow, as opposed to Earth observation, in which the FOV may easily be $60°$ from a LEO.
- Due to the narrow FOV ($\theta \approx 0°$), in radio-astronomy the obliquity factor $(1/\cos\theta = 1/\sqrt{1 - \xi^2 - \eta^2})$ can be approximated by 1.

- In addition, due to the narrow FOV, antenna patterns are approximately constant (amplitude and phase) over the FOV.
- In radio-astronomy the brightness temperature scenes are typically quasipoint sources imaged over a cold background, this allows us to use super-resolution image reconstruction algorithms, as opposed to Earth observation, in which the brightness temperatures have very low contrast, occupy the whole FOV, or even more, and may not even exhibit the Earth−sky contrast due to aliasing during the imaging process.

After the success of ESTAR, ESA pushed for and evolved a concept that performed 2D aperture synthesis. It was the MIRAS instrument that ended up being the single payload of ESA's SMOS Earth Explorer Opportunity Mission (Font et al., 2010; Kerr et al., 2010). We illustrate this concept in Fig. 4.113.

The principles of operation of 1D and 2D synthetic aperture radiometers are quite similar, although there are some differences. The fundamentals were already presented in Section 4.6.2.2.2, but are briefly revised here, since they have implications in the selection of the array topology.

The basic observable is the so-called visibility function V_{mn}^{pq}, measured as the cross-correlation of each pair of signals $b_{m,n}^{p,q}(t)$ collected by the array antennas m and n, at p- and q-polarizations, located over the z plane (i.e., $z=0$). Since the antennas occupy specific locations in the array, the visibility function is not measured continuously, but sampled at particular wave numbers (u_{mn}, v_{mn}) determined by the antenna positions, normalized to the wavelength.

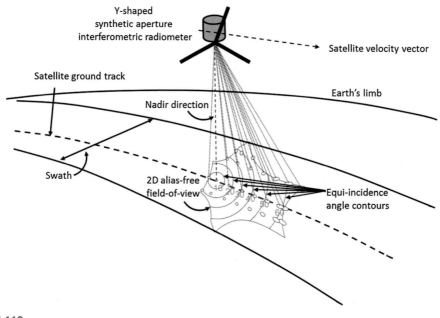

FIGURE 4.113

2D aperture synthesis radiometer imaging configuration as in microwave imaging radiometer by aperture synthesis in Soil Moisture and Ocean Salinity.

$$V_{mn}^{pq} \triangleq V^{pq}(u_{mn}, v_{mn}) = \frac{1}{k_B \sqrt{B_m B_n} \sqrt{G_m G_n}} \frac{1}{2} \langle b_m^p(t) b_n^{q*}(t) \rangle, \tag{4.124}$$

with:

$$(u_{mn}, v_{mn}) = (x_n - x_m, y_n - y_m)/\lambda_0 = (\Delta x_{mn}, \Delta y_{mn})/\lambda_0. \tag{4.125}$$

The brightness temperature

$$T^{pq}(\xi, \eta) = \frac{F_{n1}(\xi, \eta) F_{n_2}^*(\xi, \eta)}{\sqrt{\Omega_1 \Omega_2}} \frac{T_B(\xi, \eta) - T_{ph\,rec}}{\sqrt{1 - \xi^2 - \eta^2}}, \tag{4.126}$$

with the director cosines defined as:

$$(\xi, \eta) = (\sin \theta \cos \phi, \sin \theta \sin \phi), \tag{4.127}$$

can be then reconstructed in the ideal case (identical antenna patterns, negligible spatial decorrelation effects, and no antenna positioning errors), as a 1D or 2D Fourier process:

$$T^{pq}(\xi, \eta) = F^{-1}[V^{pq}(u_{mn}, v_{mn})]. \tag{4.128}$$

Due to the crucial impact of this in the array topology, antenna spacing selection, number of complex correlators, etc., this will be analyzed in more detail in the following sections for the 1D and 2D cases.

4.6.4.1.3.1 1D Synthetic Aperture Radiometers: Array Thinning.

In a full linear array of N antennas equally spaced d wavelengths, $V(u) = V(u, v = 0)$ can be measured N times at $u = 0$, $N - 1$ times at $u = 1d$, $N - 2$ times at $u = 2d$; etc. and only once at $u_{max} = (N - 1)d$. This is obviously an inefficient use of the available antennas, since most of the measurements are redundant. In 1955, Arsac found that greater efficiency can be achieved by spacing the N antennas out in such a way that the greatest multiple of d, $u_{max} = N_{max} \cdot d$, is greater than $(N - 1)d$ and all multiples up to u_{max} are also present.

The so-called "restricted" minimum-redundancy problem is to find, for a given N, a spacing pattern of this kind that has maximum efficiency (minimum redundancy). The "general" problem allows N_{max} to be greater than the number M for which all multiples of d up to $M \cdot d$ are present. The redundancy R is quantitatively defined as the number of pairs of antennas divided by N_{max} (Moffet, 1968):

$$R = \frac{1}{2} \cdot \frac{N \cdot (N - 1)}{N_{max}}, \tag{4.129}$$

For $N \leq 4$ Arsac (1955) found arrays with $R = 1$ ("zero-redundancy" linear arrays or ZRLAs), and Bracewell (1966) proved that these are the only ZRLAs. In 1956, Leech provided some optimum solutions for $N \leq 11$ and demonstrated that $1.217 \leq R \leq 1.332$ for $N \to \infty$. Although the solutions provided for $N \leq 11$ were not zero-redundant, the redundancy was the lowest possible ($R = R_{opt} > 1$: minimum-redundancy linear arrays, MRLA). For larger values of $N > 11$, the optimum solutions have not been found, but some semiempirical and numerical array patterns that approach Leech's bounds (low-redundancy linear arrays, LRLA) have been found (Leech, 1956; Ishiguro, 1980; Ruf, 1993; Trucco et al., 1997; Rossouw et al., 1997).

Table 4.8 from Camps et al. (2002) summarizes the ZRLA, MRLA, and LRLA with some new LRLA found in Camps et al. (2002). The nomenclature used to denote an array of N antennas is a bracketed list of $N - 1$ numbers $\{x; y; \ldots; z\}$ indicating the spacing between adjacent antennas. For

Table 4.8 Comparison of Existing Zero-Redundancy Linear Array (ZRLA), Minimum-Redundancy Linear Array (MRLA), and Low-Redundancy Linear Array (LRLA) With New LRLA Obtained by a Recursive Method Proposed in Camps et al. (2002)

Number of Antennas (N)	Maximum Array Spacing (N_{max})	Pattern	Redundancy (R)	Comments (Array Family)
1	0	{0}	1.00	ZRLA [1]
2	1	{1}	1.00	ZRLA [1]
3	3	{1,2}	1.00	ZRLA [1] (A)
4	6	{1,3,2}	1.00	ZRLA [1] (A)
5	9	{1,3,3,2}	1.11	MRLA [1,3] (A)
		{1,1,4,3}		(B)
6	13	{1,1,4,4,3}	1.15	MRLA [l,3] (B)
		{1,5,3,2,2}		(C$_1$)
		{1,3,1,6,2}		
7	17	{1,1,4,4,4,3}	1.24	MRLA [1,3] (B)
		{1,7,3,2,2,2}		(C$_1$)
		{1,3,6,2,3,2}		(D)
		{1,1,6,4,2,3}		
		{1,1,6,4,3,2}		
		{1,1,1,5,5,4}		
8	23	{1,1,9,4,3,3,2}	1.22	MRLA [1.3] (C$_2$)
		{1,3,6,6,2,3,2}		(D)
9	29	{1,2,3,7,7,4,4,1}	1.24	MRLA [1,3] (E)
		{1,1,12,4,3,3,3,2}		(C$_2$)
		{1,3,6,6,6,2,3,2}		(D)
10	36	{1,2,3,7,7,7,4,4,1}	1.25	MRLA [1,3] (E)
11	43	{1,2,3,7,7,7,7,4,4,1}	1.28	MRLA [1,3] (E)
12	50	{1,2,3,7,7,7,7,7,4,4,1}	1.32	LRLA [6]
		{1,1,1,20,5,4,4,4,4,3,3}		New LRLA
13	58	{1,2,3,11,3,7,8,10,4,4,4,1}	1.34	LRLA [6]
		{1,1,1,24,5,4,4,4,4,4,3,3}		New LRLA
		{1,4,3,4,9,9,9,9,5,1,2,2}		New LRLA
		{1,3,7,3,9,9,9,9,2,4,1,1}		New LRLA
		{1,1,6,7,1,10,10,10,3,4,2,3}		New LRLA
14	68	{1,1,6,6,6,11,11,11,5,5,3,1,1}	1.34	LRLA [6]
		{1,1,6,7,1,10,10,10,10,3,4,2,3}		New LRLA
15	79	{1,1,6,6,6,11,11,11,11,5,5,3,1,1}	1.33	LRLA [6]
16	90	{1,1,6,6,6,11,11,11,11,11,5,5,3,1,1}	1.33	LRLA [6]
17	101	{1,1,6,6,6,11,11,11,11,11,11,5,5,3,1,1}	1.35	LRLA [6]
18	112	{1,1,6,6,6,11,11,11,11,11,11,11,5,53,1,1}	1.37	LRLA [6]

Table 4.8 Comparison of Existing Zero-Redundancy Linear Array (ZRLA), Minimum-Redundancy Linear Array (MRLA), and Low-Redundancy Linear Array (LRLA) With New LRLA Obtained by a Recursive Method Proposed in Camps et al. (2002)—cont'd

Number of Antennas (N)	Maximum Array Spacing (N_{max})	Pattern	Redundancy (R)	Comments (Array Family)
19	121	{1,1,1,1,1,45,8,7,6,6,6,6,6,6,6,5,5,4}	1.41	LRLA [6]
	123	{1,1,6,6,6,11,11,11,11,11,11,11,11,5,5,!,1,1}	1.39	New LRLA[b]
20	133	{1,1,2,5,7,7,1,13,13,13,13,13,13,13,6,6,4,1,1}	1.43	LRLA [6]
	134	{1,1,6,6,6,11,11,11,11,11,11,11,11,11,5,5,3,1,1}	1.42	New LRLA[b]
21	145	{1,1,1,1,1,1,53,9,6,5,8,7,7,7,7,7,4,6,6}	1.45	LRLA [6]
		{1,2,1,1,1,5,5,5,14,14,14,14,14,14,14,9,4,9,3,1}		New LRLA
22	160	{1,2,6,6,8,1,13,13,8,5,13,13,13,13, 13,13,5,2,5,4,3}	1.44	LRLA [6][a]
	159	{1,2,1,1,1,5,5,5,14,14,14,14,14,14, 14,14,9,4,9,3,1}	1.45	New LRLA[b]
		{1,1,1,1,1,1,60,9,6,5,8,7,7,7,7,7,7,4,6,6}		New LRLA[b]
23	173	{1,2,1,1,1,5,5,5,14,14,14,14,14,14,14, 14,14,9,4,9,3,1}	1.46	LRLA [6]
24	188	{1,2,3,1,1,5,2,5,7,16,16,16,16,16,16,16, 16,9,9,4,7,2,2}	1.47	LRLA [6]
	191	{1,1,1,1,8,1,9,9,9,17,17,17,17,17,17,17, 8,8,8,5,1,1,1}	1.45	New LRLA[b]
25	208	{1,1,1,1,8,1,9,9,9,17,17,17,17,17,17,17,17, 8,8,8,5,1,1,1}	1.44	LRLA [6]
26	225	{1,1,1,2,3,8,8,8,17,17,17,17,17,17,17,17,17,9, 9,1,9,9,1,1,1}	1.44	LRLA [6]
27	236	{1,1,2,6,3,6,6,16,16,16,16,16,16,16,16,16,16, 16,10,3,4,3,7,3,4,1}	1.49	LRLA [6]
	242	{1,1,1,2,3,8,8,8,17,17,17,17,17,17,17,17,17, 17,9,9,1,9,9,1,1,1}	1.45	New LRLA[b]
28	257	{1,1,1,5,8,7,1,7,17,17,17,17,17,17,17, 17,17,17,17,9,9,9,6,3,1,1,1}	1.47	LRLA [6]
	259	{1,1,1,2,3,8,8,8,17,17,17,17,17,17, 17,17,17,17,17,9,9,1,9,9,1,1,1}	1.46	New LRLA[b]
29	270	{1,1,3,2,2,20,2,1,19,19,19,19,19,19, 19,19,19,19,16,1,4,3,9,1,3,8,2,1}	1.50	LRLA [6]
	276	{1,1,1,2,3,8,8,8,17,17,17,17,17,17,17,17, 17,17,17,17,9,9,1,9,9,1,1,1}	1.47	New LRLA[b]
30	287	{1,1,12,2,6,6,8,6,19,19,19,19,19,19,19, 19,19,19,19,11,2,5,6,3,2,2,3,1,1}	1.52	LRLA [6]
	293	{1,1,1,2,3,8,8,8,17,17,17,17,17,17,17,17, 17,17,17,17,9,9,1,9,9,1,1,1}	1.48	New LRLA[b]
31	306	{1,1,12,2,6,6,8,6,19,19,19,19,19,19, 19,19,19,19,19,19,11,2,5,6,3,2,2,3,1,1}	1.52	New LRLA

Continued

Table 4.8 Comparison of Existing Zero-Redundancy Linear Array (ZRLA), Minimum-Redundancy Linear Array (MRLA), and Low-Redundancy Linear Array (LRLA) With New LRLA Obtained by a Recursive Method Proposed in Camps et al. (2002)—cont'd

Number of Antennas (N)	Maximum Array Spacing (N_{max})	Pattern	Redundancy (R)	Comments (Array Family)
	310	{1,1,1,2,3,8,8,8,17,17,17,17,17,17,17, 17,17,17,17,17,17,9,9,1,9,9,1,1,1}	1.50	New LRLA[b]
32	325	{1,1,12,2,6,6,8,6,19,19,19,19,19,19,19,19, 19,19,19,19,19,11,2,5,6,3,2,2,3,1,1}	1.53	New LRLA
	327	{1,1,1,2,3,8,8,8,17,17,17,17,17,17,17,17,17,17, 17,17,17,17,9,9,1,9,9,1,1,1}	1.52	New LRLA[b]
33	344	{1,1,12,2,6,6,8,6,19,19,19,19,19,19,19,19, 19,19,19,19,19,19,11,2,5,6,3,2,2,3,1,1}	1.53	New LRLA[b]
	344	{1,1,1,2,3,8,8,8,17,17,17,17,17,17,17, 17,17,17,17,17,17,17,17,9,9,1,9,9,1,1,1}	1.53	New LRLA[b]
34	363	{1,1,12,2,6,6,8,6,19,19,19,19,19,19,19,19,19, 19,19,19,19,19,11,2,5,6,3,2,2,3,1,1}	1.55	New LRLA[b]
	361	{1,1,1,2,3,8,8,8,17,17,17,17,17,17,17, 17,17,17,17,17,17,17,17,17,9,9,1,9,9,1,1,1}	1.55	New LRLA
35	382	{1,1,12,2,6,6,8,6,19,19,19,19,19,19,19,19, 19,19,19,19,19,19,19,19,11,2,5,6,3,2,2,3,1,1}	1.56	New LRLA[b]
	378	{1,1,1,2,3,8,8,8,17,17,17,17,17,17,17,17,17,17, 17,17,17,17,17,17,17,17,9,9,1,9,9,1,1,1}	1.57	New LRLA
36	401	{1,1,12,2,6,6,8,6,19,19,19,19,19,19,19,19,19,19, 19,19,19,19,19,19,19,11,2,5,6,3,2,2,3,1,1}	1.57	New LRLA[b]
	395	{1,1,1,2,2,3,8,8,8,17,17,17,17,17,17,17,17,17, 17,17,17,17,17,17,17,17,17,9,9,1,9,9,1,1,1}	1.59	New LRLA
37	420	{1,1,12,2,6,6,8,6,19,19,19,19,19,19,19,19,19,19, 19,19,19,19,19,19,19,19,11,2,5,6,3,2,2,3,1,1}	1.59	New LRLA[b]
	412	{1,1,1,2,3,8,8,8,17,17,17,17,17,17,17,17,17,17, 17,17,17,17,17,17,17,17,17,9,9,1,9,9,1,1,1}	1.62	New LRLA

[a]There is a typographical error in (Ruf, 1993), since there are 12 antenna spacing missing in the pattern given for N = 22, N_{max} = 160.
[b]New and largest LRLAs with the least redundancies reported up to date.

example, the ZRLA {1,3,2} is a four-antenna array that looks like xxooxox (Fig. 4.114A) where crosses indicate positions occupied by an antenna, and circles are empty positions. The MRLA {1,3,3,2} is a five-antenna array that looks like xxooxooxox (Fig. 4.114B). As we can clearly see, just adding one more antenna the maximum antenna spacing passes from 6d up to 9d, adding just one redundant baseline (3d), indicated with a dashed arrow. Adding another antenna will increase the maximum antenna spacing up to 13d, which illustrates the advantages of the array thinning in terms of hardware savings.

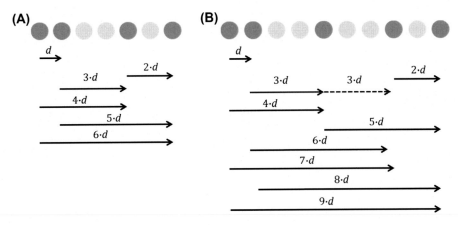

FIGURE 4.114

(A) Zero-redundancy linear array (ZRLA) {1,3,2} with 4 antennas and a maximum baseline = 6d, and (B) minimum-redundancy linear array (MRLA) {1,3,3,2} with 5 antennas, maximum baseline 9d and only one redundant baseline (3d). *Dark gray circles* indicate the antenna positions. *Light gray circles* indicate the positions where there is no longer an antenna. In a full array ZRLA {1,3,2} will have 7 antennas, while MRLA {1,3,3,2} will have 10 antennas.

It is worth noting that the effective length of a synthetic array is twice the maximum antenna separation since $\langle b_m(t) \cdot b_n^*(t) \rangle = \langle b_n(t) \cdot b_m^*(t) \rangle^*$. Therefore the angular resolution is roughly $\sim \lambda/(2 \cdot N_{max} \cdot d)$ instead of $\sim \lambda/(N_{max} \cdot d)$, at the expense of higher side lobes (twice higher in dBs).

4.6.4.1.3.2 2D Synthetic Aperture Radiometers: Array Topologies. In 2D synthetic aperture radiometers, there is not much interest in thinning the array due to the inherent low level of redundancy. For example, a Y-shaped array, such as the one in Fig. 4.115A, with N elements in each arm, and one in the center ($3N + 1$ in total), will have N_{NRV} non-Hermitian (u,v) points:

$$N_{NRV} = 3N^2 + 3N + 1, \tag{4.130}$$

while the total number of independent cross-correlations is:

$$N_{RV} = \frac{1}{2} \cdot (3N + 1) \cdot 3N. \tag{4.131}$$

As in the definition of the 1D case, it should be noted that in Eqs. (4.130) and (4.131) Hermitian visibilities $(V(-u,-v) = V*(u,v))$ have not been accounted for since $\langle b_m(t) \cdot b_n^*(t) \rangle = \langle b_n(t) \cdot b_m^*(t) \rangle^*$.

The level of redundancy is:

$$R = \frac{\frac{1}{2} \cdot (3N + 1) \cdot 3N}{3N^2 + 3N + 1}, \tag{4.132}$$

which tends to 3/2 as $N \to \infty$, as shown in Fig. 4.115B.

The shape of the array determines the sampling of the spatial frequency plane, and how the aliases will be repeated. The "aliases" are periodic repetitions of the main image that are formed during the

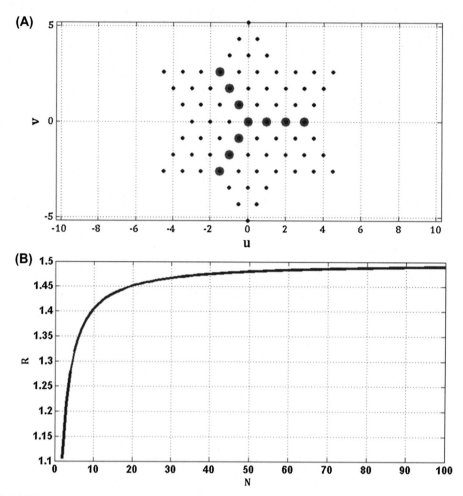

FIGURE 4.115

(A) Y-array with 10 elements, 3 elements per arm (*red circles*), leading to 73 (u,v) points, from which 36 correspond to non-Hermitian visibilities plus the zero baseline (Eq. 4.130). (B) Redundancy level (Eq. 4.132) as a function of N the number of elements per arm.

Fourier synthesis process. Arrays whose arms are oriented forming $90°$ (e.g., T- or U-shaped arrays) will lead to (u,v) points distributed over a rectangular grid, and alias images also centered over a rectangular grid (Fig. 4.116A). Arrays whose arms are oriented $120°$ apart (e.g., a Y-, or a Δ-, or a hexagonal-array) will lead to (u,v) points distributed over a hexagonal grid, and alias images also centered over a hexagonal grid (Fig. 4.116B).

Antenna spacing dictates the separation of the alias images in the (ξ,η) domain. Since the whole semispace in front of the antenna array is mapped into the unit circle, that is $0° \leq \theta \leq 90°, 0° \leq \varphi < 360°$ map into $\xi^2 + \eta^2 \leq 1$, for rectangular sampling the minimum antenna spacing to avoid aliasing is $d = \lambda/2$ (Fig. 4.116A), while for hexagonal sampling it is

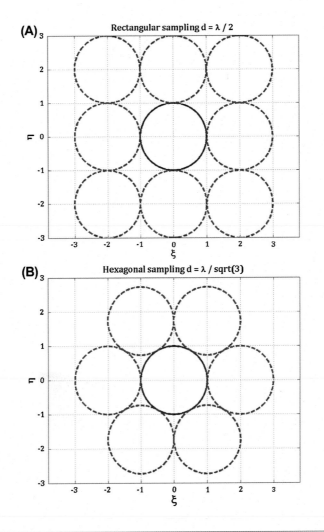

FIGURE 4.116

Unit circle (blue) where the whole brightness temperature in front of the array is mapped, and alias images of the unit circle (red), for (A) rectangular sampling ($d = \lambda/2$) and (B) hexagonal sampling ($d = \lambda/\sqrt{3}$).

$d = \lambda/\sqrt{3} = 0.577\lambda$ (Fig. 4.116B). This translates into a 15.5% savings in terms of antennas and receivers (0.577λ as compared to 0.5λ), and a 33.3% ($=1.155^2$) in correlators as compared to the rectangular sampling case. This can be intuitively understood as the savings in the space between the circles that do not correspond to any physical direction. Actually, since the physical space maps into the unit circle, it can be demonstrated that hexagonal sampling is the optimum one, and from all array topologies, the Y-shaped array is the one providing the largest (u,v) coverage, and therefore the best angular resolution (Camps et al., 1997).

In practice, the antenna spacing can be larger than the minimum required not to have aliasing, since the FOV does not have to cope with the whole unit circle, for example, in SMOS (Font et al., 2010; Kerr et al., 2010), the ground incidence angle can be smaller than 90° ($d = 0.875\lambda$, Fig. 4.117), or in

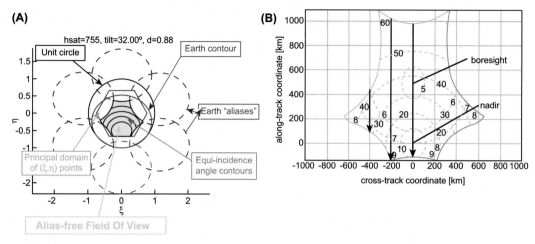

FIGURE 4.117

Aliasing in the microwave imaging radiometer by aperture synthesis instrument aboard European Space Agency's Soil Moisture and Ocean Salinity mission. (A) Antenna spacing is $d = 0.875\lambda$, which induces some level of aliasing, but despite it, (B) the swath width is ~ 1000 km and the range of incidence angles extends $0° - \sim 60°$.

PATH (https://directory.eoportal.org/web/eoportal/satellite-missions/g/geostar) or in GAS (Christensen et al., 2007) the Earth will be seen from a geostationary orbit and therefore the Earth disk occupies only a disk of radius 0.3 ($0° \leq \theta \leq 17°, 0° \leq \varphi < 360°$, that is $\xi^2 + \eta^2 \leq 0.3^2$).

Other array configurations exist, including nonuniform antenna spacing (Camps et al., 2008), circular array such as in GIMS (Zhang et al., 2013), random arrays, etc., but they have to be analyzed numerically.

4.6.4.1.3.3 Other Synthetic Aperture Radiometer Concepts. Other synthetic aperture microwave radiometer concepts involve, for example, 1D or 2D synthetic aperture radiometers such as ESTAR or MIRAS, with a certain tilt angle and with a conical scan so as to enlarge the swath width, while, potentially, reducing the range of incidence angles. These concepts have been studied by Martin-Neira and Font-Rosello (1997) and are sketched in Fig. 4.118.

Other concepts involve 1D synthetic aperture radiometry and frequency scanned antennas, as illustrated in Fig. 4.119.

4.6.4.2 Radiometer Calibration

The calibration of a synthetic aperture radiometer is a very complex process, but conceptually, it is similar to that of real aperture radiometers, involving two steps:

- An internal calibration first, prior to the image reconstruction, to correct for the relative errors of each visibility sample (cross-correlation) to prevent image aberration, and
- An external calibration to set the absolute value and the gain (equivalent to the "brightness" and "contrast" of an image), and involves the measurement of external known targets.

These procedures are explained in the next sections.

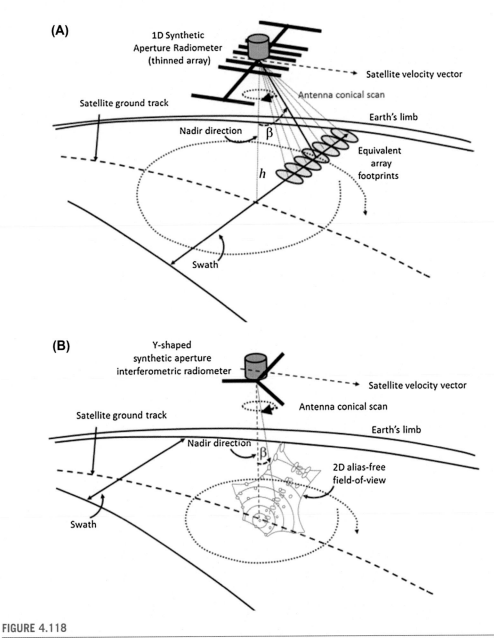

FIGURE 4.118

(A) 1D and (B) 2D synthetic aperture radiometers with a tilt angle β and a conical scan.

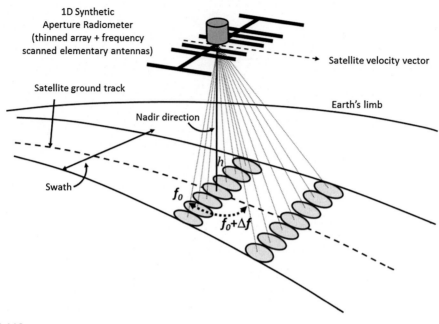

FIGURE 4.119

1D synthetic aperture radiometer: array is thinned in the cross-track direction and scanning in the along-track direction is performed by frequency scanned elementary antennas.

4.6.4.2.1 Internal Calibration

If in a real aperture radiometer (nonpolarimetric) the basic calibration equations are:

$$V_d = \frac{1}{a}(T_A - b), \tag{4.133a}$$

$$T_A = a \cdot V_d + b, \tag{4.133b}$$

the internal calibration of a synthetic aperture radiometer is a quite complex procedure and requires a detailed instrument error model. The interested reader is referred to Corbella et al. (2005) for a detailed description. For the sake of clarity, a conceptually simpler case is explained first, in which there are no quadrature errors,[12] and instrumental errors are said to be separable, that is, they can be assigned to each particular receiver forming the baseline.

The basic calibration equation (simplified case) of the baseline (complex cross-correlation) formed between the signals collected by antennas m and n of a synthetic aperture radiometer is:

$$V_{mn} = A_{mn} \cdot R_{mn} + B_{mn}, \tag{4.134}$$

---

[12]Quadrature errors refer to the error in the I/Q demodulators by which the in-phase and the quadrature components are not orthogonal. This error can be (is) frequency-dependent within the instrument's bandwidth.

where V_{mn} is the calibrated visibility sample, R_{mn} is the measured cross-correlation (after inversion of the correlator transfer function), and A_{mn} and B_{mn} are complex amplitude and offset terms. Since the observables are complex cross-correlations, the amplitude and offset can be determined when correlated and uncorrelated signals are present at receivers' inputs, respectively. Since in microwave radiometry the correlators often use just a few bits (or even just one as in MIRAS/SMOS) the measured cross-correlation is a distorted version of the cross-correlation that would be measured by an analog or a multibit correlator (e.g., Eq. 4.95 for the 1 bit/2 level case). Therefore, since the particular form of the correlation function depends on the signal statistics, it is crucial that the input signals used for calibration purposes exhibit the same statistics. This can be achieved using either true noise signals, or pseudorandom signals with a flat spectrum over the receiver's bandwidth (Ramos-Perez et al., 2009).

The offset term can be corrected for by connecting matched loads to receivers' inputs, since each matched load will be generating an uncorrelated noise signal with a power equal to $k_B \cdot T_{ph} \cdot B$.

The amplitude term can be corrected for (after proper offset correction) by injecting a common correlated noise to all receivers simultaneously. However, since the noise distribution networks or NDNs (power splitters) used to distribute the common correlated noise signal introduce correlated noise themselves as well, the common correlated noise must be of much higher level than the one generated by the NDNs themselves. This common correlated noise can be generated by a common noise source such as the solid-state hot noise sources discussed before. Since the correlated noise generated by the NDNs is additive, it can be canceled out by subtracting two measurements with different levels of the common correlated noise.

This procedure is sketched in Fig. 4.120, and summarized in the following equations assuming an ideal 1:N NDN, so that the input correlated noise signals are in-phase and have a power equal to T_N/N:

- for the uncorrelated noise injection:

$$0 = A_{mn} \cdot R_{mn}^{(3)} + B_{mn}, \tag{4.135}$$

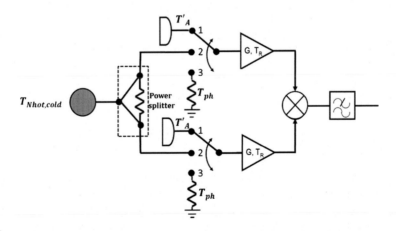

FIGURE 4.120

Basic calibration concept of a baseline of a synthetic aperture radiometer. Input switch position: (1) antenna measurement, (2) correlated noise injection either at T_{N1} or T_{N2} levels, and (3) uncorrelated noise injection.

- for the correlated noise injection:

$$\frac{T_{Nhot}}{N} + T_{offset,NDN} = A_{mn} \cdot R_{mn}^{(3,hot)} + B_{mn},$$ (4.136a)

$$\frac{T_{Ncold}}{N} + T_{offset,NDN} = A_{mn} \cdot R_{mn}^{(3,cold)} + B_{mn},$$ (4.136b)

Subtracting Eqs. (4.136a) and (4.136b), the complex amplitude term can be derived:

$$A_{mn} = \frac{(T_{Nhot} - T_{Ncold})/N}{R_{mn}^{(3,hot)} - R_{mn}^{(3,cold)}},$$ (4.137a)

and replacing Eq. (4.137a) into Eq. (4.135), the complex offset term can be derived:

$$B_{mn} = 0 - A_{mn} \cdot R_{mn}^{(3)}.$$ (4.137b)

The parallelism between the internal calibration of a baseline of a synthetic aperture radiometer and a real aperture radiometer becomes evident when comparing Eqns. (4.136a,b) and (4.137a,b) with Eqns. (4.107a,b) and (4.108a,b).

In this simplified case, the complex amplitude terms A_{mn} can be factored as the product of two terms, associated to each of the receiving channels involved:

$$A_{mn} = A_m \cdot A_n^* = \left|A_m\right| \cdot \left|A_n\right| \cdot e^{j(\phi_m - \phi_n)}.$$ (4.138)

This assumption is not completely true in the general case, and more sophisticated methods have to be used. However, it holds in many cases in radio-astronomy and allows to some particular external calibration methods that will be described afterwards.

4.6.4.2.2 External Calibration

- The external calibration of the instrument is used for the absolute calibration of the NIR by viewing known cold scenes, similar to many other real aperture radiometers (e.g., SSM/I in Section 4.6.3.5). The cold sky is obtained by inertial pointing the MIRAS instrument towards the galactic poles, so as to minimize the impact of the Galactic plane (see Section 4.2.2, Fig. 4.10), and an internal matched load placed in each NIR, and at a well-monitored physical temperature. The calibration of the NIR (actually three redundant NIRs in the case of MIRAS/SMOS) provides the absolute value of the scene:

$$T_A^{pq} \triangleq V^{pq}(0,0),$$ (4.139)

and serves as reference for the calibration of the on-board noise diodes, and the power measurement system (PMS) of each receiver (acting as a TPR), used to denormalize the visibility samples.
- The measurement of the **flat-target response (FTR)** (Martin-Neira et al., 2008), which is the instrument response to a scene as homogeneous as possible, allows us to compensate for the term $-T_{ph} \cdot \delta_{pq}$ in Eq. (4.70) and improves the instrument performance since it reduces the impact of the antenna pattern errors which are of a multiplicative nature. The measurement of the FTR must be as clean as possible from any perturbation, including the Sun, the Galaxy plane, the Earth through the antenna back lobes, attitude pointing errors from zenith, etc.

- The **ocean target transformation (OTT)** proposed by Tenerelli and Reul (2010) compensates for the residual instrument calibration errors and, mainly, the antenna imperfect characterization that, after the image reconstruction algorithm, induces ripples in the brightness temperature images. The OTT is an additive mask that is applied to each pixel over the ocean and has proven to be very effective in the accurate retrieval of the SSS.
  However, the actual nature of these errors is not additive, but multiplicative (Camps et al., 2008), so when the OTT is applied to land or ice targets errors become larger.
- To overcome this, a **multiplicative mask** derived by Wu et al. (2011) to correct at a pixel level for the residual aberrations induced during the image reconstruction process due to phase/amplitude residual errors, as well as antenna pattern errors.

Other calibration methods that do not rely on internal calibration signals can be used in principle. They have been used in radio-astronomy with success (Thompson et al., 1986), although their applicability to Earth observation is more limited (Camps, 1996), since the brightness temperatures being imaged do not consist of point sources, but are quite homogeneous instead, which translates into visibility functions more concentrated in the lower frequencies, than in radio-astronomy. These other methods are as follows:

1. the **redundant space calibration**, based on the fact that redundant baselines must measure the same visibility value, and if not, this difference can be attributed to separable amplitude and phase errors associated with each receiving chain (Thompson et al., 1986; Camps et al., 2003).
   Therefore, this technique cannot be applied to arbitrary arrays or moving arrays, since the number of redundant baselines may be very limited (if any) and
2. the **phase/amplitude closures**, based on the following two properties: the sum of the uncalibrated ("raw") phases of three visibilities forming a triangle is equal to the sum of the calibrated phases:

$$\phi_{p,q}^{raw} + \phi_{q,r}^{raw} + \phi_{r,p}^{raw} = \phi_{p,q} + \phi_{q,r} + \phi_{r,p}, \tag{4.140}$$

and the ratio of the product of the uncalibrated ("raw") amplitudes of the visibilities forming opposite sides of a quadrilateral is equal to that of the calibrated amplitudes:

$$\frac{\left|V_{m,n}^{raw}\right|\left|V_{p,q}^{raw}\right|}{\left|V_{m,p}^{raw}\right|\left|V_{n,q}^{raw}\right|} = \frac{\left|V_{m,n}\right|\left|V_{p,q}\right|}{\left|V_{m,p}\right|\left|V_{n,q}\right|}. \tag{4.141}$$

The main limitation of amplitude and phase closures in the MIRAS/SMOS case is the residual error amplification along the arms.

4.6.4.3 Image Reconstruction

The first image reconstruction algorithm for synthetic aperture radiometers was the G-matrix (Ruf et al., 1988; Tanner, 1990; Tanner and Swift, 1993), devised for the ESTAR instrument, the first airborne 1D synthetic aperture interferometric radiometer. The G-matrix relates in a matricial form the measured visibility samples to a point source located in different directions. Each measurement becomes thus a column of $\overline{\overline{G}}$:

$$\overline{V} = \overline{\overline{G}} \cdot \overline{T}. \tag{4.142}$$

FIGURE 4.121

Artist's view of European Space Agency's Soil Moisture and Ocean Salinity mission carrying the microwave imaging radiometer by aperture synthesis instrument, whose three long arms forming 120° are clearly seen.

From ESA. http://www.esa.int/spaceinimages/Images/2004/06/SMOS

As originally devised, this method cannot be directly applied to large 2D synthetic aperture microwave radiometers, because either the matrix is too large or because it is ill-conditioned and errors get amplified, but still the basic idea is the same as in Eq. (4.142). The main differences are twofold: (1) the way the $\overline{\overline{G}}$ matrix is computed, stored in memory, and inverted (Camps et al., 1998a,b; Anterrieu, 2004; Corbella et al., 2009) and (2) the preprocessing of the visibility samples to minimize the ringing in the regions with high brightness temperature contrasts, such as in the Earth's limb or in the coast lines, the way to minimize the effect of the Sun, etc. (Camps et al., 2008), and the way the mitigate RFI sources in the imagery (Camps et al., 2011a).

The interested readers are invited to consult Camps et al. (2016) for a detailed description of the instrument error model and their classification, the calibration and image reconstruction algorithms, including the random arrays.

4.6.4.4 ESA's SMOS Mission and the MIRAS Instrument

So far ESA's SMOS mission has been the only spaceborne mission using synthetic aperture radiometry. Therefore, despite the fact that there are a few other missions planned using this technique, the example must be focused on the MIRAS instrument, the single payload of the SMOS mission (Fig. 4.121). The MIRAS instrument achieves a "good"[13] spatial resolution (~ 50 km) with an array of small antennas (~ 20 cm). In principle, this solution is the lowest cost possible, while it is more easily scalable.

---

[13]In terms of other microwave radiometers at the same frequency band.

The SMOS mission was launched on November 2, 2009, and it has been a challenge itself for a number of reasons:

- First, it uses a brand new type of instrument, never before tried in space. It required revising the fundamental equation, making a new and detailed error model and correction (calibration) algorithms, and image reconstruction algorithms, as well.
- Second, it provides new types of observations at L-band, including multilook and multiangle observations, in which pixels are seen with different size and orientation, different noise level and precision, etc. and polarization mixing due to geometric and Faraday rotation effects that mix the Stokes elements.
- Third, it required the development of new L-band and multiangular ocean and soil emission models, over a wide range of incidence angles ($0°-60°$), to use them in the geophysical parameters retrievals.
- And fourth, new geophysical parameter retrieval algorithms had to be developed, taking into account the three issues above.

SMOS scientific measurements led to a Sun-synchronous (mean orbital plane inclination = 98.416°), dawn/dusk (local time ascending node = 6 a.m.), and quasicircular orbit (mean altitude = 755.5 km and eccentricity = 0.001165). The antenna spacing in the MIRAS array was selected to be $d = 0.875\lambda$, which is larger than $d = \lambda/\sqrt{3}$, required to avoid aliasing. This configuration reduces the number of antennas for a given angular resolution, while maintaining the swath width. The MIRAS array was steered 30°[14] so as to enlarge the swath width and homogenize the number of observations in the central part of the swath, and tilted 32.5° so as to enlarge the range of incidence angles observed.

The MIRAS instrument itself is described below. The most visible part of the MIRAS instrument is its large antenna array, of more than 4 m along each arm. Since the arms have to be fold so as to fit in the upper part of the launcher (Fig. 4.122A), and then deploy (Fig. 4.122B) the arms are made of three sections each with six antenna elements, as illustrated in Fig. 4.123.

The individual receiving elements are "simple" superheterodyne receivers with one down-conversion. As characteristics, each receiver has a 4:1 input switch to select between the antenna polarization (either horizontal or X-pol, or vertical or Y-pol), the correlated noise or the uncorrelated noise (internal matched load) sources, an input isolator to improve matching and minimize the noise radiated toward the antenna,[15] a slope corrector to improve the flatness of the spectrum, the 1 bit/2 level samplers, and the PMS, which is a diode detector that allows each receiver to behave as a TPR. It is needed to denormalize the "normalized" correlations (obtained using 1 bit/2 level correlators).

The NIRs have triple redundancy, since they provide the reference for any other calibration (reference noise source of calibration subsystem or CAS) and provide the average value of the antenna temperature of the scene ($V_{pq}(0,0) = T_{Apq}$). They are located in the hub and are fully polarimetric, that is, they do measure the four Stokes elements. The block diagram of the NIRs is shown in Fig. 4.125. They are composed of two LICEF receivers connected each to one antenna polarization. Correlated noise inputs from the NDN allow phase/amplitude calibration of receivers as LICEFs and the third and fourth Stokes parameters measurements (Fig. 4.124A).

---

[14]One of the arms was pointing toward the Sun.

[15]Actually the radiated term will be the one generated by the internal matched load mainly.

(A) **(B)**

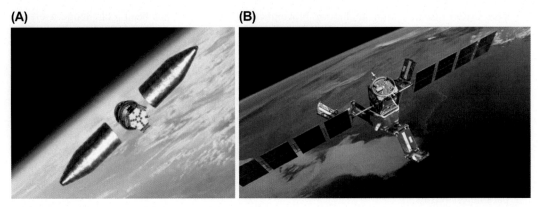

FIGURE 4.122

(A) Soil Moisture and Ocean Salinity (SMOS) in the upper stage while fairing opens, and (B) SMOS in orbit and microwave imaging radiometer with aperture synthesis antennas deploying.

Credits ESA multimedia.

The digitized signals from each LICEF are transmitted using an optical link (MOHA: microwave optical harness) to the DICOS (Digital COrrelator System) in the CCU to compute the complex cross-correlations of all signal pairs (Fig. 4.126). With 1 bit/2 level sampling, the cross-correlator reduces to an NOT-XOR gate to perform the multiplication (output equals one when the two inputs are the same, and it is zero when the two inputs are different), and an upcounter.

A final subsystem required to the correct operation of the instrument is the CAlibration System or CAS that includes all the noise sources (with double redundancy), switches and power splitters to drive the correlated noise signal to the receivers. One of the requirements of the CAS is the exquisite matching both in amplitude in phase, and stability, of the power splitters. Fig. 4.127A and B shows the CAS can be in the hub (Fig. 4.127A) or in one arm (Fig. 4.127B).

Finally, Fig. 4.128A and B show two pictures of the whole MIRAS instrument, during a deployment test at EADS-CASA premises (Madrid, Spain), MIRAS prime contractor, and prepared for functional tests at ESA/ESTEC (Noordwijk, The Netherlands).

MIRAS has two modes of operation:

- the dual-polarization one, in which all LICEFs at X-polarization or at Y-polarization at a time, and both polarizations are measured sequentially, and
- the full polarimetric one, in which LICEFs are at orthogonal polarizations. However, to measure the whole set of (u,v) points, a particular switching sequence has to be commanded. This sequence requires one arm to be at p-polarization, while the other two arms are at q-polarization. In the next two integration times the arm at p-polarization is changed. A fourth step with all arms at p-polarization is required to have the complete Stokes vector. In addition, to homogenize the integration times and have the same radiometric precision at both polarizations, the role of the p- and q-polarizations is swapped in the next four steps.

MIRAS data are geolocated using the ISEA family of grids, instead of the EASE-Grid, more popular among many of the US Earth observation missions, namely AQUA (NASA/NASDA) and

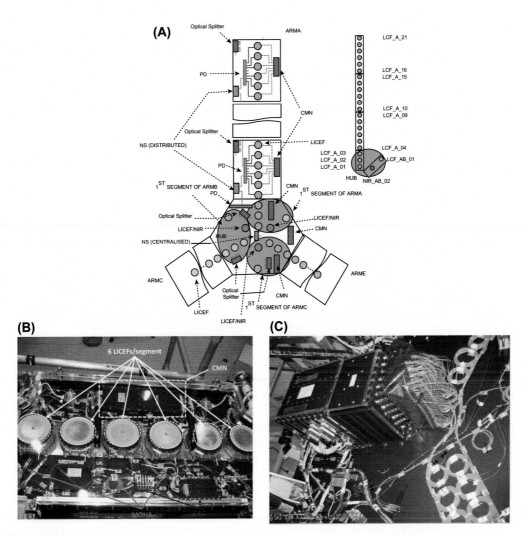

FIGURE 4.123

(A) Microwave imaging radiometer with aperture synthesis instrument array topology: 69 antenna elements (LICEF, Fig. 4.124B) are equally distributed over the three arms and the hub, with three sections of six elements each. Each section has a control monitoring node (CMN) to control the instrument state, and a (distributed) noise source. The acquired signals are transmitted optically (*blue cables*) to a central correlator unit (CCU), which computes the complex cross-correlations of all signal pairs. (B) Image of a segment with the different elements, and (C) CCU with a bunch of fiber optic cables bringing all digitized signals.

(Courtesy of Astrium EADS-CASA Espacio, Madrid, Spain; now Airbus Defence and Military).

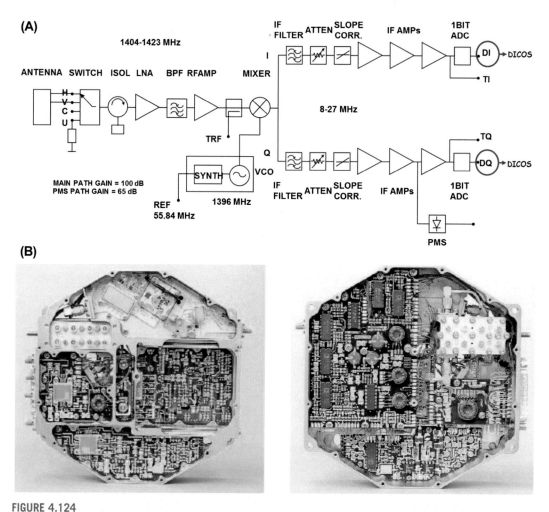

FIGURE 4.124

(A) Block diagram of a LICEF (light and cost-effective front-end) of microwave imaging radiometer by aperture synthesis, and (B) front and back pictures of a LICEF. On the left (top center) the isolator is clearly seen, while on the right (top right corner) the very selective cavity filter is located.

Courtesy of Mier Comunicaciones SA, Barcelona, Spain (now Tryo Aerospace).

Aquarius (NASA), which are particularly interesting for comparison with the SMOS products. Spatial partitioning of ISEA can be triangular, hexagonal, or diamond-based. In its hexagonal form, ISEA has a higher degree of compactness, quantizes the plane with the smallest average error, and provides the greatest angular resolution. It also possesses uniform adjacency with its neighbors, unlike the square EASE-Grid. ISEA hexagonal at aperture 4 and resolution 9 (ISEA 4-9) is made up of 2,621,442 points over the Earth spaced ∼15 km, while the EASE-Grid at ∼12 km has 3,244,518 points.

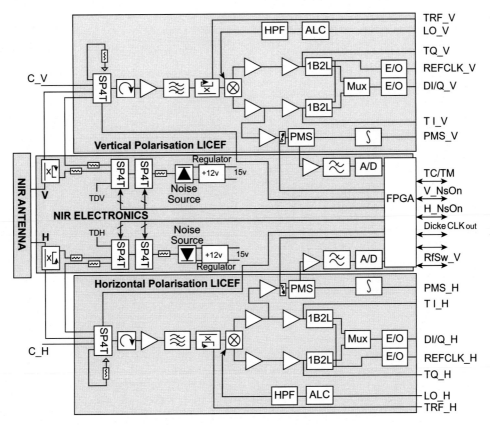

FIGURE 4.125

Block diagram of MIRAS NIRs made of two individual LICEFs and the noise injection electronics block.

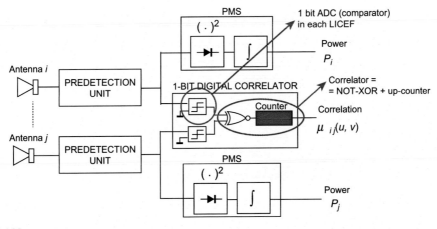

FIGURE 4.126

Block diagram of a digital correlator system.

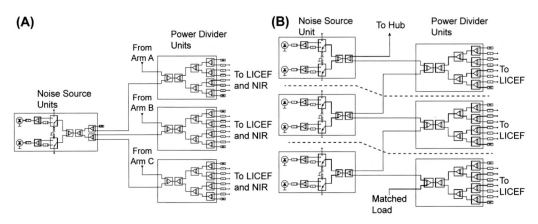

FIGURE 4.127

(A) CAlibration System (CAS) in the hub: a noise source drives correlated noise to the first six elements in each arm, from which three are in the first part of the arm and two correspond to the redundant elements, one of them being an NIR (with two LICEFs). (B) CAS in the arms: noise sources distribute noise to groups of the sections (6 + 6 receivers). All noise sources have double redundancy.

Courtesy of Astrium EADS-CASA Espacio, Madrid, Spain (now Airbus Defence and Military).

Finally, multiangular emissivity models are fit to the data aggregated for each pixel at different incidence angles to obtain the different geophysical parameters. Since ESA is only in charge of the processing up to level 2, there are two centers for processing levels 3 and 4, one in France (the CATDS: https://www.catds.fr/) and another one in Spain (the SMOS CP34: http://cp34-bec.cmima.csic.es/). A composite SSS and soil moisture map generated at the CP34 is presented here in Fig. 4.129. Fig. 4.130 shows the residual first Stokes parameters, which is associated to the excess surface roughness induced by the high winds of hurricane Danielle crossing the Atlantic and hitting the East coast of the United States.

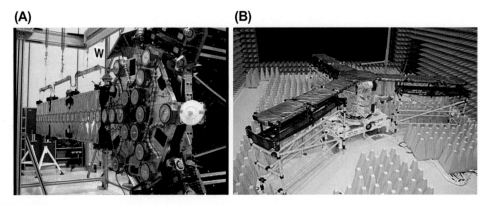

FIGURE 4.128

(A) Microwave imaging radiometer by aperture synthesis (MIRAS) instrument at Astrium EADS-CASA premises (Madrid, Spain) for deployment tests (credits Astrium EADS-CASA), (B) MIRAS at ESA/ESTEC (Noordwijk, The Netherlands) for functional tests.

Courtesy of Astrium EADS-CASA Espacio, Madrid, Spain (now Airbus Defence and Military).

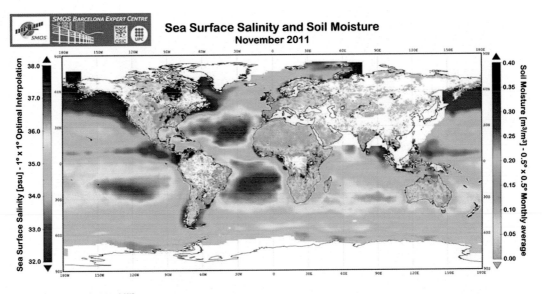

FIGURE 4.129

Sample global soil moisture and ocean salinity maps for November 2011.

Courtesy of Dr. Jordi Font, ICM/CSIC; http://cp34-bec.cmima.csic.es/.

As previously discussed, the spatial resolution that can be achieved with passive microwave sensors is quite poor, even at the highest frequency bands, because it is limited by the antenna size. This prevents one from using microwave radiometry data for regional or local applications. One might think that technology would remedy this situation in the coming years, by allowing us to manufacture and deploy in orbit larger and larger antennas, however *"the radiometer uncertainty principle"* cannot be exceeded. It states that the product of the spatial resolution and the radiometric precision is a constant for a given type of radiometer. For example, using the same receivers, if one could manufacture a 10 times larger MIRAS instrument, the spatial resolution would improve from ~ 30 to ~ 3 km, but the radiometric precision would have degraded by a similar factor ($\sim \times 10$ times larger), and instead of 2–3K, this new instrument would have 20–30K rms radiometric resolution, which is useless for many applications.

To overcome this limitation, achieving a better spatial resolution, while keeping good radiometric precision additional high quality and high resolution multispectral data can be merged with the radiometry data (Piles et al., 2011). Since June 2012, the SMOS Barcelona Expert Center (SMOS-BEC http://cp34-bec.cmima.csic.es/) is routinely producing near real time (morning and afternoon passes, if available) soil moisture maps of the Iberian peninsula[16] at 1 km, from SMOS and MODIS/AQUA and TERRA data (Fig. 4.131), that are used by the regional authorities as a proxy for forest fires prevention (Chaparro et al., 2016), or to estimate forest decline (Chaparro et al., 2017).

---

[16]SMOS Barcelona Expert Centre on Radiometric Calibration and Ocean Salinity (SMOS-BEC) website: http://www.smos-bec. icm.csic.es/smos_bec.

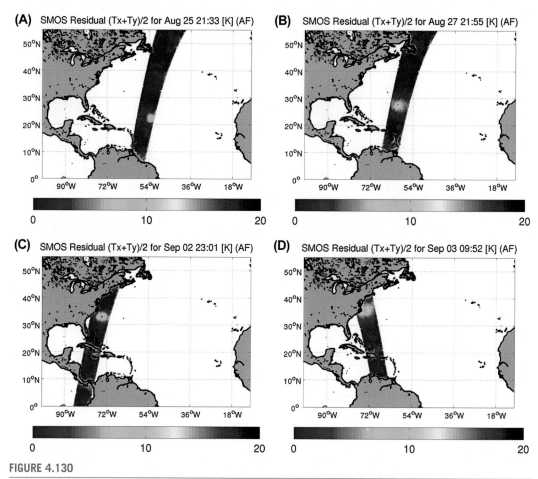

FIGURE 4.130

Wind maps over hurricanes as imaged by Soil Moisture and Ocean Salinity: (top) Hurricane Danielle, August 25 (75 kt at 982 mb) and August 27 (115 kt at 942 mb) 2010, (bottom) Hurricane Earl September 2 (100 kt at 947 mb) and September 3 (90 kt at 955 mb) 2010.

Courtesy of Dr. Nicolas Reul, IFREMER and Joseph Tenerelli, ODL.

4.6.5 FUTURE TRENDS IN MICROWAVE RADIOMETERS

Today's evolution of microwave radiometers includes (1) a much wider use of digital technology, including sampling at IF, digital I/Q demodulation, digital filtering, etc., and many techniques/technologies inherited from software-defined radio, (2) a much wider use of RFI detection and mitigation techniques, both real and synthetic aperture radiometers, to face the increasingly worrying problem of interferences, (3) highly integrated microwave and millimeter-wave circuits, that will allow low-power miniature radiometers, to be boarded in nano-satellites, and (4) a spread of synthetic aperture techniques after the success of ESA's SMOS mission. At present, three new millimeter-wave

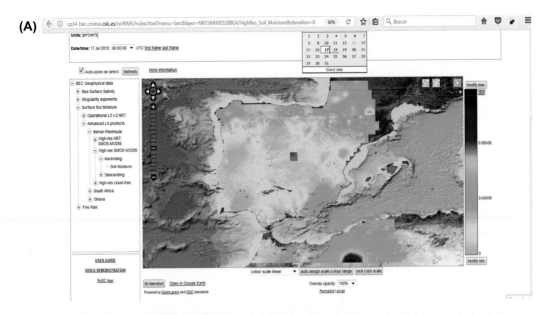

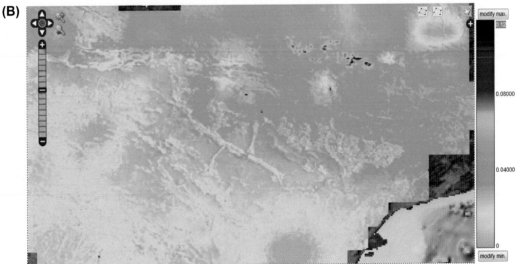

FIGURE 4.131

(A) Soil moisture map of the Iberian Peninsula at 1 km spatial resolution generated from the combination of Soil Moisture and Ocean Salinity and MODIS/AQUA and TERRA data. (B) Zoom over the Ebro river valley. (C) Zoom over the Pyrenees. In (B) the cultivated regions near the river and its tributaries are clearly wetter than the rest. In (C) the north side of the mountains is clearly wetter than the south side which is drier.

atmospheric sounders from a geostationary orbit are under study: the NASA/JPL PATH mission, the ESA GAS mission, and the China GIMS mission.

4.7 STUDY QUESTIONS

1. When the electromagnetic radiation passes through the atmosphere it may suffer two types of effects. Which ones? What is their frequency dependence?
2. What happens in the atmosphere around 22 and 60 GHz? And around 36 and 89 GHz?
3. The emissivity of a surface increases or decreases with increasing values of the dielectric constant? Why?
4. The emissivity of a surface increases or decreases with increasing surface roughness? Why?
5. From a satellite, the brightness temperature of the rain over the sea increases or decreases with increasing frequency. And, at the same frequency, increases or decreases with the rain rate? Why?
6. From a satellite, the brightness temperature of the rain over land increases or decreases with increasing frequency. And, at the same frequency, increases or decreases with the rain rate? Why?

   At 53° incidence angle:
7. The brightness temperature of the snow increases or decreases with the frequency? Why?
8. The brightness temperature of the deserts increases or decreases with the frequency? Why?
9. In vegetated regions, which is the dependence of the brightness temperature as a function of the "amount of vegetation" and the polarization? Why?
10. How does the iced sea is seen from a spaceborne microwave radiometer, for example, at 19 GHz?
11. The brightness temperature of the sea at vertical polarization increases or decreases with the wind speed? Why?
12. In the retrieval of geophysical parameters over the ocean surface, what happens when the presence of rain is detected? Why?
13. At 1.4 GHz, the sea surface brightness temperature increases or decreases with the presence of waves and salinity? Why?
14. In plain words, what is a microwave radiometer?
15. Which measurements are needed to perform the radiometer calibration?
16. In a TPR with a receiver's noise temperature of 200K, a bandwidth of 27 MHz, and an integration time of 1.2 s, compute the radiometric sensitivity when the antenna temperature is 100K.
17. If the beam width of a radiometer antenna is 2°, compute the footprint size while pointing to nadir from 800 km height.
18. Which type of scanning is performed by the SSM/I family? Which are its advantages from the electrical, mechanical, and emissivity points of view?
19. In which order must the radiometric and geometric corrections be performed?
20. Taking into account the different footprint size from 69 × 43 km at the lowest frequency band, to 15 km × 13 km at the highest one comment how the scanning of the different channels is performed.

21. How is the calibration of the SSM/I sensors performed? What happens when a calibration is wrong, and how it does look like in the brightness temperature image?

22. How is the antenna pattern correction implemented for SSM/I?

23. Comment the geometric correction technique applied to EMSR, SMMR, and SSM/I?

24. What happens if the time tag of a measurement is wrong? How it does appear in the images?

25. How can the geolocalization errors associated to imperfect orbit knowledge be reduced?

26. A spaceborne microwave radiometer at 19 GHz performs a conical scan over the sea surface at a constant incidence angle of 53.1°, at vertical and horizontal polarizations. Assuming a flat sea, will the brightness temperatures depend on the polarization? If yes, at which polarization the brightness temperature will be smaller? Which is the dependence of the brightness temperatures with the wind speed?

27. A helicopter-borne noise-injection radiometer operates at 36.5 GHz with 1 GHz bandwidth, and performs a conical scan at 200 rpm and at a constant incidence angle of 50° with respect to the vertical at a constant height of 4 km. The antenna beam width is 2.35° in both planes.

    a. Discuss the advantages and disadvantages of an NIR versus a TPR.

    b. For the nominal conditions, compute the size of the footprint (major and minor axes) over a flat surface.

    c. Compute the swath width when the helicopter flies in a straight line at the nominal height.

    d. Compute the maximum speed of the helicopter to measure all pixels over the ground.

    e. Compute the radiometric resolution if the antenna temperature is smaller than 300K, the temperature of the internal matched load is 310K, and the receiver's noise temperature is 400K.

    f. If the range of input temperatures spans from 50K to 300K, compute the minimum number of bits in the analog-to-digital converters so that the quantization noise is negligible in front of the radiometric resolution. Compute the amount of information generated.

28. A side-looking push-broom radiometer uses an antenna array and a combining network to create eight narrow beams in elevation simultaneously. Each beam is connected to a radiometer receiver. The platform height is 1 km over the surface and the ground speed is 50 m/s.

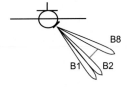

Data:
    Frequency of operation: 10.68 GHz
    Radiometer type: Dicke
    Receiver bandwidth: 150 MHz
    Reference temperature: 300K
    Receivers' noise temperature: 500K
    Antenna beam width in azimuth and elevation: 4° (each beam)
    Pointing angles B1−B8 with respect to nadir: 31°, 35°, 39°, 43°, 47°, 51°, 55°, 59° (±2°, beam width)

a. Swath width over the surface in a straight line.
   For B4 beam ($43° \pm 2°$):
b. Pixel dimensions in the along-track and cross-track directions.
c. Maximum integration time.
d. Assuming that the antenna temperature is 300K, compute the radiometric resolution, or minimum input change that can be detected.
   For the whole radiometer:
e. If the input temperature range spans from 80K to 320K, determine the amount of data generated by the radiometer assuming all eight receivers are identical.
f. Which factors can create errors in the measurements? Propose a method to compensate for them.
g. In which cases the apparent brightness temperature will significantly differ from the surface's brightness temperature?

RADAR

5

Radar has its earliest origin in observations made by Heinrich Hertz in the late 19th century that radio waves were reflected by metallic objects. Motivated by the studies of James Clerk Maxwell on electromagnetic waves, Hertz demonstrated the reflection of these waves by metallic objects. In 1887, Hertz further experimented with electromagnetic waves and found that some waves could be transmitted through some materials while others reflected these waves back to their source.

In the early part of the 20th century, another German researcher Christian Huelsmeyer built a simple ship detection device to assist in avoiding ship collisions. The term radar itself comes from the acronym radio detection and ranging (RADAR), coined by the US navy. Thus, radar must both detect a target and provide a range to it. In the years before 1934, systems either gave a range estimate and no direction, or provided an estimate of direction, but with no range information. One key development was the introduction of pulsed radar that was timed to provide range while they were sent from large antennas that could also provide directional information. Taken together these two components provide us with conventional radar, as we know it today.

The development of the wireless radio is often attributed to Guglielmo Marconi (1874−1937) although he was not the first to invent the technology. He was, however, the greatest early promoter of this new technology and its applications. Marconi also saw the potential of using radio waves for ship routing by lighthouses. He said, "I… pointed out the possibility of a lighthouse to use this technology to enable vessels to safely navigate coastal waters in foggy weather." He recognized the all-weather operational capability of radio waves over traditional optical detection methods.

In 1904, Huelsmeyer (1881−1957) demonstrated, in Germany and the Netherlands, the use of radio echoes to detect ships so that collisions could be avoided. His device consisted of a spark gap that could generate a signal that was received using a dipole antenna with a cylindrical parabolic reflector. When a signal reflected from a ship was picked up by a similar antenna, attached to a separate coherent receiver, a bell sounded. During bad weather or fog the antenna was periodically spun to scan for nearby ships. Huelsmeyer patented his invention, which he named the telemobiloscope. The name never caught on with naval authorities.

One of the many ideas that Nikola Tesla (1856−1943) included in his studies on electromagnetic waves was that from a fixed station, we could use these waves to determine the relative position or course of a moving object such as a ship at sea. He also proposed the use of pulsed radio waves to determine the relative position, speed, and course of a moving object.

In the fall of 1922, Albert H. Taylor and Leo C. Young of the US Naval Aircraft Radio Lab were conducting communications experiments when they noticed that a wooden ship in the Potomac River was interfering with their signals. In essence they had demonstrated a bistatic radar, which is a system

Introduction to Satellite Remote Sensing. http://dx.doi.org/10.1016/B978-0-12-809254-5.00005-1

where the radio signals are sent and received from two different and independent antennas. In a future system, Young suggested trying pulsing techniques to allow a direct determination of both range and direction to the target. Robert Morris Page was assigned to Young to implement these ideas. In December 1934, they used such an apparatus to detect a plane at a distance of 1 mile flying up and down the Potomac. Although the signals were weak, this demonstrated the first use of a pulsed radar system. Thus, Page, Taylor, and Young are usually credited with building and demonstrating the world's first true radar. An important subsequent development by Page was the invention of the duplexer, a device that allowed the transmitter and receiver to use the same antenna without overwhelming or destroying the sensitive receiver circuitry. This also solved the problem of synchronization of the separate transmit and receive signals that was needed for accurate position determination at long ranges.

There were parallel developments of radio ranging and detection going on in the United Kingdom, Germany, and the former USSR. Accelerated by the impending World War, radar developments were many on both sides of the war. In the United Kingdom, radar was an essential advantage for the country in scrambling its fighters against bombing attacks. Progress would not likely have come so quickly without the motivation of war and its necessities. Today, radar has advanced far beyond those early uses, which emphasized its application to detection and ranging. Today we talk of imaging radars that have been extended to use interferometry to improve their horizontal spatial resolution. We have radar altimeters to map the sea surface heights and detect significant wave heights. We look at the backscatter of the wave signal to compute ocean surface wind speed and direction. These various technologies will be discussed in the upcoming sections of this chapter.

Active radio/microwave remote sensing systems are commonly known as radars. A crucial advantage of radio/microwave sensors over optical sensors is their potential of day–night, all-weather operation, a feature of particular relevance for monitoring areas frequently covered by clouds or haze, such as tropical rain forests or polar regions. Since their development, starting at the beginning of the 20th century (Hülsmeyer, 1904), radars have been frequently used in a number of different applications, e.g., military, space and air navigation, air traffic control, ship safety, police enforcement, or planetary observations. In addition to the all-day, all-weather operation capability, radar images contain different information when compared to optical images; radars are sensitive to the complex conductivity of the targets and naturally adopt the scattering and propagation properties of their electromagnetic waves. Radars operating with shorter wavelengths are very sensitive to small geometric structures, whereas the longer wavelengths can penetrate deep into forests and vegetation.

In addition to other pertinent classifications, remote sensing radars can be divided into imaging and nonimaging. Altimeters and scatterometers are examples of nonimaging radars, usually yielding a one-dimensional (1D) profile of topography or reflectivity, respectively. Imaging radars, on the other hand, perform two-dimensional (2D) measurements resulting in reflectivity images. In remote sensing applications, synthetic aperture radars (SARs) are very popular imaging devices because of their excellent geometric resolution, on the order of a few radar wavelengths. This very fine resolution, combined with the maturity of several coherent methods, has transformed SAR remote sensing into a leading technology with applications ranging from security to disaster monitoring.

5.1 A COMPACT INTRODUCTION TO RADAR THEORY

The principle of radar remote detection and ranging is based on a send-and-receive approach, where the transmission of a signal is followed by the subsequent recording of the echoes scattered from the targets in the scene causing this backscatter in the (Skolnik, 1980a). Although radar was the first device used for remote detection and ranging, its principle is now shared with other popular systems such as sonar (acoustic waves, Urick, 1986; Le Chevalier, 1989) or LIDAR (Smullin and Fiocco, 1962).

As an example of the simplest radar measurement, let us consider the stationary radar and target depicted in Fig. 5.1.

The geometry is fairly simple, with the radar illuminating a target located at a distance. This distance is typically called the range in radar jargon. The signal transmitted by the radar is represented by $s(t)$, where t is a time variable. Since radio transmitters are band-limited and operate at certain carrier wavelengths, $s(t)$ will be, in general, a band-pass signal, i.e., $s(t)$ can be expressed as (Carlson, 1986)

$$s(t) = s_{lp}(t) \cdot \exp(j \cdot \omega_0 \cdot t), \tag{5.1}$$

where $s_{lp}(t)$ is the low-pass equivalent of $s(t)$, j is the imaginary unit $\left(\sqrt{(-1)}\right)$, and ω_0 is the central angular frequency of the radar. From classical wave theory, we know that the central angular frequency will dictate both the carrier frequency and wavelength in the following manner:

$$\omega_0 = 2\pi \cdot f_0 = \frac{2\pi \cdot c}{\lambda}, \tag{5.2}$$

where f_0 is the central or carrier frequency, c is the velocity of propagation of the transmitted waves in the medium, and λ is the central wavelength. In practical terms, however, the signal transmitted by the radar is a real signal, i.e.,

$$s_{Tx}(t) = \Re[s(t)] = \Re[s_{lp}(t)] \cdot \cos(\omega_0 \cdot t) - \Im[s_{lp}(t)] \cdot \sin(\omega_0 \cdot t), \tag{5.3}$$

where $\Re[\cdot]$ and $\Im[\cdot]$ are the real and imaginary operators, respectively. The complex model of (5.1) provides a representation of the phase of the signal. In addition, radars typically incorporate in-phase

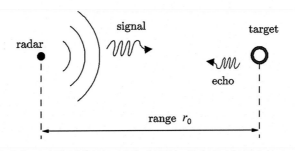

FIGURE 5.1

A sketch of a stationary radar illuminating a stationary target.

and quadrature receivers (Skolnik, 1980a; Smith, 1985) and measure the phase of the received echoes. Consequently, after demodulation, the echoes will be complex, which gives the radar accurate ranging capabilities on the order of the wavelength. After hitting the target, the transmitted energy is scattered back in the direction of the radar receiver.[1] Assuming linearity, the received echo may be approximated as

$$s_{Rx}(t) \approx \rho \cdot s(t - \tau) \cdot \exp(-j \cdot \omega_0 \cdot (t - \tau)), \tag{5.4}$$

where ρ is the complex reflectivity of the target, τ is the delay from radar to target, and the exponential accounts for the basebanding of the signal at the receiver which propagates in the opposite sense as the wave in Eq. (5.1). The value of ρ depends on the frequency of the transmitted signal and incorporates the properties of the electromagnetic interaction between wave and target. The magnitude of ρ is proportional to the square root of the ratio between the energy scattered in the direction of the radar receiver and the total amount of transmitted energy (Skolnik, 1980a). Thus, ρ depends on the carrier frequency, the observation geometry, the polarization of the incidence wave, and the geometrical and dielectric properties of the scene. For the sake of simplicity, power losses, phase rotations, or dispersion effects occurred during the propagation and reception of the radar signals has been omitted.

5.1.1 REMOTE RANGING

From the original patent, which was filed for remote detection applications (Hülsmeyer, 1904), radars soon evolved to be ranging devices. For this, the information contained in the delay of the transmitted signal (i.e., τ) becomes valuable. The variable τ measures the time elapsed between the transmission of the signal and the reception of the echo. By inspecting both (5.4) and Fig. 5.1, the delay of the signal is proportional to the distance between the radar and the target r_0 (Skolnik, 1980a), i.e.,

$$\tau = \frac{2 \cdot r_0}{c}, \tag{5.5}$$

where the factor two accounts for the signal traveling the distance r_0 back and forth. Fig. 5.2 shows an example of the reception of a single echo over the timescale of the radar.

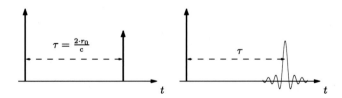

FIGURE 5.2

Ranging of the echo of a single target. In the left figure, the transmitted signal is a spike, which corresponds to an infinite bandwidth radar transmitter. In the right figure, the spike gets transformed into its band-limited version and energy spills over into adjacent delays.

---

[1] Modern radars are typically monostatic, i.e., share the same antenna for transmitting and receiving. For bistatic radars, however, the use of the term backscattering is certainly not exact, but backscattering will be used in the following pages to describe the energy scattered toward the radar receiver for bistatic cases too.

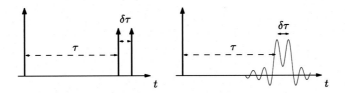

FIGURE 5.3

Ranging of the echo of two close targets. On the left figure, the infinite bandwidth case, with two spikes corresponding to the echoes of two targets. In the right figure, the energy measured by the radar receiver assuming the two targets have the same complex reflectivity.

The example contains several idealizations, i.e., no noise is depicted in the drawing, and the received signal appears as a spike, which would require an infinite bandwidth. Radars naturally have a finite bandwidth driven by resolution requirements, cost, and technological limits. In the following, we will denote the bandwidth of the radar as B_r, where the subscript r refers to ranging. The spike of the left chronogram of Fig. 5.2 becomes the band-limited echo of the right chronogram. Note that some of the echo energy is spilled into delays, i.e., ranges, not corresponding to r_0, which introduces uncertainty into the ranging measurements.

To further illustrate this effect, we might be interested in the ability of the system to discriminate two different unknown targets, located at ranges r_0 and r_1, respectively, as depicted in Fig. 5.3.

A natural question is how close can two targets be (r_1 of r_0) and still be distinguishable. The right diagram in Fig. 5.3 shows the band-limited version of the two spikes on the left. The ability to identify the two targets peaks is a function of the available bandwidth. In general, the time resolution of the radar will be inversely proportional to the signal bandwidth, i.e.,

$$\delta\tau = \frac{1}{B_r}, \tag{5.6}$$

where the proportionality factor has been set to 1 for compactness.[2] Scaling delay to range, we can express the range resolution of the radar as

$$\delta r = \frac{c}{2} \cdot \delta\tau = \frac{c}{2 \cdot B_r}. \tag{5.7}$$

Hence, good range resolution requires a large transmitted bandwidth, which usually comes at the expense of increasing hardware complexity and cost. With the progress in radio and microwave technology after World War II, the bandwidth limitations for UHF ($\sim$90 MHz) to C-band (5.4 GHz) civilian radars are basically due to international radio frequency (RF) allocations.[3] For carrier frequencies from X-band (9.6 GHz) and higher, larger bandwidths (above 10%) are available, and costs and technological limits play a more important role.

---

[2]The factor 1 roughly corresponds to a single side lobe of the response in Fig. 5.2 assuming the radar has a flat response over the frequency band.

[3]The agency in charge of the international regulations for the use of the electromagnetic spectrum is the International Telecommunication Union (ITU), belonging to the United Nations (UN).

5.1.2 DOPPLER ANALYSIS

So far, we have assumed stationarity of both the radar and the target. In practical terms, this stationarity is only required in the noninertial frame. We now consider a scenario with a relative motion between the radar and the target. For the sake of simplicity, a linear relative radial motion model (Fig. 5.4) will be used.

The delay measured by the radar is a function of time, fulfilling the following condition:

$$\frac{c \cdot \tau(t)}{2} = r_0 + v \cdot \left[t - \frac{\tau(t)}{2} \right], \tag{5.8}$$

where r_0 is the range at the start of the measurement and v the relative radial velocity between the radar and the target. Note the transmitted signal reaches the target after a time $\tau(t)/2$. After some manipulation, we obtain

$$\tau(t) = \frac{2 \cdot r_0}{c + v} + \frac{2 \cdot v}{c + v} \cdot t, \tag{5.9}$$

By substituting (5.9) into (5.4), we can express the received echo as

$$s_{Rx}(t) = \rho \cdot s_{lp}[t - \tau(t)] \cdot \exp[-j \cdot \omega_0 \cdot \tau(t)]$$

$$= \rho \cdot s_{lp} \left[\left(\frac{c - v}{c + v} \right) \cdot t - \frac{2 \cdot r_0}{c + v} \right] \cdot \exp \left[-j \cdot \omega_0 \cdot \frac{2 \cdot (r_0 + v \cdot t)}{c + v} \right], \tag{5.10}$$

which is an expression that deserves careful attention. In addition to a constant delay, we observe a scaling of the time reference by a factor $(c - v)/(c + v)$, i.e., the well-known Doppler effect. We clearly see that the scaling of the time reference introduces a frequency shift in the received echoes, corresponding to the linear term in the exponential. Fig. 5.5 shows the Doppler shifted echo of the target of Fig. 5.4, assuming s_{lp} is a constant envelope, as depicted in the left figure.

The analogy with the ranging case depicted in Fig. 5.2 is clear: in real systems, neither the available bandwidth, nor the observation time can be unlimited. As it can be seen in the spectrum of the received echo, the energy corresponding to a single target (i.e., the Doppler frequency) is spread into the

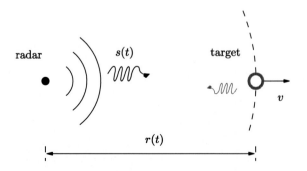

FIGURE 5.4

Radar illuminating a target with relative radial velocity v.

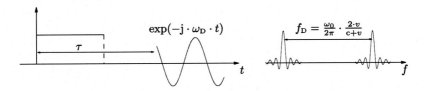

FIGURE 5.5

Doppler analysis of the echo of a single target. The transmitted signal on the left is a constant envelope of limited duration, which gets delayed and modulated by the relative motion between the target and the radar, yielding the pulsed sinusoid depicted in the figure. The right plot shows the spectra of the two pulses in the frequency domain, with energy spilling over adjacent Doppler frequencies due to the limited duration of the transmitted waveform.

adjacent frequencies introducing uncertainty in the measurement. In general, the Doppler resolution of a radar is inversely proportional to the observation time T_D, i.e.,

$$\delta f_D = \frac{1}{T_D}. \tag{5.11}$$

The velocity resolution for the particular case of Fig. 5.4 can be derived from (5.11) in a similar manner as (5.7). In other geometrical configurations, the Doppler shift can be used for the estimation of the angular or azimuth components, as it is the case of synthetic aperture radars.

Radars measuring the relative motion of the targets are known as Doppler radars. The ideal waveform in terms of Doppler resolution is a constant envelope, which becomes a tone after modulation; note that the constant envelope produces a spike in the frequency domain, an analogous situation to the one we confronted within the analysis of ranging radars.

At this point, we can draw an important conclusion from the previous analysis of ranging and Doppler radars: the requirements imposed on the radar waveform by both measurements are contradictory. This is in fact no surprise, since it reflects the dual and inverse characters of the time and frequency domains.

5.2 RADAR SCATTERING

Before we examine more properties of the radar systems, we want to consider the scattering of the radar signals from the target surfaces. The reflected radar signature is the amplitude, phase, and polarization characteristics of the radar echoes. These characteristics yield information on the geometrical and electromagnetic properties of the area imaged (Curlander and McDonough, 1991). It is only natural that this interaction is conditioned by the spectral and the electromagnetic properties of the area illuminated by the radar. In particular, the carrier frequency and bandwidth of the transmitted signal, its angular characteristics, and its observation geometry relative to the target (i.e., local incident and squint angles), along with the polarization of the radar waves, determine the complex reflectivity of the scene.

The scattering process measured by the radar depends on the morphological and dielectric properties of the scene's surface. Among the most relevant morphological properties, we identify the density and the spatial distribution of the individual scatterers in the scene, their form and orientation, the slope of the terrain, the roughness, and spatial periodicity relative to the wavelength of the radar.

Among the dielectric properties to which the radar is sensitive, we can list the soil moisture and vegetation, the characteristics of the penetration (i.e., attenuation, dephasing, depolarization) into the medium, and the conductivity of the material. The electromagnetic interaction between the radar waves and the scene is a subject of considerable complexity. Depending on the accuracy of the description of the target scene, different abstraction levels may be used to simplify the interaction with the scene. Wherever possible, we will make an effort to keep our explanations at the level of geometrical optics, which can be understood using the ray tracing analysis of the propagation phenomena (Hecht, 1997). This methodology, more stringent than the Born approximation, essentially assumes all subwavelength effects in the electromagnetic interaction can be ignored.

5.2.1 RADAR FREQUENCY BANDS

The dependence of the radar backscatter on frequency is depicted in Fig. 5.6.

This figure shows a multifrequency image acquired by the DLR airborne F-SAR system over Kaufbeuren, Germany. The different carrier frequency images are displayed in the RGB channels as follows: X (red), C (green), and L (blue). The bright white spots correspond to areas with similar backscatter in the three frequency bands. Note the images have been equalized separately and the colors might not represent with fidelity the full dynamic range of each frequency band; nevertheless, the image is illustrative.

Remote sensing radars typically operate at frequencies ranging from some tens of MHz (VHF) to some tens of GHz (W). There exist both technical and scientific reasons for this. The selection of the frequency band of the radar depends mainly on the characteristics of the electromagnetic interaction with the target surface, especially regarding its geometric and dielectric properties. As a general rule, the lower-frequency bands (e.g., roughly below C-band, 4–8 GHz) are typically used for the penetration into volumetric structures such as forests, snow, ice, dry soil, whereas higher-frequency bands (e.g., from C-band and above) typically show little to no penetration capability into natural surfaces

FIGURE 5.6

Radar image acquired by the DLR airborne F-SAR system over Kaufbeuren, Germany. Radar illumination is from the top. The different colors of the image correspond to the similarity of the radar backscatter to the different carrier frequencies.

Courtesy of DLR.

Table 5.1 Some Radar Frequency Bands and Typical Applications				
Customized Radar Suite				
Instrument	**Measurement**	**Frequency**	**Platform**	**Resolution**
MCoRDS/I	Bed topography, bed imaging, internal layering	195 MHz	DC-8, P-3	4 m
Accumulation radar	Internal layering	750 MHz	P-3	40 cm
Snow radar	Snow thickness, internal layering, topography	2–7 GHz	DC-8, P-3	5 cm
Ku band altimeter	Topography	14 GHz	DC-8, P-3	5 cm
From Tiwari, A., Airborne Radar and Applications, slide 55. https://www.slideshare.net/anupamtiwari1972/airbone-radar-applications-by-wg-cdr-anupam-tiwari.				

and are used for surface remote sensing. Likewise, the sensitivity to surface roughness also increases with increasing carrier frequency; typically, lower carrier frequencies show larger dynamic ranges in the radar signature of natural scenes than higher carrier frequencies. Table 5.1 shows a list of the different radar bandwidths with examples of airborne radars.

A further boundary condition for the usability of a given frequency band in spaceborne radar missions is the effect of the atmosphere on the propagation of the radar signals, which may introduce mild to severe attenuation, as well as delays and phase modulations in the received echoes. Lower frequencies are affected by the ionosphere (roughly up to C-band, and below), and higher frequencies are attenuated by the troposphere (from Ku band and higher).

In the previous paragraphs, the importance of the scene characteristics in the choice of the radar wavelength has only been indirectly recognized through its semitransparency, roughness, or level of detail. It is in the relationship with these properties that radar characteristics such as penetration into the media, the dynamic range of radar backscatter for a given scene, or the transmitted bandwidth for higher resolution plays a role. In general, the wavelength of the carrier shall be on the same order of magnitude as the dimensions of the structures to be detected by the radar.

Today, spaceborne radars are concentrated in the L to X bands, yielding a reasonable trade-off between penetration, sensitivity to roughness, geometric resolution, and technological maturity. In the foreseeable future, the bounds on the lower side will be extended to P-band, and on the upper side to Ku and Ka bands. All other frequency bands are generally used in airborne systems, because of the better sensitivity and the reduced influence of the atmosphere on radar wave propagation.

5.2.2 NORMALIZATIONS OF THE RADAR REFLECTIVITY

The radar cross section (RCS) of a target is the ratio of the power scattered back to the radar receiver over the incident radar power density per unit of solid angle on the target as if the radiation were isotropic (Skolnik, 1980b), i.e.,

$$\sigma_{\mathrm{RCS}} = \lim_{r \to \infty} 4\pi \cdot r_0^2 \cdot \frac{P_{\mathrm{Rx}}}{P_{\mathrm{Tx}}}, \tag{5.12}$$

where r_0 represents the distance between the radar and the target. In this expression, the limit states that the RCS is measured in the far-field. The RCS depends on the frequency and the polarization of the radar waves, and also on the form and composition of the target. The computation of the RCS of very simple targets can be approximated analytically and can be found in the classical references on radar (Skolnik, 1980b; Rihaczek, 1969). In more complex cases, however, its computation requires the use of numerical electromagnetic solvers.

The RCS describes the power efficiency of the scattering mechanism of a given target, but fails to describe homogeneous areas without clear boundaries. For these cases, the RCS can be averaged over the surface or volume where the scatterer extends. The average RCS per unit of resolution cell is denoted as sigma naught, and it can be expressed as

$$\sigma_0 = \frac{\langle \sigma_{RCS} \rangle_{cell}}{A_{cell}}, \tag{5.13}$$

where A_{cell} is the area of the resolution cell on the ground, and $\langle \cdot \rangle$ indicates spatial averaging over a multiplicity of observations. This equation is valid for imaged surfaces and can be consistently extended to the case of volumes by swapping the area in the denominator with a volume. However, a constant area on the ground might project differently on the geometry of the image depending on the local topography of the scene, which results in a terrain-dependent modulation of the scattering coefficient σ_0. The reflectivity of the scene normalized to the unit area of the slant range resolution cell is typically known as radar brightness (Raney et al., 1994), and can be approximated as

$$\beta_0 = \frac{\langle \sigma_{RCS} \rangle_{cell}}{A_{slant,cell}}, \tag{5.14}$$

where the $A_{slant,cell}$ is the size of the resolution cell in the geometry of the image. The transformation between the two different areas requires knowledge of the local topography to be able to project between a slant view and flat ground, i.e.,

$$\sigma_0 \approx \frac{\beta_0}{\sin \theta_i} = \frac{\beta_0}{\sin \left(\theta_{i,\, flat} - \alpha_{slope} \right)}, \tag{5.15}$$

where the θ_i is the local incident angle depending on look angle and the local slope, $\theta_{i,\, flat}$ is the incident angle to the flat terrain, α_{slope} is the angle of the ascending slope, and a constant resolution cell in Doppler has been assumed. Note that this equation has been derived under the assumption of coplanar slope and incidence.

Expressing the normalized reflectivity relative to the unit area of the incident wave front helps to reduce its dependence on the incidence angle, at least for a rough surface. A further discussion on the dependence of the radar scattering on slopes and incident angles is given later. The normalization to the area of the wave front, as suggested in Cosgriff et al. (1960), yields gamma naught, which is related to β_0 as follows

$$\gamma_0 \approx \frac{\beta_0}{\tan \theta_i}. \tag{5.16}$$

We illustrate the impact of the different normalizations with Fig. 5.7, which shows two sets of TerraSAR-X images scaled according to σ_0, β_0, and γ_0 from left to right, respectively.

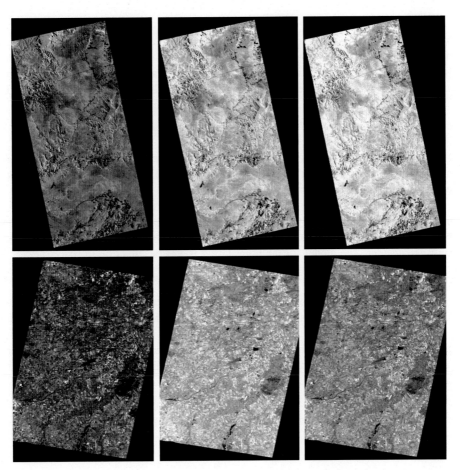

FIGURE 5.7

Two sets of geocoded TerraSAR-X images over the Brazilian rain forest (top) with an incidence angle of 46.5°, and over central Denmark (bottom) with an incidence angle of 34.7°. From left to right the normalizations are relative to σ_0, β_0, and γ_0, where σ_0 has been computed projected onto the ellipsoid; the other two take into account the local topography. As expected, σ_0 contains a significant modulation caused by the local topography, which is progressively reduced in the radar brightness and γ_0. The differences between these two are more noticeable in the bottom image due to the inhomogeneity and steeper incidence.

Courtesy of DLR.

The top image shows a rain forest area in Brazil, which is expected to yield a rather homogeneous backscatter. As expected, both β_0 and γ_0 reflect the rain forest homogeneity with more fidelity than σ_0, which in this case was normalized using the ellipsoidal height. The differences between β_0 and γ_0 are small due to this scene homogeneity and the incidence angle. The bottom image shows an image of central Denmark, where a similar trend can be observed. In this case, however, the differences between β_0 and γ_0 appear clearer due to the scene inhomogeneity and the steeper incidence angle.

5.2.3 POINT VERSUS DISTRIBUTED SCATTERERS

One of the characteristics of the scene is the density of single scatterers within each resolution cell. In radar images, single scatterers are those with dimensions on the order of the wavelength. In our single scattering approximation, the observed complex reflectivity in a resolution cell can be approximated by the coherent superposition of the reflectivities of the individual scatterers within the cell, i.e.,

$$\rho \approx \sum_i \rho_i \cdot \exp\left(-j \cdot \frac{2\pi}{\lambda} \cdot \delta r_i\right), \tag{5.17}$$

where the ρ_i are the complex reflectivities of the individual scatterers and δr_i represents the differential range of the individual targets with respect to the center of the cell. The different cases that will be analyzed in this subsection are illustrated in Fig. 5.8.

The point target is identified as a dominant single scatterer within a resolution cell. Both the amplitude and the phase observed by the radar in the corresponding image pixel will be dominated by the response of the point target, which spreads over consecutive resolution cells following the impulse response of the system. The model for the point target is a Dirac function in space, i.e.,

$$\rho(r, x) \approx \sqrt{\sigma_{\text{RCS}}} \cdot \alpha \cdot \exp(j \cdot \phi) \cdot \delta(r - r_0, x - x_0), \tag{5.18}$$

where σ_{RCS} is the radar cross section of the target, α is a constant with dimensions 1/m to make the expression dimensionally compatible, and ϕ is any constant phase. This model can be used for calibration targets like trihedral reflectors. A realistic point target of opportunity, however, looks more like the bottom purple scatterer depicted in Fig. 5.8. In addition to the dominant scatterer, the cell includes contributions of a set of smaller scatterers which will cause a slight deviation with respect to the ideal point target response. In any case, a point target is characterized at any instant in time by a single

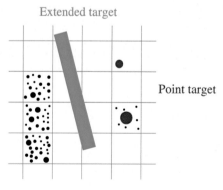

Extended target

Point target

Distributed target

FIGURE 5.8

A representation of individual scatterers in several resolution cells of the radar image. The *purple* are dominant scatterers appearing as point targets in the images. The *top purple* scatterer appears as an ideal point target, whereas the *bottom purple* is a dominant point surrounded by some additional points. The extended target appears as a deterministic target occupying several resolution cells. The distributed target is formed by a random collection of homogeneous scatterers within the resolution cells. The thickness of the individual scatterers is intended to represent the magnitude of the reflectivity of the target.

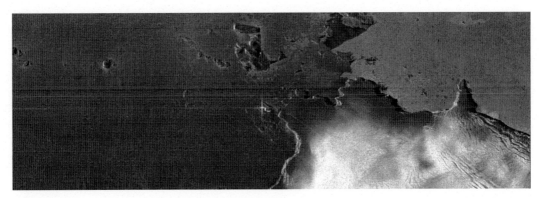

FIGURE 5.9

TerraSAR-X staring spotlight image over the DLR receiving station in the O'Higgins peninsula in Antarctica. The bright point in the center of the image shows a calibration target, appearing as an ideal point. Radar illumination is from the left. The extension in azimuth of the point target is reduced due to weighting.

Courtesy of DLR.

frequency in the Doppler spectrum, which suggests the possibility of using spectral estimation techniques for enhanced resolution imaging (Salzmann et al., 2004). An example of a point target response (e.g., calibration trihedral) in a radar image is shown in Fig. 5.9.

If the dominant single scatterer extends over several resolution cells, we will speak of an extended scatterer, as depicted in Fig. 5.8. The reflectivity model of the extended target extends in both space and Doppler domains. Exemplary extended targets in radar images are buildings, as shown in the left image of Fig. 5.10 in a neighborhood of Sevastopol, Russia.

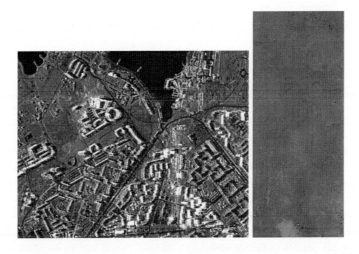

FIGURE 5.10

(Left) TerraSAR-X staring spotlight image over Sevastopol, Russia. Radar illumination is from the left. The buildings in the image appear as extended targets. (Right) TanDEM-X image of savanna area near Kweneng, Botswana. The area is homogeneous and can be interpreted as a distributed target.

Courtesy of DLR.

An extended target of a stochastic nature is a distributed target: extended because it is not small enough to fit the single-scatterer model and stochastic because it is fully characterized by a small number of statistical descriptors. Distributed targets are usually modeled by the superposition of several scatterers with similar RCSs and they are randomly placed within the resolution cell, as depicted in the black areas of Fig. 5.10. Because of the coherent superposition of the backscattered echoes, the amplitude and phase of the cell reflectivity appear to be random. Depending on whether the coherent signal of the distributed scatterer is the result of constructive or destructive interference, the subject cell will appear bright or dark in the radar image, an effect commonly referred to as speckle. The spectrum of the echoes of a distributed scene covers the entire instantaneous Doppler range and basically appears as noise in the radar images. The right image of Fig. 5.10 shows a savanna area in Botswana imaged by TanDEM-X; the image shows little contrast as expected from an ideal distributed area.

5.2.4 SPECKLE, MULTILOOK, AND RADIOMETRIC RESOLUTION

If the size of the resolution cell is much larger than the wavelength, the assumption of a large number of scatterers within the cell (i.e., distributed target) is plausible. Under these circumstances, the complex reflectivity can be assumed to be described as a circular Gaussian process with a zero-mean and a probability density function (pdf).

$$p_S(s) = \frac{1}{\pi \cdot \sigma_0} \cdot \exp\left[- \frac{\Re(s)^2 + \Im(s)^2}{\sigma_0} \right], \tag{5.19}$$

where $\Re[\cdot]$ and $\Im[\cdot]$ are the real and imaginary parts of the radar image, each having a standard deviation equal to $\sqrt{\sigma_0/2}$.[4] The intensity of the image is computed as the square of the magnitude of the complex reflectivity, i.e.,

$$I = |s|^2, \tag{5.20}$$

which for an ideal distributed target follows an exponential distribution with pdf

$$p_I(I) = \frac{1}{\sigma_0} \cdot \exp\left(- \frac{I}{\sigma_0} \right). \tag{5.21}$$

It follows from the previous expression that both the mean value and standard deviation of I will be σ_0. Fig. 5.11 shows the pdfs of the components of the complex reflectivity and the intensity of the radar images of an ideal distributed scatterer.

The statistics of the distributed scatterers can be used to separate them from point and extended scatterers in radar images. As an example, Fig. 5.12 shows a color-coded TerraSAR-X image over Berlin city center. The "distributed" areas appear bluish in the image, whereas the "urban" and "deterministic" areas appear yellowish. The condition of detection of the distributed areas was based on the assumption that the local average of the reflectivity equals its standard deviation, as predicted from the exponential distribution above.

As a consequence of (5.19), the phase of the radar image is uniformly distributed. The coherent nature of radar images results in bright and dark intensity pixels due to constructive and destructive

---

[4]The normalization with respect to the ground resolution cell is ignored in this derivation for the sake of simplicity.

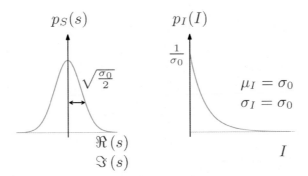

FIGURE 5.11

Probability density functions (pdfs) of the real and imaginary parts of the complex reflectivity (left) and the intensity (right) for the radar image of an ideal distributed scatterer.

FIGURE 5.12

TerraSAR-X image over Berlin city center, Germany. The image shows the Brandenburger Tor and the Central Railway Station. The distributed areas appear *blue*, whereas the urban areas are *yellow*.

interference between the echoes of the single scatterers within the resolution cell. The effect is shown in the left and right diagrams of Fig. 5.13.

The situation is exemplified in Fig. 5.13, where the different pixels of the intensity image (bottom) appear brighter or darker depending on the constructive or destructive interference of the scatterers of the resolution cells (above). Note that point targets have no speckle, and speckle tends to be reduced with increasing resolution, since a lesser number of scatterers is likely to appear in the resolution cells.

Speckle has been historically considered as a noise source, with a multiplicative character, i.e., increasing the transmitted power does not reduce its effect. This consideration is of course unfair, since the coherent character of the system is what gives radars the potential to achieve geometrical

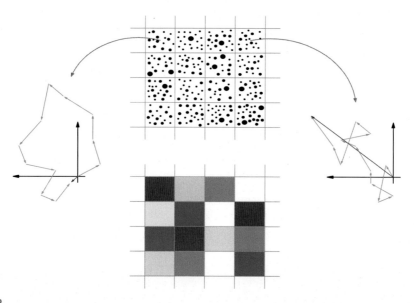

FIGURE 5.13

Speckle is a consequence of the coherence of radar echoes. The constructive and destructive interference (left and right) of the scatterers in the different resolution cells (top) results in brighter and darker pixels in the image intensity (bottom).

resolutions on the order of the wavelength. However, the intensity as the estimator of the sigma naught is responsible for giving speckle all its bad reputation. To better estimate the backscatter of the scene, we can compute a local average of homogeneous pixels, i.e.,

$$\bar{I} = \left\langle |s|^2 \right\rangle. \tag{5.22}$$

If the averaging is done over N independent pixels, the distribution of $\bar{I}$ can be shown to be a Gamma function (Papoulis, 1965), with a pdf of the form

$$p_{\bar{I}}(\bar{I}) = \frac{N^N}{(N-1)!} \cdot \frac{\bar{I}^{N-1}}{\bar{I}_0^N} \cdot \exp\left(-N \cdot \frac{\bar{I}}{\bar{I}_0}\right). \tag{5.23}$$

The variance of $\bar{I}$ is reduced by a factor N under the assumption that the samples are statistically independent, i.e.,

$$\sigma_{\bar{I}}^2 = \frac{\sigma_0^2}{N}, \tag{5.24}$$

a condition fulfilled by ideally distributed targets provided there is no oversampling, nor weighting. The local averaging of the intensity image is incoherent. The averaged intensity $\bar{I}$ will then be a much better estimator of sigma naught than the full resolution intensity. This local averaging is known as multilooking, with N the number of looks. Multilooking can be effected in both the time and frequency domains (Curlander and McDonough, 1991). The improvement in the variance of the estimation after multilooking is only achieved if two conditions are fulfilled: (1) statistical independence of the samples, and (2) the different samples correspond to the same distributed target. The compromise is a

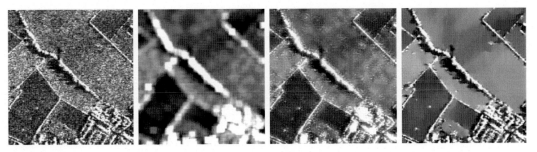

FIGURE 5.14

Original intensity image (left) and the results after several methods of speckle filtering. Multilooking with a 11 × 11 window (mid-left), 11 × 11 Lee filter (mid-right), and iterative "simulated annealing" filter after 200 iterations, with a 3 × 3 box (right). The improvement in the radiometry without loss of resolution is apparent.

reduction in the geometrical resolution of the image roughly equal to the gain in the variance of the estimator. This effect is illustrated in the left part of Fig. 5.14.

The image on the left shows the original intensity image with noticeable speckle in the distributed areas. The mid-left image shows the multilooked intensity with a 11 × 11 window. The sigma naught values appear to be significantly smoothed at the expense of a reduced geometric resolution. A number of speckle filters have been designed over the years, all of them with the similar purpose of averaging the homogeneous areas while respecting the edges in the images (Shi and Fung, 1994). Fig. 5.14 shows two examples of these space-variant approaches on the right-hand side: Lee filter (middle right), and an iterative simulated annealing filter (right).

The sensitivity of the radar system to changes in the received power determines its radiometric resolution, which is typically proportional to the ratio of the standard deviation of the image divided by its expected value, i.e.,

$$\delta\sigma_0 = 10 \cdot \log_{10}\left(1 + \frac{\sigma_{\overline{I}}}{\mu_{\overline{I}}}\right), \tag{5.25}$$

where $\mu_{\overline{I}}$ is the mean value of the multilooked intensity, $\sigma_{\overline{I}}$ is its standard deviation, and the radiometric resolution is expressed in decibels. Under the presence of noise, the standard deviation of the image is degraded, and the previous expression can be approximated as follows:

$$\delta\sigma_0 \approx 10 \cdot \log_{10}\left(1 + \frac{\gamma \cdot \frac{\sigma_0}{\sqrt{N}}}{\sigma_0}\right) = 10 \cdot \log_{10}\left(1 + \frac{1 + \text{SNR}}{\sqrt{N} \cdot \text{SNR}}\right), \tag{5.26}$$

where the γ is a decorrelation factor, well known in the SAR interferometry field. In this case, it is due only to receiver and quantization noise. The previous equation is one of the fundamental expressions driving the design of any remote sensing radar.

5.2.5 RADAR EQUATION

The idealized power budget of the radar link (transmitter-target-receiver) is usually known as the radar equation. In the case of remote sensing radars, scenes can be assumed to be distributed and not

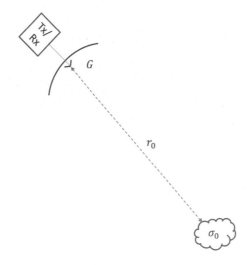

FIGURE 5.15

Schematic view of a monostatic radar illuminating an area with backscattering cross section σ_0.

composed of isolated targets, and consequently, no integration gain should be assumed. A sketch of the monostatic radar link is shown in Fig. 5.15.

The radiated power by the antenna is known as the equivalent isotropic radiated power (i.e., EIRP) and can be expressed as

$$P_{\text{Tx}} = G \cdot P_{\text{peak}}, \tag{5.27}$$

where G is the gain of the transmitter antenna in the look direction and P_{peak} is the peak power delivered by the power amplifier applied after the data are received by the antenna. Considering the free-space losses in the one-way path from radar to the scene, the surface power density at the vicinity of the scene can be expressed as

$$\xi_{\text{Tx}} = \frac{P_{\text{Tx}}}{4\pi \cdot r_0^2}, \tag{5.28}$$

where r_0 is the distance from radar to scene. The signal power at the receiver can be computed by integrating over the area on the ground, which contributes to a single sample in the radar echoes, extending in range the length of the transmitted pulse and the footprint of the antenna in azimuth, as shown in Fig. 5.16.

The shaded area of Fig. 5.16 can be approximated by

$$A_{\text{illum}} \approx \frac{c \cdot \tau_p}{2 \cdot \sin \theta_i} \cdot \frac{r_0 \cdot \lambda}{L_a}, \tag{5.29}$$

where τ_p is the duration of the transmitted pulse and L_a is the length in azimuth of the antenna. The first term of (5.29) is the ground range resolution due to the pulse envelope, and the second term is the projection of the azimuth antenna pattern on the ground. The scene is characterized by its scattering

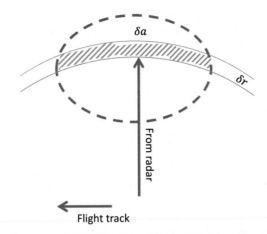

FIGURE 5.16

Area on the ground contributing to a single echo sample over which the integration of the received power of the radar shall be carried out.

cross-section sigma naught value (σ_0). Under these circumstances, the surface power density at the vicinity of the radar antenna can be computed as

$$\xi_{Rx} = \xi_{Tx} \cdot \frac{A_{\text{illum}} \cdot \sigma_0}{4\pi \cdot r_0^2}. \tag{5.30}$$

The received power is a function of the effective antenna area of the receiver $A_{e,Rx}$, which after some recollection of terms becomes

$$P_{Rx} = \xi_{Rx} \cdot A_{e,Rx} = \xi_{Tx} \cdot \frac{A_{\text{illum}} \cdot \sigma_0}{4\pi \cdot r_0^2} \cdot \frac{\lambda^2 \cdot G}{4\pi} \approx \frac{\lambda^3 \cdot G^2}{(4\pi)^3 \cdot r_0^3} \cdot \frac{c \cdot \tau_p}{2 \cdot \sin \theta_i} \cdot \frac{\sigma_0}{L_a} \cdot P_{\text{peak}}. \tag{5.31}$$

The thermal noise power at the end of the receiver chain can then be expressed as

$$N = k_B \cdot T_0 \cdot B_{eq} \cdot F_{Rx}, \tag{5.32}$$

where $k_B = 1.38064852 \cdot 10^{-23}$ J/K is the Boltzmann constant, $T_0 = 290$ K is the reference temperature, B_{eq} is the receiver's equivalent noise bandwidth, and F_{Rx} is the receiver's noise figure. Rearranging terms and identifying the noise bandwidth as the available transmitted bandwidth of the radar, the signal-to-noise ratio (SNR) at the receiver becomes

$$\text{SNR} = \frac{P_{Rx}}{N} = \frac{\lambda^3 \cdot G^2}{(4\pi)^3 \cdot r_0^3 \cdot k \cdot T_0 \cdot F_{Rx}} \cdot \frac{\delta r_g}{L_a \cdot \text{PRF}} \cdot \sigma_0 \cdot P_{\text{avg}}, \tag{5.33}$$

where the pulse repetition frequency (PRF) allows us to scale the peak power to the average power transmitted by the radar. The sensitivity of a remote sensing radar is typically measured by the noise equivalent sigma zero (NESZ) and corresponds to the value of σ_0 for an SNR of 0 dB, i.e.,

$$\text{NESZ} = \sigma_0(\text{SNR} = 0 \text{ dB}) = \frac{(4\pi)^3 \cdot r_0^3 \cdot k \cdot T_{eq} \cdot F_{Rx} \cdot L_a \cdot \text{PRF}}{\lambda^3 \cdot G^2 \cdot P_{\text{avg}} \cdot \delta r_g}. \tag{5.34}$$

As expected from this definition, the lower this value is, the better the sensitivity of the system. The NESZ is one of the driving requirements of any remote sensing radar. High values of NESZ will yield noisy radiometry and location inaccuracy, hence decreasing the quality of the final measurements. Typical values of NESZ in spaceborne radar systems are better (lower) than -20 dB.

5.2.6 RADAR WAVES AT AN INTERFACE

The surface of the Earth can be interpreted as a resistive dielectric, whose dielectric constant depends on factors such as the material, the moisture, and the radar carrier frequency. In the case of a simple carrier wave, the refractive index of the soil can be expressed as

$$n_2 = \sqrt{\varepsilon' - j \cdot \frac{\sigma_{\text{cond}}}{2\pi \cdot f \cdot \varepsilon_0}}, \tag{5.35}$$

where ε' is the relative permittivity of the soil, σ_{cond} its conductivity,[5] f is the frequency, and ε_0 is the permittivity of a vacuum. The soil will behave as a good conductor when the relative permittivity is negligible with respect to the value of the second term inside the square root, i.e.,

$$\frac{\sigma_{\text{cond}}}{2\pi \cdot f \cdot \varepsilon_0} \gg \varepsilon', \tag{5.36}$$

a condition which is naturally more easily fulfilled at lower frequencies.

Let us first consider a simple example: an incident radar wave impinging on a flat surface separating the air from a dielectric soil. A surface will be flat whenever the length of the surface is much larger than the wavelength, i.e., $L \gg \lambda$. As predicted by geometrical optics, the wave will be reflected in the specular direction (i.e., $\theta_r = \theta_i$), as depicted in Fig. 5.17.

When the material is a good conductor, all incident energy is reflected. In the opposite case, some of the energy of the incident wave penetrates and is refracted into the dielectric, with the geometry of refraction following Snell's law.

The amount of energy reflected and refracted can be computed with help of the Fresnel's coefficients, derived from the boundary conditions of the electrical and magnetic fields at the interface. The Fresnel coefficients vary depending on the polarization of the incident wave. The reflection coefficients for vertically (V) and horizontally (H) polarized waves take the following forms (Hecht, 1997)

$$\rho_{\text{Vr}} = \frac{E_{\text{Vr}}}{E_{\text{Vi}}} = \frac{n_2 \cdot \cos \theta_i - n_1 \cdot \cos \theta_t}{n_1 \cdot \cos \theta_i + n_2 \cdot \cos \theta_t},$$

$$\rho_{\text{Hr}} = \frac{E_{\text{Hr}}}{E_{\text{Hi}}} = \frac{n_1 \cdot \cos \theta_i - n_2 \cdot \cos \theta_t}{n_1 \cdot \cos \theta_i + n_2 \cdot \cos \theta_t}. \tag{5.37}$$

When $|n_2| > |n_1|$, $|\rho_{\text{Hr}}| > |\rho_{\text{Vr}}|$, this is a characteristic that will be relevant in the analysis of the scattering by rough surfaces. Note that ρ_{Hr} and ρ_{Vr} can take negative values depending on the sign of the numerator, which corresponds to a phase shift of $180°$. This is the particular case of reflection

---

[5]Not to be confounded with the RCS (σ_{RCS}) or sigma naught (σ_0) frequently mentioned throughout this chapter.

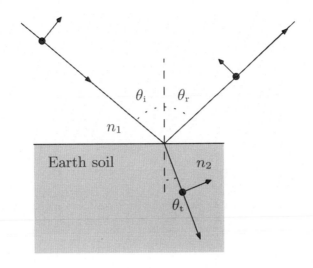

Reflection and refraction of an incident wave onto a flat surface, $L \gg \lambda$. The angle of reflection θ_r is equal to the incident angle θ_i. The angle of refraction is subject to Snell's law. The amplitude and phase balance of the reflection and refraction effects are described by the Fresnel coefficients.

from a good conductor, with $\rho_{Vr} = 1$, and $\rho_{Hr} = -1$. The refraction coefficients for the V and H components are

$$
\begin{aligned}
\rho_{Vt} &= \frac{E_{Vt}}{E_{Vi}} = \frac{2 \cdot n_1 \cdot \cos \theta_i}{n_1 \cdot \cos \theta_t + n_2 \cdot \cos \theta_i}, \\
\rho_{Ht} &= \frac{E_{Ht}}{E_{Hi}} = \frac{2 \cdot n_1 \cdot \cos \theta_i}{n_1 \cdot \cos \theta_i + n_2 \cdot \cos \theta_t}.
\end{aligned}
\tag{5.38}
$$

Eqs. (5.37) and (5.38) encompass all the necessary information needed to understand the scattering processes occurring in typical Earth observation radars. Reflection will be the relevant phenomenon in the analysis of simplified targets, such as urban areas and land and ocean surfaces. Refraction will play a role in the penetration of the radar waves and the imaging of volumetric structures such as forests, ice, or dry soil.

5.2.7 MULTIPLE REFLECTIONS: DOUBLE BOUNCE, TRIPLE BOUNCE, AND URBAN AREAS

So far, we have considered a simple radar interaction with a dielectric. Sticking to geometrical optics, let us turn to more complicated scenarios that are particularly relevant in artificial targets and urban areas: double and triple bounces. The double bounce is characterized by the double reflection happening at a dihedral (e.g., L-formed) target as the one depicted in the left panel of Fig. 5.18.

In the case of urban areas, the double bounce is combined with the already discussed layover effect. As a consequence, not only do buildings appear laid over in the radar images, but their signatures

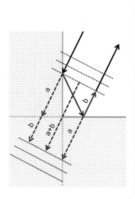

FIGURE 5.18

(Left) Double bounce scattering occurs when two reflective surfaces are perpendicular to each other. (Right) TerraSAR-X image of Guangdong; note the base of the buildings appears much brighter than the rest due to the double bounce.

Courtesy of DLR.

present bright areas at the tops and the bottoms of the structures, as shown in the radar image of Fig. 5.18. In this figure, the horizontal bright lines along the buildings correspond to the double (or triple) reflections at the vertical edge of the building aligned with the radar line of sight (LOS), i.e., the radar signature is capable of discriminating the orientation of the building with respect to the satellite orbit. Double bounce reflections are also typical in sparsely forested areas.

As an example of the sensitivity of backscattering to the orientation of the targets, Fig. 5.19 shows a monostatic (magenta) and bistatic (green) TanDEM-X image over Brasilia, Brazil.

The images are acquired with an angular separation of 2°, enough to gain sensitivity to the orientation of buildings in the city center. The magenta areas show the buildings oriented relative to the monostatic observation, whereas the green areas show an orientation favorable to the bistatic measurement. A consequence of the sensitivity to orientation is the loss of the double bounce in the general bistatic configurations, as illustrated in Fig. 5.20.

The square in the middle of the monostatic radar image (left), corresponding to a metallic fence is nowhere to be found in the bistatic image. Moreover, the solar panels at the bottom of the crop field, appear very bright in the monostatic geometry and very dark in the bistatic one. The sensitivity to the orientation of the targets can also be observed in natural areas such as crops (rows) or ocean waves.

More complex targets usually show a more elaborate reflection structure, including three or more bounces. A typical example is that of bridges, as the one shown in the left panel of Fig. 5.21. The right image shows a multitemporal image of the Sydney Harbor Bridge acquired by TerraSAR-X in spotlight mode.

The direct reflection corresponds to the thin arch and deck structures of the bridge. The double bounce causes the deck, pillars, and arch to appear a bit thicker to the left. The thicker lines correspond to the triple bounce of deck and arch, respectively. A further, better known, example of a triple bounce

FIGURE 5.19

TanDEM-X overlay of monostatic (*magenta*) and bistatic (*green*) images over Brasilia, Brazil. The angular difference between the monostatic and bistatic acquisitions is about 2°, enough to separate the monostatic from the bistatic scattering in the city area due to target orientation.

Courtesy of DLR.

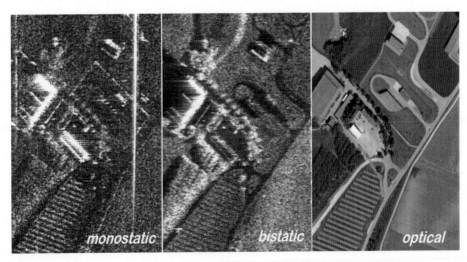

FIGURE 5.20

Monostatic (left) and bistatic (center) radar images, and corresponding optical (right) image of some buildings of Kaufbeuren airfield, Germany. The *white square* in the middle of the monostatic image corresponds to the double bounce of a metallic fence, which disappears completely in the bistatic geometry.

Courtesy of DLR.

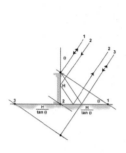

FIGURE 5.21

Schematic description of single, double, and triple bounces occurring in pedestal structures like bridges (left). TerraSAR-X spotlight image on the right over the bay of Sydney, Australia, showing the single, double, and triple bounce reflections of the Harbor Bridge.

Courtesy of DLR.

is the one occurring in trihedrals that are rather insensitive to the azimuthal angle with which they are observed. Trihedral targets, also known as corner reflectors, are commonly used for the calibration of radar systems.

So far, our discussions have dealt with the radar signatures of simplified isolated targets. In the case of a city, everything becomes much more complicated because of the more complex reflection schemes, the superposition of the signatures of buildings and structures and because of the geometrical and dielectric complexity of the individual targets. The previous TerraSAR-X images of Figs. 5.19 and 5.21 exemplify this complexity.

5.2.8 BACKSCATTERING OF SURFACES

The previous sections introduced the radar signatures of some simplified targets (e.g., dihedrals and trihedrals), which are linked to the analysis of multiple reflections and serve to help interpret scattering mechanisms in urban areas. We now turn our attention to more homogeneous areas, where the scatterers in the resolution cells are on a surface, and no penetration beyond the interface occurs. The dominant phenomenon in the interaction of radar waves and surfaces is reflection, with a spatial sensitivity on the order of the wavelength.

We will focus our consideration of the backscatter on the reflections from the observed surface, and start with a set of simple examples. The basic behavior is described by the reflection model discussed earlier under the assumption of an infinite layer boundary. The infinite layer situation is again illustrated by the left panel of Fig. 5.22, where the incident angle is now assumed to be zero, and the reflected energy is scattered back only in this direction.

If the layer has a finite length (relative to the wavelength), the situation is best described by the three figures on the right of this figure, where in all cases the incident angle is assumed to be zero.

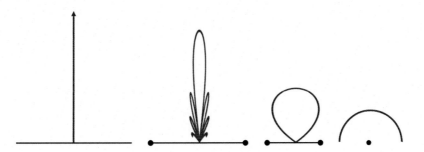

Backscatter patterns for facets of different lengths assuming a zero-degree incidence angle. The shorter the facet relative to the wavelength, the broader the backscatter pattern.

Let us start with the one on the right, which represents an isolated infinitely small scatterer. In this case, the target behaves isotropically and scatters the energy in all directions. If a short surface is assumed, as it is the case of the two figures in the middle, the surface behaves by showing directivity, the greater the larger the surface. As a consequence, most of the reflected energy follows along the reflection angle but some is scattered in other directions. In general, the aperture angle of the scattering pattern can be approximated by

$$\theta_s \approx \frac{\lambda}{L_s},$$ (5.39)

where L_s is the length of the segment, in the following facet. The former equation is analogous to that of the aperture of an antenna, and the backscattering patterns reminds one of antenna radiation patterns. For simplicity, the scattering patterns depicted have been set to sinc-forms, basically a fundamental assumption of homogeneity. The model can be assumed to be rotation-invariant; i.e., if the angle of incidence of the incoming wave is not zero, the described patterns shall be rotated accordingly. The power return of the segment can be assumed to be a function of the local refractive index of the soil, as in the reflection Fresnel coefficients.

We can represent any surface as the superposition of rotated and shifted facets of different lengths. The refractive index of the soil under a segment may be allowed to vary, hence modulating the local scattering pattern of the facet. Under the geometrical optics approximation, the scattering pattern of the scene will then be the superposition of the patterns of the facets, as shown in Fig. 5.23.

The facet model is obviously an approximation that may break down in cases where the microscopic scale of the surface is relevant or a more elaborate electromagnetic interaction between the radar waves and the surface is required. The approximation, however, serves our purposes well and helps to illustrate the relationship between the roughness of the surface and the brightness in the radar image. As a general rule, the smoother the surface, the more specular it will be. This trend is illustrated in the right plot of Fig. 5.23. Another direct consequence is that the radar backscatter will depend on the primary slope of the surface. Ascending slopes will appear brighter than flat areas, and these brighter than descending slopes, especially for smooth areas. This effect can be clearly seen in radar images over mountainous areas, such as that in Fig. 5.24. The image shows the σ_0 normalized using the ground resolution cell on the ellipsoid. Normalizing the image to the area of the wavefront

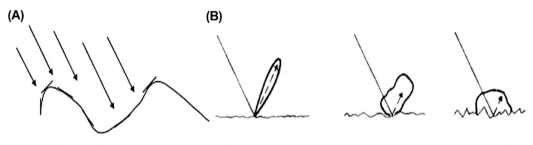

FIGURE 5.23

The left panel shows the approximation of a surface by a collection of shifted and rotated facets. The right panels show the resulting scattering diagram for different surface roughness. Increasing roughness results in a more isotropic scattering and increases the amplitude of the radar backscatter.

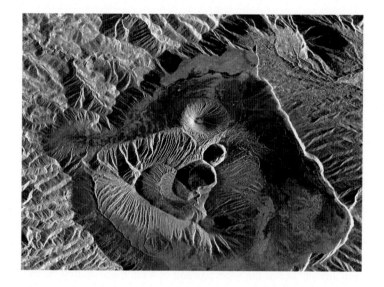

FIGURE 5.24

TerraSAR-X staring spotlight image over Mount Bromo, East Java. Radar illumination is from the left. Note that the brighter areas correspond to ascending slopes, whereas descending slopes appear darker.

Courtesy of DLR.

using the information of the local slopes results in a more homogeneous estimate of the radar backscattering.

As a summary of the previous results, Fig. 5.25 shows a plot of the radar backscatter as a function of incidence angle for three different surface roughness.

As expected, the radar backscatter decreases for increasing incident angles, more steeply for smooth surfaces, which behave in a very specular manner. Rough surfaces show brighter returns for shallow incident angles. Conversely, smooth surfaces show much brighter returns for steep incident

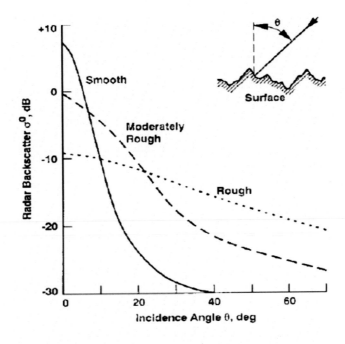

FIGURE 5.25

Radar backscatter as a function of the incidence angle for three different values of surface roughness.

angles. It is important to remember that the roughness of the surface is of course relative to the radar wavelength. A simplified model for the definition of rough assumes a small phase variation in the radar carrier due to the small height variation within the resolution cell, i.e.,

$$\frac{4\pi}{\lambda} \cdot \cos\theta_i \cdot \Delta h < k \cdot \pi. \tag{5.40}$$

where the left part of the equation approximates the phase difference between the bottom and top of the rough surface and k is a constant which can be tuned depending on the sensitivity criterion. The case where $k = 0.5$ is known as the Rayleigh criterion, whereas the more stringent $k = 0.125$ is known as the Fraunhofer criterion.

Since we have defined the directivity of a facet as a function of wavelength, the same level of geometrical roughness shall be observed differently by radars with different carrier frequencies. In particular, higher frequencies will appear more sensitive to roughness than lower ones. In Fig. 5.26, we observe a large range of radar backscatters for smooth and moderately rough surfaces. The direct consequence may be that a larger dynamic range can likewise be expected for lower carrier frequencies imaging surface areas. As an example, Fig. 5.26 shows two airborne radar images acquired by the DLR F-SAR system over Wallerfing, Germany, in X (top) and L-bands (bottom). The bottom image shows a larger dynamic range despite the equalization of the histograms prior to the display of the files.

To further illustrate the dependence of the backscatter on the roughness of the surface, Fig. 5.27 shows a calm sea (left) surrounding the island of Montserrat appearing very dark in the radar image.

FIGURE 5.26

X (top), and L (bottom) band radar images over Wallerfing, Germany, acquired by the DLR F-SAR airborne system. Radar illumination is from the top. The larger dynamic range in the bottom image is noticeable despite the equalization of the histograms performed in the generation of the images.

Courtesy of DLR.

A rougher sea, however, such as the one shown off the coast of Keflavik, Iceland (right), appears much brighter in the image. Note the runway of Keflavik airport appears very dark due to the smoothness of the asphalt. Fig. 5.28 shows fields in the state of Victoria, Australia, showing very different backscatter due to the varying equivalent roughness of the surfaces. Note the lake at the bottom of the image appears very dark, which suggests the absence of wind during the acquisition.

5.2.9 PERIODIC SCATTERING: THE BRAGG MODEL

We have so far approximated rough surfaces as a collection of translated and rotated facets. The scattering diagram of the resulting surface can be well approximated by the superposition of the scattering diagrams of the individual facets, which assumes that their combination may be somewhere between constructive and destructive. We will now turn to the analysis of periodic surfaces which yields a resonant (i.e., constructive) scattering mechanism known as Bragg scattering. This scattering process will be very helpful in the analysis of some agricultural areas, as well as ocean surfaces covered with wind generated waves. This is especially useful at lower frequencies (e.g., L band; Caponi et al., 1988).

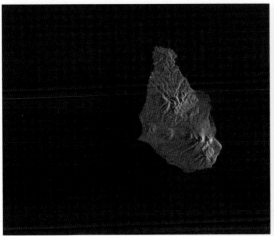

FIGURE 5.27

Calm sea surface surrounding Montserrat (left) as observed by TerraSAR-X. Rough sea surface near Keflavik, Iceland (right), acquired by TerraSAR-X. Note the reflectivity of the sea increases and a definite pattern can be observed. Note the runway of the airport appears black in the image due to the smoothness of asphalt.

Courtesy of DLR.

FIGURE 5.28

Varying backscattering in fields in the State of Victoria, Australia, as observed by TerraSAR-X.

Courtesy of DLR.

In Fig. 5.29 (top left), we see the geometry of the scattering process, where the returns of the two points on ground add coherently in the received echoes. Note the similarity of the Bragg model with the previous case of an array of antennas.

The spatial wavelength for Bragg scattering can be approximated as a function of the incident angle by

$$\lambda_B \approx \frac{\lambda}{2 \cdot m \cdot \cos \theta_i}, \tag{5.41}$$

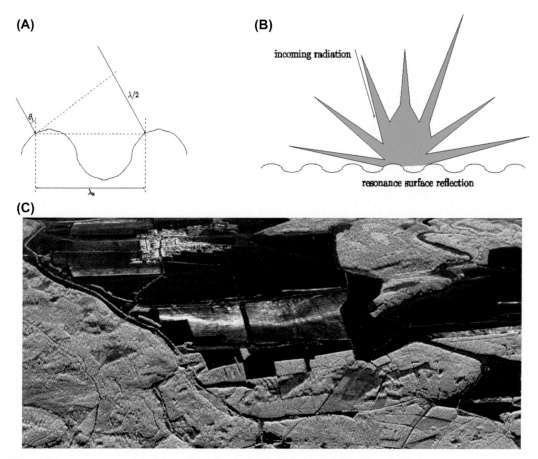

FIGURE 5.29

The top left figure presents the basic geometry of resonant or Bragg radar scatter. The top right figure shows the resulting scatter diagram of a Bragg scatterer, showing an inhomogeneous form with energy being scattered in a few different directions. The bottom figure shows an airborne radar image acquired by DLR F-SAR system at L-band over the Ammer river area, Germany. Bragg scattering occurs in the field in the center showing the white geometric pattern. Note the field appears much brighter than the forest, which suggests the presence of a resonant mechanism.

Courtesy of DLR.

where *m* is an integer. The previous condition ensures the coherent interference of the scattered echoes. The scattering diagram of a resonant surface shows an inhomogeneous pattern with periodic peaks, as presented here in Fig. 5.29 (top right). If the scene is observed with an appropriate incidence angle, the scattering of the scene appears very bright; otherwise, it remains at a lower level. In the case of ocean surfaces, Bragg scattering mainly occurs at shallow incidence angles, whereas the steep incidence angles (e.g., smaller than 20°) are dominated by specular reflection.

The bottom panel of Fig. 5.29 shows an example of Bragg scattering in an airborne radar image acquired by DLR F-SAR system over the Ammer river area, Germany. Bragg scatter appears as the white pattern in the field in the middle of the image. All other fields appear very dark[6] and even the forests are darker than the Bragg field, which clearly suggests the presence of a resonant mechanism. In addition to the surface roughness, the dielectric properties of the surface play a role in the effective brightness of the surface. In particular, changes in the permittivity of the soil will be reflected in changes in the brightness of the surface in the radar images. A typical example is humidity and water content. For the same level of roughness, wet soil appears brighter than dry soil.

5.2.10 BACKSCATTERING OF VOLUMES

Whenever the scatterers within a resolution cell are not confined to the surface, we speak of a volume scatterer. This is the typical case of the radar waves penetrating into a medium, i.e., besides reflection, a certain amount of refraction also occurs, especially in the case of homogeneous volumes. Volume scattering can occur in two different kinds of media (e.g., inhomogeneous forest and homogeneous soil) as illustrated in Fig. 5.30.

In both cases the radar is assumed to be flying on a line normal to the paper illuminating along the arrow. The purple circular corona represents the extension of a slant range resolution cell through the volume, and a fair amount of scatterers can be assumed to be present within each cell. The left plot

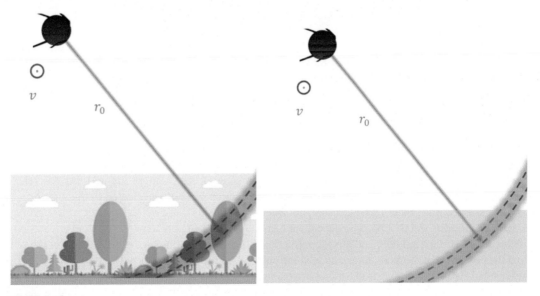

FIGURE 5.30

In volumes, the scatterers within the same resolution cell are distributed over a volume. Note the curvature of the radar wave entering into the second medium has been neglected in the figure for simplicity.

---

[6]This is a consequence of the scaling of the image due to the increase in the dynamic range due to the presence of the resonant mechanism.

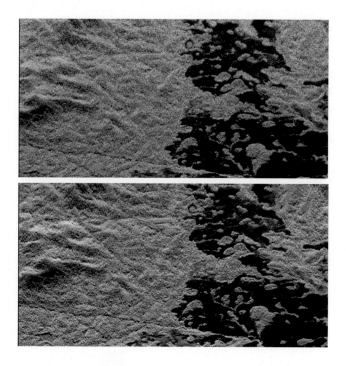

FIGURE 5.31

Fully polarimetric P band (top) and L band (bottom) airborne radar images acquired by the DLR F-SAR over Lopé National Park, Gabon. The F-SAR illumination is from the top. The color coding represents the different polarizations on the Pauli basis. The *red channel* represents the double-bounce surface trunks. Note the P band image appears more reddish due to the deeper penetration of the P band into the forest.

Courtesy of DLR.

shows a forest, a good example of an inhomogeneous volume scatterer; other examples of inhomogeneous volume scatterers are natural vegetation and crops. The right panel of the figure shows a homogeneous volume scatterer, which corresponds to snow or ice, sand, or dry soil. For the sake of simplicity, the curvature of the radar waves due to refraction has been neglected in this figure.

As expected, the scattering patterns are not as easily characterized as in the case of simple artificial targets or surfaces, and they become dependent on parameters such as heterogeneity, local orientation of dielectric properties, humidity, or varying refraction indices. Depending on the wavelength of the radar and the properties of the medium, the attenuation suffered by the radar waves will vary. Typically, longer wavelengths penetrate more than shorter wavelengths, and dense media attenuate more than diffuse media. Penetration into volumes gives the radar a rather unique capability, i.e., the potential to identify the 3-D distribution of a scatterer, allowing for structure characterization. This property is expected to play a paramount role in future radar missions in which the information of biospheric and cryospheric structures will provide insight into the evolution of climate processes. As an example, the volume backscatter of forests has been suggested to have a direct relationship to biomass (Wang et al., 1995), which will be operationally measured by the P band Earth Explorer Mission of the European Space Agency called Biomass, to be launched in 2020. For this purpose, wavelengths below S band, in

FIGURE 5.32

Fully polarimetric DLR airborne E-SAR C band (left), L band (middle), and hyperspectral AVIS instrument (right) images over Ben Gardane, Tunisia. The hyperspectral image has the RGB channels at 659, 550, and 477 nm, respectively. The color coding of the radar images represents the different polarizations in the lexicographic basis with the green channel representing the cross-pol. The dotted area on the top left side represents an area with olive trees, where the C band image indicates volumetric scattering and the L band image indicates surface scattering. The right area contains dry soil, and the features repeat, which suggests the penetration of the radar waves into the dry soil. The data were acquired in 2005 in an ESA-ESRIN project: 18746/05/I-LG. AVIS hyperspectral data by Ludwig Maximilian University in Munich.

Courtesy of DLR.

particular L band, appear to be particularly well suited. As an example, Fig. 5.31 shows a fully polarimetric airborne radar image acquired with the DLR F-SAR system over Lopé National Park, Gabon. The color coding represents the polarimetric information on the Pauli basis. Significant changes in the reflectivity of the forest can be recognized. The red channel roughly reflects the terrain. As expected, the red component is stronger in the P band image, which suggests the deeper penetration of the radar waves.

A further example over dry soil area is presented in Fig. 5.32, which shows fully polarimetric C band (left) and L band (middle) images acquired with the DLR airborne system F-SAR over Ben Gardane, Tunisia. The right-hand panel shows the corresponding hyperspectral image acquired using the AVIS instrument of the Ludwig Maximilian University (LMU) in Munich, with the color channels of 659, 550, and 447 nm. The color coding of the radar images corresponds to the lexicographic channels, with the cross-pol depicted in green. The top left area shows a plantation of olive trees, which appear strikingly different in the C band and L band images. In the C band signature, the volumetric component is stronger, while the L band image is a much clearer response of solid terrain. The right panel shows an area that contains dry soil, which appears as volume scattering in the C band image. The L band image shows a surface scattering probably due to the solid terrain underneath the sand.

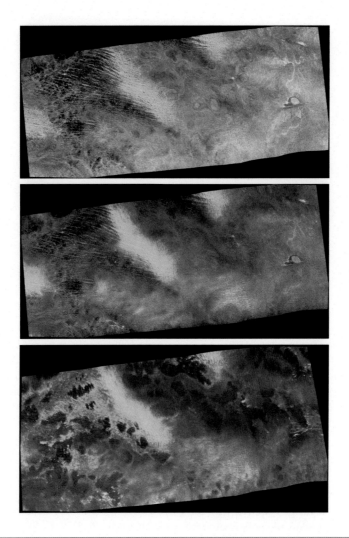

FIGURE 5.33

Fully-polarimetric geocoded DLR airborne F-SAR X-band (top), C band (middle), and L-band (bottom) images over an area close to Qeqertarsuup tunua, Greenland. The color coding represents the different polarizations in the Pauli basis. The surface and its detailed structures are clearly visible in the X band image. Some level of detail is lost in the C band image, and some subsurface structures start to become apparent due to the limited penetration of the radar signals. In the L band image, the subsurface structure of ice is more clearly depicted because of the deeper penetration of the radar waves.

Courtesy of DLR.

In Fig. 5.33 we present an even more illustrative result, with three airborne images acquired over ice in Qeqertarsuup tunua area, Greenland, by the DLR airborne F-SAR system. The figure shows X band (top), C band (middle), and L band (bottom) geocoded images, again with the color coding representing the Pauli components of the RGB channels and northing corresponding to the horizontal. Finer details of the surface appear in the X-band image and get progressively mixed with the structure of the subsurface

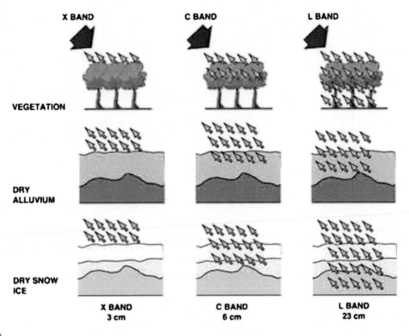

X BAND   C BAND   L BAND

VEGETATION

DRY
ALLUVIUM

DRY SNOW
ICE

X BAND   C BAND   L BAND
3 cm   6 cm   23 cm

FIGURE 5.34

The effect of increasing radar wavelength on its ability to sense through a tree canopy.

in the C and L band images. In the L band image the structure of the subsurface is clearly apparent, which illustrates the potential of longer wavelengths for imaging volumetric structures.

In Fig. 5.34 we summarize the effects of radar wavelength on the ability to sense various target surfaces.

5.2.11 OVERALL SUMMARY OF RADAR BACKSCATTER

In the previous sections, we have introduced the relevant aspects of radar backscatter mechanisms in urban areas, land and ocean surfaces, and inhomogeneous and homogeneous volumes. We have further illustrated these effects with real radar images. As a summary of these previous sections, Table 5.2 shows examples of backscatter of artificial surface, surface scatterers, and volume scatterers.

5.2.12 DEPOLARIZATION OF RADAR WAVES

Different polarizations of the radar waves are affected differently in their interactions with surfaces and volumes. Under the Born approximation, we have accepted this depolarization process to be only dependent on the geometrical and dielectric properties of the target. Conversely, the use of polarimetric radar relies on the implicit assumption that the depolarization of the radar waves yields a better understanding of the local scattering mechanisms of the scene.

Since depolarization is a relative process, all quantitative polarimetric information is retained in the amplitudes and phases between the different polarizations (Maitre, 2008). The descriptor of the scattering process in polarimetric radar needs to record the interactions between the polarizations

Table 5.2 Exemplary Values of Radar Backscatter for Typical Artificial Surface, Surface Scatterers, and Volume Scatterers

Radar Brightness	σ_0 (dB)	Target Examples
Very high backscattering	>0	Artificial objects and urban areas, ascending terrain slopes, very rough surfaces, very steep incident angles
High backscattering	−10 to 0	Rough surfaces (e.g., dense vegetation, forests, open water with wind), volumes (e.g., multiyear ice, marginal sea ice)
Moderate backscattering	−20 to −10	Moderately rough surfaces (e.g., vegetation, crops, first-year ice, open water with little wind)
Low backscattering	>−20	Smooth surfaces (e.g., calm water, roads, very dry soil, deserts)

Note the values are just representative and will change as a function of several parameters of the acquisition, such as wavelength and/or polarization.

unambiguously. In the case of a single scatterer, this can be done with the help of a complex scattering matrix, i.e.,

$$S = \begin{bmatrix} S_{HH} & S_{HV} \\ S_{VH} & S_{VV} \end{bmatrix}, \tag{5.42}$$

where the elements of the matrix represent the response of the scatterer for all combinations of input−output polarizations, i.e., the S_{HV} describes the attenuation and dephasing suffered by the incident horizontal component when scattered into the vertical component. The elements of the scattering matrix can be interpreted as the Fresnel coefficients of the target, and need to be squared to observe the power efficiency of the scattering mechanism. Assuming transparent atmospheric conditions, the scattered electrical field received by

$$E_{Rx} \approx \frac{\exp\left(-j \cdot \frac{2\pi}{\lambda} \cdot r_0\right)}{\sqrt{2 \cdot \pi \cdot r_0^2}} \cdot S \cdot \vec{E}_{Tx} = \frac{\exp\left(-j \cdot \frac{2\pi}{\lambda} \cdot r_0\right)}{\sqrt{2 \cdot \pi \cdot r_0^2}} \cdot \begin{bmatrix} S_{HH} & S_{HV} \\ S_{VH} & S_{VV} \end{bmatrix} \cdot \vec{E}_{Tx}, \tag{5.43}$$

where $\vec{E}_{Tx}$ is the transmitted electrical field. This expression characterizes all possible phase and amplitude changes due to co-polarized (HH, VV) and cross-polarized (HV, VH) scattering.[7] In monostatic observations, the reciprocity theorem for electromagnetic waves results in the symmetric assumption of the cross-pol channels

$$S_{VH} = S_{HV}. \tag{5.44}$$

While this symmetry assumption is generally valid, there are a few cases when it breaks down, notably in bistatic measurement geometries. The propagation through the ionosphere, as already

---

[7]The terms will be in the following abbreviated as co-pol and cross-pol, respectively.

hinted, may also affect the measurements of the polarization state of the radar echoes, especially in the lower frequencies (e.g., from C band downward).

Radars measuring any two elements of E_{Rx} are called dual-polarized (i.e., dual-pol) radars. Radars measuring all elements of the E_{Rx} matrix are called quad-polarized (i.e., full-pol) radars. Dual-polarized radars can be implemented by installing an antenna capable of separately receiving H and V polarizations. This architecture is likely to also provide different combinations of transmit polarizations, hence allowing the measurement of the following co-pol/cross-pol combinations VV/VH and HH/HV. Full-pol radars are typically identical to dual-pol radars from a technological point of view, only they need to be able to switch the transmit polarizations in consecutive pulses. Among the full-pol radar satellites which have been launched in the past years, we can identify TerraSAR-X/TanDEM-X, ALOS1/2, Radarsat-2, and now Sentinel-1A/B. Full-pol operation is expected to be standard in future radar missions.

Since phase and amplitude variations are descriptors of the scattering process, radar polarimetry is particularly sensitive to errors in the system. An accurate, relative calibration of the polarimetric channels is required to be able to effectively exploit the potential of radar polarimetry. In particular, channel imbalances and cross-talk effects (i.e., the contamination of a channel by an adjacent one) must be very accurately characterized in fully polarimetric radars. However, the radar system is not the only element affecting the fidelity of the recorded polarimetric channels. In their propagation through the ionosphere, the received echoes might appear rotated due to the electronic content of the atmospheric layer. This phenomenon is known as Faraday rotation and changes the distribution of the received polarimetric channels. Faraday rotation is typically estimated under the symmetry assumption of the cross-pol channels.

Under the symmetry of the crosspolar channels the scattering matrix becomes redundant, and a three-element vector suffices to describe the scattering properties of the point target. The most straightforward description is the lexicographic one, i.e.,

$$
k_L = \begin{bmatrix} S_{HH} \\ \dfrac{1}{\sqrt{2}} \cdot (S_{HV} + S_{VH}) \\ S_{VV} \end{bmatrix},
\tag{5.45}
$$

which coincides with the physical channels recorded by the radar. The lexicographic descriptor is therefore particularly well suited for the identification of residual radiometric calibration errors. The display of the received images of the polarimetric channels with an RGB-coding will allow for a visual interpretation of the dominant scattering mechanisms. For the sake of interpretation, the co-pol channels can be associated to first-order scattering mechanisms, both odd and even bounce (e.g., reflection, double and triple bounces), with the VV polarization being in general more sensitive to roughness than HH. The cross-pol channels, on the other hand, can be related to more complex scattering processes (e.g., volumes).

The choice of the lexicographic basis seems rather natural, but of course we can express the polarimetric information in many different ways. Since the information in the polarimetric descriptor does not change, the use of a different basis should have no impact on the exploitation of the

FIGURE 5.35

Examples of fully polarimetric DLR F-SAR airborne radar images of Wallerfing, Germany, in lexicographic (top) and Pauli (bottom) color coding. Radar illumination is from the top. According to the definition of the bases in and, the channel distribution is RGB for lexicographic and BRG for Pauli.

Courtesy of DLR.

polarimetric radar information. One example of particular significance, with a stronger link to the transfer coefficients of simple scattering mechanisms is the Pauli basis, i.e.,

$$k_P = \frac{1}{\sqrt{2}} \cdot \begin{bmatrix} S_{HH} + S_{VV} \\ S_{HH} - S_{VV} \\ S_{HV} + S_{VH} \end{bmatrix}. \tag{5.46}$$

A comparison between the lexicographic and Pauli basis descriptions is provided here in Fig. 5.35.

The lexicographic and Pauli representations are the most popular in the radar polarimetric community, but of course other representations may be helpful in the visual interpretation of radar images. As an

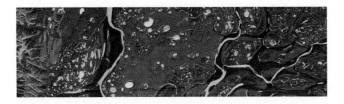

FIGURE 5.36

Delta of the Lena river in Siberia, Russia, as acquired by TerraSAR-X in dual-pol mode. The RGB channels are the HH, VV, and HH-VV components, respectively.

example, Fig. 5.36 is a dual-pol image (HH, VV) acquired by TerraSAR-X over the delta of the Lena river, Russia. The image shows the HH, VV, and HH-VV components in the RGB channels.

As discussed, the scattering matrix can be used to describe the scattering of a point target. For distributed targets, the interactions between the different scatterers within a cell require a statistical description of the scattering processes. Under the assumption of stationarity, this statistical description is fully characterized by the second-order statistics of the received polarimetric components in the desired basis. If the lexicographic basis is used, we speak of the lexicographic covariance matrix, i.e.,

$$
C_L = \frac{1}{\sqrt{2}} \cdot
\begin{bmatrix}
\left\langle |s_{HH}|^2 \right\rangle & \left\langle s_{HH} \cdot (s_{VH} + s_{HV})^* \right\rangle & \left\langle s_{HH} \cdot s_{VV}^* \right\rangle \\
\left\langle (s_{VH} + s_{HV}) \cdot s_{HH}^* \right\rangle & \dfrac{\left\langle |s_{VH} + s_{HV}|^2 \right\rangle}{\sqrt{2}} & \left\langle (s_{VH} + s_{HV}) \cdot s_{VV}^* \right\rangle \\
\left\langle s_{VV} \cdot s_{HH}^* \right\rangle & \left\langle s_{VV} \cdot (s_{VH} + s_{HV})^* \right\rangle & \left\langle |s_{VV}|^2 \right\rangle
\end{bmatrix},
\tag{5.47}
$$

where the elements of the matrix are no longer the scattering coefficients but the polarimetric channel values. In the case that the second-order statistics are expressed in the Pauli basis, we speak of the Pauli-based covariance (or coherency) matrix, i.e.,

$$
T_P =
\begin{bmatrix}
\dfrac{\left\langle |s_{HH} + s_{VV}|^2 \right\rangle}{2} & \dfrac{\left\langle (s_{HH} + s_{VV}) \cdot (s_{HH} - s_{VV})^* \right\rangle}{2} & \dfrac{\left\langle (s_{HH} + s_{VV}) \cdot (s_{VH} + s_{HV})^* \right\rangle}{2} \\
\dfrac{\left\langle (s_{HH} - s_{VV}) \cdot (s_{HH} + s_{VV})^* \right\rangle}{2} & \dfrac{\left\langle |s_{HH} - s_{VV}|^2 \right\rangle}{2} & \dfrac{\left\langle (s_{HH} - s_{VV}) \cdot (s_{VH} + s_{HV})^* \right\rangle}{2} \\
\dfrac{\left\langle (s_{VH} + s_{HV}) \cdot (s_{HH} + s_{VV})^* \right\rangle}{2} & \dfrac{\left\langle (s_{VH} + s_{HV}) \cdot (s_{HH} - s_{VV})^* \right\rangle}{2} & \dfrac{\left\langle |s_{VH} + s_{HV}|^2 \right\rangle}{2}
\end{bmatrix}.
\tag{5.48}
$$

Note that the elements of the covariance matrix contain the information of the local power transfer and dephasing between polarizations. As already discussed, the elements in the diagonal of the covariance matrices are distributed according to the Gamma distribution (Papoulis, 1965). The joint distribution of the covariance matrices can be shown to be of a Wishart class (Maitre, 2008).

A common practice in radar polarimetry consists of decomposing the previous matrices into their principal components and to associate the resulting values with elementary (orthogonal) scattering processes within the resolution cell, which helps enhance the performance of classical radar applications such as classification, target identification, and physical parameter estimation (Maitre, 2008).

5.3 RADAR SYSTEMS

We now return to a description of radar technologies concentrating on their various applications for remote sensing of the Earth.

5.3.1 RANGE-DOPPLER RADARS

Range-Doppler radars measure the delay and frequency shift of the received echoes. However, the processing of radar waveforms is confronted with an apparently unsolvable physical problem, which is the impossibility of providing optimal range and Doppler resolutions. This result, known as the Heisenberg-Gabor limit, is true for any radar waveform (Skolnik, 1980a; Le Chevalier, 1989).

The natural approach to overcome this limitation is to formally split the time coordinate t into two separate scales, one short term to measure delays, and the other one used with a longer timescale to measure Doppler shifts. In radar terminology, these two separate variables are called fast and slow times, respectively. The fast or range time, denoted as t_r, measures the time elapsed between the transmitted signal and the received corresponding echo. The slow or Doppler time, denoted as t_D, senses the Doppler shifts of the radar echoes. In reality, the slow timescale is discrete and artificially generated by the repetition of the transmitted waveform at a specific rate. The time between consecutive transmit events is called the pulse repetition interval (PRI), and its inverse the pulse repetition frequency (PRF).[8] We can rewrite the slow timescale (s_{lp}) as a train of pulses in the form

$$s_{lp}(t) = \sum_{i=0}^{I-1} p_{lp}(t - i \cdot \text{PRI}) \,,$$

$$\text{with } p_{lp}(t) = p_{lp}(t) \cdot \prod\left(\frac{t}{\text{PRI}}\right),$$

(5.49)

where the $\prod(\cdot)$ is a rectangular envelope of duration PRI, and $p_{lp}(t)$ is a pulsed complex signal of duration T_p and bandwidth B_r. From (5.49), the total observation time T_D is

$$T_D = I \cdot \text{PRI}.$$

(5.50)

---

[8]To avoid range ambiguities, all echoes must be received between two consecutive pulses, which means that the PRI must be larger than twice the difference between the maximum and minimum ranges, divided by the speed of light.

If the duration of the transmitted signal (p_{lp}) is smaller than the PRI, we will speak of a pulse radar. The duty cycle of a pulse radar is defined as

$$\eta = T_p \cdot \text{PRF}, \tag{5.51}$$

which takes values between 0 and 1. If the duty cycle equals one, i.e., the duration of p_{lp} is equal to the PRI, which refers to continuous-wave (CW) radars.[9] The essential conditions a range-Doppler radar waveform should fulfill are (1) the bandwidth of the signal must be contained within one PRI and (2) the envelope of the signal along the consecutive pulses must be constant. For convenience, (5.49) can be expressed as the discrete-time version of a 2-D continuous-time signal in the new time basis as

$$s_{lp}(t_r, t_D) = p_{lp}(t_r) \cdot \prod\left(\frac{t_D}{T_D}\right). \tag{5.52}$$

Obviously, the division of time into two separate scales is not without cost, and some intrinsic limitations are introduced into the measurement. These limitations are the maximum range and Doppler values that can be unambiguously sensed with this approach. These Doppler values can be expressed as

$$\Delta R \leq \frac{c \cdot \text{PRI}}{2} = \frac{c}{2 \cdot \text{PRF}}, \tag{5.53}$$

$$\Delta f \leq \text{PRF} = \frac{1}{\text{PRI}}, \tag{5.54}$$

where the time and frequency are again inverses of each other. We can express the 2-D impulse response of the range-Doppler radar in the new coordinates as

$$h(t_r, t_D) = p_{lp}(t_r - \tau(t_r, t_D)) \cdot \exp[-j \cdot \omega_0 \cdot \tau(t_r, t_D)] \cdot \prod\left(\frac{t_D}{T_D}\right), \tag{5.55}$$

which equals the echo of a unitary target with a delay $\tau(t_r, t_D)$ and describes the behavior of the range-Doppler radar as a linear system. Note that the previous impulse response does not necessarily correspond to a time-invariant system (Oppenheim et al., 1983; Papoulis, 1968). In typical radar cases where $v \ll c$, we can approximate (5.55) as

$$h(t_r, t_D) \approx p_{lp}\left(t_r - \frac{2 \cdot r(t_D)}{c}\right) \cdot \exp\left[-j \cdot \omega_0 \cdot \frac{2 \cdot r(t_D)}{c}\right] \cdot \prod\left(\frac{t_D}{T_D}\right), \tag{5.56}$$

where $2 \cdot r(t_D)/c$ is the delay history of the target and describes the time elapsed between transmission and reception of the echoes from a particular target as a function of the slow time. Whenever this assumption is violated, the more general expression of (5.55) should be used.

---

[9]The technological implications of this condition are large. Pulse and continuous-wave radars have often been depicted as separate technologies. This approach is in our opinion slightly confusing from the radar engineer point of view, since they can be treated in a similar way. In any case, continuous-wave radars offer a better power budget using similar peak power, whereas pulse radars offer better isolation between transmitter and receiver.

Eq. (5.56) has been derived under the assumption that the effect of the instantaneous Doppler on the low-pass pulse p_{lp} can be neglected. This condition can be expressed generally by stating that the change of τ within the PRI is negligible, i.e.,

$$d\tau(t_r, t_D) = \frac{\partial \tau}{\partial t_r} \cdot dt_r + \frac{\partial \tau}{\partial t_D} \cdot dt_D \approx \frac{\partial \tau}{\partial t_D} \cdot dt_D. \tag{5.57}$$

Besides this algebraic simplification, the previous approximation has a further geometrical interpretation: between two consecutive transmission events; the model assumes that no relative motion between the radar and the target occurs. This assumption is known in the literature as stop and go, start–stop, or stop and hop, and it is very common in Earth observation radars (Curlander and McDonough, 1991). In the example of Fig. 5.5, the stop-and-go approximation can be reduced to the expression

$$B_r \cdot T_p \ll \frac{c + v}{2 \cdot v} \approx \frac{c}{2 \cdot v}, \tag{5.58}$$

Eq. (5.58) helps identify the variables in the validity of the stop-and-go approximation. These are the transmitted bandwidth, the duration of the transmitted pulses, and the speed of the platform. Both the bandwidth (B_r) and platform speed (v) are bounded by the physical phenomena the radar is designed to observe. These limits are imposed by the geometrical resolution and observational geometry. The impact of the pulse duration has not been addressed so far. A smaller value of this impact (T_p) reduces the impact of the instantaneous Doppler, at the expense of reducing the overall transmitted energy and sensitivity of the system. A larger T_p, however, increases receiver mismatch which may degrade the SNR of the detected peaks.

The echo signal received by the radar can be seen as the superposition of the echoes over the integrated surface, which can be expressed as the 2-D integral of (5.56) times a field of complex reflectivity, all embedded in noise, i.e.,

$$e(t_r, t_D) = n(t_r, t_D) + \iint d\tau_r \cdot d\tau_D \cdot p(\tau_r, \tau_D) \cdot h(\tau_r, \tau_D; t_r, t_D), \tag{5.59}$$

where $e(t_r, t_D)$ models the radar data for an idealized continuous-time acquisition, $n(t_r, t_D)$ is a stationary random process modeling the receiver noise, $\rho(t_r, t_D)$ is the map of complex reflectivity of the illuminated scene in the radar geometry, and τ_r and τ_D are dummy integration variables, which should not be mixed up with the previous delay function $\tau(t)$.

We can define range-Doppler radar processing of the received echoes as the process of estimating the complex reflectivity map of the imaged scene from the received radar echoes, i.e.,

$$\widehat{\rho}(t_r, t_D) = R^{-1}[e(t_r, t_D)], \tag{5.60}$$

where $R^{-1}[\cdot]$ is, the operator effecting the inversion, commonly known as the radar processing algorithm. The development of efficient radar processing techniques in the Fourier domain largely depends on the possibility to express Eq. (5.60) as a convolution.

5.3.2 OPTIMAL RECEIVER FOR A SINGLE ECHO: THE MATCHED FILTER

We have assumed that the received echo of a single target can be approximated as a delayed, attenuated, and phased-shifted replica of the transmitted signal. This model has also been described as a simplification, which neglects further attenuation and dispersive terms due to propagation or reception, as well as any instantaneous Doppler spread, it is nonetheless helpful for the purpose of the analysis. Radar echoes are very weak signals, due to power losses in the two-way propagation path and to the typically small values of the reflectivity magnitude of the targets. As a sensible extension, we will be embedding the received signal in noise, i.e.,

$$s_{Rx}(t) = \rho \cdot p_{lp}(t - \tau) + n(t), \tag{5.61}$$

where $n(t)$ is assumed to be a stationary random process with a Gaussian distribution which accounts for the thermal noise in the radar. We are interested in deriving an optimal filter for the received echoes. In Fig. 5.37 we present the output of the matched filtered echo that fulfills the conditions discussed earlier to achieve good range resolution.

Without loss of generality, the maximum of the matched filtered echo is assumed to be located at $t = \tau$. The problem arises by considering as optimum a system that maximizes the SNR of the received signal (Carlson, 1986; Papoulis, 1965; North, 1963), which at delay τ takes the form

$$SNR = \frac{|\rho|^2 \cdot \left| \int_{B_r} df \cdot H(f) \cdot P_{lp}(f) \right|^2}{\int_{B_r} df \cdot |H(f)|^2 \cdot S_n(f)}, \tag{5.62}$$

where the Parseval identity has been used (Oppenheim et al., 1983). In this equation f is the frequency, P_{lp} is the Fourier transform of p_{lp}, H is the transfer function of the matched filter, and S_n is the power spectral density of noise. Using the Cauchy-Schwarz inequality (Bounjakowsky, 1859; Schwarz, 1888) and a careful identification of terms, it can be shown that the optimum solution to the problem is given by a filter with transfer function (Carlson, 1986)

$$H_{mf}(f) = K \cdot \frac{P_{lp}^*(f)}{S_n(f)}, \tag{5.63}$$

where K is any complex constant, and the symbol $*$ indicates the complex conjugate. The magnitude of the matched filter is thus directly proportional to the magnitude of the spectrum of the transmitted

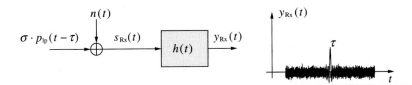

FIGURE 5.37

Optimal radar receiver for maximum signal-to-noise ratio (i.e., matched filter), with system model and sample output for a received echo.

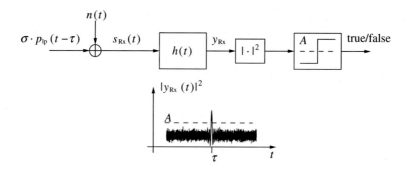

Optimal maximum likelihood radar detector with the appropriate output for a received echo.

signal, and inversely proportional to the noise power spectral density, which emphasizes the frequency bands with higher SNR and deemphasizes those with lower SNR. The impulse response of the matched filter can be computed as the inverse Fourier transform of (5.63). Particularly for the case of white noise, the impulse response of the matched filter takes the form (Oppenheim et al., 1983)

$$h_{mf}(t) = F^{-1}\left[\frac{K \cdot B_r}{\sigma_n^2} \cdot P_{lp}^*(f)\right] = K' \cdot p_{lp}^*(-t),\tag{5.64}$$

where σ_n^2 is the total noise power, $F^{-1}[\cdot]$ is the inverse Fourier transform operator, and K' is another complex constant, which can be ignored in the subsequent analysis. The matched filter has been designed to provide the best possible SNR at delay τ, which maximizes the probability of detection of the target. Using the same optimization approach, the optimal detector is depicted in Fig. 5.38.

After the matched filter the instantaneous power is computed and compared to the transmitted signal at a given threshold. This threshold is a function of the expected SNR of the scene. The key parameters playing a role in the detection capabilities of the radar are the transmitted signal energy and the sensitivity of the radar receiver, so that the filtered echo needs to exceed the noise floor. The peak after the matched filter will be proportional to

$$\langle s_{Rx}^2 \rangle = |\rho|^2 \cdot \int_{T_p} dt \cdot |p_{lp}(t)|^2.\tag{5.65}$$

The value of $|p_{lp}(t)|^2$ is bounded by the peak power of the transmitter, which emphasizes the advantage of using constant envelope waveforms to maximize power efficiency. Longer pulses (i.e., higher values of T_p) will increase the SNR of the received echoes and hence the sensitivity of the radar.

5.3.3 MATCHED FILTER VERSUS INVERSE FILTER

Let us now consider an ideal radar of infinite bandwidth transmitting a short pulse of duration T_p. The output of the matched filter will be a triangular pulse of duration $2 \cdot T_p$, as depicted in red in the left plot of Fig. 5.39.

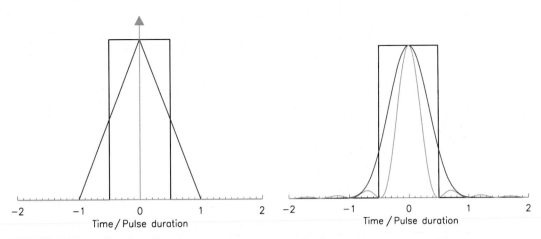

FIGURE 5.39

Output of the matched filter (red) and of the inverse filter (orange) for a transmitted pulse (black). The left plot shows the ideal waveforms after filtering. The right plot shows the real waveforms after detection. The amplitudes of the waveforms have been normalized for illustration purposes.

Note that the amplitudes of the different waveforms have been normalized for illustration purposes. The triangular waveform after idealized matched filtering becomes in a band-limited radar detector the red waveform of the right plot of Fig. 5.39.

The inversion or compression filter is defined as the filter that concentrates the maximum amount of signal energy around the delay τ, hence achieving the best resolution in the delay measurement. Consequently, the inverse filter should be able to detect the rising edge of the pulse with a precision proportional to its bandwidth, a situation which is depicted in the orange curves of Fig. 5.39. In the band-unlimited case, the filter provides the orange spike shown on the left. Note the matched filter does not concentrate so much signal energy around τ. This characteristic is independent of the bandwidth and does not change if a band-limited radar is considered. The detected band-limited spike, depicted in orange in the right plot of Fig. 5.39, still presents a narrower shape than the output of the matched filter (in red). Note that for the particular waveform used in the figure the compression filter does not exist.

The inverse filter can also be interpreted as equalizing all spectral components of the transmitted signal, i.e., the spectrum of the output signal is flat. All things considered, the transfer function of the inversion filter is

$$H_{\mathrm{if}}(f) = K \cdot \frac{1}{P_{\mathrm{lp}}(f)}, \tag{5.66}$$

where K is any complex constant. From (5.66) it is easy to see that the filter does not exist for frequencies where the signal has no energy, as is the case of many pulsed waveforms. In cases where the spectrum of the signal has low values, the inverse filter amplifies the noise energy, hence destroying the SNR and ruining the sensitivity of the system.

There exists, however, a class of radar waveforms for which matched filters are also inverse filters, i.e., best detectability and resolution are achieved simultaneously at no extra cost. Assuming a white noise environment, the waveforms whose matched and inversion filters coincide have a spectrum of the form

$$H_{\mathrm{if}}(f) = H_{\mathrm{mf}}(f) \rightarrow P_{\mathrm{lp}}(f) = K' \cdot \prod\left(\frac{f}{B_r}\right) \cdot \exp[j \cdot \Phi(f)], \tag{5.67}$$

where K' again is any complex constant, and $\Phi(f)$ is a real function. Radar waveforms fulfilling this condition are considered to be spectrally efficient.

5.3.4 OPTIMAL RECEIVER FOR RANGE-DOPPLER RADAR ECHOES: THE BACKPROJECTION OPERATOR

As just discussed, the matched filter adapts itself to the transmitted waveform, and its impulse response equals the complex conjugate of the time-inverted signal itself. In the general range-Doppler case, the construction of this matched filter is subject to two conditions: (1) the relative motion of the targets is known and (2) the transmitted signal is not affected by the instantaneous Doppler shift. The former condition can only be fulfilled in the case of stationary scenes. In many other cases, e.g., a radar measuring ocean currents, fast geophysical motions, rain Doppler changes, or moving targets, the lack of knowledge of this relative motion may cause filter mismatch. The latter condition imposes a further restriction on the transmitted radar waveform, typically known as Doppler robustness, which can be formulated as follows: besides having the full bandwidth within a PRI, the output of the matched filter should be as little unaffected as possible by instantaneous Doppler shifts (Skolnik, 1980a; Woodward, 1980).

When the transmitted signal is Doppler robust, the impulse response of the 2-D matched filter for the range-Doppler radar echoes can be expressed as the time-inverted version of the complex conjugate of (5.56), i.e.,

$$h_{\mathrm{mf}}(t_r, t_D) \approx p_{\mathrm{lp}}^*\left(-t_r - \frac{2 \cdot r(-t_D)}{c}\right) \cdot \exp\left[j \cdot \omega_0 \cdot \frac{2 \cdot r(-t_D)}{c}\right] \cdot \prod\left(\frac{-t_D}{T_D}\right). \tag{5.68}$$

Under these circumstances, we can split the 2-D matched filtering operation into two subsequent 1-D matched filtering stages, one in fast time usually known as range compression, and a second in slow time, usually known as Doppler processing.

In the general case, the matched filter after range compression, which typically shows a sinc form in the fast-time coordinate, is not time-invariant. Note this is not a consequence of the range compression, since this only happens if the filter in (5.68) is also time-variant. The time-invariance property allows for the efficient implementation of the matched filter correlation in the Fourier domain, with help of the fast Fourier transform (FFT) algorithms (Frigo and Johnson, 1998). If the matched filter is variant in both fast and slow times, radar processing requires a 2-D convolution known, in the field of coherent imaging theory, as a backprojection operator, i.e.,

$$\widehat{\rho}(t_r, t_D) = \iint d\tau_r \cdot d\tau_D \cdot e(\tau_r, \tau_D) \cdot h_{mf}(\tau_r, \tau_D; t_r, t_D), \tag{5.69}$$

where again the τ are dummy integration variables. The computational burden of the matched filter assuming 2-D time variance is roughly $O(N^4)$, where N is the number of samples in one direction.

In many cases, however, the surveys are invariant in slow time (e.g., SAR), which typically reduces the burden of the optimal receiver to $O(N^3)$. In many other cases, e.g. weather profilers, time-invariance can be used to further reduce the burden to $O(N^2)$.

5.3.5 RADAR WAVEFORMS

Recollecting the information provided on radar waveforms through the previous subsections, we can list a number of properties a good radar waveform fulfills, i.e.,

- Power efficiency: improved sensitivity and detectability require a constant envelope in the time domain, i.e.,

$$p_{\mathrm{lp}}(t) = p_0 \cdot \prod\left(\frac{t}{T_p}\right) \cdot \exp[j \cdot \phi(t)], \tag{5.70}$$

where p_0 is a real constant and $\phi(t)$ a real function. This condition can be sacrificed in case the sensitivity of the system is extremely good. As an example, waveform tapering is frequently used in sounding radars to reduce side lobes.
- Large bandwidth: the bandwidth of p_{lp} needs to be contained within one PRI. Whenever $B_r > 1/T_p$, as is usually the case, this imposes a condition on the derivative of $\phi(t)$, i.e.,

$$\frac{d^2\phi}{dt^2} \neq 0. \tag{5.71}$$

This condition is generally known in radar theory as pulse compression. Pulse compression has a dual character. For a given bandwidth, it allows the improvement of the sensitivity of the system. For a given pulse duration and sensitivity, it allows improvement of the system's range resolution.
- Spectral efficiency: good time-localization capabilities and maximal SNR, i.e., the inverse filter is also the matched filter. This condition requires a constant envelope in the frequency domain, i.e.,

$$P_{\mathrm{lp}}(f) = P_0 \cdot \prod\left(\frac{f}{B_r}\right) \cdot \exp[j \cdot \Phi(f)], \tag{5.72}$$

where P_0 is a constant and $\Phi(f)$ a real function. Further weight on the compression filter is typically used to reduce side lobes.
- Doppler robustness: filter mismatch due to instantaneous Doppler does not affect the SNR of the detected peak. Within small PRI values and large transmitted bandwidths, the approximation that instantaneous Doppler is basically a frequency shift of the received echoes is typically good.

5.3.6 A PARADIGMATIC EXAMPLE: LINEAR FREQUENCY MODULATED PULSES (CHIRPS)

Linear frequency modulated signals, commonly known as chirps, have been very popular among radar engineers due to their advantageous temporal and spectral properties. In particular, chirps fulfill the conditions just listed above. The instantaneous frequency of a chirp can be expressed as

$$f_i(t) = \pm \beta_r \cdot t, \tag{5.73}$$

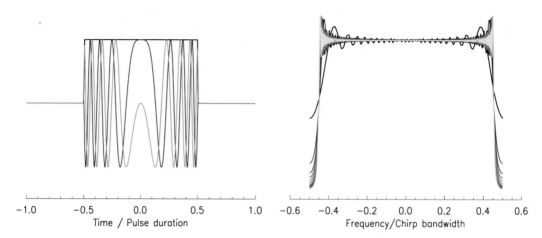

Time / Pulse duration   Frequency/Chirp bandwidth

FIGURE 5.40

The left plot shows the real (*red*) and imaginary (*orange*) parts of a chirp pulse. The right plot shows the spectral envelopes of 16 chirp signals with time-bandwidth products increasing linearly from 100 (black) to 10,000 (yellow). The spectrum of chirps with a large time-bandwidth product, tend to be asymptotic.

where the subscript i refers to the instantaneous, and β_r is the chirp rate. Depending on whether the frequency variation is positive or negative, refers to up- and down-chirps, respectively. Up- and down-chirps show identical time and spectral qualities. For simplicity and without loss of generality, in the following we will use down chirps in the derivations. The chirp bandwidth can be straightforwardly computed as the product of β_r times the duration of the pulse, i.e., the chirp rate can be expressed as

$$\beta_r = \frac{B_r}{T_p}, \tag{5.74}$$

where B_r is the transmitted bandwidth and T_p is the duration of the pulse. A linear instantaneous frequency results in a quadratic phase, i.e.,

$$s(t) = \prod\left(\frac{t}{T_p}\right) \cdot \exp\left[-j \cdot 2\pi \cdot \int dt' \cdot f_i(t')\right] = \prod\left(\frac{t}{T_p}\right) \cdot \exp\left[-j \cdot \pi \cdot \beta_r \cdot t^2\right], \tag{5.75}$$

where t' is a dummy variable and the integration constant has been neglected for compactness. The left part of Fig. 5.40 shows the real (red) and imaginary (orange) parts of a chirp pulse.

The exact Fourier transform of (5.75) cannot be computed analytically using conventional calculus. An approximate expression can nonetheless be derived using an asymptotic expansion known as the principle of stationary phase (Bleistein and Handelsman, 1975; Papoulis, 1968). Valid for oscillatory integrands, the principle of stationary phase assumes that the contribution of the high-frequency parts of the oscillation cancels out during integration, i.e., the value of the integral can be approximated by evaluating the integrand at its zero-frequency crossing, also known as the point of stationary phase. Using the principle of stationary phase, the spectrum of (5.75) can be approximated by

$$S(f) = F[s(t)] \approx C \cdot \prod\left(\frac{f}{\beta_r \cdot T_p}\right) \cdot \exp\left(j \cdot \pi \cdot \frac{f^2}{\beta_r}\right), \tag{5.76}$$

where C is a complex constant, and $F[\cdot]$ is the Fourier transform operator. As expected, the factor $\beta_r \cdot T_p$ reduces to the bandwidth of (5.75), i.e., B_r. Due to its asymptotic character, Eq. (5.76) works better for larger time-bandwidth products. The right plot of Fig. 5.40 shows the envelope of the spectrum of 16 chirp pulses with time-bandwidth products going linearly from 100 (black) to 10,000 (yellow). The magnitude oscillations are caused by the harmonic approximation of the time-domain pulse envelope, i.e., the well-known Gibbs effect (Oppenheim et al., 1983). Larger time-bandwidth products tend to be asymptotic in (5.76). Since long pulses improve the detectability of the targets and wideband signals improve the ranging capabilities of the radar, the approximation in (5.76) is valid in most typical radar cases.

Another advantage of chirp waveforms is their robustness with respect to instantaneous Doppler shifts, i.e., those that can be analyzed in terms of receiver mismatch, or segmentation of the receiver (Curlander and McDonough, 1991). To simplify the analysis, we will assume the delay to a target takes the form

$$\tau(t) \approx \tau_0 + q_D \cdot t, \tag{5.77}$$

where q_D is a dimensional and expresses the ratio between the radial relative motion radar target and the velocity of propagation of the radar waves. We have seen that (5.77) is accurate in the case of relative linear radial motion of the target, but it is also a first-order (and good) approximation within one PRI for even more complicated motions. We further neglect contributions due to τ_0 (which the radar measure anyway) and any additional attenuation terms. The received chirp echo can be expressed as

$$s(t) = \prod \left[\frac{t \cdot \left(1 - q_D \right)}{T_p} \right] \cdot \exp\left[-j \cdot \pi \cdot \beta_r \cdot \left(1 - q_D \right)^2 \cdot t^2 \right] \cdot \exp\left(-j \cdot \omega_0 \cdot q_D \cdot t \right), \tag{5.78}$$

assuming (5.77) holds and $q_D \ll 1$, the output of the matched filter in (5.78) can be expressed as

$$y(t) \approx F^{-1} \left\{ \prod\left(\frac{f}{B_r} \right) \cdot \exp\left[-j \cdot \pi \cdot \frac{\left(f + q_D \cdot f_0 \right)^2 - f^2}{\beta_r} \right] \right\}$$

$$= C \cdot \exp\left(j \cdot \frac{\omega_0^2 \cdot q_D^2}{4\pi \cdot \beta_r} \right) \cdot \text{sinc}\left[B_r \cdot \left(t - \frac{\omega_0 \cdot q_D}{2\pi \cdot \beta_r} \right) \right], \tag{5.79}$$

where C is a complex constant and $F^{-1}\{\cdot\}$ is the inverse Fourier transform. The conclusion is clear: other than a constant phase term and a slight delay shift, the form of the signal at the output of the matched filter remains unchanged. This robustness with respect to instantaneous Doppler means its detectability capabilities remain unchanged under certain circumstances. Moreover, chirp waveforms are also power efficient, and easy to generate at large bandwidths, i.e., its matched and inverse filters coincide.

There is an additional reason why chirp signals are the best waveforms for Earth and planetary observation radars. The radar data have been approximated as the superposition of the individual echoes of many scatterers in the scene. The scene, consisting of these many scatterers, is randomly distributed, and typically have a wideband character. The radar data as a sort of convolution of the reflectivity map and the radar waveform will fill the full spectral support of the system, both in range and Doppler, i.e., no information gap will be available within the bandwidth of the system. As a result, under the reasonable assumption of a wide scene spectrum, as it is customary for homogeneous and complex scenes, chirp signals are the optimal waveforms for remote sensing radars.

5.3.7 GEOMETRICAL DIALECTICS OF REMOTE SENSING RADARS

An essential requirement for the correct reconstruction of radar data is the accurate evaluation of the platform, which naturally plays a fundamental role in understanding the operation, analysis, and design of radar instruments in Earth and planetary missions. We present in this section some basic geometric definitions, combined with a discussion of the several levels of abstraction in the representation of orbital observations. The essential part of the analysis is the identification of the advantages and shortcomings of different approaches. We choose to introduce these models in a dialectic way to better exemplify their range of applications.

Let us start by considering the lateral view of the spaceborne radar shown in Fig. 5.41. The geometry is simplified, but helps to illustrate the basic elements of the radar acquisition.

The radar is assumed to be flying in the plane of the paper following the velocity vector described by the arrow and the velocity v, the s standing for spacecraft, illuminating targets on the horizontal black line below, which depicts the surface of the illuminated target. The shaded triangle in the figure represents the azimuth antenna beam of the radar, with aperture Θ_{az}. The thick gray line on the surface represents the illuminated area or instantaneous footprint of the radar antenna.

As depicted in the figure, the radar trajectory defines the along-track dimension, which is measured with the help of the slow (e.g., Doppler or along-track) timescale of the radar. A new sample is acquired with every new

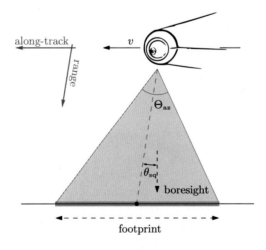

FIGURE 5.41

Simplified lateral view of a spaceborne radar. In this simplified case, the trajectory of the spacecraft defines the along-track coordinate, measured with the slow timescale of the radar. The pointing of the antenna defines the LOS coordinate, or slant range, measured with the fast timescale of the system.

transmission of the radar pulses. The dashed line between the radar and the center of the footprint represents an average LOS direction of the acquisition. The plane whose normal vector is the velocity vector $\vec{v}_s$ is called boresight or broadside. The boresight vector results from the intersection between this plane and the azimuth beam of the antenna, explicitly indicated in the figure. The angle between this vector and the LOS is called squint, denoted as θ_{sq} in the figure. This average LOS forms the (slant) range direction. Delays are measured on the fast timescale during the time elapsed within two consecutive reception events. *Stricto sensu*, the range component perpendicular to the along-track direction forms the cross-track direction of the survey. For small squint angles, range and cross-track dimensions are usually confounded.

5.3.8 PROFILER VERSUS IMAGING RADARS

A further classification of remote sensing radars refers to the dimension of the radar measurements. Radar profilers yield 1-D measurements (profiles) of range or Doppler maps, as it is the case of radar altimeters. Fig. 5.42 shows a topographic profile acquired by ESA's CryoSat altimeter over Cuba and the Caribbean Sea.

The vertical scale in this figure corresponds to the height of the surface and is measured by evaluating the position of the spacecraft, the range values of the received echoes, and the form of the reference surface. The horizontal axis is the along-track position of the spacecraft projected on the Earth's surface. Radar sounders are similar to altimeters, only the transmitted waves penetrate in the surface and image the subsurface. The profiles generated at different depths result in 2-D images as the one shown in Fig. 5.43.

Again, the vertical scale is measured through ranging from the satellite, whereas the horizontal axis corresponds to the along-track positions of the sensor. Note that the vertical thickness is a consequence of the penetration of the radar waves into the ice, allowing the imaging of the subsurface structure, using

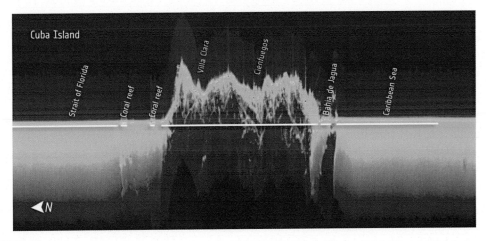

FIGURE 5.42

CryoSat altimeter view of sea level and topography over the Caribbean Sea and Cuba. The image shows radar reflections that differ in intensity between the water and elevated land. Near the edges of the island, points of high radar reflections are depicted in red. This is due to the higher reflectivity of calm waters of the bay and over coral reefs.

From ESA-AOES Medialab, http://www.esa.int/spaceinimages/Images/2012/12/Altimeter_reading_over_Cuba.

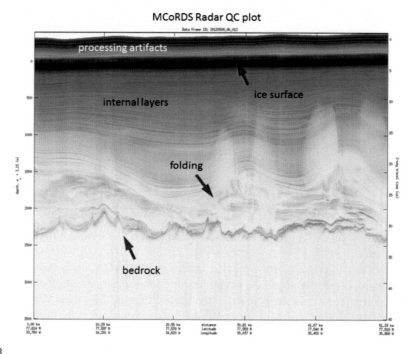

FIGURE 5.43

Example of a sounder over ice.

From NASA.

dedicated processing of the Doppler returns of the radar. In both altimeters and sounders, Doppler processing (e.g., synthetic aperture) can be used to improve the resolution in the long-track dimension.

Other radars provide 2-D images exploiting the fast-time and angular or slow-time components of the radar acquisitions. This is the case of weather radars or slide-looking radars, both with real and synthetic apertures. In Fig. 5.44 we see a radar image acquired by TerraSAR-X over Mount Okmok, in the Aleutian Islands.

In this case, the vertical side coincides with the along-track dimension, with the radar illuminating from the right.

5.3.9 NADIR-LOOKING VERSUS SIDE-LOOKING RADARS

Radars can also be classified depending on their look direction: nadir and side-looking radars. Nadir-looking radars are looking directly downward, as depicted in the left panel of Fig. 5.45. Side-looking radars are looking sideways, as shown in the figure.

Altimeters, scatterometers, and sounders are generally operated in nadir-looking geometries. Nadir-looking radars usually receive very strong echoes and typically have a much more favorable power budget. As can be inferred from the figure, in the nadir-looking configuration the left and right areas are indistinguishable for the radar. The solution of the left-right ambiguity in nadir-looking

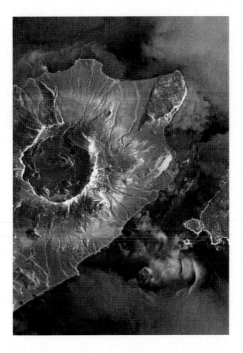

FIGURE 5.44

Radar image of Mount Okmok, United States, acquired by TerraSAR-X. Radar illumination is from the right as evidenced by the bright returns. Note also the wave trains, and the interference in the upper right corner of the image.

Courtesy of DLR.

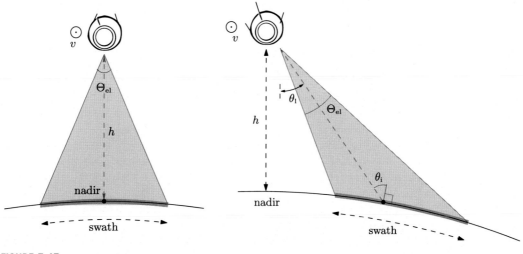

FIGURE 5.45

Along-track cuts of a nadir-looking (left) and side-looking (right) spaceborne radars.

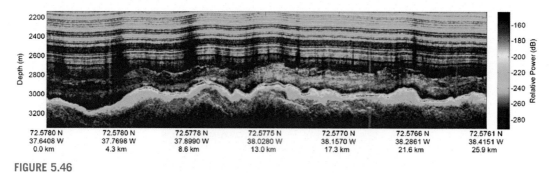

FIGURE 5.46

Ice sounder image over Greenland.

Credits: University of Kansas.

systems requires the use of two separate receive beams. These two beams can be achieved either with two separate antennas or with a single one supported by the use of analog or digital beamforming techniques (Van Trees, 2002). Downward-looking radars with a squint are usually called forward or backward-looking.

Fig. 5.46 shows an image of the ice subsurface acquired by the Multichannel Coherent Radar Depth Sounder (MCoRDS) from the University of Kansas. The vertical scale shows the penetration of the radar waves into the media.

As depicted in the right drawing of Fig. 5.45, side-looking radars point the antenna beam sideways, hence avoiding the left-right ambiguity of nadir-looking systems. The backscattered energy in side-looking observations is typically lower than in the nadir-looking radars, i.e., side-looking radars require in general higher transmit power. Real aperture radars (RARs) and SARs are typically operated in side-looking geometries.

5.3.10 DISTORTIONS OF THE RADAR SIDE-LOOKING GEOMETRY

Side-looking imaging radars exhibit a cylindrical symmetry. The real scenes appear mapped on the slant range plane, approximately formed on the surface of a family of cylinders around the radar trajectory, all with different radii. To illustrate the range projection of the side-looking radar, Fig. 5.47 (left) shows the mapping of three targets a,b,c placed on a common ground plane in the radar coordinates. The radar is placed on the gray spot of the figure and a flat-Earth geometry has been assumed for the sake of simplicity. The targets are mapped at slant ranges a', b', c' in the radar image.

It is easy to see that the radar mapping is nonlinear and the distances between points are not maintained. The natural consequence is the following: the radar images appear distorted with relative to their natural projections. To better illustrate this effect, Fig. 5.47 presents (right) an image pattern with several geometric features in the ground plane (top) and in side-looking radar coordinates (right bottom). We can see that the distances at near-range shrink, which is the reason for the curvature of the green line and the deformation of the red circle. In Fig. 5.48 we present an airborne radar image acquired by the DLR airborne F-SAR system over Kaufbeuren, Germany. The incidence angle of the near-range area (right top) is roughly 20°. Note the shrinking of the near ranges due to the side-looking geometry of the system.

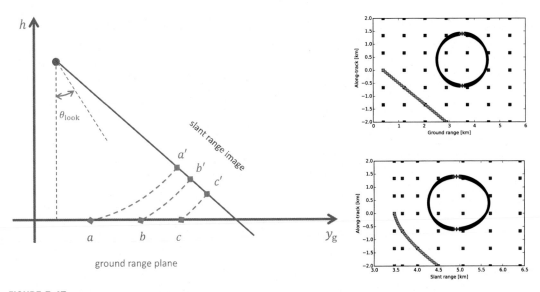

FIGURE 5.47

On the left is shown the mapping of targets placed on the ground plane onto the radar slant range plane. (right top) Mapping of an exemplary scene on a flat surface (top) to side-looking radar coordinates (right bottom). The example corresponds to an airborne geometry, with a large change in the incident angle from near to far range.

FIGURE 5.48

Airborne radar image acquired by DLR F-SAR system over Kaufbeuren, Germany. Radar illumination is from the top. Note the shrinking of the near ranges due to the side-looking observation geometry.

Courtesy of DLR.

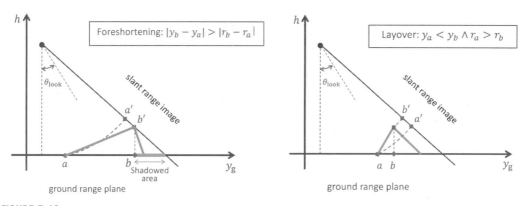

FIGURE 5.49

Sample side-looking geometry illustrating foreshortening, shadowing effects (left), and the layover effect (right).

Of course, real scenes are rarely viewed on a ground plane. In 3-D scenarios, the side-looking geometry causes several well-known distortions in the representation of the ground resolution. These distortions are foreshortening, shadowing, and layover, all illustrated in Fig. 5.49.

In the foreshortening case (left), we assume the target scene has a mountain represented by the red triangle, and we wish to map the base, denoted as a, and the top, coinciding with b', on the ground plane $x_g y_g$. In the radar image, however, the base appears mapped at a'. The distance $a'b'$ is smaller than ab, i.e., the radar image appears foreshortened. Foreshortening is typical of mountainous areas with moderate heights. Shadowing is also represented in the left panel of Fig. 5.49 by the blue stripe behind the mountain. Due to the side-looking geometry, all this area will be invisible to the radar. The shadowed area appears in the radar image having zero reflectivity and is full of noise, similar to the reflectivity of roads, calm oceans, and lakes. In Fig. 5.50 we show an example of a TanDEM-X image over Réunion

FIGURE 5.50

TanDEM-X bistatic image over the Piton de la Fournaise, on Réunion Island. The dark areas are due to shadowing. The bright areas correspond to the ascending slopes of mountains illuminated by the radar and appearing foreshortened in the radar image.

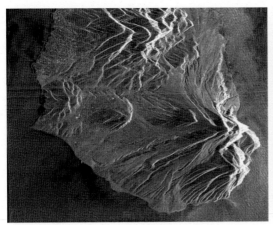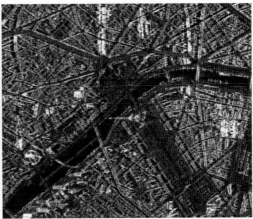

FIGURE 5.51

Examples of layover in TerraSAR-X spotlight images in mountainous and urban areas. The left image shows the Island of Montserrat; the mountains appear bent toward the radar, a typical sign of layover. The right image shows part of a Paris metropolitan area, France. The Eiffel tower shown in the middle appears laid over toward the radar.

Courtesy of DLR.

Island, close to the Piton de la Fournaise. The black areas in the image are due to shadowing. The bright areas show the ascending slopes of mountains, which appear foreshortened in the image.

Layover is another distortion in which the order of two targets is swapped in the radar image with respect to their positions on the ground plane. The effect is illustrated in the right panel of Fig. 5.51 and can be interpreted as excessive foreshortening. The situation is analogous to the previous one in the sense that the ground range coordinate of a is smaller than that of b. However, in this case, the slant range to the top of the mountain is in effect smaller than the range to the point at its base, i.e., the mountain appears laid over in the radar image. The mathematical condition is explicitly stated in the figure. Layover is typical of mountainous areas as shown in the left image of Fig. 5.51. The radar illumination is from the right, and the mountains of Montserrat appear bent toward the radar.

Since layover is a consequence of the height of the targets, layover also happens in urban areas. An example of urban layover is shown in the right panel of Fig. 5.51, showing a TerraSAR-X image over Paris city center. The radar illumination is from the left and the Eiffel tower appears bent toward the radar.

5.3.11 FLAT EARTH VERSUS CURVED SURFACE

The difference between flat-Earth and curved surface exists between airborne and spaceborne systems. A flat-Earth geometry is fairly uncomplicated in that the imaged scene lies in the horizontal plane, and the trajectory of the radar follows rectilinear trajectory.[10] This geometry is typically valid for airborne radars, which fly low enough that the Earth's curvature can be neglected. For spaceborne radars, the

---

[10]The flat-Earth approximation only assumes the first condition; the second is usually known as the linear tracks approximation. They are usually combined because their order of accuracy is similar. In the case of airborne radars, the real trajectories are usually not rectilinear (even if desired) due to turbulences and aircraft deviations with respect to the reference trajectory.

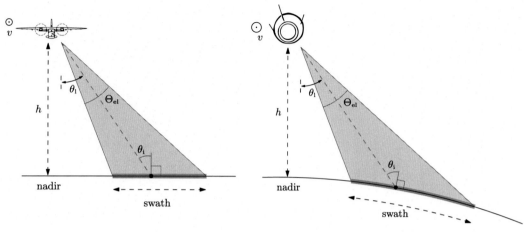

FIGURE 5.52

Flat-Earth representation of a side-looking airborne survey (left) versus a curved Earth representation (right). Flat-Earth geometries are very good approximations for airborne radar observations, as depicted in the left figure.

flat-Earth assumption is only used for preliminary and simplified analyses. Fig. 5.52 (left) shows an along-track cut of an aircraft over a flat-Earth geometry.

In Fig. 5.52 the aircraft flies at a height h over its nadir on the ground, and the antenna beam is shown in gray. The elevation aperture of the radar antenna is explicitly written in the drawing, Θ_{el}. The extension on the ground of the elevation footprint is called the swath. In a flat-Earth geometry, the look angle θ_l, between the pointing direction of the antenna and the nadir vector, coincides with the incident angle at the center of the scene, θ_i, which corresponds to the angle between the direction of propagation and the normal to the scene. In a flat-Earth geometry, slant and ground ranges are always part of a rectangular triangle and the geometry can be solved using Pythagoras' theorem, i.e.,

$$y_{gr}(p) \to r_0(p) = \sqrt{h^2 + y_{gr}(p)^2} = \frac{h}{\cos \theta_i(p)}, \tag{5.80}$$

where p indicates a point on ground, y_{gr} and r are the ground and slant range coordinates, respectively. The swath of the radar can be expressed as

$$S_w = h \cdot \left[\tan\left(\theta_l + \frac{\Theta_{el}}{2}\right) - \tan\left(\theta_l - \frac{\Theta_{el}}{2}\right) \right] \approx \frac{h}{\cos \theta_i \cdot \sin \theta_i} \cdot \Theta_{el}, \tag{5.81}$$

which, as expected, is larger for higher altitudes.

In many cases, however, the use of the flat-Earth geometry for spaceborne cases gives only a fair approximation and a more elaborate model is required. A typical example is that of the computation of coverage, where the flat-Earth approximation will result in erroneous values of orbit cycles to cover the globe. Thus, a local spherical surface is use to approximate the imaged scene, for a radar flying in a Keplerian orbit. This model is not necessarily used for real mission analysis and operation, where typically the real topography of the planet and real orbits are used, but it is definitely useful for an approximate coverage and performance analysis.

Fig. 5.52 (right) shows an azimuth cut of a spacecraft in spherical geometry.

The first obvious difference is that the equality between look and incident angles disappears, with the incident angles increasing faster for increasing slant ranges. As a consequence, the slant ranges appear compressed with respect to the flat-Earth geometry. The slant range is part of an obtuse triangle which can be solved with help of the cosine law, i.e.,

$$r_0 = R_P \cdot \cos\theta_i - \sqrt{R_P^2 \cdot \cos^2\theta_i + h\cdot(2\cdot R_P + h)}, \tag{5.82}$$

where R_P is the radius of the planet (e.g., the Earth) at the latitude of the scene. In the same manner, the incident angle can be solved as a function of the slant range

$$\theta_i = \arccos\left[\frac{(R_P + h)^2 - r_0^2 - R_P^2}{2\cdot r_0 \cdot R_P}\right]. \tag{5.83}$$

The swath can be approximated in the following manner:

$$S_w \approx \left[R_P \cdot \cos\theta_i - \sqrt{R_P^2 \cdot \cos^2\theta_i + h\cdot(2\cdot R_P + h)}\right] \cdot \frac{\Theta_{el}}{\sin\theta_i}. \tag{5.84}$$

5.3.12 GROUND VELOCITY

An additional consequence of Earth curvature and rotation is that the velocity of the footprint on ground, typically known as ground velocity, is smaller than the velocity of the spacecraft (Raney, 1986; Cumming and Wong, 1979). This is natural consequence, since both the radar and its footprint cover simultaneously similar angular spans if no steering of the antenna is required. The situation is better depicted in Fig. 5.53, where the diamonds represent both the radar and its footprint on ground.

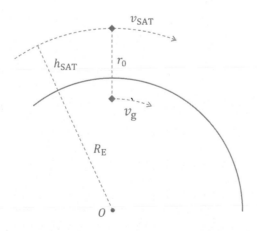

FIGURE 5.53

Illustration of the spacecraft and ground velocity in a quasicircular orbital geometry.

The ratio between the ground and spacecraft velocities decreases with higher orbits, increasing incident angles, and increasing orbit inclinations. For quasicircular orbits, the ground velocity can be approximated by

$$v_g(r_0) \approx v_s \cdot \frac{R_P}{R_P + h} \cdot \cos(\theta_i - \theta_l) \pm \frac{2\pi}{T_{\text{rot}}} \cdot |R_P \cdot \cos \theta_{\text{lat}}| \cdot \sin \varphi_{\text{incl}}, \tag{5.85}$$

where v_s is the velocity of the spacecraft, T_{rot} is the rotation period of the Earth, θ_{lat} is the latitude of the point on ground, and φ_{incl} is the inclination of the orbit. The second term approximates the effect of the rotation of the planet, and the plus or minus sign discriminates between descending and ascending passes, respectively. It is important to note that the ground velocity is changing with range (and decreasing for a flat scene), one crucial difference with respect to the flat-Earth model depicted in Fig. 5.53, where both the ground and spacecraft velocities are the same.

5.3.13 LOCAL VERSUS GLOBAL COORDINATE SYSTEMS

The choice of an appropriate coordinate system to describe the motion of the remote sensing radars is typically influenced by the application of the radar measurements and the personal preference of the user. To analyze the geometry of radar missions, where the computation of distances and angles from different points with respect to different planes is required, the use of a Cartesian coordinate system as a reference appears to be a sensible choice. Further rotations to move within the platform for changing the reference (e.g., phase centers of the antenna of the instrument, the global navigation satellite system (GNSS), or the communication link) can be easily accommodated.

A local coordinate system, as the one shown on the left of Fig. 5.54, can be either centered on the instrument or on the scene. This choice may help to simplify the analysis of the geometry of the radar survey.

These simplified analyses are only typically valid for local scales and start losing either significance or illustrative power for national or continental computations. Local coordinate systems are therefore typically only valid for the analysis of airborne surveys. For satellite surveys, they are at best suited for preliminary computations.

In the case of orbital geometries, a global coordinate system with an origin at the center of the Earth, as the one depicted in Fig. 5.54 right is preferable. Now we can distinguish between two different solutions depending on whether we focus on the analysis of the orbits or on the radar geometries. The orbits can be assumed to be confined to an orbital plane and do not rotate together with the planet. This suggests the use of the nonrotating (i.e., inertial) version of the reference frame shown in Fig. 5.54 right for propagating the state vectors of the orbit considered. For the analysis of the radar geometry, however, all computations will be done with respect to scenes rotating with the planet, and the noninertial reference frame of the figure will be preferred (Fig. 5.55). Both the trajectory of the radars and the points of the target scene will be preferably expressed in this reference frame, called Earth-centered Earth-fixed (i.e., ECEF).

The transformation of the scene into ECEF coordinates is done from a natural geographic coordinate system, typically in the form of latitude, longitude, and elevation. The elevation is usually given with respect to a well-characterized reference, linked to the Earth, which must describe unambiguously the distance between the point and the center of the Earth. Under this assumption, the transformation of the geographic coordinates ($\varphi_{\text{lat}}, \varphi_{\text{lon}}, h$) into the ECEF frame can be expressed as

$$x = [R_P(\varphi_{\text{lat}}, \varphi_{\text{lon}}) + h] \cdot \cos \varphi_{\text{lat}} \cdot \cos \varphi_{\text{lon}},$$

$$y = [R_P(\varphi_{\text{lat}}, \varphi_{\text{lon}}) + h] \cdot \cos \varphi_{\text{lat}} \cdot \sin \varphi_{\text{lon}}, \tag{5.86}$$

$$z = [R_P(\varphi_{\text{lat}}, \varphi_{\text{lon}}) + h] \cdot \sin \varphi_{\text{lat}},$$

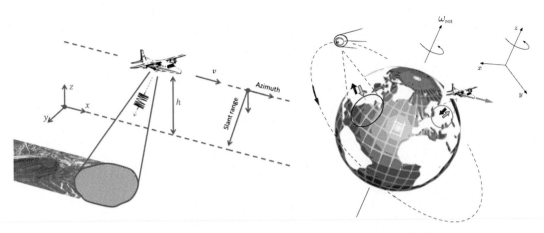

FIGURE 5.54

Local reference frame (left) and global reference frame (right). The noninertial reference frame shown on the right is typically used in the analysis of the geometry of radar missions. Note the local model might have problems accommodating Earth or orbit curvature. Conversely, the linear-track model does not seem appropriate to represent orbital geometry.

FIGURE 5.55

A geocoded airborne radar image acquired by the DLR F-SAR system over Juelich, Germany. The colors show an RGB representation of the polarimetric channels.

where the dependence of the radius of the Earth with latitude and longitude accounts for its nonspherical form. The transformation between ECEF coordinates and geographic coordinates is done by inverting the previous system of equations.

5.3.14 THE RADAR COORDINATES

As already discussed, the measurement scales of the radar are the fast and slow times. The fast-time coordinate system represents the two-way travel time of the transmitted signal when the target is placed at the center of the beam, whereas the slow-time coordinate measures the time elapsed from the start of the acquisition to the instant where the target is at the center of the beam. The previous two coordinates can be expressed with arbitrary accuracy by means of the following implicit expression

$$|\boldsymbol{p}_R(\tau_a) - \boldsymbol{p}_T| + |\boldsymbol{p}_R(\tau_a - \tau_r) - \boldsymbol{p}_T| = c \cdot \tau_r, \tag{5.87}$$

where $\boldsymbol{p}_R(\cdot)$ is the parameterized trajectory of the radar, $\boldsymbol{p}_T$ is the position of the target, and τ_r and τ_a are the fast- and slow-time coordinates of the target. Note the slow-time coordinate has been defined at the reception of the radar echo, but could have been defined as at the transmission point. Under the stop-and-go assumption, i.e., the radar remains stationary between the transmission and reception of the echoes, and the previous expression is typically written as

$$\tau_r \approx \frac{2 \cdot |\boldsymbol{p}_R(\tau_a) - \boldsymbol{p}_T|}{c} = \frac{2 \cdot r_0}{c}, \tag{5.88}$$

where r_0 is the beam-center range, or the equivalent range from the radar to the target when the latter is placed at the center of the beam, and the factor 2 accounts for the two-way propagation time of the radar signal. Making the analogy of the considered survey with a linear-track trajectory, the slow-time coordinate can be approximated by

$$\tau_a \approx \frac{x_a}{v_g}, \tag{5.89}$$

where x_a is an effective distance from the start point at range r_0 at the start of the acquisition and the considered target. Eq. (5.89) implicitly assumes the ground velocity is locally constant for the radar survey.

In a flat-Earth geometry, the projection between the ground range and the slant range depends only on the ground range, i.e., the ground range projected onto the radar coordinates becomes the slant range plane. The term is also used in curved geometries, where its validity holds only as an instantaneous approximation.

5.3.15 GEOCODING

Geocoding is the operation of mapping the radar image from radar coordinates to geographical coordinates on the surface of the Earth (Curlander and McDonough, 1991). This mapping involves the computation of the transformation between the two sets of coordinates followed by the interpolation of the image itself to fit these map coordinates. Mathematically, geocoding consists of solving a set of nonlinear equations, which corresponds to the range and Doppler coordinates measured by the radar and the point positioned on the surface of the Earth. As already discussed, these radar coordinates are related to the beam-center range and squint of the acquisition. Hence, the geocoding problem is equivalent to solving the following set of equations:

$$|\boldsymbol{p}_R(\tau_a) - \boldsymbol{p}_T| + |\boldsymbol{p}_R(\tau_a - \tau_r) - \boldsymbol{p}_T| = c \cdot \tau_r,$$

$$(\boldsymbol{p}_R(\tau_a) - \boldsymbol{p}_T) \cdot \boldsymbol{v}_R(\tau_a) + (\boldsymbol{p}_R(\tau_a - \tau_r) - \boldsymbol{p}_T) \cdot \boldsymbol{v}_s(\tau_a - \tau_r) = 2 \cdot v_R \cdot r_0 \cdot \sin\theta_{sq}, \quad (5.90)$$

$$\boldsymbol{p}_T \in \Omega_{\text{Earth}},$$

where $\boldsymbol{v}_s$ is the velocity vector of the radar, and Ω_{Earth} is a function mapping the surface of the Earth, which is assumed known; the unknown is of course the position of the target with the given coordinates $\boldsymbol{p}_T$. The previous description is a nonlinear system, which is usually solved numerically. Neglecting the motion of the satellite between transmission and reception[11] and a locally circular planet, the previous set of equations can be further simplified into

$$|\boldsymbol{p}_R(\tau_a) - \boldsymbol{p}_T| = r_0,$$

$$(\boldsymbol{p}_R(\tau_a) - \boldsymbol{p}_T) \cdot \boldsymbol{v}_s(\tau_a) = v_R \cdot r_0 \cdot \sin\theta_{sq}, \quad (5.91)$$

$$|\boldsymbol{p}_T| = R_P,$$

which also have to be solved numerically. Geometrically, the previous set of equations can be interpreted as the intersection of two spheres, one the planet and the other the iso-range surface centered at the radar, and a cone representing the LOS vector. Errors in the knowledge of the observation geometry (e.g., uncalibrated range delays, orbit inaccuracies, errors in our knowledge of the Earth's surface) all cause range and azimuth shifts which end up decreasing the geolocation accuracy of the radar. Analogously, the operation from regular coordinates to radar coordinates, i.e., the inverse operation to geocoding, is known as backgeocoding. It is easy to see that the nonlinear system is the same, the only unknowns now ar (r_0, f_{DC}).

We present in Fig. 5.55 a geocoded airborne radar image formed with multiple image strips acquired by DLR F-SAR system over Juelich, Germany. The horizontal line corresponds to the north direction in a georeferenced frame, and the tilt accounts for the inclination of the trajectory of the airplane.

5.3.16 REAL VERSUS SYNTHETIC APERTURE

A real aperture radar (RAR) is the one whose geometrical azimuth resolution is directly derived from its footprint on the ground. The azimuth resolution of a RAR is equivalent to the length on ground of the azimuth footprint and inversely proportional to the antenna length, i.e.,

$$\delta x_{RA} \approx r_0 \cdot \Theta_{az} \approx r_0 \cdot \frac{a \cdot \lambda}{L_a}, \quad (5.92)$$

where r_0 is the slant range to the scene, Θ_{az} is the azimuth aperture of the radar antenna, L_a is the azimuth length of the radar antenna, and a is a constant, which relates the aperture to the length of the antenna and is typically close to unity.[12] This value roughly approximates the footprint shown earlier. Typical values of azimuth resolutions for current low Earth orbit (LEO) spaceborne RARs are on the order of several

---

[11]This motion is fairly space-invariant during the orbit and can be easily compensated using a range-dependent along-track shift as explained in (Prats-Iraola et al., 2012).

[12]The value of $\alpha = 0.89$ corresponds to the normalized 3 dB width of a sinc pattern, which is usually approximated by 1.

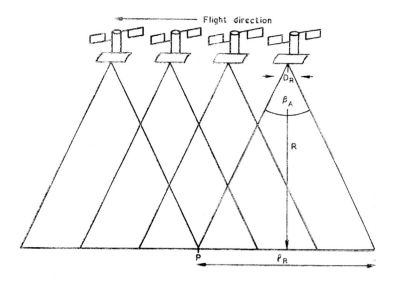

FIGURE 5.56

Synthetic aperture radar (SAR) geometry showing the SAR antenna synthesis due to the motion of the spacecraft.

hundreds to thousands of kilometers, which is acceptable to measure large-scale physical phenomena such as oceans. Medium- or small-scale phenomena, as well as details of artificial or natural surfaces, require unpractically large antennas, and can therefore only be observed with airborne RAR systems.

The concept of SAR uses the relative motion of the radar to synthesize a larger antenna size via coherent processing of the received radar echoes. Besides the relative motion of the radar, the extension of the footprint on ground is fundamental for the formation of the synthetic aperture. As depicted in Fig. 5.56, the length of the synthetic aperture is approximately equal to the extension of the footprint on ground.

This length is, however, scaled by a factor of two due to the modulation introduced by the two-way travel of the transmitted signal. Since the synthetic aperture results in a larger effective antenna size, the azimuth resolution appears improved (Curlander and McDonough, 1991), i.e.,

$$\delta x_{SA} \approx r_0 \cdot \frac{k \cdot \lambda}{L_{syn}} \approx \frac{L_a}{2}, \tag{5.93}$$

where $L_{syn} = 2 \cdot \delta x_{RA}$ is the length of the synthetic aperture. Note that the previous expression is not dependent on the range, since the length of the synthetic aperture increases linearly with it. The resolution enhancement achieved can be shown to be directly proportional to the ratio between L_{sa} and the azimuth length of the physical antenna (Curlander and McDonough, 1991), i.e.,

$$\eta_{SA} = \frac{\delta x_{RA}}{\delta x_{SA}} \approx \frac{2 \cdot r_0 \cdot \lambda}{L_a^2}. \tag{5.94}$$

In state-of-the-art SAR systems, the values of η_{SA} achieved after SAR processing are easily larger than 30 dB.

5.3.17 THE RADAR AS A COMMUNICATIONS SYSTEM

As an active remote sensing device, the radar has the structure of a classical electrical communication system, including a transmitter for the transmission of the radar waves and a receiver to receive and record the echoes backscattered by the imaged scene. Fig. 5.57 shows a schematic view of a radar illuminating a scene.

The analogy between the scene observed by the radar and the channel of a communications system is unmistakable. In formal terms, the estimation of the reflectivity of the scene is analogous to the characterization of a time-varying communication channel. As already discussed, the received signal can be expressed, locally and for a given coherence time, as the 2-D convolution of the scene reflectivity and the radar impulse response specific for the considered acquisition, i.e.,

$$s_{Rx}(t_r, t_D; \tau_r, \tau_D) = \rho(\tau_r, \tau_D) \ast h(t_r - \tau_r, t_D - \tau_D; \tau_r, \tau_D),$$ (5.95)

where $\rho(\tau_r, \tau_D)$ has been assumed to be projected onto the geometry of the radar. The radar data can be assumed to be the superposition of all these local convolutions. As expounded, the radar behaves like a linear system that is not necessarily invariant in time.

The present section is divided into three main blocks, which help understand and analyze radar systems. The first subsection presents the block diagram of the radar from a communication electronics perspective. The remaining two subsections are dedicated to antennas and propagation of radar signals.

5.3.17.1 Block Diagram

A schematic diagram of a typical Earth observation radar is presented here in Fig. 5.58. The system abstraction in the figure is to some extent arbitrary and might have been done differently; in its present form, however, it serves well the purpose of the subsection.

The figure depicts a transmitter, a receiver, and a digital unit (central electronics) responsible for the operation of the radar and the sampling and storage of the radar echoes. These blocks work synchronously with the same oscillator reference, which guarantees the coherent nature of the system. Radar oscillators are usually quartz, due to their good short-time stability, which helps in maintaining the low phase noise figure of the system.

5.3.17.2 Radar Transmitter

The particular structure of a radar transmitter changes with the specifics of the system design and is subject to constant technological improvements. Fig. 5.59 shows, as an example, a block diagram of a state-of-the-art radar transmitter. As depicted in the figure, the transmitter consists of a waveform generator synchronous to an oscillator reference, an up-converting stage, a power amplifier, and an antenna.

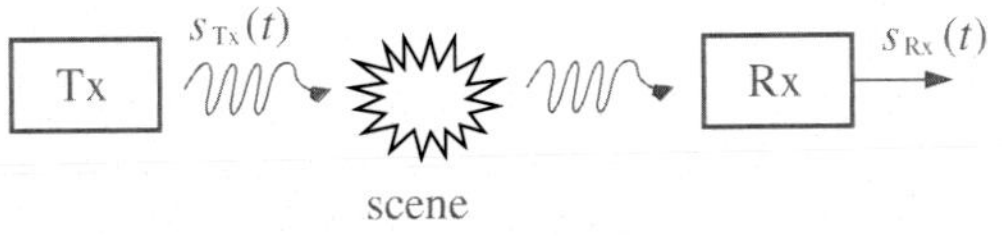

FIGURE 5.57

Schematic view of a radar measurement of a scene.

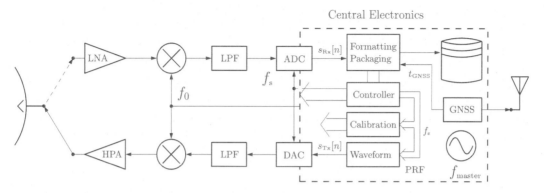

FIGURE 5.58

Schematic view of a radar system. Besides the antenna, a transmitter, a receiver, and a digital unit responsible for the generation of the radar signal, the storage of the radar echoes, and the synchronous operation of the system are all depicted.

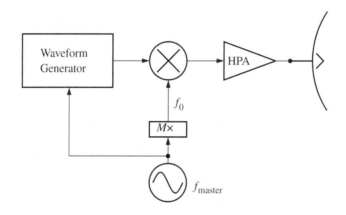

FIGURE 5.59

Schematic view of a radar transmitter.

Modern radar waveforms are generated digitally, as shown in Fig. 5.60, which suggests the idea that this block is shared between the transmitter and the central electronics. The digital generation of radar waveforms has the advantage of being flexible, accurate, and simple. A band-pass generation may simplify the up-converting stages of the receiver, thereby reducing cost and improving the linearity of the transmitter.

A waveform generator basically consists of a memory where the samples of the waveform are stored, a digital-to-analog converter, and a low-pass/band-pass filter to reduce the generation of spurious harmonics at the nonlinear stages of the transmitter (i.e., mixer, power amplifier).

Typically, direct band-pass generation of radar waveforms is possible up to frequencies around 1 GHz (P-L bands). For higher frequencies (from S band upward), an up-converting stage is necessary

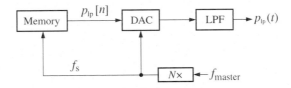

FIGURE 5.60

Schematic view of a radar waveform generator.

to bring the radar waveform to the frequency band of the radar. The band-pass signal goes through a transmission line to the power amplifier, which delivers the signal to be radiated by the antenna. In case the radar uses an active antenna, these two stages are combined. This is the case of many spaceborne past radar missions, e.g., SRTM, Envisat ASAR, ALOS1/2, CosmoSkyMed, TerraSAR-X, among others.

Power amplifiers can be solid-state, or traveling wave tubes (TWTs). The former are cheap, but deliver lower power with good linearity and are preferably used in CW or pulse radars with high-duty cycles. Active antennas use solid-state high-power amplifiers. TWTs are expensive, nonlinear, and with reduced duty cycles, but typically deliver very high peak powers. TWTs are typically used only in pulsed radars.

5.3.17.3 Radar Receiver

As in the case of the transmitter, radar receivers are subject to technological change and to the specific design of the particular system. The general structure presented here is in any case representative and instructive for the subsequent analysis. Because of the low SNR of the radar echoes, radars have typically used superheterodyne receivers (Smith, 1985; Pozar, 2000). The technological trend is to allow direct digitalization of the signal at higher frequencies, which of course reduces the cost of the electronics and improves the performance and flexibility of the receiver.

A general structure of a typical radar superheterodyne receiver is shown in Fig. 5.61.

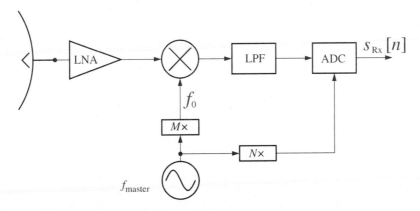

FIGURE 5.61

Schematic view of a radar receiver, including a down-conversion stage for low-pass sampling.

The master oscillator controls the radar receiver too and is used for the demodulation of the radar echoes. In monostatic radars, both transmitter and receiver share the same master oscillator, which ensures perfect synchronization of the radar system. Bistatic radars, however, with spatially separated transmitter and receivers, usually operate with different master clocks, and the synchronization of the system and the received echoes poses a fundamental challenge for their operation (Auterman, 1984; Willis, 1991).

The first stage after the antenna is a low-noise amplifier (LNA) followed by an RF filter. This structure allows to keep a low-noise figure at the receiver (Willis, 1991) and protects the SNR of the receive echoes after demodulation. Low-noise amplifiers are solid-state devices and are incorporated into the antenna structure whenever the radar uses an active antenna. Some radars, especially airborne, adapt the gain of the receiver gradually depending on the distance to the scene. This approach, known as automatic gain control, allows the use of the full dynamic range of the receiver (Skolnik, 1980a). A downconverting stage is used to baseband the received echoes followed by a filter and an analog-to-digital converter (ADC).

5.3.17.4 Central Electronics

The central electronics system is the digital unit responsible for the generation of the synchronous signals controlling the system, the operations of transmitter and receiver, and the digitization and storage of the received echoes. A block diagram of a simplified central electronics unit with some of its functionalities is found in Fig. 5.62.

The main signals generated by the central electronics are the PRF sequence including the start and end of the acquisition, the envelope and coding of the transmitted signal, the echo window, and the sampling clock. Other ancillary signals and registers like the record of the external GNSS time and

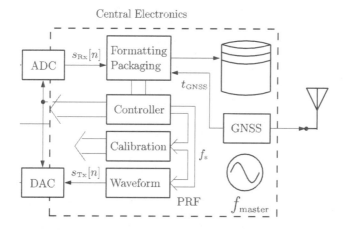

FIGURE 5.62

Schematic view of a radar digital unit. The unit includes a digital waveform generation stage, the digitizer, data packaging and storage, the controller for the synchronous operation of the system, an internal calibration network, and a global navigation satellite system (GNSS) receiver for external time and positioning references. *ADC*, analog-to-digital converter; *DAC*, digital-to-analog converter; *PRF*, pulse repetition frequency.

positioning reference, dedicated weighting coefficients for the radar antenna, or the controlling of the calibration sequences are also produced by the central electronics.

With the development of digital electronics, onboard processing of the radar signals is becoming feasible. State-of-the-art onboard processing is roughly limited to data compression algorithms (e.g., Block Adaptive Quantization or BAQ) to reduce the data volume and improve the data-link budgets of the missions. In the future, onboard processing will allow for the processing and calibration of radar data. This feature is of paramount importance in real-time applications (e.g., traffic and disaster monitoring) and planetary observation.

5.3.17.5 Radar Antennas

Antennas convert the electrical energy from the radar electronics into electromagnetic waves propagating in space and capture the electromagnetic energy scattered back by the scene. Antennas are basically conductors of a particular form through which a time-varying electrical current is flowing. As predicted by Maxwell (Ramo et al., 1994), time-varying currents generate time-varying magnetic and electric fields which propagate as electromagnetic waves in space.

The dimensions of the antenna are set proportional to the central radar wavelength. These antennas are usually small when compared to the distances over which the antenna must radiate the radar energy. Under this situation, we can identify the antenna as being the origin of a spherical coordinate system, such as the one depicted in Fig. 5.63. This coordinate system will be shown in the following to be particularly suitable for the characterization of the electrical properties of the antenna.

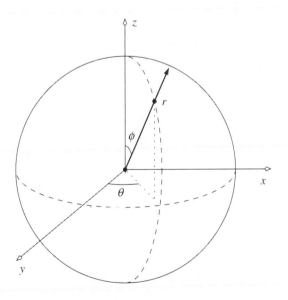

FIGURE 5.63

Spherical coordinate system for the analysis of antennas. The horizontal plane xy is commonly known as the azimuthal plane, and the vertical plane yz is commonly known as the elevation plane.

The antenna is assumed to be placed at the origin of the coordinate axes. The two angular coordinates define the planes of the antenna. The horizontal plane (xy) is known as the azimuthal plane, and the coordinate θ is known as the azimuth angle. The vertical plane (yz) is known as the elevation plane, and, consequently, the angle ϕ is known as the elevation angle. These terms are commonly used to describe the geometry of remote sensing radars.

In the region close to the antenna, i.e., the Fresnel region or near-field, the wave fronts can be assumed to be spherical, whereas farther away, i.e., the Fraunhofer region or far-field, the radar signal propagates as plane waves. The near-field region is defined as a sphere centered at the origin of coordinates, with radius proportional to the antenna size and operating wavelength, i.e.,

$$r \approx \frac{2 \cdot D^2}{\lambda},$$
(5.96)

where D is the maximum linear direction from the antenna. The far-field region is naturally the space outside the near-range sphere. In far-field, the plane wave approximation simplifies the computation of the radiated fields.

A technical boundary associated with the carrier of the radar is the size of the antennas, typically scaled with the carrier wavelength (Skolnik, 1980a; Thourel, 1971; Cardama et al., 1998). As an example, the surface of the antenna of a spaceborne L-band radar may be about 70 times larger than the surface of the antenna of an X-band radar with similar coverage. Analogously, the size of the microwave electronic components can also be roughly scaled with the wavelength, i.e., in general, radars in higher frequencies tend to be more compact than lower-frequency radars. In terms of available bandwidth, the higher-frequency bands have typically larger available bandwidths. For technological reasons, larger radar bandwidths tend to be more expensive than moderate bandwidths. However, the developments in RF electronics in the past years have made the previous statement to some extent obsolete, at least in the case of the lower radar frequency bands. In these cases, the strict limit to available bandwidth of remote sensing radars is set by international regulations (e.g., in the bands ranging from VHF to C).

The use of the RF bands is coordinated at an international level by several supranational organizations. The most notable one is the International Telecommunication Union (ITU), a United Nations (UN) agency responsible for the allocation of the RF bands for different applications. The ITU reserves a bandwidth for remote sensing in a particular frequency band, which can vary from the 6 MHz available in P band to the almost 4 GHz in Ka band. A further restriction arises in the case of airborne radars due to national regulations, especially in sensitive areas close to airports or military facilities, which typically use wideband surveillance radars. The other side of the coin is the presence of RF interference caused by artificial transmitters in the radar frequency range. This interference jams the recorded echoes and typically blurs partially or totally the relevant radar images. As an example, a P band mission like ESA's Biomass will be affected by long-range surveillance radars operating in the same frequency; its use over North America, Europe, and Russia is expected to be severely restricted for this reason. In L band missions such as JAXA's ALOS-PALSAR and ALOS-PALSAR2 typical urban radio interference has been frequently observed. Interference also happens in C and X-band systems, as reported by ESA, ASI, or DLR. In Fig. 5.64 we see an example of RF interference in a TerraSAR-X staring spotlight image over Guangdong, China. The interference reduces the contrast of the radar image for a given area hence affecting the applications based on the image analysis.

FIGURE 5.64

A TerraSAR-X staring spotlight image of Guangdong, China. The middle-left side of the image shows radio frequency interference appearing as stripes over the water.

5.3.17.6 Electromagnetic Radiation

In the far-field regions, the electrical field radiated by the antenna can be computed as the superposition of the coherent contributions of the current density over all the surface of the antenna. To exemplify this result, let us consider an antenna aperture as that depicted in Fig. 5.65. Horns have been among the most popular basic radar antennas because of their simplicity of construction, moderate gain, and bandwidths.

In the case of an antenna aperture significantly larger than the radar wavelength (λ), a narrowband approximation of the radiated field in the Fraunhofer region is given by the integral of the individual contributions from the current density with a phase term introduced by the path difference along the LOS, i.e.,

$$
\begin{aligned}
E(\theta, \phi) &= \left| E_\theta \cdot \overrightarrow{u_\theta} + E_\phi \cdot \overrightarrow{u_\phi} \right| \\
&= \int \int_{-\infty}^{\infty} dx \cdot dy \cdot A(x, y) \cdot \exp\left[-j \cdot \frac{2\pi}{\lambda} \cdot (x \cdot \sin\theta \cdot \cos\phi + y \cdot \sin\theta \cdot \sin\phi) \right],
\end{aligned} \tag{5.97}
$$

where the $A(x,y)$ is the complex current density, which is of course limited to the area of the aperture, and $\overrightarrow{u_\theta}$ and $\overrightarrow{u_\phi}$ are the director vectors of the plane defined by the direction of propagation. Note that the intensity of the radiated electric field as defined in Eq. (5.97) is the Fourier transform of the current density at the antenna.

Antennas operate in transmission and reception under the far-field assumption as reciprocal elements. Consequently, the far-field radiation pattern in reception is exactly the same as the one in transmission, which simplifies their analysis, synthesis, and characterization.

From the previous expression, one could infer one of the fundamental characteristics of antennas: the ability to concentrate or capture the radiated power within a particular part of space, a property known as directivity. The radiated power density in a given angular direction can be expressed as

$$
\xi(\theta, \phi) = \frac{|E_\theta|^2 + |E_\phi|^2}{120 \cdot \pi}, \tag{5.98}
$$

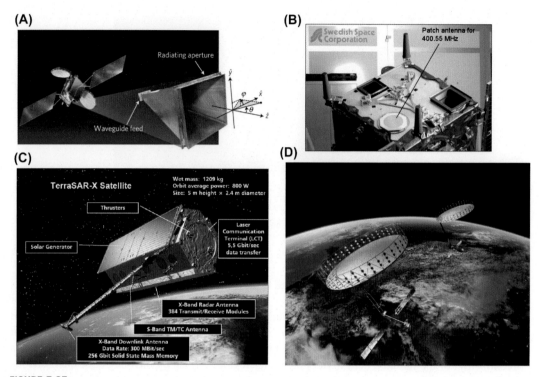

FIGURE 5.65

Picture of a horn, patch on Mango satellite, TerraSAR-X (image credit: DLR, EADS Astrium GmbH), reflector Tandem-L.

From http://www.dlr.de/hr/en/desktopdefault.aspx/tabid-8113/.

where $\eta = 120\pi = \sqrt{\mu_0/\varepsilon_0}$ is the wave impedance of free space. The radiated power density is a vector pointing radially [parallel to the Poynting vector (Ramo et al., 1994)], i.e., the total radiated power can be computed by integrating $\xi(\theta,\phi)$ on the surface of a sphere centered at the phase center of the antenna. The special case of an antenna that does not prefer any angular direction is called isotropic, which shows a radiated power density at a distance r_0 of the form

$$\xi_{\mathrm{iso}} = \frac{p_{\mathrm{rad}}}{4\pi \cdot r_0^2}, \tag{5.99}$$

where P_{rad} is the power radiated by the antenna.

The previous expressions describe the angular behavior of the antenna, which is a function of its form and of the current density distribution with which it is fed. The radiation intensity pattern of an example antenna is presented here in Fig. 5.66. Without loss of generality, only a 1-D cut of the previous function is shown, corresponding to any of the angular components (θ, ϕ). The angular component is noted off-axis to explicitly consider that the antenna might be pointing in a nonzero angle. Antenna patterns are commonly given in decibels due to their large dynamic ranges.

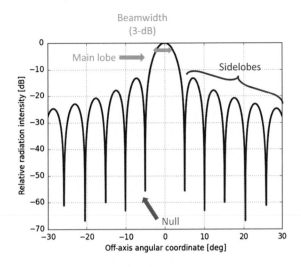

FIGURE 5.66

Relative radiation pattern in decibels for one of the antenna planes as a function of the angular coordinate. Most of the energy is radiated or captured through the main lobe of the antenna, typically characterized by its 3-dB angular extension called beam width. Spillover energy is also transmitted through the side lobes, though it is usually highly attenuated (tens of dB) with respect to the main lobe. The nulls of the antenna are angular blind spots through which almost no energy is radiated.

Most of the energy is radiated or captured back through the main lobe, marked in green, which is typically characterized by its beam width. The beam width of the antenna, also called aperture, is the angular extension where the antenna pattern falls 3 dB with respect to its maximum (called the half-power point) and is in general inversely proportional to the effective length of the antenna L_θ, i.e.,

$$\theta_a = a \cdot \frac{\lambda}{L_\theta}, \tag{5.100}$$

where the use of θ is arbitrary and does not only refer to the azimuthal plane, and a is a constant close to unity, which varies as a function of the amplitude of the current distribution. For an ideal constant amplitude of A, $a \approx 0.89$ as noted earlier.

The pattern in Fig. 5.66 has been normalized and its maximum is at 0 dB. Some energy, however, is radiated in angular directions outside the main lobe. These signals are very much attenuated, on the order of tens of decibels, relative to the energy radiated through the main lobe; this energy is transmitted through the side lobes of the antenna. As depicted in the figure, all lobes are separated by nulls of the antenna pattern, corresponding to blind angle through which almost no energy is radiated.

5.3.17.7 Polarization of Antennas

Radar waves in the far-field can be assumed to be transverse electromagnetic waves, which means the electrical field vector is orthogonal to the direction of propagation. In this plane containing the electric field vector, we can define two perpendicular axes coinciding with elevation and azimuth (or vertical

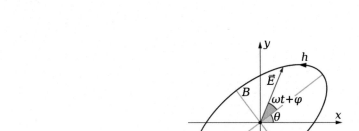

FIGURE 5.67

A polarization ellipse, which is the general polarization state. Both circular and linear polarizations can be described as elements of the elliptical case. The wave shown is right-hand polarized, assuming the direction of propagation is emerging from the paper.

and horizontal) axes of the radar antenna. The polarization of the wave is given by the shape that the instantaneous value of the electrical field projects on this plane.

Due to the periodic character of the electromagnetic fields, this shape can be linear, circular, or elliptical, depending on the phase and ratios between the amplitudes of the vertical and horizontal components of the field, e.g., E_V, E_H. The polarization of the antenna refers to the polarization of the electrical field vector of the radiated wave. Typically, antennas are either linearly or circularly polarized. Many remote sensing radars (e.g., ERS-1/2, SRTM, TerraSAR-X, COSMOSkyMed, Sentinel-1, Radarsat-2) have antennas that are able to radiate and receive in both V and H polarizations. The general case of elliptical polarization is shown in Fig. 5.67.

The condition of linear polarization can be expressed as follows:

$$\mathrm{Arg}\left(E_H \cdot E_V^*\right) = m \cdot \pi \tag{5.101}$$

where m is any integer number and the condition holds at any time. If the previous condition is not met, the polarization will be elliptical, with the special circular case if $E_H = E_V$. Whether the electric field vector moves clockwise or counterclockwise (as seen from an advanced position) will indicate, respectively, left and right-hand polarizations.

The polarization state of the radar waves is expected to change in its interaction with the target surface and consequently, the phase and amplitude differences between the different polarizations of the echoes provides relevant information of the scattering process. For this purpose, radar antennas are usually designed to provide a specific polarization of the radiated waves, which will be more sensitive to the intended measurement of a target surface.

5.3.17.8 Characterization of Antennas

Often, the electromagnetic properties of antennas are ignored and antennas are treated as an element of a transmitter-receiver chain. The classical literature on antennas identifies several parameters, which help to characterize its impact on the power budget of the radar link for both transmission and reception.

Due to the mismatch in the electronics, and to ohmic losses in the conductors, the power radiated by an antenna is less than the power delivered to it. The efficiency of the antenna is the ratio of the power radiated by the antenna to the power entering the antenna, i.e.,

$$\eta = \frac{P_{\text{out}}}{P_{\text{in}}},$$

(5.102)

which is by definition smaller than unity. The P_{out} in transmission refers to the power radiated by the antenna, and in reception to the power at the input of the line leading to the LNA. The P_{in} in transmission refers to the power delivered by the amplifier at the end of the reception chain; it is also the power scattered by the scene and captured at the antenna.

The directivity of the antenna is defined as the ratio between its radiated power density, $\xi(\theta,\phi)$, and the power density of an isotropic antenna radiating the same amount of power over a distance r_0, i.e.,

$$D(\theta, \phi) = \frac{\xi(\theta, \phi) \cdot 4\pi \cdot r_0^2}{P_{\text{rad}}},$$

(5.103)

where P_{rad} is the power radiated by the antenna. If the direction is not explicitly expressed, the directivity is assumed to be given as the direction of maximum radiation. The maximum of the directivity can be approximated for very directive antennas as

$$D = \frac{4\pi}{\Omega_{3 \text{ dB}}} \approx \frac{4\pi}{\theta_a \cdot \phi_a},$$

(5.104)

where $\Omega_{3 \text{ dB}}$ is the solid angle of the antenna aperture, and θ_a and ϕ_a are the half-power beam widths of the antenna in elevation and azimuth, respectively. The gain of the antenna is the product of its efficiency and its directivity, i.e.,

$$G(\theta, \phi) = \eta \cdot D(\theta, \phi).$$

(5.105)

As in the case of the directivity, if the direction is not explicitly stated, the gain refers to the direction of maximum radiation and the antenna is basically characterized as an amplifier. The area over which an antenna captures its power density is called the effective area. Due to its reciprocal character, the value of the effective area can be expressed as

$$A_e(\theta, \phi) = \frac{\lambda^2}{4\pi} \cdot G(\theta, \phi),$$

(5.106)

The previous parameters enable the computation of the contributions of the antennas to the power budget of the radar link. At the terminals of the antenna, the radar echoes are heavily embedded in noise. Assuming a noise power of the Nyquist form, the noise temperature of the antenna can be expressed as

$$T_{\text{ant}} = \frac{N}{k \cdot B_{\text{ant}}},$$

(5.107)

where N is noise power (usually caused by thermal and background noise radiated into the antenna), and B_{ant} is its bandwidth, which typically coincides with the bandwidth of the radar. The noise figure of the antenna can be shown to be (Cardama et al., 1998)

$$F = 1 + \frac{T_{\text{amb}}}{T_{\text{ant}}} \cdot \frac{1 - \eta}{\eta},$$

(5.108)

where T_{amb} is the temperature (in Kelvin) of the environment of the antenna.

5.3.17.9 Antenna Basics

Due to their usually adverse radar link power budgets, remote sensing radar antennas usually have moderate to high gains. The basic antenna technologies are the following:

- Wire antennas, basically wire dipoles or monopoles, with a length proportional to the central wavelength, are low-gain antennas only used for radars at lower frequencies such as UHF, VHF. The use of these antennas is limited to airborne or ground-based systems such as Carabas-II.
- Aperture antennas, basically the section of a waveguide with a specific form (e.g., rectangular horns, circular horns), achieve larger gains than wire antennas. Aperture antennas have been used extensively in airborne radars, e.g., Ramses, E-SAR (Horn et al., 1992; Dubois-Fernandez et al., 2005). Slotted waveguides are a particular case of aperture antennas, currently used in spaceborne radars such as TerraSAR-X and TanDEM-X (Werninghaus et al., 2010).
- Patch antennas made by forming a conductor pattern printed on a substrate (e.g., microstrip). These antennas usually have low gains, e.g., the ones used in the PAZ instrument (https://directory.eoportal.org/web/eoportal/satellite-missions/p/paz).
- Reflector antennas, consisting of a feed illuminated by a large reflector (e.g., a parabola). Reflector antennas usually have large gains but are difficult to deploy in space for remote sensing satellites. Reflectors will be used in future missions such as ESA's Biomass (https://earth.esa.int/web/guest/missions/esa-future-missions/biomass), DLR's Tandem-L (http://www.dlr.de/hr/en/desktopdefault.aspx/tabid-8113/), or NASA-JPL's NISAR (https://nisar.jpl.nasa.gov/nisarmission/).
- Antenna arrays are a group of antennas located with a given spatial distribution working together for transmission or reception of the radar signals. The purpose of an array of antennas is to form a more complex radiation pattern resulting from the coherent combination of the single patterns of the antenna elements of the array. Antenna arrays usually have a better directivity than a single antenna. In Fig. 5.68, we show a 1-D array of antennas, where the basic elements are depicted as thick points on the horizontal line.

Let us assume that the azimuth plane of the antenna coincides with the page, i.e., the antenna shows a cylindrical symmetry around the x-axis. Under these circumstances, the radiation pattern of the array will not depend on the elevation angle ϕ. Using the parallel rays (far-field) and the narrowband assumptions, the pattern of the antenna array of Fig. 5.68 can be expressed as

$$\xi_{\text{array}}(\theta) = \sum_{n=0}^{N-1} \xi_n(\theta) \cdot \exp\left(-j \cdot \frac{2\pi}{\lambda} \cdot \delta r_n\right) = \sum_{n=0}^{N-1} \xi_n(\theta) \cdot \exp\left(-j \cdot \frac{2\pi}{\lambda} \cdot \delta x_n \cdot \sin\theta\right), \tag{5.109}$$

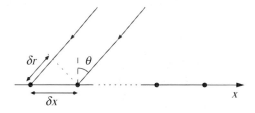

FIGURE 5.68

Depiction of a 1-D antenna array. The thick dots correspond to the positions of the antenna elements in the array.

where the ξ_n are the azimuth radiation patterns of the single antenna elements, δr_n is the difference in the range path of the n-th element with respect to a reference, and δx_n is the azimuth spacing of the n-th element with respect to the same reference. Under the condition of equidistant identical elements, the previous expression becomes separable and the array pattern becomes the product of the element pattern and a sum of the dephasing terms of (5.109), denoted in the following as array factor. The pattern of the array becomes the product of the single element, now denoted ξ, and this array factor, i.e.,

$$\xi_{\text{array}}(\theta) = \xi(\theta) \cdot \sum_{n=0}^{N-1} \exp\left(-j \cdot \frac{2\pi}{\lambda} \cdot n \cdot \delta x \cdot \sin \theta\right) = \varrho(\theta) \cdot \text{AF}(\theta). \tag{5.110}$$

Note that the summands of the array factor coincide with the kernels of a discrete Fourier transform (DFT) with a wavenumber $\sin\theta/\lambda$. Eq. (5.110) exemplifies the advantage of using antenna arrays. The shape of the pattern can be modified by changing the geometry of the array and the relative weighting of its elements. If the elements of the array have phase and amplitude setting capabilities, the array factor becomes

$$\text{AF}(\theta) = \sum_{n=0}^{N-1} w[n] \cdot \exp\left(-j \cdot \frac{2\pi}{\lambda} \cdot n \cdot \delta x \cdot \sin \theta\right), \tag{5.111}$$

where the $w[n]$ are the excitation coefficients in the array. Eq. (5.111) coincides with the DFT of $w[n]$. In this case, the analogy with apertures is clear, and arrays appear as a sampled version of the antenna aperture. The ability to excite the elements of the array is essentially the ability to change the pattern of the antenna, a quality known as beamforming.

As an example, a constant weighting of the elements sharpens the antenna pattern by a factor of $\text{sinc}\left(\frac{N \cdot \delta x}{\lambda} \cdot \sin \theta\right)$. We show the effect of this sharpening (Fig. 5.69) for two different classes of arrays. In the left plot (Fig. 5.69), the array factor (green) shows a single peak, since the spacing between the

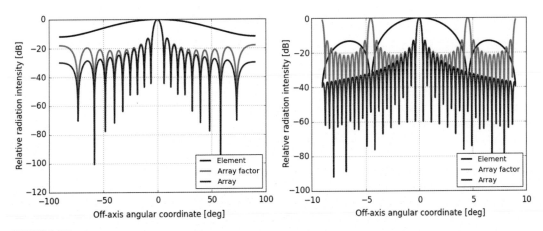

FIGURE 5.69

Component antenna (blue) and normalized array (red) patterns for the two different antenna array geometries listed in Table 5.3. The normalized array factor is plotted in green for illustration purposes.

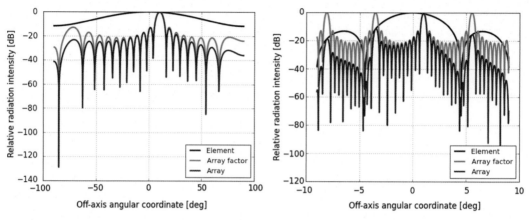

FIGURE 5.70

Electronically steered element (blue) and normalized array (red) patterns for two different array geometries. The normalized array factor is plotted in green for illustrative purposes. The left plot corresponds to the nonaliased array with a 10-degree steering. The right plot corresponds to the aliased array of Table 5.3 with a 1-degree steering. Note the ambiguous peaks of the array factor (i.e., grating lobes) are no longer attenuated by the nulls of the element pattern as they were in Fig. 5.69.

Table 5.3 Geometry of the Antenna Arrays Used for the Simulations Presented in Figs. 5.69 and 5.70

Type	Spacing (cm)	Total Length (m)	Number of Elements	Wavelength (cm)
Array 1	1	4	32	3.31
Array 2	30	4.7	12	3.31

elements is shorter than $\lambda/2$. Note that we have normalized the array factor to match the dynamic range of the component antenna pattern for illustration purposes. We also see that the pattern of the component (blue) modulates the array factor to generate the pattern of the array (red). The right plot shows the case where the spacing of the elements is larger than $\lambda/2$. The array factor (green) in this case shows several peaks in the angular span, as a consequence of aliasing in the sampling of the dephasing terms. In the array pattern (red), however, these peaks coincide with the nulls of the pattern of the component antenna (blue) and appear therefore attenuated (Fig. 5.69).

We have already seen that the array factor is the Fourier transform of the weighting coefficients of the component antennas. The properties of the Fourier transform provide us with useful information for understanding the relationship between the coefficients and the resulting patterns. In particular, the following property will be used extensively in radar antenna arrays: a linear phase in one domain transforms into a shift in the transformed domain, which will provide us with a simple way to create an electronic steering of the radiation pattern. The steering of the beam allows radars to scan different portions of space. In particular, tracking radars were the driving force in the development of antenna array technologies in the middle of the 20th century (Skolnick, 1980). For small angles, the linear phase in the spatial domain to steer the pattern an angle $\delta\theta_0$ takes the form

$$\varphi_{st}(\delta\theta_0) \approx \frac{2\pi}{\lambda} \cdot n \cdot \delta x \cdot \sin \delta\theta_0. \tag{5.112}$$

Electronic steering, unlike the mechanical one, is monochromatic, which can have an impact on the transmission of wideband signals. Fig. 5.70 shows analogous plots to those of Fig. 5.69 with two different steering laws of the radiation patterns computed after (5.111). In the left plot, a steering of $10°$ has been performed. A small increase in side lobe energy (from -30 to -25 dB) can be noticed. In the right plot, a steering of $10°$ has been carried out. The situation now is dramatically changed, and the ambiguous peaks of the array factor are no longer suppressed by the nulls of the antenna element pattern. Moreover, the peak of the main lobe of the array pattern is attenuated by the element pattern. All things considered, the main-to-side lobe ratio is significantly worsened with respect to the no steering scenario. The ambiguous peaks of the array pattern are commonly known as grating lobes.

The array model used so far does not make any assumption about the reception of the signal. In the case where the received signal is digitized and stored for every single element (or group of them), leads to a further advantage: the ability to synthesize a posteriori different receive pattern, which allows for the estimation of the direction of arrival and the use of digital beamforming techniques (Van Trees, 2002).

In an analogous way to the 1-D case, the general expression of the array factor of a 2-D array is (Cardama et al., 1998; Thourel, 1971),

$$\xi_{array}(\theta, \phi) = \sum_{n=0}^{N-1} \sum_{m=0}^{M-1} \xi_{nm}(\theta, \phi) \cdot w[n, m] \cdot \exp\left[-j \cdot \frac{2\pi}{\lambda} \cdot (x_{nm} \cdot \sin\theta \cdot \cos\phi + y_{nm} \cdot \sin\theta \cdot \sin\phi) \right]. \tag{5.113}$$

In the case of equidistant spacing in the xy plane, identical elements yield an array factor of the form

$$AF(\theta, \phi) = \sum_{n=0}^{N-1} \sum_{m=0}^{M-1} w[n, m] \cdot \exp\left[-j \cdot \frac{2\pi}{\lambda} \cdot (n \cdot \delta x \cdot \sin\theta \cdot \cos\phi + m \cdot \delta y \cdot \sin\theta \cdot \sin\phi) \right], \tag{5.114}$$

which is again a 2-D Fourier transform of the weighting function of the component antennas of the array. If the excitation coefficients are separable, i.e., $w[n, m] = w_1[n] \cdot w_2[m]$, then the array factor can be expressed as the product of the array factors of the horizontal and vertical 1-D subarrays, i.e.,

$$AF(\theta, \phi) = AF_1(\theta, \phi) \cdot AF_2(\theta, \phi)$$

$$= \sum_{n=0}^{N-1} w_1[n] \cdot \exp\left[-j \cdot \frac{2\pi}{\lambda} \cdot n \cdot \delta x \cdot \sin\theta \cdot \cos\phi \right] \cdot \sum_{n=0}^{N-1} w_2[n] \cdot \exp\left[-j \cdot \frac{2\pi}{\lambda} \cdot m \cdot \delta y \cdot \sin\theta \cdot \sin\phi \right]. \tag{5.115}$$

As expected, the steering capabilities and limitations of 1-D arrays are maintained in the case of 2-D arrays.

5.3.17.10 *Propagation of Radar Waves*

Radar waves propagate through one or several media on their two-way path from radar to target and back, suffering both attenuation and delays, which affect the detectability of the echoes and their location accuracy. This section describes the effects on the propagation in free space and through the atmosphere. The analysis of the propagation through the atmosphere can be further divided into two separate cases: (1) the nondispersive troposphere and (2) the dispersive ionosphere.

5.3.17.10.1 Propagation Through the Troposphere

As discussed earlier, the troposphere is the lowest part of the Earth's atmosphere, up to a height of about 16 km at middle latitudes. The thickness of the layer depends on the latitude of the area of interest. The two main effects of the troposphere on the propagation of the radar signals are attenuation and refraction.

Radar waves propagating through the troposphere suffer from attenuation caused by two phenomena: (1) molecular absorption by tropospheric gases, mainly oxygen and water vapor, and (2) attenuation (scattering and absorption) by hydrometeors, especially rain. Molecular absorption is negligible for carrier frequencies under 10 GHz (i.e., X-band) and typically increase for higher frequency bands, particularly at resonance peaks for water vapor and oxygen at 22.3 and 60 GHz, respectively. Typical hydrometeors are rain, snow, haze, or hail. Rain causes the most dramatic attenuation, especially for frequencies higher than 1 GHz (L band), and this attenuation depends on factors such as the type or intensity of the rain, and the size and speed of the raindrops. Due to the impact of air friction on raindrops, raindrops are flattened and the attenuation of the vertical polarization component is smaller than that for the horizontal polarization.

An example of the attenuation of the radar signals in the troposphere is shown in Fig. 5.71, corresponding to a radar image acquired by TerraSAR-X in ScanSAR mode over the Amazons in northwestern Brazil. The black spots in the image correspond to the attenuation of the radar echoes below the noise level due to heavy rain.

The propagation speed of the radar signal in the medium is inversely proportional to its refractive index n_m, i.e.,

$$c(n_m) = \frac{c_0}{n_m}.$$

(5.116)

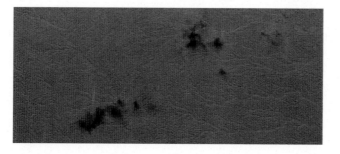

FIGURE 5.71

TerraSAR-X ScanSAR image over the Amazon rain forest in Brazil. The black spots within the image are due to the attenuation of the radar waves in heavy rain and clouds.

Courtesy of DLR.

where c_0 is the propagation speed in a vacuum. The atmospheric refractive index is a function of pressure, temperature, and water vapor content. It decreases with increasing altitude. The velocity of propagation of the radar signals in the troposphere depends on the stratification of the layer, which introduces an unknown delay in the measurement of the radar echoes. This delay changes dynamically with the stratification of the atmosphere in time and space and results in a few meters error in typical nadir and side-looking acquisitions (e.g., zenith and slant tropospheric delays). In typical cases, the inhomogeneity of the tropospheric layer introduces variable delays and phase changes in the acquired radar echoes, which need to be accounted for in applications involving accurate radar ranging (e.g., SAR interferometry).

The refractive index in the troposphere changes with altitude. We look at a model of a wave being refracted at the boundary layer of two isotropic media (e.g., air, water) with different refractive indexes. The geometry of the incident and refracted rays is depicted in Fig. 5.72.

The refractive index of the bottom layer is assumed to be higher than the refractive index of the top layer ($n_2 > n_1$); as a consequence, the angle with respect to the normal of the refracted wave, θ_{i2}, is smaller than the angle of the incident wave θ_{i1}. The geometry of the process follows Fermat's principle of the least time and its corollary which is Snell's law, i.e.,

$$n_1 \cdot \sin \theta_{i1} = n_2 \cdot \sin \theta_{i2}. \tag{5.117}$$

whereas the previous model describes the interaction at a discrete interface, the change of the refractive index in the troposphere is continuous, and the radar waves are continuously being refracted. A reference value for the refractive index in the troposphere as provided by the International Telecommunications Union (ITU) is shown in the left plot of Fig. 5.73, where the curve shows an exponential decay.

As a consequence of the decreasing refractive index with altitude, radar waves do not propagate through the troposphere along a straight line and become slightly curved while approaching the Earth's surface. The horizontal offset relative to the straight propagation for an incident angle of $35°$ and the reference refractive index is shown in the previous figure is shown in the right plot of Fig. 5.73 as a function of altitude. Note, a horizontal offset of about 1 m between the curved and the straight paths can be observed on the Earth's surface.

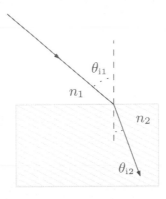

FIGURE 5.72

Refraction occurring at the boundary layer of two isotropic media with different refractive indexes.

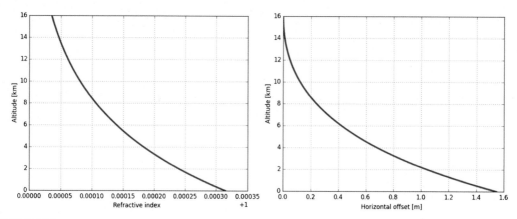

FIGURE 5.73

On the left the reference refractive index (provided by International Telecommunications Union) as a function of the altitude in the troposphere. (Right) Horizontal offset for an incident ray at 35° between the curved and straight paths assuming the reference refractive index profile on the left.

5.3.17.10.2 Propagation Through the Ionosphere

The ionosphere is the layer of the Earth's upper atmosphere between 60 and 1000 km. The ionosphere is an ionized layer with a high concentration of ions and free electrons. The ionization of this layer is mainly due to high-energy solar (X-rays and UV) and cosmic radiation. The maximum ionization occurs at the middle of the layer at altitudes of about 250–400 km. In Fig. 5.74, we show the vertical distribution of electron density in the ionosphere separated between day and night and low to the difference in solar activity.

Since the main cause of ionization in the atmosphere is solar radiation, the behavior of the ionosphere follows the solar cycles: daily, yearly, and 11-yearly, a period related to sunspot activity on the sun.

The integrated electron density along the path where the radar waves are propagating is called the total electron content (TEC) and can be obtained by integrating the density of electrons N along the path of the radar signal, i.e.,

$$\text{TEC} = \int_L N \cdot dl . \tag{5.118}$$

TEC is measured in the number of electrons per unit area. The basic unit, called TECU, can be defined as 10^{16} electrons per square meter. Depending whether the TEC is expressed in its vertical form or on the LOS of the radar, refers to either vertical or slant TEC. The available TEC maps of the Earth are typically provided by GPS measurements, wherefrom several empirical models can be adjusted.

The TEC is assumed to have two components with different spatial behavior: (1) a large-scale background value, that is rather stable and (2) a turbulent part with small-scale variations that produce

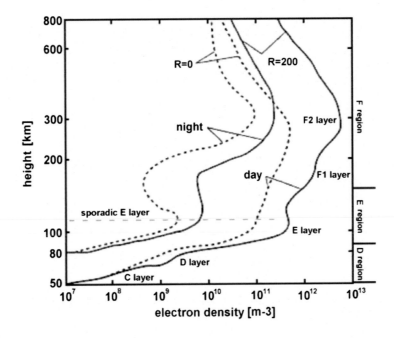

FIGURE 5.74

Electron density profiles for day and night ionospheric activity as a function of height for low and high solar activity.

the phase and amplitude scintillations of these rapid variations. The background ionosphere exhibits TEC values about one order of magnitude larger at the Equator than the poles. Scintillations, on the other hand, are due to changes in the plasma density of the layer. They are typically localized in time after the Sunset at the Equator and also occur frequently at the poles.

The plasma frequency is a function of the electron density, i.e.,

$$f_p = \frac{q_e}{2\pi} \cdot \sqrt{\frac{N}{m_e \cdot \varepsilon_0}} = \kappa \cdot \sqrt{N}. \tag{5.119}$$

where q_e is the charge of the electron, m_e is the mass of the electron, ε_0 is the permittivity in a vacuum, and κ is a proportionality constant. The plasma frequency of the ionosphere is usually on the order of a few megahertz. Radio enthusiasts will surely remember that these are the typical frequencies of the shortwaves used for continental radio communications, which benefit from ionospheric reflections. The wave propagating through the medium causes acceleration of the free charges. Part of the energy generated is radiated back to the wave and another is partially absorbed by inelastic collisions. The effect of these collisions is negligible for typical spaceborne radar frequency bands, i.e., the attenuation caused by the ionosphere can be ignored.

5.3.17.10.3 Delays, Phase Offsets, and Depolarization Caused by Inhomogeneity

The delay and dephasing of the radar signals propagating through the ionosphere is dictated by the dispersive character of the medium, which means that the impact on the radar propagation will be different for different frequencies. Neglecting both collisions and the influence of the Earth's magnetic field, the refractive index in the ionosphere is a function of the frequency, i.e.,

$$n(N,f) = \sqrt{1 - \frac{f_p^2}{f^2}}, \tag{5.120}$$

with f_p defined in Eq. (5.119). As in the case of the troposphere, the change in the refractive index of the layer will have the effect of curving the path of the radar waves, adding in this case to the dispersive character of the medium. The phase velocity of the radar wave can be expressed as

$$c_{\text{phase}}(N,f) = \frac{c_0}{\sqrt{1 - \frac{f_p^2}{f^2}}}, \tag{5.121}$$

which always takes higher values higher than c_0. Alternatively, the group velocity can be expressed as

$$c_{\text{group}}(N,f) = c_0 \cdot \sqrt{1 - \frac{f_p^2}{f^2}}. \tag{5.122}$$

The previous values dictate that the phase offset, introduced in the two-way propagation of the radar signals, can be approximated by

$$\delta\phi_{\text{iono}} \approx \frac{4\pi}{\lambda_0} \cdot \int_L \text{Re}[n-1] \cdot dl \approx -\frac{4\pi}{c_0 \cdot f_0} \cdot \kappa^2 \cdot \text{STEC}, \tag{5.123}$$

where the STEC is TEC expressed in the slant geometry of the radar, and a monochromatic approximation has been used. The additional delay in the envelope of the received echoes introduced by the ionosphere can be computed as

$$\delta\tau_{\text{iono}} \approx 2 \cdot \int_L \frac{dl}{c_{\text{group}}} - 2 \cdot \int_L \frac{dl}{c_0} \approx \frac{2 \cdot \kappa^2}{c_0 \cdot f_0^2} \cdot \text{STEC}, \tag{5.124}$$

where the monochromatic approximation has been used. Unlike the troposphere, the delay and dephasing of the radar echoes is not the same, which may have an impact on the Doppler processing of the echoes. Similarly, inhomogeneity in the TEC structure will introduce filter mismatch both in range and Doppler, causing signal energy loss, which affects the local accuracy of the system.

A consequence of the combination of ionosphere and the Earth's magnetic field is that the phase velocity of a circularly polarized wave propagating through the ionosphere will depend on its orientation. Consequently, any polarized wave propagating through the ionosphere will suffer a rotation of its polarization state due to the difference between the phase velocities of its left-hand and right-hand basic components.[13] This effect is known as Faraday rotation and can be approximated as (Gray et al., 2000),

$$\Omega_F \approx \frac{\kappa^2}{m_e \cdot c_0 \cdot f_0^2} \cdot \text{STEC} \cdot \left(\vec{B} \cdot \vec{u_k}\right), \tag{5.125}$$

where $\vec{B}$ is the magnetic field of the Earth, and $\vec{u_k}$ is the unitary wave number vector. The previous expression neglects the curvature of the radar waves in their path through the ionosphere.

5.4 SYNTHETIC APERTURE RADAR

The SAR principle,[14] suggested by Carl Wiley in 1951 (Wiley, 1985), allows an enhancement in the along-track resolution of a conventional RAR by combining the Doppler information of the echoes in a coherent manner (Brown, 1967). While still working for Goodyear, Wiley was able to demonstrate the experimental feasibility of Doppler beam sharpening with the L band DOUSER system flying onboard a Douglas DC-3 (Willis, 1991). Simultaneously and independently, Sherwin et al. (1962) developed the same idea and promptly proposed a method for SAR image formation. Using an X band radar developed at the University of Illinois and operated onboard a US Air Force Curtiss C-46 aircraft, Kovaly et al. (1952) produced the first SAR image on July 8, 1953 (Skolnik, 1980a). One year later, Cutrona et al. (1960), from the University of Michigan, started developing fast analog optical SAR processors (Cutrona et al., 1961). Using the AN/UDP-1 system, developed in cooperation with Texas Instruments, they were able to produce SAR imagery by optical means in 1959 (D'Aria et al., 2004). At that time, optical computers offered the only practicable solution for SAR processing, and remained so until the end of the 1970s. With the advent of digital computing optical SAR processors started being replaced by the slower (but more accurate) off-line digital processors (Kirk, 1975a,b). Airborne SAR remained mainly a military research tool until the 1980s, when the development of scientific civilian airborne systems started. Due to its low cost and flexibility, airborne SAR has always been the ideal test bed for the development of new SAR modes, techniques, and applications, some of which have been eventually incorporated into spaceborne missions. Airborne SARs can provide inexpensive multicarrier, multichannel, multipolarization SAR systems while still maintaining decent swath widths, very good resolution, and high sensitivity.

In 1978, the first spaceborne SAR mission was flown onboard NASA's Seasat satellite (Evans et al., 1988). Seasat's SAR was an L band instrument conceived for the imaging of ocean surfaces, polar ice caps, and coastal regions. Based on the success of the Seasat mission, NASA started a program to fly SARs onboard the Space Shuttle, known as Shuttle Imaging Radar (SIR). All SIR radars used the same carrier frequency as Seasat, orbiting at an altitude of around 250 km. The first and second generations, SIR-A and SIR-B, were launched late 1981 and 1984, respectively (Elachi et al., 1982; Cloude and Pottier, 1997). The two Soviet spacecraft, Kosmos-1870 and Almaz-1a, launched in 1987 and 1991, respectively, provided S band SAR imagery of land and oceans for scientific and commercial

---

[13]Any polarization state can be expressed as a linear combination of left-hand and right-hand circularly polarized components.

[14]Carl Wiley's name for the principle of synthetic aperture for side-looking radar was Doppler beam sharpening (DBS).

applications. Russia flew years later the L/S-band SAR system Travers onboard the Priroda module of the Mir orbital station. Shortly after the start of the Almaz-1a, ESA launched the Earth observation satellite ERS-1, which included a C band SAR among other instruments (Attema, 1991). ERS-1 was followed in 1995 by a duplicate satellite ERS-2, ensuring two decades of consistent C band SAR observations which very much helped to consolidate now mature SAR techniques such as interferometry. The European contribution to SAR remote sensing was continued with ENVISAT's C band ASAR instrument, launched in early 2002, which offered dual-pol sensing capabilities (Desnos et al., 2000). In 1992, the Japan Aerospace Exploration Agency launched an L band SAR instrument onboard the JERS-1 satellite, mainly conceived for land-oriented applications. JERS-1 was followed by ALOS-1, started in 2006, which included the fully polarimetric L band PALSAR instrument.

Between JERS-1 and ERS-2, the third SIR mission, known as SIR-C/X-SAR, was flown twice; this mission, a joint NASA/DLR/ASI project, consisted of three radars, operating at L/C, and X-bands, fully polarimetric at L/C-bands, which provided the first simultaneous multifrequency, multipolarization SAR observations. Reusing hardware from the previous mission, the same three partners of SIR-C/X-SAR carried out a single-pass C/X-band SAR interferometric mission in early 2000 (Werner, 2000); the Shuttle Radar Topography Mission, commonly known as SRTM (Farr et al., 2007), provided a digital elevation model (DEM) of the Earth's land surface between 60° north and 56° south with a vertical accuracy better than 10 m. At the end of 1995, the Canadian Space Agency (CSA) launched the C band Radarsat-1, which was followed a decade afterward by the fully polarimetric, C band Radarsat-2. In 2007, the first satellite of the Agenzia Spaziale Italiana (ASI) X band SAR constellation COSMO-SkyMed was launched. In the same year, the German X band radar satellite TerraSAR-X was started as a public–private partnership between DLR and Astrium GmbH.

Three years later, a duplicate satellite TanDEM-X was launched. These two satellites operating cooperatively are the core of the TanDEM-X mission, in which they form a single-pass bistatic X-band SAR interferometric system, with the main goal of providing a global DEM with a vertical accuracy better than 2 m. The Seasat imagery had a resolution of 25 m, with a horizontal image size of 100 km. TerraSAR-X in high-resolution mode yields images with a spatial resolution of around 1 m, and a swath width of 5 km. Between 1983 and 1984, the Polyus-V SAR onboard Soviet missions Venera-15/16 mapped the Venusian surface from the north pole to about a latitude of 30° north, which corresponds to 25% of the surface of the planet. Between 1990 and 1994, the NASA Magellan SAR mission mapped 98% of Venus with a resolution of about 100 m, including repeat-pass interferometric acquisitions.

With increasing performance and flexibility, spaceborne SAR sensors have grown to become very powerful tools for both Earth and planetary science, with specific applications in a variety of fields, i.e., agriculture, forestry, cartography, oceanography, urban planning, risk and disaster monitoring, axiological defense, geology, or transportation monitoring.

SAR interferometry, suggested by Graham in 1974 as a tool to retrieve the topographic information of the scene, measures the phase difference between two SAR images acquired from different positions. Besides topography, SAR interferometry can also be used to measure motion (e.g., moving target indication, ocean currents). Using a two-antenna modification of the NASA AIRSAR system, the first practical demonstration of (single-pass) SAR interferometry was carried out at JPL by Zebker and Goldstein in 1986 (Bamler, 1992). Two years later, Gabriel and Goldstein (1988) demonstrated spaceborne, crossed orbit, repeat-pass interferometry using L-band data acquired by SIR-B. Gabriel et al. (1989) also demonstrated differential interferometry shortly afterward. This technique, capable of detecting small topography changes, can be used to measure natural or artificial deformations with an accuracy of a fraction of the wavelength. Using differential SAR interferometry (DInSAR), Massonnet et al. detected and validated in 1993 the signature of a large earthquake using ERS-1 data; roughly at the same time, Goldstein et al. (1993) used SIR-B data to map the flow velocity of the Rutherford ice stream in Antarctica. Over the past years, SAR interferometry has become a fundamental remote sensing technique with numerous applications in many scientific and commercial fields (e.g., glaciology, oceanography, topographic mapping, etc.).

5.4.1 A COMPACT INTRODUCTION TO SYNTHETIC APERTURE RADAR THEORY

Considering the classifications presented earlier in this chapter, SARs are side-looking imaging range-Doppler radars with enhanced along-track resolution achieved through synthetic aperture processing. A very long antenna, proportional to the distance to the scene, is synthesized by the motion of the platform, which can be airborne or spaceborne. The typical geometry of the SAR is shown in Fig. 5.75, with the along-track and slant range directions of the observation as described earlier.

As already discussed, the angles with respect to the boresight of the SAR observation in elevation and azimuth are called look and squint angles, respectively. The area illuminated on the ground by the antenna is called the footprint. The extension of the footprint as the satellite moves is called the swath. This movement of the footprint in azimuth allows the illumination of the targets from different positions (angles), a necessary condition for the formation of the synthetic aperture using coherent processing. The length of the synthetic aperture can be shown to be proportional to the distance to the scene, i.e.,

$$L_{\text{syn}} \approx 2 \cdot r_0 \cdot \frac{\lambda}{L_{\text{a}}}, \tag{5.126}$$

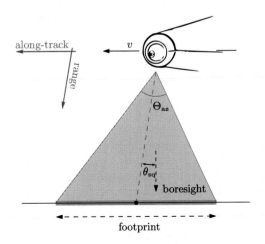

FIGURE 5.75

Across-track cut of a side-looking radar observation geometry.

where L_a is the length of the azimuth antenna, and the factor 2 is due to the two-way travel of the signal. The maximum achievable azimuth resolution is then roughly half the length in azimuth of the antenna:

$$\varphi_{syn} \approx r_0 \cdot \frac{\lambda}{L_{syn}} = \frac{L_a}{2}. \tag{5.127}$$

5.4.1.1 Range and Azimuth Resolutions

The range resolution of the SAR is inversely proportional to the transmitted bandwidth of the system, i.e.,

$$\delta r = \frac{c}{2 \cdot B_r} \tag{5.128}$$

This resolution is given in the slant range direction, which does not, in general, coincide with the orography of the scene. The projection from the LOS onto flat ground is depicted in Fig. 5.76. Note the range resolution on ground is coarser than the slant range resolution.

The ground range resolution can be approximated as (Cantalloube and Koeniguer, 2008)

$$\delta r_g \approx \delta r \cdot \frac{\nabla r}{|\nabla r|^2}, \tag{5.129}$$

where the ∇r represents the gradient of the slant range between the radar and the scene. For the simple case of a flat area, Eq. (5.129) can be approximated by

$$\delta r_g \approx \frac{c}{2 \cdot B_r \cdot \sin \theta_i}, \tag{5.130}$$

where the θ_i is the incident angle. Typical values of range resolutions are in the metric scale for frequency bands up to C band (~ 5.5 GHz), and decimetric scales for X band, and higher bands.

The derivation of the azimuth resolution in SAR presented earlier was based on a geometrical argument. Let us use now present a derivation based on the Doppler frequency analysis of the SAR

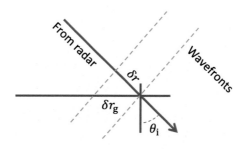

FIGURE 5.76

Slant and ground range resolutions for a side-looking imaging geometry. Note the ground range resolution is coarser than the slant range resolution.

measurement. As discussed, the Doppler frequency is proportional to the radial relative velocity between radar and target, which in the SAR case becomes

$$f_{Dop}(\alpha_{sq}) = -\frac{2 \cdot v_s}{\lambda} \cdot \sin \alpha_{sq}, \tag{5.131}$$

where v_s is the spacecraft velocity, α_{sq} is the instantaneous squint angle and the factor 2 is due to the two-way propagation of the signal. For small squint angles, the previous expression is roughly linear with α_{sq}. The Doppler support of the observation corresponds to an angular extent equal to the azimuth aperture of the antenna θ_a around the average squint angle of the observation θ_{sq}, i.e.,

$$B_a = \left| f_{Dop}\left(\theta_{sq} + \frac{\theta_a}{2}\right) - f_{Dop}\left(\theta_{sq} - \frac{\theta_a}{2}\right) \right| \approx \frac{2 \cdot v_s \cdot \theta_a}{\lambda} \approx \frac{2 \cdot v_s}{L_a}, \tag{5.132}$$

where the approximation is again valid for small θ_a and θ_{sq}. The azimuth resolution projected on the ground can now be expressed as

$$\delta x = \frac{v_g}{B_a} \approx \frac{L_a \cdot v_g}{2 \cdot v_s}, \tag{5.133}$$

which can be greatly simplified whenever the speed of the platform equals the ground velocity v_g. Typical values for the azimuth resolution are again in the metric scale for lower-frequency bands and in the decimetric scale for higher-frequency bands.

There are some interesting conclusions that can be drawn from (5.133). The first and most counterintuitive, is that the smaller the antenna of the radar, the better the resolving power of the SAR. This is of course due to the larger Doppler effect given by the wider the antenna beam. Second, the SAR azimuth resolution is independent of the distance at which the radar observes the scene. Last, but not least, the azimuth resolution appears to be independent of the wavelength. This is of course only partially true, since the value of L_a will be proportional to some extent to the wavelength. A further limitation to the achievable resolution is linked to the sensitivity of the system. As discussed during the derivation of the radar equation, larger antennas yield better gains, effective areas, and sensitivity values.

The limit for the SAR azimuth resolution is obtained after evaluating Eq. (5.132) when θ_a approaches π, i.e.,

$$\delta x_{SAR} \geq \frac{\lambda}{4}, \tag{5.134}$$

which is the theoretical limit of the integration over an infinitely linear survey or a circular one for an isotropic target. Approaching such a limit, in any case, would require the targets in the scene to remain coherent for a wide angular span (hence a longer illumination time), a very restrictive condition in natural targets. As with any other signal, the Nyquist sampling of the SAR data accommodates the transmitted bandwidth and the Doppler effect of the received echoes, which sets the minimum value for the sampling frequency and the PRF of the SAR as

$$f_s \geq B_r,$$

$$PRF \geq B_a \approx \frac{2 \cdot v_s}{L_a}, \tag{5.135}$$

where again both f_s and PRF tend to be higher for systems operating in higher-frequency bands.

5.4.1.2 Ambiguities and Doppler Centroid

SARs transmit pulses at a rate (i.e., PRF) high enough to sample the imaged scene. These phase-encoded pulses are received by the radar and can be through processing using their corresponding range and Doppler coordinates. As discussed earlier, the unambiguous range and Doppler bands, which can be scanned by the SAR, are limited. Beyond these bands, targets appear warped both in time and frequency, i.e., replicated at the wrong radar coordinates. Depending on whether or not the echoes correspond to returns in time or frequency, yields ambiguities in range or azimuth. Ambiguities are rarely a problem in airborne SAR, but pose one of the major challenges in the design of spaceborne SAR systems.

Range ambiguities are returns from previous or future[15] pulses into the desired echo window, typically arriving through side lobes of the SAR antenna. The left side of Fig. 5.77 shows a transverse cut of the SAR acquisition with the main beam in yellow and the ambiguous areas coming through the purple beams pointing at different incident angles and corresponding to far- and near-range ambiguities. As depicted, the ambiguous signal corresponds to the echoes of previous and posterior pulses transmitted by the SAR.

Assuming a reference target placed at range r_0, the ambiguity of order k arrives at the radar from range

$$r_{\mathrm{amb}}(k) = r_0 + m \cdot \frac{c}{2 \cdot \mathrm{PRF}}, \tag{5.136}$$

where m is an integer number describing the distance in pulses of the ambiguous echoes. Typical range ambiguities arrive from several tens of kilometers away from the illuminated swath area and effectively reduce the contrast in the image. Since there is a direct correspondence between slant range and look angle, range ambiguities are typically suppressed through an optimization of the elevation antenna pattern, and the PRF of the system (Curlander and McDonough, 1991). We present such an optimization in Fig. 5.78, which is an image of Naples, Italy.

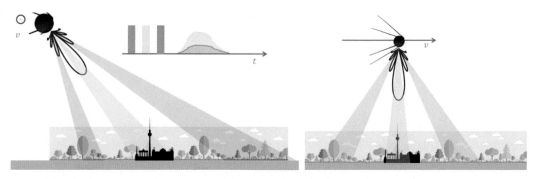

FIGURE 5.77

Range (left) and azimuth (right) ambiguities in gray. Range ambiguities are caused by echoes of previous or future pulses overlapping in time and illuminated through the elevation side lobes of the synthetic aperture radar (SAR) antenna. Azimuth ambiguities are caused by echoes overlapping in frequency, which are illuminated through azimuth side lobes of the SAR antenna.

---

[15]In spaceborne observations, the travel time of the radar signals is typically in the order of several PRIs.

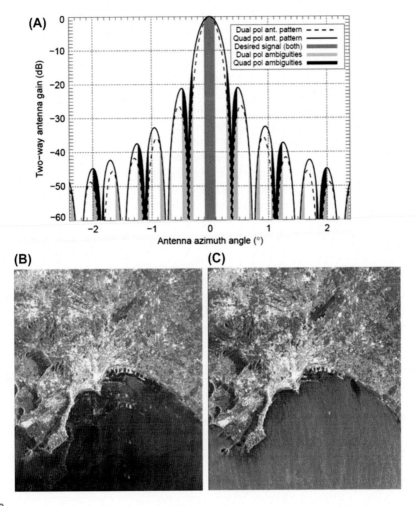

FIGURE 5.78

Suppression of ambiguities by pulse repetition frequency (PRF) optimization. Azimuth antenna pattern and ambiguous bands (A) TerraSAR-X quad-polarized acquisitions over Naples acquired with PRF of 5267 Hz (B) and optimized PRF of 6395 Hz (C). Ambiguities are clearly visible in (B).

Courtesy of DLR.

The figure of merit that describe the sensitivity of the radar to range ambiguities is the range ambiguity to signal ratio (RASR), which can be expressed as the integrated ambiguous power over the useful signal power, i.e.,

$$\text{RASR} \approx \frac{\int_{\theta_{e,amb}} \sigma_{0,amb}(\theta_e) \cdot r_0(\theta_e)^{-3} \cdot G_e(\theta_e) \cdot d\theta_e}{\int \sigma_0 \cdot r_0(\theta_e)^{-3} \cdot G_e(\theta_e) \cdot d\theta_e}, \tag{5.137}$$

where G_e is the gain pattern of the antenna in elevation and the integration is carried out over the elevation angles that correspond to the extension of the radar signal on the ground. The cube of the range is analogous to the one obtained in the derivation of the radar equation, due to the combined effect of the two-way travel and the extension of the azimuth footprint of the antenna. By construction, smaller RASR values are better. A good RASR should be consistent with the dynamic range of the received SAR echoes, i.e., typically larger in lower-frequency bands. Good ambiguity to signal ratios are commonly better than −25 dB.

Range ambiguities appear unfocused in the SAR images due to the range mismatch of the processing kernel and appear attenuated by the suppression of the elevation pattern. As a consequence, range ambiguities may only be visible to the eye in low backscatter areas (e.g., coastal), but definitely degrade the radiometric and interferometric information of the images.

Similarly, azimuth ambiguities result from echoes in ambiguous frequency bands aliased into the Doppler effect of the measurement and arriving through azimuth side lobes of the antenna, as depicted in the right drawing of Fig. 5.77. The figure shows an elevation cut of a SAR acquisition with the footprint beam in yellow and the angular areas contributing to left and right azimuth ambiguities in purple.

To derive the location of azimuth ambiguities, we need to first introduce the Doppler centroid (acronym DC), or the Doppler frequency of the mean squint angle of the acquisition, i.e.,

$$f_{\text{DC}} = -\frac{2 \cdot v_{\text{s}}}{\lambda} \cdot \sin \theta_{\text{sq}}. \tag{5.138}$$

The acquired energy in the Doppler spectrum is concentrated around f_{DC}. Azimuth ambiguities are returns at Doppler frequencies offset from the DC of the acquisition a whole number of PRFs, i.e., azimuth ambiguities of order k appear around Doppler frequency

$$f_{\text{Dop}}(k) = f_{\text{DC}} + n \cdot \text{PRF}, \tag{5.139}$$

where again n is an integer number. The previous expression can be used to compute the angular location of azimuth ambiguities, i.e.,

$$\theta_{\text{sq,amb}}(k) \approx \theta_{\text{sq}} - n \cdot \frac{\lambda \cdot \text{PRF}}{2 \cdot v_{\text{s}}}, \tag{5.140}$$

which can be translated into a shift in azimuth of the form

$$\Delta x_{\text{amb}}(k) \approx n \cdot \frac{\lambda \cdot r_0 \cdot \text{PRF}}{v_{\text{s}}}. \tag{5.141}$$

Azimuth ambiguities appear shifted on the order of several kilometers for state-of-the-art SAR systems and, like range ambiguities, reduce the contrast in the image. However, azimuth ambiguities are only slightly defocused and perfectly visible in medium to low backscattering areas (e.g., coastal). Azimuth ambiguities not only degrade but also bias the radiometric and interferometric information of the images. Fig. 5.79 shows an example of azimuth ambiguities in a TerraSAR-X image contaminating the backscattering signature of a lake in Russia.

As shown in Fig. 5.77, azimuth ambiguities are received through the side lobes of the azimuth pattern, which should be designed so as to suppress these side lobes. The figure of merit to evaluate this ambiguity is the ratio of the azimuth ambiguity to signal ratio (AASR), which gives the ratio between the ambiguous power and the signal power received through the main lobe of the antenna, i.e.,

$$\text{AASR} \approx \frac{\int_{\theta_{\text{a,amb}}} \sigma_{0,\text{amb}}(\theta_{\text{a}}) \cdot G_{\text{a}}(\theta_{\text{a}}) \cdot d\theta_{\text{a}}}{\int \sigma_0 \cdot G_{\text{a}}(\theta_{\text{a}}) \cdot d\theta_{\text{a}}}, \tag{5.142}$$

FIGURE 5.79

TerraSAR-X image of a lake in Russia. The white area on the lake is due to azimuth ambiguities. Radar illumination is from the right.

Courtesy of DLR.

where the integration is carried out over the azimuth angles corresponding to the extension of the radar signal on ground. Since the ranges of the azimuth ambiguities are very similar to the ranges of the valid echoes we can neglect them, including the integral of the free-space attenuation. As in the case of the RASR, AASR values should be similar to the dynamic range of the radar backscatter, e.g., typically values greater than -25 dB are used in the specifications of SAR missions. The total ambiguity to signal ratio (TASR) is the sum of the range and azimuth ratios, i.e.,

$$TASR = RASR + AASR. \tag{5.143}$$

As in the case of the azimuth and range ambiguities, good values of TASR are typically better than -25 dB.

5.4.1.3 An Important Synthetic Aperture Radar Choice: Swath Versus Azimuth Resolution

The time and frequency measurements of a SAR impose conflicting requirements on the radar: range ambiguities lessen as azimuth ambiguities increase and vice versa. The relationship between the available swath width and the Doppler bandwidth is related to the size of the SAR antenna. A small azimuth antenna size requires a high PRF, which dictates a small swath width and thus a large antenna. This relationship can be approximated by

$$S_w \approx r_0 \cdot \frac{\lambda}{L_e} \leq \frac{c}{2 \cdot PRF} \approx \frac{c \cdot L_a}{4 \cdot v_s}, \tag{5.144}$$

where the swath width has only been coarsely approximated and the influence of the pulse duration has been neglected. Similarly, a small elevation antenna illuminates a large swath, which requires a small PRF requiring a large antenna in azimuth. Eq. (5.144) illustrates the trade-off between the swath width and azimuth resolution. In general, the better the azimuth resolution of the SAR, the narrower the swath it covers. Conversely, the wider the swath, the poorer the azimuth resolution. This trade-off is valid even when the antenna is steered in elevation or azimuth. By further rearranging Eq. (5.144), we can derive the minimum size of the SAR antenna as the product of L_a and L_e in the following form (Freeman, http://citeseerx.ist.psu.edu/viewdoc/download?doi=10.1.1.424.2154&rep=rep1&type=pdf):

$$A_{SAR} = L_e \cdot L_a \geq \frac{4 \cdot \lambda \cdot r_0 \cdot v_s}{c}. \tag{5.145}$$

Larger antennas have reduced swath widths, or azimuth resolutions, but of course they improve the overall sensitivity of the system.

5.4.1.4 Synthetic Aperture Radar Imaging Modes

So far, we have assumed that the antenna of the radar is not steered during the data acquisition, which corresponds to the natural SAR acquisition mode called stripmap. The stripmap mode has some advantageous properties such as an invariant azimuth of the survey, which typically extends for a few seconds along the orbit, antenna pattern, and radar mapping coordinates. Other SAR imaging modes allow by steering in azimuth or elevation of the antenna pattern a selectable trade-off between azimuth resolution and scene size.

5.4.1.4.1 High Azimuth Resolution Modes: Spotlight

In the spotlight modes, the antenna is linearly steered in azimuth to illuminate a spot on the ground for a longer period of time (Carrara et al., 1995). The situation is depicted in Fig. 5.80.

The simplest spotlight mode is called staring, the center of the rotation of the antenna pattern coincides with the center of the imaged scene, hence maximizing both illumination time and azimuth resolution. The size of the scene roughly coincides with the footprint of the radar antenna, as depicted in Fig. 5.80A. As a way to overcome the short extension of the staring spotlight scenes, the sliding

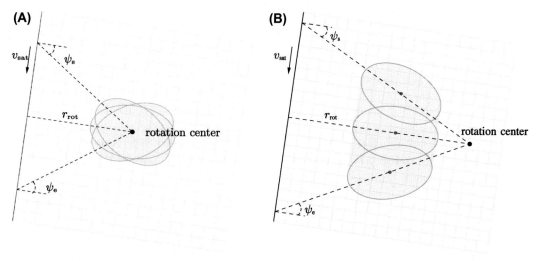

FIGURE 5.80

Steering laws, footprints, and scene size for staring (A) and sliding (B) spotlight acquisition modes. In the sliding configuration, a longer azimuth scene size is traded off for a coarser illumination time and azimuth resolution.

spotlight mode was developed (Fig. 5.80B). The rotation center is placed at a further range, usually outside the imaged scene, i.e., the steering rate is smaller than in the staring case. The targets are illuminated for a shorter time than the duration of the acquisition, which allows for a larger extension in azimuth of the imaged scene. The larger size of the scene is traded off in the sliding spotlight mode for a decrease in azimuth resolution, as depicted in the right drawing of Fig. 5.80.

As a consequence of azimuth steering, the DC of the targets changes linearly with the along-track position within the scene. This variation is in any case smaller than the azimuth bandwidth of the scene, but needs to be accommodated by the image formation kernel (Carrara et al., 1995). The spotlight mode allows the improvement of the azimuth resolution without reducing the size of the antenna. As a consequence, the PRF of the spotlight acquisitions is only marginally higher than the stripmap PRF, with the slight increase used to accommodate the migration of the DC along the scene.

TerraSAR-X was the first operational civilian SAR mission to acquire images in both staring and sliding spotlight modes, achieving azimuth resolutions of 25 and 50 cm, respectively. Many examples of both sliding and staring spotlight images can be found throughout this chapter. Fig. 5.81 (top) shows an example of two geocoded TerraSAR-X images over the Atacama Desert, Chile, in stripmap (A) and sliding spotlight (B) modes, respectively.

5.4.1.4.2 Wide-Swath Modes: ScanSAR and TOPS

Wide-swath modes are operated in bursts shorter than the stripmap illumination time of the SAR antenna, i.e., the available (integration) time is multiplexed for the scanning of different swaths accessed with a short number of steering in elevation of the SAR antenna beam (Cumming and Bennet, 1979; Moreira and Huang, 1994). In Fig. 5.82A, we show an example of a ScanSAR acquisition with three subswaths.

The consecutive ellipses represent the footprints of the antenna at different acquisition times. The number of subswaths can be extended, at the cost of each having a smaller portion of the available illumination time, hence degrading the azimuth resolution of the final images. As in the spotlight case, this degradation does not imply a decrease in the PRF as compared to stripmap, since the instantaneous Doppler bandwidth is the one imposed by the size of the antenna. A further condition of the duration of the bursts is imposed by the continuity of the images of consecutive bursts, as shown in the figure.

Depending on their azimuth position on the ground, the targets are observed with different portions of the azimuth pattern, which has two main consequences: (1) the DC of the targets changes linearly with the azimuth position (Cumming and Bennet, 1979; Moreira and Huang, 1994) and (2) a modulation of the amplitude of the images corresponding to the integrated pattern, with which the targets are observed, appears in the image. This modulation known as scalloping, is usually small, on the order of 1−2 dB, but introduces an inhomogeneity which affects both the radiometric and interferometric quality of the images. In Fig. 5.83, we present an image which shows a noise scaling effect in lake Qinghai, China, due to scalloping. In the image, acquired by TerraSAR-X in ScanSAR mode, azimuth ambiguities are also visible in the lake. The ScanSAR mode in TerraSAR-X has a swath of about 100 km with an azimuth resolution of 15 m.

ScanSAR was first used in ERS-1, and it now plays a fundamental role in the popularization of repeat-pass interferometric applications.

A way to overcome the azimuth inhomogeneity of the pattern, the TOPS (i.e., terrain observation by progressive scans) mode was suggested. The discrete scanning in elevation and burst operation of

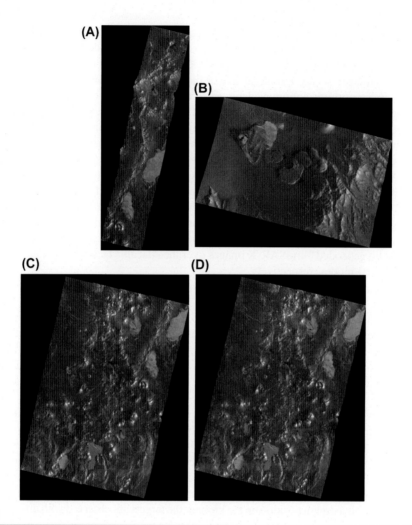

FIGURE 5.81

Geocoded TerraSAR-X images over the Atacama Desert, Chile. Stripmap (A), sliding spotlight (B), ScanSAR (C), and terrain observation by progressive scans (TOPS) modes (D). The vertical direction corresponds to north.

ScanSAR are maintained, but a linear steering in azimuth of the antenna is done at the burst level to equalize the tapering of the azimuth pattern for all targets. The steering, as depicted in Fig. 5.82B is inverse with respect to the spotlight steering. The scalloping is reduced, typically to about 0.5 dB. A more homogeneous performance is achieved, at the expense of higher DCs within the acquisitions as it can be observed in Fig. 5.82. The Terrain Observation with Progressive Scans (TOPS) mode was demonstrated for the first time on-orbit with TerraSAR-X (F. De Zan, http://elib.dlr.de/61807/1/dezan.pdf) and has become standard for Sentinel-1 Interferometric Wide Swath (IWS) and Extra Wide Swath (EWS) modes. The first SAR images acquired by TerraSAR-X (left) and Sentinel-1 (right) in the TOPS acquisition mode are included here in Fig. 5.84.

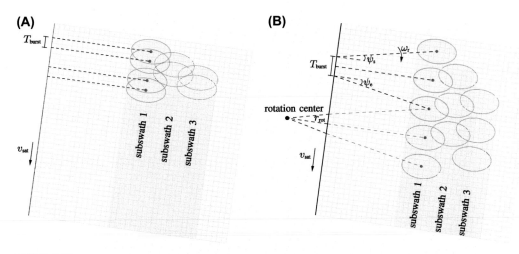

FIGURE 5.82

Steering laws, footprints, and scene size for ScanSAR (A) and terrain observation by progressive scans (TOPS) (B) acquisition modes. The azimuth steering of the TOPS mode allows for the illumination of the scene with more homogeneous radiometric qualities.

FIGURE 5.83

TerraSAR-X ScanSAR image over Qinghai Lake, China. The periodic horizontal white stripes in the lake are due to a noise scaling effect due to scalloping. The distance between the stripes corresponds to the duration of the radar bursts. Note azimuth ambiguities can also be observed in the lake.

Courtesy of DLR.

(A)   **(B)**

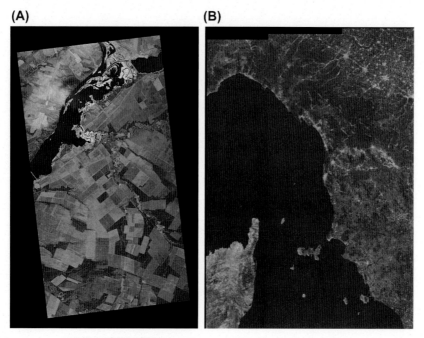

FIGURE 5.84

First synthetic aperture radar images acquired by (A) TerraSAR-X and (B) Sentinel-1 in the terrain observation by progressive scans (TOPS) acquisition mode. TerraSAR-X image acquired over the south Russian Steppes about 500 km northeast of the Black Sea and about 50 km west of Volgograd. Sentinel-1A mosaic of the first S-1A TOPS interferogram (two slices) in IW mode around the Gulf of Genoa, Italy, acquired on August 7 and 19. Range extension, 250 km; azimuth extension, 340 km.

Courtesy of DLR.

5.4.1.4.3 Circular Synthetic Aperture Radar

One specific airborne SAR acquisition mode consists of surveys with circular flight paths. The targets in the scene are observed over an angular span of 360°, which allows for both very high-resolution imaging and angular discrimination. Over isotropic targets, the impulse response of circular SAR shows a polar symmetry with the best achievable SAR resolution of $\delta x \leq \lambda/4$. This resolution appears to be independent of the transmitted bandwidth. Such a theoretical result would require an idealized isotropic target infinitely thin, not found in real life. Nevertheless, circular SAR images show a very high-resolution character with reduced speckle and a higher level of detail, as it can be seen in Fig. 5.85.

The swath shows a circular shape, generated after rotation of the footprint on the ground. The survey is space-variant and very sensitive to topography and motion compensation. This sensitivity to topography is at the core of the tomographic potential of circular SAR surveys. In Fig. 5.86, we show an example of a complete circular SAR image over Kaufbeuren, Germany, with the color coding corresponding to the different L band polarimetric channels in the Pauli basis is presented here in Fig. 5.85. The SAR image is overlaid on the Google Earth image of the area.

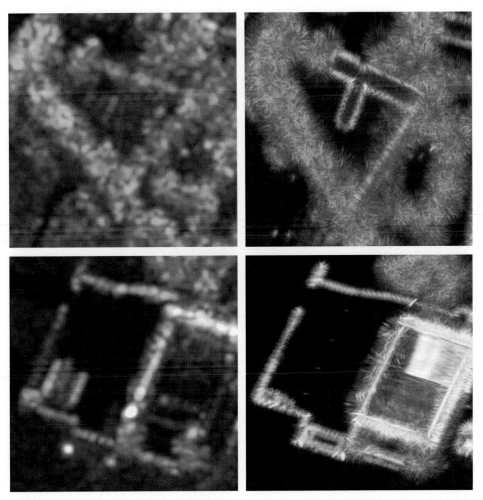

FIGURE 5.85

Comparison of stripmap (left) and circular (right) synthetic aperture radar (SAR) images acquired with DLR airborne system F-SAR in L-band over Kaufbeuren, Germany. The circular SAR images show a higher level of detail and reduced speckle.

Courtesy of DLR.

5.4.1.4.4 Synthetic Aperture Radar Image Calibration

Calibrated SAR images show radiometric values close to the actual radar brightness of the scene, with a high-fidelity location of the scatterers, and an accurate representation of the complex ratios between the polarimetric channels over the entire image. Furthermore, the times of the images in both range and azimuth accurately represent the radar observation geometry. These aspects are discussed in the following paragraphs.

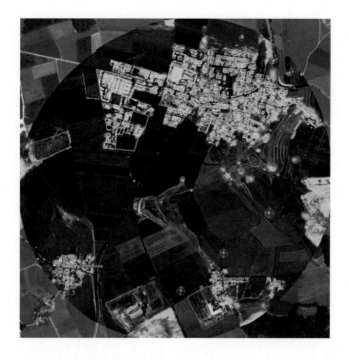

FIGURE 5.86

L-band polarimetric circular synthetic aperture radar (SAR) image acquired over Kaufbeuren, Germany, with DLR's airborne F-SAR system. The full-resolution image has a Cartesian sampling of 6 cm in both directions and is overlaid over the Google Earth image of the area.

Courtesy of DLR.

Since the propagation losses can be, on average, estimated with a very good accuracy, the radiometric fidelity of the SAR system depends on a good characterization of the antenna patterns, SAR pointing, and the payload electronics of the transmitter and the receiver (e.g., transmitted power and transfer functions, receiver gain, anomalies and drifts). For the radiometric calibration of SAR systems, external targets with high radar cross sections, typically transponders, are used. Observations over homogeneous areas (e.g., rain forest for higher-frequency bands) are used to observe the average elevation patterns of the system. Typical values for the absolute radiometric fidelity of a SAR system have values better than 30%, i.e., 1 dB.

The second aspect mentioned above is the accurate locations of the image pixels, which strongly depends on the knowledge of the observation geometry, and also on the delays of the radar signals and echoes occurring in the atmosphere and in the payload electronics. A good knowledge of the radar observation geometry depends on a precise knowledge of the orbit ephemeris data, and of the pointing attitude of the radar antenna, which includes spacecraft attitude knowledge. The delays in the propagation of the radar signals through the atmosphere can be estimated from meteorological models. The characterization of the internal delays of the payload electronics can be accomplished with the help of external targets, typically trihedrals, whose positions are very accurately known. Typical values of geolocation accuracies are better than the range resolution of the system. A further step in the accurate location of the scatterers is the relative accuracy between several acquisitions, which will be crucial for SAR interferometric applications. The relative accuracy typically depends on the separation between the available ephemeris (i.e., baseline), which should be typically kept better than a fraction of the wavelength.

Table 5.4 List of All Spaceborne Synthetic Aperture Radar Missions Over the Past 40 years

Sensor	Frequency Band/ Polarization	Agency
Seasat	L/HH	NASA/JPL
ERS-1/2	C/VV	ESA
JERS-1	L/HH	JAXA
SIR-C/X-SAR	L/quad, C/quad, X/VV	NASA/JPL, DLR, ASI
Radarsat-1	C/HH	CSA
SRTM	C/HH + VV, X/VV	NASA/JPL, DLR, ASI
ENVISAT/ASAR	C/dual	ESA
ALOS/PalSAR	L/quad	JAXA
TerraSAR-X/TanDEM-X	X/quad	DLR/Astrium
Radarsat-2	C/quad	CSA
COSMOSkyMed-1/4	X/dual	ASI/MiD
RISAT-1	C/quad	ISRO
HJ-1C	S/VV	
Kompsat-5	X/dual	KARI
PAZ	X/quad	CDTI
ALOS-2	L/quad	JAXA
Sentinel-1a/1b	C/dual	ESA
Radarsat constellation-1/2/3	C/quad	CSA
SAOCOM-1/2	L/quad	CONAE/ASI

The third element, which outlined above, is the accurate representation of the polarimetric ratios, which suggests accurate relative radiometric and interferometric information between the channels. The polarimetric calibration requires the accurate characterization of transmitter and receiver, including the antenna patterns and both horizontally and vertically polarized receive channels in terms of geometry, imbalances, and cross-talk. Moreover, the Faraday rotation, caused by the propagation of the radar signals through the ionosphere, needs to be estimated and compensated. Typical values for these ratios are below 30% in terms of amplitude deviation and a few degrees in phase.

5.4.2 SYNTHETIC APERTURE RADAR SYSTEMS AND MISSIONS

Table 5.4 shows a list of all spaceborne SAR missions over the past 40 years since the flight of NASA/JPL Seasat.

5.4.3 FUNDAMENTALS OF SYNTHETIC APERTURE RADAR PROCESSING

The difference between real and SARs lies in the coherent processing of the received echoes. Processing plays a fundamental role in SAR. SAR data acquisition is illustrated in Fig. 5.87 along with the SAR image formation process. The satellite is depicted flying from right to left and represents different

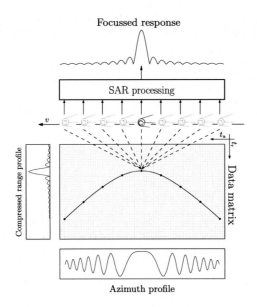

FIGURE 5.87

Synthetic aperture radar (SAR) data acquisition model and image formation process. The satellites show the different positions of the radar during the synthetic aperture. The data matrix shows the migration of the radar echoes, and the bottom and left figures show the real part and envelope of the echoes of a single scatterer in azimuth and range after pulse compression. The top image shows the envelope of the azimuth signal after SAR processing.

illuminations of the radar. The data matrix shows in black the collected echoes of a single scatterer placed at the center of the survey where all dashed lines converge. The delay of the received echoes, measured in fast time, represents the instantaneous distance to the target, which is larger at the edges of the survey. The envelope of the SAR echo migrates along the convex black line shown in the figure. After pulse compression, the echoes show envelopes similar to the one depicted on the left side of the image.

In slow or azimuth time, the slant range is shown as the phase of the received signal, which possesses the roughly quadratic (e.g., chirp) form of the bottom signal in the figure. The modulation of the amplitude of the azimuth signal is mainly due to the antenna pattern. The SAR image formation process basically retrieves the energy corresponding to a small area on the ground of the corresponding pixel, which in the case of a single scatterer shows the band-limited shape of the focused envelope at the top of the figure. As discussed earlier, the focused SAR image will be the coherent superposition of the responses of the individual scatterers in the scene.

This process can be represented in a compact way. The SAR impulse response is the 2-D signal describing the echoes received from a single unitary point target, and it can be approximated as

$$h(t_r, t_a) \approx p_{lp}\left(t_r - \frac{2 \cdot r(t_a)}{c}\right) \cdot \exp\left[-j \cdot \omega_0 \cdot \frac{2 \cdot r(t_a)}{c}\right] \cdot w_{ant}\left(\frac{t_a}{T_{int}}\right), \qquad (5.146)$$

where the Doppler time t_D has been replaced by the azimuth time t_a, and the rectangular antenna pattern has been substituted by a real one denoted as $w_{ant}(\cdot)$.[16] The signal p_{lp} is typically a chirp. Eq. (5.146) represents a fairly general expression of the SAR impulse response, which can in many practical cases be written for the linear-track case by using hyperbolical range histories, i.e.,

$$r(t_a) = \sqrt{r_{bc}^2 + v_e^2 \cdot (t_a - t_{bc})^2}, \tag{5.147}$$

where t_{bc} is the time at which the Doppler history equals the f_{DC}, and the subscript bc stands for beam center. The v_e is the effective velocity of the survey, which can be approximated by the geometric mean of the spacecraft and ground velocities, i.e.,

$$v_e^2(r_0) \approx v_s \cdot v_g. \tag{5.148}$$

As with the ground velocity, the effective velocity also varies along the swath. The model of hyperbolical surveys is very helpful in both airborne and spaceborne SAR, since it provides an exact analytical model for the SAR transfer function. In the spaceborne case, the hyperbolic model is only locally valid in range and needs to be updated along the orbit. The hyperbolic model breaks down among others in very high squint, very high-resolution, or bistatic SAR systems.

Fig. 5.88 shows the real part of the SAR raw data of four different target responses with different DCs embedded in noise.

For simplicity, a rectangular antenna pattern has been assumed. The vertical axis describes near to far ranges from top to bottom. The left side of Fig. 5.88 shows the responses of two targets acquired in a zero-Doppler geometry. The center of the circles represents the phase center in range and azimuth of (5.146). The right side shows the responses of the same point targets acquired in a squinted geometry. As expected, the phase center no longer appears symmetrical.

The Doppler history of the target is the instantaneous Doppler with which the target is seen at any time instant. It can be computed as

$$f_{Dop}(t_a) = -\frac{2}{\lambda} \cdot \frac{dr}{dt_a} = \frac{2 \cdot v_e^2 \cdot (t_a - t_{bc})}{\lambda \cdot \sqrt{r_{bc}^2 + v_e^2 \cdot (t_a - t_{bc})^2}}, \tag{5.149}$$

where the first equation is general and the second is particular to the hyperbolic case. For a short amount of time ($r_0 \gg v_{SAT} \cdot (t_a - t_{bc})$), the Doppler history can be linearized, and the proportionality factor can be expressed as

$$\beta_a \approx \frac{2 \cdot v_e^2}{\lambda \cdot r_{bc}}, \tag{5.150}$$

which is usually referred to as the Doppler rate. Essentially, this linearization is at the basis of the quadratic approximation of SAR range histories, which allows the expression of the azimuth modulation

---

[16]The stop-and-go approximation will be maintained in the entire derivation, which suggests small to moderate values of the product chirp bandwidth pulse duration (cf. Section 5.1). In the cases where this approximation does not hold (e.g., very high-resolution SAR or continuous-wave SAR), this approximation leads to defocusing, positioning, and phase errors of the resulting SAR images. Prats-Iraola et al., 2014 shows an easy way to model the effect (and correct for) of the continuous motion of the spacecraft in case the reader is interested.

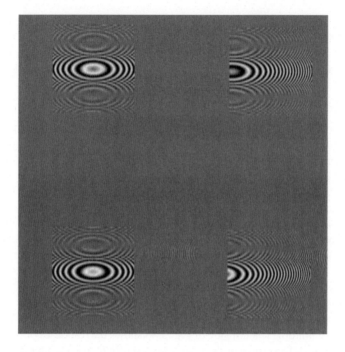

FIGURE 5.88

Real part of the synthetic aperture radar raw data of four targets with different Doppler centroids embedded in noise. The bottom corresponds to the far range of the swath. The left side shows the responses of two targets acquired in a zero-Doppler geometry. The right side shows the same two targets acquired in a squinted geometry.

of the SAR impulse response as a chirp. This approximation, while rarely used in reality, is very useful to understand the fundamentals of SAR. As an example, we can rewrite Eq. (5.133) with help of the Doppler rate as

$$\delta x_g \approx \frac{v_g}{\beta_a \cdot T_{int}} = \frac{\lambda \cdot r_{bc}}{2 \cdot v_{SAT} \cdot T_{int}}, \tag{5.151}$$

where the T_{int} is the illumination or integration time. Eq. (5.151) provides an estimate of the azimuth resolution for the cases in which the integration time does not coincide with the full illumination time of the antenna, such as the high-resolution or wide-swath modes (e.g., spotlight, scanSAR, TOPS).

5.4.3.1 Exact Synthetic Aperture Radar Image Formation: The Backprojection Integral

The optimal radar receiver, in terms of maximizing the SNR of the received echoes, is the matched filter. For the shape of the SAR impulse response discussed above, the matched filter coincides with the inversion filter, and hence achieves the compression of the SAR echoes. In terms of system theory, SAR image formation is effected via a range-variant filter with an impulse response time-inverted and conjugated using Eq. (5.146).

The most general expression of exact SAR image formation is the direct backprojection algorithm (DBP). This algorithm can be interpreted as a range and azimuth—variant correlation in the time domain based on the backprojection integral, i.e., which can be expressed as

$$i(r_0, x_0) = \int_{T_{int}} s_{RC}(r_0, t_a) \cdot \exp\left[j \cdot \frac{4\pi}{\lambda} \cdot \delta r(t_a; r_0, x_0)\right] dt_a, \tag{5.152}$$

where $s_{RC}(r_0, t_a)$ are the range-compressed data, i.e., the raw data after pulse compression. The backprojection algorithm is the most flexible SAR image formation algorithm since it accommodates in an exact way, unlike its Fourier-domain counterparts, any space-variance present in the range history of the survey and can accommodate nonlinear, circular, or bistatic surveys.

The DBP algorithm is an attractive option for low to medium resolution experimental systems, and it is very appropriate for parallelized implementations with moderate memory storage. It is usually implemented using an echo-based approach, discarding the echoes once they have been backprojected onto the final image. The computational complexity of the algorithm is $O(N_r \cdot N_a^2)$, i.e., all samples in azimuth need to be backprojected to the final image. This computational burden is acceptable for small integration times but becomes impracticable in most general SAR cases, where efficient implementations in the time and frequency domains are preferred.

5.4.3.2 Spectral Properties of Synthetic Aperture Radar Images

SAR signal properties are illustrated in different frequency and time domains in Fig. 5.89.

Assuming there is no steering in azimuth of the antenna, the spectral support of the raw data has a square shape as shown in the top left quadrant of Fig. 5.89, where range is the transmitted bandwidth,

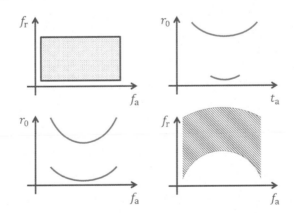

FIGURE 5.89

Synthetic aperture radar (SAR) signal properties in different frequency and time domains. The top left figure shows the raw data in wave number (range frequency-Doppler) domain. The top right figure shows two target responses of the raw data in the time (range-azimuth) domain. Note the extent of the azimuth of the near-range target is shorter to represent the same angular extent as seen from the radar. The bottom left figure shows the same two targets in their range-Doppler domain; note the Doppler support of the two targets is the same. The bottom right figure shows the spectrum of the SAR image (i.e., after SAR image formation) in the wave number domain. Note the shape of the spectrum is distorted in range with respect to the spectral support of the raw data in the top left figure.

B_r, and for azimuth it is the bandwidth of the antenna B_a around the DC, f_{DC}. The top right figure shows the envelopes of the raw data of two target responses in the time (range-azimuth) domain. The azimuth extension of the near-range target is shorter to represent the same angle as seen by the radar. Moreover, the curvature of the near-range targets is greater, as expected from Eq. (5.150). The bottom left figure shows the same two targets in the range-Doppler (time, azimuth frequency) domain. Note the Doppler effect of the two targets is now the same, and the curvature of the envelope (i.e., range cell migration) of the far-range target is greater.

After SAR image formation, the spectrum on the top left becomes the one shown on the bottom right, corresponding to a focused SAR image. The curvature of the spectrum is due to the range-variance of the SAR focusing kernel. The 2-D frequency domain is usually known in the SAR literature as the wave number domain.

5.4.3.3 Synthetic Aperture Radar Transfer Function

The SAR transfer function can be computed by transforming the SAR impulse response in the 2-D frequency (i.e., wave number) domain. Eq. (5.146) is transformed into range frequency by using the delay property of the Fourier transform (Oppenheim et al., 1983; Papoulis, 1965), i.e.,

$$H_r(f_r, t_a; r_0) = w_{ant}\left(\frac{t_a}{T_{int}}\right) \cdot P_{Tx}(f_r) \cdot \exp\left[-j \cdot \frac{4\pi}{\lambda} \cdot (f_r + f_0) \cdot r(t_a)\right], \qquad (5.153)$$

where P_{Tx} is the Fourier transform of the transmitted signal, f_r denotes the range frequency, and $r(t_a)$ is the range history of the target. The suffix r in H_r is used to explicitly state that the Fourier transform has been computed only in the range variable.

The Fourier transform in azimuth is computed using the principle of stationary phase (Papoulis, 1968; Bleistein and Handelsman, 1975; Cumming and Wong, 1979), according to which only the low-frequent parts of the time-domain signal contribute to the value of this Fourier transform. For the previous expression, the principle of stationary phase can be mathematically expressed as

$$-\frac{c}{2} \cdot \frac{f_a}{f_r + f_0} = \frac{dr(t_a)}{dt_a} \rightarrow t_a^* = -r_0 \cdot \frac{c \cdot f_a}{2 \cdot v_e^2 \cdot \sqrt{(f_r + f_0)^2 - \left(\frac{c \cdot f_a}{2 \cdot v_e}\right)^2}}, \qquad (5.154)$$

where f_a is the azimuth frequency, t_a^* is the slow time of the stationary phase and the right part of (5.154) already assumes the hyperbolic range history model in (5.147). By evaluating the kernel of the azimuth Fourier transform in t_a^*, the SAR transfer function can be approximated by

$$H(f_r, f_a; r_0) \approx C \cdot w_{ant}\left(\frac{t_a^*}{T_{int}}\right) \cdot P_{Tx}(f_r) \cdot \exp\left[-j \cdot \frac{4\pi}{\lambda} \cdot (f_r + f_0) \cdot r(t_a^*) - 2\pi \cdot f_a \cdot t_a^*\right]$$

$$= C \cdot w_{ant}\left(\frac{t_a^*(f_a)}{T_{int}}\right) \cdot P_{Tx}(f_r) \cdot \exp\left[-j \cdot \frac{4\pi}{\lambda} \cdot r_0 \cdot \sqrt{(f_r + f_0)^2 - \left(\frac{c \cdot f_a}{2 \cdot v_e}\right)^2}\right], \qquad (5.155)$$

where C is a complex constant, and the argument of $w_{ant}(\cdot)$ has been maintained for compactness. Note that the azimuth antenna pattern modulates, conveniently scaled, the Doppler spectrum of the received echoes. The matched filter used in the SAR image formation process is the complex conjugate of Eq. (5.155). Note the SAR matched filter changes with range, which constitutes a paramount challenge for the derivation of efficient implementations.

Eq. (5.155) is a very accurate approximation of the SAR transfer function for the linear-track model and is the basis of efficient SAR image formation algorithms in the Fourier domain. In general, (5.155) can be typically used in both airborne and spaceborne radar with moderate to moderately high resolution. In very high-resolution and bistatic systems, however, the range history model should preferably be approximated differently, which leads to different results in the final parts of (5.154) and (5.155). The first elements of these equations, however, are general and approximate well high-resolution and bistatic sampling.

5.4.3.4 Efficient Synthetic Aperture Radar Image Formation

As already pointed out, SAR image formation can be interpreted as the range-variant implementation of the complex conjugate of Eq. (5.155). Efficiency in the SAR image formation process is gained via a fast convolution in azimuth, be it in the Fourier frequency or in the time domain. The improvement in the computational efficiency of SAR image formation algorithms with respect to DBP is analogous to the jump between DFT and FFT. The efficiency improvement in SAR image formation follows the rule

$$O(N_a^2) \rightarrow O(N_a \cdot \log_2 N_a). \tag{5.156}$$

There are a number of efficient SAR image formation algorithms available in the literature, which are in principle equivalent for low to medium resolutions and small to medium swath widths. We will divide them into two main categories. Depending on whether or not the matched filter carries its corrections to the full transmitted bandwidth or only to the carrier frequency, we will speak about polychromatic and monochromatic algorithms, respectively.

The paradigm of monochromatic SAR image formation algorithms is the range-Doppler (Cumming and Wong, 1979; Cumming and Bennett, 1979), with its fast, interpolation-free version of chirp scaling (Raney et al., 1994; Moreira et al., 1996; Cumming and Wong, 1979). Polychromatic algorithms can be symbolized by two conjugate approaches in the time and frequency domains: the fast-factorized back-projection (FFBP) and the range-migration or $\Omega\kappa$ algorithm (Cafforio et al., 1991; Milman et al., 1993). In general terms, monochromatic algorithms, especially chirp scaling, are faster than polychromatic algorithms, naturally at the expense of higher approximation errors. In general terms, time-domain approaches are insensitive to temporal and spectral folding and are better suited for space-variant corrections such as motion compensation. They also allow for dynamic memory management.

The final choice of a SAR image formation algorithm requires a formal evaluation of the system requirements. However, most state-of-the-art civilian spaceborne systems use monochromatic approaches such as range-Doppler or chirp scaling. With the increasing performance of spaceborne SAR systems of the next generation, both the $\Omega\kappa$ and FFBP algorithms are expected to be playing a more significant role in the future.

5.4.3.5 Monochromatic Synthetic Aperture Radar Image Formation

Monochromatic SAR image formation is based on a Taylor expansion of the phase of (5.155) around the carrier frequency, i.e.,

$$H(f_r, f_a; r_0) \approx W_{\text{ant}}\left(\frac{f_a}{B_a}\right) \cdot P_{\text{Tx}}(f_r) \cdot \exp(-j \cdot \phi_{\text{AC}}) \cdot \exp(-j \cdot 2\pi \cdot f_r \cdot t_{\text{RCM}}) \cdot \exp(-j \cdot \phi_{\text{SRC}}), \tag{5.157}$$

where ϕ_{SRC} is a polynomial containing the higher-than-order-two terms of the expansion (SRC = secondary range compression). The different phase terms can be interpreted in the following manner:

- an azimuth compression phase, ϕ_{AC}, responsible for the modulation in azimuth of the echoes;
- a range cell migration delay, t_{RCM}, responsible for the spread of the phase center of the scattered echoes in the range-Doppler, domain;
- a secondary range compression term,[17] ϕ_{SRC}, which modifies the modulation of the received chirps for higher Dopplers.

For hyperbolic surveys, the expressions of the terms can be shown to be (Carrara et al., 1995; Curlander and McDonough, 1991; Cumming and Wong, 1979),

$$\phi_{AC}(f_a; r_0) = -\frac{4\pi}{\lambda} \cdot r_0 \cdot \sqrt{1 - \left(\frac{\lambda \cdot f_a}{2 \cdot v_e}\right)^2},$$

$$t_{RCM}(f_a; r_0) = \frac{2 \cdot r_0}{c} \cdot \left[1 - \left(\frac{\lambda \cdot f_a}{2 \cdot v_e}\right)^2\right]^{-1/2}, \qquad (5.158)$$

$$\phi_{SRC}(f_a; r_0) = -\frac{2\pi \cdot v_e^2}{\lambda \cdot r_0} \cdot \left[1 - \left(\frac{\lambda \cdot f_a}{2 \cdot v_e}\right)^2\right]^{-3/2} \cdot f_r^2 + \sum_{i \geq 3} \gamma_{SRC}(f_a; r_0; i) \cdot f_r^i.$$

The range-Doppler algorithm is the standard SAR image formation monochromatic algorithm. Developed at the time of NASA/JPL's Seasat mission (Cumming and Bennett, 1979), the range-Doppler algorithm remains the most popular SAR image formation algorithm due to its simplicity and performance, which is acceptable for moderate to moderately high resolutions and swaths. A further advantage of range-Doppler is that it is capable of accommodating range-varying effective velocities with good accuracy, something required for spaceborne SAR systems. In Fig. 5.90 the left panel shows the block diagram of the range-Doppler algorithm.

After a 2-D DFT, the data are bulk focused to a reference range, usually in the middle of the scene. The bulk focus accounts for an exact SRC and the range compression measurement (RCM) correction for the reference range. After bulk focusing, the data are brought into the range-Doppler domain, where a residual RCM correction (RCMC) is carried out. This RCMC is computed using an interpolation, which can be done with arbitrary accuracy. After RCMC, the curvatures of the envelopes in the range-Doppler domain are straight, as shown in the figure. The last stage is the azimuth compression, which compensates the azimuth modulation of the signal and restores the interferometric phase. After an inverse DFT, the image is focused.

The primary error of the algorithm is due to the range-invariant SRC approximation, which basically accounts for its monochromatic character. Of smaller order is the RCMC, which can be tuned within the limits of the Taylor approximation to any arbitrary accuracy. This RCMC is, however, the computational bottleneck of the algorithm. Typically, it reduces the efficiency by about 10% or 15%

---

[17]The term secondary range compression is due to the historical expression of the term as $\phi_{SRC} \approx \pi \cdot \beta_{SRC} \cdot f_r^2$. In this case, the chirp rate of the received echoes is clearly modified for higher Doppler frequencies.

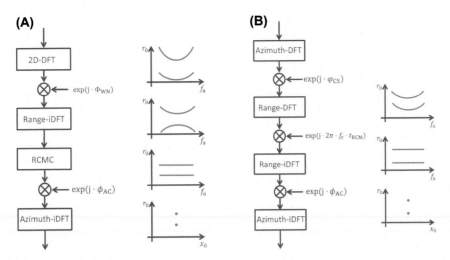

FIGURE 5.90

Block diagrams of the range-Doppler (A) and the chirp scaling (B) algorithms. *DFT*, discrete Fourier transform; *RCMC*, residual RCM correction.

with respect to its interpolation-free version: the chirp scaling algorithm, whose block diagram is shown Fig. 5.90B.

The algorithm is a simplified version of the range-Doppler algorithm with the following approximations:

- quadratic SRC, which means slight filter mismatch even at the reference range;
- quadratic interpolation for RCMC using chirp signals (Papoulis, 1968);
- phase-center correction of the scaling function.

In practical terms, however, the algorithm performs almost as well as range-Doppler in many cases where monochromatic approximation is acceptable.

The principle of the algorithm is to perform an equalization of the RCM for all ranges through chirp scaling, a multiplication in the range-Doppler domain with a phase function, as shown in the figure. In the case of the linear-tracks model, this function can be approximated by

$$\varphi_{CS}(f_a) = \pi \cdot \beta_r \cdot \frac{\beta_{SRC}(f_a; r_{0,\text{ref}})}{\beta_r + \beta_{SRC}(f_a; r_{0,\text{ref}})} \cdot \left[t_r - \frac{2 \cdot (r_0 - r_{0,\text{ref}})}{c \cdot \sqrt{1 - \left(\frac{\lambda \cdot f_a}{2 \cdot v_e}\right)^2}} \right]^2, \qquad (5.159)$$

where β_r is the chirp rate of the transmitted signal, and $r_{0,\text{ref}}$ is the reference range typically set as the middle of the swath. After chirp scaling, all ranges show the same RCM and it can be compensated for without the need of interpolation by multiplying by a linear phase in the range frequency domain. After this, the azimuth compression is common in the flow of the range-Doppler algorithm, with a residual phase compensation step to correct for the effect of the previous scaling (Raney et al., 1994; Moreira et al., 1996).

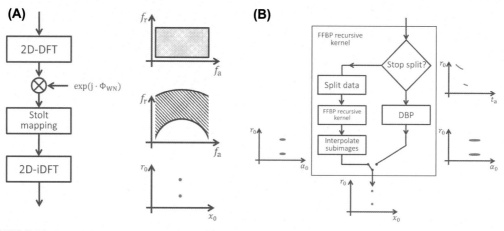

FIGURE 5.91

Block diagrams of the $\Omega\kappa$ (A) and the fast-factorized backprojection (FFBP) (B) algorithms. *DFT*, discrete Fourier transform; *DBP*, direct backprojection.

5.4.3.6 Polychromatic Synthetic Aperture Radar Image Formation

As already discussed, polychromatic SAR image formation reproduces the exact matching of the SAR impulse response or matched filter over the entire available bandwidth. The most popular polychromatic algorithms are range-migration, also known as $\Omega\kappa$, in the wave number domain and FFBP in the time domain. The block diagram of the algorithms is shown in Fig. 5.91.

5.4.3.7 The Range-Migration Algorithm

The $\Omega\kappa$ algorithm was imported from geophysical imaging techniques in seismic applications by Cafforio et al. (1991) and Milman (1993). The algorithm is based on an interpolation in the wave number domain, known as Stolt mapping, which transforms the spectrum of the raw data into the spectrum of the final image thus achieving wideband RCMC and SRC. The flow of the algorithm is shown on the left part of Fig. 5.89.

For linear tracks, the Stolt mapping takes the following form:

$$\sqrt{(f_r + f_0)^2 - \left(\frac{c \cdot f_a}{2 \cdot v_e}\right)^2} \to f_r + f_0, \tag{5.160}$$

which already hides the main weakness of its canonical version for spaceborne applications. The interpolation is not able to accommodate the variation of v_e with range in a precise manner. There are ways to overcome this limitation for high-resolution or wide-swath cases in which a decomposition of the SAR transfer function is achieved via a single-value approach. In terms of computational cost, $\Omega\kappa$ shows a similar behavior as range-Doppler, which as the Stolt interpolation is the computational bottleneck. If space-variant corrections are required, e.g., motion compensation, $\Omega\kappa$ typically requires additional Fourier transformations with respect to range-Doppler, making it the less efficient of the classical Fourier-based solutions.

5.4.3.8 Fast-Factorized Backprojection

The FFBP algorithm can be seen as a time-domain counterpart of $\Omega\kappa$. The algorithm possesses, however, a natural ability to accommodate any spatial variance in the sample (e.g., range-varying effective velocities or DC, motion compensation) without a significant modification of flow or computational efficiency. The price to pay is a larger computational burden than in $\Omega\kappa$ due to the use of 2-D (instead of 1-D) interpolations. The algorithm was developed for low-frequency SAR with large integration times where very accurate motion compensation was required.

As depicted in the right panel of Fig. 5.91, FFBP is based on a divide and conquer approach analogous to the butterfly FFT algorithm. The full aperture of the data is recursively split in two (or another factor) until it reaches a sufficiently small number of pulses. The small subapertures are processed into coarse resolution images in a convenient polar geometry which minimizes the Nyquist sampling requirements. Once the images are computed, the algorithm goes back one step in recursion and computes a finer resolution image (e.g., twice the subaperture) through interpolation, as shown in Fig. 5.91. The accommodation of the spatial variance is accommodated with finer resolution at every step of recursion. At the final stage, the full-resolution images may be interpolated to the more natural radar coordinates.

5.5 SYNTHETIC APERTURE RADAR INTERFEROMETRY

Radars measure the amplitude and delay of the backscattered echoes. As discussed earlier, the delay information is stored in the phase of the received signal with subwavelength accuracy. Unfortunately, the ranging accuracy of the phase measurement can hardly be exploited in regular SAR images because it appears mixed with many other components of the measurement, in particular, with the phase pattern resulting from the coherent combination of the scatterers distributed within the resolution cell (as we have seen, in a uniformly distributed phase). All systematic components of the phase of the SAR images cancel out if the measurement is performed differentially, e.g., by combining two SAR complex images acquired at different times or from different positions. This unique differential ranging potential of SAR systems, known as SAR interferometry, yields access to a number of applications including 3-D realization, motion detection, or vertical deformation monitoring.

To remove the geometrical phase caused by speckle—and other systematic components of the measurement—we need to access the differential phase of the two SAR images acquired from different times or from different positions. These two images are known as an interferometric pair, and their temporal or geometrical spacing is called the baseline. The two images of the interferometric pair are typically known as master and slave. A very deficient terminology, since both images are needed equally to generate the desired output. We will speak about first and second acquisitions. The two images are combined as follows:

$$i(r,x) = s_1(r,x) \cdot s_2^*(r,x), \tag{5.161}$$

where i is called the interferogram of images s_1 and s_2. If both individual observations are simultaneous, we speak about single-pass interferometry. Single-pass interferometric systems are expensive since at least two receive channels are needed, a very stringent requirement in space. The only two spaceborne single-pass interferometric SAR missions so far have required very sophisticated configurations: (1) the use of a boom on the Space Shuttle Endeavour (e.g., SRTM) and a two-spacecraft constellation

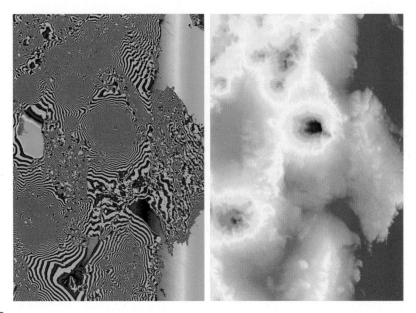

FIGURE 5.92

Wrapped (left) and unwrapped (right) interferometric phases of an interferometric pair acquired with TanDEM-X over the Atacama Desert, Chile.

Courtesy of DLR.

(e.g., TanDEM-X or COSMOSkyMed). The alternative to the two-receive channel payload is to have the second image acquired by the same radar a number of orbit cycles afterward. This scenario is known as repeat-pass interferometry, which is of course affected by temporal changes in the imaged scene. A geometric baseline in this case is naturally achieved by the tolerances in both the orbit control of the spacecraft and our knowledge of this orbit.

The phase of the interferogram is called the interferometric phase, i.e.,

$$\phi_i(r, x) = \angle\{i\} = \angle\left\{s_1(r, x) \cdot s_2^*(r, x)\right\}, \tag{5.162}$$

where the $\angle\{\cdot\}$ extracts the angle between 0 and 2π of the input complex number. As already hinted, the interferometric phase contains the subwavelength information of the differential ranging measurement between the two observations. However, it is not directly observable due to the wrapping of ϕ_i, which poses a major problem in the unambiguous recovery of the phase changes caused by the different observation geometry. The wrapped (left) and unwrapped (right) phases of a TanDEM-X interferogram over the Atacama Desert, Chile, are presented here in Fig. 5.92.

The jumps between 0 and 2π, called fringes, typically correspond to areas of phase continuity, appearing wrapped in the interferometric phase in the left figure. In this case, the fringes represent the slopes (rather than sudden jumps) due to the topography of the scene. This continuity is shown in the right unwrapped interferometric phase of Fig. 5.92, where the color scale has been changed with respect to the figure on the left.

5.5.1 GEOMETRICAL MODELS

As already discussed, SAR interferometry is a differential ranging technique with subwavelength accuracy. Depending on the observation geometry, this property will allow for the estimation of different magnitudes of the imaged scene, exemplified in the across-track cuts of Fig. 5.93.

The left panel of the figure shows the cross-track interferometric configuration for a stationary scene. The two images are acquired from slightly different positions with different observation angles. The triangulation of the delays to the top of the TV tower from the two observation centers allows its localization in space, i.e., the height information of the scene can be estimated. The phase difference between the two images can be expressed as a function of the height of the scatterer in the following form:

$$\phi_i = m \cdot \frac{2\pi}{\lambda} \cdot \left(r_{0,2} - r_{0,1} \right), \tag{5.163}$$

where m is a constant which describes whether the same transmitter path is shared by both interferometric channels. If it is like a single-pass system, $m = 1$; if it is not and instead is similar to a

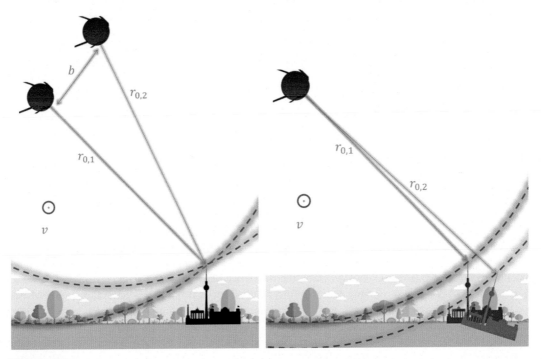

FIGURE 5.93

Typical configurations for synthetic aperture radar interferometry. The spacecraft are flying along an orbit normal to the page. The drawing on the left shows a stationary scene; the different ranges are due to the change in the observation geometry. The drawing on the right shows the observation from a stationary position of a moving scene.

repeat-pass system, $m = 2$; and in single-pass cases operated in ping-pong mode, the value of m is also 2. Under the parallel-ray approximation,[18] the differential range can be approximated as

$$\phi_i \approx m \cdot \frac{2\pi}{\lambda} \cdot b \cdot \sin(\alpha - \theta_l), \tag{5.164}$$

where the last two terms are the projection of the baseline on the LOS, also called the parallel baseline. The height of the scatterer can be straightforwardly computed as

$$h = H \cdot (1 - \cos \theta_l), \tag{5.165}$$

The previous expression is accurate but rarely useful. In real systems, a true evaluation of the actual ranges of the scene via geocoding/backgeocoding is preferable. For a fast analysis, however, the angular information in (5.164) is usually too detailed to be interpreted by eye. By further simplifying the geometry of Fig. 5.93 to the flat-Earth approximation, we can express the interferometric phase as a function of the height of the scatterer as follows:

$$\phi_i \approx m \cdot \frac{2\pi}{\lambda} \cdot \frac{b_{\text{perp}}}{r_0 \cdot \sin \theta_i} \cdot h, \tag{5.166}$$

where the b_{perp} is the projection of the baseline perpendicular to the LOS, following the purple vector of Fig. 5.93, θ_i is the incident angle, and h is the height of the scatterer. The sensitivity of the interferometric phase to topography can be straightforwardly computed from the previous expression, i.e.,

$$\frac{\partial \phi_i}{\partial h} \approx m \cdot \frac{2\pi}{\lambda} \cdot \frac{b_{\text{perp}}}{r_0 \cdot \sin \theta_i}, \tag{5.167}$$

where the $b_{\text{prep}} \approx b \cdot \cos(\alpha - \theta_l)$ is the projection of the baseline perpendicular to the LOS, following the purple vector of Fig. 5.93, and θ_i is the incident angle. In the expression above, the simplification of the geometry of Fig. 5.93 to the flat-Earth case has been implicitly made. From Eq. (5.167) it becomes apparent that larger perpendicular (hence cross-track) baselines give more sensitivity to topographic changes. The former statement is true, but limited by two other effects. With increasing cross-track baselines (1) the quality of the phase of the interferogram degrades due to the increasing dissimilarity of the images (i.e., geometric decorrelation) and (2) the unwrapping of the interferometric phase becomes more intricate due to an increase in the frequency of the fringes. As an example of the latter, Fig. 5.94 shows two interferograms over the Turrialba volcano, Costa Rica, acquired by TanDEM-X in an experimental bistatic "ping-pong" mode with an along-track separation of 20 km (Rodriguez-Cassola et al., 2012).

The left interferogram corresponds to the monostatic (or full) baseline, whereas the right interferogram corresponds to the bistatic (or half) baseline. Note the frequency of the fringes is reduced (about halved) in the bistatic interferogram due to the shorter effective baseline of the acquisition.

---

[18]The parallel-ray approximation is valid whenever the range is several orders of magnitude larger than the baseline, a typical scenario in space borne SAR interferometric systems. Note the rays in Fig. 5.91 are not depicted parallel because of an intentioned distortion of the geometry for illustration purposes.

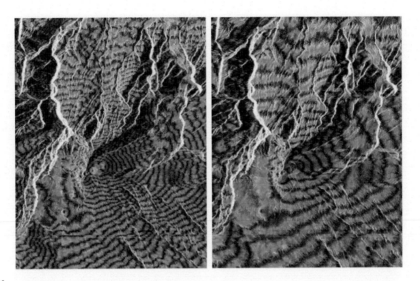

FIGURE 5.94

TanDEM-X monostatic (left) and bistatic (right) interferograms (over the corresponding reflectivity image) acquired in an experimental bistatic "ping-pong" mode over the Turrialba volcano, Costa Rica (Rodriguez-Cassola et al., 2012). Note the frequency of the topographic fringes is reduced in the bistatic interferogram due to the reduced size (about half) of the effective baseline. The interferograms shown correspond to the first bistatic interferometric acquisition of TanDEM-X.

Courtesy of DLR.

Eq. (5.168) provides the solution to the linearization of the interferometric phase as a function of the height h of the considered resolution cell, i.e.,

$$\phi_i \approx \frac{\partial \phi_i}{\partial h} \cdot h = 2\pi \cdot \frac{h}{h_{amb}} = m \cdot \frac{2\pi}{\lambda} \cdot \frac{b_{perp}}{r_0 \cdot \sin \theta_i} \cdot h. \tag{5.168}$$

The h_{amb}, called the height of ambiguity, is a descriptor of any cross-track interferometric observation and approximates the topographic height, which causes a cyclic change in the interferometric phase. As can be inferred from the previous discussion, the height of ambiguity is related to both the sensitivity of the interferometer and the difficulty in unwrapping the phase. Higher values of h_{amb} result in lesser sensitivity to topography and a reduced challenge in phase unwrapping and vice versa.

In the right panel of Fig. 5.94, the situation is inverted, and the two images are acquired from the same position. In this case, the differential path is due to a motion of areas or targets within the scene. The two images must be obviously acquired at different times if motion is to be captured. The interferometric phase in this case can be expressed as

$$\phi_i = m \cdot \frac{2\pi}{\lambda} \cdot \delta r_{LOS} = m \cdot \frac{2\pi}{\lambda} \cdot v_{LOS} \cdot \delta t, \tag{5.169}$$

where δr_{LOS} is the displacement in LOS, v_{LOS} can be interpreted as an equivalent radial velocity describing the motion of the scene as linear, and δt is the time interval between the two radar acquisitions, typically denoted as temporal baseline.

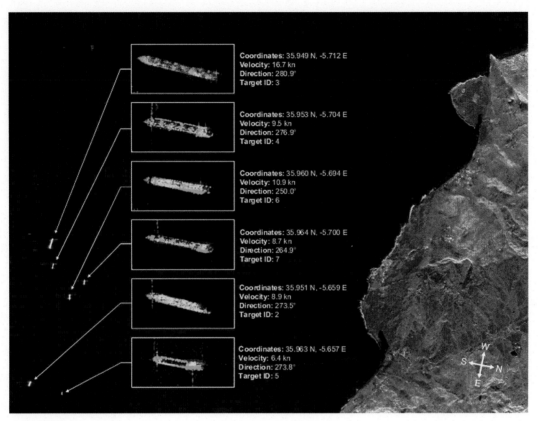

Coordinates: 35.949 N, -5.712 E
Velocity: 16.7 kn
Direction: 280.9°
Target ID: 3

Coordinates: 35.953 N, -5.704 E
Velocity: 9.5 kn
Direction: 276.9°
Target ID: 4

Coordinates: 35.960 N, -5.694 E
Velocity: 10.9 kn
Direction: 250.0°
Target ID: 6

Coordinates: 35.964 N, -5.700 E
Velocity: 8.7 kn
Direction: 264.9°
Target ID: 7

Coordinates: 35.951 N, -5.659 E
Velocity: 8.9 kn
Direction: 273.5°
Target ID: 2

Coordinates: 35.963 N, -5.657 E
Velocity: 6.4 kn
Direction: 273.8°
Target ID: 5

FIGURE 5.95

Single-channel TerraSAR-X image over the Strait of Gibraltar. The ships in the image have been detected using along-track synthetic aperture radar interferometry.

Courtesy of DLR.

Moving targets (e.g., vehicles, ships, airplanes) or sea ice and ocean currents must be observed within short-time intervals up to a few seconds (and shorter), which typically require two (or more) antennas mounted on the same platform with an along-track separation between them. Such systems are obviously single-pass interferometers commonly described in the community as along-track interferometric systems. In Fig. 5.95, we show an example of the detection and estimation of the velocity of ships in the Strait of Gibraltar using along-track interferometry with TanDEM-X.

Along-track interferometry can also be used for the estimation of fast movements of natural surfaces. As an example, Fig. 5.96 shows the estimation of the 2-D Doppler component, related to the wind, of the sea surface in the Arctic regions (Lopez-Dekker et al., 2014). The data are part of an experiment conducted with TanDEM-X in a bistatic, bidirectional (Bi-Di) mode (Mittermayer et al., 2013) to emulate the measurements suggested in Frasier and Camps (2001).

A further example of the use of SAR interferometry for measuring motion is presented in Fig. 5.97. The figure shows the interferometric signature of rotating sea ice in the Arctic regions as observed by

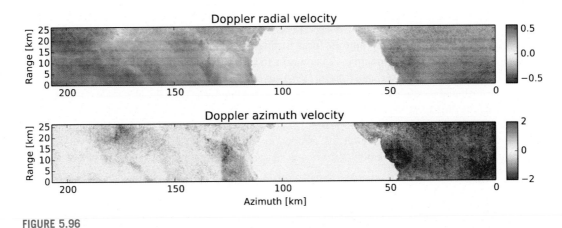

FIGURE 5.96

Estimated line-of-sight (LOS) Doppler velocity (top) and equivalent azimuth velocity (bottom). Positive LOS and azimuth velocities imply motions away from the radar and in the along-flight direction (i.e., right to left), respectively.

Courtesy of DLR.

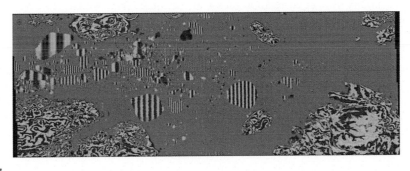

FIGURE 5.97

Rotating sea ice in the Arctic observed with TanDEM-X in ScanSAR pursuit monostatic mode with a temporal baseline of roughly 3 s. Radar illumination is from the bottom. The rotating ice can be identified by vertical fringes, whose frequency is proportional to the rotational speed of the ice. The nonrotating ice (bottom and top areas) show the usual topographic fringes.

Courtesy of DLR.

TanDEM-X in ScanSAR pursuit monostatic mode with a temporal baseline of 3 s. The rotating ice pieces show vertical fringes with spacing proportional to the velocity of rotation (Zebker and Rosen, 1994). The nonrotating parts show conventional topographic fringes.

Slower motions such as those of glaciers, ice fields, or lava flows require time intervals of several hours or even several days and are typically observed by repeat passes of the same satellite. In repeat-pass scenarios, the cross-track baseline between acquisitions is usually nonzero, which adds an additional topographic component to the interferometric phase. This topographic component is a nuisance for the estimation of motion within the images. A standard technique for the removal of the topography consists of using two consecutive interferograms to form a differential interferogram in which the systematic topography disappears. This principle is known in the literature as differential SAR interferometry, acronym DInSAR (e.g. Zebker and Rosen, 1994). In the cases of very slow motions such as terrain subsidence, artificial structure deformations, seismic and volcanic activities, the observation time intervals typically require days to years, and more than two images of the scene should be available. A set of images to that will be combined in an interferometric manner is called an interferometric stack. The processing of an interferometric stack can exploit the information of several image pairs to derive very accurate information of the temporal evolution of the observed scene. Differential interferometry will be analyzed later in greater detail.

5.5.2 COHERENCE, EFFECTIVE NUMBER OF LOOKS, AND DECORRELATION SOURCES

The model of SAR interferometry discussed above implicitly assumes that s_1 and s_2 only differ in a phase term proportional to the ranging difference between the two observations. In reality, both the instrument and the response of the scene to the two observations introduce further differences in the data, which reduce the resemblance of the images and degrade the quality of the interferometric phase. A measurement of the similarity of the interferometric pair is given by the interferometric coherence, essentially a local correlation factor of the image pair, which can be estimated in the following manner:

$$\gamma = \frac{E\left[s_1 \cdot s_2^*\right]}{\sqrt{E\left[|s_1|^2\right] \cdot E\left[|s_2|^2\right]}} \approx \frac{\left\langle s_1 \cdot s_2^* \right\rangle}{\sqrt{\left\langle |s_1|^2 \right\rangle \cdot \left\langle |s_2|^2 \right\rangle}}, \qquad (5.170)$$

where the mathematical expected values are approximated by spatial averages or any other averaging over a multiplicity of observations, an explicit acceptance of ergodicity.[19] The magnitude of the coherence approaches one when the variance of the interferometric phase is small, and tends to zero when it becomes large. If N independent samples are averaged, the variance of the estimation is naturally reduced by a factor N. Multilooking interferograms is a standard procedure to improve the quality of the interferometric phase, at the expense of decreasing the geometrical resolution. Averaging over N samples only means a factor N improvement in the variance if samples are independent. Factors such as the spectral weighting or oversampling introduce correlation between samples, and the effective number of independent samples N_e may be smaller than N. The standard deviation of the estimate of the multilooked interferometric phase can be approximated as (Miller and Rochwarger, 1972):

$$\sigma_\phi \approx \frac{1}{|\gamma|} \cdot \sqrt{\frac{1 - |\gamma|^2}{2 \cdot N_e}}, \qquad (5.171)$$

---

[19]In the case of spatial averages, typically five to seven independent pixels are used.

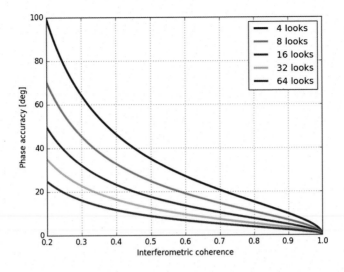

FIGURE 5.98

Approximation of the Cramér–Rao bound given by Eq. (5.171) of the phase estimator as a function of the interferometric coherence for different number of effective looks.

which provides good estimates for values of N_e higher than four.[20] The previous expression plotted against the interferometric coherence for different values of the effective number of looks is presented here in Fig. 5.98.

The sources of decorrelation degrade the coherence of the scene and the quality of the interferometric measurement. If we assume that the useful signals are formed by the sum of two statistically independent components, let us call them coherent and incoherent, respectively, the coherence can be expressed in the following manner:

$$\gamma = \frac{E[(s_{1c} + s_{1i}) \cdot (s_{2c} + s_{2i})^*]}{\sqrt{E[(s_{1c} + s_{1i})^2] \cdot E[(s_{2c} + s_{2i})^2]}} = \frac{E[s_{1c} \cdot s_{2c}^*] \cdot E[s_{1i} \cdot s_{2i}^*]}{\sqrt{E[s_{1c}^2] \cdot E[s_{2c}^2] \cdot E[s_{1i}^2] \cdot E[s_{2i}^2]}} = \gamma_c \cdot \gamma_i, \qquad (5.172)$$

where the subscripts c and i represent the coherent and incoherent components of the images, and by construction $E[s_{\cdot c} \cdot s_{\cdot i}^*] = E[s_{\cdot i} \cdot s_{\cdot c}^*] = 0$. Note the coherence in this case can be expressed as a product of coherent and incoherent components. The term $|\gamma_i|$, always smaller than one, can be seen as a source of decorrelation, which degrades the coherence of the interferometric pair. Analogously, we can write the interferometric coherence as the product of several decorrelation factors in the following manner:

$$\gamma \approx \gamma_{SNR} \cdot \gamma_{amb} \cdot \gamma_{proc} \cdot \gamma_{geo} \cdot \gamma_{temp}, \qquad (5.173)$$

---

[20]The exact expression can be numerically computed using the pdf of the phase estimator as discussed in Oppenheim and Schafer (1975).

where the separation of the contributions is linked to different physical sources (Bamler and Hartl, 1998). The suffixes describe the effects of noise, ambiguities, processing errors, as well as geometrical and temporal changes.

The first obvious cause of decorrelation is noise, which is independent in the two images. Its main contributor is the thermal noise of the receiver. A smaller contribution is due to quantization noise, mainly caused by different effective dynamic ranges, DC bias and the nonlinearity of the ADC, and data compression distortions. Under the assumption of identical SNRs for s_1 and s_2, the value of the SNR decorrelation term can be approximated as follows:

$$\gamma_{\text{SNR}} \approx \frac{\text{SNR}}{1+\text{SNR}}, \tag{5.174}$$

where the SNR may include additional contributions than those above described earlier.[21]

The second term in Eq. (5.173) describes the impact of range and azimuth ambiguities. Provided they are not coherent in the two images,[22] ambiguities appear uncorrelated in the interferogram. In these circumstances, they have a similar effect to noise, and the decorrelation factor can be approximated likewise as

$$\gamma_{\text{amb}} \approx \frac{1}{1+\text{TASR}}, \tag{5.175}$$

where again a similar TASR for both observations has been assumed.

The third source of decorrelation listed above is processing, which accounts for phase and localization errors introduced in the interferometric processing. In airborne SAR, decorrelation due to processing may be caused by significant residual motion errors of the platform. In spaceborne missions, processing errors arise from coregistration errors induced by errors in pointing, orbit knowledge, and the topographic information of the scene. For a homogeneous scene, coregistration errors introduce a decorrelation factor approximated by (Zebker and Villasenor, 1992)

$$\gamma_{\text{proc}} \approx \text{sinc}\left(\frac{\Delta r_{\text{proc}}}{\delta r}\right) \cdot \text{sinc}\left(\frac{\Delta x_{\text{proc}}}{\delta x}\right), \tag{5.176}$$

where Δr_{proc} and Δx_{proc} are the range and azimuth coregistration errors introduced during processing, respectively, and no weighting of the images has been assumed.

The fourth term in Eq. (5.173) describes the impact of the change of the observation geometry, which results in a different result of the coherent superposition of the scatterers on ground. This geometric decorrelation factor can be decomposed in two surface and one purely volumetric components (Moreira et al., 2013), i.e.,

$$\gamma_{\text{geo}} = \gamma_r \cdot \gamma_D \cdot \gamma_{\text{vol}}, \tag{5.177}$$

---

[21]In the case the SNR of the two images is different, the value of γ_{SNR} can be straightforwardly derived from (5.172) as $\gamma_{\text{SNR}} \approx [(1+\text{SNR}_1^{-1}) \cdot (1+\text{SNR}_2^{-1})]^{-1/2}$, which collapses to (5.174) when $\text{SNR}_1 = \text{SNR}_2$ (Kay, 1993).
[22]Azimuth ambiguities might be correlated if the PRF and the pointing of the acquisition is the same, and the baseline is small, but decorrelate rapidly resulting in small changes in the PRF.

where the subscripts r, D, and vol stand for range, Doppler, and volume, respectively. Neglecting the instantaneous Doppler scaling, the carrier wavelength of the radar gets projected onto the different geometrical resolution dimensions of the SAR images. The received echoes can be interpreted as the sample of the scene spectrum in the projected wave number (Gatelli et al., 1994). In elevation, the difference between the projected wave numbers of the interferometric pair is known as spectral shift (Gatelli et al., 1994), which can be approximated by

$$\Delta k_r = \frac{2\pi}{\lambda} \cdot \left(\frac{1}{\sin \theta_{i1}} - \frac{1}{\sin \theta_{i2}} \right),$$
(5.178)

where k_r is the range wave number, and the angles θ_{i1} and θ_{i2} refer to the local incident angle for the considered pixel. The previous expression can be interpreted as a shift in the range frequency domain of the form

$$\Delta B_r = c \cdot \frac{\Delta k_r}{2\pi},$$
(5.179)

which typically becomes larger for larger (perpendicular) baselines and defines the portion of the spectrum appearing as noise in the interferogram. Having exact knowledge of the observation geometry, the decorrelation factor of the spectral shift can be kept at one by filtering the common part with a band-pass filter, at the expense of reducing the spatial resolution of the interferogram. The Doppler term is analogous to the spectral shift, only in the azimuth plane, where the wave number shift due to two different squint angles θ_{sq1} and θ_{sq2} is

$$\Delta k_a = \frac{4\pi}{\lambda} \cdot (\sin \theta_{sq2} - \sin \theta_{sq1}),$$
(5.180)

where k_a is the azimuth wave number, and the 4π is due to the two-way propagation of the signal. Again, this expression can be scaled in the frequency domain, yielding the difference between the DCs of the two acquisitions. As in the case of the spectral shift, the Doppler decorrelation term can be set to one—within the pointing accuracy of the system—by common band filtering of the spectral support, at the expense of reducing the available resolution of the resulting interferograms. In the case of ScanSAR and TOPS interferometry, with an azimuth-varying DC and typically small Doppler support, the orbit control and the pointing accuracy of the system need be controlled very tightly, since any error in the long-track position of the data-take results in spectral decorrelation of the interferometric pairs.

The volumetric decorrelation is due to the change in the signatures of volumetric structures when imaged from different observation angles (Kay, 1993). Volumetric decorrelation increases with the penetration into the media and with increasing cross-track baselines. Volumetric decorrelation cannot be neutralized through processing and can only be minimized by carefully choosing (typically reducing) the baseline of the acquisition. To illustrate the impact of volume decorrelation in the pursuit monostatic and one single-pass bistatic acquisitions, Fig. 5.99 presents an example acquired with TanDEM-X during its pursuit monostatic commissioning phase in a forest area North of Turrialba volcano, Costa Rica.

FIGURE 5.99

Forest area North of Turrialba volcano, Costa Rica, as acquired in the first bistatic interferometric acquisition of TanDEM-X during its pursuit monostatic commissioning phase (Rodriguez-Cassola et al., 2012). Radar illumination from the left. The left plot shows the bistatic intensity image of the area. The middle and right plots show the pursuit monostatic ($m = 2$) and single-pass bistatic ($m = 1$) interferometric coherences, respectively. Note the decrease in the coherence of the middle crop with respect to the right crop, caused by volumetric decorrelation.

Courtesy of DLR.

The image corresponds to the first bistatic interferometric acquisition of the mission, in which simultaneously full-baseline ($m = 2$) pursuit monostatic and half-baseline ($m = 1$) interferograms were acquired (Rodriguez-Cassola et al., 2012). The figure shows the reflectivity image (left) and interferometric coherences (middle-pursuit monostatic; right-single-pass bistatic). The forested area in the middle of the crop shows a lower coherence in the middle crop (pursuit monostatic, full-baseline) due to volumetric decorrelation.

The last decorrelation term of (5.173) is due to temporal effects. Temporal decorrelation is caused by the dynamics of the scene with time, due to changes in the structure, motion, or the dielectric properties of the scene (Zebker and Villasenor, 1992). Temporal decorrelation poses a fundamental limit to interferometry using repeat passes and can only be minimized by minimizing the temporal baseline between the acquisitions. As in the case of the geometric decorrelation, temporal decorrelation also scales with the carrier frequency, affecting more the shorter wavelengths. Although temporal

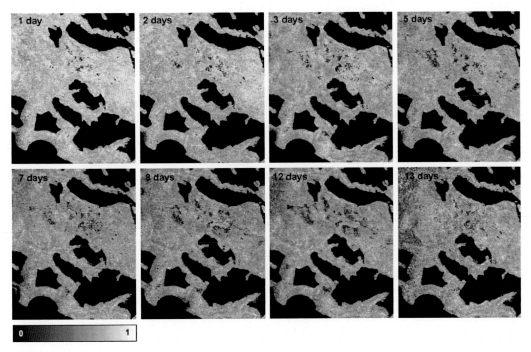

FIGURE 5.100

Interferometric coherence computed with L-band repeat-pass interferograms for temporal baselines between 1 and 13 days over Traunstein, Germany. The data have been acquired with DLR airborne E-SAR system.

decorrelation is in many cases seen as noise in the interferometric measurement, it may also give valuable information as the signature of a physical change. A series of interferograms computed with temporal baselines ranging from 1 to 13 days using L-band data acquired by DLR airborne E-SAR systems are shown in Fig. 5.100. Note the coherence decreases for increasing temporal baselines as expected.

5.5.3 INTERFEROMETRIC PROCESSING

The block diagram of an interferometric SAR processor taking the observation geometry of s_1 as a reference is shown in Fig. 5.101.

The first step of the flow is the coregistration of image s_2 in the geometry of s_1, a step typically done in both airborne and spaceborne systems using the trajectories, timing values, and an external topographic map of the scene (e.g., SRTM) (Sansosti et al., 2007). Fig. 5.102 shows an airborne bistatic cross-platform interferometric pair before (left) and after (right) coregistration.

The red represents the monostatic SAR image acquired by DLR's E-SAR system over Garonnes, France. The blue and green represent the bistatic channel acquired by ONERA's RAMSES. Note the pixels of both images coincide after coregistration.

The second step is the band-pass filtering of both images to the common range and azimuth bands of the scene, to reduce the spectral decorrelation, i.e., $\gamma_r \cdot \gamma_D \rightarrow 1$, and hence increase the overall coherence.

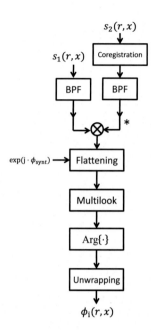

FIGURE 5.101

Block diagram of a synthetic aperture radar interferometric processor in the reference observation geometry of s_1.
BPF, band-pass filter.

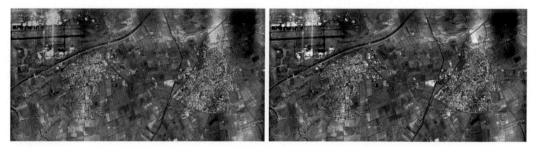

FIGURE 5.102

Airborne bistatic cross-platform synthetic aperture radar (SAR) interferometric pair over Garonnes, France,
before (left) and after (right) coregistration. The *red channel* shows the monostatic SAR image acquired by DLR's
E-SAR system, whereas the *blue and green* channels show the bistatic image acquired by ONERA's RAMSES.
The darker areas are due to echo power fading caused by abrupt antenna deviations caused by turbulence during
the data collection flight (Dubois-Fernandez et al., 2005).

After this, the interferogram is computed. As hinted earlier, the resolution of the interferogram is now
proportional to the common bandwidth of the observation. The next step is typically the removal of a
synthetic phase component computed with whichever a priori information we may have on the scene.
In the case of a cross-track interferometric acquisition, flattening typically consists of removing the

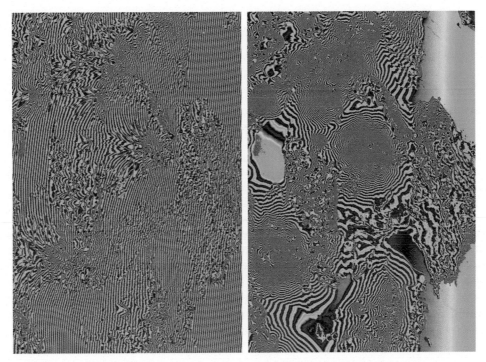

FIGURE 5.103

Interferometric fringes computed with a TanDEM-X interferometric pair over the Atacama Desert in Chile. Radar illumination is from the left. The left interferogram shows the flat-Earth fringes along the range dimension. The right interferogram shows the fringes with the topographic information of the scene after removal of the flat-Earth fringes. The remaining fringes show the topographical information in Atacama.

Courtesy of DLR.

interferometric phase expected from the ellipsoid or the available DEM.[23] Two interferograms computed with a TanDEM-X interferometric pair over the desert of Atacama, Chile, can be found in Fig. 5.103.

The periodic fringes of the left interferogram are the ellipsoidal fringes caused by the increase of look angle within the swath. The topographic fringes appear on top of them, typically visible to the trained eye. The right interferogram shows the topographic fringes after removal of the ellipsoidal (flat Earth) synthetic phase. As shown in the figure, the removal of the synthetic phase usually reduces the frequency of the fringes, hence simplifying subsequent averaging and phase unwrapping.

Once flattened, the interferogram is "multilooked" to the desired averaging factor. As already discussed, multilooking consists of a low-pass filtering followed by a downsampling to accommodate the final bandwidth of the resulting interferograms. The low-pass filtering stage can be done either in

---

[23]This step has been typically known in the community as (flat) Earth removal. The term flat Earth obviously refers to the removal of the fringes introduced by the Earth's ellipsoid and can be extensively used to describe the removal of any a priori expected interferometric phase.

the time or in the frequency domain. The step prior to unwrapping is the computation of the phase of the interferogram as shown in (5.162). After this operation, the interferometric phase is wrapped in a 2π interval.

Phase unwrapping consists of finding the most likely solution to a matrix of multiples of 2π varying for every pixel of the interferogram. Spatial continuity and energy preservation can be imposed on the solution, but it is in general an ill-conditioned problem. From vector field theory, the solution must fulfill the following condition:

$$\nabla \times \nabla \phi_i = 0, \tag{5.181}$$

where ϕ_i represents the unwrapped phase of the interferogram. The most popular phase unwrapping algorithms are branch-cut, region growing, and least squares estimation. Errors in unwrapping typically occur due to the high frequency of the fringes, due to low SNR, or to spatial discontinuities in the interferogram. An example of the latter is shown in Fig. 5.104.

Fig. 5.104 shows three interferograms generated from F-SAR airborne data over Jade Bight, Germany. The top and bottom interferograms show the wrapped and correctly unwrapped phases. The interferogram in the middle shows an unwrapping error on the left side. The unwrapping algorithm (in this case of the region growing class) has failed to ensure the spatial continuity of the phase due to the diagonal water structure in the middle of the interferogram. Unwrapping errors can be identified and corrected with the help of redundancy, i.e., a second (or more) observation(s) of the same interferometric phase with different sensitivity, be it baselines or wavelengths (Kay, 1993). With enough difference in sensitivity, phase jumps will distribute spatially in a different manner, which provides

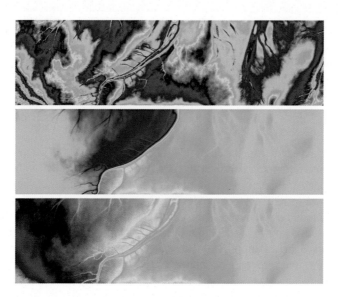

FIGURE 5.104

Wrapped (top) and unwrapped (bottom) interferometric phases of an interferometric pair acquired with the airborne F-SAR system of DLR over the Jade Bight, Germany. The middle figure shows an unwrapping error on the left of the image, where the phase continuity at both sides of the river is not maintained.

Courtesy of DLR.

reliable information for robust estimation. The unwrapping error in Fig. 5.104 (middle plot) has been detected and corrected with the help of a dual-frequency, dual-baseline technique.

5.5.4 DIFFERENTIAL SYNTHETIC APERTURE RADAR INTERFEROMETRY

As already discussed, the interferometric phase between two SAR images acquired at different times encompasses information about the dynamics of the scene itself. The general approach of DInSAR will be the combination of the two effects described in Fig. 5.93, and the interferometric phase will have contributions from any difference in the observation of the scene, be it cross-track baseline or motion within the scene. Without loss of generality, the interferometric phase of any repeat-pass acquisition can be expressed as the sum of a topographic and a differential component, i.e.,

$$\phi_i = \phi_{topo} + \phi_{diff}, \tag{5.182}$$

where the topographic component will increase for increasing cross-track baselines, and the differential component will account for any dynamic change in the observation. In the simplified model of Fig. 5.93, the differential phase was entirely attributed to displacements of the scene. This is hardly the case in reality, and even under error-free calibration, unwrapping, and topography removal, the differential interferometric phase is contaminated by atmospheric propagation and noise, i.e.,

$$\phi_{diff} = \phi_{disp} + \phi_{atm} + n, \tag{5.183}$$

where n is a phase noise process. The atmospheric term is due to uncompensated delays and defocusing introduced by the propagation of the radar waves through the ionosphere and the troposphere. As already discussed, the troposphere tends to affect shorter wavelengths (typically from C-band and above). Residual atmospheric fringes are typically caused by changes in the local water vapor pressure of the layer, which introduces residual delays in the centimetric scale. For lower-frequency bands (e.g., L and P bands), changes in the TEC introduce typical delays in the centimeter scale and high-frequent variations within the aperture during scintillations, which may account for further atmospheric phase components.

To illustrate the previous discussion, Fig. 5.105 shows repeat-pass interferometry over Mexico City using two Sentinel-1 images with a temporal separation of about 6 months from August 2015 to April 2016. All images are geocoded and the vertical component of the bottom interferograms corresponds to the north direction. Note the top images are rotated 90° with respect to the bottom ones.

The two top images show the reflectivity of the scene (left) and interferometric coherence (right). The bottom images correspond to the repeat-pass interferogram (left) and the differential interferogram (right) after subtraction of the topographic fringes estimated using the SRTM model. Note the areas with low coherence (darker) correspond to noisier fringes in the interferograms. When comparing the two interferograms, one can readily identify most of the topographic signature disappearing in the center, left part, and bottom of the images. Most of the remaining (low-pass) fringes are due to changes in the water vapor of the troposphere between the two acquisitions. The band-pass fringes at the top of the interferogram are due to subsidence caused by water extraction in Mexico City.

Another example of the applications of DInSAR is shown in Fig. 5.106.

The image shows the fringes caused by the Illapel earthquake of September 16, 2015, Chile, as seen by Sentinel-1 images. The images were acquired in August and September of 2015. The top area of the image depicting the Pacific coast, where the epicenter of the earthquake was located, appears as noise

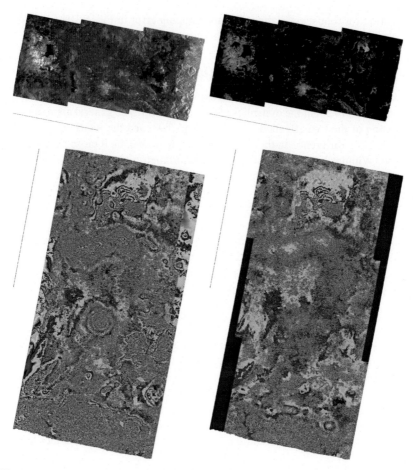

FIGURE 5.105

Repeat-pass interferometric acquisitions over Mexico City observed by Sentinel-1. The north direction corresponds to the vertical axis of the interferograms (bottom). The top images, representing the reflectivity (left) and interferometric coherence (right) are rotated 90° with respect to the interferograms to save space. The bottom images show the repeat-pass (left) and differential (right) interferograms showing the different contributions of the interferometric phase. The low-pass fringes in the differential interferogram correspond to the differential tropospheric delays. The high-pass fringes at the top of the interferogram are caused by subsidence in Mexico City caused by water extraction. The repeat-pass interferogram shows the topographic fringes in addition to the atmospheric and subsidence components.

in the interferogram. The interferometric fringes appear roughly centered around the epicenter of the earthquake, located at about 46 km offshore of Illapel.

The differential interferometric fringes follow the displacement measured in the LOS of the radar. If several acquisitions with different lines of sight are used, the different projections of the displacements can be used to estimate more than one component of the motion of the scene. The situation is illustrated in Fig. 5.107 for the cases of one (left) and two (right) lines of sight, respectively.

FIGURE 5.106

Geocoded differential interferogram computed using two Sentinel-1 acquisitions in the vicinity of Illapel, Chile. The horizontal dimension from left to right shows the north direction of the scene. The fringes are due to the deformation caused by the earthquake occurred in September 16, 2015. The geometric disposition of the interferometric fringes shows the epicenter of the earthquake.

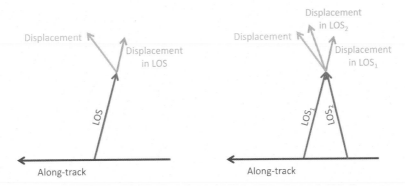

FIGURE 5.107

Line-of-sight (LOS) displacements as measured in two cases. In the second LOS case (right), 2-D displacements within the scene can be estimated. The accuracy of the estimation will depend on signal-to-noise ratio, the variance of the atmospheric perturbations, and the angular separation between the LOS.

    The accuracy of the estimation of the 2-D components will depend on the accuracy of the differential interferometric phase, the thermal and atmospheric noise variance, and the angle between the LOS vectors. The closer they are to being orthogonal, the less noise scaling occurs during the inversion.[24] The two different LOSs can be realized in many different ways, the most common being the use of

---

[24]The estimation of the displacement is done on the equivalent slant range planes of the acquisitions. The angle of incidence can be used to further decompose the motion in ground (e.g., northing and easting) and vertical components.

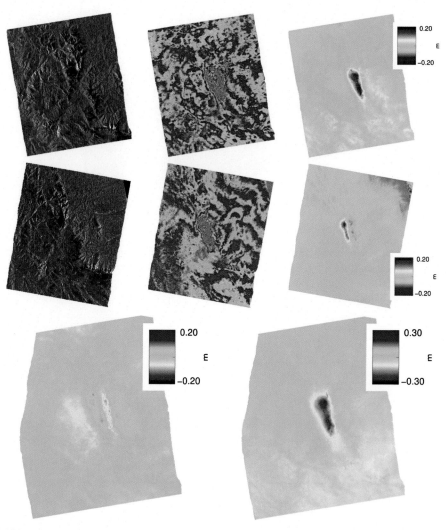

FIGURE 5.108

2-D displacement measurements (easting bottom left, northing bottom right) caused by the Central Italy earthquake of August 2016 using differential interferometry with ascending (top) and descending (mid) passes of Sentinel-1. The first two rows show the reflectivity images (left), differential interferograms (center), and line-of-sight displacements (right).

ascending and descending passes, left and right looking acquisitions, diversity in squint, or along-track separation in bistatic acquisitions.

Fig. 5.108 shows an example of the measurement of the 2-D displacement caused by the Central Italy Earthquake in August 2016 computed using differential interferometry with ascending (top row) and descending (middle row) passes of Sentinel-1 over the Accumoli area.

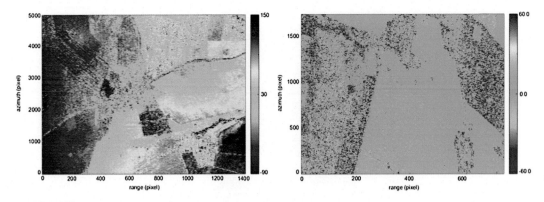

FIGURE 5.109

Differential interferometric phase (left) and closure phase (right).

All images are geocoded, with the vertical dimension oriented towards the north. The images of the first two rows correspond to the radar reflectivity (left), differential interferograms (center), and LOS displacements (right) for the ascending (top) and descending (mid) passes, respectively. The bottom images correspond to the 2-D displacement measurements, eastward (left) and northward (right), inverted using the information of the two passes. As expected from the configuration, the sensitivity in the eastward measurement is significantly better due to the high inclination of the Sentinel-1 orbit.

Sensitive to deformations on the order of the wavelength, DInSAR can also be used to detect changes in the dielectrical properties of volumes via changes in the phase center of the return of the volumes. An example of this is the model for the relationship between the differential interferometric phase and the differential moisture of the soil. These changes can also be due to changes in the humidity of forests, in the structure of snow and ice. Such an example is presented here in Fig. 5.109, which is a differential interferometric measurement of differential moisture done with DLR E-SAR system over the Demmin region, Germany.

The left figure shows an L band differential interferogram of the area. The right figure shows the residual phase (i.e., closure phase) after combining the interferograms generated with three images acquired at different weeks in the year. The combination was done in such a way as to expect a zero phase for unchanged terrain. The red and blue areas of the right figure show changes in the moisture of the terrain.

DInSAR has been extensively used for over two decades for observing strong deformation signatures such as landslides, volcanic eruptions, or earthquakes. However, as illustrated in the previous figures, errors in the removal of the topographic and atmospheric information limit the accuracy of DInSAR with only two images. The effects contaminating the DInSAR measurement, e.g., atmospheric influences and DEM errors, can be estimated if a large stack of interferograms is available. The basic idea is that the atmospheric phase errors are expected to be uncorrelated between consecutive passes and will be averaged out if a large number of acquisitions are combined.

Stack-based DInSAR is an analysis of areas that remain coherent through the stack. Depending on the content of the scene, the use of point-like targets or distributed areas will be used. The former are particularly prevalent in urban areas. Another target may be very stable natural surfaces (e.g., rocky areas). The persistent scatterers (PS) technique allows for a full-resolution amplitude-based detection

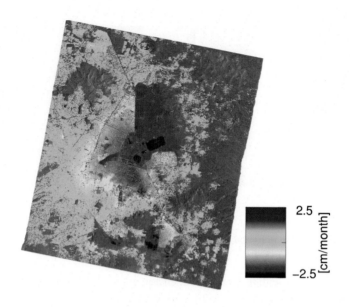

FIGURE 5.110

Geocoded mean deformation velocity map obtained with 45 Sentinel-1 images acquired from October 3, 2014 until May 1, 2016 over Mexico City using the persistent scatterers technique. The vertical dimension corresponds to the northing of the image.

of these targets. Distributed areas are typically detected using interferometric coherence, which average adds to the robustness of the detection at the expense of reducing the geometrical resolution of the areas (Franceschetti and Lanari, 1999; Graham, 1974). A combination of both amplitude and coherence-based detections, as suggested in Barnes et al. (1971), may extend the area where the measurements can be effected. The subsequent unwrapping of the phase of the detected areas helps identify false alarms. This unwrapping is effected on a 3-D sparse grid (i.e., azimuth, range, time) and can be addressed as the 2-D unwrapping of every sparse interferogram (Basu and Bresler, 2000; Franceschetti and Lanari, 1999), using phase difference between the coherent areas as in Graham (1974) and Bauck and Jenkins, (1989), or with help of 3-D unwrapping techniques as in Bleistein and Handelsmann (1975) and Braubach and Voelker (2007). Once the differential unwrapped phase is available, the different components of the differential phase can be estimated in a pixel-wise manner, yielding values for the error of the DEM, the atmospheric phase screen, and the linear and nonlinear components of the deformation. The example in Fig. 5.110 shows the geocoded mean deformation velocity map obtained with 45 Sentinel-1 images acquired from October 3, 2014 until May 1, 2016 over Mexico City using the PS technique. The deformation pattern is caused by ground water extraction.

5.5.5 SYNTHETIC APERTURE RADAR TOMOGRAPHY

The model of 3-D localization presented in Fig. 5.111 implicitly assumes that there exists only one scatterer within a resolution cell. This is typically not the case in surface areas with layover (e.g., urban areas) or volumetric areas with semitransparent media such as forests, snow areas, and ice sheets.

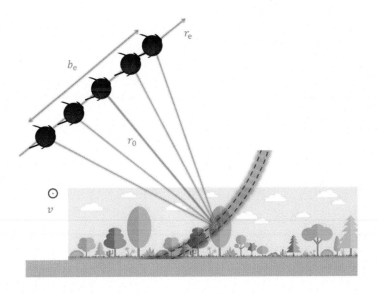

FIGURE 5.111

Geometrical model of a tomographic synthetic aperture radar acquisition. Resolution along r_e is achieved via construction of the tomographic aperture. Height resolution is the projection of δr_e onto the vertical direction.

In these cases, several scatterers are usually present in the resolution cells as depicted in Fig. 5.111, and the interferometric phase is the result of a weighted mean height of the scatterers within the slant range resolution cell.

   This limitation can be overcome by using different passes over the same area to build a synthetic aperture in elevation achieving resolution within the slant range resolution cell following the direction of r_e, as shown in Fig. 5.111. Assuming the maximum cross-track baseline in the tomographic aperture is b_e, the resolution in elevation gained via tomographic SAR processing is (Evans et al., 2005; Ender et al., 2006)

$$\delta r_e \approx r_0 \cdot \theta_e \approx \frac{\lambda \cdot r_0}{2 \cdot b_e}, \tag{5.184}$$

where again the factor of 2 is due to the two-way propagation path of the SAR signals. As expected, the tomographic resolution improves with increasing cross-track baselines, which should typically take values one to two orders of magnitude smaller than the orbit height to yield resolutions of a few meters. Projected onto the vertical direction, the vertical resolution of the tomographic SAR can be approximated by

$$\delta h \approx \frac{\lambda \cdot r_0}{2 \cdot b_e \cdot \cos \theta_i}. \tag{5.185}$$

   The minimum value of the cross-track baseline between passes defines the sampling of the tomographic aperture. To fulfill Nyquist, the minimum cross-track baseline between passes shall be smaller than the following value (Evans et al., 2005; Ender et al., 2006):

$$b_{\min} \leq \frac{\lambda \cdot r_0}{2 \cdot h_V \cdot \cos \theta_i}, \tag{5.186}$$

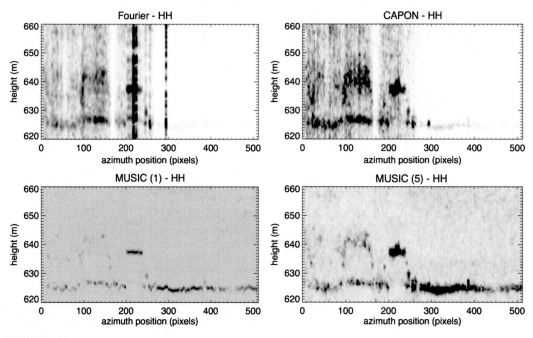

FIGURE 5.112

DLR airborne E-SAR tomograms over Oberpfaffenhofen, Germany. The plots show the results of four different tomographic imaging techniques, e.g., matched filter or Fourier (top left), Capon, (top right), MUSIC with the assumption of a single scatterer in the resolution cell (bottom left), and MUSIC with the assumption of five scatterers in the resolution cell (bottom right). As expected, the superresolution approaches exhibit improved resolution and reduced side lobes, at the expense of a less predictable energy distribution.

Courtesy of DLR.

where h_V is the maximum expected height of the scene. If the true maximum height exceeds the value of h_V, aliasing will occur. On the other hand, a conservative value of h_V results in expensive over-sampling of the tomographic aperture. In the frame of SAR tomography, a better resolution implies more passes, hence increasing the complexity of the whole acquisition geometry if a cubic volume resolution is to be obtained without undesired artifacts.

As an example, the first experimental demonstration of SAR tomography (Ender et al., 2006) used 13 passes with an equidistant separation of 20 m, reaching a resolution in elevation of 2.9 m at midrange. The data were acquired in 1998 by DLR's airborne SAR system E-SAR. Some of the resulting tomograms are shown in Fig. 5.112.

These tomograms have been computed using four different processing methods: Fourier or matched filter (top left), Capon (top right), MUSIC with the assumption of one scatterer in the resolution cell (bottom left), and MUSIC with the assumption of five scatterers in the resolution cell (bottom right). The tomograms show a forest area on the left, followed by a building (whose roof appears as a horizontal line in all four cases, and a surface area including one corner reflector to the right of the building. The matched filter approach has a limited resolution and strong side lobe contributions.

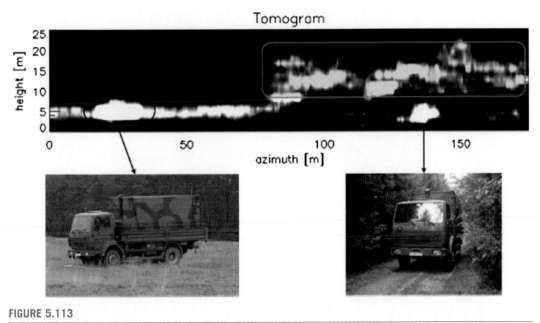

FIGURE 5.113

A tomogram containing different realizations of a covered truck either from the side or from the front. The truck shows the proper orientation in the optical images.

Courtesy of DLR.

In Fig. 5.113, we present a tomogram with some optical images of the targets being captured in the tomogram.

In the use of state-of-the-art missions to generate tomography, especially of urban areas, super-resolution approaches like Capon, MUSIC, and compressive sensing may benefit from a multidimensional signal framed in the sparse information paradigm (Aguttes, 2003; Attema, 1991).

All three methods are super-resolution approaches: Capon is nonparametric and MUSIC is parametric, which causes the results to be sensitive to the models assumed at the focusing stages. As an example, the building, the surface, and the corner reflector appear better focused in the bottom left tomogram, whereas the forest appears better characterized by the bottom right tomogram.

Also, both Capon and MUSIC require the estimation of the covariance matrix, implying a resolution loss due to the averaging operation. On the other hand, compressive sensing (CS) works at full resolution and can reconstruct nonuniformly sampled sparse signals, the latter meaning that the elevation profile to be estimated must be discrete. If the signal of interest is indeed sparse, the CS theory guarantees the possibility to obtain it at a rate significantly below the Nyquist (Moreira et al., 1996).

The use of CS in the frame of urban monitoring in combination with PS has been a topic of research in recent years. Indeed, differential SAR tomography (Cantalloube and Koeniguer, 2008) allows the discrimination of multiple scatterers in layover, e.g., ground and building facades, and at the same time the retrieval of their respective deformation velocities. These approaches have been mainly exploited using current high-resolution spaceborne sensors (e.g. Horn, 1996), becoming a powerful tool for urban monitoring (Fig. 5.114). However, in the frame of forest monitoring, CS as such does not apply

FIGURE 5.114

Three-dimensional absolutely positioned TomoSAR point clouds in 3-D (top) and 2-D (bottom). The absolute height values are color-coded and range between 70 m and 110 m. Clearly, the fusion of multitrack point clouds allow for a very detailed representation of the city where most of the structures can be easily recognized.

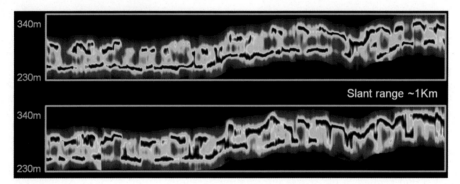

FIGURE 5.115

P-band (top) and L-band (bottom) tomograms acquired with DLR airborne F-SAR system in a forest area close to Lope, Gabon. Note the P-band tomogram shows more bottom signature due to the higher penetration of the radar waves.

Courtesy of DLR.

as well, as the elevation profile is indeed not sparse on the Fourier basis. One possible solution is to make use of a wavelet transform to obtain a sparse representation of the vertical structures, hence allowing the use of CS also for the imaging of forested areas.

One of the most appealing applications of SAR tomography is the imaging of volumetric structures, such as forests or ice. Typically, lower frequencies appear to be particularly well suited due to their penetration capabilities. The following two figures show two tomographic examples with DLR's airborne system F-SAR. Fig. 5.115 shows two tomograms in P (top) and L (bottom) bands computed with 10 baselines in a forest area over Lopé, Gabon.

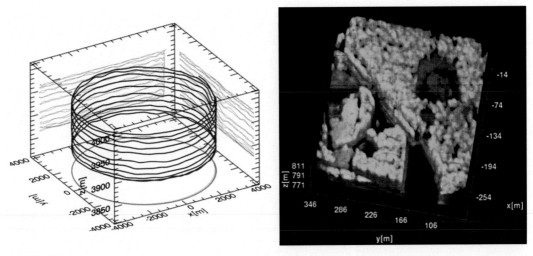

FIGURE 5.116

Circular tomographic acquisition trajectory (left). The geometry allows for the angular generation of tomograms, which can be combined in holographic representations of semitransparent media, such as the forest depicted in the right plot.

Courtesy of DLR.

The structure of the forest is clearly recognizable, with a more transparent signature in P band than in L band, supported by the stronger contribution of the ground.

By combining circular SAR with a tomographic aperture, as shown in the left plot of Fig. 5.116, the possibility of combining tomographic imaging with the multiangular observation of circular SAR allows for holographic SAR. The image on the right of Fig. 5.116 shows a holographic SAR image of a forest acquired at L band by DLR's airborne F-SAR system as the result of the coherent combination of different tomograms acquired from the different angles of the circular trajectory.

A further example over ice is shown in Fig. 5.117, acquired with a similar acquisition as the one shown in Fig. 5.116. The figure shows the hologram from different angles acquired at L band over K transect, Greenland, with the blue channel showing the co-pol HH data and the white channel showing the cross-pol HV data.

The tomographic apertures have been processed using a compressive sensing approach. The structure of the subsurface is clearly recognizable from the different observation angles.

5.6 FUTURE SYNTHETIC APERTURE RADAR SYSTEMS

Future SAR systems are expected to become more cost-effective and capable of offering a higher degree of performance and flexibility when compared to today's systems. We list in this subsection, without a claim on being comprehensive, some of the technological trends which are expected to play a significant role in future Earth observation missions.

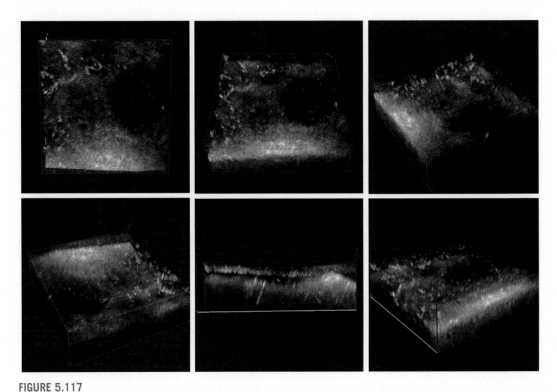

FIGURE 5.117

Holographic, i.e., circular tomographic image acquired at L band over the Ktransect, Greenland, by DLR airborne system F-SAR. The blue channel shows the co-pol HH data, whereas the white channel shows the cross-pol HV data

5.6.1 HIGH-ORBIT (MEDIUM EARTH/GEOSYNCHRONOUS) SYNTHETIC APERTURE RADAR

As already discussed, SAR systems have been flying in LEO over the past few decades. The use of higher orbits such as medium Earth (MEO) or geosynchronous (GEO) opens the door to increased coverage and reduced revisit times, which can be of paramount importance in interferometric, tomographic and polarimetric applications. MEO SAR was first suggested by Edelstein et al. (2005) for tectonic applications. With orbital altitudes between 1500 and 30,000 km, MEO offers a wide range of advantages for SAR remote sensing ranging from global coverage with one- or two-day revisit periods to continental coverage with daily multirevisit. The idea of GEO SAR first appeared in the literature in 1983 (Tomiyasu and Pacelli, 1983). GEO SAR orbits offer persistent continental coverage. With integration times on the order of several hundreds of seconds, GEO SAR systems are very much affected by atmospheric propagation due to their much higher orbital altitude.

In Fig. 5.118 we present the swaths (between 20° and 60° incidence) for one revolution of LEO (blue), MEO (green), and GEO (red) orbits.

Note the significant increase of the swath with the orbit height, which helps reduce the revisit times of these satellites. Note also, that GEO does not offer global coverage. In both MEO and GEO cases,

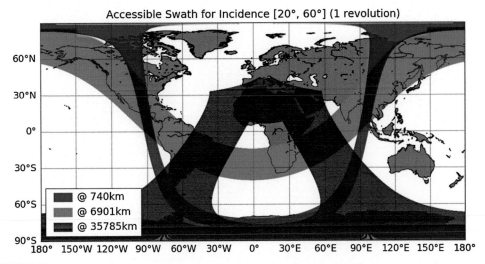

FIGURE 5.118

Swaths (between 20° and 60° incidence) for one revolution of low Earth (blue), medium Earth (green), and geosynchronous (red) orbits.

the power budget of the radar link becomes less favorable, due to their higher altitudes, and larger antennas and transmit powers are required. Moreover, integration times become larger on the order of tens and even hundreds of seconds, which make the systems extremely sensitive to environmental changes within the synthetic aperture, be it atmospheric turbulences, radio frequency interference (RFI), or changes in the reflectivity of the scenes (e.g., ocean surfaces).

5.6.2 MULTICHANNEL SYNTHETIC APERTURE RADAR SYSTEMS

Multichannel SAR systems allow for the simultaneous acquisition of several samples at the same time allowing for additional beam shaping (e.g., digital beamforming) techniques. The main advantage of multichannel systems is that they help relax the timing constraints of single-channel SAR systems, and consequently, decouple the available swath and Doppler requirements for the radar to observe phenomena unambiguously. In this manner, the number of independent pixels of the SAR images can be increased by one or two orders of magnitude.

Depending on the distribution of the channels, we can differentiate between systems with multiple channels in elevation and azimuth. As a rule of thumb, the multiple channels in elevation help improve the sensitivity of the system by digitally narrowing the receive elevation beams, whereas the multiple channels in azimuth help improve the synthetic resolution of the system by digitally narrowing the receive azimuth beams.

This narrow beam can point to different areas of the wide swath or can be steered to follow the returns of the transmitted chirp on the ground. This approach was first suggested by Krieger et al. (2008) and has been known in the literature as scan on receive (i.e., ScORe). The equivalent approach

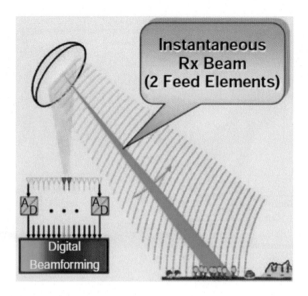

FIGURE 5.119

Digital beamforming in elevation using a reflector array, i.e., SweepSAR. The narrower receive beam follows the propagation of the transmitted signal on the ground, hence improving the sensitivity of the system.

for reflector arrays, depicted in Fig. 5.119, is known in the literature as SweepSAR the standard operation mode foreseen for both JPL's NISAR and DLR's Tandem-L.

One problem of ScORe in single platform systems is the impossibility of receiving echoes during the transmission events, which results in gaps (e.g., blind ranges) in the resulting images. Beamforming in elevation requires a precise calibration of the antenna pattern in which some pointing accuracy can be traded off for a bit of gain of the receive beams, which makes it a reasonably robust technique in terms of implementation. One clear drawback of elevation digital beamforming (DBF) techniques is the comparatively high levels of signal to interference ratios in environments with large RFI.

DBF in azimuth follows a similar pattern and is based on an effective sharpening of the azimuth receive beam after synthetic aperture processing. The basic idea is based on the possibility of reducing the PRF of the system, hence widening the unambiguous accessible swath, while collecting enough samples through the different azimuth channels for Nyquist sampling the Doppler bandwidth of the antenna. The solution is little more than a system concept in which the physical separation of the channels is such that the samples of the different channels roughly correspond to the original sampling of a single-channel system with N times the PRF. A generalization of the approach using a dedicated DBF algorithm is presented in Kim et al. (2013). Fig. 5.120 shows the basic channel distribution of a system example including DBF in azimuth and elevation, the high resolution wide swath (HRWS) system.

In the original HRWS concept, the increase of the number of independent pixels with respect to TerraSAR-X is about a factor 20.

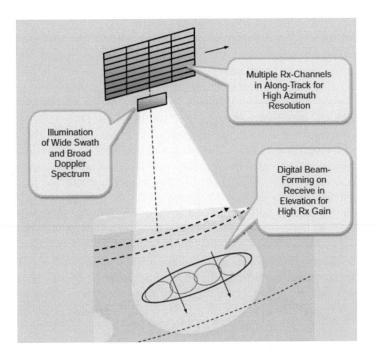

FIGURE 5.120

Basic channel distribution of the high resolution wide swath (HRWS) system for high-resolution and wide-swath imaging. Digital beamforming (DBF) in elevation for increased sensitivity and DBF in azimuth for high-resolution imaging.

5.6.3 ONBOARD PROCESSING FOR DATA REDUCTION IN EARTH AND PLANETARY SYNTHETIC APERTURE RADAR MISSIONS

With increasing performance (e.g., wider swaths and better geometric resolution) and increasing orbit duty cycles, the data volume of SAR missions shall be increasing proportionately. Traditionally, the processing of the SAR data has been conducted on ground, after the downlink of the radar raw data, and all necessary ancillary information (e.g., instrument settings, internal calibration). If things in the future are expected to remain the same, the capacity of the downlinks must grow to accommodate the increasing data volumes. However, this need not be the only solution to the problem. In many cases, the increase in the downlink capacity can be offset by the use of dedicated onboard systems for data reduction (including the SAR retrieval), which are expected to largely benefit from the fast developments in digital electronics.

Excluding the original optical SAR processors, onboard processing in spaceborne SAR has been traditionally limited to the bit adaptive quantization (BAQ) of the raw data. BAQ algorithms are adaptive quantizers capable of locally accommodating the number of bits to the apparent SNR of the received echoes. Typical BAQ ratios for current SAR missions (e.g., TerraSAR-X, Sentinel-1) yield effective reductions in the number of bits slightly better than 50% (e.g., 8:4, 8:3).

The first obvious field for the use of spaceborne onboard SAR processing techniques is planetary observation, in which the effective data transfer capacity is very limited due to the lack of direct visibility of the ground stations and the delay of the radio link over great distances. In these cases, full SAR/InSAR onboard processing followed by multilooking of the reflectivity images and interferograms may yield data reduction rates larger than one order of magnitude, as was the baseline for Veritas, a JPL-ASI-DLR mission proposal for the interferometric mapping of Venus. The use of further image compression techniques for images and interferograms may help relax the data volume bottlenecks.

The argument for planetary observations is also valid for the Earth. In the case of single-pass interferometric and tomographic systems, the additional baselines do not yield, in general, significant changes in the image reflectivities, and the changes of the observation geometry appear all packed in the phases of the images. SAR/InSAR processing may be used onboard, and then only one raw data set and the corresponding multilooked-interferograms need to be transferred to ground. As long as no significant decorrelation is artificially introduced at the processing stages the residual errors in the observation geometry and the system may be expected to be accurately calibrated on ground. Another case in favor of the use of onboard processing for data reduction is multichannel SAR, where the combination of the channels onboard may result in a reduction in the number of effective data streams. Regarding the implementation of onboard DBF algorithms, the calibration requirements (e.g., channel balancing, lever arms, antenna patterns) are in general more stringent if effected in azimuth than in elevation.

5.6.4 BISTATIC AND MULTISTATIC SYNTHETIC APERTURE RADAR CONSTELLATIONS

As previously defined, bistatic SAR has separate antennas for transmitting and receiving, which provides spatial flexibility and the possibility of achieving high isolation in the payload electronics. These attributes make them particularly well suited for CW operation which improves somewhat (e.g., 4 or 5 dB) the power budget of the radar link. In practical terms, bistatic SARs also have different master clocks in the transmitter and the receiver, which changes the time and phase references of the system and becomes a challenge for precise SAR imaging and interferometry. Fig. 5.121 shows the color composite of three airborne bistatic SAR images acquired in 2003 with the airborne systems of DLR (E-SAR) and ONERA (RAMSES) over Garonnes, France.

The colors represent different bistatic elevation angles: 0 (red), 10 (blue), and 20 (green) degrees, which reveal different characteristics of the scene in both urban and natural areas. Accurate bistatic radiometry depends on the accurate characterization of the bistatic joint patterns and the combined effect of transmitter and receiver, which raises interesting challenges in the calibration of the systems. This figure is a very good example of the potential of bistatic SAR for applications involving change detection and image classification techniques.

Arguably, more important than the additional observation angle, bistatic SAR systems offer a straightforward path to generate sufficient angular diversity in the coherent combination of two or more observation geometries. This has been, in fact, the main feature of the first bistatic SAR in space, the TanDEM-X constellation that will also be used in the future Tandem-L mission. In fact, bistatic SAR becomes an essential method for the generation of single-pass interferometric, tomographic systems, as well as for systems with distributed radar systems. Such systems pave the way for applications such as

FIGURE 5.121

Color composite of three bistatic airborne synthetic aperture radar (SAR) images acquired with DLR's E-SAR and ONERA's RAMSES systems over Garonne area, France. The colors represent different bistatic angles in elevation, i.e., 0 (red), 10 (blue), and 20 (green) degrees, showing clear relative variations within the scene.

Credits: DLR and ONERA

the generation of digital elevation and surface models, fast moving target detection and tomographic imaging, together with very high-resolution imaging. These applications require different levels of time and phase coherence between the data. As an example, Fig. 5.122 shows two enhanced resolution images in range (top) and azimuth (bottom) computed by merging the monostatic and bistatic TerraSAR-X and TanDEM-X data, respectively.

The images show two urban areas in Sydney, Australia (top), and Neustrelitz, Germany (bottom), and have been generated by coherently combining the spectral supports of the different images.

The coherent combination of bistatic SAR data requires, in general, an accurate synchronization of time and phase references of the raw data, which typically requires accuracies of a few picoseconds and degrees, respectively. An example of the lack of phase synchronization on the bistatic data is shown here in Fig. 5.123.

The figure shows the first bistatic repeat-pass interferograms acquired by TanDEM-X over the Brasilia area, Brazil, with (left) and without (right) clock phase errors. The horizontal fringes of the left interferogram, which should be interpreted as a horizontal slope in a single-pass scenario, are caused by a difference of about 1 Hz in the carrier offsets between the two acquisitions.

The time and phase synchronization of bistatic SAR data requires the incorporation of dedicated hardware and precise algorithms in the ground software. The current options encompass a hardware-driven and a software-driven one:

1. A dedicated synchronization link between the transmitter and receiver, like in TanDEM-X. Within the accuracy of the orbits and baselines, the direct or synchronization link must provide absolute time and phase referencing between transmitter and receiver, at the cost of dedicated hardware, and constraints in the angular relative position between the spacecraft and in the number of synchronization events. A duplex (e.g., two-way) link may be more robust to reciprocal errors in the system, but its accuracy is limited by the accuracy in knowledge of the baseline and the lever arms of the system.

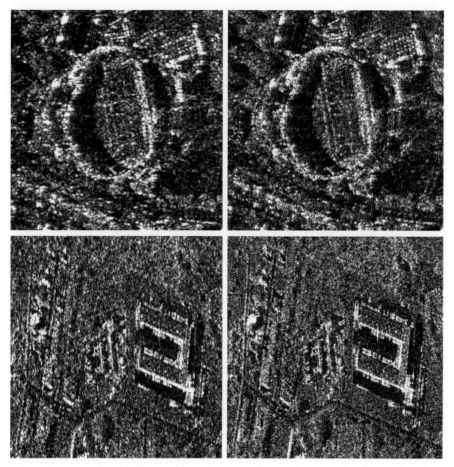

FIGURE 5.122

Enhanced resolution in range (top) and azimuth (bottom) computed after merging the monostatic and bistatic images acquired by TerraSAR-X and TanDEM-X satellites over Sydney, Australia, and Neustrelitz, Germany, respectively.

Courtesy of DLR.

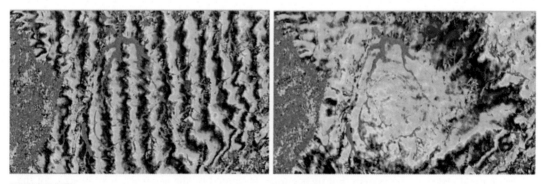

FIGURE 5.123

First bistatic repeat-pass interferograms computed with TanDEM-X over Brasilia, Brazil. The horizontal fringes in the left interferogram correspond to a residual carrier offset of about 1 Hz between the two acquisitions. The right interferogram is the one generated after accurate phase synchronization.

Courtesy of DLR.

2. A model-based (interferometric) map-drift autofocus approach known as autonomous synchronization, e.g., AutoSync. AutoSync is based on the inversion of the relative distortion introduced in the bistatic SAR image by the local changes in the clock carrier frequency, and delivers relative time and phase referencing with an accuracy dependent on the similarity (e.g., coherence) between the bistatic and the reference images. AutoSync was used to generate the first bistatic single-pass interferograms of the TanDEM-X mission computed with the satellites separated by an along-track baseline of 20 km presented earlier in Fig. 5.94. Because of the experimental commanding of the acquisition and the long along-track baseline ($\sim$20 km), the bistatic image had to be synchronized with the help of an AutoSync algorithm.

In future systems, the radar oscillator may be merged with the navigation unit of the satellite, thus enabling a solution of the clock time and phase errors for transmitter and receiver as a direct product of the standard precise orbit determination (POD) algorithms. Effective L band accuracies of a few degrees with a second rate may be expected, which suggests validity for the approach up to X-band systems.

We finish this subsection with a few comments on bistatic SAR image formation. In the general case, bistatic SAR, unlike monostatic surveys, have the following properties: (1) they are in general nonhyperbolical, (2) they show a higher dependence on topography, and (3) they do not follow an azimuth-invariant model, especially in the case where the baseline is changing with time. Bistatic SAR image formation algorithms need to accommodate the previous characteristics and must be implemented in both time and Fourier domains. The same algorithms described earlier for monostatic radars are available in the bistatic case, with the difference that the processing functions in the general bistatic case are best computed numerically. Most of the bistatic TanDEM-X SAR images shown throughout this chapter have been computed with numerical bistatic versions of the range-Doppler and chirp scaling algorithms. For more demanding scenarios, a numerical $\Omega\kappa$ approach based has already been suggested for the SAOCOM-CS mission proposal.

Whenever the baseline is changing with time, the bistatic, fast-factorized, backprojection (BFFBP) model appears as the most natural solution. Finally, Fig. 5.124 shows a bistatic airborne SAR image acquired by DLR's E-SAR and ONERA's RAMSES over Garonnes, France, processed using the BFFBP algorithm.

Bistatic SAR can also be seen as the basic unit of multistatic SAR constellations, which multiply the potential capabilities of bistatic systems. Multistatic constellations offer an increase in the coverage and revisit times with respect to single-receive channel systems, as well as the potential to adapt the observation geometry to several specific applications. Multistatic SAR constellations are configurable, scalable, and offer interesting options for the simplification of hardware and the improvement in the performance. Potential applications encompass single-pass tomography or multibaseline interferometry, very high geometric and radiometric resolution imaging, or multiaspect reconstruction.

5.7 RADAR ALTIMETERS

Radar altimeters are typically nadir-looking, ranging radars used for the measurement of the surface's height to centimetric accuracy. Initially conceived to measure the surface topography of the ocean, radar altimeters typically operate at higher-frequency bands (Ku and Ka bands) and have also been

FIGURE 5.124

Bistatic airborne synthetic aperture radar (SAR) image over Garonnes, France, acquired by DLR's E-SAR and ONERA's RAMSES. The bistatic angle of the observation was about 20°, and the baseline was kept stable during the flight. The image was generated using a dedicated bistatic, fast-factorized, backprojection kernel.

Courtesy of DLR and ONERA.

used to measure the marine geoid, wave heights and wind speeds, the topography of ice polar caps, surface roughness, as well as low-incidence radar backscatter. As already mentioned, sounders are a particular case of radar altimeters operating in lower-frequency bands and imaging the vertical profiles of volumetric structures such as ice or snow.

The history of space borne radar altimetry starts in the 1970s with two experimental missions, Skylab and Geos-3, conceived as technology demonstrations and not designed with systematic observation capabilities and hence were not sufficiently accurate to enable a scientific use of the data. The first operational radar altimeter in space was designed as part of the Seasat mission in 1978. The Seasat-ALT instrument, even for the short lifetime of the mission, demonstrated the value of sea level measurements obtained through radar altimetry. Years went by until the mid 80s and Geosat was launched, initially conceived as a classified military mission to measure the marine geoid, making the data unavailable for the public use. After completing its goals over the year and a half year geodetic mission, US scientists Jim Mitchell and George Born convinced the US. Navy to put Geosat in the Seasat orbit and allow it to measure ocean surface variability. This move provided useful data for many research projects and set the stage for future dedicated altimeter missions such as TOPEX/Poseidon.

The TOPEX/Poseidon mission was dedicated to observing the ocean surface variability and using these observations to study variations in geostrophic ocean surface currents, waves and tides. This mission operated for a total of 13 years long past its design life of 7 years. It provided increasingly accurate measurements of the ocean's surface as the altimetry science community developed new calibration procedures and refined the processing of the data to achieve higher accuracy in the measurements. One year prior to the launch of TOPEX/Poseidon (TP), the European Space Agency (ESA) launched its first radar altimeter (e.g., RA) onboard ERS-1, followed a few years later by its twin ERS-2. TOPEX/Poseidon managed to continue operating into the lifetime of the follow-on mission Jason-1. In fact, both radar altimeters were operated in the same orbit some 90 s apart for 6 months in 2001, revealing a centimetric agreement between the measurements of the two instruments (Haines et al., 2003a,b). After this calibration period (TP) was moved over between the orbits of Jason-1, which remained in the same orbit as the previous 13-years of TP measurements. This increased the spatial coverage of the combined Jason-1, TP measurement pair.

The launch of ERS-2 was followed shortly by another US navy mission known as Geosat Follow-On (GFO), which started a golden decade for spaceborne radar altimetry. The launch of ESA's large ENVISAT satellite in 2002, with its RA-2 altimeter, provided continuity for ERS-1 and 2. Jason-2, launched in 2008, continued the mission of Jason-1 flying again in the 10-day repeat Seasat orbit. In mid-2012 Jason-1, almost at the end of its lifetime, was moved from its exact repeat orbit to a geodetic one with a very long repeat, thus compromising its ability to measure ocean variability, but putting it in a better position to measure ocean bottom effects. Slightly later, ENVISAT suddenly ceased to communicate with the ground, thus losing two instruments for ocean observations over a short period of time.

In 2010, the SIRAL-2 instrument onboard ESA CryoSat-2 mission introduced the first delay-Doppler, or SAR altimeter as suggested in (Raney, 1998). With its far better along-track resolution, CryoSat-2 has been used for monitoring the polar ice caps and for tracking the changes in the ice thickness, yielding valuable data for understanding the warming of the planet.

The Chinese National Space Administration (CNSA) launched in 2011 the HY-2A satellite including the RA altimeter, still in operation in 2016. In another launch in 2013 by the French Center National d'Études Spatiales (CNES) and the Indian Space Research Organization (ISRO) the Saral satellite carries the radar altimeter ALtiKa, which is the first altimeter operated at Ka band, which improves the accuracy of the system by a factor of three over most of its predecessors. In addition, the improvement in the spatial resolution better resolves coastal areas, continental rivers and lakes. The ESA mission Sentinel-3, launched in 2015, includes a SAR altimeter (SRAL) derived from CryoSat's and Poseidon for ocean observations. Jason-3, launched in early 2016, shares with its predecessors the same instrument, orbit and objectives.

Future planned missions include Jason-CS/Sentinel 6 and the surface water ocean topography (SWOT), all to be launched around 2020. The former will fly a SAR altimeter derived from Sentinel-3 flying in the Jason orbit at 1336 km height. The latter uses a wide-swath altimetry concept with low-incidence side-looking interferometric observations. Table 5.5 is a list of radar altimetry missions over the past 40 years.

5.7.1 GEOMETRICAL MODELS

Altimeters cannot be classified in the same way as SAR systems and other satellite radars. Most satellite altimeters are nadir pointing (Fig. 5.125).

In the nadir-looking geometry, topography is derived from the measurement of the delay of the radar echoes, the position of the radar, and the calibration of the range of the system. The resolution on the ground is given by the footprint of the altimeter antenna in the real aperture case. Early suggested by Raney (1998), delay-Doppler or SAR altimeters benefit from the synthetic aperture principle to improve the along-track resolution of the system, like in the case of ESA CryoSat-2, CNES ALtiKa, and Copernius Sentinel-3.

A newer generation of wide-swath altimeters like SWOT (Fig. 5.126) (Fu et al., 2009), deviate from the strictly nadir-looking to a low-incidence side-looking observation geometry and provide a

Table 5.5 List of Spaceborne Radar Altimetry Missions Over the Past 40 years

Sensor	Operation	Frequency Band	Orbit Height (km)	Agency
Skylab (S193)	1973	Ku	435	NASA/JPL
Geos-3	1975–1978	Ku	845	ESA
Seasat (ALT)	1978	Ku	800	NASA/JPL
Geosat	1985–1989	Ku	800	US Navy
ERS-1 (RA)	1991–1996	Ku	785	ESA
TOPEX/Poseidon	1992–2005	S, Ku	1336	NASA/CNES
ERS-2 (RA)	1995–2011	Ku	785	ESA
GFO	1998	Ku	800	US Navy
Jason-1 (SSALT)	2001	C, Ku	1336	NASA/CNES
ENVISAT (RA-2)	2002–2012	S, Ku	785	ESA
Jason-2 (SSALT)	2008-	C, Ku	1336	NASA/CNES
Cryosat-2 (SIRAL-2)	2010-	C, Ku	717	ESA
HY-2A (ALT)	2011-	C, Ku	963	SOA
Saral (ALtiKa)	2013-	Ka	785	ISRO/CNES
Sentinel-3 (SRAL)	2015	C, Ku	814	ESA
Jason-3 (SSALT)	2016	C, Ku	1336	NASA/CNES/Eumetsat/ NOAA

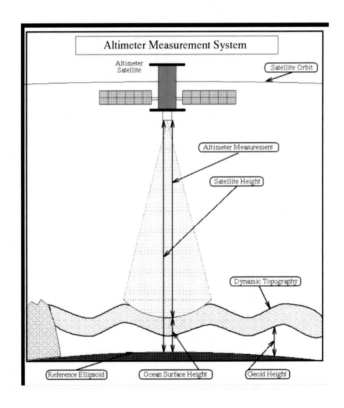

FIGURE 5.125

Nadir-looking geometry of a conventional radar altimeter.

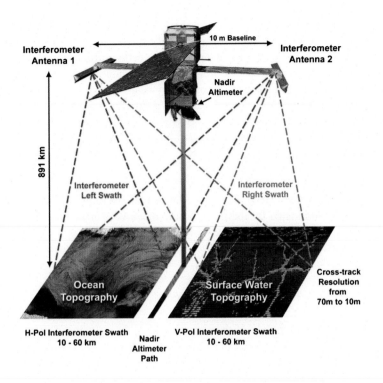

FIGURE 5.126

Surface water ocean topography measurements concept.

From http://swot.jpl.nasa.gov/gallery/galleryImages/index.cfm?FuseAction=ListPhotoGallery&CatID=1.

high-resolution scan of a wide swath on the ground (Fig. 5.126). Wide-swath altimetry is linked with the synthetic aperture principle, which allows us to obtain height accuracies typical of nadir-looking altimetry with the help of SAR interferometry.

5.7.2 ILLUMINATED AREA AND ECHO SIGNAL POWER

The illumination of a flat scene by the radar altimeter in its nadir oriented geometry is depicted in Fig. 5.127

The figure shows top-down views of the observation geometry, including the beam; the shaded rings represent the transmitted pulse radiated in all angular directions. The mid circular areas represent the area on the ground contributing to the received echo. Note in this idealized case the power of the echo increases linearly with time as the pulse illuminates a wider area, reaching its peak after the delay of first impact plus the chirp duration as shown in the top middle figure. Any radar altimeter whose footprint is smaller than the area covered by the pulse on ground is denoted as beam-limited. Conversely, if the footprint is larger than the area covered by the pulse on ground the system will be pulse-limited, and the situation is the one depicted in the top right figure, i.e., the area on ground contributing to the echo becomes an annulus, whose thickness is proportional to the duration of the transmitted chirp.

The temporal distribution of the area contributing to the received echoes is depicted in the bottom of the figure. Under the assumption of an isotropic scene, the power of the received echoes is proportional to this area. The transmitted signal hits the surface at delay t_0. The echo power increases linearly as described earlier covering a circular area until $t_0 + \tau$, with τ being the duration of the transmitted signal. This area of linear increase corresponds to the one covered by beam-limited altimeters. Afterward, the area remains constant until it is finally limited by the antenna beam. In terms of power, the energy decreases slowly in this area due to several causes, e.g., antenna patterns, increasing free space losses, etc.

For an arbitrary surface, the illuminated area can be characterized statistically. The average value of the illuminated area can be approximated as the result of the following convolution (Brown, 1977).

$$A_{\text{eff}}(t_r) \approx p_H(t_r) \cdot h_{\text{radar}}(t_r) \cdot A_{\text{flat}}(t_r), \tag{5.187}$$

where the $p_H(t_r)$ is the pdf of the surface, $h_{\text{radar}}(t_r)$ is the normalized impulse response of the instrument, and $A_{\text{flat}}(t_r)$ is the area illuminated by the pulse on the mean flat surface. As expected, the previous expression collapses to $A_{\text{flat}}(t_r)$ under the assumption of a flat surface, i.e., $p_H(t_r) = \delta(t_r - t_0)$. For generic rough (flat) surfaces, the central limit theorem allows the approximation of the surface height using a Gaussian variable with a pdf

$$p_H(t_r) \approx \frac{1}{\sqrt{2\pi} \cdot \sigma_H} \cdot \exp\left[-\frac{c_0^2 \cdot t_r^2}{2 \cdot \sigma_H^2} \right], \tag{5.188}$$

where σ_H^2 is the variance of the height of the surface, and c_0 the velocity of propagation of the radar waves. If no weighting is assumed, the impulse response of the instrument typically has a sinc form. Once weighting and the frequency response of the instrument are incorporated, h_{radar} can also be approximated with a Gaussian, i.e.,

$$h_{\text{radar}}(t_r) \approx \exp\left[-\frac{B_r^2 \cdot t_r^2}{2} \right], \tag{5.189}$$

where the B_r is the available bandwidth of the system. The power of the received echo can be approximated by integrating the power density multiplied by the effective illuminated area over the iso-ranges (Fig. 5.127) which for an ideal nadir-looking geometry can be further approximated as

$$P_{\text{alt}}(t_r) \approx \frac{P_{\text{Tx}} \cdot \lambda_0^2 \cdot G(t_r)}{(4\pi)^3 \cdot r^4(t_r)} \cdot \sigma_0 \cdot A_{\text{eff}}(t_r), \tag{5.190}$$

where P_{Tx} is the transmitted peak power, $G(t_r)$ is the antenna gain projected on the fast time, r is the slant range, and σ_0 is the average normalized RCS over the footprint (Fig. 5.127).

To compute the received echo power, t_r is the fast-time impulse response of the radar altimeter. Actually, we never deal with flat surfaces and the measurement is an average. The previous discussion is of course only valid in the case of an isotropic surface. The altimeter measures the mean surface height.

5.7.3 RADAR ALTIMETRY OVER THE OCEAN

The principle of measurement of the radar altimeter over the sea surface is sketched in Fig. 5.128.

This figure shows the different surfaces relevant to the altimetric measurement. The quantity of prime interest to oceanographers is the sea level with respect to the reference ellipsoid, the horizontal

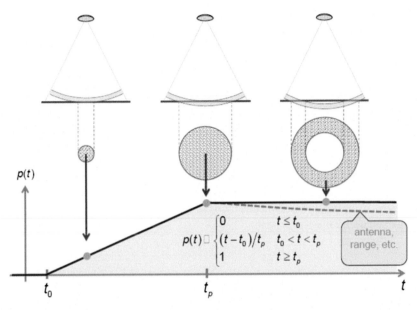

FIGURE 5.127

Power of received echoes over a flat surface (left) shortly after first reflection (mid), once the complete pulse has impacted the illuminated scene and (right) the pulse is illuminating an area toward the edges of the footprint. The dashed beam in the middle figure represents a beam-limited altimeter. The right figure represents a pulse-limited altimeter.

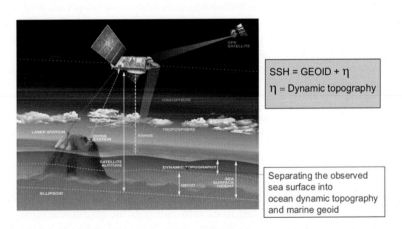

FIGURE 5.128

A schematic of the radar altimeter measurement scheme. Also shown are orbit reference elements of GPS ranging, laser tracking, and DORIS radio tracking. The contributions to the altimeter measurements are also included in the figure.

From NASA website, 2009.

gradient of which is proportional to the geostrophic ocean current. The sea surface topography measured by a satellite altimeter can be expressed as

$$\mathrm{SSH}(x, t_\mathrm{a}) = |\boldsymbol{p}_\mathrm{A}(t_\mathrm{a}) - \boldsymbol{p}_\Omega(x,\ t_\mathrm{a})| + |\boldsymbol{p}_\mathrm{A}(t_\mathrm{a} - \tau_\mathrm{r}) - \boldsymbol{p}_\Omega(x,\ t_\mathrm{a})| - c_0 \cdot \tau_\mathrm{r} = h_\mathrm{G}(x) + \eta(x, t_\mathrm{a}) + \epsilon(x, t_\mathrm{a}), \tag{5.191}$$

where SSH is the surface height, x is the along-track nadir coordinate, t_a is the azimuth time, $\boldsymbol{p}_\mathrm{A}$ is the ephemeris of the altimeter, $\boldsymbol{p}_\Omega(x, t_\mathrm{a})$ is the parametrization of the ellipsoidal surface in the altimeter coordinates, τ_r is the two-way delay of the radar signal, h_G is the height of the marine geoid, η is the dynamic topography, and ε are measurement errors. Let us ignore the latter for the present discussion, since it will be addressed in further detail later (Le Traon and Morrow, 2001b).

The marine geoid is the local surface of the planet in absence of marine perturbations; the latter are basically attributed to the dynamic topography, which describes the variation of the sea surface due to dynamic processes in the ocean. The dynamic topography can be further decomposed as the superposition of a permanent and a variable component,

$$\eta(x, t_\mathrm{a}) = \eta_\mathrm{perm}(x) + \eta_\mathrm{t}(x, t_\mathrm{a}), \tag{5.192}$$

where the former is due to permanent circulation and the second is caused by variable geostrophic ocean currents (tides, mean currents, mesoscale eddies, etc.).

The estimation of η requires the knowledge of h_G, which is not available with sufficiently high-resolution on a global scale and with sufficient accuracy. An estimation of the variable part of the dynamic topography is however possible by combining the data of several altimetric surveys to compute a mean sea surface to use as a reference. The average over several realizations of the sea surface height will be defined as the mean sea surface, i.e.,

$$\mathrm{MSSH}(x) = \langle \mathrm{SSH} \rangle_{t_\mathrm{a}} \approx h_\mathrm{G} + \eta_\mathrm{perm}(x), \tag{5.193}$$

where the operator $\langle \cdot \rangle_t$ refers to temporal averaging over a representative period. The mean sea surface height can be combined with models of h_G to compute a low-pass estimate of η_perm. The model derived from the GOCE mission (i.e., Gravity field and steady-state Ocean Circulation Explorer), provides centimetric accuracy with 100 km sampling (Fig. 5.129).

The estimation of the variable part of the dynamic topography can be straightforwardly expressed as

$$\eta_t(x, t_\mathrm{a}) \approx \mathrm{SSH}(x, t_\mathrm{a}) - \langle \mathrm{SSH} \rangle_{t_\mathrm{a}}. \tag{5.194}$$

Near the coast the altimeter data becomes less useful. First the passive microwave radiometer used for tropospheric water vapor radar path delays, has inherently large spot sizes, which are compromised when they encounter land eliminating the water vapor path length correction. In addition, the altimeter waveform is altered by the presence of land in the altimeter footprint. Here it is important to point out how the SSH and other parameters are retrieved from the altimeter waveform. The waveform (Fig. 5.130) indicates the SSH from the midpoint of the leading edge at least for the open ocean retrieval (Brown, 1977). The y-axis offset from zero indicates the instrument noise level while the waveform maximum is a measure of the σ_0 backscatter, which is related to the wind speed. The slope of the leading edge is inversely proportional to the significant wave height (SWH) and the slope of the trailing edge reflects and antenna mispointing.

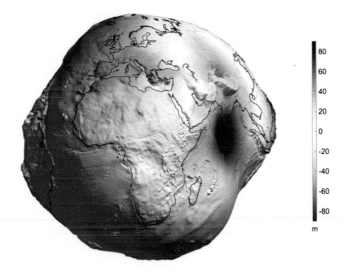

FIGURE 5.129

3-D representation of the Gravity field and steady-state Ocean Circulation Explorer geoid with variations from −100 to + 60 m.

(http://www.esa.int/Our_Activities/Observing_the_Earth/GOCE)

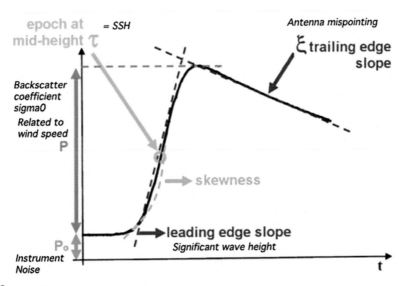

FIGURE 5.130

Radar altimeter return echo. The altimeter waveform is labeled for geophysical parameter retrievals (http://www. altimetry.info/html/alti/principle/waveform/ocean_en.html).

Credits AVISO+/CNES 2017.

5.7.4 ERROR CORRECTION AND CALIBRATION

The calibration of radar altimetry data encompasses the following terms: (1) instrument, (2) atmospheric propagation, (3) sea state bias, (4) tides, and (5) inverse barometer corrections. In this section, we will discuss the contributions of every component of the list.

Instrument errors are mainly due to internal delays of the instrument, but can also be caused by attitude errors. These can be calibrated out using measurements of external targets. Other contributions from the instrument include (long-term) oscillator drifts and Doppler shifts. Another source of errors is the tracking system of the spacecraft for the improvement of the ephemeris determination on the ground. Orbital errors remain the largest contribution to the range error budget of radar altimeters. Orbital errors can be approximated using a sinusoidal model with a fundamental frequency being the instantaneous argument of latitude of the orbit plus some high-frequency noise. Currently POD algorithms supported by accurate error models and differential GNSS measurements achieve accuracies in the order of a couple of cm (~ 2 cm). By supporting POD with external tracking systems (e.g., GPS, radio, and laser) these accuracies are consistently 1 or 2 cm, as was the case of TOPEX/Poseidon. TOPEX/Poseidon provided a huge improvement relative to the ± 10 cm orbital accuracy associated with the ERS-1,2 satellites.

As discussed earlier in this chapter, the propagation through the atmosphere introduces additional delays on the radar echoes which bias the range measurements. Ionospheric variability can also introduce range errors from 1 to 20 cm, which have been traditionally calibrated out exploiting the dispersive character of the volume. Dual-frequency altimeters benefit from the frequency dependence of the group delays to correct the ionospheric perturbations down to 0.5 cm. The troposphere introduces biases of about 2–3 m, due to the hydrostatic and wet contributions. Each modern altimeter also carries a bore-sighted multichannel microwave radiometer to expressly correct for tropospheric water vapor (Brown et al., 2004, 2009). The accuracy of the correction of the wet tropospheric delay can be assumed to be in the order of 1 cm. If the data from the radiometer are missing, the wet tropospheric delay can be estimated from atmospheric climate models.

Atmospheric variations have a further impact other than the additional delays on the propagation path. Spatial and temporal variations of the surface pressure result in changes of the sea level on the order of 1 cm/mbar. In addition, wind and pressure forcing creates a higher-frequency barotropic response at periods shorter than 20 days (Le Traon and Morrow, 2001b). Tidal variations can result in sea surface height changes in open ocean up to 60 cm, and take even larger values in coastal areas where tides become highly nonlinear.

Probably the most obscure bias to compensate for in correcting an altimetry measurement is the one due to the sea state. The sea state bias is due to the response of the wave covered sea surface to the spherical wave fronts of the radar waves, much better reflected in wave troughs than in wave crests. The sea state tends to shift the sea surface height away from the mean sea surface toward the troughs and can take values of about 2 cm (Fu et al., 2010). We present the error budget for some radar altimetry missions (Le Traon and Morrow, 2001b) in. Fig. 5.131.

In their review of these errors Le Traon and Morrow (2001b) also presented these errors as a function of altimeter mission stratified with time (Fig. 5.131), which clearly demonstrates the decrease of the overall budgets with time resulting in making satellite altimetry a measurement that can be used operationally.

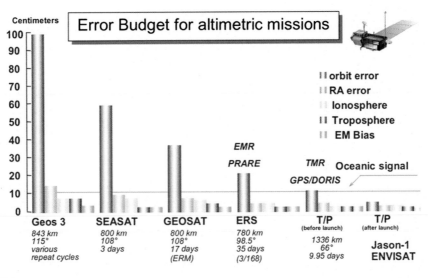

FIGURE 5.131

Error budget for altimetric missions from Geos-3 to Jason-1.

From the mid-1990s to early 2012 there were multiple altimeter satellites operating providing the real golden age of altimeter data availability. During this period researchers were able to merge these different altimeter data together to improve the space-time coverage. To merge multisatellite altimeter data one must first create homogeneous and intercalibrated data sets. A method to achieve this is to use the most precise mission as a reference for the other satellites (Le Traon and Ogor, 1998). During this period, the best references were TOPEX/POSEIDON or Jason-1,2. Further to compute the SLA it is best to use a common reference surface to get the SLA relative to the same ocean mean.

5.8 RADAR SCATTEROMETRY FOR OCEAN WIND VECTOR OBSERVATIONS

Scatterometers are unique among satellite remote sensors in their ability to determine wind speed and direction over water. These wind observations have a wide variety of applications including weather forecasting, marine safety, commercial fishing, El Nino prediction and monitoring of long-term climate fluctuations (http://coaps.fsu.edu/scatterometry/about/overview.php).

A radar scatterometer does not directly measure the wind, but rather the radar backscatter (Bragg scatter) from the ocean's surface that is caused by the small capillary waves that form when the wind begins to blow across the sea surface. These capillary waves have the surface tension of water as their restoring force that balances the initial displacement by the wind. These are the very first responses to the wind forcing on the water and hence can be used as a very good proxy for the wind itself.

To measure this backscatter the scatterometer transmits a pulse of RF and measures the backscattered power from this pulse as compared with the transmitted signal. Using the radar equation, it is possible to compute the normalized radar cross-section of the ocean's surface. This estimate is then converted into the near-surface wind speed using a geophysical model function, which relates the backscatter intensity to the wind stress. Due to the biharmonic dependence of the model function on wind direction, multiple colocated measurements from different azimuth angles are required to determine the wind vector. Even then, there is usually a 180° ambiguity that needs to be resolved with external wind or atmospheric pressure measurements.

The wind direction is found by determining the angle that is most likely to be consistent the backscatter observed from multiple angles. In roughly 5 min, a satellite in a low polar orbit will move far enough to view a point on water surface from angles spanning 90°. The mathematical function describing the fit of the observed backscatter (as a function of the wind direction) usually has multiple minima (ambiguities). Ideally, the best fit corresponds to the true direction of the wind. Typically, the next best fit is in approximately the opposite direction, and the next two minima are in directions roughly perpendicular to the wind direction. The process of selecting the direction from among the multiple minima is called ambiguity selection. Noise in the observations can change the quality of fit and thereby cause incorrect directions (also known as aliases) to be chosen. NSCAT ambiguity selection has proven to be much better than previous scatterometers, with roughly 90% successful selection of the correct ambiguity. Most of the problems with ambiguity removal occur for low wind speeds, where the signal is weak and easily confounded by noise. For wind speeds greater than 8 ms^{-1} successful ambiguity removal is near certain (http://coaps.fsu.edu/scatterometry/about/overview.php).

The backscatter measurements can also be used to map the extent and motion of sea ice, track icebergs, monitor snow melt-accumulation, global rain measurement (at Ku band), and track global change. It should be noted that unlike other space borne instruments, scatterometers observe the same location at multiple azimuth angles. Required for ocean vector wind retrieval, this capability can be explored for the use of scatterometer data in the mapping of the Earth's surface in regions such as the cryosphere, deserts, and tropical vegetation.

5.8.1 BRIEF HISTORY OF SCATTEROMETRY

Historically weather data and in particular wind data could only be observed over land. Geostrophic winds computed from atmospheric pressure fields could only be computed for areas where atmospheric pressure was measured and that was primarily over land. There are occasional reports from ships at sea, but these are very sparse when compared with the network of weather stations scattered over the land's surface. Scatterometry has its very early origins in the radar developed and used in World War II. Early radar measurements over the oceans were corrupted with sea clutter particularly in the region closest to the radar. It was not realized at this time that this "clutter" was the radar's response to winds over the ocean. Radar backscatter was directly related to wind in the late 1960s.

The first real scatterometers flew on the Skylab missions in 1973 and 1974, which demonstrated that spaceborne scatterometers were indeed feasible. As discussed above the Seasat-A satellite scatterometer (SASS, http://nasascience.nasa.gov/missions/seasat-1/?searchterm=seasat) was the first instrument dedicated to a wind vector measurement, which lasted over the short life of Seasat. The subsequent NASA scatterometer (NSCAT, http://winds.jpl.nasa.gov/missions/nscat/index.cfm) was launched aboard Japan's ADEOS-Midori satellite in August 1996. It was the first dual-swath, Ku-band scatterometer to fly since Seasat although the European Space Agency (ESA) flew a single-swath scatterometer on the ERS-1 and ERS-2 missions. It operated well over the short 10-month life of the ADEOS satellite providing comprehensive mapping of the ocean wind vector over the globe. The success of NSCAT prompted the acceleration of the follow-on mission which was SeaWinds carried on the QuikSCAT satellite launched in June 1999. This was only 2 years after the failure of the ADEOS satellite and the termination of the NSCAT observations, but QuikSCAT carried the new NASA SeaWinds scatterometer.

5.8.2 SCATTEROMETER ANTENNA TECHNOLOGY

Different scatterometers have employed various types of antenna systems to collect the information needed. The original Seasat scatterometer (SASS) used a very simple pair of "fan-beam" antennas (Fig. 5.132), which were both crosspolarized. This pair of fan beams gave the required two looks at a surface location, but could not scan more locations than were possible to view with the orientation of these antennas. Fan beam antennas were also used on ERS-1,2 satellites (Fig. 5.132), but the antennas were oriented in a slightly different configuration from SASS. The NASA NSCAT also used a fan beam antenna system much like SASS, but it added another beam to provide a better geometrical solution to the direction resolution problem by resolving the 180° ambiguity mentioned earlier when having only two fan beam antennas.

	SASS	ERS-1/2	NSCAT	SeaWinds
FREQUENCY	14.6 GHz	5.3 GHz	13.995 GHz	13.6 GHz
AZIMUTHS	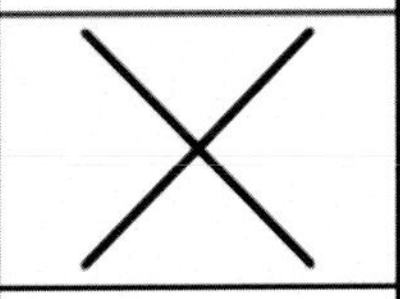	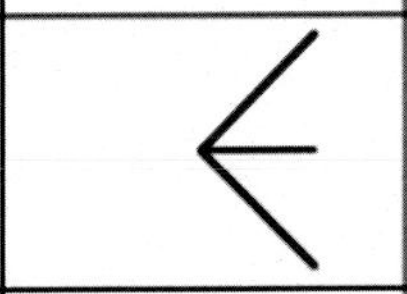	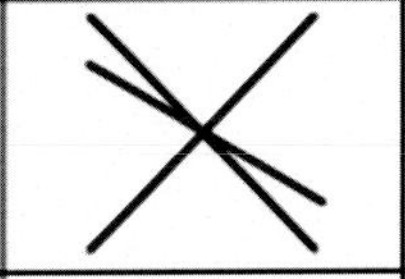	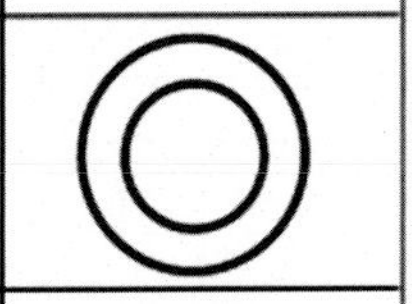
POLAR.	V-H, V-H	V ONLY	V, V-H, V	V-OUTER/H-INNER
BEAM RESOLUTION	FIXED DOPPLER	RANGE GATE	VARIABLE DOPPLER	PENCIL-BEAM
SCI. MODES	MANY	SAR, WIND	WIND ONLY	WIND/HI-RES
RESOLUTION	50/100 km	25/50 km	25/50 km	25 km/6x25km
SWATH	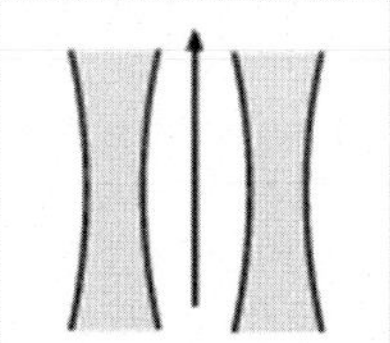	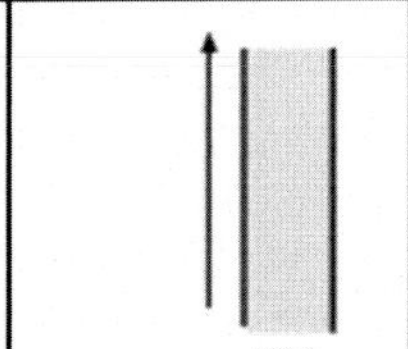 500	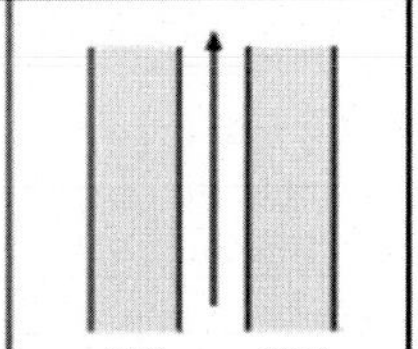 600 600	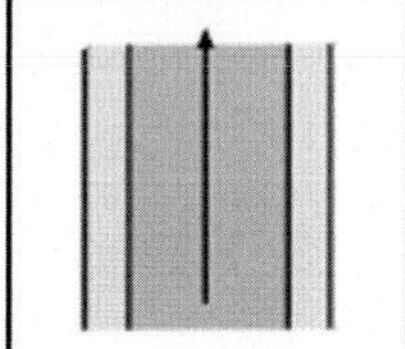 1400,1800
INCIDENCE ANGS	0° - 70°	18° - 59°	17° - 60°	45° & 54°
DAILY COVERAGE	VARIABLE	< 41 %	78 %	92 %
DATES	6/78 – 10/78	92–96 & 96–	8/96 – 6/97	5/99 & 11/01

FIGURE 5.132

Various scatterometer antennas and their coverages.

Finally, NASA SeaWinds used a completely new concept, which was a scanning pencil beam that then retrieved ocean surface wind vectors for a rather wide swath beneath the satellite. SeaWinds used a rotating dish antenna (Fig. 5.130) with two spot beams that sweep in a circular pattern. The antenna radiates microwave pulses at a frequency of 13.4 GHz across broad regions on Earth's surface. The instrument collects data over ocean, land, and ice in a continuous, 1800-km-wide band, making approximately 400,000 measurements and covering 90% of Earth's surface in 1 day (http://winds.jpl. nasa.gov/missions/quikscat/index.cfm).

5.8.3 SEAWINDS A SCATTEROMETER EXAMPLE

As the scatterometer that produced the longest global record we will consider NASA's SeaWinds scatterometer flown on the QuikSCAT satellite as an example of overall scatterometer technology. The SeaWinds scatterometer consists of three major subsystems: the electronics subsystem (SES), the antenna subsystem (SAS), and the command and data subsystem (CDS, http://winds.jpl.nasa.gov/aboutScat/index.cfm).

The SES is the heart of the scatterometer and it contains a transmitter, receiver, and digital signal processor. It generates and sends high RF waves to the antenna. The antenna transmits the signal to the Earth's surface as energy pulses. When the pulses hit the surface of the ocean, it causes a scattering affect referred to as backscatter. A rough ocean surface returns a stronger signal because the waves reflect more of the radar energy back toward the scatterometer antenna. A smooth ocean surface returns a weaker signal because less of the energy is reflected. The echo or backscatter is routed by the antenna to the SES through waveguides (rectangular metal pipes that guide RF energy waves from one point to another). The SES then converts the signals into digital form for data processing.

The CDS is essentially a computer housing the software that allows the instrument to operate. It provides the link between the command center on the ground, the spacecraft and the scatterometer. It controls the overall operation of the instrument, including the timing of each transmitted pulse and collects all the information necessary to transform the received echoes into wind measurements at a specific location on Earth. To locate the precise position on Earth at which the echo was collected, the CDS samples (for each pulse) the antenna rotational position, spacecraft time, and an estimate of the spacecraft position. The CDS also collects instrument temperature, operating voltages and currents, so that the overall health of the instrument can be monitored. It is through the CDS that the other two subsystems receive the commands that control all of their functions.

The SAS consists of a 1-m parabolic reflector antenna mounted to a spin activator assembly, which causes the reflector to rotate at 18 rpm's (revolutions per minute; Fig. 5.133).

The activator assembly provides very accurate spin control and precise position or pointing information to the CDS. Optical encoders, glass disks with small patterns printed on the surface, tell the CDS exactly where the antenna is pointing to about 10/1000 of a degree. The antenna spins at a very precise rate, and emits two beams about 6° apart, each consisting of a continuous stream of pulses. The two beams are necessary to achieve accurate wind direction measurements. The pointing of these beams is precisely calibrated before launch so that the echoes may be accurately located on the ground from space.

The SeaWinds radar operated at 13.4 GHz transmitting 110W pulses at 189 Hz PRF. It weighed 200 kg and consumed 220 W of power during steady-state operation. The average data rate was 40 kbits/s. It has an 1800 km swath, which provided nearly 90% coverage of the world's ocean each day. Wind

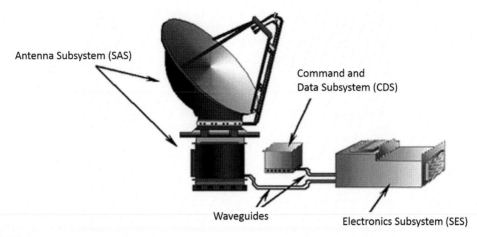

FIGURE 5.133

Block diagram of the SeaWinds scatterometer.

speeds between 3 and 20 m/s (for wind speeds less than 3 m/s the scatterometer winds were unreliable) were accurate to 2 m/s in magnitude and ±20° in direction. The wind vector spatial resolution was 25 km. These accuracies are similar to the wind vector accuracies measured in situ by a moored buoy. The QuikSCAT satellite was a small spacecraft built by Ball Aerospace which was 3-axis stabilized using a star tracker and RU reaction wheels and a C/A Code GPS. This resulted in a pointing accuracy <0.1° absolute per axis. Pointing knowledge was even better at <0.05° per axis. The satellite weighed 970 kg and could generate an average power of 874 W.

5.8.4 SCATTEROMETER LIMITATIONS

While scatterometer wind vector retrievals are a unique source of ocean wind vector information, they are not without problems. In the presence of rain wind vector estimation reliability is diminished. As discussed earlier the wind direction estimates are not unique and require some independent method of selecting the right direction in the face of multiple direction possibilities. Also, as discussed before, for wind speeds less that 3 m/s the scatterometer wind vectors are not reliable. In fact, the lower the wind speed, the less reliable the wind direction estimate.

5.8.5 EXAMPLES OF SCATTEROMETER MEASUREMENTS

A good example of global SeaWinds is shown here in Fig. 5.134 where the colors indicate the wind speed (as indicated by the legend on the right) and the lines denote the atmospheric circulation or the ocean wind pattern. This is a very early result of the SeaWinds mission. Later SeaWinds images and data became very important for the mapping and monitoring of hurricanes, typhoons and tropical storms. One very well-known event was hurricane Katrina that was responsible for significant devastation in the states bordering the northern Gulf of Mexico. In Fig. 5.135 from Liu et al. (2008) we present the QuikSCAT image for Hurricane Katrina for August 28, 2005. Here the color indicates the wind

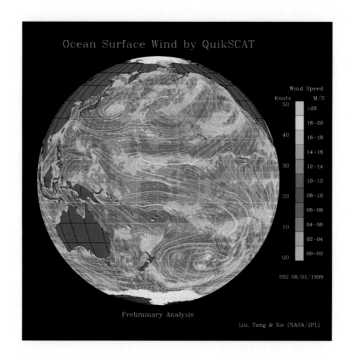

FIGURE 5.134

Ocean surface wind from QuikSCAT.

Courtesy of NASA/JPL. From https://www.jpl.nasa.gov/spaceimages/details.php?id=PIA01346.

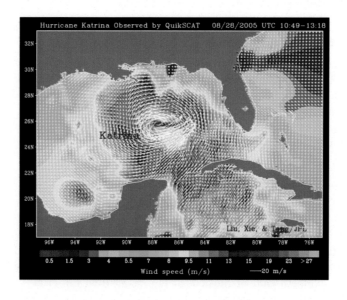

FIGURE 5.135

Hurricane Katrina seen in QuikSCAT data on August 28, 2005 (Liu et al., 2008).

Credits: NASA/JPL.

FIGURE 5.136

QuickSCAT wind vectors overlain on a visible advanced very high resolution radiometer image. Vector colors indicate wind speed as given in the color bar at the top.

magnitude as indicated by the color scale at the bottom of the image. Wind direction is indicated by the white barbs superimposed on the colors of the wind magnitudes. The eye of the hurricane is clearly evident both as a change in color and as the centroid of all the wind barbs in the core region of the storm.

A combination of NSCAT wind vectors and a visible image from the advanced very high resolution radiometer (AVHRR) is shown in Fig. 5.136 for typhoon Violet off the southern coast of Japan. NSCAT wind directions are indicated by the wind vectors and wind magnitudes are revealed both by the size of the vectors and by their color. There is an excellent correspondence between the NSCAT wind vectors and the AVHRR cloud image.

Finally, the image in Fig. 5.137 is a combination of various sampling periods of the SeaWinds scatterometer. Over the ocean, the colors indicate wind speed with orange as the highest wind speeds and blue as the lowest speeds. The white streamlines indicate the coincident wind direction. These wind measurements were made by SeaWinds on September 20, 1999. The large storm in the Atlantic off the coast of Florida is Hurricane Gert and the high wind region in the Gulf of Mexico is the expression of Tropical Storm Harvey. Further west in the eastern Pacific one can see Tropical Storm Hillary. There is also a very strong storm system in the south Atlantic approaching Antarctica.

The land portions of this image were made from four days of SeaWinds data with the aid of a resolution enhancement algorithm developed by Dr. David Long at Brigham Young University (BYU). The lightest green areas correspond to the highest radar backscatter and indicate dense vegetation. You can compare the bright Amazon and Congo rain forests to the dark Sahara desert. The Amazon River can be seen as a dark line running horizontally through the bright South American rain forest. Cities appear as bright spots on the image particularly in the US and Europe. Greenland and the north polar ice cap were generated from SeaWinds data collected on a single day. Here white corresponds to the largest radar return while purple is the lowest. These changes in color reveal variations in the local ice sheet and snow cover conditions.

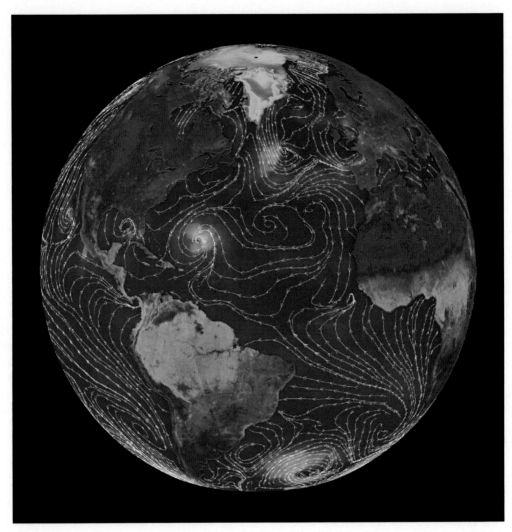

FIGURE 5.137

Composite SeaWinds scatterometer image.

5.9 STUDY QUESTIONS

A SAR operating at a central frequency of 1257.5 MHz, with a bandwidth of 80 MHz. In the ascending passes, the SAR is pointing to the west of the satellite's ground-track, and the swath is determined by incidence angles from 12° up to 45°.

1. Compute the maximum PRF so that there are no ambiguities (take into account Earth's curvature).
2. Compute the antenna dimensions (length l_x and height l_y).
3. Compute the achievable spatial resolution in the azimuth direction.
4. Compute the spatial resolution in the range direction at the edges of the swath.
5. Assuming that σ^0 is not dependent with the incidence angle, which will be the difference in decibels (ratio in linear units) between the SNRs at the edges of the swath?
6. To have roughly square pixels, a number of consecutive pixels in the range direction can be averaged. How many of them can be averaged at the edge of the swath, and by which factor the SNR will improve?
7. To form a 400-km long image, how long the data must be recorded so that all pixels in the image can be focused to achieve the maximum azimuth resolution?
8. If the complex I/Q samples are acquired at the Nyquist rate (no oversampling), and they are stored in 1 byte each, please compute the amount of data to be stored to form the image.
9. Please provide the name and comment each of the following phenomena that occur in SAR imagery, making a drawing on how some representative points will appear in the SAR image.

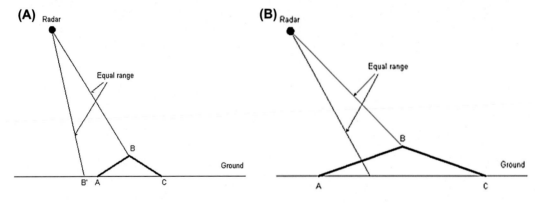

10. The scattered power is stronger in a nadir looking, a side-looking, or in a bistatic configuration. Why? Does it depend on the surface roughness? And on the dielectric constant of the surface? If there is vegetation on top of the surface, which phenomena will take place?

REMOTE SENSING USING GLOBAL NAVIGATION SATELLITE SYSTEM SIGNALS OF OPPORTUNITY

6.1 BRIEF HISTORICAL REVIEW

Radio occultation (RO) measurements have been used in planetary atmospheric studies since 1965 (Fjeldbo and Eshleman, 1965), and until 1995 14 bodies of the Solar system had been explored using RO techniques, the first ones being Mars and Venus as Mariner 4 and 5 flew past these planets (see Table 6.1, adapted from Yakovlev, 2002).

Although its use to study Earth's atmosphere dates back to 1965, it was not until 1990 when the first large-scale study was conducted from the Mir space station and a geostationary satellite. It required, however, the advent of the GPS satellite constellation to provide a suitable source of radio signals for global and continuous monitoring. Some of the most important missions carried aboard GPS-RO nowadays are GPS/MET, CHAMP, COSMIC, and the GRAS instrument onboard MetOp:

- The GPS/MET experiment managed by UCAR demonstrated the potential of using RO to sound the Earth's atmosphere using the Microlab-1 minisatellite (1995−97).
- The Challenging Minisatellite Payload (CHAMP) was a German polar orbiting satellite mission (2000−10) managed by the GeoForschungsZentrum (GFZ) Potsdam, intended for atmospheric and ionospheric research, including a GPS RO experiment (Heise et al., 2008; Pavelyev et al., 2011).
- The Constellation Observing System for Meteorology, Ionosphere, and Climate (COSMIC) is a joint-venture between the United States and Taiwan, and it is the first operational mission constellation carrying GPS RO payloads for space weather. Launched in 2006, it consisted of six satellites in different orbital planes at 700−800 km height, 72° inclination, 30° spacing between them in terms of the right ascension of ascending node, and 45° spacing in mean anomaly between adjacent orbital planes (Fong et al., 2009). COSMIC can be considered as the first GNSS-RO operational mission. Due to the success of COSMIC-1, US agencies and Taiwan have decided to move forward with a follow-on RO mission (called FORMOSAT-7/COSMIC-2a) that will launch six satellites into low-inclination orbits in 2017, and another six satellites (COSMIC-2b) into high-inclination orbits after 2020.
- MetOp is a series of three polar orbiting meteorological satellites (MetOp-A, launched in 2006; MetOp-B, launched in 2012; and MetOp-C scheduled for 2018) operated by the European Organization for the Exploitation of Meteorological Satellites (EUMETSAT). They form the

Introduction to Satellite Remote Sensing. http://dx.doi.org/10.1016/B978-0-12-809254-5.00006-3

Table 6.1 Celestial Bodies Explored Using Radio Occultations

Celestial Body	Spacecraft	Years
Mars	Mariner 4, 6, 7	1965, 1969
	Mars 2, 4, 6	1971, 1974
	Mariner 9	1971
	Viking	1976
Venus	Mariner 5, 10	1967, 1974
	Venera 9, 10	1975
	Pioneer Venus	1978
	Venera 15, 16	1983
Earth	Mir—Geostationary satellite	1990
	GPS/MET	1995
Jupiter	Pioneer 10, 11	1974
	Voyager 1, 2	1979
Saturn	Pioneer 11	1976
	Voyager 1, 2	1980
Titan	Voyager 1	1980
Uranus	Voyager 2	1986
Neptune	Voyager 2	1988
Saturn and Uranus rings	Voyager 2	1980
Moon	Luna 19, 22	1973, 1974
Mercury	Mariner 10	1974
Io	Pioneer 10	1974
Halley's comet	Vega 1, 2	1986

Adapted from Yakovlev, O.I., 2002. Space Radio Science, CRC Press.

space segment component of the EUMETSAT Polar System (EPS), which is the European contribution to the EUMETSAT/NOAA Initial Joint Polar System (IJPS). Among the 11 scientific instruments, MetOp carries the Global Navigation Satellite System Receiver for Atmospheric Sounding (GRAS) (Bonnedal et al., 2010b; Carlstrom et al., 2012).

The "COSMIC" program [http://www.cosmic.ucar.edu/] has initiated the acquisition and processing of data from additional missions beyond the COSMIC mission: the USAF C/NOFS-CORISS (Communication/Navigation Outage Forecast System-C/NOFS Occultation Receiver for Ionospheric Sensing and Specification) mission, the Argentinian SAC-C mission (as of August 2014 suspended due to a spacecraft power anomaly), the US—German GRACE mission, the German TerraSAR-X mission, and the Korean KOMPSAT-5 mission.

Finally, it is worth mentioning that all these experiments and missions use right-hand circularly polarized (RHCP) antennas to receive the direct GPS signal. The upcoming Spanish PAZ mission (launch pending) will include the first polarimetric radio occultations receiver to study the depolarization effects in the propagation of the GPS signal due to the presence of hydrometeors (Cardellach et al., 2015).

The origin of GNSS-Reflectometry (GNSS-R) dates back to 1988, when Hall and Cordey proposed the concept of multistatic scatterometry using GPS signals (Hall and Cordey, 1988). In July 1991 an incident with a French Alpha jet aircraft testing a GPS receiver showed that GPS navigation signals scattered off the sea surface could be collected and tracked (Aubert et al., 1994). In 1993, the concept of reflectometry using GNSS-R signals was proposed for mesoscale altimetry to reduce the revisit time (Martin-Neira, 1993) (Fig. 6.1). Later, in 1996 GNSS-R was proposed as a way to correct for the ionospheric delay errors in ocean altimetry (Katzberg and Garrison, 1996).

In the 90s the first GPS-R observations from an aircraft were collected using an ad hoc receiver (Garrison and Katzberg, 1997, 2000), the first GPS-R data were "found" in segments of SIR-C data acquired in 1994 when radar returns were not present (Lowe et al., 2002a) (Fig. 6.2), and in 2005 GPS-R

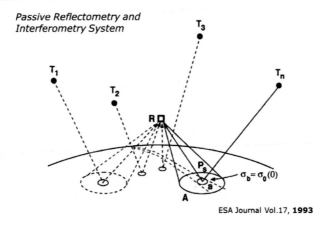

FIGURE 6.1

PARIS concept (Passive Reflectometry and Interferometry System) (Martín-Neira, 1993).

From ESA, The figure 15: "Overall system configuration realising the PARIS concept." from Martin-Neira, M., 1993. A passive reflectometry and interferometry system (PARIS): application to ocean altimetry. ESA J. 17 (4), 331–355.

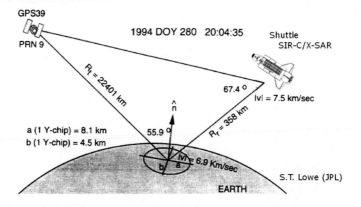

FIGURE 6.2

First GPS-R data collected from space in segments of SIR-C data (Lowe et al., 2002a).

data were acquired from the UK-DMC satellite using the first dedicated space-borne instrument (Gleason et al., 2005).

In addition, since 1993, many studies and experimental activities have been performed, consolidating the understanding of these techniques for other Earth observation applications. Just to cite a few milestones, the ESA sponsored OPPSCAT 1 and 2 and PARIS-α, -β, and -γ projects laid the foundations for GNSS-R scatterometry, ocean altimetry, and radio occultations (GNSS-RO). Other applications include ice monitoring, ionospheric TEC sounding, etc.

More recently, in 2012, NASA selected the CYGNSS mission, a constellation of eight micro-satellites in low inclination orbit for hurricane monitoring (https://directory.eoportal.org/web/eoportal/satellite-missions/c-missions/cygnss), and ESA selected the GEROS experiment on board the International Space Station to perform polarimetric conventional and interferometric GNSS-R and GNSS-RO (Wickert et al., 2011).

In 2016, around 90 operational GNSS satellites from several systems are in orbit including GPS, Glonass, Beidou, Galileo, etc., but in 5 years, this number will reach more than 120, including a number of Space Based Augmentation Systems (SBAS), such as Japan's Quasi-Zenith Satellite System (QZSS) and India's Regional Navigation Satellite Systems (IRNSS). With so many satellites transmitting signals of opportunity, the number of simultaneous signals being received is very large (Fig. 6.3). Therefore, the distinctive feature of remote sensing using these highly precise and continuous signals is in an unbeatable inherent wide swath ($\geq$1000 km, as illustrated in Fig. 6.1), and short revisit time.

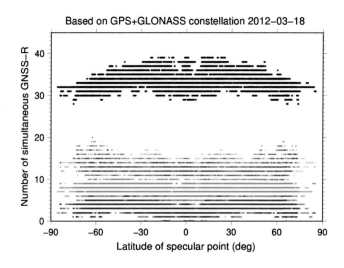

FIGURE 6.3

Number of simultaneously reflected GPS + GLONASS satellites as a function of the latitude coordinate of their specular point on the Earth surface, as computed from a receiver orbiting at 800 km altitude and a 72° inclination (GPS and GLONASS constellation as in March 18, 2012). Statistic of 24 h. Black for all reflected signals, red, green, and blue after applying an elevation cut-off at 30°, 45°, and 60° respectively.

From Jin, S.G., Cardellach, E., Xie, F., 2014. GNSS Remote Sensing: Theory, Methods and Applications, Springer, Dordrecht, Netherlands, ISBN: 978-94-007-7481-0.

Before entering into the details of GNSS-RO and GNSS-R systems, it is important to revise the main properties of the GNSS signals to understand their advantages and "limitations" as compared to the radar systems presented in the previous chapter.

6.2 FUNDAMENTALS OF GLOBAL NAVIGATION SATELLITE SYSTEM SIGNALS

The GPS was the first fully operational GNSS. Its signal structure was designed to allow multiple transmitters using the same frequency band and to have a certain tolerance to multipath and jamming, a serious issue for military applications, and more and more often to civilian applications as well. It was also conceived to have a low power spectral density to avoid mutual interference with other microwave systems, and to allow the estimation of the ionospheric delay for accurate range determination. These features are achieved by means of spread spectrum techniques. In short, this requires a spread the bandwidth of the navigation signal (a biphase modulation with a symbol rate of 50 Hz) by mixing it with a pseudo-random rectangular pulse train that has a much higher frequency than the data. The higher the spreading frequency the higher the power spectral density decrease for a given total radiated power. The spreading sequences are known as pseudo-random noise (PRN), since they have auto-correlation and cross-correlation properties similar to those of Gaussian noise, but with the advantage that they can be precisely generated and regenerated, since they are in essence deterministic. Each GPS satellite has its own PRN code that not only allows discriminating between transmitters, but also achieves the required jamming and multipath resilience and provides range estimations to determine the user position by triangulation.

To understand the structure and properties of the PRN codes, it is useful to consider first the case of a pure random sequence of pulses of width τ_c (Fig. 6.4 and Eq. 6.1):

$$P(t) = \sum_{n=-\infty}^{+\infty} x_n \cdot \Pi\left(\frac{t - n \cdot \tau_c}{\tau_c}\right), \tag{6.1}$$

where x_n takes the values ± 1 with equal probability.

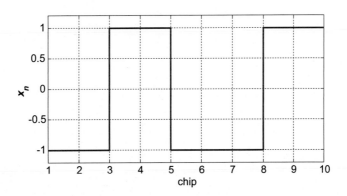

FIGURE 6.4

Sample random sequence of pulses $x_n = \{-1, -1, +1, +1, -1, -1, -1, +1, +1 \ldots\}$. Chip duration $= \tau_c$.

Each individual pulse that composes the sequence is known as a "chip" in opposition to a "bit", since it does not carry any information. The autocorrelation of P(t) is approximately a triangle function given by:

$$R_p(\tau) \approx \wedge_{T_c}(\tau) = \begin{cases} 1 - \dfrac{|\tau|}{T_c}, & |\tau| < \tau_c \\ 0, & \text{elsewhere} \end{cases}, \tag{6.2}$$

where τ is the time lag. For the L1 C/A code $\tau_c = 0.977$ μs, that corresponds to 293 m (146.6 m round trip). As compared to conventional radar altimeters, this value is too large to ensure a satisfactory range resolution for altimetry applications. Higher signal-to-noise ratios (SNR), and larger bandwidth codes, with narrower autocorrelation functions (ACFs) would then be required. The actual PRN codes cannot be strictly speaking random, since it is necessary to regenerate the spreading sequence used by the transmitter at the receiver to decode the navigation signal and retrieve the pseudo-range observable. Therefore, the PRN codes will be deterministic and periodic sequences, but with autocorrelation properties similar to those of a pure random sequence.

The coarse acquisition (C/A) codes are used for the open-access civil service. They have a period of 1 ms to allow quick signal acquisition and a length of 1023 chips. This implies a chip rate of 1.023 MHz, and a bandwidth 2.46 MHz. C/A codes are obtained as the product of two 2^n-1 long "maximal length sequences" (MLS) G1 and G2 generated from n-stage linear feedback shift register (LFSR), so that the cross-correlation properties of the single MLS are improved. Both G1 and G2 are generated by LFSR of 10 stages driven by a 1.023 MHz clock. The actual satellite ID is determined by the relative delay between G1 and G2. This delay is determined by the position of the two connectors of the cells that compose the G2 LFSR. For example, PRN 1 is generated when taps 2 and 6 are selected, and PRN 31 with taps 3 and 8. There are only 37 delay combinations: 32 of them are reserved for the satellites, and 5 are used for other applications, such as ground transmission. The code generation is summarized in Fig. 6.5.

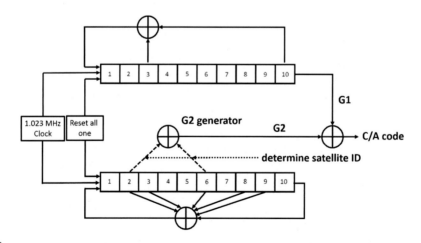

FIGURE 6.5

Generation of the coarse acquisition code as the product of two maximal length sequences.

Adapted from Tsui, J.B.-Y., 2000. Fundamentals of Global Positioning System Receivers, Wiley-Interscience.

The resulting C/A codes have high autocorrelation peaks to clearly identify an acquired satellite and low cross-correlation peaks so that the satellites do not interfere between them (Fig. 6.6). To discriminate a weak signal surrounded by strong ones it is necessary for the autocorrelation peak of the weak signal to be higher than the cross-correlation peaks of the stronger signals. In the ideal case of using random sequences the codes would be orthogonal and the cross-correlations zero. The used PRN codes are almost orthogonal, and the cross-correlation values are as low as $-65/1023$ (12.5% of the time), $-1/1023$ (75% of the time), or $63/1023$ (12.5% of the time).

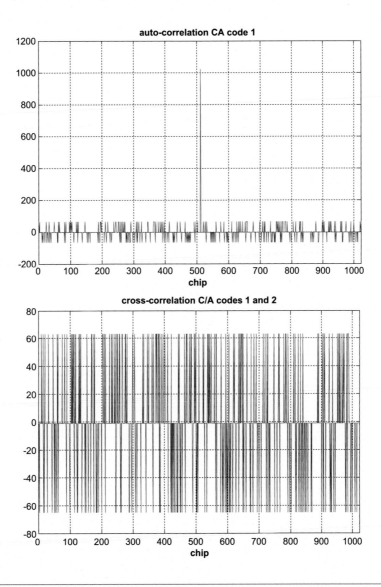

FIGURE 6.6

Autocorrelation of coarse acquisition (C/A) code 1 (top) and cross-correlation of C/A codes 1 and 2 (bottom).

There is also a precise code (P) used for the restricted military signal. It has a chipping rate 10 times faster than the C/A code (10.23 MHz) that results in a 10-fold increase of the pseudo-range observable accuracy. The code period is of 1 week, so that the direct acquisition of the code (i.e., the estimation of the code offset) is pretty cumbersome. Therefore, to acquire the P code special data fields of the navigation frames are used [Z-count and Time of Week]. To increase even more the code robustness, it is possible to switch the system operation to use an encrypted version of the P code, noted as P(Y). The C/A and P codes are modulated in-phase and quadrature on the L1 carrier as shown in Eq. (6.3):

$$S_1(t) = \sqrt{2 \cdot P_{C/A_1}} D(t) \cdot CA(t) \cdot \cos(\omega_1 \cdot t + \phi_1) + \sqrt{2 \cdot P_{P_1}} D(t) \cdot P(t) \cdot \sin(\omega_1 \cdot t + \phi_1), \qquad (6.3)$$

where $S_1(t)$ is the signal transmitted by a given GPS satellite, P_{C/A_1} is the transmitted power for the civil signal at L1, and P_{P1} is the transmitted power for the restricted signal at L1. On L2, for a long time only the P code was broadcast (Eq. 6.4):

$$S_2(t) = \sqrt{2 \cdot P_{P_2}} \cdot P(t) \cdot \cos(\omega_2 \cdot t + \phi_1), \qquad (6.4)$$

but on April 28, 2014, the US Air Force began broadcasting civil navigation (CNAV) messages on the L2C and L5 signals. Prior to that, L2C and L5 provided a default message (Message Type 0) containing no data. The L2 signal began with the launch in 2005 of GPS Block IIR(M). As of June 25, 2014, 13 GPS satellites are broadcasting this signal, and by 2018 all 24 GPS satellites will be broadcasting it.

The LC2 sequence has the same chip rate of the C/A signal, but it is composed of two PRN codes of different length. On one side the moderate length code (CM) is 10.230 chips long, repeats every 20 ms, and it is modulated with navigation data. On the other side, the long code (CL) has 767.250 chips, repeats every 1.5 s and has no data modulation. Both CM and CL codes are generated using the same 27-state LFSR, which is restarted every one CM or CL period. The initial state of the LFSR determines the ID of the satellite the generated code belongs to. Each code is generated at 511.5 kHz and then multiplexed on a chip-by-chip basis to obtain the composite signal at a rate of 1.023 chips/s. A detailed description can be found in Fontana et al. (2001).

Navigation information such as ephemeris, almanacs, or corrections and constellation health are conveyed by the 50 Hz biphase code D(t). All the bit/chip transitions in the C/A, P, and D codes are synchronous, since they are all driven by the same clock. These various signals are broadcast at L-band, thus suffering low atmospheric and rain attenuations. The carrier frequencies are all multiples of 10.23 MHz: $f_{L1} = 154 \cdot 10.23$ MHz $= 1575.42$ MHz; $f_{L2} = 120 \cdot 10.23$ MHz $= 1227.60$ MHz; and $f_{L5} = 115 \cdot 10.23$ MHz $= 1176.45$ MHz.

The frequency spacing between L1 and L2 allows to estimate the ionospheric delay as:

$$\Delta t_1 = \frac{f_2^2}{f_1^2 - f_2^2} \cdot \delta(\Delta t), \qquad (6.5)$$

where Δt_1 is the time delay at the frequency L1 due to the ionosphere, f_1 and f_2 are the L1 and L2 frequencies, and $\delta(\Delta t)$ is the measured time difference between frequencies f_1 and f_2.

The minimum received power (P_R) is computed using the standard propagation equation:

$$P_R = P_T \cdot G_T \cdot \frac{1}{4\pi R^2} \cdot A_{eff} \cdot L, \qquad (6.6)$$

where P_T is the transmitted power, G_T is the gain of the transmitting antenna, $1/4\pi R^2$ are the propagation losses, A_{eff} is the effective area of the receiving antenna

$$A_{eff} = \frac{\lambda^2}{4\pi} \cdot G_R, \tag{6.7}$$

and L are the losses in the atmosphere and other media between transmitter and receiver. Assuming an isotropic receiving antenna ($G_R = 0$ dBi), Table 6.2 summarizes the received power and SNR.

The L1 P signal is -133 dBm for a 0 dBic RHCP antenna. At L1 the C/A signal is 3 dB higher than the P signal (minimum received power of -130 dBm). At L2 the P code is transmitted 3 dB below the L1 P signal. The SNR for the direct signal ranges between 39 dBHz and 52 dBHz, depending on the geometry, the actual transmitted power, and the instrumental and propagation losses. Within the GPS satellite antenna field of view (FOV), the different signal attenuation due to different propagation losses and atmospheric absorption is compensated by the pattern itself of the transmitting antenna. More specifically, the edge of the Earth is $14°$ off the antenna boresight, and therefore the pattern maximum is located at this angle. The transmitted signal is RHCP, and so it is immune to the atmospheric Faraday rotation and the receiving antenna does not have to be pointing to the transmitting satellite to avoid polarization mismatch.

The new L5 signal designed for Safety of Life applications is broadcast in a radio band reserved exclusively for aviation safety services. It features higher power, larger bandwidth, and an advanced signal design including two in-phase and quadrature multiplexed signals: a navigation data channel and a data-free channel to allow a more robust carrier phase tracking. It began in 2010 with GPS Block IIF, as of June 25, 2014 six GPS satellites are broadcasting this signal, and it will be available for all 24 GPS satellites by 2021 (http://www.gps.gov/).

Other satellite navigation systems such as Galileo share the same frequency bands, as illustrated by Fig. 6.7, but the PRN sequences are not necessarily generated using shift registers, but instead

Table 6.2 GPS Typical Received Signal Level and C/N$_0$

Parameter	Symbol	Value	Units
Transmitter power (C/A code)	P_T	14.3	dBW
Transmitting antenna gain	G_T	10.2	dB
Atmospheric loss	L	-2	dB
Space loss @ R = 20,200 km		-157.1	dBm^{-2}
Power density at receiver	PD	-134.6	dBWm^{-2}
Effective area of isotropic antenna		-25.4	dBWm^2
Receiving antenna gain	G_R	0	dB
Received power	P_R	-160.0	**dBW**
Noise spectrum	N_0	-204	dBW/Hz
Carrier-to-noise ratio	C/N_0	**44**	**dB-Hz**

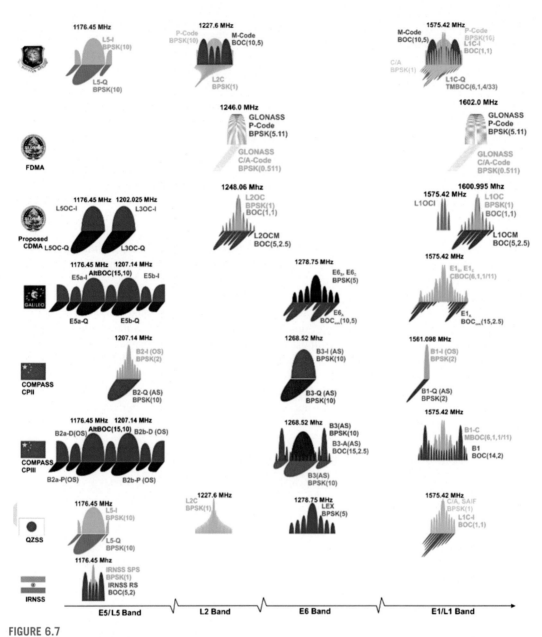

FIGURE 6.7

Frequency bands used by the different GNSS systems (http://www.navipedia.net/index.php/GNSS_signal).

look-up tables. Also, advanced modulation techniques (Binary Offset Carrier Signals or BOC) are used to increase the achievable accuracy with the same bandwidth (http://www.gsc-europa.eu/gnss-markets/segments-applications/os-sis-icd). The BOC modulation is the result of multiplying the PRN code with a subcarrier, which is equal to the sign of a sine or a cosine waveform, yielding so-called sine-phased or cosine-phased BOC signals respectively as shown in J.W. Betz (2001). The BOC signal is commonly referred to as BOC(m,n) where $f_s = m \cdot 1.023$ and $f_c = n \cdot 1.023$, and unless indicated in a different way, when talking about BOC signals it will always be understood the sine-phased variant.

For the sine-phased BOC signals (i.e., L1M, E1B, and E1C), the ACF can be expressed as an addition of triangles (Pascual et al., 2014):

$$R_{BOC_s(\alpha m,m)}(\tau) = \wedge_{t_c}(\tau) * \sum_{k=-\alpha+1}^{\alpha-1} (\alpha - |k|)[2 \cdot \delta(\tau - 2k) - \delta(\tau - 2k - 1) - \delta(t - 2k + 1)], \quad (6.8)$$

where $\alpha = n/m$ is the symbol ratio and $t_c = \tau_c/\alpha$. For the cosine-phased BOC signals (i.e. E1A), the ACF is given by:

$$R_{BOC_c(\alpha m,m)}(\tau) = \wedge_{t_c}(\tau) * \sum_{k=-\alpha+1}^{\alpha-1} (\alpha - |k|)[6\delta(\tau - 2k) + \delta(\tau - 2k - 1) + \delta(\tau - 2k + 1)$$

$$- 4\delta(\tau - 2k - 1/2) - 4\delta(\tau - 2k + 1/2)]. \quad (6.9)$$

Finally, the ACF for the E5 signal can be closely approximated using the general expression of a CDBOC modulation:

$$R_1(\tau) = \wedge_{T_{B_{12}}}(\tau) * \sum_{i=0}^{N_1-1} \sum_{k=0}^{N_2-1} \sum_{i_1=0}^{N_1-1} \sum_{k_1=0}^{N_2-1} (-1)^{i+i_1+k+k_1} \cdot \delta(\tau - (i - i_1)T_{B_1} - (k - k_1)T_{B_{12}}), \quad (6.10)$$

$$R_2(\tau) = \wedge_{T_{B_{12}}}(\tau) * \sum_{l=0}^{N_3-1} \sum_{m=0}^{N_4-1} \sum_{l_1=0}^{N_3-1} \sum_{m_1=0}^{N_4-1} \sum_{p=0}^{N_{res}-1} \sum_{p_1=0}^{N_{res}-1} (-1)^{l+l_1+m+m_1} \cdot \delta(\tau - (l - l_1)T_{B_3}$$

$$- (m - m_1)T_{B_{34}} - (p - p_1)T_{B_{12}}), \quad (6.11)$$

$$R_{AltBOC\left(a\frac{N_1}{2}, a\right)}(\tau) = R_1(\tau) + R_2(\tau), \quad (6.12)$$

where $T_{Bi} = \tau_c/N_i$, $T_{Bij} = \tau_c/N_iN_j$, and $N_{res} = N_1N_2/N_3N_4$.

In Fig. 6.8 we present the ACFs absolute value for infinite bandwidth signals. As an example, Fig. 6.8A is the composition of three functions: a triangle of base $[-1, +1]$ C/A code chips (corresponding to the L1 C/A signal), another triangle of base $[-0.1, +0.1]$ C/A code chips (corresponding to the P code), and two aside peaks (corresponding to the M code, if available, depending on the satellite). Table 6.3 below shows the main signal characteristics for the GPS L1 and L5 and for the Galileo E1 and E5 bands. The carrier frequency is denoted by F_c and the chipping rate and the BOC subcarrier frequency are f_c and f_s, respectively. The bandwidths and the received powers are the ones defined in the GPS Interface Specification (IS) documents [NAVSTAR IS-GPS-200, 2011; NAVSTAR IS-GPS-705, 2011], and the Galileo Interface Control Document (Galileo OS-SIS-ICD, 2010), except for the L1M power, for which Martín-Neira et al. (2011) is used as a reference, and for the E1A power that is assumed to be equally distributed within the E1 band. These documents describe the minimum

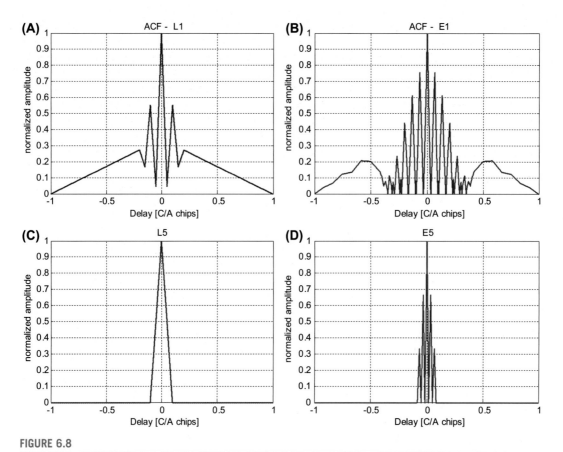

FIGURE 6.8

Autocorrelation functions absolute value for infinite bandwidth signals (L1, L5, E1, and E5).

received power, and therefore may lead to pessimistic performance estimates. It is known that, actually, typical power values for the GPS signals are about 3 dB higher than the specified minimums.

6.3 GLOBAL NAVIGATION SATELLITE SYSTEM—RADIO OCCULTATIONS
6.3.1 BASIC PRINCIPLES

GNSS-RO relies on the detection of a change in a radio signal between a GNSS satellite and a receiver placed on a Low Earth Orbit (LEO), as the radiation passes through the Earth's atmosphere.

According to ITU-R recommendations P.531-11 on "ionospheric propagation" [ITUR-R 531-11; 2012] and P.618-7 [ITU-R 618-7] on "propagation data and prediction methods required for the design of Earth-space telecommunication systems", when the electromagnetic radiation passes through the atmosphere and ionosphere, it is affected by absorption in atmospheric gases; absorption, scattering, and depolarization by hydrometeors; atmospheric refraction; phase decorrelation across the antenna

Table 6.3 GNSS Signals Main Parameters (Pascual et al., 2014)

GNSS	Band (MHz)[a]	Service	Component	Modulation	f_c (MHz)	f_s (MHz)	Power (dBW)[c,d,g]	Main Lobe Bandwidth (MHz)[h]
GPS	L1 [1] Fc = 1575.42 BW₁ = 20.46 BW₂ = 30.69	P(Y)[b]	DATA	BPSK-R10	10.23	–	Min: –161.5 Typ: –158.5 Max: –155.5	20.46
		C/A	DATA	BPSK-R1	1.023	–	Min: –158.5 Typ: –155.5 Max: –153	2.046
		M[b]	N/A	BOCs(10,5)	5.115	10.23	Min: –157[e] Typ: –154 Max: –150	30.69
	L5 [2] Fc = 1176.45 BW = 24	SoL	DATA (L5I)	BPSK-R10	10.23	–	Min: –157.9 Typ: –154.9 Max: –150	20.46
			DATA (L5Q)	BPSK-R10			Min: –157.9 Typ: –154.9 Max: –150	20.46
GALILEO [3]	E1 Fc = 1575.42 BW = 24.552 BW_assumed = 32	PRS[b]	DATA (E1A)	BOCc(15,2.5)	25.575	15.345	Min: –157[f] Typ: –154 Max: –150	35.805
		OS,SoL,CS	DATA (E1B)	CBOC(6,1,1/11) BOCs(1,1)	1.023	1.023	Min: –157 Typ: –154 Max: –150	4.092
			PILOT (E1C)	BOCs(6,1)		6.138		14.322
	E5 Fc = 1191.795 BW = 51.15 Fca = 1176.45 Fcb = 1207.14	OS	DATA (E5aI) PILOT (E5aQ)	AltBOC(15,10) + constant envelope	10.23	15.345	Min: –155 Typ: –152 Max: –148	51.15
		OS,SoL,CS	DATA (E5bI) PILOT (E5bQ)				Min: –155 Typ: –152 Max: –148	

[a] IS defined RF bandwidths.
[b] Restricted codes.
[c] GPS signals: Minimum received RF signal strength on Earth's surface when space vehicle (SV) is above 5° user elevation angle with 3 dBi linearly polarized antenna.
[d] Galileo signals: Minimum received RF signal strength on Earth's surface when SV is above 10° user elevation angle with an ideally matched and isotropic 0 dBi linearly polarized antenna and lossless atmosphere.
[e] From Martín-Neira, M., D'Addio, S., Buck, C., Floury, N., Prieto-Cerdeira, R., June 2011. The PARIS ocean altimeter in-orbit demonstrator. IEEE Trans. Geosci. Remote Sens. 49 (6), 2209–2237. http://dx.doi.org/10.1109/TGRS.2010.2092431.
[f] Equal power distribution assumed.
[g] For GPS signals, typical value is 3 dB above minimum.
[h] For BOC signals, bandwidth is defined between outer nulls of largest spectral lobes.

aperture, caused by irregularities in the refractive-index structure (only important in large apertures); slow fading due to beam-bending caused by large-scale changes in refractive index and more rapid fading (scintillation) and variations in angle of arrival, due to small-scale variations in refractive index; Faraday rotation; dispersion due to differential time delay across the bandwidth of the transmitted signal; excess time delay; ionospheric scintillation.

Among all these effects, the one that is exploited by GNSS-RO is the bending of the direction of propagation due to refractive index gradients (Snell's Law). The magnitude of the refraction depends on the refractivity gradient normal to the path, which in turn depends on the density and water vapor content gradients. This effect is more important when the radiation traverses a long atmospheric limb path.

This effect at optical frequencies is illustrated in Fig. 6.9. When the Sun is on the horizon, or close to it, refraction in the lower atmosphere is more important and it appears to flatten the solar disk. Sunlight directed from the bottom portion of the Sun passes through a slightly denser atmosphere than do rays coming from the Sun's upper portion and refraction is more important.

At radio frequencies the bending angle (α), or total change in ray direction with respect to a signal path trajectory through free space (Fig. 6.10), is $\sim 1°-2°$ at the surface, falling exponentially with height, and it cannot be measured directly. The amount of bending can be related to the refractive index by using an Abel transform on the formula relating bending angle to refractivity (Section 6.2.3). In the case of the neutral atmosphere (below the ionosphere) information on the atmosphere's temperature, pressure, and water vapor content can be derived, leading to RO data applications in meteorology, with good vertical resolution ($\sim 100-200$ m), broad horizontal resolution ($\sim 70\%$ of the bending occurs over a ~ 450 km section of the ray-path, centered on the tangent point), quasi-instantaneous observations (a radio-link with tangent point at ~ 60 km altitude, to a link with tangent point at the surface level takes only ~ 2 min, during this time, the tangent point has drifted horizontally ~ 150 km).

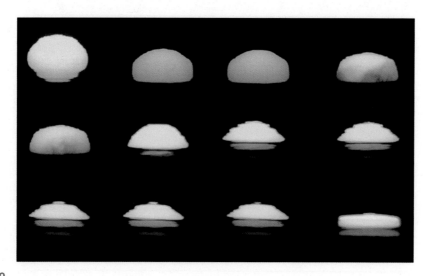

FIGURE 6.9

Images of the Sun distorted due to refraction.

https://commons.wikimedia.org/wiki/File:Sunset_mirage_and_green_flash_9-28-10.jpg.

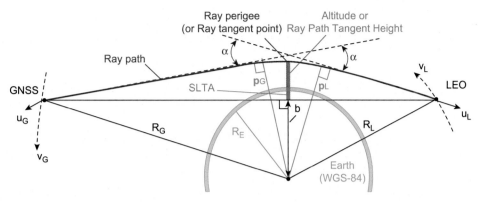

FIGURE 6.10

Geometry of GNSS radio occultations. Shown are the following: the bending angle (α), the GNSS and Low Earth Orbit (LEO) side impact parameters (radial distances to the tangent point of the virtual straight ray paths across the free space, defined at the incident p_G or exit p_L virtual ray, but $p_G = p_L$ if spherical symmetry applies), the GNSS and LEO coordinate vectors (R_G, R_L), the ray path (solid red line), the straight line tangent altitude, altitude or ray path tangent height (in orange), and the satellite side asymptotes of the ray path (dashed).

From eoPortal Directory – MSG.

According to Fig. 6.10, and applying Snell's law at each infinitesimal layer of the atmosphere, at any point along the ray with index of refractivity $n(r)$, being r the radial distance, and $\theta(r)$ the ray angle with respect to the radial, Eq. (6.13) is satisfied:

$$n(r)\cdot r\cdot\sin(\theta(r)) = p, \tag{6.13}$$

and at the tangent point $r = r_{min}$, and $\theta(r_{min}) = 90°$, so:

$$n(r_{min})\cdot r_{min} = p_{G,L}. \tag{6.14}$$

A setting occultation occurs when a transmitter initially above the horizon gradually sets down below the Earth's limb. A rising occultation occurs when a transmitter signal blocked by the Earth gradually comes up above the limb [from (https://directory.eoportal.org/web/eoportal/satellite-missions/m/metop-sg)].

The so-called Level-1 data consist of a time series of "**excess phase delay**" and the "**SNR**" of the radio-link. The higher the index of refraction ($n > 1$), the slower the propagation speed and the greater the path length, which produces an "**excess phase**" or difference between the total measured phase delay of the bent radio-link and the delay corresponding to a straight signal propagation between the transmitter and the receiver.

The bending of the path also modifies the apparent Doppler frequency, which produces an "**excess Doppler**" frequency shift that is obtained by differentiating the excess phase delays ($\Delta\phi$) with respect to time:

$$f_D = -\frac{f_0}{c}\left[v_r^T\cdot\cos\left(\theta^T\right) + v_\theta^T\cdot\sin\left(\theta^T\right) + v_r^R\cdot\cos\left(\theta^R\right) - v_\theta^R\cdot\sin\left(\theta^R\right)\right], \tag{6.15}$$

and because of the spherical symmetry assumption:

$$r^R \cdot \sin(\theta^R) = r^T \cdot \sin(\theta^T), \tag{6.16}$$

$$\alpha = \theta^T + \theta^R + \theta - \pi, \tag{6.17}$$

and

$$p = r^R \cdot \sin(\theta^R). \tag{6.18}$$

The observables time series $\{\Delta\phi(t_k), SNR(t_k)\}$; and $\{p(t_k)$ and $\alpha(t_k)\}$ are obtained from all GNSS' available subbands (e.g., L1 and L2 for GPS). In addition, linear combinations of different band observations can also be performed (Fig. 6.11). In particular, the "ionosphere-free" combination allows

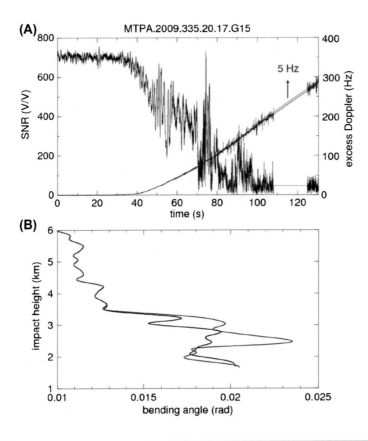

FIGURE 6.11

Example of Metop/GRAS occultation with data gap. (A) SNR and excess Doppler filled with the model inside the gap. (B) Retrieved bending angles. Shifting the Doppler model inside the gap by 5 Hz (A) results in the difference in retrieved bending angles (lower panel).

From Schreiner, W., Sokolovskiy, S., Hunt, D., Rocken, C., Kuo, Y.H., 2011. Analysis of GPS radio occultation data from the FORMOSAT-3/COSMIC and Metop/GRAS missions at CDAAC. Atmos. Meas. Tech. 4. http://dx.doi.org/10.5194/amt-4-2255-2011.

us to remove the contribution of the ionospheric electron content into the bending, and the corrected bending becomes the following:

$$\alpha_{\text{ionosphere-free}}(p) = c \cdot \alpha_{L_1}(p) - (c - 1) \cdot \alpha_{L_2}(p), \tag{6.19}$$

$$c = \frac{f_{L_1}^2}{f_{L_1}^2 - f_{L_2}^2}. \tag{6.20}$$

From Eq. (6.13), the differential equation of the ray path can be derived:

$$\frac{d}{ds}\left\{ n \cdot \frac{d\vec{r}}{ds} \right\} = \vec{\nabla}n, \tag{6.21}$$

with $d\vec{r} = \vec{s} \cdot ds$. The rate of change in the direction along the ray-path is given by:

$$\frac{d\vec{s}}{ds} = \frac{\vec{\nabla}n}{n} - \vec{s}\frac{1}{n} \cdot \frac{dn}{ds} = \frac{\vec{\nabla}_{\perp s}\, n}{n}, \tag{6.22}$$

and the change in the bending angle by:

$$d\alpha = \frac{ds}{n} \cdot \sqrt{\left(\frac{\partial n}{\partial r} \cdot \sin(\theta) + \frac{\partial n}{\partial x} \cdot \cos(\theta) \right)^2 + \left(\frac{\partial n}{\partial y} \right)^2}, \tag{6.23}$$

which under spherical symmetry, becomes:

$$d\alpha = dr \cdot \left(\frac{d\,\ln(n)}{dr} \cdot \frac{p}{\sqrt{n^2 \cdot r^2 - p^2}} \right), \tag{6.24}$$

and the total refractive bending is given by:

$$\alpha(p) = 2 \cdot \int_{r_t}^{\infty} d\alpha = 2p \cdot \int_{r_t}^{\infty} \frac{d\,\ln(n)}{dr} \cdot \frac{p}{\sqrt{n^2 \cdot r^2 - p^2}} \cdot dr. \tag{6.25}$$

An Abel transformation can be used to invert Eq. (6.25) and extract the refraction index:

$$n(r) = \exp\left\{ \frac{1}{\pi} \cdot \int_{P_t}^{\infty} \frac{\alpha}{\sqrt{p^2 - p_t^2}} \cdot dp \right\}. \tag{6.26}$$

Finally, some atmospheric thermodynamic variables can be inferred from the following equation that relates the coindex of refraction N(*h*):

$$N = (n - 1) \cdot 10^6 = N_{\text{dry}} + N_{\text{wet}} + N_{\text{iono}} = 77.6\frac{P}{T} + 3.732 \cdot 10^5\frac{P_W}{T^2} - 40.3\frac{n_e}{f^2}, \tag{6.27}$$

to the total pressure P [hPa], the water vapor pressure P_W [hPa], the absolute temperature T [K], the frequency of the GPS carrier signal f [Hz], and the electron density n_e [number of electrons per cubic meter] [http://www.navipedia.net/index.php/Ionospheric_Delay#cite_note-4].

GNSS-RO by itself cannot separate these contributions, therefore some assumptions or ancillary information is required. For example, if the wet term is neglected, using the ideal gas law:

$$P \cdot V = n \cdot R \cdot T, \tag{6.28}$$

or

$$P = \rho \cdot R_{specific} \cdot T, \tag{6.29}$$

where P is the pressure [Pa], V is the volume [m^3], n is the number of moles [mol], R is the universal gas constant (R = 8.314 J/K mol), T is the absolute temperature [K], ρ is the density (ρ = m/V) [g/m^3], and P = $\rho \cdot R_{specific} \cdot T$ is the specific constant of a gas (or mixture of gases), and it is given by the molar gas constant (R), divided by the molar mass (M) [g/mol] of the gas (or mixture), the refractivity (N) is linearly proportional to density and using the hydrostatic equation the pressure profile can be derived:

$$P(z) = P(z_{top}) - \frac{1}{77.6 \cdot R} \int_{z}^{z_{top}} gN(z) \, dz, \tag{6.30}$$

and

$$T(z) = 77.6 \frac{P(z)}{N(z)}. \tag{6.31}$$

6.3.2 GNSS-RO INSTRUMENTS

GNSS-RO instruments are basically very sensitive GNSS (or GPS only) receivers that can provide very high sampling rates, up to 50 Hz, for accurate phase measurements.

The development of GPS-RO instruments starts in the US with the GPS/MET mission in the mid 90's. In 1998, JPL started developing a new class of high-precision GPS space science receivers, called BlackJack, a configuration of the NASA/JPL dual-frequency codeless TurboRogue Space Receiver (TRSR) to perform radio occultations that could take precise measurements of how GPS signals are delayed along their way to the receiver. BlackJack GPS flight instruments have been used in 2000 on SRTM, SAC-C, CHAMP, JASON-1, etc., in 2001 on FEDSat, ICESat, and GRACE. In 2006, the Constellation Observing System for Meteorology, Ionosphere, and Climate (COSMIC) was launched through a collaboration among Taiwan's National Science Council and National Space Organization, the US National Science Foundation, NASA, the National Oceanic and Atmospheric Administration (NOAA), and other Federal entities. COSMIC was the first demonstration of a GPS RO constellation, and its primary payload was the so-called "integrated GPS Occultation Receiver" or IGOR, based on NASA/JPL BlackJack receiver (see Fig. 6.12). Pyxis is the next generation of GNSS-RO instruments based on the evolution of IGOR. It includes L2C and L5 frequencies as well, and eventually Galileo frequencies provide increased occultation data and improved PVT resolution. The next generation of US GNSS-RO instruments (Tri-G) will have the capability to track both the legacy GPS L1CA and L1 + L2 semicodeless and the new L2C/L5 signals from GPS, as well as additional new GNSS signals, such as GLONASS (CDMA) and Galileo (E1, E5), and will have digital beam steering antenna capability, more channels and memory, and "wider" open loop tracking function.

In Europe, Saab Ericsson Space (now RUAG), ESA, and EUMETSAT developed the GRAS (GNSS-RO Receiver for Atmospheric Sounding) for Metop A to C satellites, and THALES developed ROSA (radio occultation sounder for atmosphere) for OCEANSAT (India), and SAC-D (Argentina).

FIGURE 6.12

Dual-frequency (L1 + L2) receiver, IGOR is available in two versions for Low-Earth orbit: for precise orbit determination (POD) only, or as a receiver for POD and GNSS-RO (http://spinoff.nasa.gov/Spinoff2011/er_4.html).

These receivers are based in the AGGA (Advanced GPS/GLONASS and Galileo ASIC) device developed by ESA within the Earth Observation Preparatory Program (EOPP) to support the Earth observation applications of navigation signals. The AGGA is available to all European space industries, and it is flying or will fly in a large number of ESA missions (e.g., GRAS instrument in Metop A to C for GNSS-RO, or GOCE, Swarm, EarthCare, GMES Sentinels 1, 2, 3 for POD), and non-ESA missions (e.g., Radarsat-2 and Cosmo-Skymed for precise orbit determination or POD, and ROSA in Oceansat-2, SAC-D, and MegaTropiques for GNSS-RO). The next generation of GNSS-RO instruments will benefit from open loop processing techniques, already demonstrated in MetOp since 2007, for improved range and Doppler modeling, better open loop with new open CDMA GNSS signals (GPS, Galileo, Glonass, Beidou), and no need for codeless techniques, and larger coverage up to ~2600 observations per satellite and day.

Table 6.4 summarizes the past, current, and planned GNSS-RO instruments, including the name of the mission where they were flown, the instrument full name, and a short description of the instrument heritage.

6.3.3 GNSS-RO APPLICATIONS

6.3.3.1 Atmospheric Profiles of Temperature, Pressure, and Water Vapor

Solving the aforementioned equations, atmospheric excess phases and atmospheric profiles of bending angle, refractivity, dry temperature, temperature, and water vapor are inferred, and routinely distributed. Fig. 6.13 shows sample retrievals using COSMIC data.

6.3.3.2 Numerical Weather Forecast Contributions

GNSS-RO data are intrinsically self-calibrated, while they also have a long-term stability, and good vertical resolution. The contribution to the forecast error reduction of different sources of Earth

Table 6.4 Summary Table of GNSS-RO Instruments Acronyms and Full Name, Missions, and Heritage

Instrument Acronym (Mission/Satellite)	Instrument Full Name
Legacy Instruments	
BlackJack (CHAMP)	BlackJack (configuration of the NASA/JPL TurboRogue Space Receiver to perform radio occultations)
GOLPE (SAC-C)	GPS Occultation and Passive reflection Experiment (adaptation of the NASA/JPL BlackJack)
GPS/MET (OrbView-1, formerly MicroLab)	Global Positioning System/Meteorology
IGOR (TacSat-2)	Integrated GPS Occultation Receiver (based on NASA's BlackJack)
Current Instruments	
AOPOD (KOMPSAT-5)	Atmosphere Occultation and Precision Orbit Determination
BlackJack (GRACE)	BlackJack
CORISS (Communication/Navigation Outage Forecasting System)	C/NOFS Occultation Receiver for Ionospheric Sensing and Specification (dual-frequency, tracking during rising, and setting)
GNOS (Feng-Yun - 3C)	GNSS Radio Occultation Sounder
GRAS (MetOp A to C)	GNSS Receiver for Atmospheric Sounding
IGOR (COSMIC, TanDEM, TerraSAR)	Integrated GPS Occultation Receiver
ROSA (SAC-D, OceanSat)	Radio Occultation Sounder of the Atmosphere
TRSR (Ørsted)	TurboRogue Space Receiver (Ørsted)
Future Instruments	
GNSS-RO (CLARREO)	Global Navigation Satellite System-Radio Occultation
Radiomet (Meteor-3M)	Radio Occultation sounder
RO (GRAS follow-on of MetOp A to C)	Radio Occultation sounder (follow-on of GRAS flown on MetOp A to C)
ROHPP (PAZ)	Radio Occultations and Heavy Precipitation with PAZ (PAZ)
Tri-G (COSMIC-2, GRACE-FO, JASON-CS)	Triple G: GPS, Galileo, GLONASS (COSMIC-2 and GRACE-2)

observation data is shown here in Fig. 6.14. The largest impact is from AMSU-A, the primary microwave sounder on NOAA's polar orbiting satellites. GNSS-RO (actually GPS-RO) is essentially tied for second with two infrared (IR) sounders. However, it should be noted that AMSU and IR performance is significantly improved by the calibration using GNSS-RO data. Without GNSS-RO, bars will be smaller, therefore the true impact of GNSS-RO is much larger than this plot induces to think, and this improvement comes from a system that is at least two orders of magnitude less expensive than microwave sounders.

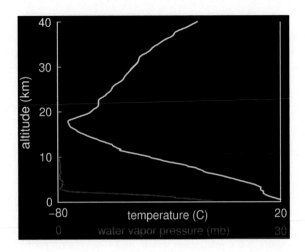

FIGURE 6.13

Sample water vapor and temperature height distributions as derived from COSMIC.

From http://www.cosmic.ucar.edu/data.html.

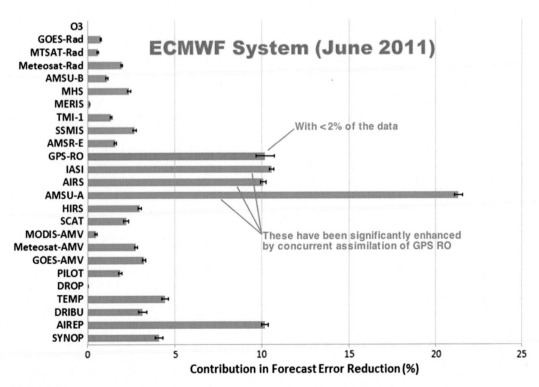

FIGURE 6.14

Forecast contribution to error reduction from different sensors: GPS RO is essentially tied for second with two infrared sounders (http://geooptics.com/?page_id=9).

6.3.3.3 Ionosphere

The ionospheric total electron content (TEC) can be inferred from the phase delay L_i of the occulting radio link at frequency band f_i [Hz], which depends on the integrated path length of the link (s_i), and the TEC [e^-/m^3] to be determined (Ruffini et al., 1998):

$$L_i = s_i - \frac{40.3082 \frac{m^3}{s^2}}{f_i^2} \cdot TEC_i. \tag{6.32}$$

Combining two observations L_1 and L_2 at two different frequencies the TEC along the signal path can be extracted:

$$TEC = \frac{f_1^2 \cdot f_2^2 \cdot (L_1 - L_2)}{40.3082 \frac{m^3}{s^2} \cdot (f_1^2 - f_2^2)}. \tag{6.33}$$

Ionospheric excess phases, S4 scintillation index, and ionospheric profiles of electron density can also be inferred and are currently distributed (Fig. 6.15).

6.4 GLOBAL NAVIGATION SATELLITE SYSTEM-REFLECTROMETRY

GNSS Reflectrometry, or GNSS-R in short, can be understood in radar terminology as a "passive bistatic radar": passive because it operates with transmitters of opportunity, and bistatic (actually multistatic) because transmitter(s) and receiver are separated by a considerable distance. Although the very first radars were already bistatic, because of the difficulty to implement the duplexers to separate the transmitted and received signals, most of the existing radar technology is for monostatic radars. Bistatic radars have experienced, though, a number or revivals, in the 50s for semiactive homing missiles…, in the 70s–80s with the sanctuary project … and since the mid 90s using opportunity signals such as GNSS or TV signals.

The GNSS-R concept is sketched in Fig. 6.16 showing a receiver in an LEO satellite collecting the reflected (actually scattered) signal (marked in red) of four different navigation satellites, and (eventually) the direct signals as well (marked in orange, dashed line). Depending on the application, GNSS-R instruments can be operated in different "modes":

- as a scatterometer: when the observable is the amplitude of the (forward) scattered power,
- as an altimeter: when the observable is the time delay between the direct and the scattered signals arriving to the receiver, or
- as a synthetic aperture radar or SAR (actually an unfocused SAR due to the short coherent integration time that can be achieved): when the observable is the whole delay Doppler map (or DDM) that can be deconvolved to derive a radar cross-section (RCS) density (σ^0) image or to perform Doppler altimetry.

Although based on the same principles, depending on the application, the instruments have different requirements. For example, for scatterometric purposes, bandwidth is not an issue, and a single frequency receiver suffices. On the other hand, for altimetric applications, bandwidth is an issue,

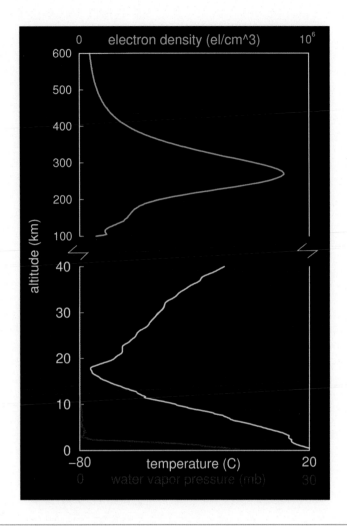

FIGURE 6.15

Sample electron density height distribution as derived from COSMIC.

From © 2013 University Corporation for Atmospheric Research / COSMIC.

since it determines (together with the SNR) the achievable altimetric accuracy, and dual frequency instruments are needed to compensate for the ionospheric delay. Instruments measuring DDMs will be a lot more complex, since they will require a large number of complex correlators for each delay-Doppler bin, while for scatterometry and altimetry applications just a few correlators are needed to measure a delay cut along the central Doppler frequency.

Before discussing the specifics of GNSS-R techniques, instruments, and applications, it is convenient to review briefly some basic concepts of bistatic radar, which are needed to better understand GNSS-R.

FIGURE 6.16

GNSS-R concept: receiver collects the reflected signal (red solid line) and (eventually) the direct (orange dashed line) from navigation satellites.

6.4.1 BASIC PRINCIPLES: GNSS-R AS A MULTISTATIC RADAR

6.4.1.1 Isodelay and Iso-Doppler Contours

A sketch of a bistatic radar consisting of a transmitter (T_x) and receiver (R_x) is shown in Fig. 6.17. Both are separated with a baseline L that lies in the same plane as the target: the x-y plane or the "bistatic

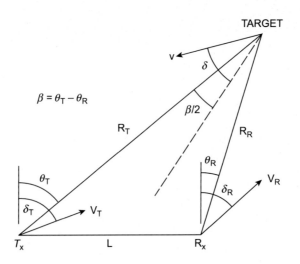

FIGURE 6.17

Geometry for bistatic Doppler.

Modified from Fig. 25.5 from Skolnik, M.I. (Ed.), 1990. Radar Handbook. McGraw-Hill, New York.

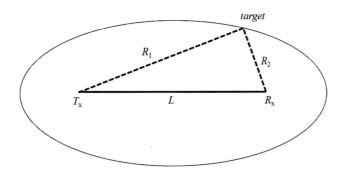

FIGURE 6.18

Geometry of the isodelay contours.

plane." The angles θ_T and θ_R are, respectively, the transmitter and receiver look angles, or angles of arrival, and the bistatic or scattering angle β is defined as $\beta = \theta_T - \theta_R$, and it will be used in the calculations related to the target.

The "isodelay" contours:

$$\tau = \frac{R_1(t) + R_2(t)}{c}, \tag{6.34}$$

are ellipses with the transmitter and receiver in the two foci of the ellipse (Fig. 6.18). In Eq. (6.34) $R_1(t)$ is the distance between transmitter and target, and $R_2(t)$ is the distance between receiver and target.

Taking as a reference the position of the receiver, and defining $R_{tot}(t) = R_1(t) + R_2(t)$, the equations of the ellipses are given by:

$$R_2(t) = \frac{R_{tot}^2(t) - L^2}{2(R_{tot}(t) + L \cdot \sin\theta_R)}. \tag{6.35}$$

The bistatic "iso-Doppler" contours are given by:

$$f_D = -\frac{1}{\lambda} \frac{\partial}{\partial t} \{R_1(t) + R_2(t)\}, \tag{6.36}$$

where λ is the wavelength. In the particular case of a stationary (or quasi-stationary) target (e.g., the Earth's or the ocean surface), Eq. (6.36) becomes:

$$f_D = \frac{V_T}{\lambda} \cdot \cos(\delta_T - \theta_T) + \frac{V_R}{\lambda} \cdot \cos(\delta_R - \theta_R), \tag{6.37}$$

where $\delta_T - \theta_T$ and $\delta_R - \theta_R$ are the angles formed by the transmitter (T_x) and receiver (R_x) velocity vectors and the unitary vectors from transmitter and receiver to the target ($\hat{n}_{Tx-target}$ and $\hat{n}_{Rx-target}$). Eq. (6.37) can be alternatively written as:

$$f_D = \frac{\overrightarrow{V_R} \cdot \hat{n}_{Tx-target}}{\lambda} + \frac{\overrightarrow{V_R} \cdot \hat{n}_{Rx-target}}{\lambda}. \tag{6.38}$$

For a flat Earth, the iso-Doppler contours become hyperbolae. Fig. 6.19 shows the isodelay (or isorange) contours together with the iso-Doppler contours. As it can be noticed, each pixel in the (x,y) plane has a pair delay-Doppler (τ, f_D) associated, but this correspondence is not unique, since two pairs

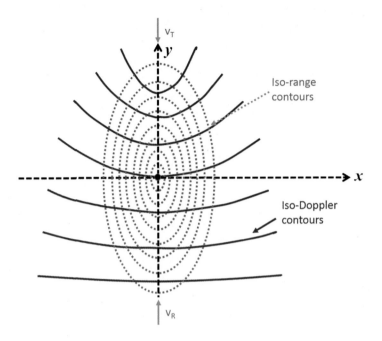

FIGURE 6.19

Geometry of the isodelay and iso-Doppler contours overlaid: two pairs of (x,y) points have the same (τ, f_D) values.

of (x,y) points have the same (τ, f_D) values. In a SAR, this ambiguity is solved by steering the antenna toward an unambiguous zone away from the y axis, but in GNSS-R, in principle, this is not the case, since the antennas point toward the specular direction point (0,0). This issue will be addressed in a subsequent section when dealing with GNSS-R as imagers.

6.4.1.2 Received Power, Signal-to-Noise Ratios, and Ovals of Cassini

The received power (P_R) in a bistatic radar can be computed in a similar way as for a monostatic radar, but taking into account that R_1 and R_2 are different, and so the antenna gains of the transmitting (G_T) and receiving antennas (G_R), that must be evaluated in the direction of the target, and that the monostatic RCS (σ) must be replaced by the bistatic one (σ_b):

$$P_R = \underbrace{\underbrace{\frac{P_T \cdot G_T}{4\pi R_1^2} \cdot \sigma_b \cdot \frac{1}{4\pi R_2^2}}_{\left.\overrightarrow{|P|}\right|_{target}} \cdot \underbrace{\frac{G_R \cdot \lambda^2}{4\pi}}_{A_{eff,R_x}} \cdot L.}_{\left.\overrightarrow{|P|}\right|_{R_x}} \quad (6.39)$$

Additional terms in Eq. (6.39) are P_T the transmitted power, and L which includes all losses in the transmitter, receiver, and propagation. In the second term of Eq. (6.39) the first part corresponds to the

Poynting vector [power density, units W/m^2] at the target, the product of the first two to the power actually being intercepted by the target, the product of the first three to the Poynting vector at the receiving antenna, and $G_R \cdot \lambda^2/4\pi$ corresponds to the effective area of the receiving antenna. Note that in the case of a surface target σ_b must be replaced by the RCS density times the surface associated to the (τ, f_D) cell (see Fig. 6.19).

The calculation of the SNR is usually computed by dividing Eq. (6.39) by the thermal noise introduced by the receiver:

$$N = k_B \cdot T_{sys} \cdot B, \tag{6.40}$$

where k_B is the Boltzmann's constant ($k_B = 1.38 \cdot 10^{-23}$ J/K), $T_{sys} = T_A + T_R$ is the system temperature, which includes the noise collected by the antenna (T_A), and the noise generated internally by the receiver (T_R) (see Chapter 4 for a detailed discussion on the calculation of these terms). However, in GNSS-R, the background noise (Eq. 6.40) is increased by the power within the band of other interfering signals, including other GNSS signals: I_{Intra}, or intrasystem interference that is due to the signals coming from satellites belonging to the same system, $I_{Interop}$ that is coming from a satellite of a different constellation, but with the same signal structure as that of the desired signal, and I_{Inter} that is coming from signals with a different signal structure, belonging or not to the same system (Navipedia, GNSS Interference Model, 2014; Wu and He, 2011). If we are now interested in finding the contours of constant SNR, $R_1 \cdot R_2$ can be isolated from the ratio of Eqs. (6.39) and (6.40), leading to:

$$\left.(R_1 \cdot R_2)\right|_{max} = \sqrt{\frac{P_T \cdot G_T \cdot G_R \cdot \lambda^2 \cdot \sigma_b \cdot L}{(4\pi)^3 \cdot k_B \cdot T_{sys} \cdot B \cdot SNR_{min}}} = \kappa, \tag{6.41}$$

or:

$$SNR = \frac{\kappa}{R_1^2 \cdot R_2^2}, \tag{6.42}$$

with $\kappa \triangleq \kappa^2 \cdot SNR_{min}$.

Finally, the equation of the ovals of Cassini, or contours of constant SNR on any bistatic plane, can be derived:

$$R_1^2 \cdot R_2^2 = \left(r^2 + \frac{L^2}{4}\right)^2 - r^2 \cdot L^2 \cdot \cos^2\theta. \tag{6.43}$$

The contours of constant SNR, or ovals of Cassini, are presented in Fig. 6.20 as derived from Eq. (6.43).

6.4.1.3 Considerations on Bistatic Scattering

Depending on the scattering configuration three different regimes must be considered to evaluate how well both individual targets and the background (clutter in radar terminology, but main target in GNSS-R applications) scattering influences the incoming signal toward the receiver:

Pseudo-monostatic region. At very high frequencies (wavelength tending to zero) the bistatic RCS (σ_b) of a sufficiently smooth, a perfectly conducting target is equal to the monostatic RCS measured on the bisector of the bistatic angle β (Crispin and Siegel, 1968). For spheres of radius up to ~ 0.4 λ, the pseudo-monostatic region extends to $\beta \approx 40°$, but for complex targets (assembly of discrete scattering centers) the extent of the pseudo-monostatic region is considerably reduced: for small bistatic angles ($\beta \leq 5°$), the bistatic RCS is equal to the monostatic RCS measured on the bisector of the bistatic angle

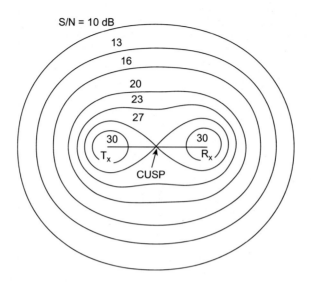

FIGURE 6.20

Contours of constant SNR, or ovals of Cassini, where the baseline $= L$ and $k = 30$ L. Ovals of Cassini define three distinct operating regions for a bistatic radar: receiver-centered region, transmitter-centered region, and receiver-transmitter—centered region, or simply the cosite region.

From Fig. 25.2, Skolnik, M.I. (Ed.), 1990. Radar Handbook. McGraw-Hill, New York.

at a frequency lower by a factor of cos ($\beta/2$). At $\beta > 5°$ the change in the radiation properties from discrete scattering centers will likely dominate over any cos ($\beta/2$) frequency reduction term (Kell, 1965).

Bistatic RCS Region. Changes in the relative phase between discrete scattering centers, radiation changes from individual scattering centers, and appearance/disappearance of scattering centers produce a bistatic RCS pattern more complex than the monostatic one that, in general, is of lower amplitude.

Forward-scattering region. When the bistatic angle approaches 180° the RCS can be very large, even for perfectly absorbing targets since, according to Babinet's principle, a perfectly absorbing target of area A will produce the same RCS as a target-shaped hole in an infinite ground plane. The bistatic RCS is $\sigma_b = 4\pi A^2/\lambda^2$, and the roll off away from $\beta = 180°$, is a uniformly illuminated antenna aperture, with an angular width of the order of λ/d radians, being d the maximum dimension of the target.

As a rough approximation (depends very much on the target) the **pseudo-monostatic region** is valid for $\beta \leq 5°$ and can be applied for near-nadir applications, the **bistatic region is valid in the range from $5° < \beta \leq 175°$ and will usually cover most of the angular range used in GNSS-R** altimetry/scatterometry applications, and the **forward-scattering region** is valid for $175° < \beta \leq 180°$ and will be valid in scattering at near-grazing angles, such as those occurring in GNSS-RO. The bistatic RCS of the clutter (background) is defined as $\sigma_c = \sigma_b^0 \cdot A_c$, where σ_B^0 [m²/m²] is the scattering coefficient. In the **pseudo-monostatic region, it can be approximated by:**

$$\sigma_B^0 \approx e^{-\left(\frac{\beta_c}{\sigma_s}\right)^2}, \tag{6.44}$$

where $\beta_c \triangleq |90° - (\theta_i + \theta_s)/2|$ and σ_s is the rms surface slope.

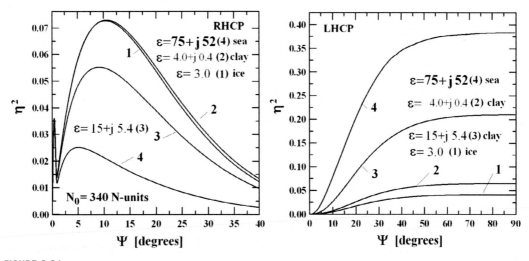

FIGURE 6.21

The reflection coefficient as a function of the grazing angle ψ for the case of the right hand (RHCP) and left hand (LHCP) circular polarizations of a GNSS-RO receiving antenna.

From Pavelyev, A.G., Zhang, K., Matyugov, S.S., Liou, Y.A., Wang, C.S., Yakovlev, O.I., Kucherjavenkov, I.A., Kuleshov, Y., 2011.
Analytical model of bistatic reflections and radio occultation signals. Radio Sci. 46. http://dx.doi.org/10.1029/2010RS004434.

In the **bistatic region** the constant-γ model is used:

$$\sigma_B^0 \approx \gamma \cdot (\sin(\theta_i) \cdot \sin(\theta_s))^{1/2}. \tag{6.45}$$

Finally, in the **forward-scattering region** the scattering is nearly specular. Actually, the ratio of the reflected to the direct powers is one of the observables in GNSS-RO. In Pavelyev et al. (2011), a study is performed to predict the GNSS reflections and ROs in GNSS-RO scenarios. Results in Fig. 6.21 show the reflection coefficients at near grazing angles for RHCP and left hand circular polarizations (LHCP), when the incident wave is a pure RHCP wave. Note that for LHCP, the reflection coefficient vanishes for $\psi = 0°$.

6.4.1.4 Woodward Ambiguity Function

The Woodward ambiguity function (WAF) (Woodward, 1980) shows the point target response of the waveform as a function of delay τ and Doppler frequency f_D:

$$|\chi(\tau, f_D)| = \left| \int_{-\infty}^{+\infty} u(t) \cdot u^*(t - \tau) \cdot e^{j2\pi f_D t} \cdot dt \right|, \tag{6.46}$$

where $u(t)$ is the complex envelope of the transmitted signal.

In the bistatic case, the shape of the ambiguity function gets distorted due to the nonlinear relationships between the Doppler frequency and the speed, and the delay and the range (Tsao et al., 1997). However, in the GNSS-R case, the definition in Eq. (6.46) suffices since the Doppler is dominated by the movement of the transmitter and/or receiver, while the target is still or moves slowly (e.g., ocean waves). We will focus on two relevant cases: a rectangular pulse and a chirp pulse.

- The voltage rectangular pulse of duration T as given by Eq. (6.47):

$$s(t) = \frac{1}{\sqrt{T}} \Pi\left(\frac{t}{T}\right), \tag{6.47}$$

has a WAF as given by Eq. (6.48):

$$|\chi(\tau, f_D)| = \left|\left(1 - \frac{|\tau|}{T}\right) \cdot \text{sinc}\left(T \cdot f_D \cdot \left(1 - \frac{|\tau|}{T}\right)\right)\right|, \tag{6.48}$$

which is plotted in Fig. 6.22A normalized for T = 1. It should be noted, that the width along the τ direction for $\Delta f = f_D = 0$ is 1 unit (1/T), while in frequency, along $\tau = 0$, it is a sin c function. Therefore, a rectangular pulse can achieve high temporal resolution (range) if the pulse is short (T → 0), but then the frequency resolution degrades (1/T → ∞).

- The chirp pulse modulates the amplitude of the rectangular pulse with a linear frequency modulation (LFM) with $\gamma = \pm B/T$, as in Eq. (6.49):

$$s(t) = \frac{1}{\sqrt{T}} \Pi\left(\frac{t}{T}\right) \cdot e^{j\pi \cdot \gamma \cdot t^2}, \tag{6.49}$$

has a WAF as given by Eq. (6.50):

$$|\chi(\tau, f_D)| = \left|\left(1 - \frac{|\tau|}{T}\right) \cdot \text{sinc}\left(T \cdot (f_D + \gamma \cdot \tau) \cdot \left(1 - \frac{|\tau|}{T}\right)\right)\right| \quad \text{for } |\tau| \leq T, \tag{6.50}$$

which is plotted in Fig. 6.22B normalized for T = 1. It can be noted, that now the widths along the τ direction ($\Delta f = f_D = 0$) and Δf directions are similar. This means that good performance is achieved in range and Doppler, although not for targets along the diagonal in Fig. 6.22B right.

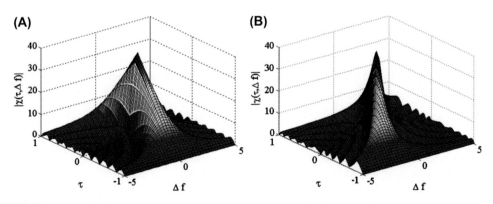

FIGURE 6.22

Woodward ambiguity function for: (A) rectangular pulse, and (B) chirp pulse.

6.4.2 GNSS-R PARTICULARITIES

As in most navigation receivers, GNSS reflectometers correlate coherently during T_c seconds (typically ~ 1 ms) the received reflected signal $s_R(t)$ with either a locally generated replica of the transmitted signal $a(t)$ (publicly available codes only, conventional GNSS-R) or with the direct signal itself ($s_D(t)$) or a reconstructed version of it ($\widehat{s}_D(t)$) after proper Doppler frequency (f_D) and delay adjustments (interferometric GNSS-R or reconstructed GNSS-R):

$$Y^c(t, \tau, f_d) = \int_t^{t+T_c} s_R(t')a^*(t' - \tau)e^{-j2\pi(F_c+f_d)t'}dt', \tag{6.51a}$$

$$Y^i(t, \tau, f_d) = \int_t^{t+T_c} s_R(t')s_D^*(t' - \tau)e^{-j2\pi(F_c+f_d)t'}dt', \tag{6.51b}$$

$$Y^r(t, \tau, f_d) = \int_t^{t+T_c} s_R(t')\widehat{S}_D^*(t' - \tau)e^{-j2\pi(F_c+f_d)t'}dt', \tag{6.51c}$$

where t is the time when the coherent integration starts, and F_c is the carrier frequency. Since the reflected signal is of even weaker amplitude than the direct signal (the one of interest for navigation purposes), the signal-to-(thermal) noise ratio is even poorer. In addition, since the scattered signal comes from multiple scatterers over the surface with different phases within the resolution cell, it also suffers from speckle noise,[1] and a large number of incoherent averages (N_i) are required to improve the SNR of $Y^{c,i,r}(t, \tau, f_d)$:

$$\left\langle \left| Y^{c,i,r}(\tau, f_d) \right|^2 \right\rangle \approx \frac{1}{N_i} \sum_{n=1}^{N_i} \left| Y^{c,i,r}(t_n, \tau, f_d) \right|^2. \tag{6.52}$$

The analysis of the noise properties is deferred to a later stage.[2] At present, we will focus on the expected values. A detailed analysis of $\left\langle |Y(\tau, f_d)|^2 \right\rangle$ (superscripts not shown unless required from now on) was performed by Zavorotny and Voronovich (2000):

$$\left\langle |Y(\tau, f_d)|^2 \right\rangle \frac{P_T \cdot \lambda^2}{(4\pi)^3} T_c^2 \iint G_T(\overrightarrow{r}) G_R \frac{(\overrightarrow{r})}{4 \cdot R_1^2(\overrightarrow{r}) \cdot R_2^2(\overrightarrow{r}) \left| \chi\left(\tau - R_1(\overrightarrow{r}) + R_2\frac{(\overrightarrow{r})}{c}, f_D(\overrightarrow{r}) - f_c\right) \right|^2} \sigma^0(\overrightarrow{r}) \cdot d^2r. \tag{6.53}$$

The physical interpretation of $\left\langle |Y(\tau, f_d)|^2 \right\rangle$ in Eq. (6.53) is simply the received power by a bistatic radar (to be compared to Eq. 6.39) from all the points in the surface where the incident wave is scattered. The following considerations must be made.

---

[1]Speckle noise is a "multiplicative noise" also found in other imaging instruments with a coherent illumination source, such as SARs. It is caused by the coherent sum of the scattered waves in many different "facets" of the surface with different phases. Speckle noise is especially significant in ocean and it can only be reduced by incoherent averaging.
[2]Increasing the coherent integration time (T_c) increases the SNR, but it cannot be increased indefinitely, due to the temporal decorrelation of the surface where the reflection takes place. This will be explained later in more detail in Section 6.4.3.

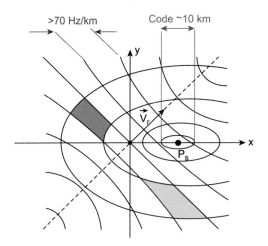

FIGURE 6.23

Isodelay and iso-Doppler contours for a bistatic configuration. Two delay-Doppler bins with the same delay and Doppler are marked with different gray levels.

From ESA, The figure 7: "Iso-range and iso-Doppler lines in a bi-state configuration." from Martin-Neira, M., 1993. A passive reflectometry and interferometry system (PARIS): application to ocean altimetry. ESA J. 17 (4), 331–355.

6.4.2.1 The Woodward Ambiguity Function

The WAF $|\chi|^2$ is maximum around surface points that satisfy that the delay and Doppler are:

$$\tau \approx \frac{R_1(\vec{r}) + R_2(\vec{r})}{c},\tag{6.54a}$$

and

$$f_D \approx f_c.\tag{6.54b}$$

Away from these points, $|\chi|^2$ vanishes and it is almost zero, although there are residual contributions away from them. The resolution cells or "delay-Doppler" bins are then the regions where the WAF is larger than half its peak value (see Fig. 6.22). The isodelay and iso-Doppler contours for a bistatic configuration are depicted in Fig. 6.23. Two "ambiguous" delay-Doppler bins with the same delay and Doppler are marked with different gray levels.[3] The shape of these contours is dominated by the dynamics of the LEO satellites (much faster than the MEO ones), or by the MEO ones (where most of the GNSS are) in airborne or ground-based instruments.

The shape of the WAF for the GNSS signals can be computed using Eq. (6.46). However, in the computation of Eq. (6.53), the error committed is negligible if the time and frequency dependencies are separated. In this case, the WAF can be approximated by the product of their ACFs (see

---

[3]Since most GNSS-R instruments point to the specular reflection point, this ambiguity cannot be resolved. The only way to resolve it is by steering the beam off the specular reflection point direction, and attenuate the alias pixel with the antenna pattern or by a multilook processing.

Fig. 6.9 and Section 6.2) and the sinc($f \cdot T_c$). In the case of the GPS C/A code, this approximation leads to:

$$|\chi(\tau,f_D)| \approx \left(1 - \frac{|\tau|}{T_c}\right) \cdot \text{sinc}(f \cdot T_c), \tag{6.55}$$

Using Eqs. (6.8)–(6.12) (or Fig. 6.9) the WAFs for other GPS, Galileo, or any other navigation signals can be readily computed.

6.4.2.2 The Bistatic Scattering Coefficient

The term $\sigma^0(\overrightarrow{r}) \cdot d^2r$ in Eq. (6.53) corresponds to the bistatic RCS (σ_b) computed as the area of the surface d^2r times the bistatic RCS density σ^0. A number of models exist to predict σ^0 over the ocean surface more accurately than the simple qualitative expressions given in Section 6.4.1.3.

6.4.2.2.1 Kirchhoff Model Under the Stationary Phase Approximation

This approximation is also known as the tangent plane. This means that scattering occurs only for points on the surface for which there are specular reflection, and local diffraction effects are excluded.

$$\sigma^0(\overrightarrow{r}) = \pi \cdot k^2 |\Re|^2 \cdot \frac{q^2}{q_z^4} \cdot P(Z_x, Z_y), \tag{6.56}$$

where $k = 2\pi/\lambda$ is the electromagnetic wave number, $\Re$ is the Fresnel reflection coefficient (actually it is $\Re_{pq}$ to indicate the incident polarization "q" and the scattered polarization "p"), $\overrightarrow{q} = q_x \cdot \hat{x} + q_y \cdot \hat{y} + q_z \cdot \hat{z} \triangleq k \cdot \left(\hat{k}_s - \hat{k}_i\right)$, $q = |\overrightarrow{q}|$, and P (Z_x, Z_y) is the probability density function of the surface slopes along the x- and y-directions (Z_x and Z_y) and can be obtained from the sea surface wave spectrum or from the slopes statistics (e.g., Cox and Munk, 1954).[4]

Since navigation systems use RHCP to avoid ionospheric effects and antenna pointing requirements at receiver level, the Fresnel reflections coefficients to be used in Eq. (6.56) are given as follows (Zavorotny and Voronovich, 2000):

$$\Re_{RR} = \Re_{LL} = \frac{\Re_{VV} + \Re_{HH}}{2}, \tag{6.57a}$$

$$\Re_{RL} = \Re_{LR} = \frac{\Re_{VV} - \Re_{HH}}{2}, \tag{6.57b}$$

where:

$$\Re_{VV} = \frac{\varepsilon_r \cdot \sin(\theta) - \sqrt{\varepsilon_r - \cos^2\theta}}{\varepsilon_r \cdot \sin(\theta) + \sqrt{\varepsilon_r - \cos^2\theta}}, \tag{6.58a}$$

---

[4]At microwave frequencies a "reduced pdf" must be used. Wilheit (1979) proposed a reduction of Cox and Munk slopes below 35 GHz. At L1, this factor is equal to 0.33 at L1. Katzberg et al. (2006) proposed a direct fit to GNSS-R measurements in hurricane conditions, differentiating three regimes (below 3.5 m/s, from 3.5 to 46 m/s, and above 46 m/s).

$$\mathfrak{R}_{HH} = \frac{\sin(\theta) - \sqrt{\varepsilon_r - \cos^2\theta}}{\sin(\theta) + \sqrt{\varepsilon_r - \cos^2\theta}}. \tag{6.58b}$$

6.4.2.2.2 Kirchhoff Model Under the Physical Optics Approximation

The Kirchhoff model under the physical optics approximation involves the integration of the scattered fields over the entire rough surface, not just the facets contributing to the specular reflection. Unlike the geometric optics approximation, the physical optics approximation predicts a coherent component given by:

$$\sigma_{pq}^{0,\,coh} = \pi k^2 |a_0|^2 \delta(q_x)\delta(q_y)e^{-q_z^2 \cdot \sigma^2}, \tag{6.59}$$

which decreases with increasing rms surface roughness (σ). In the specular direction $a_{0,VV} = +2\mathfrak{R}_{VV}(\theta_i)\cdot\cos(\theta_i)$, $a_{0,HH} = -2\mathfrak{R}_{HH}(\theta_i)\cdot\cos(\theta_i)$, and $a_{0,VH} = a_{0,HV} = 0$. However, this analysis is limited to surfaces with small slopes (Ticconi et al., 2011).

6.4.2.2.3 The Small Perturbation Method

The Small Perturbation Method (SPM) is based on finding a solution in terms of a series of plane waves that matches the surface boundary conditions, i.e., the tangential component of the field must be continuous across the boundary. The zeroth order term in the expansion corresponds to the surface field if the surface was flat. The SMP is a good model if the surface height standard deviation is much less than the incident wavelength, and the average surface slope is comparable to or less than the surface standard deviation times the wave number. To first order, the SPM does not account for multiple scattering, but it does to some extent in the higher order solutions. It is the most appropriate model to account for Bragg scattering and polarimetric effects.

6.4.2.2.4 The Two-Scale Model

The Two-Scale Model decomposes the surface roughness in two scales: the large scale roughness, which is modeled using the Kirchhoff approximation under the geometric optics (KAGO), and the small scale roughness, which are modeled using the SPM averaged over the statistics of the tilt of the large-scale roughness (Bass and Fuks, 1979; Valenzuela, 1978).

6.4.2.2.5 The Integral Equation Model

The Integral Equation Model (IEM) is a unifying method proposed in the mid 80s (Fung and Pan, 1986) that closes the gap between the Kirchhoff method and the SPM, being able to reproduce results of these two methods in appropriate limits. The method is accurate, but very intensive computationally, and it is used as a reference to check the accuracy of other methods.

6.4.2.2.6 The Small Slope Approximation

In the mid-1980s, Voronovich (1985) proposed the Small Slope Approximation (SSA) that is valid for any wavelength, and arbitrary roughness provided that the surface slopes are smaller than the angles of incidence and scattering. As the IEM, the SSA closes the gap between the Kirchhoff approximation and the SPM. In 1996 (Voronovich, 1996), the nonlocal SSA was proposed as an extension to situations in which multiple scattering from points situated at significant distance becomes important.

Calculations have shown a good agreement between the KAGO and the SSA methods for the LHCP scattered around the nominal specular direction. **Because of its simplicity, the Kirchhoff model in the geometric optics limits is the one most often used in GNSS-R.** However, the SSA calculations for the RHCP scattered signal show that Bragg scattering has to be accounted for, even at the specular direction. Therefore, **for polarimetric studies**, the KAGO method is not valid, and **the SSA must be used instead.** Table 6.5 summarizes the limitations and applicability of different scattering models revised.

6.4.3 THERMAL NOISE, SPECKLE, AND COHERENCE TIME

6.4.3.1 Simplified Approach

Since the two applications of GNSS-R are scatterometry and altimetry, our study will be focused in the determination of the SNR of the observables and in the determination of the delay estimation error.

The SNR (thermal noise only) at the input of the correlator is given by:

$$\text{SNR} = \frac{\left[P_R(\tau, f_d) * |\chi(\tau, f_d)|^2\right]}{k_B \cdot T \cdot \min\{B,\ B_{IS}\}}, \tag{6.60}$$

where P_R is the total received reflected power, χ is the WAF of the transmitted signal, k_B is the Boltzmann's constant, T is the equivalent system's noise temperature, and B and B_{IS} are the baseband bandwidths of the receiver's filter and the transmitted signal accordingto the IS documents.

For a conventional GNSS-R (cGNSS-R) instrument, in which the scattered signal is cross-correlated with a locally generated replica of the transmitted signal, the SNR at correlator's output is also given by Eq. (6.60), but replacing min $\{B, B_{IS}\}$ by the inverse of the coherent correlation time $(1/T_c)$.

Table 6.5 Limitations and Applicability of Different Scattering Models	
Method	**Limitations**
KA	• Surface correlation length larger than the electromagnetic wavelength, and • Surface mean radius of curvature larger than the electromagnetic wavelength
KGO	• Large standard deviation of the surface height compared to the electromagnetic wavelength (high-frequency limit)
KPO	• Small vertical-scale roughness and • Small slope statistics
SPM	• Standard deviation of the sea surface height smaller than electromagnetic wavelength, and • Surface correlation length smaller than electromagnetic wavelength
TSM	• Difficulty to define the limit between large and small scales
IEM	• Computationally expensive
SSA	• Slopes of the roughness small compared to the incidence and scattering angles

Adapted from Cardellach, E., Rius, A., Martin-Neira, M., Fabra, F., Nogués-Correig, O., Ribó, S., Kainulainen, J., Camps, A., D'Addio, S., 2014. Consolidating the prescision of interferometric GNSS-R ocean altimetry using airborne experimental data. Trans. Geosci. Rem. Sens. 52. http://dx.doi.org/10. 1109/TGRS.2013.2286257.

For an interferometric GNSS-R (iGNSS-R) instrument, the SNR at correlator's output is given by (Martin-Neira et al., 2011):

$$SNR = \frac{SNR_{cr}}{1 + \dfrac{1 + SNR_R}{SNR_D}}, \tag{6.61}$$

where SNR_{cr}, SNR_R, and SNR_D are the SNRs of the clean-replica cross-correlation (as in cGNSS-R), reflected and direct signals, respectively. If $SNR_D \gg 1$, then $SNR \rightarrow SNR_{cr}$, but this requires antennas with very large directivities. Table 6.6 summarizes these values for the PARIS IoD instrument, different bands, and signals. It can be noted that for moderate SNR_D the SNR loss as compared to SNR_{cr} is ~ 2 dB, to be traded-off with the increase of β.

However, the SNR is not only affected by thermal noise, but speckle noise can be dominant. A rough approximation that allows us to assess the effect of speckle noise is by adding it to the thermal noise as:

$$NSR'_{R/cr} = NSR_{D/cr} + NSR_{speckle}, \tag{6.62}$$

where $NSR_{speckle} = 1/3.63 \, (-5.6$ dB$)$, and it is independent of bandwidth. Finally, the SNR is improved with incoherent averaging of N_i.

6.4.3.2 Realistic Approach

In the previous section a coarse approximation to the effect of noise was presented. A more precise estimation of the delay is presented here using the Cramér-Rao bound (CRB) including the covariance matrix $\left(\overline{\overline{C}}\right)$ of the data:

$$\sigma_\tau^2 \geq \frac{1}{\sum\limits_{k,l} \overline{\overline{C}}^{-1}_{k,l} \, s\prime(t_k - \tau) s\prime(t_l - \tau)}, \tag{6.63}$$

where s' is the waveform's derivative. In the case of AWGN, uncorrelated from sample to sample, Eq. (6.63) reduces to $\sigma_\tau^2 \geq \sigma_n^2 / \sum\limits_l \{s\prime(l)\}^2$, which is equivalent to $,\sigma_\tau^2 \geq 1/(SNR \cdot \beta^2)$. but in the (discrete) time domain.

In reality, the ultimate achievable scatterometry or altimetry performance depends on the snapshot SNR (no incoherent averaging or $N_i = 1$) and bandwidth, and on the cross-correlation of the noise present in consecutive lags $Y^{c,i,r}(t_n, \tau_m, f_d)$ and $Y^{c,i,r}(t_n, \tau_{m+1}, f_d)$ (fast time) as well. Additionally, the

Table 6.6 PARIS IoD Predicted Minimum Signal-to-Noise Ratios (No Speckle) for $\theta_i = 0°$				
	SNR_D (dB)	**SNR_R (dB)**	**SNR_{cr} (dB)**	**SNR (dB)**
L1	+2.50	−22.40	+3.97	+2.02
L5	−0.84	−25.73	−2.37	−5.83
E1	+5.36	−19.53	+8.92	+7.80
E5	+2.16	−22.83	+8.31	+6.25
Extracted from Camps, A., Pascual, D., Park, H., Martin, F., 2012b. PARIS IOD: ID-16A Contribution to Performance and Error Budgets Report, PARIS-PhA-IEEC-UPC-TN-008, Rrev 3.0.				

GNSS-R observable was estimated as the average of the square of N_i complex cross-correlations of $Y^{c,i,r}$ (t,τ,f_d) as the only way to reduce speckle noise. However, the amount of reduction depends on the correlation between the noise in the same lag τ in consecutive observables $Y^{c,i,r}$ (t_n,τ,f_d) and $Y^{c,i,r}$ (t_{n+1},τ,f_d) (slow time). The physical interpretation is presented in Fig. 6.24.

The analysis of the noise correlation between consecutive lags is performed using the covariance matrices defined as:

$$C(\tau_1, \tau_2) = \left\langle Y^{c,i,r}(t, \tau_1, f_d = 0) \cdot Y^{c,i,r\,*}(t, \tau_2, f_d = 0) \right\rangle, \tag{6.64}$$

which can be understood as the sum of two terms: one corresponding to the signal, and the other one corresponding to the noise. Fig. 6.25 illustrates this for a space-borne cGNSS-R instrument. The interested reader can refer to a simulation study for the cGNSS-R case in (Garrison, 2012), a detailed analytical study for the cGNSS-R case in Martín et al. (2014), and for the iGNSS-R case in Martín-Neira et al. (2011).

As can be observed from Fig. 6.25A and B, the noise component is present in all the delays, while clearly this is not the case for the signal term (since it is dependent on the backscattered signal). The covariance noise term follows the shape of the ACF, and the covariance signal term is dependent on the complex multiplication of the ACF at delays τ_1 and τ_2. This analysis is fundamental to estimate the achievable SNR and the ultimate instrument performance, as well as to specify the instrument in an optimum way in terms of bandwidth (Pascual et al., 2014), sampling frequency, width, and central

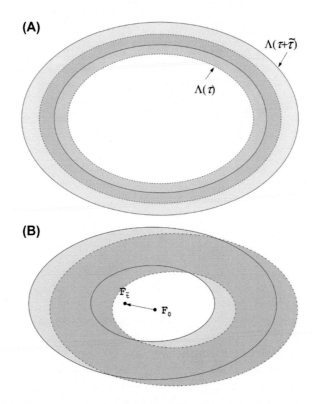

FIGURE 6.24

Physical interpretation of the noise correlation (A) between consecutive lags of the same waveform, and (B) between the same waveform lag in consecutive waveforms (You et al., 2004).

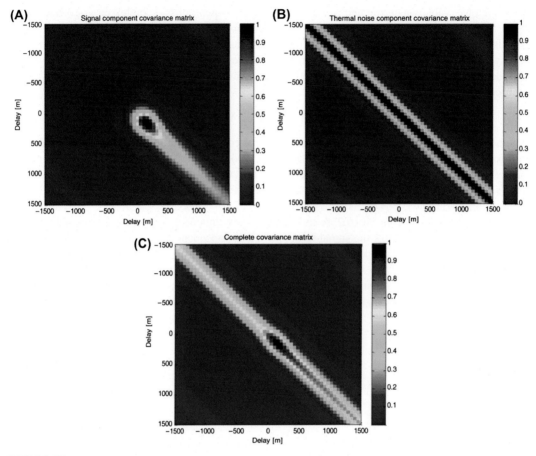

FIGURE 6.25

(A) Signal component covariance matrix (signal statistics) for cGNSS-R. (B) Thermal noise component covariance matrix (noise statistics) for cGNSS-R. (C) Complete covariance matrix (complex cross-correlation statistics, including both signal and noise terms). Simulation parameters: h = 700 km, T_c = 1 ms, N_i = 12.000.

From Martin, F., Camps, A., Park, H., d'Addio, S., Martín, M., Pascual, D., 2014. Cross-correlation waveform analysis for conventional and interferometric GNSS-R approaches. IEEE J Select. Top. Appl. Earth Observ. Remote Sensing 7 (5), 1560–1572.

position of the tracking window (Martín et al., 2015). In addition, the correlation between consecutive lags is also related to the achievable data compression that can be achieved, for example, using the wavelet transform (Camps et al., 2012a).

Consecutive waveforms (cGNSS-R) are plotted versus the correlation lag (x-axis) for up to 1000 snapshots (T_c = 1 ms, N_i = 1000) in Fig. 6.26. As it can be seen, the amplitude fluctuations are quite strong, and the correlation between the noise in the same lag τ in consecutive observables ("bin-to-bin" correlation) limits the effectiveness of the incoherent averaging. The speed of these fluctuations depends on two factors: (1) the properties of the surface under observation (i.e., how fast it changes, if it changes at all, for example, the ice, the land, or the variable ocean surface under different wind speeds), and

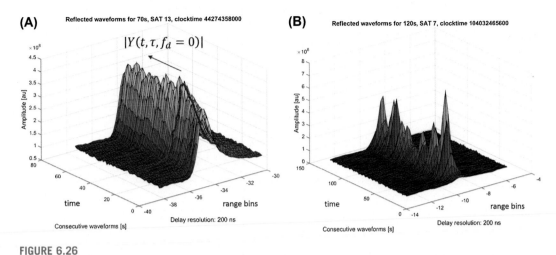

FIGURE 6.26

Temporal evolution of consecutive waveforms over (A) the ocean, and (B) land measured with the PAU instrument.

Courtesy of UPC.

(2) the relative movement between the transmitter and receiver (i.e., how fast the two coronae move away one from the other).

This effect is better illustrated in Fig. 6.27, which shows the progress of the waveform estimation using the PIR-A airborne instrument (iGNSS-R) in a field experiment on the Baltic sea (date November 11, 2011). The coherent integration time is $T_c = 1$ ms, and the number of incoherent averages increases from 1 up to 10,000. As can be seen, longer integration times are required to achieve a clear waveform due to higher noise of the interferometric processing. The zoom shows that individual waveforms are formed by the scattering on a few facets only (in particular the zoom shows one around the specular reflection point), and how the impulse response to that facet is simply the ACF squared.

In Fig. 6.28A we present the standard deviation of each correlation lag as a function of the incoherent integration time (as in Fig. 6.27). It can be appreciated that:

- the standard deviation is higher where the waveform amplitude is higher (speckle noise or "multiplicative" noise), and
- the standard deviation does not decrease as the squared root of N_i, because the corresponding area of the lags associated with the higher peaks is smaller and contains fewer scatterers. Before the leading edge, and in the tail, thermal noise dominates and the SNR goes as the squared root of N_i (Fig. 6.28B). This effect can be interpreted as an "effective" number of incoherent averages $N_{i,eff}$ that depends on the lag, and that unfortunately, is smaller where the waveform carries the information. The plot in Fig. 6.28C shows the variation of $N_i/N_{i,eff}$ versus the lag position for this data set. This value is actually related to the ratio of the coherence time of the sea surface t_{coh}:

$$t_{coh} = \frac{\lambda}{2 \cdot v} \sqrt{\frac{h}{2 \cdot c \cdot \tau_c \cdot \sin(\gamma)}}, \tag{6.65}$$

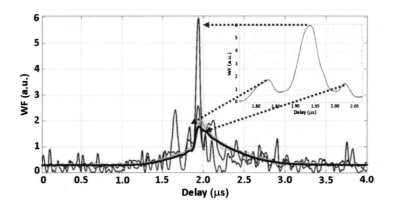

FIGURE 6.27

Sample interferometric waveforms obtained with PIR-A instrument for $T_c = 1$ ms, and $N_i = 1$ (red), 10 (purple), 100 (orange), 1.000 (green), and 10,000 (blue) waveforms. Zoom for red plot ($N_i = 1$) around the main peak compares well with the autocorrelation function of the C/A and P codes signals squared (Fig. 6.8A) both in the relative amplitude of the side lobes, and in their location at ± 1 P-chips of the main peak (Camps et al., 2014).

WF, *waveform*.

and the coherent integration time (T_c), which determines roughly the number of consecutive correlated waveforms. In this experiment $\lambda = 0.19$ m, $v = 237$ km/h, $h = 3.000$ m, $\tau_c = 97,7$ ns (P-chips), $\gamma \sim 70-77°$, and $T_c = 1$ ms, so $t_{coh} = 10,6$ ms, and $t_{coh}/T_c = 10,6$.

Fortunately, for a space borne instrument $t_{coh} = 1-2$ ms, and $N_i/N_{i,eff}$ is much closer to 1, so noise reduction by incoherent averaging becomes more effective.

In summary, as a rule of thumb, the coherent integration time should be on the order of 1 ms, and the number of incoherent averages to obtain a "clean" DDM is on the order of 1 s ($N_i = 1000$). This is illustrated graphically in Fig. 6.29 with the only publicly available data from UK-DMC (Gleason and Gebre-Egziabher, 2009) over the ocean, for $T_c = 1$ ms, N_i from 10 to 800. Although the DDM starts to be distinguishable above $N_i = 100$, the SNR is still quite poor, and much longer integration times are required.

It is worth to note that the properties of the speckle noise are, however, strongly dependent on the target. The situation shown in Fig. 6.29 is quite different over land and ice, shown in Fig. 6.30 for $T_c = 1$ ms, $N_i = 200$, together with the ocean one for better intercomparison. As we can see, the DDM in the case of the ice corresponds very closely to the WAF, that is, the product of a triangle squared in the delay domain, times a sin c function in the Doppler domain. This suggests that the reflection is nearly specular, despite the fact that the geometry is a bistatic one (not forward-scattering or near-grazing angle). Over land, the shape is not as clean as in the ice case, but a strong coherent reflection is still present, although with some blurring due to changing topography effects during the integration time, and some Doppler spread is also visible.

Finally, to illustrate the combined effects of type of target and observation geometry, Fig. 6.31 shows the griPAU instrument measurements of the peak of the delay-Doppler map over: (1) the ocean from a 382 m height cliff (Valencia et al., 2010) and (2) a bare soil from ~3 m height scaffolding

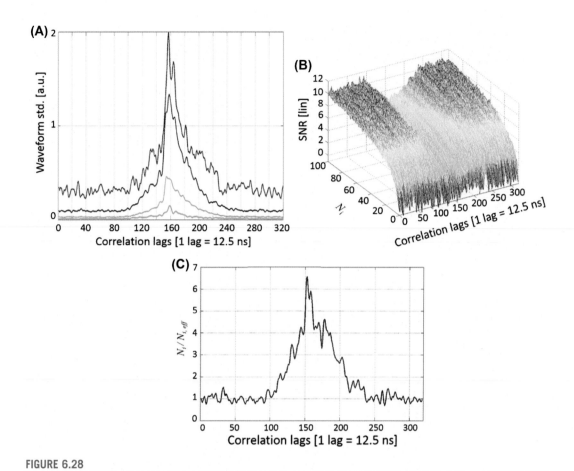

FIGURE 6.28

(A) Standard deviation of each waveform lag as a function of the number of incoherent averages, (B) evolution of the signal-to-noise with the number of incoherent averages as a function of the correlation lag, and (C) ratio of the number of incoherent averages and the "effective number" of incoherent averages.

From Camps, A., Park, H., Valencia, E., Pascual, D., Martin, F., Rius, A., Ribó, S., Benito, J., Andres-Beivide, A., Saameno, P., Staton, G., Martín, M., d'Addio, S., Willemsen, P., 2014. Optimization and performance analysis of interferometric GNSS-R altimeters: application to the PARIS IoD mission. IEEE J. Selected Top. Appl. Earth Observations Remote Sens. 7 (5), 1436–1451.

(Valencia et al., 2013b). In the first case, peak fluctuations are very fast due to the changes in the ocean's surface, while in the second case the fluctuations are much slower and are only due to the movement of the GPS satellites themselves. In both cases the estimated SNR is ~ 6.5 dB, not too far from the expected SNR for pure speckle noise (5.6 dB). However, in the second case, much longer integration times will be required to reduce speckle noise, since it is strongly correlated from sample to sample.

A last remark is that in the above analyses it is implicitly assumed that delay and Doppler tracking and retracking are properly performed, otherwise the DDMs and waveforms get blurred, the peak amplitude

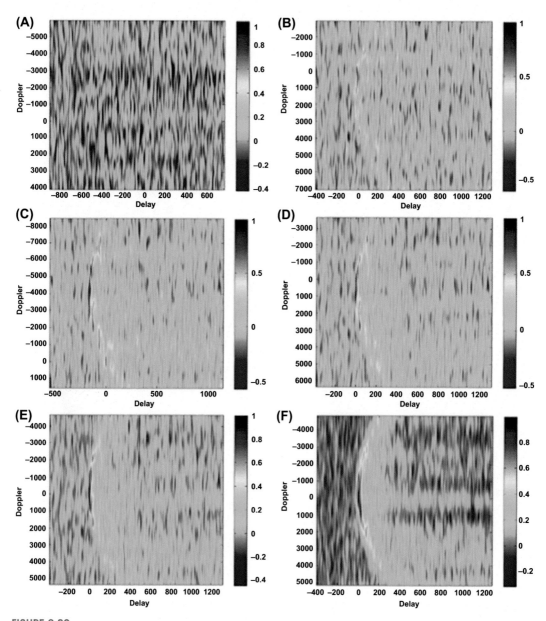

FIGURE 6.29

Open UK-DMC GPS-R data (Gleason and Gebre-Egziabher, 2009) processed with UPC software for different incoherent integration times: $T_c = 1$ ms, (A) $N_i = 10$, (B) $N_i = 40$, (C) $N_i = 80$, (D) $N_i = 120$, (E) $N_i = 200$, and (F) $N_i = 800$.

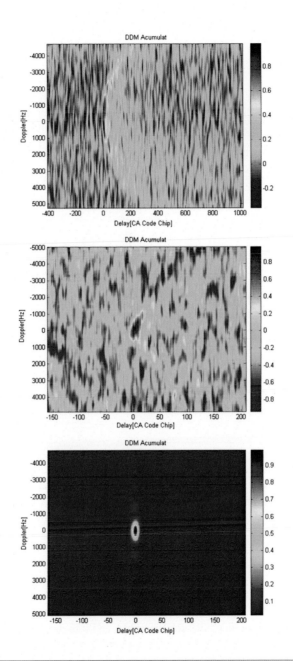

FIGURE 6.30

Open UK-DMC GPS-R data processed with UPC software sample DDMs from UK-DMC $T_c = 1$ ms and $N_i = 200$, over: (left) Ocean: DDM obtained by 12 s GPS [space vehicle (SV) #22] reflected data over the ocean, November 16, 2004, average wind speed = 7.8 m/s; (center) land: DDM obtained by 20 s GPS (SV #15 and 18) reflected data over land (Iowa and Nebraska), December 7, 2005; and (right) ice: DDM obtained by 7 s GPS (SV #13) reflected data over ice (Kuskowkwim Bay region, Alaska). February 4, 2005.

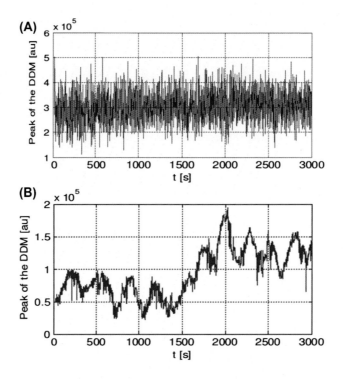

FIGURE 6.31

Peak of the delay-Doppler map over: (A) the ocean surface (Valencia et al., 2010); and (B) land (Valencia et al., 2013b).

decreases, and the altimetric accuracy degrades. For a detailed study of these effects, and how to compensate them, the interested reader is directed to Park et al. (2012a,b, 2013, 2014a). The effect of tracking or nontracking in the DDM formation is illustrated in the nontracking case (https://www.youtube.com/watch?v=flI6_bPxsEE) and tracking case (https://www.youtube.com/watch?v=gQJdsEBXoSo).

Fig. 6.32 (from Park et al., 2014a) illustrates these effects using the publicly available data from UK-DMC over the ocean.

6.4.4 GNSS-R INSTRUMENTS

6.4.4.1 Observables
GNSS-R instruments can measure different observables:

- The DDM is the most complete GNSS-R observable, and it is simply Eqs. (6.52) and (6.53) for a variable number of delays and Doppler cuts around the cross-correlation peak. It contains all the information to derive, for example, wind speed and direction over the ocean.
- The waveform (WF) is the cut of the DDM in the delay direction through the peak. This observable contains the minimum information required to derive the altimetric information from the peak of its derivative.

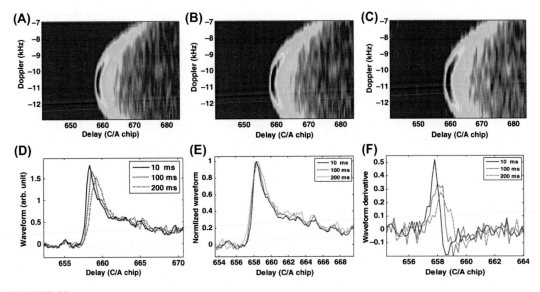

FIGURE 6.32

Delay tracking impact using UK-DMC data. Delay-Doppler maps obtained by the incoherent integration for 3 s ($T_c = 1$ ms, $N_i = 3000$), with the tracking refresh periods of (A) 10 ms; (B) 100 ms; (C) 200 ms; (D) comparison of the waveforms at the Doppler frequency; (E) comparison of the normalized waveforms; and (F) comparison of the waveform derivative.

From Park, H., Pascual, D., Camps, A., Martin, F., Alonso-Arroyo, A., Carreno-Luengo, H., 2014a. Analysis of spaceborne GNSS-R delay-Doppler tracking. IEEE J. Selected Top. Appl. Earth Observation Remote Sens. 7 (5), 1481–1492.

- The waveform peak amplitude (WPA) is the maximum value of the WF. This observable contains the minimum scattering information required, for example, to derive wind speed, soil moisture, or vegetation biomass. This observable can be either absolute or differential (relative amplitude between the peaks measured by an up-looking and a down-looking antenna). It can also be measured in reflection or in transmission.
- The ratio of the WPAs at two polarizations, e.g., LHCP and RHCP. This observable contains the scattering information required, for example, to derive wind speed, although it depends on the polarization purity of the transmitted signals, and the cross-polar level of the receiving antennas.

Finally, it is worth mentioning that other techniques exist, such as the "SNR Technique" (Larson et al., 2008a) and the "**Interference Pattern Technique**" (IPT) (Rodriguez-Alvarez et al., 2011) that include the measurement of the fading produced by the interference between the direct and reflected signals. While the SNR Technique uses GNSS geodetic receivers with RHCP antennas pointing to the zenith, the IPT uses ad-hoc receivers with linear polarization antennas (vertical and horizontal) pointing to the horizon so as to pick the direct and the reflected signals simultaneously. In the IPT, the angular position where the fading disappears, its peak-to-peak amplitude as a function of the elevation angle, and/or the relative phase between the interference patterns at vertical and horizontal polarizations are measured. The constructive/destructive interference between the direct and the reflected signals can also be used to infer geophysical information such as the soil moisture or the vegetation/ snow height (Rodriguez-Alvarez et al., 2009, Rodríguez-Alvarez, 2011a; Alonso-Arroyo et al., 2014a). For a vertically polarized antenna, the fading disappears at the Brewster angle, which depends on the dielectric constant, and the amplitude of the fading fluctuations depend on the magnitude of the reflection coefficient, which depends on the dielectric constant as well. If the relative phase of the fading patterns at vertical and horizontal polarizations is observed, a 90° phase shift occurs at the Brewster angle, which is almost insensitive to roughness effects.

The instruments required to measure the first four observables are basically the same, except for their computing capabilities in terms of number of complex correlators, antenna types, and tracking capabilities. They will be explained in the next section. The instruments required to measure the IPT can be static standard GNSS receivers with the appropriate antennas and data loggers.

6.4.4.2 Techniques

In Section 6.4.3 the most typical GNSS-R types were mentioned: the conventional GNSS-R (cGNSS-R), the (partial) interferometric GNSS-R [(p)iGNSS-R], and the reconstructed GNSS-R (rGNSS-R). The diagrams in Figs. 6.33−6.36 present the block diagrams of these different instruments.

For scatterometric applications, the cGNSS-R approach is enough, since the peak of the waveform does not depend much on the bandwidth of the (public) signals (compare Fig. 6.37A and B). However, the slope of the leading edge of the iGNSS-R waveform is much steeper than that of the cGNSS-R waveform, which should lead to a higher range resolution (depending on the SNR).

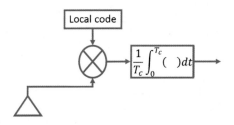

FIGURE 6.33

Basic concept of a conventional GNSS-R instrument.

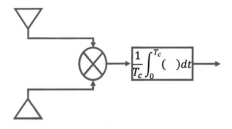

FIGURE 6.34

Basic concept of an interferometric GNSS-R instrument.

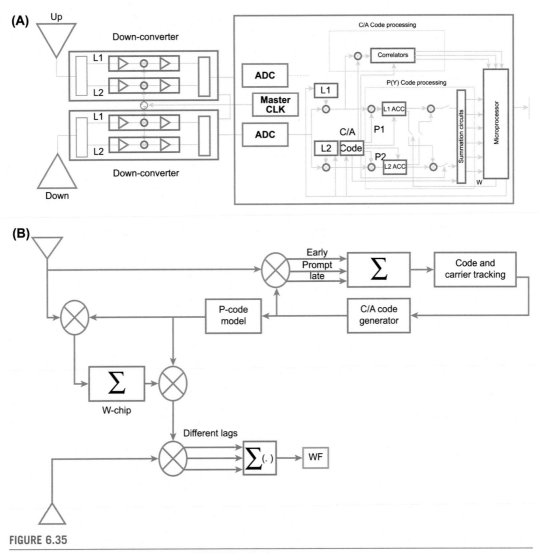

FIGURE 6.35

Basic approaches of the reconstructed GNSS-R technique.

Adapted from Carreño-Luengo, H., Camps, A., Ramos-Pérez, I., Rius, A., 2014. Experimental evaluation of GNSS-reflectometry altimetric precision using the P(Y) and C/A signals. IEEE J. Select. Top. App. Earth Obs. Rem. Sens. 7, http://dx.doi.org/10.1109/ JSTARS.2014.2320298; Lowe, S.T., Meehan, T., Young, L., May 2014. Direct signal enhanced semicodeless, processing of GNSS surface-reflected signals. IEEE J. Select. Top. Appl. Earth Observ Remote Sens. 7 (5), 1469–1472.

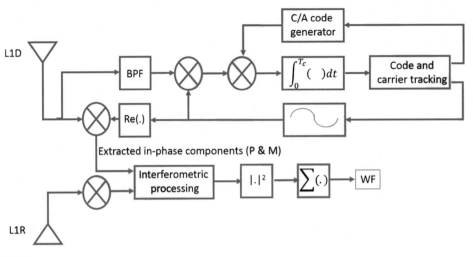

FIGURE 6.36

Basic approach of the partial interferometric GNSS-R technique.

Adapted from Li, W., Yang, D., D'Addio S., Martín-Neira, M., 2014. Partial interferometric processing of reflected GNSS signals for ocean altimetry. IEEE Geosci. Remote Sens. Lett. 11 (9), 1509–1513.

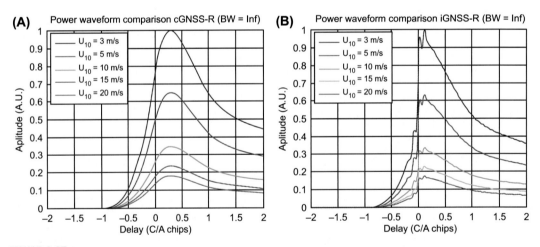

FIGURE 6.37

Normalized power waveforms for different wind speeds normalized to $U_{10} = 3$ m/s, for h = 700 km and $\theta_i = 0°$, for (A) iGNSS-R, and (B) cGNSS-R.

From Martín, F., 2015. Interferometric GNSS-R Processing: Modeling and Analysis of Advanced Processing Concepts for Altimetry (Ph.D. Dissertation). Universitat Politècnica de Catalunya. http://hdl.handle.net/10803/316583.

However, for altimetric applications,[5] the width of the ACF matters, since it determines the best achievable range resolution. Under the assumption of uncorrelated additive white Gaussian noise (AWGN) the delay estimation error is given by the[6] CRB:

$$\sigma_\tau^2 \geq \frac{1}{\text{SNR} \cdot \beta^2}, \tag{6.66}$$

where SNR is the signal-to-noise ratio, and β is the so-called rms bandwidth, defined as:

$$\beta^2 \triangleq \frac{\int_0^{\min\{B, B_{IS}\}} f^2 |S(f)|^2 df}{\int_0^B |S(f)|^2 df}, \tag{6.67}$$

where B and B_{IS} are the baseband bandwidths of the receiver's filter and the transmitted signal according to the IS documents (NAVSTAR IS-GPS-200, 2011; NAVSTAR IS-GPS-705, 2011; Galileo OS-SIS-ICD, 2010), and $|S(f)|^2$ is its spectrum.

One way to overcome the limitation of the bandwidth (B_{IS}) of the public codes, is the so-called interferometric GNSS-R processing (Martin-Neira, 1993), or "iGNSS-R" in short, in which the reflected signal is cross-correlated with the direct signal itself $s_D(t)$ after proper Doppler frequency and delay adjustment, as sketched in Fig. 6.34, and formulated in Eqs. 6.51b and 6.52.

The intercomparison between cGNSS-R and iGNSS-R is not straightforward since there are pros and cons for each method. In cGNSS-R the code replica is generated locally: it allows us to separate different satellites by their codes, it inherently has a high (infinite[7]) SNR, and smaller size (directivity) antennas can be used to track the reflected signals. However, only the public codes can be used, which exhibit a limited bandwidth, and limited range resolution. Also, the delay and Doppler frequency dynamics are larger, and these values must be adjusted more frequently for proper operation. In iGNSS-R there is no need to know the code, since the direct signal itself is used instead. It allows not only to use GNSS signals, but satellite radio, satellite television, or any other sources of opportunity with larger transmitted power, larger bandwidth, and better SNR, leading to potentially improved range resolution. In addition, the differential processing produced in the cross-correlation leads to slower delay and Doppler frequency dynamics, which are—in principle—easier to track. The main drawbacks are the large antenna size (directivity) required for the up-looking antenna, even when satellite television signals are used, which leads to the use of beam steering techniques, and eventually multibeam antennas if several reflection points are to be tracked, the need to separate different

---

[5]The ultimate performance depends not only on β, but on the SNR as well, the noise correlation, the width of the tracking window etc., as discussed previously. However, the value of β computed from the waveforms shown in Fig. 6.37A is $\beta \approx 4.8$ MHz, a much smaller value than the receiver's and the signal's bandwidth, and it will ultimately limit the achievable altimetry resolution improvement by a factor of ~ 4.8 MHz/2.2 MHz (rms bandwidth of WF signal with composite signal/bandwidth of C/A code) = 2.18 (Camps et al., 2012a,b, 2014). Similar results have been obtained by Rius et al. (2012), D'Addio and Martin-Neira (2013) and Cardellach et al. (2014). The physical explanation of this result is the smoother shape of the waveform resulting from the convolution of the ACF with the scatterers on the surface, as compared to the ACF itself. The best altimetry performance could only be achieved for quasi-specular reflections, when the WF looks like the ACF.
[6]For low SNRs (<5 dB) the Ziv-Zakai bound (Kay, 1993) provides a better indication of the magnitude of the estimation errors, which can be actually much larger than the CRB ones.
[7]Except for mismatches in the frequency responses of the transmitter filters that smooth the signals in a slightly different way for each satellite.

satellites from their signature ("location") in the delay-Doppler map, and the higher susceptibility to radio frequency (RF) interference.

To overcome some limitations of the previous techniques, newer approaches have been developed, namely the reconstructed GNSS-R (rGNSS-R) (Carreño et al., 2012, 2014; Lowe et al., 2014) and the partial interferometric GNSS-R (piGNSS-R) (Li et al., 2014). The rGNSS-R is similar to the cGNSS-R technique, but semicodeless techniques are used to reconstruct the P(Y) code which is then correlated with the reflected signal. The piGNSS-R is similar to the iGNSS-R technique, but the P and M codes components of the direct signal are extracted from the reference signal (direct signal) by coherent demodulation, and the interferometric approach is then applied to the reflected signal.

In Fig. 6.35A the correlation approach used in the downward-looking channel (slave: shown) instrument provides P-code processing of encrypted GPS signals without knowledge of the encrypted code, in addition to the C/A code for cGNSS-R, while the upward-looking channels (master: not shown) use a similar correlation approach and feed the information to the down-looking channel (slave) (Carreño et al., 2012, 2014). In Fig. 6.35B the direct L1-C/A signal is processed with typical DLLs and PLLs. The locked C/A code model is used to form a L1P model, which is then applied to the direct signal (center left), and after integration over ~ 0.5 MHz W-chips to estimate their signs, it is combined with the P-code model to form an L1 Y-code model that is used to correlate with the downward-looking channel. The advantages of this technique rely mainly on the larger bandwidth of the P(Y) codes, as compared to the C/A ones, and the large SNR, despite the losses of the semicodeless approach.

The basic approach of the piGNSS-R technique is presented here in Fig 6.36. The advantages of this technique are an even better range resolution as compared to the iGNSS-R one, but at the expense of a 3 dB signal loss (C/A code has been removed), which needs to be compensated by a 3 dB larger antenna directivity.

Relative altimetry or scatterometry observations can also be performed by applying the cGNSS-R techniques shown in Fig. 6.5 to the direct signal as well. This approach is intrinsically more insensitive to errors than absolute measurements performed with the basic scheme shown in Figs. 6.35 or 6.36. Alternatively, the second receiving chain can be replaced by mechanical or electrical beam steering to alternate between the direct and reflected signals. Table 6.7 summarizes the pros and cons of the different GNSS-R approaches.

6.4.4.3 Hardware Considerations

GNSS-R instruments share many commonalities with standard navigation receivers. In the following sections the different subsystems will be reviewed and the main characteristics pointed out.

Antennas pick up the GNSS signals and deliver them to the receiver. While most low-cost GNSS receivers worry more about the antenna size, without paying much attention to the antenna pattern shape, polarization purity, and ohmic losses, and geodetic receivers worry more about multipath and interference mitigation and phase center stability, GNSS-R antennas must be carefully designed and characterized to fulfill the instrument requirements. The following is a quick review of the main antenna parameters that have to be revised while defining a GNSS-R instrument.

6.4.4.3.1 Operating Frequencies and Bandwidths

Precise GNSS-R instruments will make use of the full available bandwidth of GNSS signals, either using the (p)iGNSS-R or the rGNSS-R techniques, and—for altimetric applications—at least at two frequency bands. On the other hand, most low-cost GNSS receivers only pick the reduced bandwidth

Table 6.7 Advantages and Disadvantages of the Different GNSS-R Approaches

GNSS-R Type	Advantages	Disadvantages
cGNSS-R	• Code replica generated locally ⇒ high SNR, smaller antennas • Transmitters separable by code	• Only public codes can be used • Limited bandwidth ⇒ limited resolution • Large dynamics of τ and f_d
iGNSS-R	• Any opportunity signal can be used (e.g., TV): larger bandwidth and transmitted power ⇒ better SNR • Full bandwidth can be used ⇒ improved resolution • Differential processing reduces dynamics of τ and f_d	• Direct signal has poor SNR ⇒ larger up-looking antennas required • Transmitters have to be separated by different τ, f_d in delay-Doppler domain
piGNSS-R	• Improved resolution as compared to iGNSS-R	• Poorer SNR than iGNSS-R ⇒ +3 dB larger directivity required
rGNSS-R	• Large bandwidth code replica generated locally using pseudo-correlation techniques ⇒ high SNR (smaller than in cGNSS-R), smaller antennas • Transmitters separable by code	• Large dynamics of τ and f_d

codes at L1. Table 6.8 summarizes the frequencies and bandwidths of the signals used by different systems. Designing a multiband large-bandwidth and well behaved is a challenge, especially is size and losses matter.

6.4.4.3.2 Gain Pattern and Polarization

Antennas for GNSS receivers must have an azimuthally symmetric antenna pattern at RHCP with a null[8] at 90° from boresight to mitigate interferences. For GNSS-R instruments the downward-looking antenna must be LHCP, and possibly RHCP as well for polarimetric instruments. In addition, while the polarization purity is not an issue in navigation receivers and an axial ratio above 3 dB is enough, for GNSS-R instruments the cross-polarization must be higher than ~25−30 dB within the FOV (not necessarily the full half-power beamwidth), because of the low level of the expected depolarization effects (Schiavulli et al., 2015).

6.4.4.3.3 Multipath Mitigation and Interference Suppression

Multipath and interference are mitigated by having a highly attenuated antenna pattern below the cutoff angle θ_C: choke rings, resistive loading, conducting ground plane, or metamaterials are placed at the rim of high-performance GPS antennas. While these are not typical requirements for GNSS-R antennas, these techniques also improve the isolation between neighbor antennas, and therefore, the antenna pattern similarity both in amplitude and phase.

---

[8]Actually it is not a null, but the pattern below a given "cut-off angle" (θ_c), typically ~5°, must be "low enough".

Table 6.8 Signals and Constellations of Major GNSS Systems: GPS, GLONASS, Galileo, and Compass

System	Country	Constellation	Coding	Carrier/Center Frequency (MHz)	Maximum Bandwidth (MHz)
GPS	IJSA	24 MEO (32 satellites in orbit in 2011 and under modernization)	CDMA	L1/L1C: 1575.420 L2/L2C: 1227.600 L5: 1176.45	30.69
GLONASS	Russia	24 MEO (24 satellites in orbit in 2011 and 6 active spares to be added)	FDMA CDMA FDMA CDMA CDMA CDMA	L1: $1602.000 + k \times 0.5625^a$ Li: 1575.420 L2: $1246.000 + k \times 0.4375^a$ L2: 1242.000 L3: 1202.025 L5: 1176.450	40.96
Galileo	European Union	27 MEO plus 3 spares (2 in operation in 2011)	CDMA	El: 1575.420 E6: 1278.750 E5b: 1207.140 E5: 1191.795 E5a: 1176.450	40.96
Compass	China	27 MEO plus 5 GEO and 3 IGSO (4 MEO, 5 GEO, 5 IGSO in, 2011)	CDMA	B1: 1559.052−1591.788 B2: 1162.220−1217.370 B3: 1250.618−1286.423	30.69

$^a k = 0$ to 24; each satellite having an FDMA channel.
From Wang, J.J.H., July 2012. Antennas for global navigation satellite system (GNSS). Proc. IEEE 100 (7), 2349−2355.

6.4.4.3.4 Phase Center Stability

The stability of the phase center depending on the direction is an important requirement for positioning arrays, and it is so for GNSS-R arrays whenever phase matters (e.g., phase altimetry). Tables 6.9 and 6.10 summarize the typical antenna performance requirements depending on the application, and the basic antenna types and their characteristics.

Low noise amplifiers and front-ends. Since the received signals are very weak, the required performances for the front-ends and low noise amplifiers (LNAs) are those of a microwave radiometer: very low noise figure (NF $\approx$ 1 dB), high gain (G $\geq$ 30 dB) to make the overall noise figure independent on the following stages, and high IP3 to avoid nonlinearities that may cause out-of band signals desensitize the receiver. A number of LNAs, or complete GPS receiver chips are available on the market that can be used, including in-phase and quadrature outputs already sampled, such as the MAX2769B, the MAX2745, or the former ZARLINK GP2015/2010 front-end and the GP2021 12 channel correlator. This last family of chips was the one used in NASA's DMR (the first "delay mapping receiver" GNSS-R receiver ever designed and built).

Table 6.9 GNSS Antennas Classified Based on Intended Platform, With Their General Characteristics and Anticipated Changes

Platform	Applications	Bands[a]	Instant. Bandwidth[a]	Gain Pattern	Multipath Rejection[a]	Interference Rejection[a]	Phase Center Stability[b]	Size	Weight	Cost
Large	Geodetic, ships, etc.	2 or more	>40 MHz	Very strict[a]	High	High	Good	Diameter >15 cm	Heavy[a]	High[a]
Medium	Car, truck, train, aircraft	1–2	>10 MHz	Somewhat strict[a]	Medium	Medium	Fair	Diameter >3 cm	Medium	Medium
Small	Body-wearable, laptop	1	>2 MHz	Not strict	Low	Low	Poor	Small and con formal	Light	Low
Handheld	Cell phone, GNSS receiver	1	>2 MHz	Ignored	None	None	Very poor	Very small	Very light	Very low

[a]Large changes in a few years are expected.
[b]Some will have serious degradation when bands and bandwidth are expanded.
From Wang, J.J.H., July 2012. Antennas for global navigation satellite system (GNSS). Proc. IEEE 100 (7), 2349–2355.

Table 6.10 Basic Antenna Types Used in GNSS and General Characteristics

Basic Antenna Type	Applications	Inherent Bandwidth	Gain Pattern[a]	Multipath Rejection	Interference Rejection	Phase Center Stability	Size	Profile	Weight	Cost[a]
Patch antenna	Most	Narrow	Strict	Medium	Medium	Medium to poor	Small to medium	Low	Medium	Low if single band
Quadrifilar helix	Handheld and small platform	Narrow	Somewhat strict	Medium	Medium	Poor	Small to medium	High	Small	Medium
Cross-slot dipole	Medium-large platform	Narrow	Strict	Medium	Medium	Medium	Medium	Low-	Medium	Medium
Planar spiral[b]	All but handheld applications	Very- wide	Strict	Medium to high	Medium	Very good	Medium	Low to medium	Medium	Medium
4-Element ring array	All but handheld applications	Narrow to wide	Strict	Medium to high	Medium ro high	Poor to food	Generally large	Low to high	Medium to heavy	Medium to high
Traveling-wave (TW) antenna[b]	All applications	Very wide	Strict	Medium to high	Medium	Good to very good	Medium to small	Low	Medium	Medium
Adaptive 2-element array	Handheld and small platforms	Narrow	Not strict	Medium	Medium	Poor	Small to medium	Low to high	Light	Low to medium
Adaptive multielement circular array	Medium to large platforms, mostly military	Narrow to wide	Very strict	High	Very high	Medium	Medium to large	Low to medium	Heavy	Very high

[a]Large changes in a few years are expected.
[b]Spiral antenna and TW antenna overlap each other considerably.
From Wang, J.J.H., July 2012. Antennas for global navigation satellite system (GNSS). Proc. IEEE 100 (7), 2349–2355.

Table 6.11 Table of Sample Commercial SAW Filters

		Standard Models			
Model Number	Center Frequency (MHz)	Bandwidth (MHz)	Ins Loss (dB)	Package	Application
TA0675A	1176.45	20	4.1	3.1x3.1	Wireless Corners
TA0730A	1178.12	40	5	3.1x3.1	Wireless Corners
TA1521A	1207	20	4	3.1x3.1	Remote Control
TA0SS3A	1210	60	5	3.1x3.1	Wireless Corners
TA1617A	1224	108	5.5	3.1x3.1	GPS
TA0425A	1542.5	35	4.0	2.5x2.0	GPS
TA15S4A	1568	18.32:	2.2	3.1x3.1	GPS
TA1703A	1570	80	4.2	3.1x3.1	GPS
TA1153A	1575	20	4.0	3.1x3.1	GPS
TA1575IG	1575.42	2.40	2.0	3.1x3.1	GPS
TA0531A	1575.42	10	3	2.5x2.0	GPS

From http://www.golledge.com/docs/products/saw/gsrf_list.htm.

Filters eliminate out-of-band signals that may interfere with the GNSS signals, or induce nonlinear effects in the amplifiers. SAW filters are typically used because of their small size (a few squared millimeters) and acceptable losses. Table 6.11 shows some commercial SAW filters, with sample GPS filters for L1 and L2 bands, with bandwidths from 2 to 20 MHz, and insertion losses from 1.5 to 4 dB (at L1).

However, if the power in adjacent bands is too strong, filters' rejection cannot be enough and may desensitize the GNSS receiver. A dramatic example of this was the Lightsquared communications system that intended to combine satellite communications services with a ground-based 4G-LTE network that would transmit on the same radio-band as its satellites (right next to L1).[9] Fig. 6.38 presents the spectra of the Lightsquared system, Inmarsat GPS, and Glonass, and their relative power levels, as well as the filters' response of different GPS receivers, illustrating the interference that would be created by these strong nearby signals.

However, filters cannot eliminate harmonics from services at lower frequency bands, falling inside the navigation bands (Table 6.12), self-interference from clocks and memories from the same GNSS-R instrument (Fig. 6.39 from Querol et al., 2014), or deliberate emissions from jammers, which are becoming more and more a problem because of the widespread of the so-called PPDs (Personal Privacy Devices). The performance degradation of a few GPS receivers as a function of the power of

---

[9]In February 2012, following extensive testing and analysis, the FCC announced it would not allow LightSquared's terrestrial operations and planned to withdraw LightSquared's ATC authorizations (http://www.gps.gov/spectrum/lightsquared/, August 25, 2014).

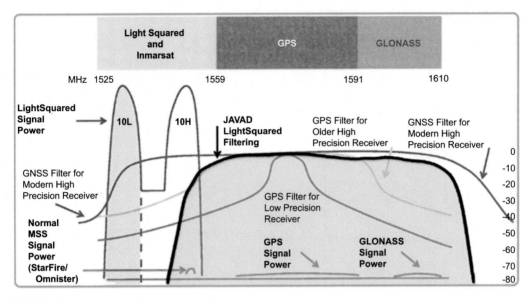

FIGURE 6.38

Spectra of Lightsquared, Inmarsat, GPS, and Glonass and filters' response for different receivers.

From June 2012. http://www.gpsworld.com/professional-oem/survey/news/javad-gnss-announces-partnership-with-
lightsquared-12096.

Table 6.12 Lower Frequency Harmonics Falling into the GPS L1 Band

Order	Band (MHz)	Usage
L1	1571.42–1579.42	GPS C/A
2nd	785.71–788.71	TV-UHF
3rd	523.807–526.472	TV-UHF
4th	392.855–394.855	Mobile/station
5th	314.284–315.884	Mobile/station
6th	261.903–263.237	Mobile/station
7th	224.488–225.631	Broadcast
8th	196.427–197.428	TV-VHF
9th	174.602–175.491	TV-VHF
10th	157.142–157.942	VHF maritime
11th	142.856–143.584	VHF military
12th	130.952–131.618	VHF Com
13th	120.878–121.494	VHF Com
14th	112.224–112.816	VOR/ILS
15th	104.761–105.295	Broadcast FM
16th	98.214–98.714	Broadcast FM

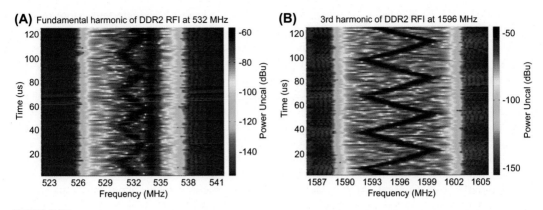

FIGURE 6.39

Measured interference caused by (A) a DDR2-SDRAM at 533 MHz, generating (B) a third harmonic at L1 navigation frequency band.

From Querol, X., Alasfuey, A., Pandolfi, M., Reche, C., Pérez, N., Minguillón, M.C., Moreno, T., Viana, M., Escudero, M., Orio, A., Pallarés, M., Reina, F., 2014. 2001–2012 trens on air quality in Spain. Sci. Total Environ. 15. http://dx.doi.org/10.1016/j.scitotenv.2014.05.074.

the jammer (Querol et al., 2014) is presented here in Fig. 6.40. At the time of writing this book, counter-measurements were mostly limited to setting nulls of the antenna pattern in the direction of the interferences, such as the GAJT (http://www.novatel.com/products/gnss-antennas/gajt/), a seven antenna array, that can set up to six nulls, or the GINCAN (http://www.roke.co.uk/resources/datasheets/

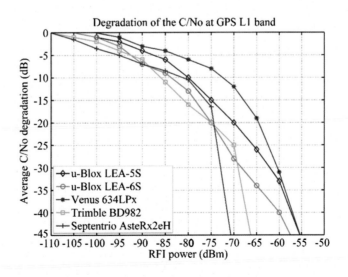

FIGURE 6.40

Performance degradation in terms of C/N_0 as a function of the interference power for different receivers.

From Querol, X., Alasfuey, A., Pandolfi, M., Reche, C., Pérez, N., Minguillón, M.C., Moreno, T., Viana, M., Escudero, M., Orio, A., Pallarés, M., Reina, F., 2014. 2001–2012 trens on air quality in Spain. Sci. Total Environ. 15. http://dx.doi.org/10.1016/j.scitotenv.2014.05.074.

02248-GINCAN.pdf) that has two antenna connectors, and can insert a single null. More recently, other counter-measurements, similar to those used in microwave radiometry are being undertaken (see Section 4.6.4.4), with the main difference that the signal cannot be "blanked," otherwise the signal tracking is lost. An example of such an experimental system, named FENIX, was developed by Querol Borràs and Camps Carmona (2017).

Down-converters. The main approaches for the down-conversion are analog down-conversion to zero IF ($f_{IF} = 0$) using I/Q demodulators, nonzero IF using a single mixer at a $f_{IF} \geq B_{RF}$, and direct sampling using band-pass sampling, a digital technique consisting in picking the appropriate replica generated during the sampling. In the first case there are two channels to be sampled at, at least, $f_s \geq B_{RF}$, while in the second one there is just one channel to be sampled at $f_S \geq 2 \cdot B_{RF}$. In practice, more than one down-conversions are needed to properly reject the image frequencies.

Automatic Gain Control (AGC) compensates the fluctuations of the input power in order to optimize the dynamic margin of the **Analog-to-digital converter (ADC)**. The combined response of the AGC and ADC determine the capability to reject RFI, since it provides a natural immunity to pulsed RFI. Although most GNSS navigation and GNSS-R receivers still use 1-bit sampling, advances in ADCs and digital processing allow nowadays to use a larger number of bits (e.g., 3 bits) that facilitates the implementation of digital filtering, down-conversion, RFI detection/mitigation algorithms etc. The increase in the number of bits also reduces the so-called quantization losses, a degradation of the SNR created by the increased noise levels coming from the sampling aliases. Table 6.13 presents the number of levels, N half the number of levels, ε the spacing between adjacent input levels, and the quantization losses (in linear units). It is clear that the quantization loss for 1 bit/2 levels is -1.96 dB ≈ -2 dB, while for 3 bits/8 levels is just -0.16 dB, which is almost negligible.

The **Local Oscillator (LO)** impacts the down-converters' performance, since an LO error translates into phase and/or frequency errors (apparent Doppler frequency). For example, a 10 Hz error in a 10 MHz reference clock leads to 1.5 kHz at 1.57542 GHz. LO errors are nonstationary and are characterized by the Allan's variance (or standard deviation) also known as two-sample variance is a

Table 6.13 Number of Levels, N (Half the Number of Levels), ε (Spacing Between Adjacent Input Levels), and Quantization Loss [lin] (Thompson et al., 2007)

Number of Levels	N	E	Quantization Loss [lin]
2	–	–	0.636620
3	1	1.224	0.809826
4	2	0.995	0.881154
8	4	0.586	0.962560
9	4	0.534	0.969304
16	8	0.335	0.988457
32	16	0.188	0.996505
64	32	0.104	0.998960
128	64	0.0573	0.999696
256	128	0.0312	0.999912

measure of frequency stability in clocks, oscillators, and amplifiers, as for the microwave radiometers sensitivity. The definition of the Allan's variance is:

$$\sigma_y^2 = \frac{1}{2(N-1)} \sum_{n=1}^{N-1} (y[n+1] - y[n])^2, \tag{6.68}$$

where y[.] is the average of M consecutive measurements. A typical Allan's standard deviation plot versus the time interval τ, in which for short τ averaging reduces the standard deviation as $1/\tau$ due to the averaging of the white phase noise, and for long τ averaging degrades the standard deviation as τ due to random walk is plotted here in Fig. 6.41.

In GNSS-R instruments all clocks must be synchronized, and the choice of the LO (or clock) will be determined by the maximum coherent integration time (the longer the better the clock stability), but also the choice of the IF.

Correlator's architectures have evolved as the digital technology has been capable integrating more correlator units per chip. Originally most of the correlations were performed serially, because of the few hardware resources required and easy implementation, but the performance was very slow. In opposition if all correlations could be performed in parallel, it will be very fast, but it will require many hardware resources, which will be used during a very low duty cycle. Finally, taking advantage of the fact that today's clocks run much faster than the navigation (GPS) signals, a semiparallel approach can be implemented where the same correlator blocks in the chip are used to compute different delay-Doppler bins within the coherent integration time (typically) ~1 ms.

6.4.4.4 Past, Present, and Future of GNSS-R Instruments

The first GNSS-R instrument was the **NASA delay mapping receiver** (http://ccar.colorado.edu/dmr/receiver/). It was developed by Garrison and Katzberg to study GPS surface reflections. A software-configurable GPS receiver was modified to measure signal postcorrelation power at a series of discrete

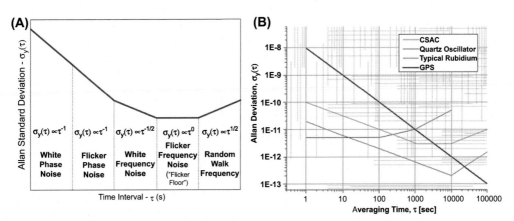

FIGURE 6.41

(A) Evolution of Allan's standard deviation versus time interval and causes, and (B) sample measurements of Allan's standard deviation.

From http://www.eetimes.com/document.asp?doc_id=1279697; (B) From John Vig.

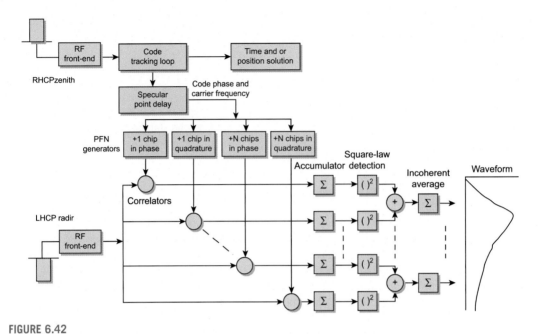

FIGURE 6.42

Delay mapping receiver block diagram (http://ccar.colorado.edu/dmr/receiver/).

code-delay steps and Doppler frequencies. This "delay-Doppler-mapping receiver" concept consists of an array of correlator pairs (Fig. 6.42) that accumulate the results of the correlation between a locally generated PRN code replica (at some defined delay and Doppler frequency) and the I/Q components of a down-converted, digitized reflected signal. The postdetection power in each bin is computed from the sum of the squares of the I and Q components as $I^2 + Q^2$.

The receiver was implemented using a COTS GPS-development kit with the MITEL (Zarlink) GP 2021 chip with 12-channel correlators spaced ~ 0.5 μs (2 MHz sampling frequency), fed by two RF front ends one connected to an RHCP zenith-looking antenna, and the other one to an LHCP nadir-looking antenna, to produce waveforms of $T_c = 1$ ms, and $N_i = 100$. The DMR instrument (Fig. 6.43; left) and the LHCP nadir-looking antenna (Fig. 6.43; right) (Garrison and Katzberg, 1997) performed the first ever GPS-R scatterometry measurements.

The **NASA/JPL delay/Doppler-mapping software receiver systems** (Lowe et al., 2002b) originally consisted of a modified TurboRogue GPS receiver that computed 1 s position measurements from direct GPS signals in the RF input and stored the results in a flash card, with raw L1 pseudo-range and carrier-phase observables sampled at 20.456 MHz recorded onto an AIT tape with a Sony SIR-1000 for off-line processing (Fig. 6.44). This instrument provided the first ocean-altimetry measurement made using reflected GPS signals, from a Cessna airplane flying off the coast of Santa Barbara, California, at 1.5 km height and ~ 50 m/s (Fig 6.45).

An upgrade of the single channel instrument shown before was developed later, which included four TurboRogue front-ends modified for synchronous operation, 16 channel in-flight data recording system, and L1 I/Q and L2 I/Q signals from all four front-ends recorded with two Cybernetics

FIGURE 6.43

(Left) DMR instrument and (right) LHCP nadir-looking antenna.

From Garrison, J.L., Katzberg, S.J., April 12–14, 1997. Detection of ocean reflected GPS signals: theory and experiment. In: Southeastcon '97. 'Engineering new New Century'., Proceedings. IEEE, pp. 290–294.

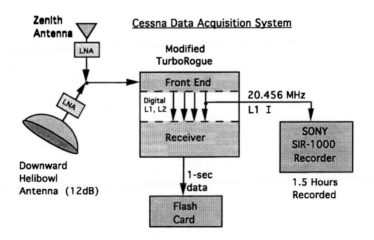

FIGURE 6.44

Data acquisition system of the experiment.

From Lowe, S.T., Zuffada, C., LaBrecque, J.L., Lough, M., Lerma, J., Young, L.E., 2000. An ocean-altimetry measurement using reflected GPS signals observed from a low-altitude aircraft. In: IGARSS 2000. IEEE 2000 International Geoscience and Remote Sensing Symposium. Taking the Pulse of the Planet: The Role of Remote Sensing in Managing the Environment. Proceedings (Cat. No.00CH37120), vol. 5. Honolulu, HI, pp. 2185–2187. http://dx.doi.org/10.1109/IGARSS.2000.858350.

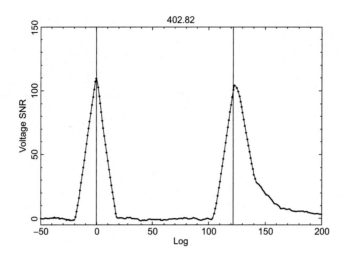

FIGURE 6.45

One second waveform showing voltage SNR as a function a lag (48.9 ns samples). Peak on the left is the direct satellite-to-Cessna signal, and the right peak is the ocean-reflected signal. Vertical lines show the best fit to the direct and specular reception, respectively.

From Lowe, S.T., Zuffada, C., LaBrecque, J.L., Lough, M., Lerma, J., Young, L.E., 2000. An ocean-altimetry measurement using reflected GPS signals observed from a low-altitude aircraft. In: IGARSS 2000. IEEE 2000 International Geoscience and Remote Sensing Symposium. Taking the Pulse of the Planet: The Role of Remote Sensing in Managing the Environment. Proceedings (Cat. No.00CH37120), vol. 5. Honolulu, HI, pp. 2185–2187. http://dx.doi.org/10.1109/IGARSS.2000.858350.

CY-DDR recorders onto a 600-GB Medea Corporation "VideoRack RTX" hard drive array, consisting of two 300-GB partitions (Fig. 6.46).

The **ESA delay/Doppler-mapping software receiver system** consisted of two TurboRogue GPS scientific receivers used as front ends (I component only), whose outputs at 1 bit, after base-band conversion, were stored in two separate Sony Recorders, using two different magnetic tapes to store the data (Germain et al., 2003). The block diagram is shown in Fig. 6.47.

The **UPC/IEEC DOppler DElay software RECeiver (DODEREC)** was designed and manufactured in 2002 by the Universitat Politècnica de Catalunya-Barcelona Tech (UPC) under a contract for the IEEC/ICE. It contained three chains delivering the in-phase and quadrature components at zero-IF, sampled at 1 bit (Fig. 6.48). It included a common LO generation system at 20.46 MHz, and an original (not commercial) data acquisition system that stored the data at 20 MB/s up to 1 h directly into a commercial IDE/ATA-100 hard disk drive. The channels could be connected in any arbitrary configuration, for example, an RHCP up-looking antennas, and a pair of RHCP and LHCP down-looking antennas (Nogués et al., 2003). An external PPS signal from a Novatel receiver was recorder synchronously with the GNSS-R data for off-line software processing.

The **OCEANPAL GNSS-R** instrument has been commercialized by Starlab since 2002. It consists of two GPS L1/Galileo E1 capable receiving chains connected to an up-looking (RHCP) and a down-looking (LHCP) antennas, and processing was performed off-line in software (http://spacetech.starlab. es/oceanpal/).

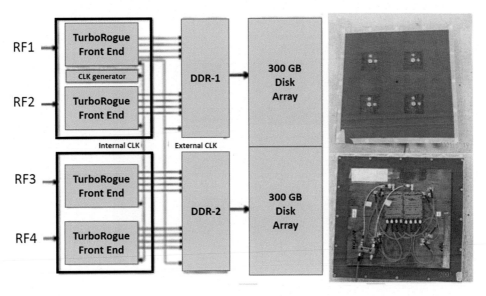

FIGURE 6.46

Schematic diagram of full 16-channel in-flight data recording system and antenna system.

From Lowe, S.T., Kroger, P., Franklin, G., LaBrecque, J.L., Lerma, J., Lough, M., Marcin, M.R., Muellerschoen, R.J., Spitzmesser,
D., Young, L.E., May 2002b. A delay/Doppler-mapping receiver system for GPS-reflection remote sensing. IEEE Trans. Geosci.
Remote Sens. 40 (5), 1150–1163.

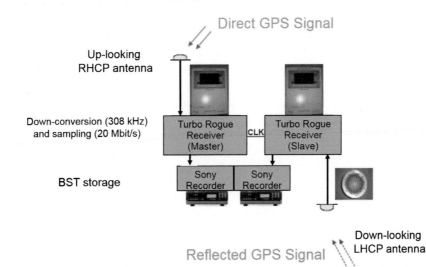

FIGURE 6.47

Block diagram of ESA's Delay/Doppler Mapping Software Receiver system. Data from two TurboRogues are
stored in separate magnetic tapes for off-line processing.

From Germain, O., Ruffini, G., Soulat, F., Caparrini, M., Chapron, B., Silvestrin, P., July 2003. The GNSS-R Eddy Experiment II:
L-band and optical speculometry for directional sea-roughness retrieval from low altitude aircraft. Paper Presented at 2003
Workshop on Oceanography with GNSS Reflections, Starlab, Barcelona, Spain. Available at: http://arxiv.org/abs/physics/0310093.

(A)

(B)

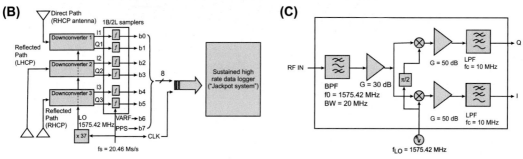

FIGURE 6.48

(A) Picture, (B) block diagram, and (C) detail of the RF front-end of the Doppler Delay Receiver.

From Nogues, O., Sumpsi, A., Camps, A., Rius, A., 21–25 July 2003. A 3 GPS-channels Doppler-delay receiver for remote sensing applications. In: 2003 IEEE International Geoscience and Remote Sensing Symposium, vol. 7, pp. 4483–4485.

In September 2003, SSTL launched the **UK-DMC satellite**, carrying aboard the first dedicated GNSS-R experiment from space. It included the **SSTL SGR-10 GPS receiver** (http://www.sstl.co.uk/Products/Subsystems/Navigation/SGR-10-Space-GPS-Receiver), a dual antenna L1 C/A code GPS receiver that weights about 1 kg and consumes 5.5 W. It was connected to a 12 dBi LHCP downward-looking antenna (Fig. 6.49) and a recorder of the IF sampled outputs, which stored bursts of 20 s of raw data.

Since 2003 **IEEC** has developed the **GOLD-RTR** (GPS Open Loop Differential Real-Time Receiver) instrument. It contains three receiving chains, driven by the same LO at 300 kHz below the L1 central frequency. Real time correlators were implemented in Altera Stratix FPGA to produce 64 lag complex waveforms at 20 MS/s, which provided a temporal spacing of 50 ns. The system clock at 40 MHz was also used to generate the LO and the FPGA clock, while a 1 pps (pulse per second) signal was generated by a Novatel Navigation receiver. Fig. 6.50A shows an image of the GOLD-RTR instrument and Fig. 6.50B its block diagram.

The Purdue GPS Instrument Multistatic and Occultation Sensing for atmospheric, land, and ocean remote sensing (**GIZMOS**) is an instrument to measure the water vapor distribution in the troposphere

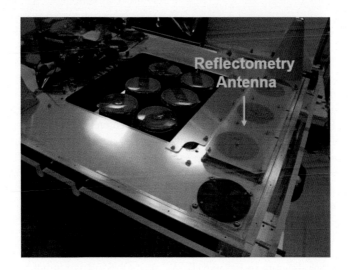

FIGURE 6.49

Picture of the three patch antenna used for the GNSS-R payload on board the UK-DMC satellite (http://digital-transmissions.co.uk/2003-11surreysat.htm]).

(A)

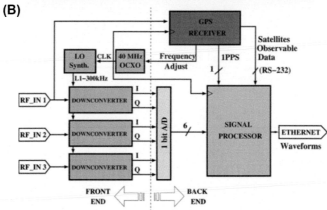

(B)

FIGURE 6.50

(A) Picture and (B) block diagram of IEEC's GOLD-RTR instrument.

From Nogués-Correig, O., Cardellach-Galí, E., Sanz-Campderrós, J., Rius, A., January 2007. AGPS-reflections receiver that computes Doppler/delay maps in real time. IEEE Trans. Geosci. Remote Sens. 45 (1), 156–174.

by inversion of the bending angle profile produced during occultation of a GPS satellite by the limb of the Earth. The instrument also uses GPS signals reflected from land or ocean to determine properties of those surfaces (soil moisture and surface roughness, respectively). Occultation samples represent distances of ca. 300 km and so are complementary to the single-point measurements obtained by conventional soundings. It is a passive remote sensing system that records the full bandwidth (intermediate frequency) of the GPS signals and stores the measurements on hard disks for subsequent analyses. The instrument started to be developed in 2006 (https://www.eol.ucar.edu/instruments/gnss-instrument-system-multi-static-and-occultation-sensing).

The **griPAU** (GNSS-R instrument for PAU) was developed by UPC (Marchan-Hernandez et al., 2007; Valencia et al., 2010) as a precursor of the **PAU** space-borne instrument (Camps et al., 2011b; Alsonso et al., 2012), a secondary payload of INTA Microsat-1. The griPAU instrument is shown in Fig. 6.51A and its block diagram in Fig. 6.51B. As most GNSS-R instruments, it had two receiving chains connected to an RHCP up-looking antenna and an LHCP down-looking antenna mounted on an antenna positioner. Its main feature was that, thanks to the implementation of hardware reuse techniques, this instrument was able to include a VHDL module in a Virtex-4 FPGA to compute in hardware, and in real time, the complete DDM, not by slices, as a sequence of waveforms. The DDM

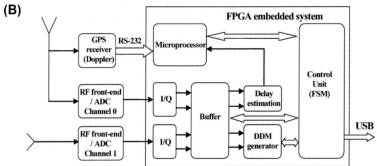

FIGURE 6.51

(A) Picture and (B) block diagram of the griPAU instrument.

From Valencia, E., Camps, A., Marchan-Hernandez, J.F., Bosch-Lluis, X., Rodriguez-Alvarez, N., Ramos-Perez, I., 2010. Advanced archititures for real-time delay-Doppler map GNSS-reflectometers: the GPS reflectometer instrument for PAU (gruPAU). Adv. Space Res. 46. http://dx.doi.org/10.1016/j.asr.2010.02.002.

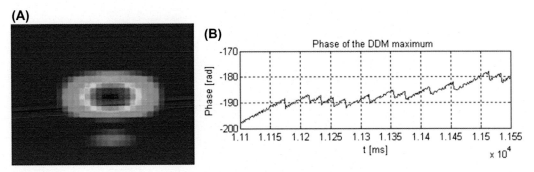

FIGURE 6.52

(A) Sample delay Doppler map ($T_c = 1$ ms, $N_i = 1$) over the ocean, (B) phase of the (complex) DDM peak changes due to the navigation bit changes, in multiples of 20 ms, suggests the presence of a coherent component in the scattered signal.

From Valencia, E., Camps, A., Marchan-Hernandez, J.F., Bosch-Lluis, X., Rodriquez-Alvarez, N., Ramos-Perez, I., 2010. Advanced architetures for real-time delay-Doppler map GNSS-reflectometers: the GPS reflectometer instrument for PAU (griPAU). Adv. Space Res. 46. http://dx.doi.org/10.1016/j.asr.2010.02.002.

size was 32 × 24 delay-Doppler bins, in programmable steps, and the coherent integration and integration times were adjustable as well, with default $T_c = 1$ ms and $N_i = 1000$. It also compensated automatically and dynamically the Doppler frequency shift and the delay (1 estimation every 5 ms). Sample Delay-Dopler measurements and phase of the peak of the complex waveform are shown in Fig. 6.52.

The **flight model of the PAU instrument** is shown here in Fig. 6.53. PAU is a hybrid total power radiometer (with Kurtosis RFI detector) and GNSS-R receiver (Camps et al., 2007). It features two cold redundant receivers and processing boards based on Virtex-4 FPGAs with in-orbit reconfiguration capability. DDMs' size is 4096 samples in delay times, 16 samples in Doppler and are computed in near real time (1 s of data acquisition and ∼1 s of data processing).

The **IEEC PIR and PIR-A (PARIS Interferometric Receiver—Airborne)** was developed in 2010 by the IEEC (Nogues-Correig et al., 2010). It implemented the iGNSS-R technique to validate the concept of the PARIS IoD concept (Martin-Neira et al., 2011). A picture of the Zeeland Brugg experiment carried out in July 2010 is presented here in Fig. 6.54A. The up- and down-looking antennas are clearly visible (directivity 15 dB, gain 9 dB, side lobes < −35 dB). Two RF channels with a bandwidth of 24 MHz each, are sampled at 80 MHz, and processed in real time to produce 320 lags interferometric waveforms. The block diagram is shown in Fig. 6.54B.

The **SSTL SGR-ReSI** is the GNSS-R payload on board the UK satellite TechDemoSat launched on July 8, 2014. SGR-ReSi is based on COTS RF front-ends and LNAs. The L1 front-end is based on the MAX2769 reconfigurable GNSS front-end, and the L2c front-end on the reprogrammable satellite tuner MAX2112, plus an A/D converter. It can also be programmed to E6, L5, and E5ab bands. The LNA is the MAX2659, and an appropriate RF matching circuitry is used to suit both L1 and L2C bands. The antenna is a 2 × 2 patch array, as seen in Fig. 6.55.

The **³Cat-2** mission (Carreno-Luengo et al., 2016) is a Spanish mission developed by UPC-Barcelona Tech based on a 6 unit (3 × 2) CubeSat structure. Its main payload is the PYCARO (**P(Y)** and **C/A R**eflect**O**meter) GNSS-R instrument (Carreno-Luengo et al., 2014; Olive et al., 2016), which implements the reconstructed GNSS-R technique (Fig. 6.35A). It has a 3 × 2 antenna array of

FIGURE 6.53

Picture of the flight model of the PAU instrument.

Courtesy of UPC.

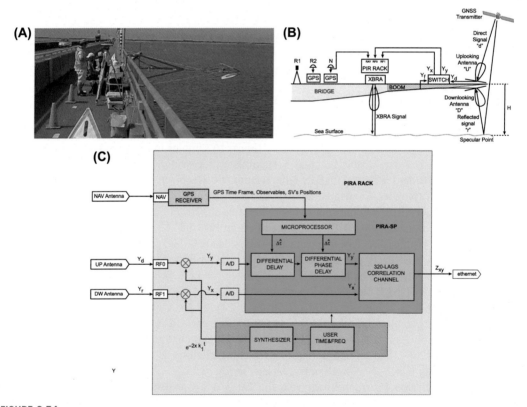

FIGURE 6.54

(A) Picture of the Zeelang Brugg experiment: first demonstration of the PARIS concept. The up- and down-looking antennas at the tip of the mast are clearly visible, (B) experiment set-up, and (C) block diagram of the PIR instrument used.

FIGURE 6.55

Picture of UK TechDemoSat with the 2 × 2 patch array of spiral antennas (http://www.sstl.co.uk/Missions/TechDemoSat-1–Launched-2014/TechDemoSat-1/TechDemoSat-1–The-Mission).

dual frequency (L1 and L2) and dual polarization (RHCP and LHCP) antennas that produce beam with a directivity better than 13 dB (Fig. 6.56). It was launched on August 15, 2016.

The **NASA CYGNSS** (Cyclone GNSS) was selected on June 18, 2012, and the launch date was December 15, 2016. This mission is the first full mission award in NASA's low-cost venture-class Earth science program. It consists of a constellation of eight small satellites in LEO at 35° inclination and each of them carries on board a modified version of the SSTL SGR-ReSI DMR to perform up to four simultaneous scatterometric observations of GPS L1 signals using the C/A code to infer wind speed maps (Fig. 6.57). This point will be discussed in more detail in the ocean applications section. Fig. 6.58A shows an artist view of a CYGNSS satellite and its parts are shown here in Fig. 6.58B. The green rectangles are two 3 × 2 patch array with ∼14 dB gain (CYGNSS Handbook, 2016).

Finally, in the 2019 time frame the **ESA GEROS ISS experiment (GNSS REflectometry, Radio Occultation and Scatterometry onboard the International Space Station)** (Wickert et al., 2016) should be installed in the Columbus module of the International Space Station (Fig. 6.59). This experiment is an answer to ESA call in 2011 to perform "Climate change related research aboard ISS". GEROS-ISS goals are as follows:

- Primary: To measure and map altimetric sea surface height of the ocean using reflected GNSS signals to allow methodology demonstration, establishment of error budget and resolutions and comparison/synergy with results of satellite-based nadir pointing altimeters. This includes POD of the GEROS payload.
- Secondary: To retrieve scalar ocean surface mean square slope (MSS), which is related to sea roughness, wind speed, with a GNSS space-borne receiver to allow methodology testing, establishment of error budget, and resolutions. In addition, it would be desirable to obtain 2D MSS (directional MSS, related to wind direction).

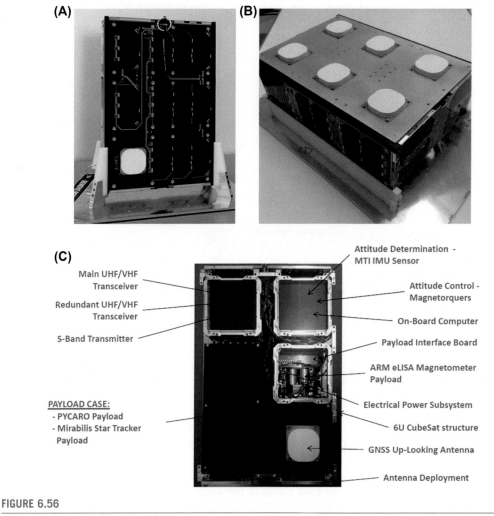

FIGURE 6.56

Picture of (A) zenith side of 3Cat-2, (B) nadir-looking antenna, and (C) satellite assembly.

Courtesy of the Universitat Politècnica de Catalunya, NanoSat Lab.

- Additional: To assess the potential of GNSS scatterometry for land applications and in particular to develop products such as soil moisture, vegetation biomass, and mid-latitudes snow/ice properties and to further explore the potential of GNSS RO data (vertical profiles of atmospheric bending angle, refractivity, temperature, pressure, humidity, and electron density), particularly in the Tropics, to detect changes in atmospheric temperature and climate relevant parameters (e.g., tropopause height) and to provide additional information for the analysis of the reflectometry data from GEROS (several new aspects: precipitation, low inclination, Multi-GNSS).

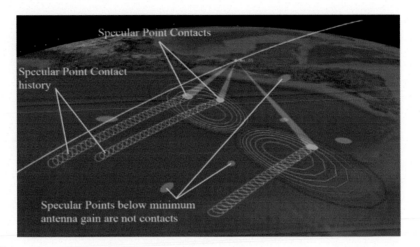

FIGURE 6.57

CYGNSS signal retrieval: Geometry of observation of the CYGNSS satellites. Only GPS reflections within the antenna footprint are used as valid observables.

Image courtesy of CYGNSS Project (http://spaceflight101.com/cygnss/cygnss-instrument-overview/).

At the time of finishing this book the GEROS ISS payload has been proposed as the G-TERNS Earth Explorer 9 mission. The final selection by ESA will be made in June 2017.

Table 6.14 summarizes of **some existing GNSS-R instruments** and their main parameters.

6.4.5 APPLICATIONS

GNSS-R applications extract the information from any of the observables explained in Section 6.4.4.1, namely: the DDM, the WF, the WPA, the ratio of the WPAs at two polarizations, e.g., LHCP and RHCP, or the IPT. In what follows, the geophysical information will be extracted from one or another GNSS-R observable, although in some cases, the techniques could not be applied to an LEO satellite. In the coming sections the following applications are presented: Ocean (wind speed mapping, σ^0 and altimetry applications), ice and dry snow monitoring, and land (soil moisture and vegetation biomass/water content).

6.4.5.1 Ocean Winds

Ocean wind retrieval was the first application of GNSS-R, and it was demonstrated with NASA's DMR instrument. A few techniques have been developed to infer wind speed (and eventually direction) from GNSS-R observables.

6.4.5.1.1 Parameter Estimation

Parameter Estimation is based on finding the best fit of a model to the measured waveform $|Y(\tau, f_D)|^2$. Recalling from Eqs. (6.53) and (6.56) that the wind speed information lies in σ^0 and that this parameter depends on the slopes pdf, a simple expression of the slopes pdf must be used in the inversion, for example, the up/cross-wind variances, or the wind speed dependence as in the Cox and Munk model, properly scaled at L-band.

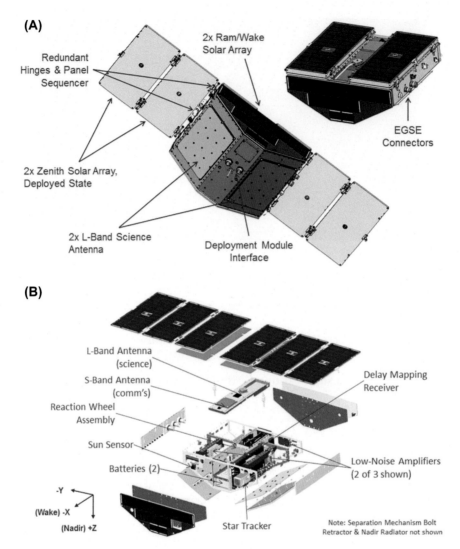

FIGURE 6.58

(A) Artist view of a CYGNSS satellite, and (B) satellite assembly.

Courtesy of Keith Smith, Southwest Research Institute; reproduced from CYGNSS Handbook: Cyclone Global Navigation Satellite System — Deriving Surface Wind Speeds in Tropical Cyclones, Michigan Publishing, April 1, 2016. http://clasp-research.engin. umich.edu/missions/cygnss/reference/cygnss-mission/CYGNSS_Handbook_April2016.pdf.

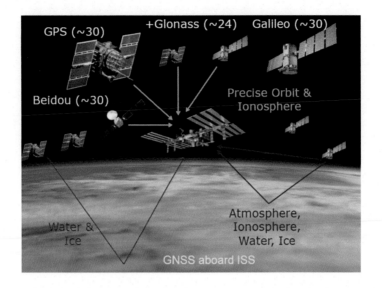

FIGURE 6.59

Artist view of the GEROS-ISS mission to assess the potential of new GNSS-R and GNSS-RO techniques for Earth observation.

From Wickert, J., Cardellach, E., Bandeiras, J., Bertino, L., Andersen, O., Camps, A., Catarino, N., Chapron, B., Fabra, F., Floury, N., Foti, G., Gommenginger, C., Hatton, J., Høeg, P., Jaggi, A., Kern, M., Lee, T., Li, Z., Martin-Neira, M., Park, H., Pierdicca, N., Ressler, G., Rius, A., Rosello, J., Saynisch, J., Soulat, F., Shum, C.K., Semmling, M., Sousa, A., Xie, J., Zuffada, C., 2016. Geros-iss: GNSS reflectometry, radio occultation and scatterometry onboard the International space Station. IEEE J. Select. Top. Appl. Earth Observ. Remote Sens.

Garrison et al. (2002) developed a technique in which the waveforms were adjusted by just three parameters: the slopes mss, the peak delay, and a scaling factor, and the model was approximated by a series of exponentials, as shown in Eqs. (6.69) and (6.70):

$$J(U_{10}, \tau_c, S) = \sum_{\tau} \sum_{f_D} \left[|Y(\tau, f_D)|^2 - |Y_{model}(\tau - \tau_c, f_D, U_{10})|^2 \right]^2, \qquad (6.69)$$

where:

$$|Y_{model}(\tau, \tau_c, U_{10}, S)|^2 \approx S \cdot \exp \left[\sum_{m=0}^{M} \sum_{n=0}^{N} a_{m,n} \cdot U_{10}^m \cdot (\tau - \tau_c)^n \right]. \qquad (6.70)$$

Komjathy et al. (2004) derived wind speed and direction minimizing the total rms residual of waveforms from two different satellites at different azimuths (Fig. 6.60). Wind speed errors are 1.2 m/s rms if only one satellite is used, and it decreases down to 0.7 m/s rms if multiple satellites are used.

Table 6.14 Summary of Some Existing GNSS-R Instruments

ID:	HW/SW:	Number of RF Ports:	Frequency Bands:	BB Bandwidth (MHz):	Sampling Rate (MHz):	Output Rate (Hz):	Receiver Technique:	GNSS Constellations:
GOLD-RTR	HW	3	L1	8	20	1000	cGNSS-R (C/A)	GPS
PIR/A	HW	3	L1	12	80	1000	iGNSS-R	ANY at L1
GORS-1(2)	HW	2(4)	L1 + L2				CGNSS-R (C/A, L2C)	GPS, Galileo
TR	SW	2	L1 + L2	18	20	20 MHz	RAW	GPS
BJ	SW	4	L1 + L2				RAW	GPS
TriG (extended)	HW	8(16)	Any 4 within L-band	2–40 Configurable	20/40	0.1–1000	ANY: SW Configurable	GPS, Glonass FDMA, Galileo, other 1–2 GHz
OceainPal/S AM	SW	2	L1	4	16.367	1000	RAW	GPS
PAD	HW							
OoenGPS	HW	2	L1		5.7	<100	cGNSS-R (C/A)	GPS
COMNAV	SW	1	L1		5 7		RAW	GPS
NordNAV R30(Quad)	SW	1(4)	L1	2	16.4		RAW	GPS
GRAS	HW	3	L1+L2	20	28.25	1000	cGNSS-R (C/A, P-semicodeless)	GPS
DMR	HW		L1+L2					GPS
Polito- GNSS-R	SW	1	L1		8.1838		RAW	GPS
SPIR	SW	16	L1	80	40	40 MHz	RAW	ANY at L1
Ublox LEA-4-T	HW	1	L1	2	4		cGNSS-R (C/A)	GPS
(gri)PAU	HW	(1)2	L1	2.2	(5.745) 16.384		cGNSS-R (C/A)	GPS
SMIGOL	HW	1	L1	2.2	5.745	1	cGNSS-R (C/A)	GPS
PYCARO[a]	HW	2	L1 + L2	20		20	cGNSS-R (C/A), rGNSS-R (P-semicodeless)	GPS
SPIR-UAV[b]	SW	8	L1	80	40	40 MHz	RAW	ANY at L1
GRIP[c]	HW	2	L1 + L5.E1 + E5	52	<=150	1	cGNSS-R	GPS arid Galileo

[a] Spaceborne system developed in the E-GEM EU FP7 607126 project.
[b] Airborne system developed in the E-GEM EU FP7 607126 project.
[c] Ground-based system developed in the E-GEM EU FP7 607126 project.
From Cardellach, E., May 2014. EU FP7 Project E-GEM, WP4 Report *"Assessment of Technologies and Solutions"*.

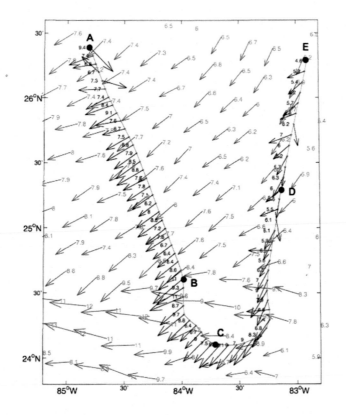

FIGURE 6.60

GPS-derived wind vector estimates at the end of the flight for hurricane Keith on October 1, 2000, overlaid on QuickSCAT wind field measurements.

From Komjathy, A., Armatys, M., Masters, D., Axelrad, P., Zavorotny, V., Katzberg, S., March 2004. Retrieval of ocean surface wind speed and wind direction using reflected GPS signals. J. Atmos. Oceanic Technol. 21, 515–526.

Finally, Germain et al. (2004) performed a full DDM inversion using data obtained with a GNSS-R software receiver. The cost function to be minimized was:

$$J(DMSS_\lambda, \tau_c, f_c) = \sum_\tau \sum_{f_D} \left[|Y(\tau, f_D)|^2 - |Y_{model}(\tau - \tau_c, f_D - f_c, DMSS_\lambda)|^2 \right]^2, \qquad (6.71)$$

where DMSS is the Directional Mean Square Slope.

6.4.5.1.2 Interferometric Complex Field

For a static receiver. It can be demonstrated that the measured waveforms are random processes, that exhibit the following functional relationship (Soulat et al., 2004):

$$\langle Y(t, \tau, f_C) \cdot Y^*(t + \tilde{t}, \tau, f_C) \rangle \propto \exp\left\{ -4k^2 \cdot \sigma_z^2 \cdot \frac{\tilde{t}}{2 \cdot \tau_z^2} \cdot \sin^2\gamma \right\}, \qquad (6.72)$$

therefore, measuring the correlation time of $Y(\tau_F, \tau, f_C)$ provides an estimation of τ_F:

$$\frac{\tau_F}{\lambda} \cdot \pi \cdot \sin\gamma = \frac{\tau_z}{SWH} \cdot \tag{6.73}$$

Numerical simulations with Elfouhaily's spectrum show good results with the following semi-empirical fit:

$$\tau'_F = \frac{\lambda}{\pi} \cdot \left(\frac{a + b \cdot SWH}{SWH}\right), \tag{6.74}$$

where $\tau'_F = \tau_F \cdot \sin\tau'_F$ is the effective correlation time, MWP is the mean wave period, SWH is the Significant Wave Height, and a and b are constants.

6.4.5.1.3 Identification of Waveform/Delay Doppler Map Features

Inversion by parameter estimation requires the computation of scattering models and gradients that are computationally very expensive. An alternative approach consists of the identification of features in the waveform or the DDM that can be related to the geophysical parameter.

Simple observables of the waveform are the position and amplitude of the peak used to perform scatterometric measurements, the position of the maximum derivative point with respect to the peak (ρ_{scatt}), the value of the maximum derivate of the waveform, or the length of the tail (τ_{tail}) (see Fig. 6.61).

When the sea surface roughness increases, the DDM spreads in the delay and Doppler domains. However, the enlargement of the boomerang shape of the DDM (Figs. 6.29 and 6.30-top) is not necessarily symmetric, since it depends on the relative velocities of transmitter and receiver and the wind direction. If the volume under the normalized DDM (peak amplitude equal to one) is computed for the delay-Doppler bins that are above a given threshold, larger than the noise level ($|Y_{th}|^2$):

$$V_{DDM} = \iint_{\overline{|Y(\tau, f_D)|^2} \geq |Y_{th}|^2} \overline{|Y(\tau, f_D)|^2} d\tau \cdot df_D, \tag{6.75}$$

this observable can be linked to the sea state. This is the approach proposed in the PAU instrument (Camps, 2003) to perform the sea state correction of L-band radiometric measurements, to improve the accuracy of the sea surface salinity retrievals. If the whole DDM is not available, the area under the normalized waveform, above a given threshold can be used instead,

$$A_{WF} = \int_{\overline{|Y(\tau, f_{D,max})|^2} \geq |Y_{th}|^2} \overline{|Y(\tau, f_{D,max})|^2} d\tau. \tag{6.76}$$

The approach of Eq. (6.75) was simulated in a space-borne scenario in Marchan-Hernandez et al. (2010), and with real data from the ALBATROSS field experiment (Valencia et al., 2011a), and the approach in Eq. (6.76) was tested in Valencia et al. (2011b) with waveforms acquired using the IEEC GOLD-RTR instrument. Fig. 6.62 shows the flight tracks from inner water body of fresh water to the Baltic sea. Fig. 6.63 shows the changes in the first Stokes parameter associated to changes in the sea surface salinity and temperature, and the sea state (rougher in the open Baltic sea), the changes in the area of the normalized waveforms and their relative scatter plot. Finally, Fig. 6.64 shows the impact of the GNSS-R corrections in the brightness temperatures and in the retrieved sea surface salinities.

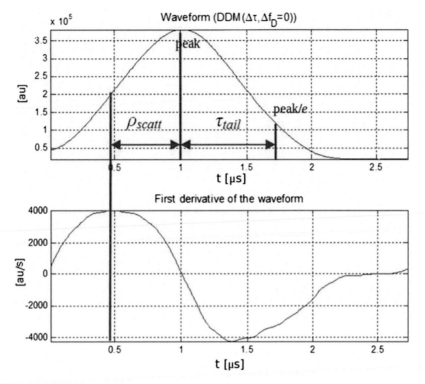

FIGURE 6.61

Some waveform observables.

From Valencia, E., Camps, A., Rodriquez, N., Ramos-Perez, I., Bosch-Lluis, X., Park, H., 2011. Improving the accuracy of sea surface salinity retrieval using GNSS-R data to correct the sea state effect. Radio Sci. 46. http://dx.doi.org/10.1029/2011RS004688.

Other approaches studied in the literature are derived from the DDM image resulting from $I(\tau, f_D) = 10 \cdot \log_{10}(\text{DDM}_{\text{norm}}(\tau, f_D))$, if $I(\tau, f_D) \geq 10 \cdot \log_{10}(e^{-1})$ have been analyzed:

1. The weighted area:

$$S_{\text{DDM}} = \sum_{\tau} \sum_{f_D} I(\tau, f_D) \cdot d\tau \cdot df_D \tag{6.77}$$

2. The 2-norm Euclidean from the center of mass $\left(CM_\tau, \ CM_{f_D} \right)$ to the maximum value of the DDM:

$$CM_{f_D} = I_0^{-1} \sum \sum f_D \cdot I(\tau, f_D) \cdot d\tau \cdot df_D, \tag{6.78a}$$

$$CM_\tau = I_0^{-1} \sum \sum \tau \cdot I(\tau, f_D) \cdot d\tau \cdot df_D, \tag{6.78b}$$

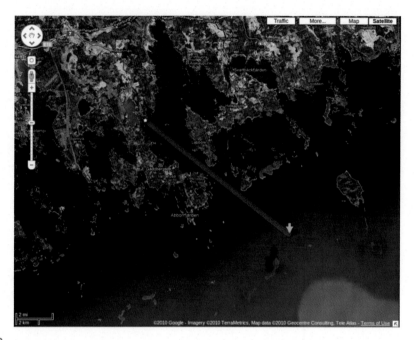

FIGURE 6.62

Track of the flights performed (CoSMOS-2007, Gulf of Finland).

From Valencia, E., Camps, A., Rodriguez-Alvarez, N., Ramos-Perez, I., Bosch-Lluis, X., Park, H., 2011b. Improving the accuracy of sea surface salinity retrieval using GNSS-R data to correct the sea state effect. Radio Sci. 46, RS0C02. http://dx.doi.org/10.1029/2011RS004688.

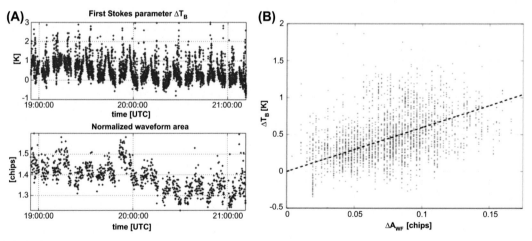

FIGURE 6.63

(A) Time evolution of the changes in the brightness temperature ($\Delta T_B(t)$) and normalized waveform area ($A_{WF}(t)$) and (B) scatter plot of $\Delta T_B(t)$ versus $A_{WF}(t)$.

From Valencia, E., Camps, A., Rodriguez-Alvarez, N., Ramos-Perez, I., Bosch-Lluis, X., Park, H., 2011b. Improving the accuracy of sea surface salinity retrieval using GNSS-R data to correct the sea state effect. Radio Sci. 46, RS0C02. http://dx.doi.org/10.1029/2011RS004688.

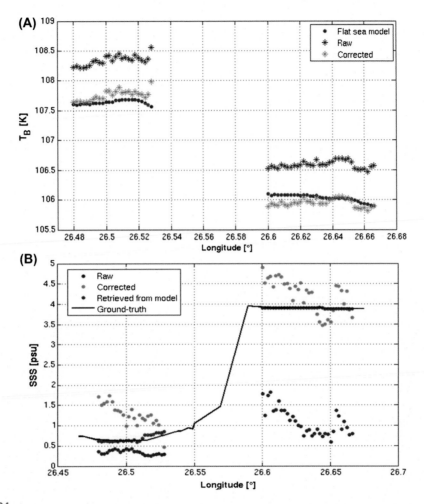

FIGURE 6.64

(A) Sea state correction of the brightness temperatures: $T_{B,flat} = T_B - \Delta T_B$ and (B) sea surface salinity ground-truth and estimated with and without GNSS-R corrections.

From Valencia, E., Camps, A., Rodriguez-Alvarez, N., Ramos-Perez, I., Bosch-Lluis, X., Park, H., 2011b. Improving the accuracy of sea surface salinity retrieval using GNSS-R data to correct the sea state effect. Radio Sci. 46, RS0C02. http://dx.doi.org/10.1029/2011RS004688.

where:

$$I_0 = \sum \sum I(\tau, f_D) \cdot d\tau \cdot df_D. \tag{6.79}$$

3. The distance from the geometric center to the maximum value of the DDM.

4. The 1-norm Euclidean distance ("taxicab" distance) $d = |x_1 - x_2| + |y_1 - y_2|$, from the center of mass to the maximum value of the DDM.

In Rodriguez-Alvarez et al. (2013) these methods were analyzed using airborne data and it was found that, after selecting the most suitable coherent integration time ($T_c = 1$ ms) and the threshold $\left(10 \cdot \log_{10}\left(e^{-1}\right)\right)$, the error of wind retrievals based on each observable was a minimum for the retrieval based on the taxicab distance from the center of the mass to the maximum position, whose mean error (bias) and standard deviation were the lowest: 0.76 and 0.79 m/s, respectively. However, as the authors acknowledge, this result is for a particular experiment carried out in the NOAA Gulfstream-IV jet aircraft on January 24, 2010, and that for a space-borne one the spatial resolution would be unacceptably large.

Using the same data set from NOAA (January 24, 2010) in Valencia et al. (2014) the DDM asymmetry (skewness) was used to infer the wind direction (i.e., receiver flying direction and wind direction). To measure the DDM skewness, the skewness angle $\left(\phi_{1,\,skew}\right)$, was properly defined to maintain sensitivity to wind direction and receiver flying direction, but otherwise to remain insensitive to the parameters that cause the DDM spreading (Fig. 6.65). For a single DDM measurement, there are four possible solutions due to the double ambiguity of the problem, although one of the solutions was always close to the ground-truth wind direction. To solve for one of the two

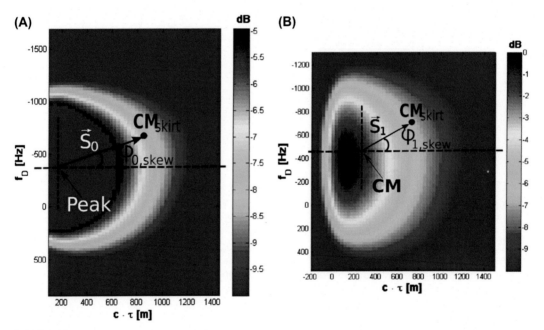

FIGURE 6.65

(A) Definition of the DDM skirt center of mass CMskirt and skewness angle $\phi_{0,\,skew}$ and (B) new definition of the skewness angle $\phi_{1,\,skew}$.

From Valencia, E., Zavorotny, V.U., Akos, D.M., Camps, A., 2014. Using DDM asymmetry metrics for wind direction retrieval from GPS ocean-scattered signals in airborne experiments. IEEE Trans. Geosci. Remote Sens. 52 (7), 3924–3936. http://dx.doi.org/ 10.1109/TGRS.2013.2278151.

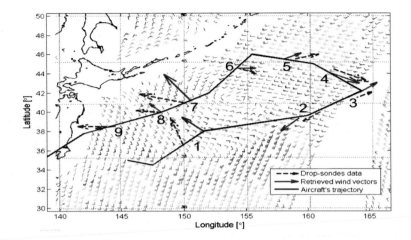

FIGURE 6.66

Wind vectors retrieved from the January 12, 2010, flight data, superimposed to an OceanSAT wind vector map as measured 10 h before the flight. The aircraft track is in a counterclockwise direction.

From Valencia, E., Zavorotny, V.U., Akos, D.M., Camps, A., 2014. Using DDM asymmetry metrics for wind direction retrieval from GPS ocean-scattered signals in airborne experiments. IEEE Trans. Geosci. Remote Sens. 52 (7), 3924–3936. http://dx.doi.org/ 10.1109/TGRS.2013.2278151.

ambiguities, simultaneous DDM measurements for different space vehicles were used. The final rmse is 22° (Fig. 6.66).

A few final comments on how the CYGNSS Level 2 retrievals will be performed are provided further (Ruf, 2016):

- From the DDMs calculated by CYGNSS only the first three delay lags, and the five first Doppler cuts can be successfully used.
- The 25-km resolution requirement limits the region to the leading edge of the waveform.
- Reflectivity measurements are calibrated as in a microwave radiometer:

$$P_g = \frac{(C - C_N) \cdot (P_B + P_r)}{C_B},$$
(6.80)

where C_N is estimated from each DDM, at delays where no signal is present, P_r is estimated using a temperature-dependent LUT, updated using Open Ocean calibration, and P_B and C_B are estimated from Black Body measurements (Ruf, 2016).

From these measurements, a number of DDM features are computed: the minimum variance, the Delay-Doppler map average (DDMA), the Delay-Doppler map variance (DD-MV), the Allan Delay-Doppler map variance (ADD-MV), the leading edge slope (LES), and the trailing edge slope (TES). These observables can be combined appropriately to obtain a generalized observable that maximizes the SNR. The predicted wind speed retrieval error for different methods, as a function of the wind speed (from Rodriguez-Alvarez and Garrison, 2016) are reproduced here in Fig. 6.67.

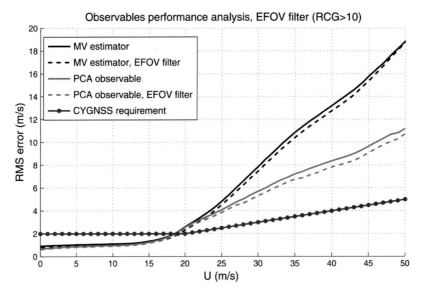

FIGURE 6.67

CYGNSS observables performance analysis: RMSE computed for different observables (*solid lines*) without time averaging and (*dashed lines*) after applying time averaging. Time averaging has been applied to minimum variance (MV) and principal component analysis (PCA). The CYGNSS 25-km resolution requirement is shown in *solid line with dot markers*.

From Fig. 11 of Rodriguez-Alvarez, N., Garrison, J.L., February 2016. Generalized linear observables for ocean wind retrieval from calibrated GNSS-R delay–Doppler maps. IEEE Trans. Geosci. Remote Sens. 54 (2), 1142–1155.

6.4.5.1.4 Delay-Doppler Map Deconvolution

One of the most recent approaches is the imaging of the bistatic RCS by deconvolving the DDM, Eq. (6.53), reproduced here for convenience:

$$
\left\langle |Y(\tau, f_d)|^2 \right\rangle = \frac{P_T \cdot \lambda^2}{(4\pi)^3} \, T_c^2 \iint \frac{G_T(\overrightarrow{r}) G_R(\overrightarrow{r})}{4 \cdot R_1^2(\overrightarrow{r}) \cdot R_2^2(\overrightarrow{r})} \left| \chi \left(\tau - \frac{R_1(\overrightarrow{r}) + R_2(\overrightarrow{r})}{c}, f_D(\overrightarrow{r}) - f_c \right) \right|^2
$$
$$
\times \sigma^0(\overrightarrow{r}) \cdot d^2 r. \tag{6.81}
$$

Eq. (6.81) can be interpreted as the two-dimensional convolution of the WAF $|\chi(\tau, f_D)|^2$ and the function $\Sigma\,(\tau, f_D)$ defined as (Valencia et al., 2013a):

$$
\left\langle |Y(\tau, f_d)|^2 \right\rangle \equiv |\chi(\tau, f_D)|^2 ** \sum (\tau, f_D), \tag{6.82}
$$

$$
\sum (\tau, f_D) = \frac{P_T \cdot \lambda^2}{(4\pi)^3} \, T_c^2 \iint \frac{G_T(\overrightarrow{r}) G_R(\overrightarrow{r})}{4 \cdot R_1^2(\overrightarrow{r}) \cdot R_2^2(\overrightarrow{r})} \, \sigma^0(\overrightarrow{r}) \cdot \delta(\tau) \cdot \delta(f_D) \cdot d^2 r, \tag{6.83}
$$

which can be written as:

$$
\sum (\tau, f_D) = \frac{P_T \cdot \lambda^2}{(4\pi)^3} \, T_c^2 \, \frac{G_T(\overrightarrow{r}) G_R(\overrightarrow{r}) \sigma^0(\overrightarrow{r})}{4 \cdot R_1^2(\overrightarrow{r}) \cdot R_2^2(\overrightarrow{r})} \cdot |J(\tau, f_D)|, \tag{6.84}
$$

where the term $|J(\tau, f_D)|$ is the Jacobian function resulting from the change of variables (x,y) to (τ, f_D) as defined in Marchan-Hernandez et al. (2009).

However, this technique suffers from aliasing since two (x,y) points map into the same (τ, f_D) bin. As in conventional SAR, the only way to eliminate this is by tilting (electronically or mechanically) the antenna away from the specular reflection point so that the antenna pattern attenuates the alias image (Park et al., 2011) (Fig. 6.68). In addition, the antenna can be more sophisticated an introduce nulls in the direction of the alias.

This technique has been tested for oil slick detection (Valencia et al., 2013a) and for hurricane wind mapping. The estimated on ground resolutions compare well with the expected values of range (Δr) and azimuth (Δa) resolutions found in bistatic SAR theory (Moccia and Renga, 2011):

$$\Delta r = \frac{c}{B \cdot \sin(\theta_i)}, \tag{6.85}$$

$$\Delta a = \frac{1}{T_c} \cdot \frac{\lambda \cdot h_R}{V_R \cdot \cos(\theta_i)}. \tag{6.86}$$

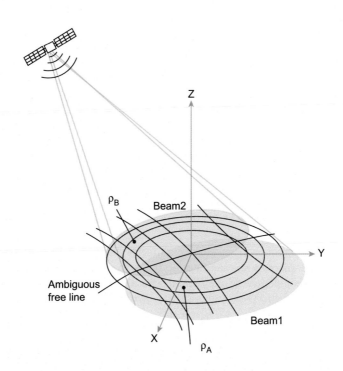

FIGURE 6.68

Spatial filtering using beamforming to avoid ambiguity problem of GNSS-R imaging.

From Park, H., Valencia, E., Rodriguez-Alvarez, N., Bosch-Lluis, X., Ramos-Perez, I., Camps, A., July 2011. New approach to sea surface wind retrieval from GNSS-R measurements. 2011 IEEE International Geoscience and Remote Sensing Symposium, Vancouver, BC, pp. 1469–1472.

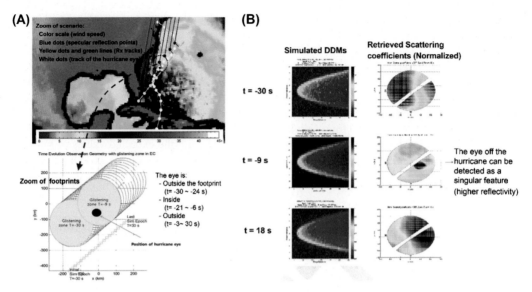

FIGURE 6.69

(A) Scenario of simulation of hurricane Sandy passing near Florida as seen by the ISS. (B) Series of three DDMs and corresponding σ^0 maps (two sides after application of the technique in Fig. 6.79).

From Park, H., Camps, A., Pascual, D., Alonso-Arroyo, A., Martin, F., Carreno-Luengo, H., Onrubia, R., 2014b. Simulation study on tropical cyclone tracking from the ISS using GNSS-R measurements. In: 2014 IEEE Geoscience and Remote Sensing Symposium, Quebec City, QC, pp. 4062–4065.

The scenario simulated in Camps et al. (2013a) and Park et al. (2014b) for mapping hurricane Sandy on October 27, 2012, from the ISS, when it went close to the coast of Florida is shown here in Fig. 6.69A while Fig. 6.69B shows the simulated DDMs and their corresponding deconvolved σ^0 maps. As in the case of oil slicks, the smoother surface in the hurricane eye enhances the forward scattering that appears as a bright area in the DDM and in σ^0.

6.4.5.2 Altimetry

The first proposal to use GNSS-R for mesoscale altimetry dates back to 1993 (Martin-Neira, 1993) with the PARIS (Passive Reflectometry and Interferometry System) concept. Much has been advanced since then, including two competitive Phase A studies for a PARIS In orbit demonstration mission, and the GEROS experiment on board the Columbus module of the International Space Station.

GNSS-R altimetry differs from conventional nadir-looking altimetry in:

- the bistatic (actually multistatic) geometry,
- the frequency bands: L-band, instead of C to Ka bands, for which atmospheric effects (mainly hydrometeors) are less important, but ionospheric effects are more important,
- the available signal bandwidths ~2 or ~20 MHz for the GPS C/A and P(Y) codes, as compared to the ~300 MHz for some radar altimeters, which result in a leading edge of 300 or 30 m, as compared to 1 m,

- the identification of the specular reflection point in the waveform shape, and
- the scattering area, which is neither beam-limited (highly directive antennas), nor pulse-limited (short pulses), but more "roughness limited," as the Sun glint over the ocean.

Altimetry observations can be performed using the group delay or the phase delay. While group delay provides (coarse) absolute range measurements, phase delays provide precise, although not accurate, differential range measurements. Moreover, as illustrated in Fig. 6.30, GNSS-R from Earth surfaces tends to be noncoherent (except for ice and—maybe—for some land smooth surfaces), so only group delay between the transmission and reception is usually feasible.

6.4.5.2.1 Phase Altimetry

Phase altimetry is estimated from the phase of the complex waveform (I and Q components). If the GNSS-R receiver includes a phase-locked loop, then phase variations are dynamically compensated, and the delay can be readily estimated from the pseudo-ranges and a model of the scattering geometry. If the GNSS-R receiver is an open-loop, then the LO phase appears at the phases of the output waveforms (direct and reflected fields, Fig. 6.70A), and needs to be stopped (Fig. 6.70B):

$$\varphi(t) = \arg\{E_r \cdot E_d^*\}, \tag{6.87}$$

before the phase unwrapping (Fig. 6.70C).

In delay altimetry, the observable (ρ_i) is the delay between transmission and reception at a given frequency band (f_i). The delay associated to the geometry (ρ_{geo}), that is the altimetry observable, is given by:

$$\rho_{geo} = |\vec{R}_{spec} - \vec{R}_T| + |\vec{R}_R - \vec{R}_{spec}|, \tag{6.88}$$

where $\vec{R}_T$, $\vec{R}_{spec}$, and $\vec{R}_R$ are the vectors indicating the position of the transmitter, specular reflection point, and receiver, respectively. However, in addition to ρ_{geo}, there are other error contributions such as:

- the ionosphere $(\rho_{iono,i})$, which—at L-band—can introduce error delays of up to ~ 40 m,
- the troposphere (ρ_{tropo}), largely driven by the "dry component" (about 2.3 m), and the "wet component", smaller, but more variable geographically,
- the electromagnetic bias $(\rho_{EM\ bias})$ that is due to the asymmetry of the surface waves, which scatter the electromagnetic radiation differently. It is a small, although not negligible contribution, on the order of 1% of the significant wave height, and
- other errors induced by multipath $(\varepsilon_{multipath})$, instrumental errors (ε_{instr}), or noise errors (ε_n).

$$\rho_i = \rho_{geo} + \rho_{iono,i} + \rho_{tropo} + \rho_{EMbias} + \varepsilon_{multipath} + \varepsilon_{instr} + \varepsilon_n. \tag{6.89}$$

In many ground-based and air-borne experiments, however, a differential measurement (reflected minus direct ranges: $\rho_{i,r} - \rho_{i,d}$) is used instead because some errors cancel out ($\rho_{iono,i}$ and ρ_{tropo}, for low altitude flights, ε_{instr}), and because—assuming a "flat Earth"—the expression becomes extremely simple:

$$\rho_{geo} = |\vec{R}_{spec} - \vec{R}_T| + |\vec{R}_R - \vec{R}_{spec}| - |\vec{R}_R - \vec{R}_T| + \rho_{antoffset}$$
$$\approx 2 \cdot h \cdot \sin(\theta_e) + \rho_{antoffset}, \tag{6.90}$$

where h is the receiver height above the "flat Earth", θ_e is the elevation angle, and $\rho_{ant\ offset}$ is an offset term associated to the fact that the phase centers of the up- and down-looking antennas are not

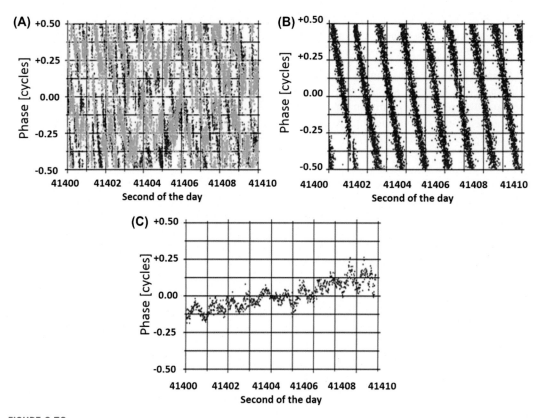

FIGURE 6.70

(A) Direct (green) and reflected (red) phases, (B) phase difference, and (C) counter-rotated phase by model.

From Fabra-Cervellera, F., 2013. GNSS-R as a Source of Opportunity for Remote Sensing of the Cryosphere (Ph.D. dissertation).

UPC, Barcelona, Spain.

physically in the same point. $\rho_{ant\ offset}$ is computed as the projection of the offset vector between the up- and down-looking antennas in the direction of the reflected signal.

The altimetry observables from a bi-static waveform can be obtained either by fitting a model of the waveform, as in Eq. (6.69), from the slope of the geometric term corrected by antenna offsets (e.g., Eq. (6.91) for a "flat Earth"[10]):

$$H = \frac{1}{2} \cdot \frac{\partial \rho_{geo}}{\partial(\sin(\theta_e))}, \tag{6.91}$$

or from the maximum peak of the derivative of the waveform as illustrated in Fig. 6.82, an approach originally proposed by Hajj and Zuffada (2003), and demonstrated later by Rius et al. (2010) (Fig. 6.71).

---

[10]If other models rather than "flat Earth" are used, it is still possible to stop the signal first using an a-priori value of H, and then estimate a correction.

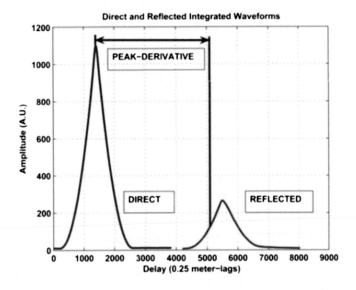

FIGURE 6.71

The altimetry delay in a bistatic altimeter can be obtained from the delay from the direct signal to the maximum peak of the derivative of the reflected waveform.

Adapted from Carreno-Luengo, H., Park, H., Camps, A., Fabra, F., Rius, A., 2013. GNSS-R derived centimetric sea topography: an airborne experiment demonstration. IEEE J. Selec. Top. App. Earth Obs. Remote Sens. 6 (3), 1468–1478. http://dx.doi.org/10. 1109/JSTARS.2013.2257990.

In the following sections the correction of the error contributions in Eq. (6.89) is discussed:

- **Ionospheric Effects** (Klobuchar, 1991)
  In ground-based and airborne instruments both the direct and scattered signals suffer the same ionospheric effects and cancel out. In a typical LEO, however, there is a residual part of the ionosphere above the satellite and the ionospheric delay suffered by the direct and the scattered signals are different, and different from the model predictions that account for the whole ionosphere.
  Ionospheric correction requires at least two pseudo-range observations ρ_1 and ρ_2 at two sufficiently spaced frequencies[11] f_1 and f_2:

$$\rho_1 = -2 \cdot h \cdot \sin\theta_e + \frac{I'}{f_1^2}, \tag{6.92}$$

$$\rho_1 = -2 \cdot h \cdot \sin\theta_e + \frac{I'}{f_2^2}, \tag{6.93}$$

---

[11]f_1 and f_2 can be L1 and L2, or L1 and L5, for example.

The altimetry observable h can then be estimated as:

$$\widehat{h}_{1,2} = -\frac{1}{2} \cdot \frac{\rho_1 + \rho_2}{2 \cdot \sin\theta_e} + \frac{1}{2} \cdot \left(f_1^{-2} - f_2^{-2} \right) \frac{\langle I'_N \rangle}{2 \cdot \sin\theta_e}, \tag{6.94}$$

where $I' \equiv 40.3$ sTEC, and sTEC is slant TEC [e^-/m^2], and I'_N is the average over N samples (moving average of three consecutive ionospheric delay estimates).

- **Tropospheric Effects** (Brunner and Welsh, 1993; Spilker, 1996; Royal Observatory of Belgium, 2012)

  The tropospheric delay induced by the troposphere above a receiver site can be modeled as the sum of two contributions: the hydrostatic zenith delay (HZD) and the wet zenith delay (WZD):

$$\rho_{tropo} = m_{hz} \cdot HZD + m_{wz} \cdot WZD, \tag{6.95}$$

  where m_{hz} and m_{wz} are the mapping functions that project HZD and WZD, from the zenith to slant directions of observation, and depend on θ_e, the geographical coordinates of the receiver, and the day of the year. The "dry component" (HZD) is larger that the "wet component" (WZD), and it can be accurately computed from atmospheric pressure data at the receiver site/level, while WZD is not. The direct and the reflected signals in ground-based GNSS-R instruments are affected in similar ways by tropospheric effects. In airborne instruments, these effects depend on the flight height, and in an LEO instrument only the reflected signal is affected. Since none of the planned space-borne GNSS-R instruments has a multifrequency radiometer (23.8 and 36.5 GHz, as a minimum) to infer the "wet delay," corrections we should rely on proper modeling.

- **EM bias**

  The EM bias is the correction for the measurement bias introduced by varying reflectivity of wave crests and troughs. It is defined as the normalized correlation between the RCS and the sea surface elevation (Elfouhaily et al., 1999):

$$\rho_{EMbias} = \frac{\langle \xi \cdot \sigma^0 \rangle}{\langle \sigma^0 \rangle}. \tag{6.96}$$

  Few studies have been performed on the EM bias at L-band and in a bistatic configuration (Picardi et al., 1998; Ghavidel et al., 2016). The predicted EM bias has been computed numerically using the fundamental equation as a function of the wind speed and incidence angle for wind direction $\phi_{wind} = 45°$ (Fig. 6.72A), and azimuth angle (Fig. 6.72B).

  As we learn from Fig. 6.72, EM bias is an important error source in GNSS-R altimetry that will need to be corrected for, taking into account the observation geometry and the wind speed conditions.

- **Multipath**

  Multipath and direct signal cross-talk can easily arise in ground-based, experiments, and even in airborne ones. Since these signals are more coherent than the scattered ones, the coherent component of the signal can be estimated and subtracted as (Martín et al., 2015):

$$\text{Var}(Y(\tau)) = \frac{1}{N} \sum_{j=1}^{N} |Y(t_{o,j}, \tau)|^2 - \left| \frac{1}{N} \sum_{j=1}^{N} Y(t_{o,j}, \tau) \right|^2. \tag{6.97}$$

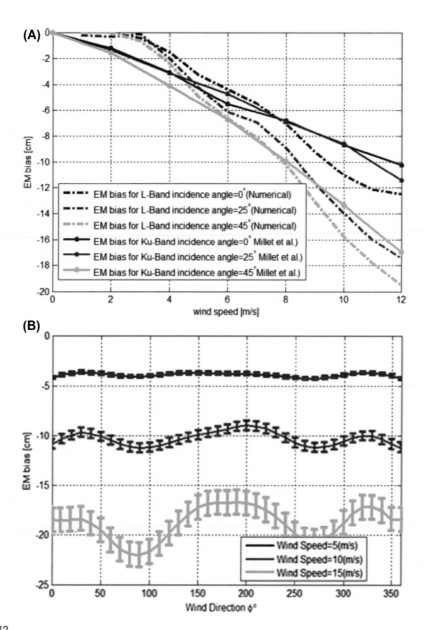

FIGURE 6.72

(A) Electromagnetic bias as a function of wind speed for two θ_e, (B) as a function of the azimuth angle.

From Ghavidel, A., Schiavulli, D., Camps, A., January 2016. Numerical computation of the electromagnetic bias in GNSS-R altimetry. IEEE Trans. Geosci. Remote Sens. 54 (1), 489–498.

If the complex waveforms are fully uncorrelated, the second term in Eq. (6.96) vanishes and the variance of the complex waveforms yields the incoherently averaged power waveform as desired and illustrated in Fig. 6.73.

Finally, as an example, the results of a flight over the Baltic on November 11, 2011, using the GOLD-RTR and the PIR-A instruments were performed simultaneously. The GOLD-RTR had access to the GPS C/A code only, while the PIR-A has access to all. The values of $\sigma_{h,rGNSS-R} = 13$ cm, and $\sigma_{h,GOLD-RTR} = 25$ cm in 20 s (Fig. 6.74).

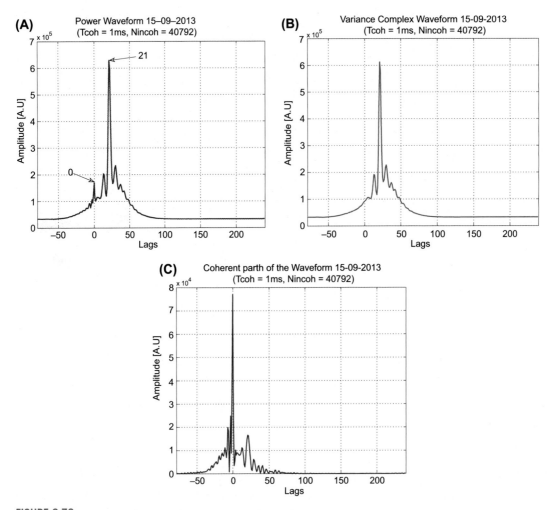

FIGURE 6.73

(A) Power waveform obtained from an incoherent average of 40792 individual power waveforms, (B) variance waveform obtained applying from 40792 individual complex waveforms, (C) coherent power waveform.

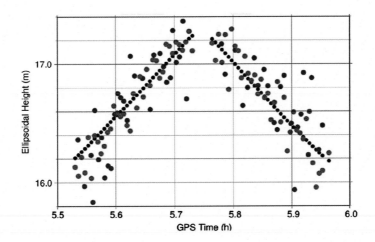

FIGURE 6.74

Evolution of the ellipsoidal height along the flight. *Blue dots* are obtained with the GOLD-RTR cGNSS-R instrument, while *red dots* are obtained with the PIR-A iGNSS-R instrument.

From Cardellach, E., Rius, A., Martin-Neira, M., Fabra, F., Nogué s-Correig, O., Ribó, S., Kainulainen, J., Camps, A., D'Addio, S., 2014. Consolidating the precision of interferometric GNSS-R ocean altimetry using airborne experimental data. Trans. Geosci. Remote Sens. 52. http://dx.doi.org/10.1109/TGRS.2013.2286257.

The interested reader is directed to Camps et al. (2012a,b), D'Addio and Martin-Neira (2013) and Carreno-Luengo et al. (2013) for further analyses on the altimetric performance of GNSS-R instruments using different techniques and types of signals.

6.4.5.3 Soil Moisture
6.4.5.3.1 Techniques Based on the "Interference Pattern"

The earliest evidences that the GPS signal fading depends, among other geophysical parameters, on the soil moisture (actually the dielectric constant: ε_r) are found in Kavak et al. (1998). The experimental set-up to measure the fading over a grass field using a GPS receiver is depicted in Fig 6.75A. By matching the experimental results to the simulated ones (Fig. 6.75B), ground electrical characteristics can be derived (roughness, dielectric constant etc.).

Later, Larson et al. (2008a) showed that the multipath present in the GPS signals collected by the network of geodetic receivers, induced a modulation of the received power that could be correlated to the soil moisture through the dielectric constant. Although the antennas of geodetic GPS receivers are specifically designed and manufactured to mitigate the antenna pattern back lobes, they are not zero, and some residual reflected signal is picked up by the receiver (Fig. 6.76A). Fig. 6.76B (top) shows a sample of L2 SNR data for satellite PRN 9 with a low frequency modulation due to the antenna pattern, and a high frequency modulation due to multipath. SNR data for a satellite with direct signal contribution is removed using a low-order polynomial fit (Fig. 6.76B bottom), showing the high frequency fluctuations associated with the multipath. Finally, the comparison of GPS multipath amplitudes, the volumetric water content for two satellites (PRN 1 open circles, PRN 9 filled circles), and the daily precipitation values as measured by a nearby airport are presented here in Fig. 6.76C.

(A)

(B)

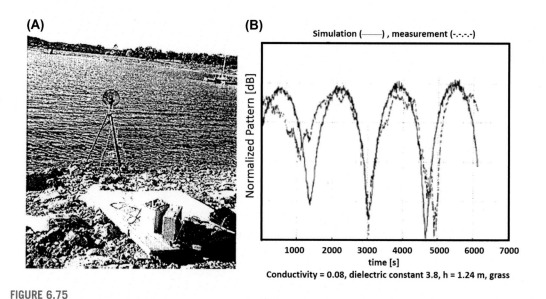

FIGURE 6.75

(A) Experimental set-up, and (B) modeled and measured signal variations received on a dry grass field.

Reproduced by permission of the Institution of Engineering & Technology, Kavak, A., Vogel, W.J., Xu, G. February 5, 1998. Using GPS to measure ground complex permittivity. Electron. Lett. 34 (3), 254–255.

This technique has started to be used today in a quasi-operational manner to study the severity of the 2014 drought in the United States (http://gpsworld.com/gps-network-shows-drought-in-the-u-s-west/; Borsa et al., 2014).

One of the issues associated with picking the signal through the antenna side lobes is the characterization of the antenna pattern itself and its cross-polarization. To solve this, and to increase the amplitude of the reflected signals, and thus the amplitude of the oscillations and the sensitivity, the technique was optimized and named the Interference Pattern Technique or IPT, for short. A dedicated instrument[12] was designed and manufactured to validate this technique. It consisted of a commercial GPS receiver (tuned) connected to a **vertical polarization** antenna with the maximum of its radiation pattern pointing to the horizon and with very good azimuthal revolution symmetry (Rodriguez-Alvarez et al., 2009). Under these conditions, the received power can be approximated as:

$$P_R \propto |E_i + E_r|^2 = F_n(\theta, \varphi) \cdot |E_{0_i}|^2 \cdot \left| 1 + |\Gamma_q(\theta, \varepsilon_r)| \cdot e^{j(\Delta\beta + \beta_{Rq}(\theta, \varepsilon_r))} \right|^2, \tag{6.98}$$

where E_i and E_r are the incident and reflected electric fields, $|E_{0_i}|$ is the amplitude of the electric field, $F_n(\theta, \varphi)$ is the antenna pattern with symmetry of rotation, $|\Gamma_q(\theta, \varepsilon_r)|$ is the reflection coefficient, θ satellite elevation angle, ε_r is the soil dielectric constant, q is the polarization (q = v), $\Delta\beta = 4\pi/\lambda h$ $\sin\theta$ is the different phase shift between the direct and reflected signals, $\beta_{Rq}(\theta, \varepsilon_r)$ is the reflection coefficient phase, h is the antenna phase center height, and λ is the electromagnetic wavelength at the frequency of operation (i.e., 19 cm for GPS L1).

---

[12]Instrument's name SMIGOL stands for Soil Moisture Interference-pattern GNSS Observations at L-band.

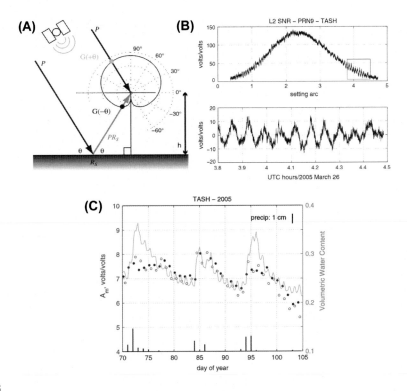

FIGURE 6.76

(A) Geometry of the observation of a geodetic antenna pattern, (B) amplitude of the collected GPS signal without and with antenna pattern removal, (C) retrieved soil moisture values for two different GPS satellites.

Larson, K.M., Small, E.E., Gutmann, E., Bilich, A., Axelrad, P., Braun, J., 2008a. Using GPS multipath to measure soil moisture fluctuations: initial results. GPS Solutions 12, 173–177.

The experimental set-up is shown in Fig. 6.77A, and Fig. 6.77B shows a sample interference pattern. A nice feature of this technique is that the amplitude fluctuations disappear when the reflected signal vanishes, and that happens at the Brewster angle, since $\Gamma_v(\theta_{Brewster}) = 0$. This observable is a very precise one of the dielectric constant, which can be related afterward to the soil moisture. In addition, away from the Brewster angle, the amplitude of the oscillations depends on the magnitude of the reflection coefficient $\Gamma_v(\theta)$ itself, which can be used to create soil moisture maps along the strips on the ground defined by the position of the reflection point.[13]

However, all GNSS-R observables, either based on the amplitude of the reflected signals, or on the interference pattern generated by multipath, are also affected by the presence of vegetation and surface roughness. In Alonso-Arroyo et al. (2014a,b) an improvement of the IPT was proposed that makes use of the differential phase shift between the interference patterns at vertical and horizontal polarizations

---

[13]Actually, despite the coherent behavior shown by the interference patterns (e.g., Fig. 6.77B), the reflection is not completely specular and the "reflected" (scattered) signal, as in the case of the ocean, comes from an area around the specular reflection point determined by the intersection of the first Fresnel zone with the surface.

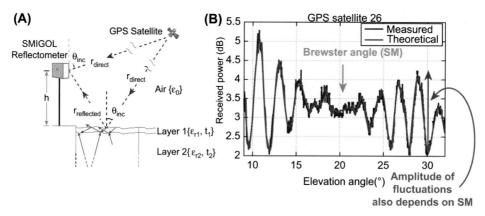

FIGURE 6.77

(A) Experiment set-up of the instrument applying the interference pattern technique, (B) sample interference pattern showing a null corresponding to the position of the Brewster angle.

Adapted from Rodriguez-Alvarez, N., Bosch-Lluis, X., Camps, A., Vall-llossera, M., Valencia, E., Marchan-Hernandez, J.F., Ramos-Perez, I., November 2009. Soil moisture retrieval using GNSS-R techniques: experimental results over a bare soil field. IEEE Trans. Geosci. Remote Sens. 47 (11), 3616–3624.

to better locate the position of the Brewster angle and therefore improve the soil moisture retrievals. Fig. 6.78A shows a sample of these interference patterns and their relative phases. As it can be seen, the presence of even low vegetation (grass) and surface roughness prevents a precise location of the Brewster angle in the vertical polarization interference pattern (red plot). However, for low elevation angles, the phase of the reflection coefficients at vertical and horizontal polarizations is the same ($\sphericalangle\Gamma_v = \sphericalangle\Gamma_h$) and both patterns are in phase (peaks and valleys go together), but as the elevation angle increases, the phase of the reflection coefficient at vertical polarization changes by 180° (peaks and valleys are in opposite phase). The optimum estimation of the Brewster angle occurs when the difference is exactly 90°. The evolution of the estimated soil moisture values using both techniques, and the ground truth as measured by the two closest stations during a field experiment conducted from July 16–31, 2013, at Yanco, New South Wales, Australia is depicted here in Fig. 6.78B.

The IPT (or any multipath technique) has two main limitations:

- The observables require the movement of the navigation satellite over a large fraction of its orbit to produce the "interference pattern", and it may take a few hours. Therefore, it is clear that this technique cannot be used in moving platforms (even ground based).
- To observe the fringes of the pattern, both the direct and the reflected signals must arrive "in phase", which means well within the same chip. In practice, this limits the receiver height to much less than half a chip, i.e., 150 m at GPS L1 using the C/A code. Using higher chipping rates codes only limits even more the maximum receiver height.

However, it has also the advantage of being able to produce maps of the local soil moisture distribution, as illustrated in Fig. 6.79.

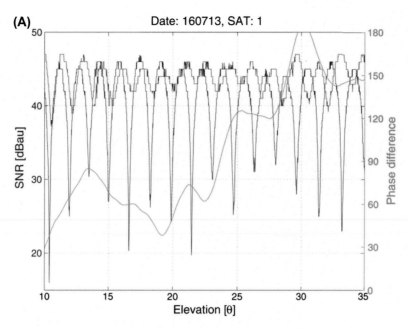

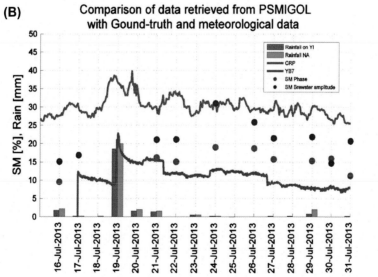

FIGURE 6.78

(A) Sample interference patterns at vertical (red) and horizontal (blue) polarizations, and phase shift among them (green). (B) Retrieved soil moisture using the position of the Brewster determined by the amplitude "notch" (*blue dots*) of by the phase difference (*red dots*), and ground truth from the closest stations.

From Alonso-Arroyo, A., Camps, A., Aguasca, A., Forte, G., Monerris, A., Rudiger, C., Walker, J.P., Park, H., Pascual, D., Onrubia, R., December 2014a. Improving the accuracy of soil moisture retrievals using the phase difference of the dual-polarization GNSS-R interference patterns. IEEE Geosci. Remote Sens. Lett. 11 (12), 2090–2094.

FIGURE 6.79

(Left) SMIGOL instrument with four sectorial antennas mounted on the tip of a 6 m tall wooden mast, (center) soil moisture map in the traces described by the specular reflection points, and (right) space vehicle paths. Note that due to the inclination of the orbital planes, the north side cannot be "covered" by an instrument located in the northern hemisphere.

6.4.5.3.2 Techniques based on "scatterometry"

The first observations showing a correlation between the scattered signal power and soil moisture were performed by Masters in 2004 using NASA's DMR instrument (Figs. 6.42 and 6.43). The ground track of specular reflections on a flight collected during flights over Northern Colorado are in Fig. 6.80A and 6.80B shows the GPS scattered SNR versus volumetric in situ soil measurements for 31 field sites containing soybean and corn crops. As we can easily see, reflectivity correlates fairly well with soil moisture.

In 2009, the griPAU instrument (Fig. 6.51) was mounted in a ~3 m height scaffolding in the Remedhus test site in Zamora (Spain) to measure the complete DDM. Due to low height, instantaneous DDMs actually looked like the ideal WAFs. The most important results were as follows: the confirmation that the scattering over land has a strong coherent component and the information is mostly in the peak of the DDM itself, which correlates well with the soil moisture (Fig. 6.81A), and that the peak amplitude fluctuations were a lot larger than over the ocean, and had a lower temporal scale, which difficult the averaging (Fig. 6.81B).

Fig. 6.82 shows some airborne results obtained with the LARGO instrument (Alonso-Arroyo et al., 2014b) during the GELOz campaigns, Yanco Region, NSW, Australia (September and October 2013). The LARGO footprint, as computed from the size of the first Fresnel zone, ranged from 15 to 60 m, depending on flight height and incidence angle. On the left column, LARGO reflectivity measurements are plotted in a particular color scale to illustrate that the lower reflectivity (red) corresponds to the drier soils, and the higher reflectivity corresponds to the wetter ones (green). The reflectivity over the ocean or inland water bodies is even higher, as expected from the higher dielectric constant. As can be seen, in some cases up to four different GNSS reflections are being tracked simultaneously. In the right

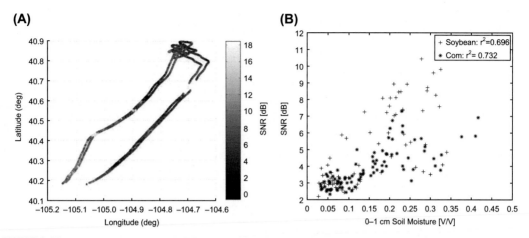

FIGURE 6.80

(A) Sample ground track of specular reflections collected during flights over Northern Colorado, (B) GPS scattered SNR versus volumetric in situ measurements for 31 fields with soybean and corn.

From Masters, D.S., 2004. Surface Remote Sensing Applications of GNSS Bistatic Radar: Soil Moisture and Aircraft Altimetry (Ph.D. thesis). University of Colorado, 189 pp.

column, the PLMR radiometer (http://prosensinginc.gravityswitch.com/crb-product/plmr/) measurements are overlaid on the LARGO ones.[14] PLMR is an L-band dual-polarization push-broom radiometer with six beams of ~15° beamwidth each, and a projected antenna footprint of ~250 m, depending on the flight height. As can be seen, the correlation between the GNSS-R and the radiometric measurements is very strong, confirming that both observables respond to the same geophysical parameters.

Finally, Fig. 6.83 (from Camps et al., 2016) shows some space borne results from the GNSS-R experiment onboard the UK TechDemoSat-1 mission over the north east of Spain to illustrate the reflectivity changes (blue: high reflectivity, red: low reflectivity) over land with different soil moisture conditions as illustrated in the boxes with the in situ data, or due to the proximity of a large water body (Fig. 6.83 top right) or to the Ebro river (green dot on the left of Fig 6.83 left), or due to the sea state from very calm water (Fig. 6.83 bottom right), to rougher (windier) conditions. A sensitivity analysis of all the UK TDS-1 data set over land, with pointing errors smaller.

An analysis by Camps et al. (2016) of all the publicly available data from UK TDS-1 (1/9/2014−5/2/2015), quality filtered for antenna gains larger than 10 dB (near-nadir $\theta_i < 15°$ observations) within the antenna half-power beamwidth and collocated with soil moisture data from SMOS level 3 (http://cp34-bec.cmima.csic.es/) and Normalized Difference Vegetation Index (NDVI) data from MODIS (http://neo.sci.gsfc.nasa.gov/view.php?datasetId=MOD13A2_M_NDVI) shows that for nearly bare soils, there is a high sensitivity to soil moisture ~38 dB/(m³/m³), and a high Pearson correlation parameter R = 0.63, in agreement with (Chew et al., 2016) where a sensitivity of ~7 dB/(0.2 m³/m³) was quoted.

---

[14]Surface roughness or vegetation corrections not applied to any data set.

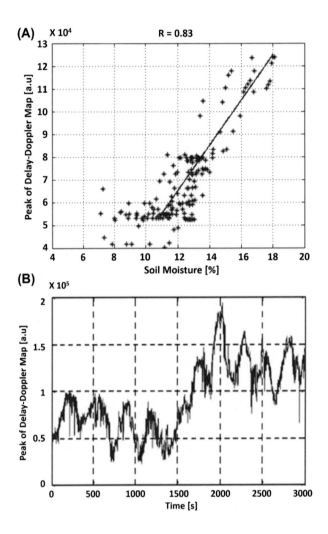

FIGURE 6.81

(A) Correlation between the amplitude peak of the DDM and soil moisture. Below 10% soil moisture sensitivity is lost due to the smaller reflection coefficient. (B) Fluctuations of the DDM peak over land (same as Fig. 6.31B). Note the slower temporal scale of fading due to the slow movement of the navigation satellites.

From Valencia, E., Camps, A., Vall llossera, M., July 2013b. GNSS-R Delay-Doppler Maps over land: preliminary results of the GRAJO field experiment. 2013 IEEE International Geoscience and Remote Sensing Symposium.

6.4.5.4 Vegetation Parameters
6.4.5.4.1 Techniques based on the "interference pattern"

The IPT can also be used over vegetated areas provided that the attenuation and scattering are not strong enough so that the ground reflection and the coherent component, respectively, are still detectable. Over vegetated areas, additional "notches" appear in the interference pattern due to "multiple" reflections in the air-vegetation and vegetation-ground interfaces. An example of the received power as a function of the elevation angle (V-pol antenna) is depicted here in Fig. 6.84A.

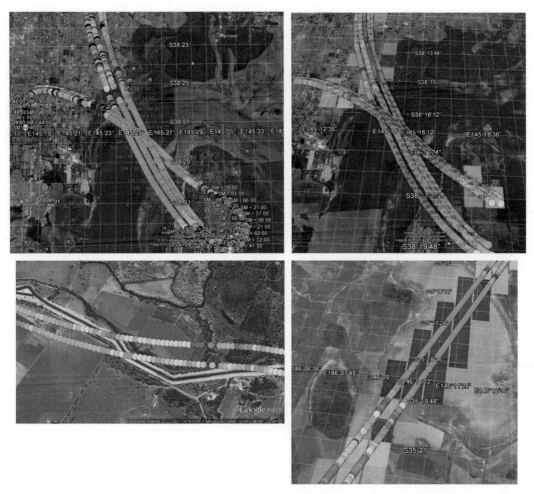

FIGURE 6.82

Sample airborne measurements from GELOz campaigns, Yanco Region, NSW, Australia (September and October 2013): Left column: LARGO geolocated GNSS-R measurements. Right column: LARGO GNSS-R and PLMR L-band radiometry geolocated measurements overlaid. Higher reflectivity and lower brightness temperatures correspond to wetter soils (or water bodies), and lower reflectivity and higher brightness temperatures correspond to drier soils.

From Alonso-Arroyo, A., Camps, A., Monerris, A., Rüdiger, C., Walker, J.P., Forte, G., Pascual, D., Park, H., Onrubia, R., July 2014b. The light airborne reflectometer for GNSS-R observations (LARGO) instrument: initial results from airborne and Rover field campaigns. In: 2014 IEEE Geoscience and Remote Sensing Symposium, pp. 4054–4057.

Despite the low frequency of the fringes, due to the low antenna height, four to five notches can be distinguished, from which the vegetation height can be inferred (Fig. 6.84B). This information can be used to determine the average vegetation height in the area surrounding the instrument, as illustrated in Fig. 6.85 over a corn field (Rodriguez-Alvarez et al., 2010).

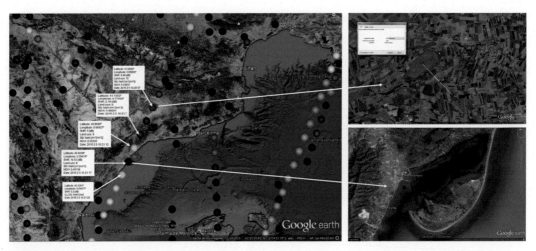

FIGURE 6.83

UK TechDemoSat-1 GNSS-R data over the north east of Spain showing reflectivity changes associated to different soil moisture values, presence of water bodies, and sea surface roughness.

From Camps, A., Park, H., Pablos, M., Foti, G., Gommenginger, C., Liu, P.-W., Judge, J., November 2016. Sensitivity of the GNSS-R spaceborne observation to soil moisture and vegetation. IEEE J. Selected Top. Appl. Earth Observations Remote Sens. (in press).

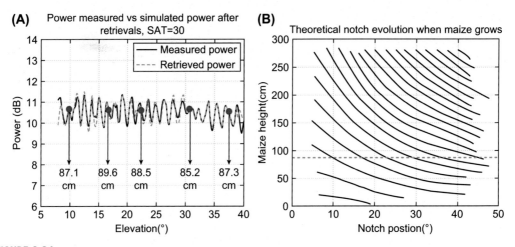

FIGURE 6.84

(A) Measured (*black solid*) and simulated (*gray dashed*) interference patterns versus elevation angle (V-pol antenna) for a ~85 cm height vegetation layer. (B) Simulated contours of the angular notch positions versus vegetation height.

From Rodriguez-Alvarez, N., Bosch-Lluis, X., Acevo, R., Aguasca, A., Camps, A., Vall-llossera, M., Ramos-Perez, I., Valencia, E., July 2010. Study of maize plants effects in the retrieval of soil moisture using the interference pattern GNSS-R technique. 2010 IEEE International Geoscience and Remote Sensing Symposium, pp. 3813–3816.

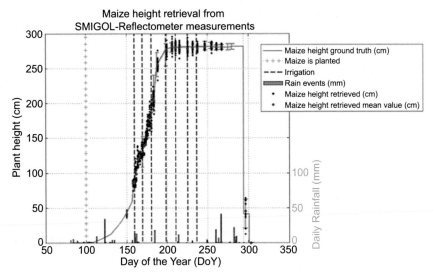

FIGURE 6.85

Evolution of vegetation height estimates, ground truth, and rain events (Rodriguez-Alvarez et al., 2010).

Similar to GNSS-R techniques, GNSS-T, where the −T stands for transmission, can be used to infer the vegetation water content (VWC). The technique is actually very simple and consists of measuring the attenuation of the GNSS signal, as it passes through the vegetation canopy. The attenuation can be related to the VWC using the well-know formula for L-band microwave radiometry:

$$L = e^{-\tau \cdot \sec(\theta_i)}, \tag{6.99a}$$

$$\tau = b \cdot VWC, \tag{6.99b}$$

where b is a vegetation-dependent factor. In the experiment shown in Rodriguez-Alvarez et al. (2012) over walnut trees, the value of b was found to depend between 0.82 and 0.88, in the elevation angle range (55°−70°), since the leaves in this range of elevation angles are not homogeneous, and the highest correlation was computed for 60° (Rodriguez-Alvarez et al., 2012). Note that the attenuation (Eq. 6.99a) in a GNSS-R scenario has to be squared since the signal attenuates in the down-welling path and in the up-welling path after the reflection takes place.

6.4.5.4.2 Techniques based on "scatterometry"

Egido (2013) performed a comprehensive study including theoretical and experimental measurements of the polarimetric scatterometry measuring Γ'_{RL} and Γ'_{RR} over different fields and forest. It was found that Γ'_{RL} shows remarkable variations in the presence of vegetation, with a variation of the specular point apparent reflectivity of more than 3 dB, when transitioning from a bare surface to a mid-biomass poplar plot with similar soil moisture and roughness conditions. Γ'_{RR} also decreased, but not so obviously as for Γ'_{RL}, which experiences a steady decrease with forest above ground biomass up to a

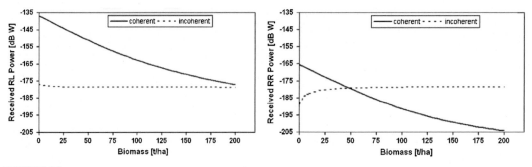

FIGURE 6.86

Received power at circular polarizations versus biomass. $\theta_i = 20°$, $\sigma_z = 1.5$ cm, SMC = 15%, receiving antenna gain G = 25 dB (left) RL polarization (right) RR polarization.

From Ferrazzoli, P., Guerriero, L., Perdicca, N., Rahmoune, R., 2011. Forest biomass monitoring with GNSS-R: theoretical simulations. 47, 1823–1832.

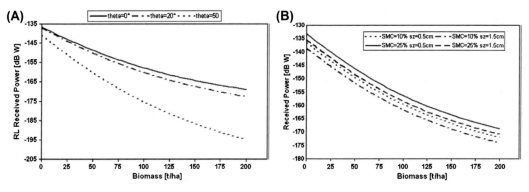

FIGURE 6.87

(A) Received power at RL polarization versus biomass of Hawaiian forest for incidence angles at $\theta_i = 0$, 20, and 50°, $\sigma_z = 1.5$ cm, SMC = 15%, receiver antenna gain G = 25 dB. (B) Received power at $\theta_i = 20°$, RL polarization, versus biomass of Hawaiian forest. Curves correspond to different soil moisture and roughness conditions.

From Ferrazzoli, P., Guerriero, L., Perdicca, N., Rahmoune, R., 2011. Forest biomass monitoring with GNSS-R: theoretical simulations. 47, 1823–1832.

biomass content of 300 t/ha. The calculated sensitivity yields 15 dB/(100 t/ha), with a correlation coefficient of R = 0.91.

These results are in relative good agreement with the simulation study by Ferrazzoli et al. (2011). Actually simulation results predict a bit higher decrease versus biomass: ~20 dB/100 ha (Fig. 6.86), except at large incidence angles (Fig. 6.87).

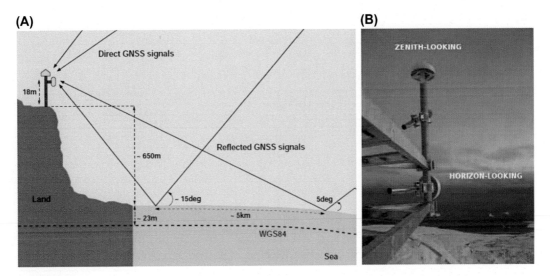

FIGURE 6.88

Antenna system mounted on the telecommunications tower, overlooking Disko Bay: (A) geometry of the experiment, and (B) pictures taken in October 2008, during the installation of the equipment for the sea ice campaign.

From Fabra-Cervellera, F., 2013. GNSS-R as a Source of Opportunity for Remote Sensing of the Cryosphere (Ph.D. dissertation). UPC, Barcelona, Spain.

6.4.5.5 Cryospheric Applications

The sea ice surface is smooth enough so as to reflect almost specularly GNSS signals even for large elevation angles, as illustrated in Fig. 6.30. The carrier phase of the reflected signals can also be tracked, and therefore centimetric altimetry solutions can be obtained. A number of applications have been developed, including the estimation of ice thickness.

6.4.5.5.1 Sea Ice Thickness

Sea ice thickness can be inferred from the altimetry solution, once the oscillations due to tides are removed. In Fig. 6.88A and B (from Fabra et al., 2012) we show the main results of an experiment performed by IEEC/CSIC. The GOLD-RTR instrument was installed on a tower above a 700-m cliff at Disko Bay, Greenland, from October 2008 to May 2009 (Arctic winter) to cover sea ice formation, its evolution, and melting. Due to experimental site constraints, the observation angles were restricted to elevation angles smaller than 20°, and since the "downward-looking" dual-polarization antenna was pointing to the horizon, the direct signal was also picked up (Fig. 6.88).

The altimetry solution is consistent with a 5 km resolution Arctic Ocean Tidal Inverse Model (AOTIM-5) (http://www.esr.org/polar_tide_models/Model_AOTIM5.html), at both LHCP and RHCP polarizations (Fig. 6.89A). Once the tides have been corrected for, the resulting mean sea ice level inversely relates to temperature (Fig. 6.89B), that is the cooler, the higher the sea ice level (thicker ice), and the warmer, the lower.

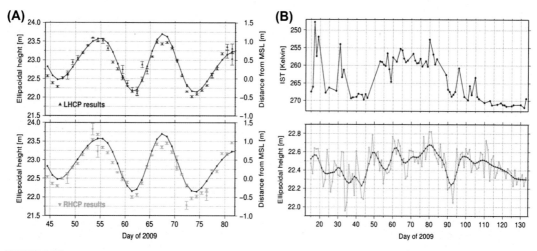

FIGURE 6.89

(A) The altimetry solution follows closely the ocean tides at both polarizations, and (B) after correction of tides, the altimetry solution inversely follows the iced sea temperature.

From Fabra-Cervella, F., Cardellach, E., Rius, A., Ribo, S., Oliveras, S., Nogues-Correig, O., Belmonte-Rivas, M., Semmling, M., D'Addio, S., 2012. Phase altimetry with dual polarization GNSS-R over sea ice. IEEE Trans. Geosci. Remote Sens. 50, 2112–2121. http://dx.doi.org/10.1109/TGRS.2011.2172797.

6.4.5.5.2 Sea Ice Permittivity

Sea ice permittivity and surface roughness determine the reflection coefficient, and therefore the peak power of the reflected waveform. Early research in this area (Komjathy et al., 2000) showed that the power of the reflected signal strongly depended on the type of ice, and therefore it can be classified accordingly (Fig. 6.90).

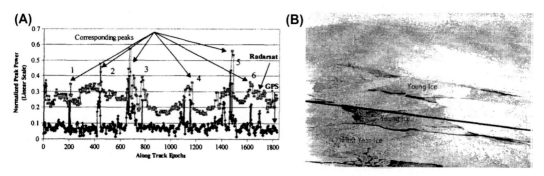

FIGURE 6.90

(A) Normalized signal power for PRN 30. (B) Photo taken from the aircraft in the vicinity of Label 5 (A) showing black and darker gray areas of newly formed leads, new and young ice.

From Komjathy, A., Maslanik, J., Zavorotny, V.U., Axelrad, P., Katzberg, S.J., 2000. Sea ice remote sensing using surface reflected GPS signals. IEEE 2000 International Geoscience and Remote Sensing Symposium, vol. 7, pp. 2855–2857.

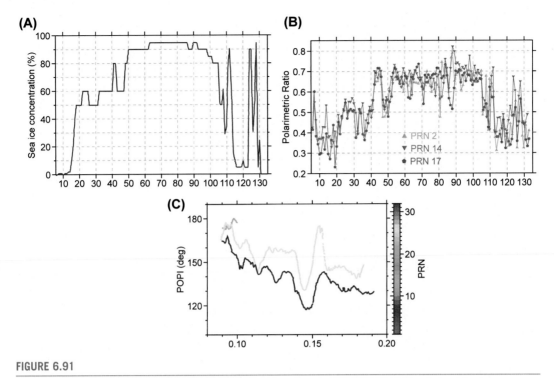

FIGURE 6.91

(A) Sea ice concentration, (B) RHCP/LHCP polarimetric ratio, and (C) RHCP/LHCP polarimetric phase
(x-axis: Day of the Year, 2009).

From Fabra-Cervellera, F., 2013. GNSS-R as a Source of Opportunity for Remote Sensing of the Cryosphere (Ph.D. dissertation).
UPC, Barcelona, Spain.

An accurate measurement of the GPS reflectivity provides an indication of the sea ice development stage (i.e., thickness), initially through the dampening of the strong bottom reflection coming off seawater and then via changes in the bulk salinity of the ice layer. Roughness is also sensitive to deformation processes that affect the ice pack (Rivas et al., 2010).

The RHCP/LHCP polarimetric ratio (amplitude) and the polarimetric phase shifts also contain information relative to the dielectric constant that can be used to infer geophysical information (Fabra-Cervella et al., 2012; Fabra-Cervella, 2013). The sea ice cover evolution (Fig. 6.91A) is linked to the example of the evolution of the polarimetric ratio for three different satellites, and how it changes from ~0.3 to ~0.7 when the sea ice concentration passes from 0% to almost 100% (Fig. 6.91B). Similarly, Fig. 6.91c shows the phase difference changes from ~180° for 0% to ~120° when the sea ice concentration passes from 0% to 100%.

6.4.5.5.3 Snow Depth

The same physical phenomena that produces the fringe pattern with multiple valleys in Figs. 6.75, 6.78, 6.79, or 6.85, can be used to infer snow thickness. The issue is, in practice, more complex, since usually the topography in mountain regions is not flat, and snow height may vary significantly from one place to a nearby one. The IPT has also been used for ocean applications to measure the sea state, by locating the angle at which the fringes (oscillations) disappear.

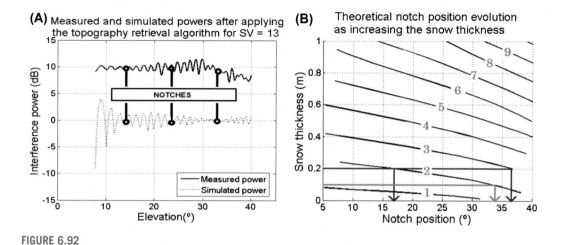

FIGURE 6.92

(A) Simulated fringe pattern for a 40 cm thick snow, (B) measured and simulated fringe patterns showing three notches.

From Rodriguez-Alvarez, N., Aguasca, A., Valencia, E., Bosch-Luis, X., Camps, A., Ramos-Perez, I., Vall-llossera, M., 2012. Snow thickness monitoring using GNSS measurements. http://dx.doi.org/10.1109/LGRS.2012.2190379.

The IPT can be applied to estimate the snow thickness by locating the position of the notches in the fringe pattern.[15] A real measurement and the associated simulations (Fig. 6.92A) show that the multinotch pattern depends on the snow thickness, and therefore thickness can be estimated from these (Figs. 6.93 and 6.94).

As for soil moisture studies, snow depth at geodetic sites has also been studied and reported from very low altitude ~1−2 m receivers, and tracking of the "direct" link. Fig. 6.95 shows the retrieved snow height profile at a geodetic site with a standard GNSS receiver (Larson and Nievinski, 2013) (see Fig. 6.95).

6.4.5.5.4 Dry Snow Substructure

In Antarctica and Greenland deep snow layers exist, up to ~3000−4000 m thick (Fig. 6.96A). In addition, there are underground lakes of liquid water (Fig. 6.96B) that have been found to affect the L-band brightness temperatures as measured from space. Since dry snow is nearly transparent to L-band, GNSS signals can penetrate down to a few hundreds of meters deep.

Since the interfaces are smooth, and the dielectric constant gradient is relatively small coherent reflections occur, and interference patterns appeared generated by reflections at multiple substructure layers giving information on the location of the most reflecting layers. A multilayer model consisting of layers of homogeneous dielectric constant (Fig. 6.97) can be used to explain the observed behavior of the reflected waveforms: a kind of superposition of a few waveforms, whose peaks are at increasing delays, and rotating at a given frequency, which is given by the residual Doppler frequency Fig. 6.98.

---

[15]The IPT has also been used for ocean applications to measure the sea state, by locating the angle at which the fringes (oscillations) disappear.

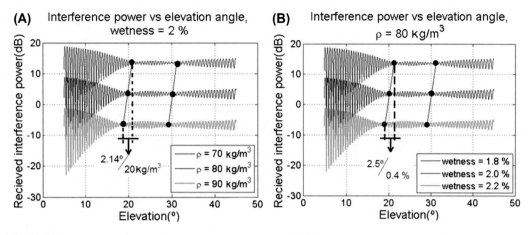

FIGURE 6.93

Simulated (A) IPT as a function of the elevation height, and evolution of notch position as a function of the snow pack density, (B) same as (A) but for changing snow wetness.

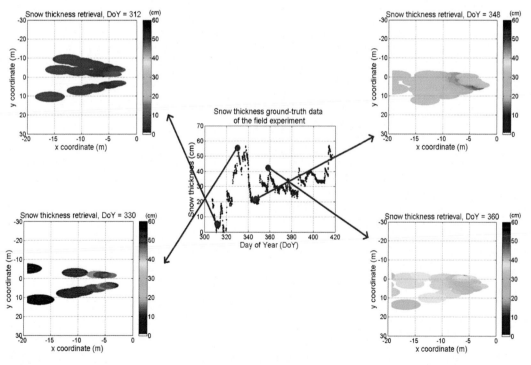

FIGURE 6.94

Retrieved snow thickness using an SMIGOL instrument at Vall d'Aran (Pyrenees mountains) during winter 2001, and snow thickness ground-truth derived from an ultrasonic range meter mounted in the meteo tower.

From Rodriguez-Alvarez, N., Aguasca, A., Valencia, E., Bosch-Luis, X., Camps, A., Ramos-Perez, I., Park, H., Vall-Ilossera, M., 2012. Snow thickness monitoring using GNSS measurements. IEEE Geosci. Remote Sens. Lett. 9 (6), 1109–1113. http://dx.doi. org/10.1109/LGRS.2012.2190379.

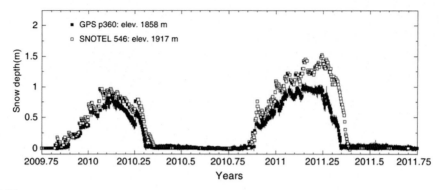

FIGURE 6.95

Snow height evolutions during 2 years showing the seasonal cycle.

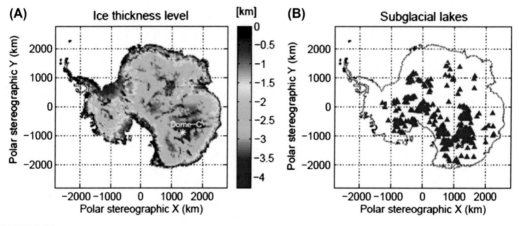

FIGURE 6.96

(A) Ice thickness and (B) subglacial lakes on Antarctica.

Modified from Pablos, M., Piles, M., González-Gambau, V., Vall-llossera, M., Camps, A., Martínez, J., September, 2014. SMOS and aquarius radiometers: inter-comparison over selected targets. IEEE J. Sel. Top. Appl. Earth Obs. Remote Sens. 7 (9), 3833–3844. http://dx.doi.org/10.1109/JSTARS.2014.2321455.

6.5 FUTURE TRENDS IN GNSS-R

Today's evolution of GNSS-Reflectometry includes the following:

- An extensive use of digital technology and software-defined radio to cope with varying signal frequencies, bandwidths, and modulations.
- Start of the use of RFI detection and mitigation techniques, due to an increased presence of jammers and other PPDs.

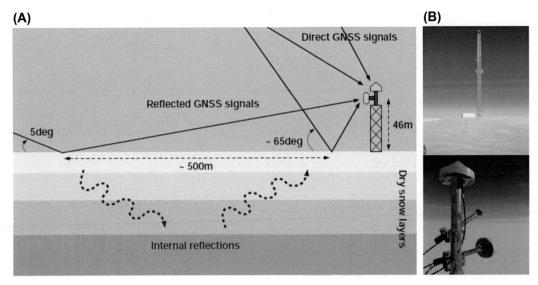

FIGURE 6.97

(A) GNSS-R experimental set-up at Dome C, Antarctica, (B) tower is ~45 m height and has two antennas: a zenith looking one at the tip and a horizon looking at the tip of the mast.

From Cardellach, E., Fabra, F., Rius, A., Pettinato, S., D'Addio, S., 2012. Characterization of dry-snow sub-structure using GNSS reflected signals. Remote Sens. Environ. 124. http://dx.doi.org/10.1016/j.rse.2012.05.012.

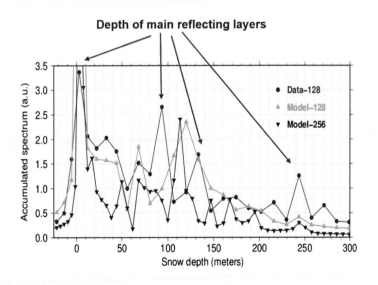

FIGURE 6.98

Sample waveforms including three reflections at deeper layers.

From Cardellach, E., Fabra, F., Rius, A., Pettinato, S., D'Addio, S., 2012. Characterization of dry-snow sub-structure using GNSS reflected signals. Remote Sens. Environ. 124. http://dx.doi.org/10.1016/j.rse.2012.05.012.

- Development of new lower power GNSS-R receivers (~ 1 W) that can be boarded in constellations of nanosatellites (Carreño-Luego et al., 2016; Olive et al., 2016). At the time of writing this chapter, the NASA CYGNSS constellation has been recently launched and scientific results will start appearing in the next months.

6.6 STUDY QUESTIONS

1. GNSS transmit circularly polarized signals, why? If the transmitted signal is RHCP, which is the dominant polarization of the reflected one? What if the elevation angle is very small (smaller than the Brewster angle)?

2. A satellite navigation system transmits pseudo-random sequences of 1023 square pulses in 1 ms. Which is the duration of the pulses, the shape of the ACF, and its length in meters? What if the signal be transmitted at a rate 10 times higher, i.e., the period is 0.1 ms?

3. Why precise satellite navigation systems require two frequencies? Which type of corrections are made and how?

4. At 1575.42 MHz the power density received by a receiver at the Earth's surface is -134.6 dBW/m^2, and the antenna gain is 4 dB. Which is the received power in dBm?

5. If the noise power spectral density is $N_0 = -204$ dBW/Hz, and the signal bandwidth is 2 MHz, which is the noise power collected by the receiver? From questions 3 and 4, which is the carrier-to-noise ratio in dB?

6. If the size of the isodelay ellipse is given by $a = \sqrt{2h\tau'}\big/\sin^{3/2}(\gamma)$, and $b = \sqrt{2h\tau'}\big/\sin^{1/2}(\gamma)$, where $\tau' = \tau - \tau_s$ (delay relative to specular), compute the size of the isodelay ellipse corresponding to $\tau' = 1$ μs as seen from a satellite at h $= 500$ km height, and with an elevation angle of $\gamma = 90°$, and $\gamma = 45°$.

7. Taking into account the power density at the surface (question 3), compute how much power is intercepted by the first isodelay ellipse (question 5), and how much power will be collected by a 15 dB gain antenna onboard the satellite in the same conditions as in question 5?

8. If the reflection occurs over a perfectly flat surface, the scattering is purely coherent. Taking into account that the ERIP (equivalent radiated isotropic power, equivalent to the product of the transmitted power times the antenna gain) is 24.5 dBW, and the transmitting satellite height is 20,200 km, compute the received power by the same satellite as in questions 5 and 6?

9. Repeat questions 5 and 6 if the transmitted pulses are 10 times faster (duration 10 times smaller) so that $\tau' = 0.1$ μs?

10. The scattering over the ocean, land, and ice is dominated by coherent or incoherent scattering? Explain why.

ORBITAL MECHANICS, IMAGE NAVIGATION, AND CARTOGRAPHIC PROJECTIONS

7.1 HISTORY

The earliest study of orbital parameters traces back to Johannes Kepler (December 27, 1571 to November 15, 1630) a German mathematician, astronomer, and astrologer. His first major astronomical work, *Mysterium Cosmographicum*, was first published in defense of the Copernican system of planetary motions. While teaching in Graz, Kepler developed a system of complex geometrical shapes which seemed to explain the orbital motions of the six known planets, which were Mercury, Venus, Earth, Mars, Jupiter, and Saturn. As suggested in the title, Kepler thought he had deduced God's geometrical plan for the universe. His enthusiasm for the Copernican system came from his theological beliefs connecting the physical and spiritual. After the publication of the *Mysterium*, Kepler began a very ambitious program to extend his work. He planned four additional books: one on the stationary aspects of the universe; one on the planets and their motions; one on the physical nature of planets and their geographical features; and one on the effects of the heavens on the Earth including meteorology and astrology.

By 1599 Kepler felt that his theoretical work was limited by the inaccuracy of the available data at a time when increasing religious tension threatened his continued employment in Graz. In December Tycho Brahe invited Kepler to visit him in Prague where Tycho had just built a new observatory. Kepler went to Prague in February 1600 and began looking at Brahe's data. While Tycho was generally very tight with his data, Kepler's theoretical understanding impressed Brahe and he allowed Kepler greater access to the data collected. Kepler planned to test his theory of the *Mysterium* using data collected on Mars.

An argument with Brahe caused a brief interruption in Kepler's work, but in June 1600 he returned to Graz to retrieve his family. Political and religious difficulties kept Kepler from returning immediately to Prague, but on August 2, 1600, after refusing to convert to Catholicism, Kepler and his family were banished from Graz and moved to Prague. Through most of 1601 he was supported by Tycho, who assigned Kepler to analyzing planetary observations. Two days after the sudden death of Tycho Brahe on October 24, 1601, Kepler was appointed as his successor as imperial mathematician with the responsibility of finishing the work that Tycho had begun. The next 11 years would be the most productive of Kepler's life.

Kepler's research in this period culminated in the *Astronomia nova* (new astronomy) including the first two laws of planetary motion. This line of study had started with the study of Mars' orbit under the direction of Tycho Brahe. With Kepler's religious view of the cosmos, the Sun (a symbol of God the Father) was the source of all motive force in the solar system causing faster or slower movements as planets move closer or farther away from the Sun.

He worked on the orbit of Mars using an egg-shaped, ovoid orbit. After 40 failed attempts, Kepler settled on an ellipse, which had previously seemed too simple a solution for astronomers to have overlooked. Finding that an elliptical orbit fit the data for Mars, he then concluded that all other planets move in ellipses with the Sun as the focus.

Politics drove Kepler back to Linz, Austria, where his wife died from Hungarian spotted fever, which she had contracted in Prague. All three of his children became ill with smallpox and one of them, Friedrich, of age 6 died. In Linz, Kepler taught at the district school and provided astronomical and astrological services to the local government.

Since the publication of *Astronomia nova* Kepler had intended to write an astronomy textbook. In 1615, he completed the first of three volumes of *Epitome astronomiae Copernicanae* (*Epitome of Copernican Astronomy*); the first volume (books I—III) was printed in 1617, the second (book IV) in 1620, and the third (books V—VII) in 1621. The title referred to heliocentrism, but the texts also involved Kepler's own ellipse-based orbital system. The *Epitome* became Kepler's most influential work. It contained all three of Kepler's laws of planetary motion and attempted to explain the movements of celestial bodies through physical means.

A spin-off from the *Rudolphine Tables* (which he had started work on in Prague) and related to a latter work known as the *ephemerides*, Kepler also published astrological calendars, which were very popular and helped to offset the costs of producing his other more scientific books. In these calendars, produced between 1617 and 1624, Kepler forecast planetary positions and weather as well as political events. The latter turned out to be cannily accurate thanks to his keen insight into present-day politics. His calendars became very controversial and the last one in 1624 was publicly burned in Graz.

Kepler was convinced that "the geometrical things have provided the Creator with the model for decorating the whole world." In publishing his next work *Harmonices Mundi*, Kepler attempts to explain the astronomical and astrological aspects of the known world in terms of music (harmonies). Unfortunately, this led to a dispute with Robert Fludd who had recently published his theory of harmonies. Among many other harmonies, Kepler explained what later became to be known as Kepler's third law of planetary motion. In this law Kepler states that the square of the periodic times are to each other as the cubes of the mean distances. He gave no indication as to how he came up with this idea, but it was later found to be consistent with gravitational theory by Isaac Newton and his contemporaries such as Robert Hooke and Edmund Halley.

In 1623 Kepler at last completed the *Rudolphine Tables* that he had started in Prague with Tycho Brahe. Politics and the "30-years war" kept them from publication until 1627 when Kepler had to have them printed at his own expense. In 1625 agents of the Catholic Counter-Reformation placed most of Kepler's library under seal, and in 1626 the city of Linz was besieged. Kepler moved to Ulm and ultimately to Regensburg. Soon after arriving in Regensburg, Kepler became ill and died there on November 15, 1630, where he was buried.

7.2 KEPLER'S LAWS OF PLANETARY MOTION

To preface our discussion of Kepler's laws of planetary motion, it needs to be observed that these laws can be derived from Newton's law of gravitation and the laws of motion when it is assumed that the

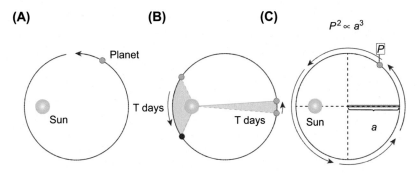

FIGURE 7.1

Graphical representation of the three Keplerian laws.

orbiting body is subject only to the gravitational force of the central attractor. The basic tool used in all orbital mechanics studies is differential calculus. Standard assumptions in astrodynamics include noninterference from outside bodies, negligible mass for one of the bodies, and negligible other forces (such as solar wind, atmospheric drag, etc.). More complicated calculations can be made numerically without these assumptions when one considers the perturbations from the classical system and studies the orbital anomalies.

7.2.1 KEPLER'S FIRST LAW

The orbit of every planet is an ellipse with the Sun at one of the two foci of the ellipse (Fig. 7.1A). The amount of "stretching" from a perfect circle is known as the eccentricity of the ellipse.[1] A circle is an ellipse with zero eccentricity. The eccentricities of the planets vary from 0.007 for Venus to 0.2 for Mercury. After the discovery of Kepler's laws, many celestial bodies have been identified with highly eccentric orbits. Heavenly bodies such as comets and asteroids can have parabolic or even hyperbolic orbits under Newtonian theory.

7.2.2 KEPLER'S SECOND LAW

A line joining a planet and the Sun sweeps out equal areas during equal intervals of time. The planet moves faster near the Sun, so the same area is swept out in a given time as at larger distances, where the planet moves more slowly. Isaac Newton had proved Kepler's second law, as described in his *Principia Mathematica*. If an instantaneous force is considered on the planet during its orbit, the vector defined from the Sun to the planet sweeps the same area in a fixed time interval (Fig. 7.1B). When the interval tends to zero, the force can be considered continuous.

---

[1]The formal definition of the eccentricity is $e = \sqrt{1 - (b/a)^2}$, where a and b are the semimajor and semiminor axes of the ellipse.

7.2.3 KEPLER'S THIRD LAW

As stated in the history section earlier, Kepler's third law is: the square of the orbital period of a planet is directly proportional to the cube of the semimajor axis of its orbit. This third law published by Kepler in 1619 captures the relationship between the distance of the planets from the Sun and their orbital periods. Using symbols, it can be written as:

$$P^2 \propto a^3, \tag{7.1}$$

where P is the orbital period of the planet in question and a is the semimajor axis of the orbital ellipse.[2] If we suppose that the planet A is four times as far from the Sun as planet B, then planet A must traverse 4 times the distance of planet B each orbit. Moreover, it turns out that planet A travels at half the speed of planet B, to maintain equilibrium with the reduced gravitational-centripetal force due to being four times farther from the Sun. Thus, in total it takes $4 \times 2 = 8$ times as long for planet A to travel an orbit in agreement with the law ($8^2 = 4^3$). The third law can also be used to estimate the distance from an exoplanet to its central star.

7.2.4 THE TWO-BODY PROBLEM

The classical problem of a satellite orbiting a planet is called the two-body problem and amounts to determining the motion of two point particles that interact with each other (Fortescue et al., 2003). The two-body problem can be reformulated as two independent one-body problems; one of which is trivial and the other that involves solving for the motion of one particle in the presence of an external force potential. Since many one-body problems can be solved exactly the corresponding two-body problem can also be solved.

Let x_1 and x_2 be the positions of the two bodies and m_1 and m_2 be their masses. We wish to determine the trajectories $x_1(t)$ and $x_2(t)$ for all times t given the initial positions $x_1(t = 0)$ and $x_2(t = 0)$ and the initial velocities $v_1(t = 0)$ and $v_2(t = 0)$. Using Newton's second law they can be written as:

$$F_{12}(x_1, x_2) = m_1 \cdot \frac{d^2 x_1}{dt^2}, \tag{7.2}$$

$$F_{21}(x_1, x_2) = m_2 \cdot \frac{d^2 x_2}{dt^2}, \tag{7.3}$$

where F_{12} is the force on mass 1 due to its interaction with mass 2 and vice versa. Adding and subtracting these two equations decouples them into two one-body problems, which can be solved independently. Adding Eqs. (7.2) and (7.3) leads to an equation describing the motion of the center of mass (called the barycenter, Fig. 7.2).

Dividing both force equations by the respective masses and subtracting the second equation from the first one and rearranging them, leads to:

$$\frac{d^2 r}{dt^2} = \frac{d^2 x_1}{dt^2} - \frac{d^2 x_2}{dt^2} = \frac{F_{12}}{m_1} - \frac{F_{21}}{m_2} = \left(\frac{1}{m_1} + \frac{1}{m_2}\right) \cdot F_{12}, \tag{7.4}$$

---

[2]In the case of a satellite orbiting the Earth, Eq. (7.1) is $P = T_E \cdot (a/R_E)^{3/2}$, where $T_E = 84.4$ minutes, and $R_E = 6378.1363$ km is the reference radius of the Earth.

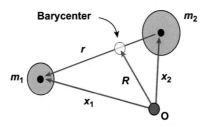

FIGURE 7.2

Coordinates for the two-body problem with the barycenter.

where we have used Newton's third law that $F_{12} = -F_{21}$ and where r is the displacement vector from mass 2 to mass 1 (Fig. 7.2). The force between the two objects, which originates in the two objects, should only be a function of their separation r and not of their absolute positions. This allows us to write Eq. (7.4) as:

$$\mu \cdot \frac{d^2 r}{dt^2} = F_{12}(x_1, x_2) = F(r),$$ (7.5)

where μ is the reduced mass given by:

$$\mu = \frac{1}{\dfrac{1}{m_1} + \dfrac{1}{m_2}} = \frac{m_1 \cdot m_2}{m_1 + m_2},$$ (7.6)

Actually, Eq. (7.5) is a set of 3 second order differential equations, that once solved require the determination of six independent integration constants. These constants are usually expressed in the form of the so-called Keplerian elements that completely determine the orbit[3]: the longitude of the ascending node on the equatorial plane (Ω), the inclination of the orbital plane with respect to the equatorial plane (i), measured at the ascending node where the orbit passes from South to North through the equatorial plane, the argument of the perigee (ω) at the ascending node (γ), or the angle from the ascending node to the closest point at which the satellite comes to the primary object around which it is orbiting, the semimajor axis of the ellipse (a), the orbit eccentricity (e), and the time of passage at the perigee (t_p), which is the reference initial time. The physical meaning of these parameters is represented in Fig. 7.3.

The general solution of the two-body problem is:

$$u(\theta) = \frac{1}{r(\theta)} = \frac{\mu}{h^2}(1 + e \cdot \cos(\theta - \theta_0)),$$ (7.7)

for any nonnegative e. The h represents the total angular momentum divided by the reduced mass. For circular orbits this becomes

$$r \cdot v^2 = r^3 \cdot \omega^2 = \frac{4 \cdot \pi^2 \cdot r^3}{P^2} = \mu.$$ (7.8)

---

[3]Real orbits depart from the ideal Keplerian orbits due to a number of effects: asymmetry in the Earth's gravity field, atmospheric drag, solar radiation pressure, Moon and Sun gravity fields, tides, Earth's magnetic field... For satellites at orbital heights above ~200 km, from all these perturbations the Earth's oblateness is by far the largest one. Real orbits can be accurately approximated by segments of ideal orbits.

FIGURE 7.3

Graphical explanation of the six Keplerian elements of an ideal orbit.

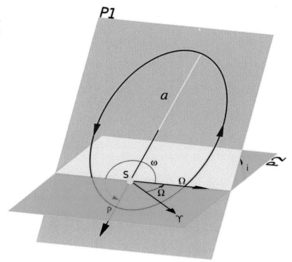

Most of the satellite orbits in Earth observation are nearly circular so this equation applies to them. There are other highly elliptical orbits (HEOs) such as the Molniya orbits, and orbits to L_1 and L_2,[4] but have a more reduced range of applications (see Fig. 7.6).

7.2.5 LOW EARTH ORBITS

A Low Earth orbit (LEO) orbit is generally defined as one with an altitude of 2000 km or less. Since the orbital decay due to atmospheric drag is substantial below about 200 km, most LEO orbits are between 160 and 2000 km. Most polar-orbiting weather satellites and some Earth mapping satellites generally operate in the 500–900 km range. Many of these weather and Earth mapping satellites are in what is known as "Sun-synchronous orbits." A Sun-synchronous orbit (Fig. 7.4) describes the orbit of a satellite that provides consistent illumination of the Earth-scan view. The satellite passes over the equator and each line of latitude at about the same time each day. For example, a satellite's Sun-synchronous orbit might cross the equator 12 to 15 times a day (depending on the orbital height) each time at about 3:00 p.m. local time. The orbital plane of a Sun-synchronous orbit must also "precess" (rotate) approximately one degree each day, eastward, to keep pace with the Earth's revolution around the Sun. Actually the regression of the orbital plane must satisfy:

$$\Delta\Omega = 2\pi \frac{P}{T_{es}} \left[\frac{\text{rad}}{\text{orbit}} \right] = 2\pi \left[\frac{\text{rad}}{\text{year}} \right], \tag{7.9}$$

where P is the orbital period, and $T_{es} = 3{,}155{,}815 \times 10^7$ s is the Earth–Sun orbital period.

---

[4]The L1 point lies on the line defined by the two masses (e.g. the Earth and the Sun), and between them, and corresponds to the point where the gravitational attraction of both masses cancels out, and the L2 point lies on the line defined by the two masses, beyond the smaller one, and corresponds to the point where the gravitational forces of the two masses compensate the centrifugal force on a body at this point.

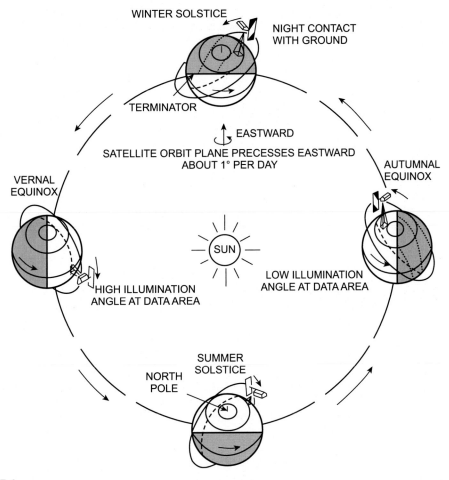

FIGURE 7.4

Seasonal changes in a Sun-synchronous orbit.

Other polar orbits require the synchronization with the Earth to overpass periodically the same geographical location. These orbits require that:

$$n|\Delta\Omega| = m \cdot 2\pi, \tag{7.10}$$

where n is the number of orbits, m is the number of Earth revolutions (days), and

$$|\Delta\Omega| = |\Delta\Omega_1 + \Delta\Omega_2| = \left| -2\pi\frac{P}{T_e} - 3\pi\left(\frac{R_E}{a}\right)^2 \cdot \frac{1}{(1-e^2)^2} \cdot J_2 \cdot cos(i) \right|. \tag{7.11}$$

In Eq. (7.11), the first term ($\Delta\Omega_1$) is the Earth's rotation in one orbit, which is the dominant term, and the second one ($\Delta\Omega_2$) is the so-called regression of the ascending node, which is a perturbing effect due to the Earth's oblateness, that creates an asymmetry of the Earth's gravitational field. Other parameters in Eq. (7.11) are $T_e \approx 86164{,}09055$ s ≈ 23 h 54 min, which is the Earth's rotation period with respect to the stars, $J_2 = 1082{,}6 \times 10^{-6}$, and i is the inclination of the orbit.

Finally, some missions require the synchronization with the Sun and with the Earth simultaneously, which requires Eqs. (7.9) and (7.10) to be satisfied at the same time:

$$n \cdot \left| -2\pi \frac{P}{T_e} + 2\pi \frac{P}{T_{es}} \right| = m \cdot 2\pi, \tag{7.12}$$

or:

$$n \cdot P \left(\frac{1}{T_e} - \frac{1}{T_{es}} \right) = m. \tag{7.13}$$

These orbits are usually referred to by the two integers $n{:}m$. For example, a 14:1 orbit will repeat exactly the first ground-track in the 15th orbit, at the expense of a separation between consecutive ground-tracks of

$$d = \frac{2\pi}{n} \cdot R_E = 2862.43 \text{ km}, \tag{7.14}$$

which is excessive for Earth observation satellites taking into account the field of view (FOV) of most of the sensors studied in previous chapters. Therefore, other combinations of n and m are selected to achieve a much narrower separation between adjacent ground-tracks, at the expense of a much longer revisit time. Following with the example aforementioned, LANDSAT 1 and 2 satellites used a 251:18 orbit, which means that they performed 251 orbits in 18 days (very close to the 14:1 orbit), with a separation between tracks about 160 km. Once these conditions are defined, the rest of the orbit parameters are defined, in this case the orbital period was 103 min, the inclination of the orbit was 99°, and an orbital height at the apogee (point in the orbit with the largest distance to the center of the Earth) of 920 km.

Usually these weather satellites "fly" in pairs with one satellite having a morning crossing which takes advantage of the fact that clouds are at a minimum before the diurnal solar heating gets going. The other satellite in this constellation has a mid-afternoon crossing designed to sample at the solar max when the land surface vegetation would be the most responsive. This is done at the expense of increased cloud cover in the afternoon pass.

The primary advantage of the LEO orbit is that the closeness of the Earth allows better resolution of surface features and gives stronger reflected and emitted signals for the satellite sensors. These polar LEO orbits have the added advantage that this resolution is maintained over the entire Earth's surface, unlike geostationary satellites whose resolution decreases dramatically toward the poles. Due to sensor swath overlap the polar-orbiting satellites generally see an improvement in satellite coverage as the sensor swaths overlap at higher latitudes.

There are LEO satellites that are not in Sun-synchronous polar orbits, and frequently these satellites are in much lower inclination orbits designed to optimally cover the tropical to mid-latitude portion of the Earth. Such a satellite was the Tropical Rainfall Mapping Mission (TRMM), which was designed to study the precipitation in the tropical region. Data from TRMM found many other applications like the passive microwave mapping of sea surface temperature (SST) from the 10 GHz channel. This SST was, however, restricted to the 40°N to 40°S orbit of the TRMM satellite. Thus, while the microwave SST would have been a great benefit to mapping SST in the polar regions, it was not possible with the TRMM radiometer.

It should be noted that receiving antennas designed to collect data from these satellites must be able to track the satellite as it moved from horizon to horizon. Early in the lives of these satellites that was accomplished by a very costly antenna system with a multihorn receiver, which could sense the satellites motion and thus track its signal. Later it was realized that much less expensive antenna systems could be used that used the orbital ephemeris data to predict the location of the satellite and could command the antenna movement with this information. This led to a great revolution in polar-orbiter antenna systems.

The next big change in weather satellite data was the realization that it was not necessary for everyone to operate their own antennas as it would be a lot more cost-effective if a central data repository was created and the satellite data were staged and delivered over the web. Such a system is the Comprehensive Large Array-*data* Stewardship System (CLASS). This system was set up to handle all of the US weather satellite data including the advanced very high-resolution radiometer (AVHRR) (HRPT, GAC, and LAC) and GOES data. Recently, CLASS has been extended to include AVHRR and other weather data from satellites operating around the world. As long as the data are requested and delivered online there is no charge for the imagery.

Most communications satellites are deployed in geostationary orbit (GEO), which provides poor coverage to higher latitudes. To overcome this limitation, the Iridium constellation of satellites was created. Initially created by Motorola Corp., this system is now operated by the US Department of Defense. The principle is that by connecting with a number of satellites in LEO can provide continuous coverage a series of polar-orbiting satellites. The Iridium constellation uses 66 LEO satellites, which are distributed in a staggered manner over six orbits (11 satellites per orbit) 30° apart. There are also some additional spare satellites that can be brought online in case of a satellite failure. Satellites are in LEO at an altitude of 781 km with an inclination of 86.4° and an orbital period of 100 min approximately. Satellites communicate with neighboring satellites via four K_a-band intersatellite links: two neighbors fore and aft in the same orbital plane, and two satellites in neighboring planes to either side. This polar design gives equal communications potential at all latitudes unlike the geostationary communications satellites. Since the communications areas of each satellite overlap this constellation these satellites provide continuous communications at every point on the globe.

Iridium NEXT is the second generation version of this system, and the first ten Iridium NEXT satellites were successfully launched on January 14, 2017, using a SpaceX Falcon 9 rocket. It will continue the 66 satellite constellation, with six on-orbit and nine on-ground spare satellites. This improved system will have new features such as data transmission which was not emphasized in the original design intended primarily for voice communications. These satellites will carry some additional payloads such as cameras and other sensors. This new constellation will provide L-band data speeds of up to 1.5 Mbps and high-speed K_a-Band service of up to 8 Mbps. The existing constellation of satellites is expected to remain operational until Iridium NEXT is fully operational with many of the satellites to remain in service until the 2020's. This new system is being built by Thales Alenia Space in a $2.1 billion deal. In June 2010, Iridium signed the largest commercial rocket launch deal ever at that time, a US$492 million contract with SpaceX to launch 10 Iridium NEXT satellites on 7 Falcon 9 rockets from 2017 onward from the Vandenberg AFB Space Launch Complex.

Iridium Communications Inc. (formerly known as GHL Acquisition Corp.) was incorporated under the laws of the state of Delaware on November 2, 2007; completed its initial public offering on February 1, 2008. On September 24, 2009, following shareholder approval of the acquisition by GHL Acquisition Corp. of Iridium Holdings LLC, GHL Acquisition Corp. began trading on NASDAQ under the IRDM trading symbol. At the closing of the acquisition on September 29, 2009, GHL Acquisition Corp. was renamed Iridium Communications Inc.

7.2.6 GEOSTATIONARY ORBITS

Since the goal of GEOs is to have the satellite constantly pointing at the same location on the Earth there is only one altitude where this is possible and that is at 35,786 km over the equator (inclination equal to zero degrees). At this altitude the satellite orbits the Earth at such a speed as to be constantly viewing the same spot on the ground. Here the satellite is in a circular orbit and rotates at the same rate as the rotation of the Earth, and therefore appears to remain stationary in the sky even while orbiting the Earth at a very high altitude. This orbit is ideal for both communications satellites and for weather satellites. In this orbit antennas do not have to track the satellite's motion, but can remain fixed oriented toward the satellite of interest. The idea of a geosynchronous satellite for communication purposes was first published in 1928 (Potocnik, 1928). It caught on much later when Clarke (1945) wrote a paper on how it would work for wireless communications.

Most commercial communications satellites, broadcast satellites, and SBAS satellites operate in GEOs. A geostationary transfer orbit is used to move a satellite from LEO into a GEO. Russian television satellites have used highly elliptical Molniya and Tundra orbits[5] due to the high latitudes of the receiving audience. The first satellite placed into a GEO was the Syncom-3, launched by a Delta-D rocket in 1964.

There is also a world constellation of geostationary weather satellites. In the United States these satellites were first spin-stabilized requiring a new mechanism to scan the Earth. In the polar-orbiters the forward movement of the spacecraft is used to increment the line-scan and thus produce an image. This problem was solved by Dr. Vern Suomi of the University of Wisconsin who invented the spin-scan camera, which later became the visible infrared spin scan radiometer (VISSR). Since atmospheric soundings are a critical part of meteorological forecasting a VISSR atmospheric sounder (VAS) was created. It suffered from the fact that since it is difficult to accurately sense vertical profiles of atmospheric temperature and moisture the VAS profiles were far less accurate than those from the TIROS Operational Vertical Sounder (TOVS). Interestingly enough when data from TOVS first became available for weather forecasting in late 1982 it was found that the atmospheric models performed worse when the temperature and moisture profiles from TOVS were assimilated into the weather forecast models. In spite of their much greater spatial coverage these data did not improve the model forecast. It was later discovered that the source of this problem was the fact that the temperature and water vapor moisture vertical profiles were very noisy and were introducing more noise in the solution than signal. This problem was overcome when it was realized that the numerical models could produce the radiances that were collected by TOVS and therefore the radiances themselves were directly assimilated into the model. This dramatically improved the weather forecast model results, and it became a regular practice to assimilate the TOVS radiances into the forecast models.

7.2.6.1 US Geostationary Operational Environmental Satellites

The Geostationary Operational Environmental Satellite (GOES) program grew out of the experimental Stationary Meteorological Satellite program where NASA tested out systems such as the spin-scan camera. Once it was established that this system worked, the GOES program was turned over to the National Oceanic and Atmospheric Administration (NOAA) for continuing operation and data delivery to users. The provision of timely weather information including the advance warning of

---

[5]Molniya and Tundra orbits have an inclination of 63.4° to have zero precession of the orbital plane. Molniya orbits have an orbital period of ~12 h, while Tundra orbits have an orbital period of ~24 h. Both spend most of the orbital period over a chosen area of the Earth, a phenomenon known as apogee dwell.

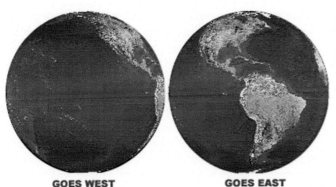

FIGURE 7.5

GOES East and GOES West satellite coverage.

GOES WEST   GOES EAST

developing and impending storms is the primary function of GOES. GOES 1—3 simply carried a VISSR imager and it was not until GOES-4 that VAS profiler was included. This second generation GOES satellite also had the Earth-pointing antennas in a nonrotating module at the bottom of the platform so that they could point continuously at the antennas on the ground.

Also in this new generation of GOES satellite a Weather Facsimile (WEFEX) capability where the GOES imagery was analyzed on the group on the ground and the analyzed product was then broadcast back through the GOES communication capabilities to ground and ship stations that were not able to receive the original image. The GOES satellites also carried a space environment monitor (SEM), which investigated solar particle emissions and helped study the effect of solar activity on Earth's telecommunications systems. The SEM detected solar protons, alpha particles, solar electrons, solar X-rays, and magnetic fields.

The United States operates two GOES satellites (Fig. 7.5) one located at 75°W and one at 135°W longitude. Why two satellites when only one can cover the entire United States? Well GOES West is clearly designed to sense all of the bad weather that hits the United States and comes out of the West. This includes storms that form in the Gulf of Alaska and enter the northwestern United States. GOES East is designed to sense hurricanes and other severe weather conditions on the eastern portion of the United States.

7.2.7 HIGHLY ELLIPTICAL ORBITS

A HEO is one with a low-altitude (about 1000 km) at perigee and a very high-altitude (greater than 35,786 km) at apogee. These very elongated orbits have the advantage of long dwell times at a point in the sky during the approach to and descent from apogee. Spacecraft visibility near the apogee can exceed 12 h with a much faster movement at the perigee. These bodies moving through the long apogee can appear almost stationary to the ground when the orbit is in the right inclination and the angular velocity of the orbit in the equatorial plane closely matches the rotation of the Earth surface beneath it. Thus, these orbits are also useful for some communications satellites. Sirius satellite radio uses HEO orbits to keep two satellites positioned above North America while another satellite quickly sweeps through the southern part of its 24 h orbit. The longitude above which the satellite dwell at apogee remains fairly constant as the Earth rotates. A special HEO orbit is the "Molniya orbit" named after a series of Soviet communication satellites, which used them.

The L2 orbit is an elliptical orbit about the semistable second Lagrange point. It is one of the five solutions by the mathematician Joseph-Louis Lagrange in the 18th century to the three-body problem. Lagrange was searching for a stable configuration in which three bodies could orbit each other yet stay in the same position relative to each other. He found five such solutions, and they are called the five Lagrange points in honor of their discoverer. In three of the solutions found by Lagrange, the bodies are in line (L1, L2, and L3); in the other two, the bodies are at the points of equilateral triangles (L4 and L5). The five Lagrangian points for the Sun—Earth system are shown in the diagram further. An object placed at any one of these five points will stay in place relative to the other two. In the case of the James Webb Space Telescope, the three bodies involved are the Sun, the Earth, and the Webb telescope. Normally, an object circling the Sun further out than the Earth would take more than one year to complete its orbit. However, the balance of gravitational pull at the L2 point means that the Webb telescope will keep up with the Earth as it goes around the Sun. The gravitational forces of the Sun and the Earth can nearly hold a spacecraft at this point, so that it takes relatively little rocket thrust to keep the spacecraft in orbit around L2.

The five Lagrange points are illustrated here in Fig. 7.6, which is enlarged to clearly show the relationships between the various elements. The Lagrange Points are positions where the gravitational pull of two large masses precisely equals the centripetal force required for a small object to move with them. This general "Three-Body-Problem" was addressed by Lagrange in his prize-winning paper (Essai sur le Probleme des Trois Corps, 1772). The L1 point (Fig. 7.6) of the Earth—Sun system provides an uninterrupted view of the Sun and is currently home to the Solar and Heliospheric Observatory Satellite (SOHO). It is also the home of LISA Pathfinder (LPF), launched on December 3, 2015, which arrived at L_1 on January 22, 2016. The L2 point of this Earth—Sun system was home to the Wilkinson Microwave Anisotropy Probe (WMAP), current home of ESA's Planck mission, and the future home of the James Webb Space Telescope. L2 is ideal for astronomy observations because the spacecraft is close enough to readily communicate with the Earth, can keep Sun, Earth, and Moon behind the spacecraft for solar power and, with appropriate shielding, it provides a clear view of deep space for space telescopes. The L1 and L2 points are unstable with periods of approximately 23 days, which requires satellites orbiting these positions to undergo regular course and attitude corrections.

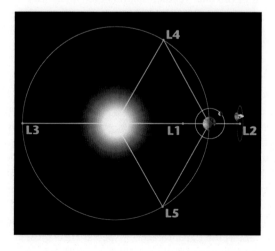

FIGURE 7.6

Lagrange Points of the Earth—Sun system (not drawn to scale).

The L3 point remains hidden behind the Sun at all times and thus is not a useful position for spacecraft sensing. The idea of a hidden Planet-X at the L3 point has been a popular science fiction topic for many years. The L4 and L5 points are locations of stable orbits so long as the mass ratio between the two large masses exceeds 24.96. This restriction is satisfied for both the Earth−Sun and Earth−Moon systems, and for many other pairs of bodies in the solar system.

7.3 MAP PROJECTIONS, IMAGE NAVIGATION, AND GEORECTIFICATION

When working with satellite images one must realize that satellites view the Earth differently depending on their orbits. As described in the previous chapter the primary orbits for Earth-oriented satellites are LEO and GEO. LEO satellites travel approximately from pole to pole as the Earth rotates beneath it. Thus, the satellite image sensor sees a curved Earth that is rotating as the satellite flies overhead. This requires geometric corrections for both these effects. These geometric corrections can be carried out using an accurate knowledge of the satellite ephemeris data, which provide the time and location where the satellite views the Earth (Emery et al., 2010; Rosbourgh et al., 1994).

Additional corrections that must be made are the compensation for spacecraft attitude that controls where the sensor is pointing to (Emery et al., 2010). Most modern satellites now perform this correction using auxiliary information from the onboard "star trackers" that calculate the actual spacecraft attitude using views of known stars. Added to the geometric corrections for curvature and rotation this "image navigation" as it is commonly called can geolocate an image pixel down to its native spatial resolution.

Prior to the wide use of star trackers, satellite orientation sensors were much more limited and were based on Sun and Earth horizon sensors. The inaccuracies inherent in these sensors led to very poorly navigated or geolocated satellite images. In response to these inaccuracies corrections were made to the images themselves. Very often these corrections were made to force significant landmarks in each image to correspond to the corresponding "ground control points" on the desired map. Emery et al. (2010) developed an automated routine to perform these attitude corrections for AVHRR 1 km imagery, creating accurately corrected base images in the thermal infrared using a maximum cross-correlation (MCC) technique to automatically calculate the attitude corrections that would produce an accurately mapped image.

One significant advantage to this method is that the attitude parameters calculated over a coastal region can be propagated out over the ocean where no landmarks are available to correct the image. This makes it possible to correctly navigate the large archive of AVHRR data that is readily available via NOAA's CLASS archival system. Here one can order AVHRR online and pick it up by FTP for free. This has led to a real change in the way weather satellite imagery can be used.

7.3.1 MATHEMATICAL MODELING OF THE EARTH'S SURFACE

Before choosing a map projection for the satellite imagery, it is required first to define how to model the Earth's surface. Human understanding on the Earth's shape has varied during history. Pythagoras (550 BC) was the first one admitting Earth's sphericity, Aristotle (384 BC) saw the Earth's shadow over the Moon (eclipses), and Eratosthenes (250 BC) was able to calculate the circumference of the Earth by measuring the Sun's angle elevation at noon in Alexandria, and knowing that at local noon on the summer solstice in Syene (modern Aswan, Egypt) the Sun was directly overhead. The method of measurement consisted of making a scale drawing of that triangle which included a right angle between a vertical rod and its shadow, as illustrated in Fig. 7.7.

FIGURE 7.7

Illustration showing a portion of the globe. The sunbeams shown as two rays hitting the ground. Angle of the sunbeams and the gnomons (*vertical sticks*) is shown at Alexandria, which allowed Eratosthenes' estimates of radius and circumference of Earth (http://www.ngs.noaa.gov/PUBS_LIB/Geodesy4Layman/80003002.GIF).

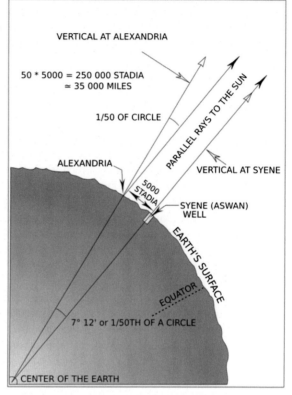

ERATOSTHENES METHOD FOR DETERMINING
THE SIZE OF THE EARTH

During the middle ages, all this knowledge was lost, and it was not until the Renaissance when these ideas were revived. Later, Sir Isaac Newton concluded that the Earth is a fluid rotating itself, so it should have the shape of an ellipsoid of revolution.

The simplest model the Earth's surface can then be is either a sphere or an ellipsoid. Spherical models are generally useful for large area maps such as world atlases and maps overlain on globes. At these scales the map error is not noticeable as it is not large relatively to the scales of the maps. For more localized and precise maps an ellipsoid (either local or global) is used to map the Earth's surface. This ellipsoid is sufficient for most topographic maps or other zoomed-in area maps.

A third model of the Earth's shape is known as the geoid, which is a complex and more accurate representation of the Earth's surface. The geoid is a hypothetical surface of constant gravity field, i.e., the gravity field vector is constant to this surface, and therefore calm water bodies appear as flat surfaces. The geoid is critically important for mapping global mean sea level from satellite altimeters, and it is computed using both terrestrial and satellite gravity measurements. The geoid itself is not used for mapping applications, but digital elevation models (DEMs) are usually referred to it, so that DEMs have the physical meaning as the height above sea level. The geoid can be approximated by a global ellipsoid with a maximum error of ±100 m. Datums are usually constructed for specific regions such as the North American Datum from 1927, or 1983 (NAD27 or NAD83), or the European Datum from 1950 or 1979 (ED50 or ED79), which have been replaced today by the European Terrestrial Reference System 1989 (ETRS89). There are also global datums such as the World Geodetic System, 1984 (WGS84), used in the global positioning system, which is optimized to represent the whole Earth with a single ellipsoid at the expense of similar accuracy in specific smaller regions. A datum is defined by seven parameters: a and b the major and minor semiaxes of the ellipsoid; Δx, Δy, Δz that are the ellipsoid center with respect to the center of mass of the Earth, and α, β, γ that are the rotations to align the ellipsoid axes with respect to the Earth axes.[6] The transformations between different datums are performed using the so-called Bursa−Wolf or Helmert transformations (Martín-Asín, 1990). Its detailed description is out of the scope of this book.

7.3.2 IMAGE GEOREFERENCING

Georeferencing is the process of scaling, rotating, translating, and deskewing an image to match a particular size and location (usually chosen to fit a selected map projection). All satellite images are raster images made up of pixels that are not naturally located at any locations on a particular map projection. Without georeferencing the image is defined by the width and height of the image in pixels each of which has its own resolution (counts per pixel). This image size alone usually has no direct relationship to the map locations corresponding to the pixel locations. What is needed is a method to assign the vector locations in a vector map file to the pixel locations in the satellite raster file. What is usually done is to find the correct "image navigation" that locates the pixel elements to the selected map projection and then resample the image to match this map projection.

7.3.2.1 THE ADVANCED VERY HIGH-RESOLUTION RADIOMETER AS AN EXAMPLE: GEOMETRIC CORRECTIONS

A good example of this need for georeferencing is the use of AVHRR data. The TIROS-N satellites that carried the AVHRR were limited in their geolocation systems to an Earth horizon sensor and a Sun sensor. Thus, it is necessary to use information on the satellite's orbit to correct the images with the appropriate orbital model. In Fig. 7.8 we show the orbit of a typical TIROS-N satellite, which is in a high-inclination, Sun-synchronous polar orbit with a nominal altitude of 850 km. This orbit has a period of 102 min with a nadir repeat period of 7 days. One can clearly see the need to correct the AVHRR image for both Earth curvature and Earth rotation. This will be done using knowledge of the satellite's orbit from ephemeris data. These data come from ground-tracking stations such as that operated by the US Air Force and the US Navy.

Another important consideration in correcting an AVHRR image for geometric distortions is the realization that the movement of the scan mirror across track also modifies the pixel size at the edge of each scan line. This is clearly seen in Fig. 7.9, which shows the size of each AVHRR pixel as a function of scan angle.

---

[6]In the above list, actually eight parameters have been defined, but in practice only seven are used, because the shape of ellipsoid (a and b) is preserved, just scaled by a factor m.

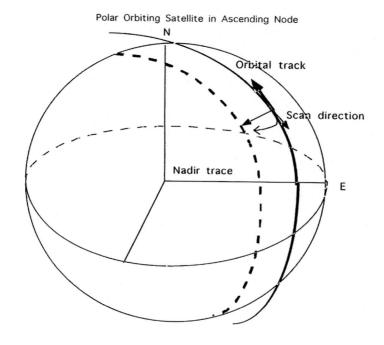

FIGURE 7.8

TIROS-N satellite orbit.

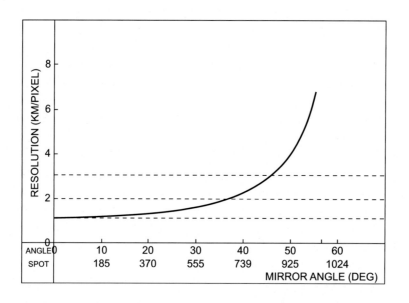

FIGURE 7.9

A very high-resolution radiometer pixel size as a function of scan angle from nadir.

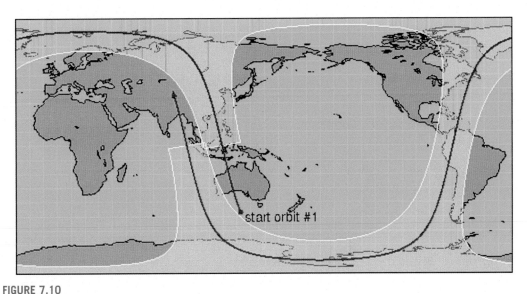

FIGURE 7.10

Advanced very high-resolution radiometer orbit and scan limits.

Here it is clear that the AVHRR retains its basic resolution of about 1 km until about 40° in scan where the pixels get much larger as the main scan mirror continues to rotate and the Earth's curvature increases. The AVHRR swath consists of 2048 pixels across each scan, which corresponds to approximately 3000 km. The instrument collects six scans per second. Beyond the 40° limit the pixel size increases from about 2 to 7 km at about 55° scan angle where the field stop limits the size of the scan swath.

To better understand the effects of Earth curvature and rotation Fig. 7.10 with the AVHRR swath overlaid as a green line showing the swath and the precession of the orbit. As an artifact of the mapping projection used, the swath width apparently increases toward the poles. The repeat period is approximately 7 days referred to the equator-crossing repeat time. The nature of a Sun-synchronous orbit is to maintain the same solar relationship established by the equator crossing time. This will vary slightly over the 7-day repeat period.

From Fig. 7.10 it is clear that satellite image data do not repeat for successive orbits. Also the geometric distortions due to curvature and rotation are apparent. Orbit ephemeris information can be used to correct for these geometric distortions. There are two basic types of georeferencing. One is called "direct referencing" that computes the latitude and longitude of the center of the instantaneous field of view (IFOV). This does not change the image picture, but merely generates the location information. In indirect referencing the FOV is calculated that contains a given latitude and longitude (Fig. 7.11). In this way, an image can be resampled to fit a given map projection. The advantage of this method is that there are no data gaps in the resultant image, but at the expense of having to repeat pixels to fill in those areas where the pixel resolution expands with increasing scan angle. Still, this is the dominant method applied in the routine georeferencing of AVHRR images (also called image navigation).

FIGURE 7.11

An example of an advanced very high-resolution radiometer image of the US West Coast in satellite perspective.

The same image as presented in Fig. 7.11 is repeated here in Fig. 7.12 using the indirect method to produce an accurately georeferenced image. In this indirect method a grid is specified by the user including geographic boundaries (coastlines, rivers, lakes, etc.), image pixel resolution, and the specific map projection. The portion of a given satellite image that corresponds to this grid is

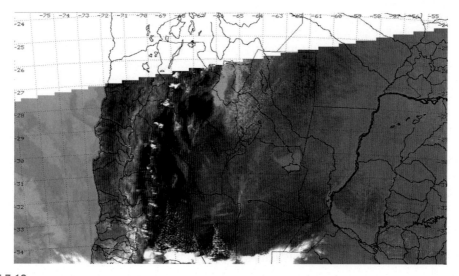

FIGURE 7.12

An example of the image in Fig. 7.11 georegistered with the indirect method.

georegistered using the orbital model. Each grid point is defined by an (i,j) position in the grid and each FOV is then selected using the orbital model and the appropriate ephemeris data. This FOV is then inserted at the appropriate location.

The steps in this indirect navigation are:

1. A vector **R** from the center of the Earth to the latitude, longitude of interest is constructed.
2. Using the orbital model at the time that this latitude, longitude pair was observed is computed. It is important to realize that time is the critical link between the ground and the satellite on orbit. This solution requires iteration with an initial choice of the spacecraft time at which the position and velocity of the spacecraft are calculated. A "look vector **d**" is then calculated that points from the spacecraft to the latitude, longitude of interest (Fig. 7.13).

Next transform **d** to **d'** a vector in a coordinate system centered on the satellite. If the time is correct, vector **d'** will be in the scan plane. Now compute the angular difference between **d'** and the scan plane. Transform this angular difference to a time correction **Δt** using the satellite's mean motion. Apply this time correction and return to calculate a new **d'**. Continue this iteration until **Δt** is less than 1/12 s (which corresponds to half a pixel on the ground). Compute the scan line number i from the solution time and the time of the first scan line. Compute the spot number **s** from the scan angle of **d'**.

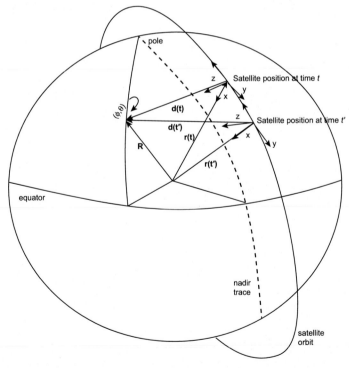

FIGURE 7.13

Orbital and scan geometries at solution time t and an intermediate time t' for the indirect image navigation.

Orbital and scanning geometry at solution time t
and intermediate time t' for indirect navigation

FIGURE 7.14

A geometrically corrected advanced very
high-resolution radiometer image.

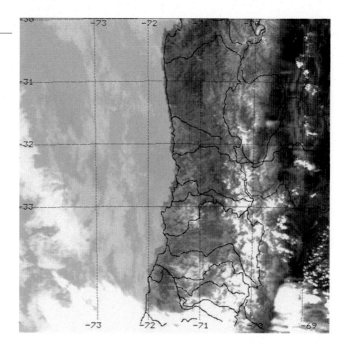

An example of the geocorrected image is given here in Fig. 7.14, which again is a portion of the US West Coast.

Note even in this geometrically corrected image there are misregistration errors that are clearly seen by offsets in coastal features and the locations of rivers and lakes. These remaining errors are due primarily to a lack of precise knowledge of the exact pointing angle of the AVHRR sensor, which depends on the attitude of the spacecraft. In the next section a method to correct for these misalignment errors will be presented. Other possible sources of registration errors are as follows: inaccuracies in the ephemeris data [data used verified to be accurate to 0.5 (IFOV)], inaccuracies in the model (validated to be accurate to within 1 IFOV for daily ephemeris data), inaccuracies in the satellite clock (a clock drift of ± 0.5 s, ~ 3 IFOV) results in registration errors of 4 IFOVs, and finally incorrect knowledge of the satellite's roll, pitch, and yaw attitude alignment. It should be noted that TIROS-N clocks are notorious for drifting, but that these drift errors can be compensated for by correcting the image for pitch, since a time error will appear as an offset due to satellite pitch.

7.3.3 ADVANCED VERY HIGH-RESOLUTION RADIOMETER ACCURATE AUTOGEOREGISTRATION USING IMAGE CALCULATED ATTITUDE PARAMETERS

In an effort to automate the attitude corrections for AVHRR imagery Emery et al. (2010) developed a method using infrared reference "base images" to compute MCCs between images that could then be converted to spacecraft attitude corrections and applied more widely to similar images along the same orbit. This MCC method is applied to widely dispersed cloud-free regions of the new and only geometrically corrected AVHRR image to find the offsets needed to bring the new image into agreement with the base image which has been very accurately georegistered. In the process these displacements are converted to attitude corrections (roll, pitch, and yaw) needed to bring the new AVHRR image into precise alignment with the AVHRR base image.

The MCC method was originally developed to calculate ocean surface currents from the displacements of SST patterns between sequential images (Emery et al., 1986). In the image navigation application there has been no movement of features between the images, but one image is distorted relative to the base image and the MCC calculation is then used to determine the displacements needed to geocorrect the new image to match the base image. This procedure is depicted here in Fig. 7.15, which shows portions of both a base image on the left and the new (and distorted) image on the right.

It should be pointed out that an infrared image is generally used to generate the base image. This is due to two effects: first infrared images are available both day and night and second an infrared image senses emitted radiation and thus has no dependence on solar illumination angle. If a visible channel had been used instead it would have a very specific dependence on solar angle and thus not be generally applicable for mapping all subsequent images. Since a critical factor is detecting cloud cover even in the infrared images both the thermal infrared and visible images are used together to define those pixels that are considered to be dominated by clouds. These regions are not used to compute the MCC against the base image.

The AVHRR autonav algorithm (Emery et al., 2010) has three means of eliminating the effects of clouds. The first is a simple threshold method where pixels having brightness temperatures less than a user specified value are ignored in the cross-correlation computation. The threshold value should be larger than the brightness temperatures of the clouds present in the images. While this technique will identify most of the cloudy pixels, it is likely that a few clouds will pass through and be used to compute

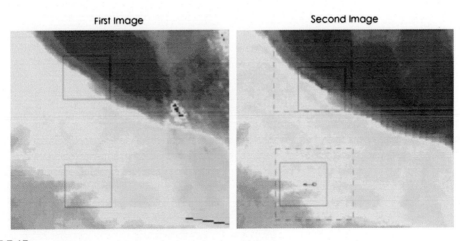

FIGURE 7.15

The maximum cross-correlations method illustrated for two grid locations. The solid boxes in the first image are the template subwindows; this is the data feature to search for within the dashed search windows in the second image. The second-image subwindows that provide the highest cross-correlations suggest the most likely displacements of the data features (*solid boxes*).

cross-correlations. Since the MCC method computes displacements by correlating data patterns between two images, each displacement vector has a correlation coefficient, which quantifies how well a pattern was matched. Displacements contaminated by clouds usually have low cross-correlations, and eliminating all vectors with cross-correlations lower than the 95% confidence level (~ 0.7) is the second means of cloud filtering. The third step, known as the coherency filter, makes use of the fact that displacements due to misregistration are fairly coherent for local regions, while those due to clouds are not. The coherency filter removes vectors that have different lengths and directions than their immediate neighbors. The three cloud filtering processes result in a set of vectors that represent the registration offsets of the target image without the influence of clouds.

An example of an AVHRR base image is included here in Fig. 7.16 for the US West Coast. Note that this image is not totally without clouds, but has a really low amount of cloud cover. The most important fact about this cloud cover is that there are open areas widely distributed over the image where cloud-free portions of the new images can be used to calculate the MCC displacement vectors. There are many coastal and inland river and lake features to match the infrared image to. Even with all of the efforts to make this image as exactly georegistered as possible, there are still some registration errors as can be seen in some of the smaller lakes.

An unregistered image of approximately the same area as the base image in Fig. 7.16 is presented here in Fig. 7.17. Note again the offsets of coastline features, displacements of rivers and lakes. There is quite a lot of clouds in this image, but fortunately the coast is almost cloud-free and enough of the rivers and lakes have relatively low cloud cover and can be used to compute the MCC offsets that will be used to calculate the satellite attitude parameters.

The displacements calculate by the MCC method for the image in Fig. 7.17 are shown here in Fig. 7.18.

FIGURE 7.16

A nighttime thermal infrared base image for the US West Coast.

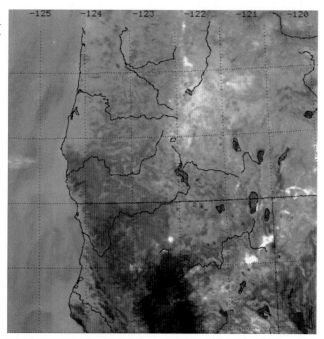

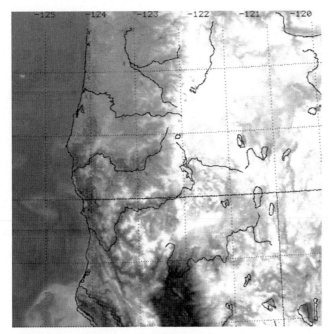

FIGURE 7.17

An advanced very high-resolution radiometer thermal infrared target image to be used for autonav georegistration. Note the misalignment of lakes, rivers, and coastline.

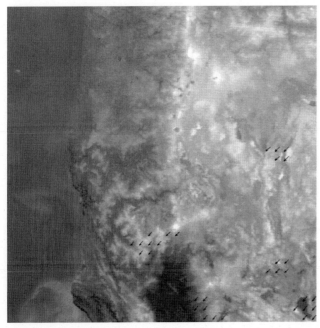

FIGURE 7.18

The advanced very high-resolution radiometer target image from Fig. 7.17 with the maximum cross-correlation displacement vectors in cloud-free regions.

It should be remembered that the autonav routine does not require human intervention to select the regions where the MCC displacement vectors are calculated. The algorithm uses the cloud identification routines outlined above to select the cloud-free regions and then computes the displacement vectors. These steps are shown as demonstrations of the method, but the algorithm was designed to operate completely autonomously without the need for human intervention in the process.

Finally, the autonav geo-registered image in Fig. 7.19 is presented. It can clearly be seen how well the coastal features, the rivers and the lakes line up. In some regions thin cloud cover makes this comparison difficult. Elsewhere the relatively low cloud cover in this image makes it easy to see how accurately autonav has geo-registered this image when compared to its original in Fig. 7.17. Lakes and rivers match very closely as do the many coastline features in the image. This confirmation of land registration features gives one confidence that the thermal infrared patterns in the coastal ocean are also well navigated in spite of the absence of ocean landmarks to confirm this accurate georegistration. It will be shown later that the attitude parameters derived from the autonav applied to a single image can be applied to images in other parts of the single orbit thus making it possible to accurately geo-locate open ocean images with no land references.

To evaluate the accuracy of autonav applied autonomously to a series of AVHRR 1 km images the x (East-West) and y (North-South) offsets for a series of nine coastal images in September of 2007 were computed (Table 7.1). The average x-offset is 0.0 km and the average y-offset is −0.4 km (West). These averages are both much smaller than the 1 km resolution of the images thus confirming the accuracy of the autonav method even when applied autonomously.

Another way to evaluate the performance of this georegistration method is to evaluate it against other AVHRR georegistration techniques. The first comparison is with the NOAA archived data from

FIGURE 7.19

The autonav georegistered target image from Fig. 7.17. Note the alignment of map and image features.

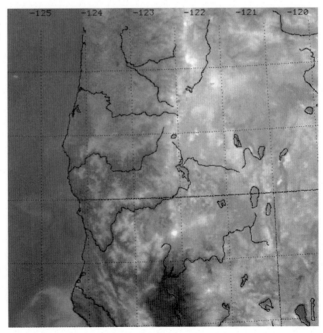

Table 7.1 Advanced Very High-Resolution Radiometer autonav Registration Errors

Satellite Pass Identification	x-Offset (km)	y-Offset (km)
3 September 2007, 06:34 GMT, NOAA-18	−1	0
3 September 2007, 10:24 GMT, NOAA-15	0	0
4 September 2007, 21:20 GMT, NOAA-15	−1	−1
5 September 2007, 15:34 GMT, NOAA-17	1	−1
6 September 2007, 15:11 GMT, NOAA-17	0	−1
7 September 2007, 14:48 GMT, NOAA-17	1	−1
8 September 2007, 02:08 GMT, NOAA-17	0	0
8 September 2007, 10:07 GMT, NOAA-15	0	0
9 September 2007, 07:13 GMT, NOAA-18	0	0

an NOAA-11 satellite in 1991. The navigation of this image was carried out using software developed by NOAA's National Environmental Satellite Data and Information Service (NESDIS) and is routinely embedded in the level 1b satellite data available online from NOAA/NESDIS through their CLASS. This navigation method has improved in recent years by updating the TIROS-N clocks more frequently and including some attitude information. As it will be shown, however, the NESDIS georegistration still results in a rather large geolocation error.

The second AVHRR image navigation method is from the European Organization for the Exploitation of Meteorological Satellites Advanced ATOVS Processing Package (EUMETSAT AAPP) software, which uses attitude parameters obtained from Moreno and Meliá (1993). It should be noted that although AAPP can compute latitudes and longitudes from level 1b AVHRR data, the software is not capable of producing a georegistered AVHRR image. Thus, for this comparison the latitude, and longitude values have had to be used, and the image has been remapped.

In Fig. 7.20 an image from the US West Coast is shown, but this time navigated with the NOAA/NESDIS software. At this scale misregistrations are not easy to see. To make this comparison easier the small portion of the image in the white square in the upper right corner of the image that contains the large Lake Tahoe has been extracted. The boundaries of this large lake demonstrate the misregistration errors of each of the three methods. First is the NOAA/NESDIS registration with about a 4 km offset with the EUMETSAT AAP registration yielding an error of about 3 km. The autonav software is clearly accurate to within the 1 km native resolution of the AVHRR image (Fig. 7.21).

Finally, we want to demonstrate how the attitude corrections calculated for one part of the orbit can be used for a completely different part of the orbit. While land references will be used in both cases to demonstrate the success of the autonav method, this success means that it is possible to use attitude corrections computed over land to accurately geolocate images over the open ocean where there are no landmarks to use for georeferencing. To demonstrate this effect, we will use attitude parameters calculated for the US West Coast to geolocate an AVHRR image of the island of Hawaii 3 h and two orbits later for the same satellite and AVHRR instrument.

For this comparison an AVHRR image of Hawaii is geometrically corrected, but unregistered for spacecraft attitude correction (Fig. 7.22).

FIGURE 7.20

A NOAA-18 channel four AVHRR image geo-registered with NOAA/NESDIS attitude-corrected level 1b data. The small *white square* in the upper right center will be used to compare geolocation software.

It can be seen that the geolocation is off by more than 4–5 pixels in this 1 km resolution image. This is particularly evident on the Southeast shore of the Big Island, which shows the highly reflective land offset from the map coastline. Using the attitude parameters calculated from an image of US West Coast 3 h and two orbits earlier the geolocation accuracy demonstrated in Fig. 7.23 can be achieved.

While there are still some small misregistration errors, most of the problems in Fig. 7.22 have been corrected. This again, means that we can use attitude parameters computed for one part of the orbit that contains landmarks to georegister images that are over the open ocean where no landmarks exist to geolocate the image data.

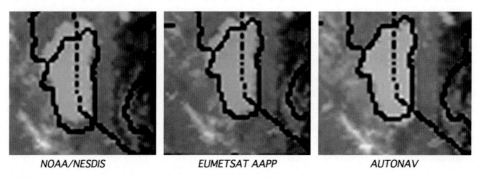

NOAA/NESDIS   *EUMETSAT AAPP*   *AUTONAV*

FIGURE 7.21

Lake Tahoe navigated with: NOAA/NESDIS software (left), EUMETSAT AAPP (center), and autonav (right).

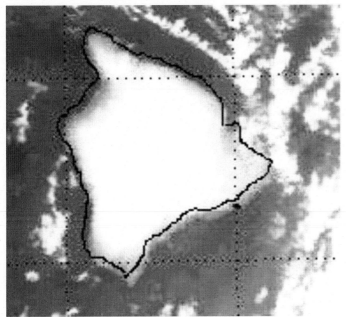

FIGURE 7.22

The Big Island of Hawaii navigated using geometric corrections without attitude corrections.

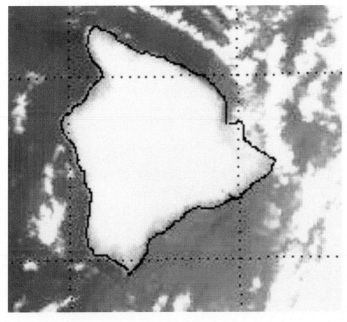

FIGURE 7.23

The image in Fig. 7.22 georegistered with attitude parameters computed from a US West Coast image 3 h and two orbits earlier for the same satellite.

7.4 MAP PROJECTIONS

A map projection is a method of representing the surface of the Earth on a two-dimensional plane. This is the process of creating a map of the Earth's surface all of which distort the Earth's surface in some way as originally proved by Gauss who said that no sphere can be represented on a plane without distortion. These distortions can be minimized depending on the specific application of the map projection. Because there are many different applications, there are a consequent large number of different map projections. Only a few types will be presented here as examples, but the interested reader can find many volumes that are addressed to subject of map projections (e.g. Martín-Asín, 1990; Kraak and Ormeling, 1996).

For simplicity it will be assumed that the Earth's surface is a sphere rather than the precise ellipsoid or geoid as discussed earlier. The purpose of a map projection is to specify the transformation between the curved elliptical surface of the Earth and the flat planar surface of the map. One example is the Albers projection shown here in Fig. 7.24.

This projection is a conic, equal area map projection that uses two standard parallels. This particular projection shows areas correctly, but to do so, it alters the shapes and scales of these areas. It is used by the US Geological Survey and the US Census Bureau.

The most widely used map project is the "transverse Mercator", which is convenient for large-scale maps as it preserves size and shape for areas within the same limited range of latitudes. The projection geometry is described here in Fig. 7.25, which reveals that this is a cylindrical projection with the tangent points of the cylinder at both poles. A particular "transverse Mercator" projection called the Universal Transverse Mercator (UTM) is probably the most recommended one from latitudes 84°N to 80°S (Fig. 7.26A). For polar regions for latitudes North of 84°N, and South of 80°S, the Universal Polar Stereographic (UPS) is widely used (Fig. 7.26B). Both are coordinate systems that use a metric-based Cartesian grid laid out on a conformally projected surface. The properties of the UTM are as follows: it is conformal (shapes are preserved), its central meridian is automecoic (distances are preserved), the equator and the central meridian intersection is the origin of the (x,y) coordinates, which are expressed in meters. The whole globe is "sliced" in 60, 6° fuses, with the Greenwich meridian lying between fuses 30 and 31. The advantage is that no point is too far away from the central meridian, so the distortions in the fuse are small, but this benefit is achieved at the expense of the discontinuities.

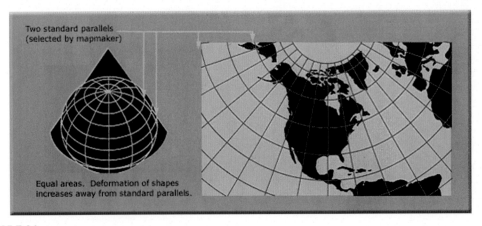

FIGURE 7.24

Albers projection of North America.

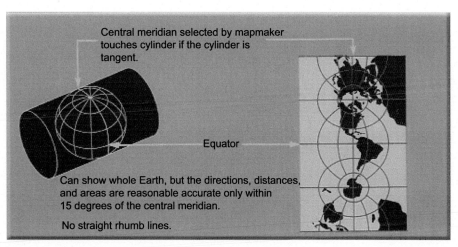

Central meridian selected by mapmaker touches cylinder if the cylinder is tangent.

Equator

Can show whole Earth, but the directions, distances, and areas are reasonable accurate only within 15 degrees of the central meridian.

No straight rhumb lines.

FIGURE 7.25

Transverse Mercator projection.

The transverse Mercator projection is mathematically the same as the standard Mercator projection except that it is oriented around a different access. The standard Mercator projection has the cylinder oriented North-South (Fig. 7.27), which results in the familiar stretching of the meridional lines as higher latitudes.

Once the choice is made between projecting on to a cylinder, a cone or a plane, the shape of the map features must be specified. This is done by specifying how the projection surface is located relative to the globe. It may be normal so that the map surface of symmetry coincides with the Earth's axis or it may be transverse that is at right angles to the Earth's axis. The map surface may be tangent in just touching the Earth's surface or it may be secant where the map plane intersects or "slices" through the globe.

A sphere or globe is the only way to depict the Earth that exhibits constant scale throughout the entire map in all directions. A flat map cannot achieve this property for any area regardless of projection method or limited size of the area of interest. Scale depends on location on the Earth, but not on orientation. Scale is constant along any line of parallel (latitude) in the direction of the latitude.

Map projections can be classified by the type of projection surface onto which the globe is projected. Examples are cylindrical (e.g., Mercator), conical (e.g., Albers), and azimuthal or planar (e.g., stereographic used frequently for polar maps). Another way to classify map projections is according to the properties that they preserve such as: preserving direction (azimuthal), preserving local shape (conformal or orthomorphic), preserving area (equal-area), preserving distance (equidistant), and preserving the shortest route (gnomonic). Because a sphere cannot be flattened, it is impossible to have a map projection that is both equal-area and conformal. There are, of course, many other mathematical formulations for map projections that do not fit easily into any of these classification schemes.

The National Atlas of the United States uses a Lambert azimuthal, equal-area projection to display the country: it is a particular mapping from a sphere to a disk, which accurately represents area in all regions of the sphere, but it does not accurately represent angles. Conformal maps are used for navigational or meteorological charts. The US Geological Survey uses a conformal projection for

(A)

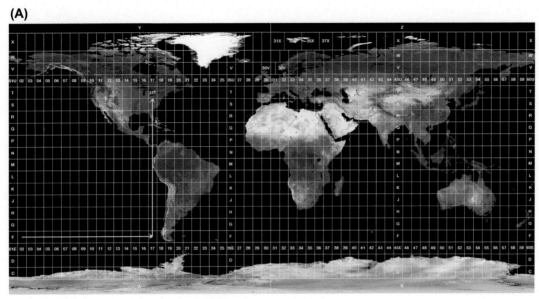

(B)

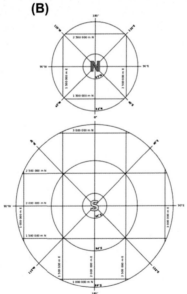

FIGURE 7.26

(A) UTM projection system: longitude zones are 6° wide, numbered from 01 at 180° West, increasing toward the east until 60 at 180° East. Latitude zones are 8° high and are labeled from C to X, omitting the letters "I" and "O", beginning at 80°S. Letters A, B, Y, and Z are used in the polar regions by the Universal Polar Stereographic (UPS) grid system. Exceptions to the system are in the west coast of Norway and the zones around Svalbard. (B) Diagrams of the UPS coordinate system whose two plane grids cover the Arctic and the Antarctic.

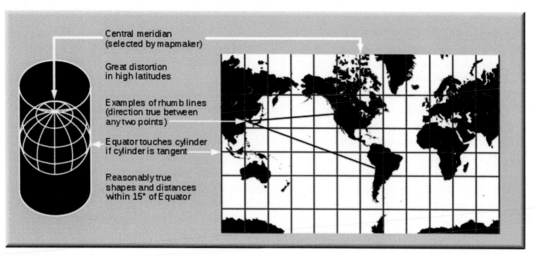

FIGURE 7.27

Standard Mercator projection.

many of its topographic maps. Equidistant projections are used for radio and seismic mapping and often for ship navigation. Aeronautical charts employ maps that preserve direction and are called again azimuthal or zenithal projections.

7.5 STUDY QUESTIONS

1. Calculate the radius of orbit for an Earth satellite in a geosynchronous orbit, taking into account that the Earth's rotational period is 86,164.1 s.
2. Compute the ground FOV over the equator of a geosynchronous satellite. How many of them will be required to view the whole Earth?
3. A circular Sun−Earth synchronous orbit produces every 3 days, 43 different ground tracks over the Earth's surface. Compute the orbital height, the orbit inclination, and the separation between two consecutive ground tracks. Note: $J_2 = 1082.6 \times 10^{-6}$
4. A satellite in a near polar Sun−Earth synchronous orbit repeats its traces every 30 days, and it has an orbital period of 100 min. If the satellite is observing a geographical location in the ascending pass at 10 a.m. local time, at which time it will observed this position 4 months later in a descending pass?
5. A space-borne remote sensor has to observe the Earth's surface in the nadir direction. Determine the orbital height if the orbit must be a perfectly circular between 800 and 1000 km height, the orbit inclination is 42° with respect to the equator, Sun-synchronous, but not Earth-synchronous, and the separation between consecutive ground-tracks must be 482.7 km.
6. A satellite in a nearly polar Sun-synchronous orbit performs 251 ground-tracks every 18 days. If the satellite embarks a high-resolution imager, which must be the instrument's swath to obtain images of the whole Earth?

7. In the previous case, which is the range of latitudes covered by the imager. Are there latitudes near the poles not covered? Which ones?

8. During the testing period a satellite is in an LEO 29:2 nearly polar orbit. After the validation period, the orbit is transferred to a 449:31 one. Compute the orbital heights if both are Sun-synchronous orbits.

9. Compute the orbital period, the altitude of the apogee and perigee of an elliptical orbit with the following parameters: semimajor axis = 26,560 km, eccentricity = 0.722, inclination = 63,4°.

10. What are HEOs intended for?

11. What is a geodesic datum, which parameters does it contain, and what it is used for?

12. If we need to represent high-resolution satellite imagery around the equator ($-10°$ to $+10°$ latitude) obtained from a LEO 0° inclination orbit, which projection will perform best? Why?

13. Make a list with the geometric and radiometric distorsions, their correction techniques, and the order in which they must be applied.

14. The following images correspond to different processing levels of an optical image of SPOT. Figure (A) shows the preprocessed L1A data. Which type of corrections have already been applied and why?

    What types of corrections are being applied in Figures (B) and (C)? Explain the shape of the image in terms of the imaging characteristics of the sensor and the satellite orbit.

(A)   **(B)**   **(C)**

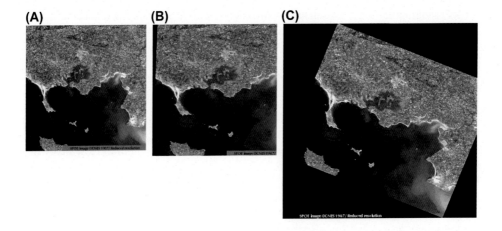

15. Explain what are the topographic surface, the geoid, and the reference ellipsoid?

16. Explain what are the local and the global ellipsoids? Please list at least one local and one global ellipsoids currently used.

17. What are the main properties of the UTM projection? In how many fuses the Earth is divided?

18. Which is the projection typically used to represent the polar regions?

ATMOSPHERE APPLICATIONS

There are many books on atmospheric remote sensing that are listed in the references so the purpose of this chapter is not to present a comprehensive review of atmospheric remote sensing. Instead a brief review of some of the primary application of satellite remote sensing to the atmosphere will be provided.

8.1 CLOUD REMOTE SENSING

One of the prime atmospheric elements that can be sensed by satellites is clouds. For many Earth surface applications clouds are merely a hindrance as reflected and emitted thermal radiation do not pass through clouds, but clouds are prime indicators of conditions of the atmosphere. Clouds are related to the atmospheric elements (gases and aerosols) that make them up, and they will be addressed separately.

8.1.1 CLOUD TOP TEMPERATURE

Accurate cloud top temperature and height information is very important for aviation forecasting and for early warning of thunderstorm development. For optically thick clouds, weather satellite imagery provides a reasonable estimate of cloud top temperature. Atmospheric absorption above the cloud may be neglected and the brightness temperature of a thermal window channel can be taken as the thermodynamic cloud top temperature. In the case of semitransparent cloudiness, the direct use of the measured thermal brightness temperature will lead to a significant overestimation of the true cloud top temperature resulting in a substantial underestimation of the cloud height.

Cloud top temperatures are first retrieved assuming that all clouds are black bodies. During the daytime when visible radiances of a cloud are available to estimate cloud optical thickness transmissive cloud are corrected for the radiance transmitted from below through the cloud deck thus decreasing cloud top temperature and subsequently increasing cloud height. At nighttime, however, semitransparent cirrus clouds (high ice clouds) may be falsely identified as mid-level clouds due to the lack of visible information to estimate cloud optical thickness.

The advanced very high-resolution radiometer (AVHRR) derived cloud heights have been compared with coincident measures of cloud top temperature from the TIROS Operational Vertical Sounder (TOVS). These cloud heights were found to agree quite well (Stubenrauch et al., 2005) in spite of their very different spatial resolutions. Both products show that the average cloud height of high clouds (cloud pressure less than 440 kPa) is largest in the tropics, due to a higher tropopause and that in these regions there are nearly no general cloud systems with the highest cloud layer being in the middle troposphere. The Southern Hemisphere mid-latitudes are mostly covered by low-level clouds. Regional and seasonal distributions of bulk microphysical properties of large-scale semitransparent cirrus clouds obtained from the revised TOVS data have been published by Rädel et al. (2003) and Stubenrauch et al. (2005).

Cloud top temperatures are indicators to forecasters if a storm is getting stronger or weaker. When cloud top temperatures get colder it means that the cloud is getting higher in the atmosphere, reflecting the uplift of warm moist air, and a stronger thunderstorm will form with a potential for cyclonic behavior, and potentially a tornado. Cloud top temperatures that warm up indicate that the cloud tops are lower than they were, reflecting a weakening of the storm.

When NASA's AQUA satellite passed over the Malacca island on September 23, 2010, the Atmospheric Infrared Sounder (AIRS) instrument measured the temperature of the cloud tops in the storm colder than 215K in the strongest parts of the storm (Fig. 8.1).

These cloud top temperatures indicate that the storm has a lot of energy. Coincident thermal infrared MODIS imagery showed an eye with thunderstorms banding around it, and convection rebuilding over the system reflecting strengthening of the system. Malacca had maximum sustained winds near 113 km/h (just under Typhoon strength).

Similar cloud top temperatures were computed from AQUA's MODIS instrument (Fig. 8.2) for a region in the Gulf of Mexico where a storm was just making landfall in Louisiana and Mississippi. Only the coldest cloud top temperatures have been colored in this image showing that there are smaller pockets of cold cloud tops following the inflow region of the storm. It should be noted that the temperature given in the temperature scale below the image are in Kelvin to provide an absolute temperature reference. This storm has the coldest cloud top temperatures in the center of the main storm with similar low temperatures in some storm cells located in the outflow streamer to the south of the main storm center. Only the lower temperature cloud tops have been colored while the rest of the infrared field is left in gray shades. This image of the cloud top temperatures gives a good indication of where the storm is stronger.

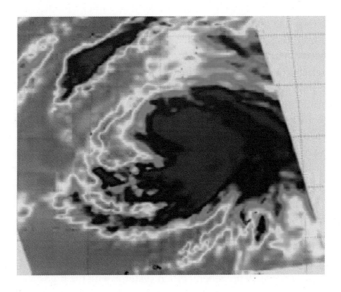

FIGURE 8.1

Tropical storm over the Malacca island from AQUA/AIRS on September 23, 2010.

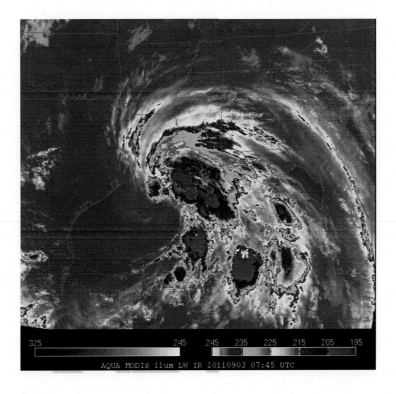

FIGURE 8.2

AQUA 11 μm MODIS cloud top temperatures from September 3, 2011, over the Gulf of Mexico.

An 8-year average cloud top temperature computed from MODIS on NASA's Terra satellite is presented here in Fig. 8.3 as generated by the MODAP system at the Goddard Space Flight Center (GSFC). From the color bar at the bottom of the image it is clear that none of these smoothed cloud tops are as cold as those seen in the storm in Fig. 8.2. The warmest clouds are located in the subtropics and mid-latitudes. The coldest cloud tops are found in the polar-regions. There is an interesting band of cooler cloud tops near the equator in the oceans. This is a fairly narrow band in the eastern Pacific and is a bit wider, but weaker in the Atlantic. In the eastern Indian Ocean and western Pacific this region is much larger meridionally and colder. This band could be associated with the Intertropical Convergence Zone (ITCZ), which is a well-known location of strong convective activity. In general cloud tops over the land are cooler than over the ocean reflecting the differential heating between land and ocean.

Another workhorse imager, the AVHRR, can also be used to estimate cloud top temperature. Here Fig. 8.4 presents the cloud top temperatures computed from an AVHRR image for hurricane Andrew. The cloud top temperatures are given by the color bar in the upper left of the image. Note that the background temperatures are all fairly uniform as shown by the red color. It is interesting that the eye of the hurricane shows the same red color and reflects the background conditions and not the storm. This is a well-known characteristic of hurricanes.

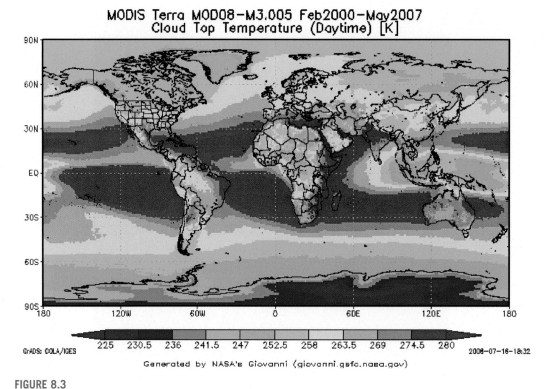

FIGURE 8.3

Terra MODIS daytime cloud top temperature (K) February 2000–May 2007.

Geostationary satellites can also be used to compute cloud top temperatures from their infrared channels. A visible image (left) and the corresponding cloud top temperatures (right) from the European Meteosat satellite are shown in Fig. 8.5. Note the row of small cloud patches along the equator that all correspond to fairly cold cloud top temperatures seen here in blue ($\sim$ 220K). Similar cold cloud tops are seen over the central part of South Africa, the Western Africa, and in the Polar Regions. Warm cloud temperatures surrounding these cold spots on the equator are shown in the figure. There are wisps of fairly warm clouds over South Africa and in the central Atlantic. Medium temperature clouds cover most of the North Atlantic. The South Atlantic exhibits very low cloud cover with many regions being completely cloud-free. Moderate cloud top temperatures dominate the area over Europe.

8.1.2 CLOUD SHAPE AND CLOUD TYPE

Clouds are classified according to their shape, height, content, and formation mechanism. A good summary of the cloud types is provided here in Fig. 8.6.

Cumulus clouds are often called "fair weather" clouds as they occur generally over land on sunny days when solar heating creates thermal convection currents. This explains the "puffy" shape of these clouds. These clouds are found typically between 0.6 and 0.9 km altitude. They occur worldwide

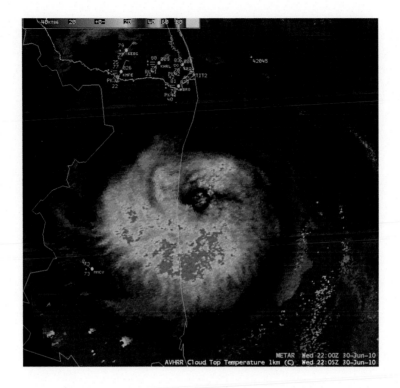

FIGURE 8.4

AVHRR derived cloud top temperatures for hurricane Andrew, June 30, 2010.

FIGURE 8.5

Visible image (left) and cloud top temperatures (right) from Meteosat.

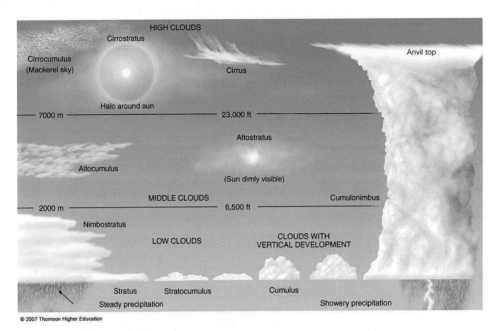

© 2007 Thomson Higher Education

FIGURE 8.6

Cloud types and their shapes, heights, and functions.

except in Antarctica where it is too cold. They are liquid water clouds and are formed by thermal convection currents.

Cumulonimbus clouds are the towering thunderclouds that strike fear into many who see them. As suggested in Fig. 8.6 the top of this tower spreads out reflecting the stronger winds at these higher elevations. To form, there are three critical conditions: (1) a ready supply of warm, moist air, which rises at speeds of up to 40–113 km/h, (2) tropospheric winds need to increase considerably with height to encourage the forward slant of the cumulonimbus tower, and (3) the atmosphere around the cloud needs to be "unstable." These clouds are common in the tropics and temperate regions, but very rare at the poles. Heavy rainfall is associated with these clouds as suggested in Fig. 8.6. These are liquid clouds throughout with ice crystals at the top. They are formed by very strong thermal convection.

Stratus clouds form as cold air flows out over warm sea temperatures and pick up both heat and moisture from the ocean. This air then flows over colder sea surface temperatures (Fig. 8.7), which forms fog that is closely associated with stratus clouds.

These clouds can be at the surface, which makes them seem like fog, and they can reach 2 km in height. Only light precipitation is generally associated with these liquid clouds. Their formation is due to advective cooling. A typical example is the marine stratus that forms off the coast of southern California. Here warm air flows out over cold water due to coastal upwelling and a dense layer of stratus clouds form a continuous layer off the coast.

Stratocumulus clouds are very similar to cumulus clouds, but they can form a semicontinuous layer much as the stratus clouds. These clouds generally form from either stratus or cumulus clouds.

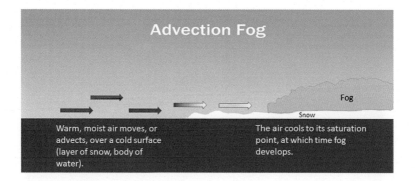

FIGURE 8.7

The formation of advective fog.

FIGURE 8.8

Altocumulus cloud bands.

Altitudes again range from 0.6 to 2 km. There is occasional light rain or snow associated with these clouds. Altocumulus clouds are the cloud bands (Fig. 8.8) that can be seen high in the atmosphere. These clouds are above the influence of thermals formed by heating of the atmosphere from the Earth's surface.

Altitudes range from 2 to 5.5 km and occasional light rain is often associated with these clouds. While most of the cloud is made up of liquid water the cloud may also contain ice crystals. They are formed by mid-level atmospheric disturbances and wave propagation (i.e., mountain waves).

Altostratus clouds are really just stratus clouds that get above 2 km. It is these clouds that make for such beautiful sunsets and sunrises. They can get as high as 7 km. They are most common in middle latitudes. They are both liquid water and ice crystal content clouds. Occasional light rain or snow can occur with these clouds. They are potentially dangerous to aircraft as they can lead to ice accumulation on the wings.

Nimbostratus clouds constitute a thick, wet blanket of clouds with a ragged based caused by continuous precipitation. Moderate to heavy rain or snow is generally steady and prolonged. These

FIGURE 8.9

Effect of cirrostratus cloud on the Sun.

clouds contain liquid water, raindrops, snowflakes, and ice crystals. They are formed by the thickening and lowering of an altostratus cloud.

Cirrus clouds are the highest of all clouds extending up to 14 km. They contain only ice crystals and are formed as streaks due to strong (161–240 km/h) upper tropospheric winds. Thus, they appear as thin streaks in the upper atmosphere. Any precipitation associated with these clouds does not reach the ground. They have a minimum altitude of 5 km. Cirrocumulus clouds are a transition phase between cirrus clouds and cirrostratus clouds. A large number of these clouds may indicate the arrival of bad weather. Cirrostratus clouds are difficult to spot as they appear as a pale, milky lighting of the sky. They never block out the Sun completely, but rather produce a variety of optical effects such as that shown here in Fig. 8.9. As the other cirrus clouds they are made completely of ice crystals and are formed through the spreading and joining of cirrus clouds.

8.1.3 REMOTE SENSING OF CLOUDS AND CLOUD PROPERTIES

Clouds play an important role in the Earth's energy budget by exerting a control on the solar illumination of the Earth's surface. High cirrus clouds reflect little or no solar radiation and thus contribute to the warming of the Earth's surface. Likewise, low-level cumulus clouds reflect a portion of the Sun's radiation back toward the outer space excluding it from the solar radiation reaching the Earth's surface. This has the net effect of cooling the Earth's surface due to reduced incoming solar radiation and persistent outward radiation from the Earth's surface. Very thick clouds block the radiation exchange in either direction resulting in no net change of Earth's surface properties.

How might clouds change in a warming world in the future. An increase in Earth's surface temperature would lead to an increase in evaporation increasing the amount of water vapor in the atmosphere. This might create more low clouds with a high water vapor content that act as the cumulus clouds discussed earlier to block incoming solar radiation and thus biasing the process to outward long radiation and cooling the Earth's surface. These clouds will necessarily contain more liquid water, which would then result in a greater frequency and intensity of rainfall events. These increased rainfall episodes would then result in a warming of the Earth's surface. Thunderstorms may also be more frequent and more intense leading again to surface warming due to the presence of ice clouds at the top of the thunderstorm clouds. If the ice particles are larger and hence fall at greater speeds the clouds will dissipate more rapidly resulting in surface cooling.

Climate model simulations show no agreement of the effect of clouds in a doubled CO_2 environment (Fig. 8.10). About half of them have clouds warming the atmosphere while another half of the models show the atmosphere cooling due to clouds in the same double CO_2 scenario.

Modelers have concluded that to improve their numerical models they need to have a better knowledge of the cloud macrophysical properties such as cloud-frequency, cloud height, cloud type, and the amount and type of mass contained in the clouds. They also need to know the microphysical properties such as cloud phase (water/ice) and the shape of the ice crystals.

Before talking about sensing the clouds two other fundamental cloud properties have to be discussed: aerosols and optical thickness. It should be pointed out here that for some, the sensing of the clouds themselves is the target of their research, while for many others the clouds represent a barricade to their research and most of their concern is to get around the clouds.

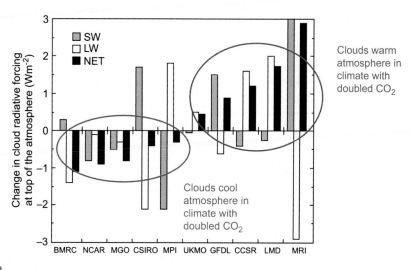

FIGURE 8.10

Climate model simulations of the effects of clouds in a double CO_2 atmosphere.

8.2 ATMOSPHERIC AEROSOLS AND OPTICAL THICKNESS

Aerosols are suspended solid or liquid particles in the gaseous atmosphere and they can contribute to clouds and to air pollution such as smog. There are several ways to characterize an aerosol. The most important in terms of environmental science and its application to health is the mass concentration, which is defined as the mass of particulate matter per unit volume. Also commonly used is the number concentration, which is the number of particles per unit volume with units such as number/m^3 or number/cm^3.

Particle size has a major influence on their properties and the aerosol particle radius or diameter is the key property used to characterize aerosols. Most aerosols have a range of particle sizes and the particles are nonuniform. While liquid aerosol particles are almost always spherical, solid particles can have a variety of shapes. As a consequence, these particles are considered in terms of an equivalent diameter. This equivalent diameter is the diameter of a regular particle that has the same value of some physical quantity of a regular particle.

For a monodisperse aerosol a single number can be used, the particle diameter, to describe the size of the particles. For a mixed polydisperse aerosol the size of the aerosol has to be described with a particle-size distribution (PSD). The PSD defines the relative amount of particles present, sorted by size. The best way to present this information is in terms of a histogram where each bar depicts the number of particles present that have sizes included in the size bins normalized by the total number of particles in the sample. Most aerosols have a skewed distribution with a long tail of larger particles. For this reason, a Gaussian population cannot be used to describe the PSD of most aerosols. It can be appropriate for some aerosols such as certain pollens and spores.

An often chosen PSD is the log-normal distribution (NIST/SEMATECH, 2016) where the size frequency is given by:

$$\mathrm{d}f = \frac{1}{\sqrt{2 \cdot \pi} \cdot \sigma} \, e^{-\frac{\left(d_p - \bar{d}_p\right)^2}{2 \cdot \sigma^2}} \cdot \mathrm{d} \, d_p, \tag{8.1}$$

where σ is the standard deviation of the size distribution and d_p is the arithmetic diameter. The log-normal distribution has no negative values, and it can cover a wide range of values and fits observed size distributions very well. It is often used to approximate the PSD of aerosols, aquatic particles, and pulverized material. Other distributions that can be used are the Weibull or Rosin–Rammler distribution (Rosin and Rammler, 1933), which applies primarily to coarsely dispersed dusts and sprays, the Nukiyama–Tanasawa distribution for sprays having extremely broad size ranges, the power function, which has been applied to atmospheric aerosols, the exponential distribution, which is applied to powdered materials (Gonzalez-Tello et al., 2008), and the Khrgian–Mazin distribution for cloud water droplets (Plank, 1991)

There are three different dynamical regimes that govern the behavior of an aerosol. These regimes can be defined by the Knudsen number of the particle which is given by

$$K_n = \frac{2 \cdot \pi}{d}, \tag{8.2}$$

where K_n is the mean free path of the suspending gas, and d is the diameter of the particle. Particles are in the free molecular regime when $K_n \gg 1$. Here the particles are small compared to the mean free path of the suspending gas. In this regime, particles interact with the suspending gas through a series of ballistic collisions with gas molecules. Thus, they behave similar to gas molecules, tending to follow streamlines and diffusing rapidly due to Brownian motion.

The continuum regime applies when $K_n \ll 1$ where the particles are large when compared to the mean path length of the suspending gas. This means that the gas can be regarded as a continuous flowing fluid around the aerosol particle. Finally, all the particles that are between the free molecular and continuum regimes and $K_n = 1$ are in the transition regime. Here the forces are a complex mixture of the interactions of particles with the suspending gas molecules.

Aerosol partitioning theory governs the condensation and evaporation of substances to and from aerosol surfaces. Condensation causes the mode of an aerosol number/size distribution to grow larger diameters as the particles add mass. Conversely, evaporation results in smaller diameters as the particles lose mass. Nucleation is the process of forming aerosol mass from the condensation of a gaseous precursor, specifically a vapor. In order for the vapor to condense, it must be super-saturated (its partial pressure must be greater than its vapor pressure). This can happen for three reasons: if the vapor pressure is lowered by lowering the temperature of the vapor, if chemical reactions increase the partial pressure of the gas or lower its vapor pressure, and if the addition of another vapor lowers the equilibrium vapor pressure, and hence the dew point, due to the Raoult effect (Kwak et al., 2008).

8.2.1 AEROSOL OPTICAL THICKNESS

Aerosol optical depth (AOD) or thickness is a measure of the transparency of an "atmosphere." Optical depth is defined as the negative natural logarithm of the fraction of radiation that is not scattered or absorbed along a path through the atmosphere. As a result, optical depth is dimensionless, and it is not actually a length as would be needed if it were a "depth." Thus, the differential optical depth can be written as

$$d\tau = -k \cdot \rho \cdot cos\theta \cdot ds, \tag{8.3}$$

where k is the opacity ($=K/\rho$ where K is the extinction coefficient), ρ is the density, θ is the angle to the normal, and ds is the optical path through the atmosphere. The optical thickness or optical depth for an atmosphere with an extinction coefficient of K can be computed by defining:

$$\mu = cos\theta, \tag{8.4}$$

$$d\tau = -k \cdot \rho \cdot \mu \cdot ds = n \cdot \sigma \cdot \mu \cdot ds, \tag{8.5}$$

where n is the number density and σ is the cross-section, and integrating for normal incidence:

$$\tau = \int_0^z K \cdot dz = N \cdot \sigma, \tag{8.6}$$

where N is the column density.

8.2.1.1 MODIS Cloud Optical Thickness

The MODIS cloud optical thickness algorithm is mainly intended for plane-parallel liquid water clouds (King et al., 1997). It is assumed that all MODIS data have been screened by the cloud mask of

Ackerman et al. (1998) with additional information regarding particle phase computed from the algorithm of Menzel and Strabala (1997). To retrieve the cloud optical thickness and effective particle radius, a radiative transfer model is first used to compute the reflected intensity field. When the optical thickness of the atmosphere is sufficiently large, numerical results for the reflection function must agree with known asymptotic expressions for very thick layers (van de Hulst, 1980).

King et al. (1997) assume that the reflection function is not dependent on the exact nature of the PSD, depending primarily on the effective radius and, to a lesser extent, on the effective variance, as first suggested by Hansen and Travis (1974). Nakajima and King (1990) demonstrated that the similarity parameter is unaffected by the variance of the cloud PSD, but the asymmetry parameter and hence the scaled optical thickness is weakly affected by the detailed shape of the size distribution.

The determination of the optical depth (τ_c) and the effective radius (r_e) from spectral reflectance measurements (from a satellite sensor such as MODIS) constitutes the inverse problem and it is typically solved by comparing the measured reflectances with entries in a lookup table and searching for the combination of τ_c and r_e that would give the best fit.

The spherical albedo as a function of wavelength for water clouds containing various values of the effective radius is presented here as Fig. 8.11. Since spherical albedo represents a mean value of the reflection function over all solar and observational zenith and azimuth angles, the reflection function itself must have a similar sensitivity to particle size.

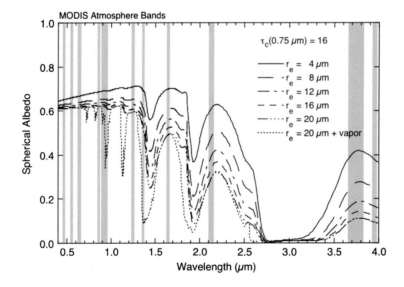

FIGURE 8.11

Cloud spherical albedo as a function of wavelength for selected values of the effective radius of cloud droplets. Results apply to water clouds having a modified gamma size distribution with an effective variance of $v_c = 0.11$, cloud optical thickness (τ_c at 0.75 μm) equal to 16, and saturated water vapor $\omega_\gamma = 0.45$ g/cm^2. The locations and bandwidth of selected MODIS atmosphere bands are also shown as dark vertical stripes in the figure.

The spectral albedo represents a mean value of the reflection function over all solar and observational zenith and azimuth angles, the reflection function itself must have a similar sensitivity to particle size. The calculation behind Fig. 8.11 (King et al., 1997) was carried out using asymptotic theory for thick layers and the complex refractive indices of liquid water including the additional contribution of water vapor. These computations only apply to clouds composed solely of liquid water and water vapor (King et al., 1990). The spherical albedo, and hence the reflection function, is sensitive to particle size at wavelengths near 1.64, 2.13, and 3.75 μm, wavelengths for which water vapor absorption is small.

Cloud properties can also be estimated from the thermal infrared bands of MODIS. A combination of visible and near-infrared (VNIR) bands provides information on both optical thickness and effective radius. In the 3.7 μm window, both solar reflected and thermal emitted radiation are present. The use of the reflectance for estimating cloud droplet size is considered to be much more sensitive than the thermal component. In either case, the thermal and solar components of the 3.7 μm channel must be separated to provide the desired estimate. CO_2 absorption is more important in the 4.3 μm channel and at wavelengths larger than 13 μm. The MODIS bands in these spectral regions can be used to infer vertical changes of atmospheric temperature.

Corrections are made for cloud phase, atmospheric water vapor content, and other atmospheric effects such as Rayleigh scattering. An algorithm for computing cloud optical depth and effective radius is shown here in Fig. 8.12.

The use of the 3.75 μm band complicates this calculation due to the presence of both reflected and emitted radiation. Cloud emission at 3.75 μm is weakly dependent on cloud effective radius unlike solar reflectance so that the relative strengths of the two components depend on particle size. A separate algorithm can be developed for using the 2.75 μm data alone. An example of the MODIS cloud optical thickness computed from the algorithm described in Fig. 8.12 is shown here in Fig. 8.13.

This image is computed from MODIS data taken by NASA's AQUA satellite over a 24-h period on October 23, 2011. The colors have been separated into water and ice contributions to the cloud optical thickness. A summary of zonal mean cloud optical thickness values is presented here in Fig. 8.14 which is a plot of the daytime cloud optical thickness values from MODIS and VIRS over land and water separately.

8.2.2 GROUND VALIDATION OF SATELLITE OBSERVED OPTICAL THICKNESS

8.2.2.1 The Beer–Lambert Law

The primary source of validation data for satellite estimates of optical thickness or cloud optical depth is ground measurements made by a Sun-photometer. Modern Sun photometers incorporate Sun-tracking into their operation along with a spectral filtering photodetector and a data acquisition system. The variable measured is referred to as direct-Sun radiance.

It has to be acknowledged that a Sun-photometer operates within the Earth's atmosphere and therefore the measured radiance does not equal the radiance emitted by the sun, which is then filtered by the atmosphere above the Sun-photometer. This atmospheric effect is given by Beer's law (Mayerhöfer et al., 2016), which states that there is a logarithmic dependence between the transmission, T, of light through a substance and the product of the absorption coefficient of the substance,

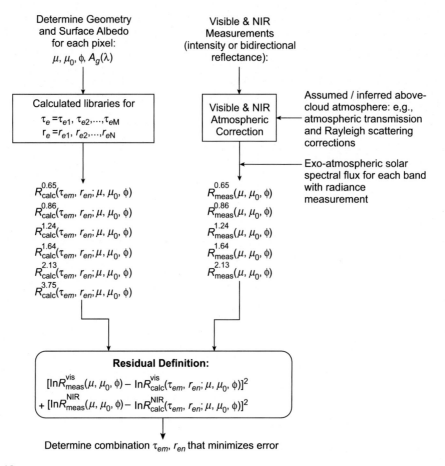

FIGURE 8.12

General cloud retrieval algorithm for determining best fit for τ_c and r_e in the 0.65, 1.64, and 2.13 μm MODIS bands (King et al., 1997).

α, and the distance the light travels through a material (i.e., the path length), l. The general Beer–Lambert law can be written for atmospheric gases as

$$T = \frac{I}{I_0} = e^{-\alpha' \cdot l} = e^{-\sigma \cdot l \cdot N},$$ (8.7)

where I_0 and I are the intensity of the incident light and that of the transmitted light respectively, σ is the cross-section of the light absorption by a single particle and N is the density (number per unit volume) of absorbing particles. Historically the Lambert Law states that the absorption is proportional to the light path length, whereas the Beer Law says that absorption is proportional to the concentration of absorbing species in the material rather than simply on the path length.

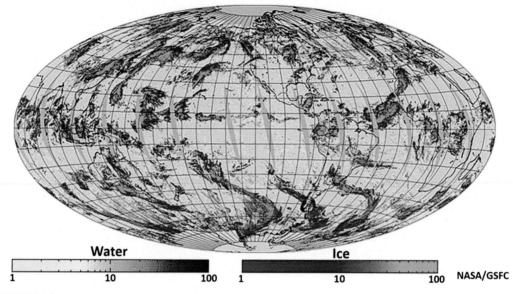

FIGURE 8.13

Aqua MODIS 24-h cloud optical thickness on October 23, 2011.

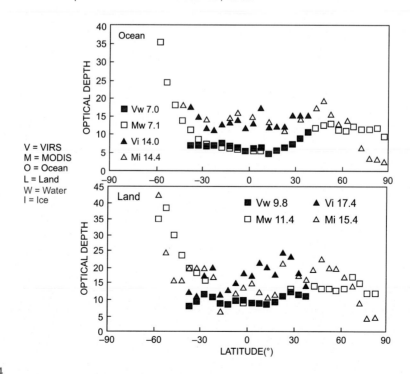

FIGURE 8.14

Zonal mean daytime cloud optical depths from MODIS and VIRS.

There are at least six conditions that must be fulfilled in order for Beer's law to be valid. These are given as follows:

1. The absorbers must act independently of each other.
2. The absorbing medium must be homogeneous in the interaction volume.
3. The absorbing medium must not scatter the radiation.
4. The incident radiation must consist of parallel rays, each traversing the same length in the absorbing medium.
5. The incident radiation should preferably be monochromatic, or have at least a width that is narrower than that of the absorbing transition.
6. The incident flux must not influence the atoms or molecules; it should only act as a noninvasive probe of the species under study. In particular, this implies that the light should not cause optical saturation or optical pumping, since such effects will deplete the lower level and possibly give rise to stimulated emission.

In the atmosphere the Beer–Lambert Law becomes a bit more complicated. There is scattering in addition to absorption that attenuates the radiation. It can be written as:

$$I = I_0 \cdot e^{-m\left\{\tau_a + \tau_g + \tau_{NO_2} + \tau_w + \tau_{O_3} + \tau_r\right\}}, \tag{8.8}$$

where each τ_x is the optical depth whose subscript identifies the source of the absorption or scattering: a refers to aerosols (that absorb and scatter), g are uniformly mixed gases [mainly carbon dioxide (CO_2) and molecular oxygen (O_2), which only absorb], NO_2 is nitrogen dioxide, mainly due to urban pollution (absorbs only), w is water vapor absorption, O_3 is ozone (absorbs only), r is Rayleigh scattering from molecular oxygen (O_2) and nitrogen (N_2) which is responsible for the blue color of the sky. m is the *optical mass* or air mass factor, a term approximately equal (for small and moderate values of θ) to $1/cos\theta$, where θ is the observed object's zenith angle (the angle measured from the direction perpendicular to the Earth's surface at the observation site).

This equation can be used to retrieve τ_a, the aerosol optical thickness, which is necessary for the correction of satellite images and also important in accounting for the role of aerosols in climate.

When the path taken by the light is through the atmosphere, the density of the absorbing gas is not constant, so the original equation must be modified as follows:

$$T = \frac{I}{I_0} = e^{-\int \alpha' \cdot dz} = e^{-\sigma \int N \cdot dz}, \tag{8.9}$$

where z is the distance along the path through the atmosphere, all other symbols are as defined earlier. This is taken into account in each τ_x in the atmospheric equation aforementioned.

8.2.2.2 The AERONET

The AERONET (AErosol RObotic NETwork) is a network of ground-based Sun photometers that measure atmospheric aerosol properties. The measurement system is a solar-powered CIMEL Electronique 318A spectral radiometer (Fig. 8.15) that measures Sun and sky radiances at a number of fixed wavelengths within the VNIR spectrum. AERONET provides continuous cloud-screened observations of spectral AOD, precipitable water, and inversion aerosol products in diverse aerosol regimes. Inversion products are retrieved from almucantar scans (circle on the celestial sphere parallel to the horizon) of radiance as a function of scattering angle and include products such as aerosol volume size distribution,

FIGURE 8.15

CIMEL Sun photometer (http://www.cimel.fr/?instrument=multi-band-sunsky-photometer&lang=en).

From CIMEL.

aerosol complex refractive index, optical absorption (single scattering albedo), and the aerosol scattering phase function. All these products represent an average of the total aerosol column within the atmosphere.

Developed by NASA under the leadership of Dr. Brent Holben, AERONET provides ground-truth calibration data for NASA and other international satellite missions. The CIMEL Sun photometer was selected for this system as it was deemed to be rather weather resistant and automatic in its operation. A sky scanning instrument provides spectral radiometry to measure atmospheric aerosol properties. These data are relayed regularly via geostationary communications satellites making it possible to continuously monitor atmospheric aerosols with this ground network (Fig. 8.16). The associated data

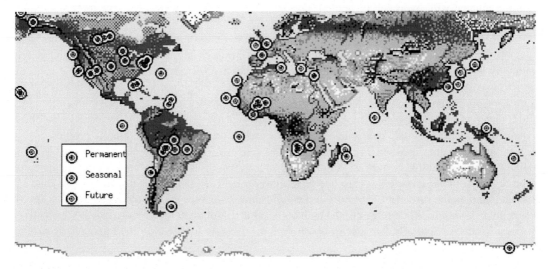

FIGURE 8.16

Approximate locations of the AERONET CIMEL Sun photometers (Holben et al., 1998).

collection system makes it possible to map approximately 90% of the Earth's surface. The resultant database is completely open access and users can directly download data from the AERONET network (Holben et al., 1998). This is enabled by a user-friendly graphical user interface making it easy for the user to select the data desired (Holben et al., 1998).

The aerosol properties are retrieved via an inversion algorithm developed by Dubovik and King (2000). Further algorithms were developed, for example, by Dubovik et al. (2006) to take into account nonspherical shapes of aerosol particles such as mineral dust. The Sun photometers are calibrated at least twice per year with a 2-m integrating sphere at NASA's Goddard Space Flight Center (GSFC). In addition, 3000 dark current measurements are made during the calibration cycle for each instrument. To judge the CIMEL's stability, the measurements made at NOAA's Mauna Loa Observatory are monitored. The location of this site above local contaminating sources and its high altitude provides a very stable aerosol and irradiance regime in the mornings (Shaw, 1963).

The results of an inversion are presented with the four channels sky radiances as a function of scattering angle, volume size distribution from 0.1 to ~8.0 μm, scattering phase function, and a table of the aerosol optical thickness and wavelength exponent from both direct Sun and the aureole (Sun's halo) measurements made by the photometer (Fig. 8.17).

The first AERONET instruments were installed in the early 1990's and since 1997 the full system of over 100 instruments has been operating. This network continues to operate providing regular global measurements of atmospheric aerosols at their homepage (http://aeronet.gsfc.nasa.gov/).

8.3 ATMOSPHERIC PROFILING
8.3.1 RADIOSONDES, RAWINSONDES, AND DROPSONDES

Traditional atmospheric profiling has been carried out using some type of sonde, which is a device that is either carried aloft by a balloon or dropped from an aircraft. Derived from a French word meaning to sound or profile the term was first applied in the early 1900s. Since the late 1930s the weather service of the National Oceanic and Atmospheric Administration (NOAA) has collected atmospheric profiles with radiosondes (Fig. 8.18), which is an instrument package suspended 25 m or more beneath a large balloon inflated with hydrogen or helium gas. Thus, these systems were given the more commonly known designation of "weather balloon."

As the radiosonde ascends at about 300 m/min, sensors on the radiosonde measure pressure, temperature, and relative humidity thus producing data on the profile of atmospheric temperature and relative humidity. These sensors are linked to a battery powered, 300 mW, or less, radio transmitter that sends the data to a sensitive ground tracking antenna on a radio frequency ranging from 1675 to 1685 MHz. Wind speed and direction aloft can be obtained by tracking the positions of sondes equipped with GPS or by a radio direction finding antenna. These latter sondes are generally referred to as rawinsondes to differentiate them from the simple radiosondes that only measure temperature and humidity profiles.

A typical weather balloon (Fig. 8.19) sounding can last longer than a couple of hours and the balloon can reach altitudes over 35 km and drift more than 300 km from their original launch site. During its ascent the sonde is exposed to temperatures as cold as −90°C and air pressure less than 1% of that found on the surface of the Earth. If the radiosonde enters the jet stream, it can travel at speeds exceeding 400 km/h. When released, the balloon is about 1.5 m in diameter and gradually expands as it rises. When it reaches a diameter of about 6−8 m, it bursts terminating the atmospheric profile.

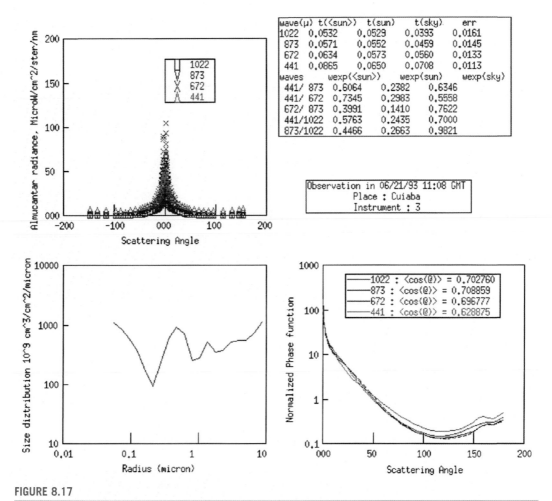

FIGURE 8.17

A successful inversion of almucantar radiances during low aerosol loading and high aerosol loading.

A small orange parachute slows the descent of the radiosonde instrument package minimizing the danger to lives and property.

Although all the data from the radiosonde ascent are used, the data from the surface to a pressure of 400 hPa (about 7,185 m in altitude for the International Standard Atmosphere for dry air) are considered to be minimally acceptable for National Weather Service operations. Thus, a flight may be deemed a failure and a second radiosonde will be released if the balloon bursts before reaching the 400 hPa pressure level or if more than 6 min of pressure and/or temperature data are missing between the surface and the 400 hPa level.

Worldwide there are over 800 upper-air observation stations and through international agreements data are exchanged routinely between countries. Most upper-air stations are located in the Northern Hemisphere and all observations are usually taken at the same time each day (up to an hour between

FIGURE 8.18

A typical radiosonde instrument package.

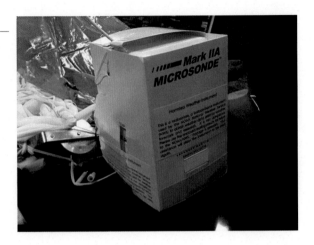

FIGURE 8.19

Typical weather balloon configuration.

00:00 and/or 12:00 UTC), 365 days a year. When severe weather is expected additional sounding may be taken during the day at a selected number of stations.

Radiosonde profiles are used as input for computer-based weather forecast models; for local forecasts of severe weather, for aviation and marine forecasts; to study weather and climate change, as input for air pollution models and as ground truth for satellite data.

Another sonde-based method of atmospheric profiling uses a dropsonde, which is a radiosonde package that is dropped from a specified altitudes. Each dropsonde also carries sensors to measure pressure, temperature, and relative humidity, and they also carry a GPS receiver as well to sense wind speed and direction. Like the radiosonde, during its descent the dropsonde is equipped with a parachute to slow its descent and to increase its vertical stability during its flight (Fig. 8.20).

During the dropsonde descent the data on pressure, temperature, and relative humidity are sent back by radio just as in the radiosonde. The advantage of the dropsonde over the radiosonde is that it provides a nearly vertical atmospheric profile when compared with the typical radiosonde profile. Each sonde is about 40 cm long, and 5.8 cm in diameter, and weighs 0.39 kg.

8.3.2 SATELLITE REMOTE SENSING ATMOSPHERIC PROFILING

8.3.2.1 The TIROS Operational Vertical Sounder

First deployed in space in 1978 the TOVS was the first system designed to measure atmospheric profiles from orbiting weather satellites. TOVS was actually not a single instrument, but rather a group of instruments. The primary sensor was the high-resolution infrared radiation sounder (HIRS), which is a discrete stepping, line-scan radiometer measuring scene radiances in 20 spectral bands to make it possible to calculate the vertical temperature profile from the Earth's surface to about 40 km altitude.

Multispectral data from a single visible channel (90.69 μm), 7 shortwave channels (3.7−4.6 μm), and 12 longwave channels (6.5−15 μm) are obtained from a single telescope, and a rotating filter wheel containing 20 filters. An elliptical scan mirror provides the cross-track scanning of 56 increments each being 1.8°. The mirror steps rapidly (<35 ms), then holds at each position while the 20 filter segments are sampled, which takes about 100 ms. The resulting field of view (FOV) for each channel is approximately 1.4° in the visible and shortwave IR bands, and 1.3° in the longwave IR band, which from an altitude of 833 km covers an area 18.9 km in diameter at nadir on the scan (http://www.class.ngdc.noaa.gov/data_available/tovs_atovs/index.htm).

The TOVS system also includes the microwave sounding unit (MSU) which is a four-channel passive microwave spectrometer operating in the 5.5 μm oxygen region of the atmosphere. The MSU has two four-inch diameter antennas each having an instantaneous FOV (IFOV) of 7.5° giving a ground resolution of 124 km at nadir for a satellite altitude of 833 km. The distance between these 124 km spots at nadir is 168.1 km. The instrument has a 12-bit digitization and the values may be converted into brightness temperatures using the calibration information, which is appended to the data stream but not automatically applied. Latitude and longitudes for each Earth FOV is included in the data stream as are time tags.

Because the MSU wavelengths can penetrate most clouds, its measurements can provide temperatures below clouds. Some of the problems with these microwave measurements come from changes in surface emissivity and the sensitivity of these measurements to changes in precipitation.

Finally, TOVS includes the stratospheric sounding unit (SSU), which is another step-scanned, infrared spectrometer with only three channels in the 15 μm carbon dioxide absorption band designed to make measurements at the top of the Earth's atmosphere. The three channels have the same

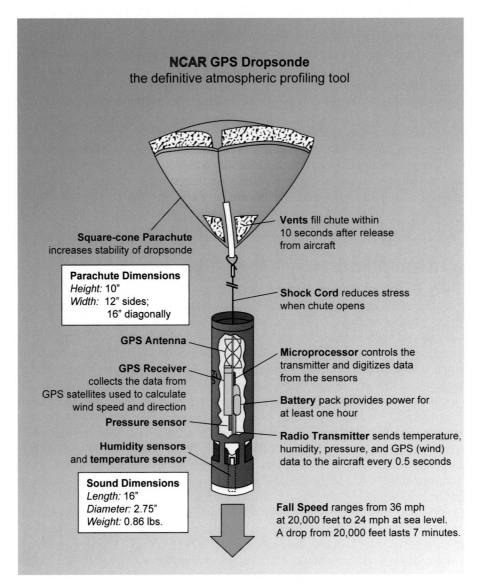

FIGURE 8.20

NCAR GPS Dropsonde (https://www.eol.ucar.edu/content/what-dropsonde).

From © 2013 University Corporation for Atmospheric Research/COSMIC.

frequency, but different cell pressures designed for different atmospheric heights. The 10° FOV results in a 147 km Earth surface resolution at nadir. A calibration sequence is initiated every eight scans and the calibration are included in the data stream, but are not automatically applied to the sensor data.

The HIRS, MSU, and SSU channels are described here in Table 8.1 with a reference to the principal atmospheric gas that they are sensing and the primary purpose of this radiance observation.

Table 8.1 Characteristics of TOVS Channels

HIRS Channel Number	Channel Central Wave Number	Central Wavelength (μm)	Principal Absorbing Constituents	Level of Peak Energy Contribution	Purpose of the Radiance Observation
1	668	15.00	CO_2	30 mb	*Temperature sounding.* The 15-μm band channels provide better sensitivity to the temperature of relatively cold regions of the atmosphere than can be achieved with the 4.3-μm band channels. Radiances in channels 5, 6, and 7 are also used to calculate the heights and amounts of cloud within the HIRS field of view.
2	679	14.70	CO_2	60 mb	
3	691	14.50	CO_2	100 mb	
4	704	14.20	CO_2	400 mb	
5	716	14.00	CO_2	600 mb	
6	732	13.70	CO_2/H_2O	800 mb	
7	748	13.40	CO_2/H_2O	900 mb	
8	898	11.10	Window	Surface	*Surface temperature* and cloud detection.
9	1028	9.70	O_2	25 mb	*Total ozone* concentration.
10	1217	8.30	H_2O	900 mb	*Water vapor sounding.* Provides water vapor corrections for CO_2 and window channels. The 6.7-μm channel is also used to detect thin cirrus cloud.
11	1364	7.30	H_2O	700 mb	
12	1484	6.70	H_2O	500 mb	
13	2190	4.57	N_2O	1000 mb	*Temperature sounding.* The 4.3-μm band channels provide better sensitivity to the temperature of relatively warm regions of the atmosphere than can be achieved with the 15-μm band channels. Also, the short-wavelength radiances are less sensitive to clouds than those for the 15-μm region.
14	2213	4.52	N_2O	950 mb	
15	2240	4.46	CO_2/N_2O	700 mb	
16	2276	4.40	CO_2/N_2O	400 mb	
17	2361	4.24	CO_2	5 mb	
18	2512	4.00	Window	Surface	*Surface temperature.* Much less sensitive to clouds and H_2O than the 11-μm window. Used with 11-μm channel to detect cloud contamination and derive surface temperature under partly cloudy sky conditions. Simultaneous 3.7- and 4.0-μm data enable reflected solar contribution to be eliminated from observations.
19	2671	3.70	Window	Surface	
20	14 367	0.70	Window	Cloud	*Cloud detection.* Used during the day with 4.0- and 11-μm window channels to define clear fields of view.

MSU	Frequency (GHz)	Principal Absorbing Constituents	Level of Peak Energy Contribution	Purpose of the Radiance Observation
1	50.31	Window	Surface	*Surface emissivity and cloud attenuation determination.*
2	53.73	O_2	700 mb	*Temperature sounding.* The microwave channels probe through clouds and can be used to alleviate the influence of clouds on the 4.3- and 15-pm sounding channels.
3	54.96	O_2	300 mb	
4	57.95	O_2	90 mb	

SSU	Wavelength (μm)	Principal Absorbing Constituents	Level of Peak Energy Contribution	Purpose of the Radiance Observation
1	15.0	CO_2	15.0 mb	*Temperature sounding.* Using CO_2 gas cells and pressure modulation, the SSU observes thermal emissions from the stratosphere.
2	15.0	CO_2	4.0 mb	
3	15.0	CO_2	1.5 mb	

Temperature and humidity profiles were inferred from the responses of the various atmospheric gases that have preferred altitudes in the atmosphere. As a result, the HIRS, MSU, and SSU channels have the weighting functions shown here in Fig. 8.21 where the y axis it given in mbars of atmospheric pressure, which is of course greatest at the surface and decreases as going up in the atmosphere.

Note that most of the channels are used to compute the atmospheric temperature profile while only four of the longwave thermal infrared channels are used to compute the water vapor profile. Since most of this water vapor is in the troposphere, these weighting functions (Fig. 8.21 lower right) go from the surface up to about 400 mbar. This set of weighting functions has a very different vertical scale than the other three groups of weighting functions. The first seven channels are well distributed over the atmosphere with channels 4–7 concentrated in the lower levels of the atmosphere. Likewise, the channels 13–16 are also concentrated near the surface. The SSU channels are all in the upper layer while the MSU microwave channels all have their temperatures below the altitude of 20 mbar (Fig. 8.21 lower left).

One of the biggest challenges in the TOVS system is deciding just how to combine the HIRS, MSU, and SSU data. For example, the HIRS and MSU sampling are dramatically different as pointed out in Fig. 8.22. The challenge of merging the much larger MSU spots with the high resolution of the HIRS data can be appreciated. This is done by averaging those HIRS spots together that are completely within an MSU spot in the final TOVS product. As indicated by Table 8.1 the HIRS channels are also used individually for other surface and atmospheric parameter retrievals.

The actual retrieval of atmospheric temperature and water vapor profiles involves an inversion method using atmospheric radiative transfer modeling. This results in a linear least squares fit of the atmospheric model to the TOVS radiances. A Taylor series expansion is used to linearize the atmospheric radiative transfer equations and boundary conditions are set using ancillary surface observations (Svensson, 1985). These many approximations resulted in rather large errors in the derived atmospheric temperature and moisture profiles.

Thus, the assimilation of the independently derived TOVS temperature and humidity profiles into numerical forecast models did not improve the model forecast (Eyre et al., 1993). In response, methods were developed to directly assimilate TOVS sensor radiances directly into the numerical model. A scheme known as "one-dimensional variational analysis" (1DVAR) was developed at the European Center for Medium-range Weather Forecasts (ECMWF) as a method for extracting information from TOVS radiances for use in their operational data-assimilation system. The 1DVAR is based on variational principles applied to the analysis of the atmospheric profile at a single location, using a forecast profile and its error covariance as constraints. This development has led to a three-dimensional approach, which is now the standard for the routine assimilation of cloud-cleared TOVS radiances into long-range weather forecast models (Derber and Wu, 1997).

The TOVS system was replaced by the Advanced TIROS Operational Vertical Sounder (ATOVS) with the launch of the NOAA-15 satellite December 15, 1998. In the ATOVS system the MSU was replaced by the Advanced Microwave Sounding Unit-A (AMSU-A).

8.3.2.2 The Advanced TIROS Operational Vertical Sounder

A change in the TOVS system took place when the satellite NOAA-15 was launched carrying the AMSU-A, the Advanced Microwave Sounding Unit-B (AMSU-B) and the High Resolution Infrared Sounder Version 3 (HIRS/3). These instruments replaced the SSU, the MSU and the HIRS/2.

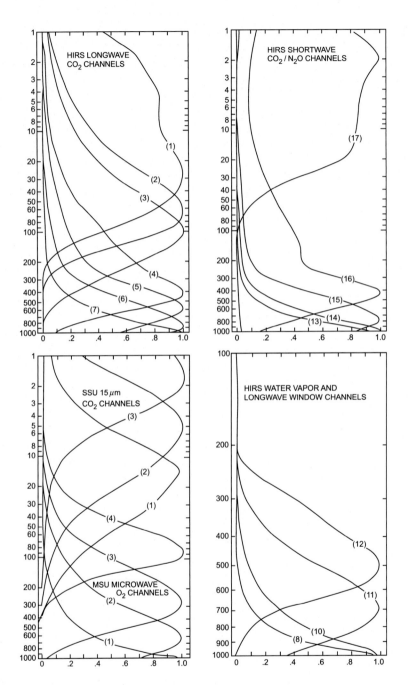

FIGURE 8.21

TOVS weighting functions (normalized).

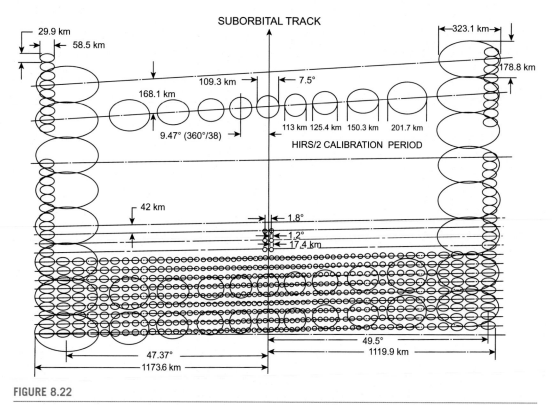

FIGURE 8.22

TOVS HIRS and MSU scan patterns projected on to the Earth's surface.

Beginning with NOAA-18 the microwave humidity sounder (MHS) replaced the AMSU-B and the HIRS/3 was upgrade to the HIRS/4. The European MetOp satellites also carry this same suite of instruments.

The characteristics of HIRS/4 are given here in Table 8.2, which briefly summarizes the parameters of this instrument.

In addition to profiles of atmospheric temperature and moisture, HIRS data are used for a number of other applications such as the derivation of surface skin temperature of the land and ocean surfaces, sensing outgoing longwave radiation, cloud fraction, cloud-top height, total integrated column ozone, precipitation and wind speed and direction. Global coverage is available for the HIRS data, which has an IFOV of approximately 1.4° in the visible and shortwave channels, and 1.3° in the longwave infrared bands. For the nominal altitude of 833 km these result in a ground IFOVs of 20.3 and 18.9 km in diameter respectively at nadir. At the edge of scan this increases to a pixel that is 68.3 km cross-track by 34.8 km along-track for the visible and shortwave infrared channels. At the equator the 99° scan equates to a swath width of 2248 km centered on the subsatellite track. As a result of this scan geometry, there is a variable underlap between steps and scans.

Table 8.2 Summary of HIRS/4 Parameters

Parameter	Value
IR calibration	Warm target and space background
Channels	20
Cross-track scan angle	±49.5°
Scan time	6.4 s
Number of steps	56
Step angle	1.8°
Step time	100 ms
Data precision	13 bits
Time between start of each scan	6.4 s
Angular field of view (FOV) (visible, shortwave IR)	1.40°
Angular FOV (longwave IR)	1.30°
At an altitude of 833 km:	
Nadir ground instantaneous field of view (IFOV) (visible, shortwave IR)	20.3 km diameter
Nadir ground IFOV (longwave IR)	18.9 km diameter
Scan-end ground IFOV (visible, shortwave)	68.3 km cross-track by 34.8 km along-track
Swath width (visible, shortwave IR)	2248.8 km to far edge of outer FOV

8.3.2.2.1 Advanced Microwave Sounding Unit-A

The AMSU-A is designed as a multichannel microwave radiometer to measure atmospheric temperature profiles and to gather information on atmospheric water in its various forms excepting small ice particles, which are transparent at microwave frequencies. It is a cross-track, line-scanned instrument that measures scene radiances in 15 discrete frequency channels. These measurements make it possible to calculate the vertical atmospheric temperature profiles from about 3 mbar (~45 km) to the Earth's surface. At each frequency the antenna beamwidth is a constant 3.3° (at the half-power point). Thirty contiguous scene resolution cells are sampled in a stepped-scan fashion every 8 s with each scan covering 50° of each side of the subsatellite path. This scan pattern and angular resolution translate to a 50 km diameter cell at nadir and a 2343 km swath width for an 833 km nominal orbital altitude.

The AMSU-A is composed of two separate modules: the AMSU-A1 and the AMSU-A2. The first has 12 channels (3 through 14 μm), and one 15 μm channel, which together provide a complete and accurate vertical atmospheric temperature profile from the Earth's surface to about 45 km altitude. AMSU-A2 contains the two lower-frequencies (K-band channel 1 and Ka-band channel 2), which are used to study the atmospheric water content excepting the small ice particles as noted earlier. A summary of AMSU-A parameter can be found here in Table 8.3.

AMSU-A replaced the functions of the MSU and the SSU in the original TOVS system. For this reason, AMSU-B data were not used in the ATOVS system to retrieve atmospheric temperature and moisture profiles. Global coverage is available for the AMSU-A data, which have an IFOV of approximately 3.3°, which for the usual 833 km altitude translates to a ground resolution of 48.05 km

Table 8.3 Summary of AMSU-A Parameters

Parameter	Value
Calibration	Internal target and space background
Channels	15
Cross-track scan angle	±48.33°
Scan time	8.0 s
Number of steps	30
Step angle	3.33°
Step time	202.5 ms
Data precision	16 bits
Time between start of each scan	8.0 s
Angular FOV	3.33°
At an altitude of 833 km:	
Ground IFOV at nadir	48.05 km diameter
Ground IFOV at center of outer FOV	149.1 km cross-track by 79.4 km along-track
Swath width	2226.8 km to far edge of outer FOV

in diameter at nadir. At scan edge the resolution reduces to a 149.1 km cross-track by a 79.4 km along-track footprint.

AMSU-B is a 5-channel microwave radiometer designed to measure radiation from different layers in the atmosphere to obtain global data on humidity profiles. AMSU-B covers channels 16–20 where channels 18, 19, and 20 span the strongly opaque water absorption line at 183 GHz to provide data on the atmosphere's humidity level. Channels 16 and 17 at 80 and 150 GHz, respectively, penetrate through the atmosphere to the Earth's surface. Beginning with NOAA-18, AMSU-B was replaced by the MHS. At each channel frequency the antenna beamwidth is a constant 1.1° (at the half-power point). The scan pattern and this geometric configuration translate to the ground resolution of 16.0 km diameter at nadir for the nominal 833 km altitude.

The MHS is a self-calibrating microwave radiometer, observing the Earth with an FOV of ±50° from nadir in five frequency channels of the millimeter-wave band (90–190 GHz). Together with AMSU-A the MHS provides the operational microwave atmospheric sounding capability for NOAA 18, 19 satellites as well as the European MetOp satellites. A summary of MHS parameters is given in Table 8.4.

8.3.2.3 The Visible Infrared Spin-Scan Radiometer Atmospheric Sounder

The Visible Infrared Spin-Scan Radiometer (VISSR) Atmospheric Sounder (VAS) was designed to operate in three distinct modes to provide parameter measurement, sampling flexibility, spectral band selection, geographic location, and variable sensitivity. In the VISSR mode it was a scanner just like the previous VISSR instruments that had been on board GOES 1, 2, 3. The visible (0.55–0.75 μm) and the thermal infrared (10.5–12.5 μm) use common Ritchey–Chrétien optical system. A single west-to-east raster scan line was collected for each revolution of the spinning spacecraft. A 20° north-to-south image frame was collected from 1821 steps of the scan mirror, one 0.192-mrad step for each spacecraft revolution. A full image took 18.2 min to complete with an additional 2 min to reset for the next image.

Table 8.4 Summary of Microwave Humidity Sounder Parameters

Parameter	Value
Calibration	Internal blackbody target and space background
Channels	5
Cross-track scan angle	±49.44°
Scan time	2.67 s
Number of steps	90
Step angle	N/A
Step time	N/A
Data precision	16 bits
Time between start of each scan	1.11 s
Angular field of view (FOV)	1.1°
At an altitude of 870 km:	
Ground instantaneous FOV (IFOV) at nadir	17 km diameter
Ground IFOV at center of outer FOV	51.6 km cross-track by 26.9 km along-track
Swath width	2348 km to far edge of outer FOV

Eight visible-spectrum detectors (0.9 km horizontal resolution) and one mercury-cadmium-telluride thermal infrared detector (6.9 km horizontal resolution) would sweep the Earth during each scan. The dwell-sounding mode used up to 12 spectral filters in a wheel covering the range 14.74 μm through 3.94 μm positioned in the optical train while the scanner occupies a single scan line. The filter wheel was programmed so that each spectral band filter dwells on a single scan line for from 0 to 255 spacecraft spins. Either the 6.9 km or 13.8 km resolution detectors can be used for the 7 filter positions in the spectral range 6.725−14.25 μm. The 13.8 km resolution detectors were used. Selectable frame size, position, and scan direction were programmable from the ground. For VAS a 10-bit reduced resolution (3.5 km) visible image provided the general information needed for the VAS data processing. In some of the spectral channels multiple-line sampling was required to enhance the signal-to-noise ratio. Typically, 20−30 satellite spins at a single north-to-south scan line were required to obtain the desired accuracy in the sounding data. This number of repeat spins per line should provide the atmospheric sounding at a 30 × 30 km resolution.

The multispectral imaging mode can provide normal VISSR-IR imagery plus data in any two selected spectral bands having a resolution of 13.8 km. This band takes advantage of the small mercury-cadmium-telluride detector offset in the north-to-south plane. Using the data from these detectors simultaneously produces a complete infrared map when they are operational every other scan line. This fact allows using the larger detectors during half of the imaging/scanning sequence to obtain additional spectral information. Unlimited north-to-south frame sizes and position detection, within the maximum north-to-south FOV scan direction, could be selected. Visible data are not available in this mode.

The VISSR output was to be digitized and transmitted to the ground. There, the signal was fed into a "line stretcher," where it was stored and time-stretched. The processed data were sent back to the

Table 8.5 VAS Channel Characteristics

| VAS Ch. No. | Spectral Filters | | | Purpose for Sounding | Main Absorbing Gas | Other Significant Effects |
| | Center | | Width | | | |
	μm	cm^{-1}	cm^{-1}			
1	14.730	0678.7	10.0	Temperature	CO_2	O_3
2	14.480	0690.6	16.0	Temperature	CO_2	O_3
3[a]	14.250	0701.6	16.0	Temperature	CO_2	O_3
4[a]	14.010	0713.6	20.0	Temperature	CO_2	O_3
5[a]	13.330	0750.0	20.0	Temperature	CO_2	O_3
6	04.525	2210.0	45.0	Temperature + cloud	N_2O	Sun
7[a]	12.660	0790.0	20.0	Moisture	H_2O	CO_2 + dust
8[a]	11.170	0895.0	140.0	Surface	H_2O	CO_2 + dust
9[a]	07.261	1377.0	40.0	Moisture	H_2O	CO_2
10[a]	06.725	1487.0	150.0	Moisture	H_2O	—
11	04.444	2250.0	40.0	Temperature + cloud	N_2O, CO_2	Sun
12	03.945	2535.0	140.0	Surface	H_2O	Sun + dust

[a]*Available at 6.9 km as well as 13.8 km instantaneous geographic field of view.*

satellite in a reduced resolution form to serve as the WEFAX for broadcast on the VHF band. VISSR and VAS data are then transferred to the Comprehensive Large Array-data Stewardship System (CLASS) where it is archived and available for distribution.

The VAS channels and their uses are described here in Table 8.5 (Montgomery and Uccellini, 1985).

These channels correspond to the weighting functions shown here in Fig. 8.23. A study of simulations of this instrument by Chesters et al. (1983) determined that: significant moisture gradients can be seen directly in image of the VAS channels, temperature, and moisture profiles can be retrieved with sufficient accuracy to delineate the major features of a severe storm environment, and the quality of VAS mesoscale sounding improves with conditioning using local weather statistics. It was also found that it was still possible to retrieve mesoscale regions of potential instability for a mostly clear prethunderstorm environment. The RMS tropospheric profile errors are $\pm 1°C$ and $\pm 25\%$ in temperature and mixing ratio, respectively. These simulations suggest that the VAS data will yield the best soundings when a human being classifies the scene, picks relatively clear areas for retrieval, and applies a "local" statistical database to resolve the ambiguities of satellite observations in favor of the most probably atmospheric structure.

A comparison between the characteristics of the VAS and HIRS are given here in Table 8.6 from Montgomery and Uccellini (1985).

This same study produced a numerical comparison between the coincident channels on VAS and HIRS for three different VAS on three different GOES satellites (Table 8.7). All of the large mean differences have the VAS value higher than that for HIRS and are generally in the higher frequencies

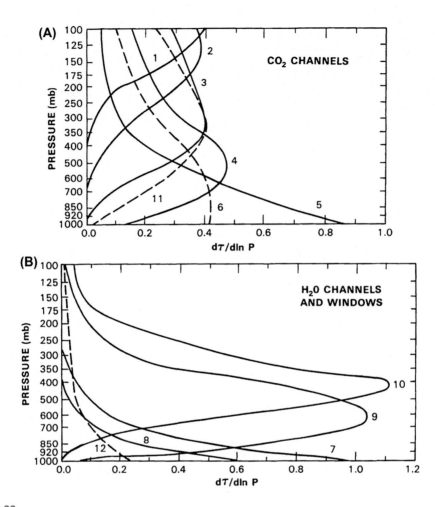

FIGURE 8.23

Standard VAS weighting functions.

used for the moisture retrieval. The large mean differences are not always associated with a large standard deviation suggesting a bias in the VAS measurements. Different VAS instruments have different large biases in different channels so this behavior is not consistent between the different VAS instruments. It must be remembered that the HIRS and VAS were not completely coincident in time and space when these comparisons were made.

Montgomery and Uccellini (1985) concluded that the agreement between the VAS and HIRS observations is surprisingly good for the CO_2 channels. The temperature calibrations of the respective instruments are accurate to only $\pm 1K$, which is about the general order of magnitude of the agreement between the channels. The surface weighting bands 7, 8, and 12 are very sensitive to good geolocation and temporal coincidence. The water vapor bands 9 and 10 are sensitive to the constancy of the water

Table 8.6 A Comparison Between the Characteristics of the VAS and HIRS		
Feature	**VAS on GOES**	**HIRS on NOAA**
Orbit	Geostationary	Polar
Altitude	30,000 km	800 km
Relative flux	1 unit	1400 units
Mirror radius	20 cm	7.5 cm
Sampling time	Programmable dwell (0–225) × 8 μs	Single glance 700 μs
IGFOV	13.8 or 6.9 km	21.8 km
Noise	±0.5K	±0.1K
Coverage	Programmable 15° latitude swath per 10 min	Suborbital 30° longitude swath per overpass
Frequency	Programmable, e.g., 15-min intervals	Fixed every 12 h
Calibration	Space/internal blackbody foreoptics corrections	Space/external blackbody
IR channels	12 total: 5 CO_2 @ 13–15 μm 2 CO_2, N_2O @ 4.4–4.5 μm 3 H_2O @ 6.7, 7.2, 12.6 μm 2 windows @ 11.2, 3.9 μm	19 total: 7 CO_2 @ 13–15 μm 5 CO_2, N_2O @ 4.2–4.5 μm 3 H_2O @ 6.7, 7.3, 8.3 μm 3 windows @ 11.1, 4.0, 3.7 μm 1 O_3 @ 9.7 μm

IGFOV, instantaneous geographic field of view.

vapor concentration in each satellite's FOV. With the 3-h time separation used here, large differences in water vapor content are likely. Earlier studies Menzel et al. (1981), and Chesters et al. (1983) have demonstrated that VAS channels 2, 3, and 4 exhibit a consistent negative bias with respect to HIRS. The VAS radiances were also compared to radiative transfer calculations of VAS radiances using radiosonde data for several hundred cloud-free FOVs. As in the HIRS comparison the VAS radiances of GOES-4 bands two to four are consistently lower than the corresponding values from radiosonde analyses. The VAS calibration algorithm appears to generate values that are 2–3°C below the radiosonde values for these bands. The relative errors for GOES-5 and GOES-6 are typically less than 1°C for the CO_2 bands, which is well within the expected accuracy of these data.

8.3.2.4 Atmospheric Infrared Sounder

AIRS is a NASA facility instrument to support climate research and improve weather forecasting by providing detailed and accurate atmospheric temperature and moisture profiles. It was launched on NASA's Aqua satellite on May 4, 2002, and is one of the six instruments carried by this satellite which is a part of NASA's Earth Observing System of Satellites. Two of the other instruments carried by Aqua are the AMSU (discussed earlier) and the Humidity Sounder for Brazil (HSB). Combined with the AIRS these instruments provide very comprehensive atmospheric column information on both temperature, cloud amount-height, spectral outgoing infrared (3.7–15.4 μm) radiation and water vapor from the top of the atmosphere to the Earth's surface where it also provides estimates of the

Table 8.7 Brightness Temperature Differences for VAS Versus HIRS for Three GOES Satellites

VAS		HIRS		T* (HIRS Est.) − T* (VAS Obs.)					
				GOES-4 vs. NOAA-6		GOES-5 vs. NOAA-7		GOES-6 vs. NOAA-8	
Ch. No.	ν (cm^{-1})	Ch. No.	ν (cm^{-1})	Mean (K)	Dev. (±K)	Mean (K)	Dev. (±K)	Mean (K)	Dev. (±K)
1	679	2	679	0.94	1.38	0.78	0.80	−1.56	1.08
2	691	3	691	0.82	1.01	−0.43	0.47	−0.30	0.32
3	702	4	704	1.65	0.77	−0.20	0.21	0.14	0.51
4	714	5	716	1.28	0.40	−1.26	0.43	0.71	0.46
5	750	7	748	−1.03	0.45	−0.88	0.52	−2.80	0.49
6	2210	14	2212	−0.33	0.60	−2.52	0.55	−1.93	0.53
7	790	10	1217	−5.61	0.61	−2.97	1.07	−3.45	1.18
8	895	8	900	−3.61	0.92	0.76	0.99	−0.89	1.02
9	1377	11	1363	−1.21	1.29	−5.99	1.88	−2.73	0.99
10	1487	12	1484	0.12	0.99	−5.51	1.50	−0.78	0.99
11	2250	15	2240	0.95	1.39	NA	NA	NA	NA
12	2535	18	2511	−0.52	1.84	−2.89	1.07	−1.82	1.99

surface emissivity. AIRS's advanced technology makes it the most advanced water vapor sensor ever built. AIRS can also measure trace greenhouse gases such as ozone, carbon monoxide, carbon dioxide, and methane. Together these data and the attendant scientific investigations are answering long-standing questions about the exchange and transformation of energy of radiation within the atmosphere and at the Earth's surface. The Joint Center for Satellite Data Assimilation, established to accelerate the assimilation of satellite observations into operational weather forecast models, announced a significant improvement in forecast skill achieved with the assimilation of AIRS data.

The AIRS instrument (Fig. 8.24) was built by BAE Systems for NASA/JPL and operates as a cross-track scanning instrument. Its scan mirror rotates around an axis along the line of flight and directs infrared energy from the Earth into the instrument with the basic swath covering approximately 800 km on either side of the nadir ground-track. The nadir spatial resolution is 13.5 km. There is also a set of VNIR detectors divided into four intermediate and broadband spectral channels. The VNIR spatial resolution is approximately 2.3 km. These channels provide a diagnostic imaging capability for observing low-level clouds. AIRS uses a cooled (155K) grating spectrometer to separate the wavelengths into fairly narrow wavebands across arrays of high sensitivity HgCdTe detectors. These IR detectors are divided into 12 modules containing 17 linear arrays distributed in a two-dimensional pattern on the cold focal plane. The detectors are of two types, photovoltaic (PV, up to 13.7 μm) and photoconductive (PC) for the 13.7 and 15.4 μm wavelengths.

The word sounder in the instrument's name means that its primary purpose is to measure temperature and water vapor as functions of atmospheric height. As noted earlier, AIRS also measures

FIGURE 8.24

The AIRS instrument (http://airs.jpl.nasa.gov/mission_and_instrument/instrument).

clouds, the abundance of such atmospheric trace components as ozone, carbon monoxide, carbon dioxide, methane, and sulfur dioxide, and it also detects suspended dust particles.

To sense these profiles and atmospheric constituents, AIRS measures the infrared brightness coming up from the Earth's surface as well as from the atmosphere for specific infrared wavelengths. Each infrared wavelength is sensitive to temperature and water vapor over a range of heights in the atmosphere from the Earth's surface up to the stratosphere. Using multiple infrared detectors, each sensing a specific wavelength, a temperature profile (or sounding) of the atmosphere can be observed. While prior profilers had 15 detectors, AIRS has 2378 of them, all of which are sampled simultaneously. This simultaneity is an essential requirement for accurate temperature retrievals under partly cloudy conditions. Calibration is carried out each scan by views of onboard blackbody radiometric calibration sources and of cold space, which greatly improves the sounding accuracy, making it comparable to measurements made by weather balloons.

The high spectral infrared resolution of AIRS dramatically improves the vertical profile resolution, but all infrared sensing degrades in the presence of clouds. The microwave AMSU and HSB instruments are insensitive to clouds and provide all-weather sounding to compliment the higher resolution AIRS soundings. Thus, a system has been developed to combine AIRS, AMSU, and HAB data to routinely provide global information on the state of the atmosphere.

8.4 RAIN RATE, ATMOSPHERIC LIQUID WATER, AND CLOUD LIQUID WATER

Rain rate can be estimated from different techniques. Microwave radiometers can provide the total averaged rain rate over the antenna footprint, and radars can provide the volumetric distribution of radar backscatter that can be related to the rain rate.

8.4.1 RAIN RATE ESTIMATION USING MICROWAVE RADIOMETRY

Due to the high emissivity of land, there is not much difference between the emission produced by the raindrops and the emission of the surface below the rain. Thus, emission-based techniques cannot be used to infer rain rate, but it can only be found from the dispersion it produces. On the other hand, the low emissivity of the oceans makes the measurements taken in the lower frequency channels much smaller, so a large dispersion would be required to obtain a positive scattering index (SI) between the 22 and 85 GHz channels. The reader is encouraged to review Figs. 4.52, and 4.58–4.60 for a visual interpretation of the processes involved.

8.4.1.1 Precipitation Detection Over Ground Surfaces

To detect the presence of rain in the atmosphere, we rely on the fact that the dispersion created by precipitation is always larger at 85 than it is at 37 GHz, as it occurs in the case of snow. It is also important to filter iced soils, cold deserts, warm deserts … which is accomplished by setting thresholds (1) between the brightness temperatures at a given polarization, typically the vertical polarization, because around 53° incidence angle (incidence angle of most conical scanning radiometers) it is less sensitive to surface roughness effects, and (2) between the brightness temperature at the same frequency, but at different polarizations (vertical minus horizontal). Once the detection of the pixels where there is rain, the rain rate can be estimated in (mm/h) using different algorithms. As an example, for illustration purposes, in the Ferraro algorithm the rain rate is estimated from the SI, which is defined in Eq. (8.10) as a combination of the SSM/I channels. The SI measures the difference between the predicted T_{85V} from the T_{19V} and T_{22V} channel in rain free conditions over land, and the measured value of T_{85V}. Note that the dependence with the 22 GHz channel is quadratic and has opposite sign than the other linear terms, to compensate for the increased attenuation due to the water vapor absorption:

$$SI = 438.5 - 0.46 \cdot T_{19V} - 1.735 \cdot T_{22V} + 0.00589 \cdot T_{22V}^2 - T_{85V}, \qquad (8.10)$$

$$RR\,(\text{mm/h}) = -2.71 + 0.362 \cdot SI. \qquad (8.11)$$

8.4.1.2 Precipitation Detection Over the Ocean

Since the ocean's emissivity is much lower than that of the land, the algorithm can be based on the rain emission. Liquid water produces a much stronger signal than ice. In addition, the microwave emission from the ocean surface is strongly polarized, while the emission from raindrops is unpolarized, which allows us to distinguish it from the ocean surface's emission using measurements at vertical and horizontal polarization. As in the previous case the rain rate is estimated from the SI computed as the difference between the predicted T_{85V} from the T_{19V} and T_{22V} channel in rain free conditions over the ocean, and the measured value of T_{85V}:

$$SI = -182.7 + 0.75 \cdot T_{19V} - 2.543 \cdot T_{22V} + 0.00543 \cdot T_{22V}^2 - T_{85V}, \qquad (8.12)$$

$$RR\,(\text{mm/h}) = -1.05 + 0.149 \cdot SI. \qquad (8.13)$$

Global rain rate maps are currently distributed by Remote Sensing Systems (http://www.remss.com/measurements/rain-rate) using the more sophisticated Hilburn and Wentz (2008) algorithm, and data from the SSM/I, SSMIS, TMI, AMSR-E, AMSR-2, and WindSat missions.

Because of the strong water vapor absorption line near 22 GHz, microwave radiometers can also measure the **atmospheric water vapor** (total column water vapor, or total precipitable water) over the oceans, which is a measure of the total gaseous water contained in a vertical column of atmosphere, and it is measured in linear units (mm), as would be the depth of the water if it condensed to liquid form.

At 18 and 37 GHz, clouds are semitransparent allowing for measurement of the total columnar absorption. After accounting for oxygen and water vapor absorption, the absorption can be related to the total amount of liquid water in the viewing path. The **cloud liquid water** is then a measure of the total liquid water contained in a cloud in a vertical column of the atmosphere, excluding solid water (e.g., snow, ice), it is a very variable measure which depends on the type of clouds, and it is measured in (g/m^3) or (g/kg), or also in (mm) as the depth of water in a column. The liquid water content is always one or two orders of magnitude lower than the content of water vapor. Sometimes it is also called liquid water in clouds, because it is the main responsible parameter for low altitude clouds, which are dominant in the marine meteorology of the first kilometer of the atmosphere. It can be computed as:

$$LWC\ (\text{kg}/\text{m}^2) = 131.95 - 39.5 \cdot log(280 - T_{22V}) + 12.49 \cdot log(280 - T_{37V}). \quad (8.14)$$

These variables are also routinely distributed by Remote Sensing Systems through their online servers (http://www.remss.com/measurements/atmospheric-water-vapor and http://www.remss.com/measurements/cloud-liquid-water-content).

8.4.2 RAIN RATE ESTIMATION USING RADAR

In Chapter 5, the basic radar equation was presented. In the monostatic case, it provides the returned power P_r from a given target of "radar cross-section" σ, as:

$$P_r = \frac{P_t \cdot G^2 \cdot \lambda^2 \cdot \sigma}{(4 \cdot \pi)^3 \cdot R^4}, \quad (8.15)$$

where P_t is the transmitted power, G is the antenna gain, λ is the electromagnetic wavelength, and R is the distance from the radar to the target. In the case of a volumetric target, the "radar cross-section" σ becomes the product of the volume observed by the radar, times the radar cross-section volumetric density (η). Since the volume observed by the radar is equal to its resolution cell, then:

$$\sigma = \frac{c}{2B} \cdot (R \cdot \Delta\theta_H) \cdot (R \cdot \Delta\theta_V) \cdot \eta, \quad (8.16)$$

where B is the bandwidth of the transmitted signal, $\Delta\theta_H$ and $\Delta\theta_V$ are the antenna beam widths in the horizontal and vertical planes. Note that, inserting Eq. (8.16) in Eq. (8.15), now the dependence with distance becomes as $1/R^2$, as opposed to $1/R^4$ for point targets. This is due to the fact that the radar resolution cell increases with the distance.

However, raindrops are not of the same size. The wavelengths used in rain radars (1−10 cm) ensure the validity of Rayleigh scattering criterion for all raindrops, so that the radar returns are proportional to the rain rate. The reflectivity measured by the radar (Z_e) varies by the sixth power of the rain droplets' diameter (D), the square of the dielectric constant (K) of the targets, and the drop size distribution. For a Marshall-Palmer (1948) distribution, this gives:

$$Z_e = \int_0^{D_{max}} |K|^2 \cdot N_0 \cdot e^{-\Lambda \cdot D} \cdot D^6 \cdot dD. \quad (8.17)$$

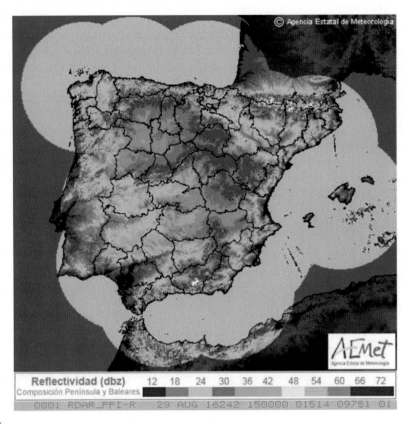

FIGURE 8.25

Rain radar composite image for the Iberian peninsula (August 29, 2016 at 17:00 h local time). Radar reflectivity in dBZ corresponds to the lowest elevation scan (0.5°) over the horizon.

On the other hand, the rain rate (RR), depends on the number of particles, their volume and their fall speed [$v(D)$] as:

$$RR = \int_0^{D_{max}} N_0 \cdot e^{-\Lambda \cdot D} \cdot \frac{\pi D^3}{6} \cdot v(D) \cdot dD. \tag{8.18}$$

Since both have similar functional expressions, the relationship between them can be expressed as:

$$Z = a \cdot RR^b, \tag{8.19}$$

where a and b depend on the type of precipitation. In rain radars, the radar reflectivity is expressed in dBZ (10 times the base 10 logarithm of the ratio of the radar return to the radar return of a standard 1 mm diameter drop filling the same scanned volume). Typical values range from 65 dBZ (extremely heavy precipitation, possible hail) to 20 dBZ (light precipitation). Modern weather radars are pulse-Doppler radars, capable of detecting the motion of rain droplets in addition to the intensity of the precipitation, and estimate its type (rain, snow, hail, etc.).

A rain radar composite image for the Iberian peninsula corresponding to August 29, 2016, at 17:00 h local time is shown here in Fig. 8.25. Radar reflectivity in dBZ corresponds to the lowest elevation scan (0.5°) over the horizon. Note the artifacts that appear as segments due to terrain clutter in two of the regional radars.

8.5 STUDY QUESTIONS

1. Discuss the mix of satellites used to operationally forecast weather and climate in the United States. How did this suite of satellite instruments develop historically? What is the future of this forecast satellite system?
2. Discuss how carbon dioxide is sensed by satellite in the atmosphere. What are the different sensors used and accuracies that each of them have? How do satellite measurements compare to ground measurements of CO_2?

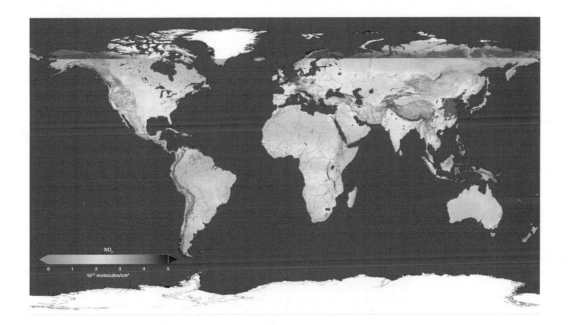

3. How do you measure atmospheric pollution? Interpret the attached slide comparing atmospheric aerosols and population density. How do you believe that the aerosols were measured? What about the population density (Kaufman et al., 2002).

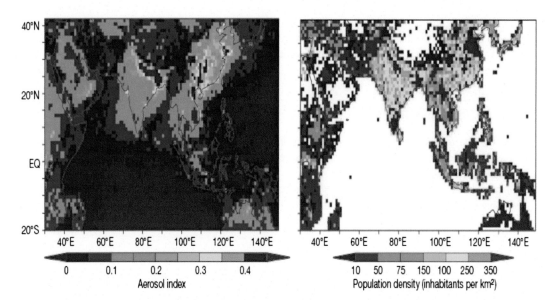

4. Explain how satellite infrared or microwave data is used to detect and map vertical atmospheric profiles of temperature and water vapor. What are the weighting functions shown in the attached figure? How do infrared weighting functions differ from passive microwave weighting functions?

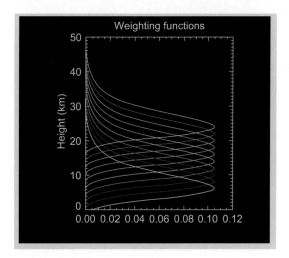

5. Which are the two phenomena that play a role in the propagation of electromagnetic radiation through a rain cell? How do they vary with frequency?
6. Modern rain radars measure the speed of the raindrops in addition to the rain intensity. Which is the underlying physical phenomena?
7. Why rain radars scan down to a minim elevation angle of $\sim 0.5°$? What happens at lower elevation angles?
8. Explain how you would estimate aerosol optical depth from AVHRR and MODIS satellite data.
9. How would you compute cloud top temperature from infrared satellite data and how would you use this information to estimate the potential for tornadoes?

OCEAN APPLICATIONS

One important reality of ocean remote sensing is that it is basically restricted to the surface or at most very near surface of the ocean. While satellites are critical components of open ocean autonomous ocean profiling systems, satellite-borne sensors cannot penetrate very far into the ocean itself. Satellite altimeters measure the ocean's surface and its variations and therefore reflect a vertical integral, but still the altimetric signal does not penetrate into the interior of the ocean. Ocean color may penetrate a few tens of meters into the upper layer of the ocean, but that is about the deepest any spaceborne sensor penetrates into the ocean.

That said there is plenty of information available from satellite sensors that is relevant to processes that link the ocean and atmosphere and contribute to the Earth's climate system. The sea surface temperature (SST) of the ocean is one of the oldest ocean measurements that has been collected, but the shift to satellite SST has changed the way we sense SST and thus its relationship to the climate puzzle. Satellite SST sensing itself varies between infrared (IR) sensing of SST and passive microwave sensing of SST. Other important ocean surface sensing parameters are ocean color to map biological activity, ocean surface wind stress, and synthetic aperture radar (SAR) images to map currents and winds.

9.1 SEA SURFACE TEMPERATURE

The first ocean parameter to be observed routinely was the SST. Originally this was a temperature measurement of a bucket sample of surface water taken on a sailing ship or a direct SST measurement from a coastal observing station. Such measurements were routinely collected in the mid-to-late 18th century. During this time shipboard measurements evolved from water samples taken in a wooden bucket (Fig. 9.1) to one taken with a smaller and folding canvas bucket. This change introduced SST errors due to different on-deck cooling conditions for the canvas bucket relative to the better insulation of the wooden bucket. As a result, shipboard SSTs measured between the late 1980s and 1940s had a cool bias introduced by this cooling effect.

In the 1950s faster powered vessels made it necessary to give up collecting water samples and shift the SST measurement to measuring the temperature of the cooling water for the ship's engines. An analog measurement taken by "eye" in a very warm engine room has introduced a warm bias in ship-based SSTs (Fig. 9.2) that had to be compensated for in creating a climatological time series. This engine intake cooling water temperature measurement was called an "injection" temperature since the temperature sensor was inserted into the cooling water circulation system. It seems only appropriate that the first parameter to be widely and routinely collected by spaceborne sensor would be SST.

Introduction to Satellite Remote Sensing. http://dx.doi.org/10.1016/B978-0-12-809254-5.00009-9

FIGURE 9.1

Wooden (A) and canvas (B) surface water sample buckets.

9.1.1 INFRARED SENSING OF SEA SURFACE TEMPERATURE

The first satellite instruments to sense SST were IR sensors originally designed to provide the meteorologist with atmospheric information during the night when reflected imagery was no longer available. These early IR images evolved into a more sophisticated system that used combinations of channels to correct for atmospheric attenuation of the IR signal from the ocean's surface (McClain et al., 1985). The first sensor used routinely for SST calculations was the scanning radiometer (SR), which provided the basic information used as input for the GOSSTCOMP system. The SSTs from this system were not found to be very realistic when compared with in situ SST measurements.

The SR flew on the now operational National Oceanic and Atmospheric Administration (NOAA) satellites, which were copies of the Improved Television InfraRed Observing Satellite (TIROS) Operational Satellite (ITOS) providing three-axes stabilization. Another instrument that flew on these satellites was the very high resolution radiometer. This instrument added more channels and a higher spatial resolution, about 1 km. While operated only as an experimental instrument it was soon apparent that they instrument provided the information that would be really needed to measure IR SST thus setting the stage for future satellites which carried something known as the advanced very high resolution radiometer (AVHRR) which then flew on the TIROS-N series of satellites. This AVHRR became a workhorse instrument for satellite SST that continued for more than 20 years.

9.1.2 THE ADVANCED VERY HIGH RESOLUTION RADIOMETER

The AVHRR is a four- or five-channel visible—thermal IR radiometer with a nadir resolution of 1.1 km which stretches out to about 6 km at the edge of scan. The channel locations and widths are shown here in Fig. 9.3 in comparison with channels from other radiometers. They are also spelled out in Table 9.1.

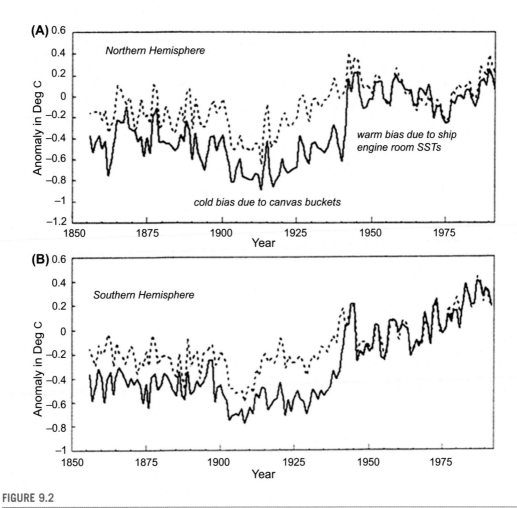

FIGURE 9.2

Time series of ship sea surface temperatures (SSTs) between 1850 and 1990. Top (A) is the time series for the Northern Hemisphere while the bottom (B) is for the Southern hemisphere.

These channels have changed slightly over time with the first AVHRRs having only four channels. The addition of channel 5 provided an additional thermal IR channel to correct the IR SST for atmospheric water vapor attenuation of the IR signal. Later channel 3 was split into a 1.6-μm channel during the day, which switched back to 3.7 μm at night. Remember in a polar orbit the satellite goes from day to night every 45 min. This additional 1.6-μm channel improved cloud property mapping during the daylight hours. The applications for these various channels are listed here in Table 9.1.

Note this list is for the latest version of the AVHRR, which has the switched channel 3. It is important to note that this channel also contributes to SST, but at nighttime when the 3.7-μm channel is available. During the daytime, it is not used in the SST formulation.

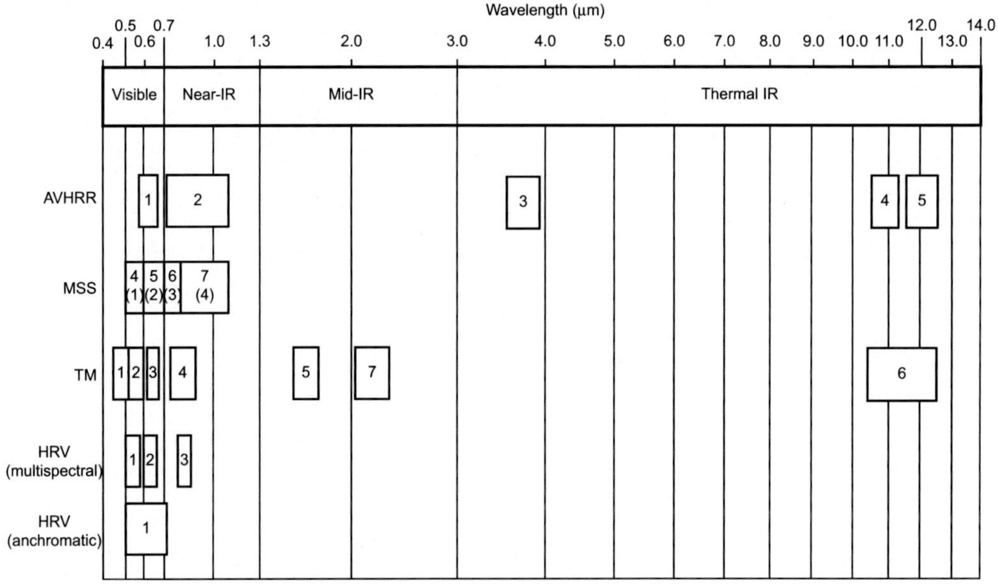
Wavelength (μm)
0.4 0.5 0.6 0.7 1.0 1.3 2.0 3.0 4.0 5.0 6.0 7.0 8.0 9.0 10.0 11.0 12.0 13.0 14.0
Visible
Near-IR
Mid-IR
Thermal IR
AVHRR
MSS
TM
HRV (multispectral)
HRV (anchromatic)
Channel 1: Detects visible light in the "green" frequencies
Channel 2: Detects radiation just outside of visible light
Channel 3: Detects radiation in the thermal frequency range but is subject to noise
Channel 4: Detects radiation in the heart of the thermal frequency range; highlights heat patterns well
Channel 5: Detects radiation also in the heart of the thermal frequency range

FIGURE 9.3

Advanced very high resolution radiometer (AVHRR) channels (1–5) in comparison to the Landsat multispectral scanner (MSS), the Landsat Thematic Mapper (TM), and the SPOT (HRV). Channel number and spectral width indicated by *boxes* in Fig. 9.3.

A picture and an exploded view of the AVHRR are presented here in Fig. 9.4, which clearly shows the radiant cooling system for the IR components of the system. The purpose of this cooling is to improve the performance of the IR detectors that are the essential components of the SST observing system.

Table 9.1 AVHRR/3 Channel Characteristics

Channel Number	Resolution at Nadir	Wavelength (μm)	Typical Use
1	1.09 km	0.58–0.68	Day-time cloud and surface mapping
2	1.09 km	0.725–1.00	Land-water boundaries
3A	1.09 km	1.58–1.64	Snow and ice detection
3B	1.09 km	3.55–3.93	Night cloud mapping, sea surface temperature (SST)
4	1.09 km	10.30–11.30	Night cloud mapping, SST
5	1.09 km	11.50–12.50	SST

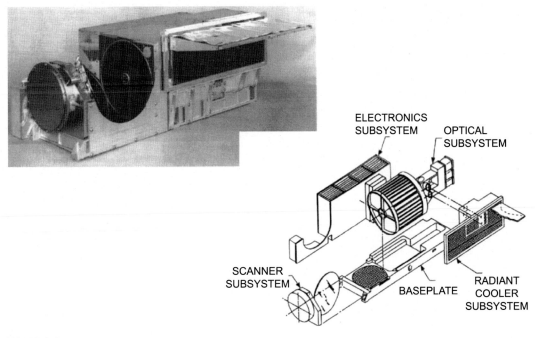

ELECTRONICS
SUBSYSTEM   OPTICAL
SUBSYSTEM

SCANNER
SUBSYSTEM   BASEPLATE   RADIANT
COOLER
SUBSYSTEM

FIGURE 9.4

The advanced very high resolution radiometer.

One real limitation of the AVHRR for calculating accurate SST is the lack of a real heated blackbody reference source in the optical system. Instead the backplane of the housing was honeycombed and painted black. It was equipped with six thermistors to measure the temperature of this pseudo-blackbody for real-time temperature calibration of the thermal IR channels. This had the negative factor that the backplane was not uniform in its temperature and therefore did not present a uniform target to the optics as an ideal blackbody. In spite of these limitations the overwhelming availability of AVHRR data made it a prime source of SST measurements.

First launched on the TIROS-N satellite in 1978 in a test configuration, the AVHRR became the operational imager on the TIROS-N series of satellites starting with NOAA-6 in 1979. The first AVHRR was a four-channel radiometer, which included channels 1, 2, 3B, and 4 of Table 9.1. This version of the AVHRR flew on NOAA satellites 6, 8, and 10. NOAA 7, 9, 11, 12, and 14 carried the AVHRR2, which added channel 5 in Table 9.1. Finally, the AVHRR3 on NOAA 15, 16, 17, 18, and 19 has the full number of channels listed in Table 9.1.

This time series is also documented here in Table 9.2, which shows the evolution of the NOAA satellites over time. It demonstrates the transition from the alpha designation given to the satellites when they were first built to a numeric designation when they were deployed on orbit. It is clear that there is not a one-to-one relationship between the alpha designations and the numeric ones. This is due to the fact that some of the NOAA satellites did not achieve orbit and thus were never commissioned as operational spacecraft. NOAA-13 however did achieve orbit was considered operational, but then

Table 9.2 Evolution of the National Oceanic and Atmospheric Administration (NOAA) Satellites Over Time

On Orbit Name	Before Launch	Launch Date	AVHRR Version	End of Mission
TIROS-N	TIROS-N	October, 1978	1	February 27, 1981
NOAA-6	NOAA-A	June 1979	1	November 11, 1986
NOAA-7	NOAA-C	June 1981	2	June 7, 1986
NOAA-8	NOAA-E	March 1983	1	October 31, 1985
NOAA-9	NOAA-F	December 1984	2	November 5, 1994
NOAA-10	NOAA-G	September 1986	1	October 2000
NOAA-11	NOAA-H	September 1988	2	September 13, 1994
NOAA-12	NOAA-D	May 1991	1	August 10, 2007
NOAA-13	NOAA-I	August 1993		Failed after launch
NOAA-14	NOAA-J	December 1994	2	October 03, 2001
NOAA-15	NOAA-K	May 1998	3	AM operational, secondary
NOAA-16	NOAA-L	September 2000	3	PM operational, secondary
NOAA-17	NOAA-M	June 2002	3	AM operational, backup
NOAA-18	NOAA-N	May 2005	3	PM operational, secondary
NOAA-19	NOAA-N'	February 2009	3	PM operational, primary
METOP-A[a]	METOP-A	October 2006	3	AM operational, primary

TIROS, televsion infrared observing satellite.
[a]Operated by EUMETSAT.

suffered a massive power failure while on orbit which killed the mission. At this time NOAA had a "no safe hold" policy for their satellites, which resulted in a complete satellite failure, which started in a single unit and then spread via the central electronics bus to all of the other satellite systems. A "failure review board" for this failure learned that the initial failure was due to an overtightened screw that shorted out a single system leading to a failure of the entire satellite.

9.1.3 ADVANCED VERY HIGH RESOLUTION RADIOMETER PATHFINDER SEA SURFACE TEMPERATURE

One of the main SST products derived from these data was the NOAA-NASA pathfinder SST (PFSST) product produced originally by the remote sensing group at the University of Miami. In this algorithm the ocean surface is assumed to emit almost as a blackbody, and without an absorbing and emitting atmosphere between the sea surface and the satellite it would be possible to estimate SST using a single IR channel observation. The reality is that the atmosphere attenuates surface-leaving IR radiance before it reaches a satellite sensor. It is therefore necessary to make corrections for atmospheric effects. Water vapor, CO_2, CH_4, NO, and aerosols are the major atmospheric constituents that cause the extinction of IR radiance (Minnett, 1990). Among these, the attenuation due to water vapor accounts for most of the needed atmospheric correction (Barton et al., 1989).

Various methods have been proposed to correct for atmospheric moisture attenuation to produce accurate retrievals of SST. One dominant solution is to use the radiance difference of two slightly different wavelength thermal IR channels to estimate the water vapor attenuation (Anding and Kauth, 1970). Using differences in brightness temperatures (BTs) measured by an earlier satellite radiometer, Prabhakara et al. (1974) were able to estimate SST to within a reasonable accuracy. Barton (1995) reviewed a number of SST algorithms all of which used this IR channel difference to correct for atmospheric moisture attenuation in the form

$$SST = a \cdot T_i + \gamma(T_i - T_j) + c, \tag{9.1}$$

where T_i and T_j are BT measurements in channels i and j and a and c are constants. The γ term is defined as

$$\gamma = \frac{1 - t_i}{t_i - t_j}, \tag{9.2}$$

where t is the transmittance through the atmosphere from the surface to the satellite. In cases of weak absorption, the transmittance can be approximated by $(1 - k \cdot u)$, where k is the absorption due to atmospheric absorbers, and u is the path length through the atmosphere (Barton, 1995).

All AVHRR SST algorithms have the same general form described in Eq. (9.1) although various modifications have been introduced through the years with changes in channels and satellites sensors. McClain et al. (1985) developed SST algorithms based on linear differences between thermal IR AVHRR channels. Known as the multichannel sea surface temperature or MCSST this algorithm assumed a constant channel difference between different AVHRR instruments. This MCSST became NOAA's first operational SST product (McClain et al., 1985) with future algorithm improvements including a correction for increased path lengths at larger satellite zenith angles (Cornillon et al., 1987). Subsequent improvements in atmospheric correction involved nonlinear formulations where the difference in the longwave IR channels was considered proportional to the BTs, as in the cross-product SST (CPSST) described by Walton (1988) and Walton et al. (1990).

That this "split-window" SST algorithm may not be the optimal correction to thermal IR SST due to atmospheric water vapor attenuation (WVSST) was discussed by Emery et al. (1994) where they use independent estimates of atmospheric water vapor from coincident measurements by the special sensor microwave imager (SSM/I) to correct for atmospheric water vapor absorption for AVHRR data. They applied their algorithm to data from the South Pacific and the Norwegian Sea and then compared the results with coincident measurements of both skin and bulk SST. In addition, a quadratic SST formulation is tested against the in situ SSTs. While the quadratic formulation provided a considerable improvement over the CPSST and the MCSST, the WVSST showed overall smaller errors when compared to both the skin and bulk in situ SST observations. Applied to individual AVHRR images the WVSST-CPSST difference reveals a water vapor pattern consistent with that from the SSMI/I itself. The quadratic SST was found to underestimate the SST corrections in the lower latitudes and overestimate the corrections in the high latitudes.

The most recent version of the channel combination algorithm is now the NOAA operational algorithm called the nonlinear SST (NLSST) where the starting SST is assumed to be proportional to a first-guess SST value (usually from the earlier MCSST). The AVHRR Oceans Pathfinder SST algorithm is based on the NLSST algorithm developed by Walton et al. (1998). This NLSST algorithm has the following form:

$$SST = a + b \cdot T_4 + c(T_4 - T_5) \cdot SST_{MCSST} + d(T_4 - T_5) \cdot sec(\rho - 1), \tag{9.3}$$

where SST is the satellite-derived SST and T_4 and T_5 are the AVHRR BTs in AVHRR channels 4 and 5, respectively, and SST_{MCSST} is an SST first guess usually computed as the MCSST described above. The satellite zenith angle is given by the last term and coefficients a, b, c, ρ, and d are estimated from regression analyses using colocated in situ and satellite measurements (called matchups). Typically, NOAA produces sets of coefficients using matchups for certain periods and the coefficients would not be changed unless there was a significant need. A good example is the need to change algorithm coefficients after the eruption of Mt. Pinatubo in June 1991, which introduced a large amount of atmospheric aerosols. Whenever a new AHVRR instrument is launched into orbit a new set of coefficients needs to be introduced so that Eq. (9.3) depends on the specific satellite sensor.

SST algorithm coefficients can be derived in two fundamentally different ways. Historically, coefficients were found by matching satellite values with nearly coincident in situ SST measurements (earlier ships and later drifting/moored buoys). This approach mixes the satellite IR radiances, which respond only to the IR emissions of skin SST with the in situ measurements of bulk SST made by the ships and buoys. The other method of SST algorithm coefficient estimation uses radiative model simulations to derive a set of atmospheric profiles to estimate the atmospheric attenuation then giving the estimate of the skin SST observed directly by the satellite BTs. This skin SST is fundamentally different from the subsurface bulk SST as discussed by Wick et al. (1992). This skin SST represents the 10-μm-thin upper layer of the ocean and is the temperature controlling the IR emissions, which are detected by the satellite sensor. An operational version of this approach has been used by the along-track scanning radiometer (ATSR, Llewellyn-Jones et al., 1993; Zavody et al., 1995), which then regressed the simulated satellite BTs against in situ measured SSTs (buoy SSTs) to derive SST algorithm coefficients. Considered as a "semiphysical" approach, this method resulted in an estimate of the skin SST.

The PFSST product used the direct regression of the satellite BTs against nearly coincident in situ measurements of bulk SST. In the PFSST the in situ matchup SSTs that were used to derive the SST coefficients are distributed together with the global SSTs. For the PFSST there are very tight space—time constraints: in situ and satellite observations are considered as coincident if they occur within ±30 min and ±0.1 degree of latitude and longitude. In addition, the pathfinder matchup database has been very carefully screened to exclude the most cloud-contaminated matchups from the SST coefficient estimation procedure. Only these relatively cloud-free matchups were used in the computation of the SST algorithm coefficients. The availability of the matchup in situ SSTs make it possible for the user to check the accuracy of the SST algorithm coefficients.

For the PFSST, some modifications to the NLSST were made. Studies of early versions of the PFSST revealed a correlation between the $(T_4 - T_5)$ temperature difference and the SST estimate itself, which suggested that this atmospheric correction was different for dry and moist atmospheres. There seemed to be low $(T_4 - T_5)$ values associated with a positive bias in SST residuals (low SST estimates) that was true for all satellite zenith angles. This is consistent with Emery et al. (1994) who studied the effects of atmospheric moisture on the SST residuals using atmospheric water vapor computed independently from microwave radiometry data. There is also a strong indication that the functional relationship between $(T_4 - T_5)$ and SST changes markedly at 0.7—1.0°C. Thus suggests that there is a fundamental change in the balance between the various radiance sources sensed by the AVHRR (emitted ocean surface radiance versus atmospheric moisture content) as a function of atmospheric moisture itself. In addition, atmospheric absorbers and scatterers other than water vapor become more important in drier atmospheres.

In an effort to capture this change in functional relationship the PFSST implemented a "piecewise" fit to the derivation of SST algorithm coefficients. SST algorithm coefficients were estimated for low, and intermediate to high $(T_4 - T_5)$ values. The boundary between these two atmospheric regimes was selected to be $(T_4 - T_5) = 0.7°C$. To avoid discontinuities in the global PFSST fields as the algorithm coefficient changed, they implemented another modification to the selection of algorithm coefficients. Early versions of the PFSST used a single set of SST algorithm coefficients, but an analysis of the resulting SST suggested a dependence on temporal trends in algorithm performance. These temporal trends included a variety of time scales ranging from seasonal [e.g., higher root mean squared (RMS) and bias in SST residuals during Northern Hemisphere summers] to interannual (differences between years). No clear cause was found for this interannual variability, which may stem from the gradual degradation of an individual AVHRR instrument. As an example during the later stages of both NOAA-9 and NOAA-11 operational lifetimes, the baseplate upon which the onboard calibration targets are mounted was operated at a much higher ambient temperature than was used in previous years (this was also higher than the temperatures used for prelaunch sensor characterization).

As stated in the PFSST guide the design goals for the AVHRR thermal IR channels were a noise equivalent differential temperature of 0.12K at 300K, and a signal-to-noise ratio of 3:1 at 0.5% albedo (Kidwell, 1997). Using the calibration technique developed by Brown et al. (1985), the overall calibration error estimate for the thermal IR channels as derived from the prelaunch thermal vacuum test is ±0.2K. In an effort to develop a consistent set of in-flight calibration algorithms for channels 4 and 5, a radiance-based correction procedure was developed to account for the nonlinear response characteristics of the detectors. This procedure resulted in a consistent algorithm applied over the entire range of the AVHRR operating temperatures. This correction constituted a significant improvement over the operational NOAA SST algorithm.

A large validation data set was also available consisting primarily of drifting and moored buoys measuring SST. The procedure for using these data is presented here in Fig. 10.5. Here the number in each box indicates the sequence with which it is carried out. The abbreviation TCAP refers to the transaction capabilities application part, which embedded in each SST report provided by a drifting or moored buoy. It verifies the status of the SST subsystem on each particular platform. Thus, it is used as a filter to remove SSTs reported from buoys, which have faulty SST subsystems.

Note that in step 5 there is a matchup criterion for the space—time window in which an in situ measurement is matched up with a satellite radiance. Finally, cloud-contaminated matchups are removed before the matchup statistics are generated. Since most of these in situ SST data come from data archives, one must perform various levels of quality control to sort out the best in situ SSTs for matchups with the satellite radiances. As a result, the pathfinder program (Fig. 9.5) did not carry out any additional quality control to further filter out erroneous buoy SST data. One exception was to correct for known errors in buoy location. These consistency checks were initially developed by Reynolds and Smith (1994). The in situ SST reports were converted to a continuous time coordinate, which became known as "pathfinder time." This pathfinder time took into account leap days, but not leap seconds. Since the matchup conditions were computed from this continuous

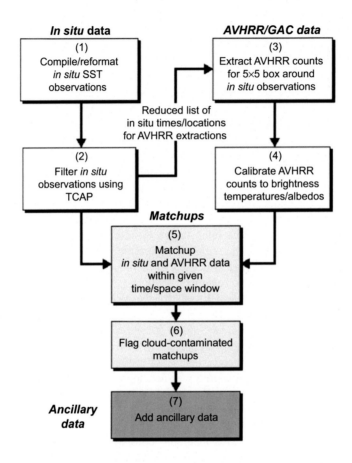

FIGURE 9.5

Advanced very high resolution radiometer (AVHRR) matchup data processing stream. *SST*, sea surface temperature; *TCAP*, transaction capabilities application part.

Pathfinder Time there may be small round off errors in the original in situ time that should not introduce errors any larger than 1 s.

The PFSST dataset is global as it is based on NOAA global area coverage (GAC) data that collected globally for each polar-orbiting satellite every day. While the instrument has a 1 km field of view (FOV) for the GAC data, four of every five samples along the scan line are averaged and the data from every third scan line are processed. The result is a spatial resolution that is nominally referred to as 4 km.

It should be noted that in spite of this global coverage persistent cloud cover results in many areas not being covered by a daily PFSST (Fig. 9.6). It can be seen here that in the polar regions clouds obscure most of the SST field while there are many other areas such as the tropical zones where cloud cover also dramatically reduces the SST coverage.

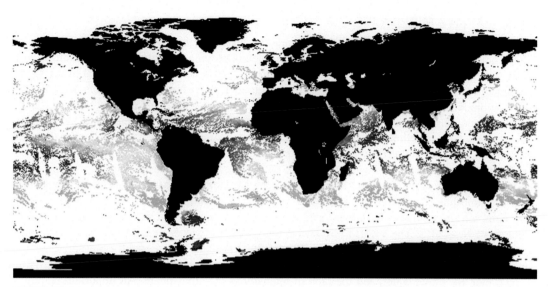

FIGURE 9.6

NOAA/NASA advanced very high resolution radiometer Oceans Pathfinder sea surface temperature; Descending Pass December 30, 1995.

Detailed patterns are retrieved where the IR SST is available, but the regular disruption of the patterns by clouds and coverage problems means that individual daily IR SST maps are rather useless. Only when these SSTs are composited over some period of time does a useful SST field emerge due to the fact that clouds move around much more rapidly than the SST field changes.

The PFSST values are given on three equal angle grids. The highest spatial resolution grid is nominally referred to as a 9 km grid and it has 4096 × 360 pixels per degree of longitude and latitude. Called the 18 km grid, values are also provided with a spatial resolution of 2048 × 360 pixels per latitude–longitude degree. Finally, the lower spatial resolution is approximately 54 km with a 720 × 360 pixel per degree resolution.

The PFSST product has made a transition from production at the Rosenstiel School of Marine and Atmospheric Science at the University of Miami to production and archival at the National Oceanographic Data Center (NODC). The goal is to have NODC provide this SST as a long-term climate data record going back to 1981 with the advent of the NOAA-7 satellite and its early AVHRR. This transition includes modernizing the PFSST processing code into a package; this is compatible with the NODC data archive architecture and easily scalable from large institutional data centers down to single individuals who wish to further improve on the PFSST in the future. This transition also requires software and analysis packages for the continued quality control of the associated matchup database. Formal documents will include a PFSST user's guide and a sustaining operations plan so that NODC personnel can continue the production of the PFSST consistent with what has been done in the past. The need to perpetuate the legacy of the AVHRR PFSST is emphasized that this sensor will continue to operate on US and European platforms well into 2024 and longer.

9.1.4 PASSIVE MICROWAVE SEA SURFACE TEMPERATURE

In their review, Gentemann et al. (2010) provide the history of microwave radiometry sensing of SST. They point out that satellite microwave radiometers provide not only SST, but simultaneously wind speed, water vapor, cloud liquid water, rain rate, and now even sea surface salinity (SSS). Designed to measure rainfall the first microwave radiometer was launched into space in December 1972 on the Nimbus-5 satellite. The first instrument specifically designed to measure SST was the scanning multichannel microwave radiometer (SMMR) carried first on the short-lived SeaSat satellite and then on Nimbus-7 (from October 1978 to August 1987). Due to some calibration problems SSTs were not terribly accurate from the SMMR.

The next microwave sensor widely used for SST was not originally designed for this purpose. The microwave imager on the Tropical Rainfall Mapping Mission (TRMM) was originally designed to measure atmospheric moisture and at 10 GHz was a bit too high in frequency to be very sensitive to SST emission. In addition, the TRMM satellite was placed in an orbit that restricted its measurement to 40°N to 40°S. The TRMM microwave imager (TMI) was designed to provide quantitative rainfall information over a wide swath width of 760 km. The TMI carefully measures the tiny amounts of microwave energy emitted by the Earth and its atmosphere, and it is able to quantify the water vapor, the cloud water, and the rainfall intensity in the atmosphere. An important feature of microwave retrievals is that SSTs can be measured through clouds, which are nearly transparent at 10.7 GHz. This is a great advantage over the IR SST observations that require a cloud-free FOV. With TMI, it is now possible to view ocean areas with persistent cloud coverage. TRMM's lower altitude (350 km) results in TMI having improved ground resolution, which is 45 km for the 10.7 GHz channel used for SST.

The TMI instrument is calibrated by viewing a "warm load" and then cold space to provide calibration values for each conical scan. The temperature of the warm load is monitored by three thermistors while the cold reflector views the cosmic microwave background at 2.7 K. TMI is capable of measuring SST using these calibration references. TMI suffered calibration problems due to an emissive sky reflector, for which corrections were developed and implemented making the TMI SSTs quite reliable.

The next most important microwave radiometer was the Japanese Space Agency's (JAXA) advanced microwave scanning radiometer (AMSR-E) which is onboard NASA's AQUA satellite, launched in May 2002. AMSR-E had 12 channels corresponding to the six frequencies 6.9, 10.7, 18.7, 23.8, 36.5, and 89.0 GHz. All of these except 23.8 GHz were both vertical and horizontally polarized (Parkinson, 2003). The 23.8 GHz channel was only vertically polarized and was intended to estimate the atmospheric water vapor content. The instrument calibration was similar to the TMI using a cold reflector pointed out to space and a hot absorber with eight thermistors. The AMSR-E hot load had large thermal gradients not well measured by the thermistors. A correction for this error in the calibration reference point had been developed and implemented. SST was primarily computed from the 6.9 GHz channel.

In an effort to demonstrate the retrieval of ocean wind speed and direction the Naval Research Laboratory launched the Coriolis satellite in January 2003, which carried the WindSat instrument. This instrument has fully polarimetric channels at 10.7, 18.7, and 37.0 GHz, and vertical and horizontal polarization channels at 6.8 and 23.8 GHz. Calibration is similar to the other two instruments with a cold space reflector and a hot absorber measured by six thermistors. Again, while this instrument was not designed to measure SST, its data can be used to map microwave SST.

Calibration is often a problem with microwave radiometers. The cold reflector is relatively trouble-free as long as problems with reflector emissions are avoided. Occasionally the reflector will not point to cold space, but to the Moon instead. These values must be removed from the data.

The hot absorber has proven to be a lot more problematic particularly since the thermistors often do not adequately measure thermal gradients across the hot absorber. A hot load correction is needed for AMSR-E because of a design flaw. The hot reference acts as a blackbody emitter and its temperature is measured by precision thermistors. Unfortunately, during the course of an orbit, larger thermal gradients develop as the satellite transitions from night into day due to solar heating making it difficult to determine the average effective temperature from the thermistors. The thermistors themselves measure these temperature gradients and may vary up to 15K between them for any given time for AMSR-E. Radiative transfer modeling simulations are used to determine the effective hot load temperature as a function of the measured hot load thermistor temperatures.

The main reflector is also a potential source of calibration error. It is generally assumed to be a perfect reflector with an emissivity of 0, but this may not always be true. A bias in the TMI measurements was attributed to a degradation of the primary reflector. At TMI's relatively low-altitude (350 km) atomic oxygen present in this part of the troposphere led to a rapid oxidization of the thin, vapor-deposited aluminum coating of the graphite primary reflector resulting in a much higher antenna emissivity than was expected from the prelaunch calibration. This emissivity was determined to be 3.5% and is constant for all of the TMI channels.

Geophysical parameters are retrieved from the microwave channel data using a radiative transfer model to derive a regression algorithm (Wentz, 1998; Meissner and Wentz, 2012). A large ensemble of ocean-atmosphere scenes is first assembled and atmospheric conditions are specified using a set of quality-controlled radiosonde profiles launched from small islands (Wentz, 1997). Half of these radiosonde profiles are used to derive the regression algorithm while the other half are withheld for testing the algorithm. A cloud layer of various columnar water densities ranging from 0 to 0.3 mm is superimposed on the radiosonde profiles. A rough sea surface is placed beneath these simulated atmospheres. With these simulated conditions SST is randomly varied from 0 to 30°C, the wind speed is randomly varied from 0 to 20 m/s, and the wind direction is randomly varied from 0 to 360 degrees.

Atmospheric BTs and transmittances are computed from these scenes and noise appropriate to the measurement noise expected (which depends on spatial resolution) is added. The noise-added simulated BTs along with the known environmental scene are used to generate multiple linear regression coefficients. The algorithm is tested by repeating this process with the withheld parameter values.

Between 4 and 11 GHz the vertically polarized BT of the ocean surface is sensitive to SST. BT at these frequencies is also influenced by the sea surface roughness and on the atmospheric temperature and moisture profiles. Fortunately, the spectral signatures of the surface roughness and the atmosphere are quite distinct from that of SST making it possible to remove the influence of these effects given the multiple frequencies and polarizations available. Both TMI and AMSR-E measure multiple frequencies that are more than adequate to remove the competing surface-roughness and atmospheric profile effects. Sea-surface roughness is usually parameterized in terms of wind speed and direction. For AMSR-E the additional 6.9 GHz channel provides an additional estimate of sea-surface roughness and improved accuracy of SST less than 12°C. All of the channels together are used to simultaneously retrieve SST, wind speed, columnar water vapor, cloud liquid water, and rain rate (Wentz and Meissner, 2000). Microwave SST retrievals are not possible in regions of sun-glitter and near land (which depends on the microwave spot size, which at the low SST frequencies is quite large). Over the rest of the ocean the ability of the microwave to view the ocean surface through cloud cover results in almost complete global coverage daily. Errors in wind speed, water vapor, cloud liquid water can contribute to errors in the retrieved SST.

Remote Sensing Systems Inc. performs the SST microwave retrievals using moored buoy SSTs from the TOGA-Tropical Atmosphere Ocean (TAO) buoy network in addition to the Pilot Research (PIRATA) moored buoy array in the tropical Atlantic. The nighttime mean differences, mean satellite minus buoy SST differences, and standard deviations (STDs) for each of the buoy arrays are presented here in Table 9.3.

While the biases are very small the STDs are about 0.5°C which is less accurate than the IR SST that often has an STD of about 0.15°C. Thus, while the microwave radiometry SST has the all-weather sensing advantage it also has the drawbacks of poor spatial resolution ($\sim$50 km), and it is a bit less accurate in terms of absolute temperature.

Near land, the side lobes of the microwave antenna pattern can result in side lobe contamination as stray signals make it in to the observational data stream. This effect is clearly going to be geography dependent as it occurs only where there is sufficient land to contaminate the microwave SST retrieval. It should be noted that, because of the larger footprint, the 10.7 GHz channel is much more affected by land emissions, which results in a warm bias and small increase in STD for both the TMI and the AMSR-E near land, but the effect is substantially greater for TMI which only has the 10.7 GHz channel that is less affected by the nearby land than is the 6.9 GHz channel.

The TRMM satellite also carries a visible infrared radiometer scanner (VIRS) from which IR SST estimates can be made, which were found to have a STD of 0.7°C when compared to Reynolds optimal interpolated SSTs (Ricciardulli and Wentz, 2004). The mean difference of SST observations away from land is about 0.12°C that is approximately the difference that would be expected between the IR skin (VIRS) SST and the subskin (TMI) SST. As the distance to land decreases, this mean SST difference increases to a maximum of about 0.72°C, indicating that the bias due to land contamination is on the order of 0.6°C. From Fig. 9.7, it is clear that this bias applies only for SST retrievals 100–150 km from land. It should be noted that these results are specific to the 10.7 GHz channel of TMI.

Table 9.3 Nighttime Satellite–Moored Buoy SST Errors, Bias, and Standard Deviation (STD) in °C						
	TOGA TAO Buoys			**PIRATA Buoys**		
Satellite	**Colocations**	**Bias**	**STD**	**Colocations**	**Bias**	**STD**
TMI	84,702	−0.09	0.67	1169	−0.09	0.60
AMSR-E	21,461	−0.03	0.41	2837	−0.00	0.35

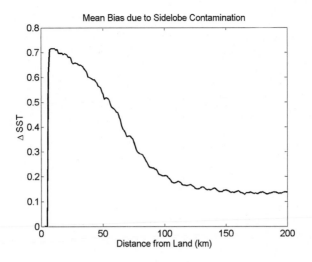

FIGURE 9.7

Land contamination bias derived from TMI—VIRS SST comparisons. This global average shows that by removing data within 100—150 km of land, side lobe contamination will be removed.

Another source of errors in microwave radiometry is radio frequency interference (RFI), which is the fastest growing source of error in microwave sensing of SST (and wind speed). The RFI impact on water vapor, cloud liquid water is less, but is growing as are the sources of RFI. These sources can generally be traced to media broadcasting activities (including television and radio) from commercial satellites in geostationary orbit. Signals from these geostationary satellites reflect off of the Earth's surface into the microwave radiometer's FOV. Other ground-based RFI comes from radio and cell phone transmission activities. Some of these sources have been identified and cataloged. Many change with time, and new sources appear all the time.

It is easy to identify and correct a microwave signal affected by very large RFI, but it is a lot harder to correct microwave signals affected by lower level RFI. The time—space changes in RFI must be carefully monitored so as to avoid their influence on measurements that will be used to infer climate change in the Earth-atmosphere system. The increasing encroachment on microwave radiometry frequency bands by people wanting to use these parts of the spectrum for their communication needs indicates that RFI will be increasing in activity and intensity in the future.

The RFI resulting from geostationary broadcast sources depends on broadcast frequency, power and direction, microwave radiometers' bandwidths, signal glint angle, and ocean surface roughness. Unfortunately, to improve the radiometric sensitivity, microwave radiometers' bandwidths are typically wider than the protected bands for microwave remote sensing meaning that these instruments can receive RFI from legal sources using nearby frequency bands allocated for communications and other commercial uses. Of the instruments discussed here AMSR-E and Windsat are the two most affected by RFI. This is due to the lower frequency bands of these two instruments. Windsat receives more RFI than AMSR-E due to wider measurement bandwidths. Wider bandwidths result in more signal by reducing noise but it also opens the bandwidth to RFI. For example, Windsat's 18.7 GHz channel receives a lot of RFI from direct TV's nationwide broadcast beams, while AMSR-E with a narrower 18.7 GHz channel is not significantly influenced by this RFI source.

To serve smaller, but geographically dispersed markets media satellites also broadcast wide, low-power beams to cover such large and less populated regions. These lower-power signals also cause RFI and they can be a lot more difficult to detect and correct for. In addition, the RFI depends heavily on how closely the satellite reflection points at the RFI source. This is called the glint angle in analogy to sunglints and optical remote sensing, where the direct reflection of the strong solar signal overwhelms the ocean surface signal.

Ruf et al. (2006) introduced a new type of microwave radiometer detector that is capable of identifying both high and low levels of RFI and of reducing or eliminating the effects of this RFI on radiometer BTs. High-level, localized RFI can be easily identified by its abnormal appearance in BTs. Low-level or persistent RFI is much more difficult to identify and filter out. Their agile digital detector (ADD) is able to discriminate between RFI and natural thermal emission signals by directly measuring higher-order moments of the signal, than just the variance that is usually measured. The ADD then uses spectral filtering methods to selectively remove RFI. ADD performance has been verified in controlled laboratory tests and in the field near a commercial air traffic control radar. A detailed description of the different RFI detection and mitigation techniques is provided in Section 4.6.3.4 of this book.

9.1.5 MERGING INFRARED AND PASSIVE MICROWAVE SEA SURFACE TEMPERATURES

As discussed in previous sections, IR and microwave SST capabilities are complimentary. While the IR sensors have higher spatial resolutions, they are subject to obscuration by clouds and attenuation due to atmospheric water vapor. Sensing of SST using microwave radiometry passes through clouds and is not affected by atmospheric moisture. It is affected by the presence of water droplets in heavy rain and by high winds. The microwave radiometry SSTs also suffer from a poor spatial resolution. It is therefore very attractive to figure out how to combine these two different SST sensing mechanisms, into a single SST field that takes best advantage of each of the individual sensors.

In a study that combines AVHRR and TMI SST estimates, Wick et al. (2004) designed an SST product that merged these two different sensing systems using an optimal interpolation scheme. They also address the complex spatial and temporal differences in these two SST products that result from the very different error sources and differences in measurement characteristics. They use existing IR and microwave radiometry SST products to generate their blended product rather than go back to the original radiances to generate the blend. The IR product is the operational AVHRR NLSST product from the US Naval Oceanographic Office (NAVO) (May et al., 1998), and the microwave radiometry SST is the TMI SST produced by remote sensing systems (Wentz et al., 2000). The IR is available as orbital files with an approximately 8 km resolution while the TMI SSTs are provided as daily ascending and descending grids at 0.25 degree resolution. To blend these two products each was mapped onto day and night grids at a 0.25 degree resolution.

The blending process of combining the two different SST products has several different components. First is to identify and account from retrieval errors in each of the products. Second is to compensate for different measurement times and effective sampling depths (IR radiances are emitted from the 10-μm-thin skin of the ocean while the microwave emissions represent the upper 1 mm of the ocean). The third step is to blend the SSTs together into a common product.

Combining SST products from two different types of retrievals requires the assumption that the SST products represent the same quantity or that there is some method to account for differences in the products. Differences between three monthly composites of these two different SST products reveal some complex space—time differences (Fig. 9.8). From these differences, it is clear that the two different SSTs cannot be simply combined without any consideration given to these space—time differences. The differences result from both retrieval errors and from different geophysical processes. To first explore the retrieval uncertainties, SST estimates from both products were compared to coincident observations from moored and drifting buoys and the differences evaluated as functions of the environmental parameters that affect retrieval accuracy (Castro et al., 2004).

This latter study concluded that bias adjustments are needed to reconcile microwave and IR SSTs to coincident buoy SST observations. Further improvements are possible by including other corrections based on climatological SST anomaly and aerosol optical depth. These additional corrections reduce the monthly RMS differences by as much as 42% and reduce the difference with the independent reference buoys by as much as 10%. The largest individual corrections are to the IR data in regions of low SST, but the microwave corrections are more geographically distributed and they are larger in number.

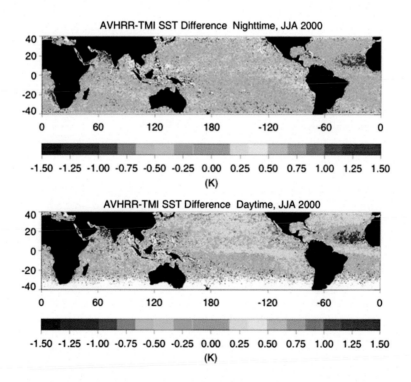

FIGURE 9.8

Difference in maps for the TRMM microwave imager (TMI) and advanced very high resolution radiometer (AVHRR) sea surface temperature (SST) products for night and day composites for June—August 2000.

For their blending efforts, Wick et al. (2004) used these results to formulate bias adjustments for the TMI SSTs as functions of wind speed, SST, water vapor, and atmospheric stability. For the AVHRR, bias corrections were computed for aerosol content, water vapor, and SST. It is important to note that coincident measures of wind speed and water vapor are provided with the TMI data stream, while the aerosol content is obtained from the NOAA/NESDIS AVHRR operational product (Husar et al., 1997), and air temperature is estimated from a climatology developed at Oregon State University (Esbensen and Kushnir, 1981). Four separate bias adjustments are computed and applied to the appropriate SST product as specified in Castro et al. (2004).

Additional compensation must be applied to each SST retrieval to account for geophysical processes, which also contribute to the differences seen in Fig. 9.8. One major contributor is the role of diurnal warming that results in complex temperature gradients just beneath the ocean's surface. As stated earlier the IR SST are in response to emissions from the 10-μm-thin skin of the ocean while the microwave emissions extend down to a 1-mm-thick ocean layer. Diurnal warming can create a vertical temperature gradient even in this very small vertical range. To compensate for this effect, the SST retrievals were referenced to the same predawn time, while the diurnal warming estimates were computed as functions of wind speed and daily mean solar insolation using the model of Fairall et al. (1996). The model was run for expected ranges of wind speed and insolation to generate a diurnal signal due only to this diurnal warming. Curves were fitted to the results to formulate a simple lookup table to apply these corrections to subsequent measurements. No further correction was made for the presence of a skin layer in their study (Wick et al., 2004). Both SST products include as a reference to buoys in their derivation and the bias adjustments discussed earlier were all made relative to buoy measurements. Thus, both SST products were effectively referenced to a subsurface temperature near 1 m depth.

Once the bias adjustments were made and the sample times were adjusted, the two different SST products could be merged. Two different methods were explored for this blending step. The first is a simple merge with no additional effort to fill in the gaps in either of the original fields. The second is the production of a final full SST field using an optimum interpolation scheme. This was done sequentially as the simply merged product is a good interim step that depends directly on the data and is needed as the input data field for the optimum interpolation.

The merged SST product is computed as the simple average of the SST products on the 0.25 degree grid wherever values are available. If a grid cell contains an SST retrieval from only one sensor at a time, the merged product takes the SST value of that sensor for that location and time. In the averaging, all of the SSTs are given the same weights. No additional attempt is made to incorporate any additional information on the uncertainty of a particular product. The optimum interpolation method of Reynolds and Smith (1994) was used to map the merged SST product and fill in all of the gaps.

The primary input to this analysis step was the 0.25 degree grid merged SST. The resulting daily mapped grids were computed by updating a first-guess SST with these daily products. The first-guess SST field was taken from a weekly averaged merged SST data product that was linearly interpolated to determine the daily-average SST field. Using this weekly averaged first guess helped to eliminate numerical instabilities in the optimal interpolation method and to fill regions void of data in the daily maps. Data increments were defined as the difference between the input data and the first guess and were weighted using a Gaussian distance. They were then added to the first guess to give the updated daily SST analysis. Both types of satellite SSTs were assumed to have equivalent amounts of correlated and uncorrelated error. The error for each input data set was defined as the ratio of the data to the first-guess error. The error ratio for the MCSST was set to unity, while the TMI error ratio had a minimum error ratio of 1.67 and this ratio was allowed to vary as a function of wind speed due to greater SST retrieval uncertainty under low (<3 m/s) and high (>10 m/s) wind speeds. As a result, the resulting error structure more heavily weights the MCSST data due to their greater overall accuracy. While buoy SST references were used to determine the individual SST bias corrections, no buoy data were used in this SST mapping analysis.

Using this process, Wick et al. (2004) created daily merged AVHRR + TMI SST fields as shown for example in the upper panel of Fig. 9.9 for August 10, 2000. While the coverage is better than for either AVHRR or TMI alone, there are still gaps in the coverage. These gaps are the consequence of a lack of satellite sampling or the inability to retrieve SST due to cloud cover or precipitation. The SST field itself is in general smooth and the SST values are consistent in regions where data from the AVHRR and TMI are known to overlap.

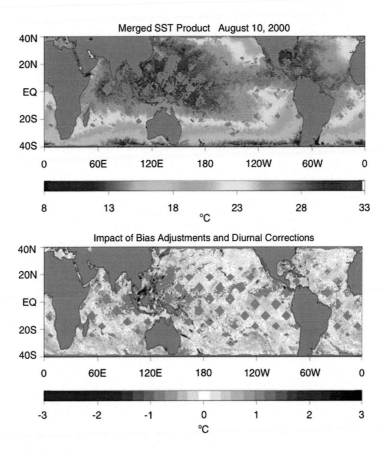

FIGURE 9.9

Sample merged sea surface temperature (SST) product (top panel) and differences between the merged SST product (lower panel) with and without the bias adjustments and diurnal corrections.

The lower panel in Fig. 9.9 shows the impact of the inclusion of the bias corrections and diurnal warming compensation as it is the difference between the merged SST product with and without these adjustments. It is evident that these adjustments have a significant impact, and corrections can be as large as 1−2°C. There are clear patterns to these corrections, which are similar to the SST difference patterns presented in Fig. 9.8 with some important differences. A region of positive difference is located to the west of central Africa where Saharan dust is known to blow out over the Atlantic Ocean strongly influencing the IR sensing of SST. Another large positive difference is also apparent over the Arabian Sea where the aerosol content is also very high.

These differences are such that the merged SST product is warmer as a result of compensating for cooling of the AVHRR SST retrievals due to the presence of the atmospheric aerosols. Likewise, regions where diurnal warming conditions have been removed show up as localized strong negative differences in the Pacific Ocean just west of Central America and east of Australia. The region near the south China Sea was an area of large TMI SST adjustments.

To provide some independent validation of this merged SST, Wick et al. (2004) compared the merged SST with drifting and moored buoy SSTs. Since all SSTs were referenced to predawn values, only nighttime buoy observations were used in these comparisons. The results are presented here as Table 9.4.

In this table it can be seen how including the corrections reduces both the bias and RMS errors in this merged product. Also included is a comparison with buoy SSTs the MCSST and TMI SST estimates. The accuracy of the merged SST is comparable to the accuracy of the TMI only SST product while it is only slightly poorer than the AVHRR only SST field.

Most people will not work with the directly merged SST product, but would rather have a smoothly interpolated field. The optimally interpolated version of Fig. 9.9 is presented here as Fig. 9.10.

All the gaps appearing in Fig. 9.9 have been filled and the result is a smooth map of SST. The advantage of merging these two different types of SSTs is to be able to retrieve this type of a smooth SST product everywhere it was mapped.

Again to test the validity of this interpolated SST maps comparisons were carried out with moored and drifting buoy SSTs. The results are presented here as Table 9.5.

Surprisingly the accuracy of the interpolated SST product was found to be better than the raw merged SST. This may be the result of a greater number of comparison samples inherent in the interpolated field. By including the bias adjustments and diurnal corrections the bias in the interpolated product is reduced, but the RMS differences decrease only slightly. The best performance in terms of RMS variability were obtained using only nighttime data, thus eliminating the need for diurnal corrections in the daytime data.

Table 9.4 Merged Sea Surface Temperature Product Accuracy Summary

Merged Product	Bias (K)	RMS (K)
All corrections	−0.08	0.59
No corrections	+0.14	0.67
MCSST only, all corrections	−0.02	0.58
TMI only, all corrections	−0.10	0.61
RMS, root mean squared.		

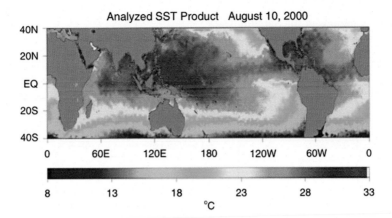

FIGURE 9.10

Optimally interpolated merged sea surface temperature (SST) for August 10, 2000.

Table 9.5 Analyzed Product Accuracy Summary

Analyzed Product	Bias (K)	RMS (K)
All corrections	−0.01	0.59
No corrections	+0.18	0.61
Nighttime data only, all corrections	+0.03	0.53
All corrections except diurnal warming	+0.05	0.56

RMS, root mean squared.

If the diurnal corrections were not applied the RMS difference was reduced, but the bias slightly increased. This indicates that the diurnal corrections are reducing the mean effects of diurnal warming, but are also adding scatter in the data due to the difficulty in predicting diurnal warming at specific times because of the warming at specific times given the limited detail on the forcing data. The other bias adjustments had a smaller impact on the optimally mapped product than on the simply merged SST. This suggests that further improvements to the mapped SST product may be achieved by a better treatment of diurnal warming.

A similar analysis was carried out by Guan and Kawamura (2004). They again used an AVHRR SST together with an IR SST from the geostationary meteorological satellite stretched−visible infrared spin scan radiometer (GMS-VISSR) and the TRMM TMI microwave radiometry SSTs together with the VIRS also on the TRMM satellite. They generated daily 0.05 degree resolution cloud-free SST products in three areas of the Pacific Ocean: the Kuroshio region, the Asia−Pacific region, and the Pacific during a 1-year period from October 1999 to September 2000. Comparison of their merged SST product with in situ SSTs from buoys operated by the Japanese Meteorological Agency (JMA) suggests an accuracy of 0.95K for the merged SST product. Guan and Kawamura (2004) summarized their data sets in Table 9.6 assigning each an SST accuracy. These range from 0.6K for the AVHRR to 0.8K for the GMS-VISSR.

Table 9.6 Specifications of the Satellite Data

Satellite Sensor	Spatial Resolution (Degree)	Temporal Resolution	Coverage	Accuracy (K)
NOAA AVHRR	0.01	Twice per day per satellite	20°N ~ 60°N 120°E ~ 160°E	0.6
GMS S-VISSR	0.05	Hourly	20°N ~ 60°N 120°E ~ 160°E	0.8
TRMM MI	0.25	3 days for full coverage	38°S ~ 38°N 0° ~ 360°E	0.7
TRMM VIRS	0.05	3 days for full coverage	38°S ~ 38°N 0° ~ 360°E	0.7

The AVHRR value was the RMS number given by Strong and McClain (1984), and McClain (1989), the GMS-VISSR around Japan from Sakaida and Kawamura (1992a,b), and the TMI RMS error from Shibata et al. (1999), Wentz et al. (2000), and Kachi et al. (2001). It is to be noted that the spatial resolution of the AVHRR is about 1 km, the GMS-VISSR about 5 km, and the TMI about 50 km.

In this study the role of diurnal variation was acknowledged, and rather than model it SSTs at different times were used to estimate the diurnal change. This was greatly enabled by using the frequent sampling of the GMS-VISSR SST product that in spite of its slightly worse spatial resolution provides frequent SST samples over a day. In addition, this study (Guan and Kawamura, 2004) ignores the difference between skin and bulk SSTs and rectifies all measurements against in situ buoy SSTs.

One important thing is that this study uses an AVHRR SST produced at Tohoku University (Sakaida et al., 2000) rather than the NAVO SST product used by Wick et al. (2004). Also the GMS-VISSR data are unique to this study and come from the Sendai Research Center (Tanahashi et al., 2000). Finally, the TMI SST product is generated by the Earth Observation Research Center of the then National Space Development Agency that included the TRMM VIRS processing. The important thing is that all of these SST sources are independent of those used in Wick et al. (2004).

They also used objective analysis (aka optimum interpolation) to merge the different SST fields. They used an analytical correlation function for their length scale in this optimum interpolation method. An example of the merged GMS-VISSR and TMI SST fields is shown here in Fig. 9.11.

For this merged product all SSTs were interpolated to a 0.05 degree grid. In their processing they used a 3-day SST centered at the current day. If the time difference was equal to 2 days, they just selected the 5-day value within the spatial window with the added condition that all the cross-correlation values were positive. For quality control, any SST value that deviated by more than 2K from the overall mean value was rejected. If several different SST values were available for any one grid point, the following priority was set for selecting the value to retain: AVHRR + VIRS, TMI, and GMS-VISSR.

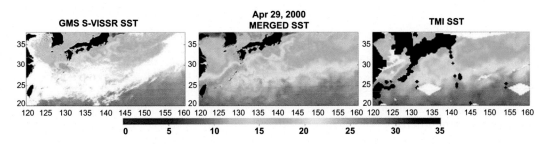

FIGURE 9.11

Comparison of geostationary meteorological satellite stretched—visible infrared spin scan radiometer (GMS-VISSR), merged, and TRMM microwave imager (TMI) sea surface temperatures (SSTs) in the Kuroshio region.

The benefit of merging TMI SST with the GMS-VISSR SST can be clearly seen in Fig. 9.11, where the white areas of the GMS-VISSR image were all filled in by the TMI SST in forming the merged SST. The primary benefit of the merged product was its ability to retain a lot of the fine structure of the Kuroshio present in the IR data but absent in the microwave TMI SST image.

The merged SST product was compared with JMA buoys for the period October 1999 to April 2000 (Fig. 9.12). Daily means buoy SSTs were calculated for each buoy and the nearest matching merged

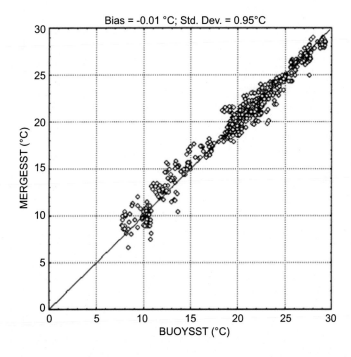

FIGURE 9.12

Comparison between merged and buoy daily-mean sea surface temperatures (SSTs).

SST value was selected for comparison yielding a total of 600 cases. In Fig. 9.12 the bias is −0.01K and the STD is 0.95K.

This accuracy estimate is higher than that found in Wick et al. (2004) in a very similar analysis. There can be many sources of this accuracy difference. To begin with, this analysis includes a new SST data product not considered in the Wick et al. (2004) study. In addition, all of the AVHRR and TMI data were processed into SST products by completely different systems than those used by Wick et al. (2004). In addition, the objective mapping scheme is a bit different which would also contribute to this difference in accuracy. Finally, a very different suite of buoys was used for comparisons and these may have been processed in a very different manner than those used in the Wick et al. (2004) study. The fact that the Guan and Kawamura (2004) study was restricted primarily to the western Pacific may have also influenced this error difference since the Wick et al. (2004) study addressed the global ocean.

Both studies, however, are encouraging in that they discuss methods of putting together very disparate SST products to take best advantage of the beneficial characteristics of each. Only in this way will it be possible to produce a regular and global SST product for related studies of air−sea interaction and global climate change.

9.2 SEA SURFACE HEIGHT AND SATELLITE ALTIMETRY

Satellite altimetry has resulted in a real change in the way physical oceanographers regard the task of mapping surface geostrophic currents. Once limited to ship-based hydrographic surveys physical oceanographers were now presented with satellite-based sea surface height (SSH) measurements that could estimate the sea surface and its variations in time and space. The ocean's surface is much like an atmospheric pressure map is to a meteorologist in that the geostrophic winds can be inferred from the spatial distribution of atmospheric pressure and changes in this pressure distribution reflect space−time changes in the winds. Similarly, the ocean's surface and its variability reflect the ocean's geostrophic surface currents. While this is not the complete picture of surface currents, it goes a very long way to providing a unique means of viewing surface currents and their variability in a way that was previously not possible.

9.2.1 RADAR ALTIMETERS

A large class of real-aperture radars is radar altimeters (RAs). Designed initially to provide an accurate measurement of the surface topography of the ocean, the RA has evolved into an instrument for studying not only ocean phenomena (including sea ice), but also terrestrial hydrology conditions. A great deal has been learned about the ocean since the late 1980s when a navy altimeter satellite called Geosat was moved from its classified geodetic orbit into a 10-day exact repeat mission (ERM) following the same orbit as the earlier and short-lived SeaSat altimeter. Most significant during this period was the Topex/Poseidon (T/P) satellites launched in late 1992 and operating for a full 13 years and overlapping with its replacement Jason-1. The real pinnacle of satellite altimeter coverage was in the late 1990s and early 2000s when multiple altimeter satellites operated by different countries and agencies provided dramatically improved coverage.

9.2.2 HISTORY OF SATELLITE ALTIMETERS

As is clearly described in Fig. 9.13, spaceborne RAs got their start with the Skylab and Geos-3 missions in the mid-1970s. These instruments were strictly exploratory and did not yield useful time-series measurements. The first real operational satellite altimeter was part of the short-lived SeaSat mission in 1979. This instrument demonstrated the value of sea-level measurements made from such a RA. Still there was a gap of more than a decade before a new mission dedicated to altimeter measurements was flown.

As it is apparent from Fig. 9.13, there was another satellite altimeter in the mid-1980s, but it was not originally designed as a monitoring mission for ocean variability. Geosat was initially conceived as a mission to measure the Earth's geoid for the US navy operations, which was a classified mission with the data unavailable to the general science community. After the 1-year classified "geodetic" mission was completed, the navy was convinced by some ocean scientists to put Geosat in the exact repeat orbit flown previously by SeaSat and allowed to operate as an altimeter to measure ocean surface variability. ThisERM phase of the Geosat program provided useful data for many research projects and set the stage for the future dedicated missions such as T/P.

The T/P satellite altimeter was the first satellite dedicated to observing the ocean surface variability and using these observations to study surface currents, waves, and tides. This satellite operated for a total of 13 years providing increasingly accurate measurements of the ocean's surface as the altimetry science community learned how to process these data and make the various corrections needed to make the results more accurate. In 1991, the European Space Agency (ESA) launched its first RA on ERS-1, which was followed in 1995 by ERS-2. T/P managed to continue operating into the lifetime of its follow-on Jason-1 satellite where for 6 months the two satellites were operated in the same orbit 90 s apart. This intercalibration period revealed that the new Jason-1 SSHs were biased high relative to T/P and the in situ measurements made at the Harvest Platform off southern California of

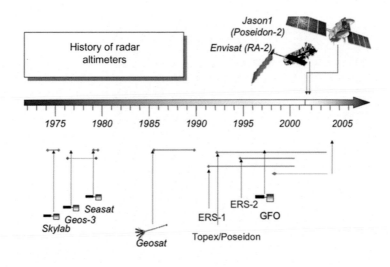

FIGURE 9.13

History of spaceborne radar altimeters (Le Traon, 2007).

138 ± 18 mm (Haines et al., 2003). Thus, while the altimeter variations are very similar, there is an offset in the Jason-1 heights relative to the T/P observations.

The launch of ERS-2 was followed shortly by another US Navy mission known as Geosat follow-on (GFO). This started a decade when multiple satellite altimeters provided the best coverage possible for these instruments. The launch of Europe's large ENVISAT satellite in 2002 continued the series started with ERS-1, 2. Thus, the best observing period for satellite altimetry was between 1999 and early 2012. In mid-2012, Jason-1 was found to have a limited lifetime and it was decided to move this satellite from its exact repeat orbit to a geodetic orbit thus compromising the ocean variability monitoring aspects of this particular altimeter. A short time later ENVISAT suddenly ceased to communicate with the ground bringing this amazing mission to an abrupt close. Also in 2011 the US Navy canceled any plans to replace the GFO satellite, which ceased to operate November 28, 2008. As a result, in mid-2012 there were only the two Jason satellite altimeters operating with one of them in a geodetic orbit.

9.2.3 PRINCIPLE OF OPERATION

As is clearly depicted in Fig. 9.14 the satellite RA measures the range between it and the sea surface relative to a reference ellipsoid from which the satellite orbit is well known. The quantity of prime interest to the physical oceanographer is the sea level relative to the marine geoid, which then indicates the ocean currents. The sea surface topography measured by a satellite altimeter can be written as:

$$\text{SSH}(x, t) = G(x) + \eta(x, t) + \varepsilon(x, t), \tag{9.4}$$

where G is the geoid, η is the dynamic topography, and ε are measurement errors (Le Traon, 2007). At present, the marine geoid is not well enough known to estimate globally the absolute dynamic height except at very long wavelengths. However, the interesting variable part of the

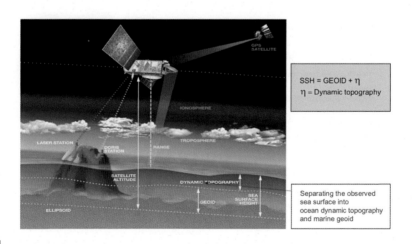

FIGURE 9.14

Satellite altimeter measurement principle.

Credits CNES and mention of Aviso+ website https://www.aviso.altimetry.fr/en/my-aviso.html.

dynamic topography $\mathrm{SSH}'(x,t)$ is easily calculated using the "repeat-track method." For a given satellite track (recall that the altimeter is a nadir sampling instrument) SSH' (x, t) is obtained by removing the mean profile over several altimeter cycles which contain both the geoid G and the mean dynamic topography. Thus,

$$\mathrm{SSH}'(x,t) = \mathrm{SSH}(x,t) - \langle \mathrm{SSH}(x) \rangle_t = \eta(x,t) - \langle \eta(x) \rangle_t + \varepsilon'(x,t), \qquad (9.5)$$

To get the absolute SSH, one needs to use a mean dynamic height climatology from historical in situ data or to use existing geoid estimates together with an altimeter mean sea surface. One can also rely on a model average. Gravimetric missions such as CHAMP (Challenging Minisatellite Payload) and Gravity Recovery and Climate Experiment (GRACE) have provided improved geoid estimates. Launched in March of 2009 the ESA's gravity field and steady-state ocean circulation explorer (GOCE) is now delivering unprecedented resolution of the geoid with an accuracy of $1-2$ cm vertically over a 100 km spatial resolution (Fig. 9.15).

The GOCE geoid can be used with an altimetric mean sea surface to derive $\langle \mathrm{SSH}(x) \rangle_t$ that can then be added to $\mathrm{SSH}'(x,t)$ to give the ocean surface elevation. This SSH is thus composed of the geoid, the dynamic topography and noise. Of these, the geoid variability contributes about 100 m, while the dynamic topography variations are about 2 m. Added to this uncertainty is the variability due to ocean tides $(1-20$ m) and other correction terms.

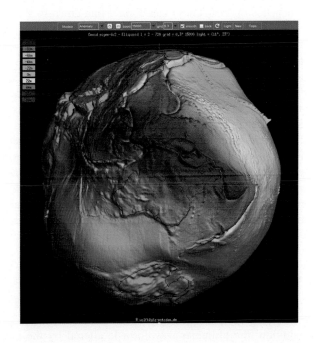

FIGURE 9.15

Gravity field and steady-state ocean circulation explorer model geoid with variations from -110 to $+88$ m (http://www.gfz-potsdam.de/typo3temp/pics/Geoid_01_0eff7212c9.png).

9.2.4 ALTIMETER ERROR CORRECTIONS

These other corrections are comprised of instrumental corrections, a sea state bias correction, an ionospheric correction, wet and dry tropospheric corrections, tidal corrections, and inverse barometer corrections. Added to these, there are errors in the orbit. Instrument errors are oscillator drift, Doppler shift, tracker response, pointing angle variability, and internal calibration.

Orbit error is a consequence of less than perfect knowledge of the spacecraft's altitude. It is actually the largest of the altimeter errors, but thanks to excellent tracking information and good orbital models, it has been possible to reduce this orbit error to ± 2 cm. This is small compared to the ± 10 cm accuracy associated with the ERS-1, 2 satellites. Instrumental in this success was the addition of GPS monitoring to the T/P satellite, which became a standard for future altimeter missions. In addition, the DORIS Doppler radio tracking system was added by CNES, which added a great many ground stations to the tracking of T/P and future altimeter satellites. Finally, a network of ground-based lasers was used with onboard laser retroreflectors to track the altimeter satellites. It is important to recognize that orbit error has a very long wavelength characteristic.

Space/time variations in atmospheric pressure result in changes in sea level on the order of a centimeter/millibar change in atmospheric pressure. In addition, wind and pressure forcing creates a higher-frequency barotropic response at period less than 20 days (Le Traon, 2007). Tidal variations can result in SSH changes between 10 and 60 cm in the open ocean, which can be larger in the coastal ocean where the tides become highly nonlinear.

Ionospheric changes can introduce range errors from 1 to 20 cm, but the dual-frequency altimeters (this range delay error depends on frequency) can correct this down to 0.5 cm. One of the largest corrections is for atmospheric path delay due to atmospheric water vapor. For this reason, most of the radiometers carry a bore-sighted microwave radiometer optimized to measure the water vapor induced path delay (Brown et al., 2004, 2009). Called the wet delay, this correction varies both spatially and temporally with magnitudes from 5 to 30 cm (highest in the tropical convergence zones where atmospheric convection is strong). The precision of the onboard water vapor correction is about 1 cm. When the microwave radiometer data are missing the water vapor content is calculated from atmospheric climate models.

One of the most difficult errors to compensate for, it is known as the sea state bias and it is due to the fact that the concave form of the altimeter wave fronts respond differently to the wave covered sea surface where wave troughs better reflect the altimeter signal, as compared to the wave crests. This shifts the SSH away from the mean sea surface to a bias that favors the troughs (Le Traon et al., 2010). The SSH error for the presently operating altimeter frequencies can be as large as 2 cm. The temporal evolution of these errors up to the T/P mission is summarized here in Fig. 9.16 (Le Traon, 2007).

9.2.5 ALTIMETER WAVEFORMS AND BACKSCATTER

Near the coast, altimeter data become less useful. First the microwave radiometer has inherently large spot sizes, which are compromised when they encounter land eliminating the water vapor path length correction. In addition, the altimeter waveform is altered by the presence of land in the altimeter footprint. Here it is important to point out how the SSH and other parameters are retrieved from the altimeter waveform. The waveform (Fig. 9.17) indicates the SSH from the midpoint of the leading edge at least for the open ocean retrieval (Brown, 1977). The y-axis offset from zero

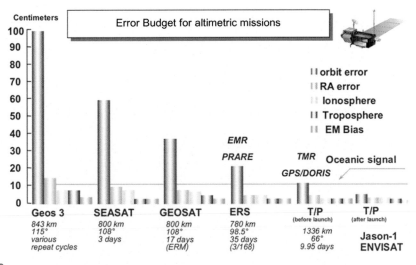

FIGURE 9.16

Altimeter overall error evolution up to Topex/Poseidon (T/P) and Jason 1 (Le Traon et al., 2010).
RA, radar altimeter; *EM*, electro magnetic; *EMR*, electro magnetic radiation; *IMR*, TOPEX microwave radiometer.

indicates the instrument noise level while the waveform maximum is a measure of the sigma-0 backscatter, which is related to the wind speed. The slope of the leading edge is inversely proportional to the significant wave height (SWH) and the slope of the trailing edge reveals the antenna mispointing.

Fig. 9.18A describes the response to an incoming altimeter pulse both in terms of the interaction with the altimeter wave front and the returned waveform if the sea surface is flat. If the sea

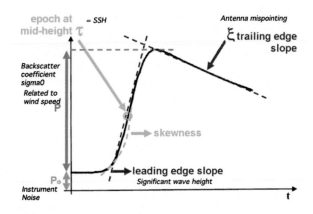

FIGURE 9.17

Altimeter waveform labeled for geophysical parameter retrievals (http://www.altimetry.info/html/alti/principle/waveform/ocean_en.html). *SSH*, sea surface height.

Credits AVISO+/CNES 2017.

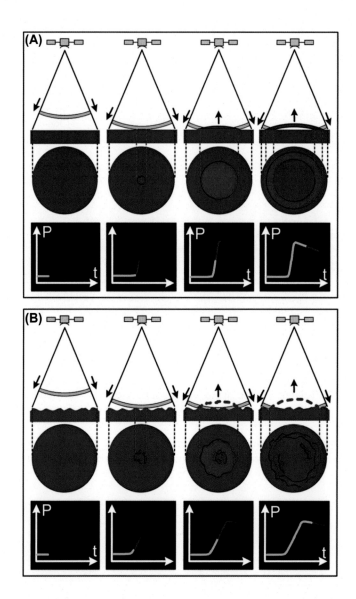

FIGURE 9.18

Altimeter echo from (A) rough seas, (B) a flat or calm water surface (Amazon river); and (C) over land. . At top is the radar pulse, in the center is the response in red and at the bottom are the waveforms that would result (https://www.aviso.altimetry.fr/en/techniques/altimetry/principle/pulses-and-waveforms.html).

Credits AVISO+/CNES 2017.

surface is rough, the response is smoother, as shown in Fig. 9.18B. In the case of a wave covered rough seas, the symmetry of the early echos is not maintained resulting in waveform distortions. Typical waveforms from real altimeters are actually quite different from the ideal waveforms in Fig. 9.19.

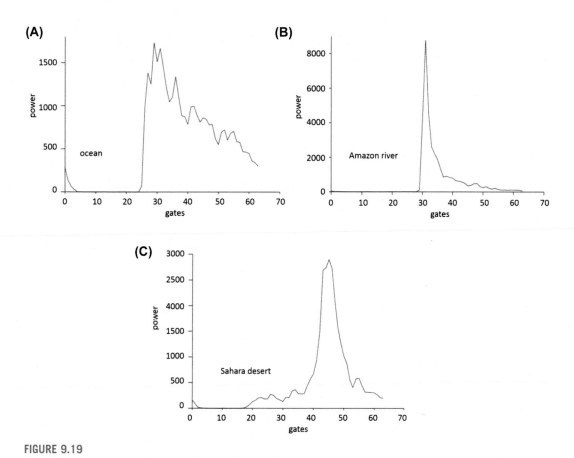

FIGURE 9.19

Sample real altimeter waveforms over (A) the ocean, (B) the Amazon river, and (C) land (Sahara desert).

Adapted from https://www.aviso.altimetry.fr/en/techniques/altimetry/principle/pulses-and-waveforms.html.

9.2.6 ALTIMETER DATA MERGING

From the mid-1990 to the early 2012, there were multiple altimeter satellites operating giving us the real heyday of altimeter data availability. During this time, it became advisable to merge these different altimeter data together to improve the space–time coverage of the combined altimeter data. To merge multisatellite altimeter data one must first create homogeneous and intercalibrated data sets. A method to achieve this is to use the most precise mission as a reference for the other satellites (Le Traon and Ogor, 1998). During this period the best references were T/P or Jason 1, 2. Further to compute the SLA, it is best to use a common reference surface to get the SLA relative to the same ocean mean.

9.2.7 SYNTHETIC APERTURE RADAR ALTIMETRY

One method of increasing the spatial resolution of an altimeter is to operate in a delay-Doppler mode (Raney, 1998). This has become known as SAR altimetry by analogy since the improved

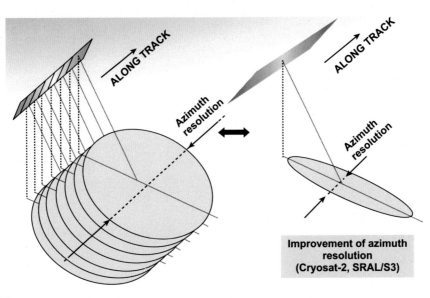

FIGURE 9.20

Synthetic aperture radar altimeter data collection.

resolution is due to multilooks of the same target. The key innovation in the delay-Doppler RA is delay compensation, analogous to range curvature correction in a burst-mode SAR. The height estimates are sorted by Doppler frequency and integrated in parallel (Fig. 9.20).

There are more looks accumulated than in a conventional altimeter and the relatively small along-track resolution is constant at about 250 m for a Ku-band altimeter. This type of altimeter is more efficient in response to the radar pulses and hence this new approach is more efficient than that for a conventional altimeter. Thus, this approach leads to a smaller altimeter that requires less power to achieve the same level of response.

The first satellite to carry one of these altimeters is the ESA's Cryosat-2 mission (http://www.esa. int/esaLP/ESAMNM1VMOC_LPcryosat_0.html). Unlike conventional altimeters where the interval between pulses is about 500 µs, the Cryosat-2 altimeters sends out a burst of pulses with an interval of only 50 µs between them. The returning echoes are correlated (SAR-like) and examining them in burst mode, the echoes are separated into strips arranged across the track by exploiting the frequency shifts caused by the Doppler effect in the forward and afterward looking parts of the beam. Each strip is about 250 m wide and the interval between bursts is arranged so that the satellite moves forward by 250 m each time step. The strips laid down by successive bursts can therefore be super-imposed on each other and averaged to reduce noise. This mode of operation is called the SAR mode. The primary goal of this mission is to evaluate the ice-sheet elevation and estimate the freeboard of sea-ice.

9.2.8 ALTIMETRY APPLICATIONS

9.2.8.1 Mapping Geostrophic Ocean Surface Currents

Geostrophic currents are by definition a balance between the Coriolis force and the horizontal pressure gradient. This pressure gradient at the sea surface is expressed by the slope of the sea surface with the strength of the geostrophic currents being proportional to the strength of the surface gradient of this elevation. This balance can be written as:

$$v_s = +\frac{g}{f} \cdot \frac{\partial \zeta}{\partial x},$$ (9.6a)

$$u_s = -\frac{g}{f} \cdot \frac{\partial \zeta}{\partial y}.$$ (9.6b)

where the sea surface is given by $\zeta(x, y)$, g is the acceleration of gravity, f is the Coriolis parameter (which represents the Coriolis force and changes with latitude), and x and y are the horizontal coordinates. Look at the x dimension only this situation (Fig. 9.21) can be depicted as in Eq. (9.6a).

The geostrophic surface current resulting from this sea surface slope is into the paper (in the northern hemisphere and due to a reversal in the sign of the Coriolis force it would be out of the paper in the southern hemisphere) as indicated by the cross-labeled by v_s and the strength of this surface geostrophic current is proportional to the gradient of the surface elevation relative to the geoid. A surface current of 10 cm/s corresponds to a horizontal elevation change of 10 cm over a 100 km. This almost imperceptible sea level change is responsible for this typical ocean current velocity. Strong currents such as the Gulf Stream and Antarctic Circumpolar Currents (ACCs) can have sea level changes as large as 20 cm over 100 km.

The geoid is by definition a level surface and therefore a level of constant geopotential (no work can be done in the vertical direction). Sea surface topography is caused by processes that cause horizontal motion in the ocean that will be in balance with the excursions of the sea surface. Such processes include tides, ocean currents, and changes in barometric pressure. Surface topographic variations due to ocean currents are considered as "dynamical changes" and are therefore called "dynamic topography." This dynamic topography is roughly one-hundredth of the geoid undulations. Thus, the shape of the sea surface is dominated by local variations of gravity.

The influence of ocean currents on the sea surface is much smaller. Typically, sea-surface topography has an amplitude of ± 1 m with typical slopes of $\partial g/\partial x \sim 1-10$ μrad for $v = 0.1-1.0$ m/s at midlatitudes. The height of the geoid, smoothed over horizontal distances greater than about 400 km, is known with an accuracy of ± 1 mm from data collected as part of the NASA GRACE satellite mission.

Because the geoid was not well known locally before 2004, satellite altimeters were flown in orbits that exactly repeat their ground tracks. This made it possible to subtract out one repeat track from an earlier one, thus computing the altimetric sea surface anomaly without accurate knowledge of the geoid. Eventually, a long-term mean altimetric height was used along this repeat ground track to

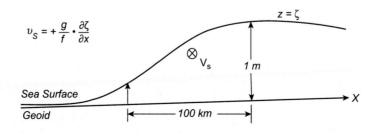

FIGURE 9.21

The sea surface associated with the geostrophic surface current.

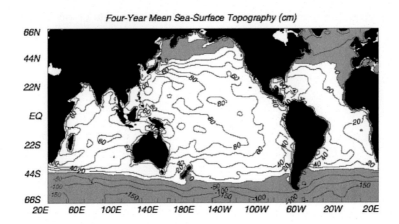

FIGURE 9.22

Global distribution of time-averaged topography of the sea surface from T/P altimeter data between 10/3/92 and 10/6/99 relative to the JGM-3 geoid.

estimate the SSH anomalies. T/P (1992−2005) was the first satellite designed to make the accurate SSH measurements needed to observe the sea surface topography and the associated geostrophic currents. It was followed by Jason-1 launched in 2001, and by Jason-2 launched in 2008. This series of satellites was instrumental in ushering in the era of altimetric mapping of mesoscale ocean currents with scales between 20 and 500 km.

It was also used to map the long-term mean average surface topography relative to some earlier geoid estimates. An example is presented here in Fig. 9.22, which is a global distribution of time-averaged topography of the ocean from T/P altimeter data relative to the JGM-3 geoid model from 10/3/92 to 10/6/99 relative to the CSR98 from the Center for Space Research of the University of Texas.

Geostrophic currents in Fig. 9.22 are parallel to the contours with higher sea levels to the right in the northern hemisphere and to the left in the southern. It can clearly be seen the parallel contours of the ACC and the gradients that mark the Gulf Stream in the North Atlantic, and the Kuroshio in the western North Pacific.

9.2.8.2 Mapping Mesoscale Ocean Dynamics With Satellite Altimetry

Observing mesoscale ocean dynamics with satellite altimetry is reviewed by Morrow and Le Traon (2012) using an 18-year record of satellite altimetric measurements. They state that "this global, high-resolution data set has allowed oceanographers to quantify the previously unknown seasonal and interannual variations in eddy kinetic energy and eddy heat and salt transports and investigate their causes." Most of these variations are in the form of mesoscale eddies that can be tracked using satellite altimetry eddy tracking methods to monitor their propagation pathways and to explore the role of Rossby waves versus nonlinear eddy dynamics in the mid-to-high latitude bands of the ocean. Satellite altimetry also makes it possible to explore how these mesoscale eddies impact the atmospheric circulation and hence the Earth's climate.

The ocean is a turbulent system and is composed of a wide variety of mesoscale and finer circulations that make up the space/time variability of the ocean's circulation. The geostrophic portion of the mesoscale variability can be mapped with satellite altimetry. This mesoscale variability generally refers to spatial scales of 50−500 km and time scales of 10−100 days. The energy of this mesoscale variability generally exceeds that of the mean flow by an order of magnitude or more. These eddies are primarily generated by instabilities of the mean flow (Stammer and Wunsch, 1999), but fluctuating ocean winds can also provide a direct forcing mechanism (Frankignoul and Müller, 1979). In addition to being formed as instabilities of the mean flow eddies feed energy back into the mean flow and drive the deep ocean circulation (Morrow et al., 1994; Lozier, 1997). They also transport heat, salt, carbon, and nutrients as they propagate in the ocean and are the principal mechanism for the poleward transport of heat across strong zonal currents such as the ACC (Karsten and Marshall, 2002; Jayne and Marotzke, 2002).

Satellite altimetry has made it possible over the past couple of decades to map these mesoscale sea levels and associated geostrophic ocean circulation variations. Early altimeter missions such as SeaSat (1978) and GEOSAT (1986–89) provided the first global coverage of the ocean's mesoscale sea level variations. This was the first time that mesoscale circulations and their spatial variations could be studied relative to turbulence theory (Le Traon, 1991). A review of early mesoscale studies with satellite altimetry is given by Le Traon and Morrow (2001a) and by Fu et al. (2010).

9.2.8.2.1 Multimission Mapping Capabilities

One thing that has made this altimetric mapping of mesoscale circulation possible is the period of overlapping satellite altimeter missions that has made it possible to retrieve SSH from many different satellites all at the same time. During the first decade of this century there were at least three different altimeters operating that could be used together. An example is shown in Fig. 9.23 from Morrow and Le Traon (2012) for the NW Mediterrancan Sea in January 2005.

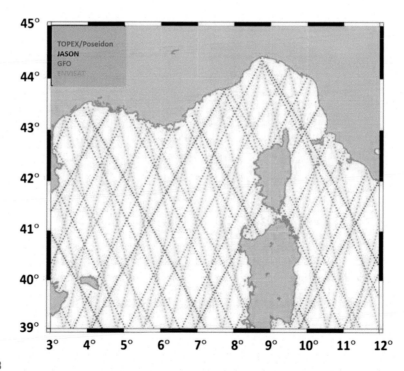

FIGURE 9.23

35-day coverage of altimeter missions in the NW Mediterranean Sea in January 2005. Topex-Poseidon (light red), Jason (dark red), ENVISAT (green), Geosat Follow-On (blue).

While these tracks overlap, merging the data together is not simple in that each satellite system has different errors and orbit constraints the effects of which need to be considered when merging the different altimeter data together. Still, the potential for increased spatial coverage is too great to ignore the opportunity to merge these different altimeter data sets. Dibarboure et al. (2011) discuss an interpolation technique used by the SSALTO/DUACS project to construct global gridded data sets from multimission altimeter data. In this merging a crossover analysis is used to minimize the errors between ground tracks, including a correction for the large-scale orbit errors. All of the available altimetric data are then mapped onto a regular 1/3 degree Mercator grid every 7−10 days, which makes it possible to properly map the mesoscale dynamics.

A summary of the mapping capabilities of the T/P, Jason-1, and ENVISAT combination is presented in Le Traon and Dibarboure (2002) using simulations from the eddy-resolving 1/10 degree Los Alamos Model (Smith et al., 2000). The modeled 1/10 degree sea level anomalies (SLAs) were added to random noise and subsampled along altimeter tracks. The aforementioned SSALTO/DUACS suboptimal interpolation technique was then applied to these subsampled data to reconstruct the 2D SLA. This reconstructed field was then compared with the full-resolution model SLA field. The difference in SLA is an estimate of the sea level mapping error inherent in altimeter sampling. In this study they found that with two altimeters in the T/P-ERS configuration, sea level can be mapped with an error less than 10% of the signal variance (depending on latitude) as shown here in Fig. 9.24. Part of the mapping error is due to high-frequency, high-wave number signals that are not properly resolved by even this combined altimeter sampling.

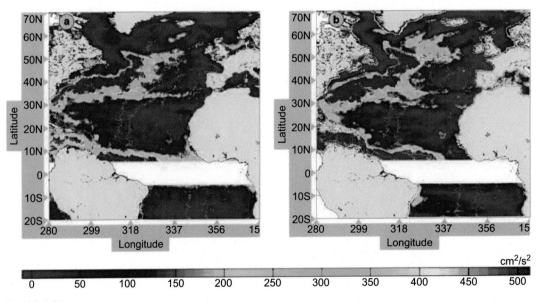

FIGURE 9.24

Comparison of eddy kinetic energy levels in the North Atlantic calculated from model sea level data subsampled along T/P + ERS ground tracks (left) and from the full 1/10 degrees resolution numerical model (right).

Another major contribution to being able to resolve the space and time scales of the mesoscale ocean circulation was the tandem mission first started when the aging T/P satellite was placed in an orbit halfway between the Jason-1 repeat ground-tracks (which continued the previous T/P ground-tracks). The tandem mission had as objectives to provide a greater spatial resolution of surface topography for mesoscale ocean circulation studies and for coastal tidal models. By comparing results from the tandem mission with independent ERS-2 data, Le Traon and Dibarboure (2004) were able to show that in regions with strong mesoscale variability, the sea level and geostrophic velocity can be mapped, respectively, with a height error of about 6% and velocity error of 20%−30% of the signal variance. Stammer and Theiss (2004) calculated along-track geostrophic velocities from the tandem mission and compared them with in situ velocities measure from acoustic Doppler (ADCP) current profiles along a ground-track near Bermuda. It was found that the tandem mission geostrophic velocity variances to be about 25% lower than the current variance observed by the ADCP, which contained both geostrophic and ageostrophic currents. To improve over the tandem mission results a spatial spacing of 10 km or less is required as is planned for the future wide swath altimetric mission.

In a classic study of nonlinear mesoscale eddies, Chelton et al. (2011) used data from two altimeters (T/P followed by Jason-1 and ERS-1 followed by ERS-2) that were merged by SSALTO/DUACS at 7-day intervals on a Mercator grid with a 1/3 degree spacing. This SSALTO/DUACS product included the removal of the 7-year mean SSH (1993−99) to eliminate the unknown geoid. These SSH fields are referred to by AVISO (French altimetry data distribution system) as the "reference series." Their analysis was based on the reference series that was available in early 2010 that included SSH fields for the period between October 14, 1992 and December 31, 2008. They found that mesoscale eddies (radius scale of O(100 km)) are readily apparent in these high-resolution SSH fields and could be identified with an automated procedure which isolated 35,891 eddies with lifetimes >16 weeks. These longer lived eddies were part of the approximately 1.15 million individual eddies identified with an average lifetime of 32 weeks and an average propagation distance of 550 km. Their mean amplitude and a speed-based radius scale defined by the automated procedure were 8 cm and 90 km, respectively.

An example of the mapped eddy field is given here in Fig. 9.25. Here it is clear that much of the SSH variability is due to energetic mesoscale features; it is also apparent that at lower subtropical and tropical latitudes (<20 degrees) in the Pacific there are long crests (middle panel Fig. 9.25) and troughs consistent with Rossby waves that have been distorted into westward pointing patterns by refraction (Chelton et al., 2011). While these features are relatively small in amplitude and are full of much more energetic mesoscale features, these patterns are identifiable across most of the South Pacific. They are less evident in the North Pacific in the middle panel of Fig. 9.25 because of the overall higher SSG in the northern hemisphere caused summertime steric heating (Fig. 9.25 is an August map). These patterns become more apparent in the eastern North Pacific when the data are spatially high-pass filtered to remove the steric effects of large-scale heating and cooling. Even then these features do not penetrate further than 2000 km westward from the eastern boundary in the North Pacific.

The tracked eddies were found to originate nearly everywhere in the World Ocean, consistent with previous conclusions that virtually all of the World Ocean is baroclinically unstable. In general, there is a preference for cyclonic eddies, while eddies with longest lifetimes and greatest propagation distances tend to be anticyclonic. There is a tendency toward larger cyclonic eddies in the southern hemisphere,

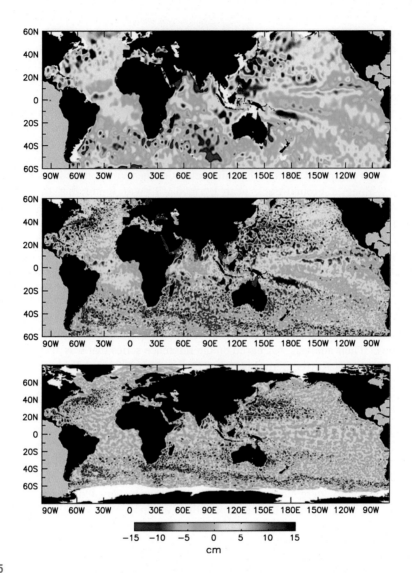

FIGURE 9.25

An example of global maps of sea surface height (SSH) on August 28, 1996 constructed from Topex/Poseidon (T/P) data only (top) and from the merged T/P and ERS-1 data (middle). The bottom panel is the SSH from merged field after spatially high-pass filtering with half-power filter cutoffs of 20 degrees of longitude by 10 degrees of latitude. The automated eddy tracking procedure identifies 3291 eddies in the bottom panel of which 2398 could be tracked for 4 weeks or longer.

while there is an overall preference for anticyclonic eddies in the northern hemisphere. In general, there is no evidence of anisotropy of these eddies of either sign. Most are approximately Gaussian in shape, but there is an indication that a quadratic function is a better approximation of the eddy profile (Chelton et al., 2011).

Chelton et al. (2011) considered that the most significant conclusion of their study was that essentially all of the observed mesoscale features outside of the tropical band (20°S–20°N) are nonlinear by the metric U/c, where U is the maximum circum-average geostrophic speed within the eddy interior, and c is the translation speed of the eddy. A value of $U/c > 1$ indicates that there is fluid trapped within the eddy interior. Many of the eddies were found to be highly nonlinear with $U/c > 10$. The probability distributions of all the measures of eddy nonlinearity were found to be skewed toward large values for cyclonic eddies in the southern hemisphere, while they are skewed toward anticyclonic eddies in the northern hemisphere extratropics. Added evidence of the nonlinear nature of these eddies is given by the fact that they propagate nearly due west with propagation speeds that are nearly equal to the long baroclinic Rossby wave phase speed which is twice as fast as that predicted by linear Rossby wave theory.

Between September 2002 and the end of the Geosat Follow-On mission in 2008, there have been four altimeter missions flying simultaneously. The resulting merged data set has greatly improved the estimate of the surface mesoscale geostrophic currents, and a number of higher resolution (1/8 degree) products have been developed for some regional applications. On average, the merged Jason-1 + ERS-2 + T/P + GO maps yield eddy kinetic energy levels 15% higher than the two-satellite configuration of Jason-1 + ERS-2 (Morrow and Le Traon, 2012). The consistency between altimetry and coincident maps of SST, tracks of Lagrangian drifters, and tide gauge sea levels was also significantly improved when four satellites were merged compared to the details available when the results from only the two-satellite constellation were available.

A number of studies have tracked the propagation of individual eddies at mid-to-high latitudes using the mapped multimission altimetric SLAs. Different automatic eddy tracking methods have been developed and tested on altimetry data such as the Okubu-Weiss parameter (e.g., Isern-Fontanet et al., 2003), the skewness of the relative vorticity (Niiler et al., 2003), the wavelet decomposition of the SLA (e.g., Lilly et al., 2003), the SLA itself (Fang and Morrow, 2003; Chaigneau and Pizarro, 2005; Chelton et al., 2011), and a geometric criteria using the winding angle approach (e.g., Souza et al., 2011). The different techniques all yield differences in the number of eddies detected, their duration, and their propagation velocities. Still, all methods do detect an abundance of mesoscale circulation features in the merged altimetry data set.

These many studies have revealed that there are definitely "eddy corridors" (Morrow and Le Traon, 2012) where eddies are more abundant. In the region south of Madagascar there is the regular formation of eddy dipole pairs resulting in a regular train of dipoles in the regions that, for example, started in December 1999 and continued into 2000. This dipole train remained coherent and triggered an unusually early retroflection of the Agulhas Current in late 2000. In the south Indian Ocean, Fang and Morrow (2003) analyzed the pathways of anticyclonic warm-core eddies from 1995 to 2000. They found long-lived warn-core eddies that could be tracked for periods longer than 6 months and over distances longer than 1500 km. The number of eddies was found to oscillate with the ENSO cycle. All of these features propagated west as they should. Still, it was possible to overturn this westward movement of the eddies in regions of very strong currents or bathymetric steering.

While this combination of multiple altimeters has made it possible to map the mesoscale eddy field, we cannot adequately sample the submesoscale. The submesoscale refers to features that surround the mesoscale eddies. They are filaments and smaller eddy structures that interact with the mesoscale eddies. To study them, the altimetric measurements have to be combined with higher resolution (1−5 km) sampling systems such as those for SST or work with large eddy simulation numerical models that can discretely measure these submesoscale features.

9.2.8.3 Application of Satellite Altimetry to Sea Level Rise

Since the advent of T/P the continuing series of satellite altimeters has made it possible to accurately estimate changes in global mean sea level. These measurements are continuously calibrated again a network of tide gauges as a reference. To calculate the mean sea level rise, the seasonal and local tidal variations must be first subtracted from the record. This time series is presented here in Fig. 9.26 from Nerem et al. (2010). A series of three satellite altimeters has been stitched together to create a longtime series of mean sea level from late 1992 to 2013. The interannual variations have been retained in this series, but a long-term line has been fit to the overall series showing the linear increase in sea level at a rate of 3.2 ± 0.4 mm per year. The seasonal signals have been removed.

This rate of sea level rise computed by the University of Colorado compares very well with the rates calculated independently by other groups. The French AVISO group gets 3.2 ± 0.6 mm per year, the Australian CSIRO get 3.2 ± 0.4 mm per year, and NOAA gets 3.2 ± 0.4 mm per year. Thus, it is clear that there is convergence in these estimates, and that sea level in increasing at this rather steady level.

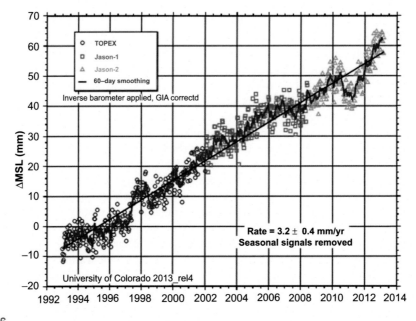

FIGURE 9.26

Global mean sea level (MSL) time series from satellite altimeters (seasonal signals removed).

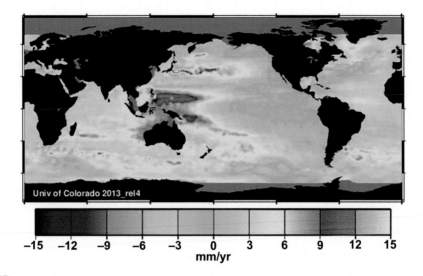

FIGURE 9.27

Map of sea level trends made by the University of Colorado.

This sea level increase has a geographical trend, which can be seen here in Fig. 9.27. Local trends have been calculated with a least-squares fit of 10-day, 0.25 degree resolution grids of sea level. A trend, bias, annual and semiannual terms are fit simultaneously (Nerem et al., 2010). These have been calculated only for the altimeter period (1993−2012). They therefore reflect the impact of decadal scale climate variability on the regional distribution of sea level rise. In addition, local SSH trends and variations are a result of many factors, including local crustal displacement, glacial isostatic adjustment, steric effects, and even local wind pattern. All of these effects and some others must be considered when interpreting these estimates of sea level trends.

To estimate mean sea level change from altimeter data, it is very important to correct for altimeter drift over time. This is done using sea level measured in situ with tide gauges. The method used for the CU sea level rise estimates is described in detail in Mitchum (2001). Briefly, the method creates an altimetric time series at a tide gauge location, and then differences these time series with the time series measured by the tide gauge. In this difference series signals due to common oceanographic phenomena will largely cancel out leaving a difference time series that is dominated by the sum of the altimetric drift and the land motion at the tide gauge site. Making independent estimates of land motion rates and by combining the difference series from a large number of global tide gauges results in a difference time series that is dominated by the altimeter drift. Since the difference series at separate tide gauge locations have been shown to be statistically almost independent (Mitchum, 1998), the final drift series has a variance that is very much smaller than any of the individual series that contribute to it. Since this results in very large degrees of freedom, this method outperforms altimeter calibrations from dedicated altimeter calibration sites although it is a relative calibration meaning that it is a relative bias and cannot compute an absolute bias. It can, however, detect a change in bias due to an altimeter drift.

Ideally, one would want to include all of the available tide gauges in the calibration. A number of the gauges, however, have a significant lag in reporting and are not available for real-time Jason calibration. In contrast some of the other stations do not extend backward in time to the T/P mission. For the sea level rise the CU group has restricted the ∼100 available gauges to a set of 64 near real-time stations that space the T/P period up to the current Jason-2 mission.

9.2.8.4 Estimating Ocean Bathymetry With Altimeter Data

The primary principle being invoked in the satellite altimetric mapping of ocean bottom topography is that the surface of the ocean reflects the shape of the ocean floor and it is a matter of teasing this information out of the satellite altimeter record (Sandwell and Smith, 1995). The bumps and dips in the ocean surface are caused by small variations in the Earth's gravitational field. For example, the extra gravitational attraction due to a large subsurface mountain on the ocean floor attracts water toward it resulting in a local bump at the ocean's surface. Even for a mountain 2000 m tall with a radius of 20 km the sea surface bump is too small to be seen with the naked eye.

It can, however, be detected by a satellite altimeter. The US Navy launched the Geosat satellite in 1985 to map the Earth's geoid height at a horizontal resolution of 10−15 km and a vertical resolution of 0.03 m. Geosat was placed in a nearly polar orbit to obtain a high-latitude coverage.

Two very precise measurements must be made to establish the topography of the ocean's surface to an accuracy of 0.03 m. First the height of the satellite above the ellipsoid must be measured by tracking the satellite from a globally distributed network of lasers and/or Doppler stations. This height is further refined using orbital dynamic calculations. Second, the height of the satellite above the closest ocean surface is measured with a microwave radar operating in a pulse-limited mode on a carrier frequency of 13 GHz. The ocean is a good reflector at this frequency. The large radar spot is narrowed using sharp radar pulses and accurately recording the two-way travel time. The footprint of the pulse must be large enough to average out the local irregularities in the ocean's surface due to ocean waves. The spherical waves of the altimeter ensure that the altitude is measured to the closest ocean surface. A high pulse repetition rates (1000 pulses per second) is used to improve the signal-to-noise ratio. Correction to the travel time is made for ionosphere and atmospheric attenuations and known tidal corrections are also applied. The difference between the height above the ellipsoid and the altitude above the ocean surface is approximately equal to the geoid height.

Thus, as the satellite orbits the Earth it collects a continuous profile of geoid height across an ocean basin. Taken together profiles from many different altimeter satellites collected over many years can be combined to make high-resolution images. Maps of geoid height measurements from 4.5 years of Geosat and 2 years of ERS-1 satellite altimeters have been used to compute the gravity anomaly, which is then mapped to a grid (Fig. 9.28). These grid values can then be compared with coincident ship-based measurements of the gravity anomaly. These estimates show agreement with ship data at a level of 5 milligal (mgal), where one mgal is about one-millionth the normal pull of gravity (which is 9.80665 m/s^2). Typical variations in the pull of gravity are 20 mgal although over the deep ocean trenches they can exceed 300 mgal.

The dense satellite altimeter measurements can be combined with sparse accurate measurements of seafloor depth to construct a uniform resolution map of seafloor topography as shown, for example, here in Fig. 9.29. These maps are sufficiently accurate for subsurface navigation, and they are useful for applications such as locating the major obstructions/constrictions to the major ocean currents, and locating shallow seamounts where fish and lobster are more abundant.

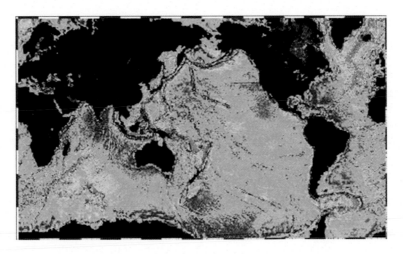

FIGURE 9.28

Global map of gravity anomaly calculated from Geosat and ERS-1 altimetry.

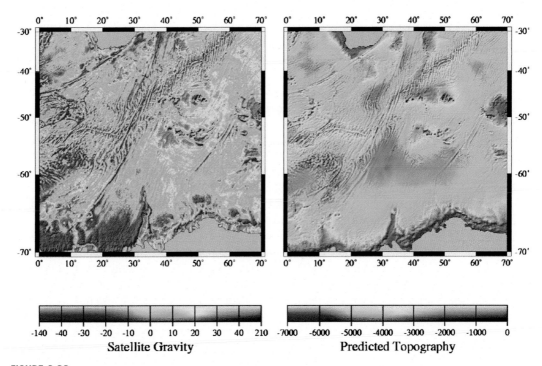

FIGURE 9.29

Maps of satellite-derived surface gravity and seafloor topography.

9.3 SYNTHETIC APERTURE RADAR OCEAN APPLICATIONS
9.3.1 MEASURING AND MAPPING OCEAN WINDS FROM SYNTHETIC APERTURE RADAR

SAR measures the backscatter from the sea surface, which is a measure of the roughness of the ocean surface. This roughness is strongly influenced by the local wind field so that the radar backscatter can be used to estimate the wind field (Horstmann et al., 1998). For their study, Horstmann et al. (1998) used SAR imagery from ERS-1 and ERS-2 satellites which both carried a C-band SAR. The advantage of using SAR to compute wind fields is that the resulting wind fields have a very high spatial resolution and the winds can be computed very close to the shoreline providing a major advantage over many other microwave techniques for computing ocean surface winds. Because of its potential for providing such high-resolution coastal wind fields, SAR-derived winds complement measurements made from wind scatterometers and microwave radiometers that yield very coarse-resolution estimates of the oceanic wind fields (Thompson and Beal, 2000). Such high-resolution wind fields are important for many coastal applications.

The spectral density of small-scale surface waves is a strong function of the surface wind stress. In addition, the intensity of the electromagnetic field that is transmitted from a radar and backscattered from the ocean surface is a clear function of this small-scale surface wave spectral density. Thus, there is a strong correlation between the wind stress (or wind speed) and the intensity of the backscattered radar return. Thus, even early researchers concluded that wind speed could be estimated from radar backscatter.

For the wind speed calculation from ERS-1 and ERS-2 SAR images, Horstmann et al. (1998) used the empirical C-band model CMOD4 (Stoffelen and Anderson, 1997) and used CMOD IFR2 (Quilfen and Bentamy, 1994) to compute wind speeds from the SAR backscatter. Both models were developed for the scatterometer on ERS-1 and 2. These models require input of the normalized radar cross section, the incidence angle of the radar beam, and the wind direction. The radar cross section and incidence angle can be derived in a straightforward manner from the details of the ERS SAR instruments. Since small changes in radar cross section result in very large wind speed changes, the radar has to be calibrated as accurately as possible. This calibration requirement is discussed further in Laur et al. (1997).

The wind direction can be easily derived if wind streaks are present in the SAR image (Fig. 9.30). In about 65% of the SAR images studied by Horstmann et al. (1998) wind streaks were found to be present from which the wind direction could be extracted. Shadowing of the wind field due to coastal topography can remove the 180 degrees ambiguity of the wind direction. For a general derivation of the wind direction a 10 km × 10 km subimage was analyzed with a fast Fourier transform (FFT) as plotted here in Fig. 9.31. Here the gray levels correspond to energies of the power spectrum, which are plotted for wavelengths between 500 and 1500 m. The resulting wind direction is perpendicular to the dashed line indicating the direction of peak power of the wind streaks.

Without wind streaks it is not possible from a single azimuth sample to accurately compute the wind direction (Thompson and Beal, 2000). Conventional wind scatterometers must sample at least two independent directions to solve for the wind direction (even then with a 180° ambiguity). An independent estimate of the wind direction is therefore needed to even formulate the wind speed retrieval.

Even the use of wind streaks due to coastal shadowing may sometimes result in erroneous wind directions that are almost orthogonal to the in situ measured wind direction (Thompson and Beal, 2000). One option is to incorporate numerical model estimates of wind direction. These models give

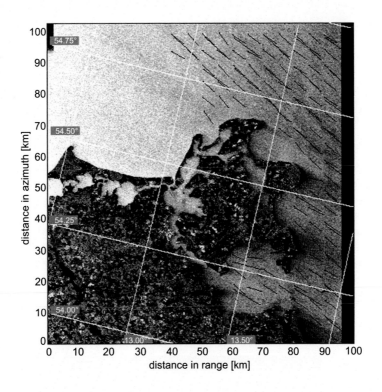

FIGURE 9.30

Synthetic aperture radar image of the island Rügen at the Baltic coast of Germany, from August 12, 1991 at 21:07 UTC from ERS-1. The *solid black lines* give the orientation of the visible wind streaks (Horstmann et al., 1998).

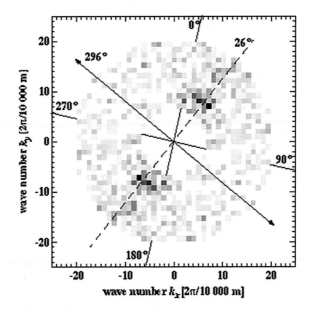

FIGURE 9.31

Power spectrum of an area in the open water north of Rügen. Wavelengths between 500 and 1500 m are taken for the estimation of wind direction. The main energy is along an orientation of 26 degrees from north; the resulting main direction is oriented perpendicular to this line.

wind directions estimates that are sufficiently reliable that wind speed can be retrieved from the C-band models and the SAR image.

Horstmann et al. (1998) used the wind fields from the two ERS satellites to investigate the spatial variability of the ocean wind field. The ERS SAR-derived winds are a snapshot of the ocean surface winds for an area up to 100 km × 100 km. An example of the SAR wind field on December 1, 1992 northwest of the Shetland Islands is shown here in Fig. 9.32. This image exhibits the expression of range traveling waves with approximately 300 m wavelengths. Wind streaks are used to define the solid lines, which represent the wind directions. Shadowing off the coast of the Shetlands was used to remove the 180 degrees directional ambiguity.

The mean wind direction was used to compute the wind speed with a resolution of 100 m × 100 m. The resultant mean wind speed was 14 m/s. The wind-speed spectrum was computed from the wind speeds with a two-dimensional FFT, which was then integrated over all directions to produce the one-dimensional wave number spectrum shown here in Fig. 9.33. Between wavelengths of 56 and 2 km the spectral density decreases in an almost linear fashion. For wavelengths shorter than 2 km the spectral density increases. This same behavior was found by Horstmann et al. (1998) to occur in several

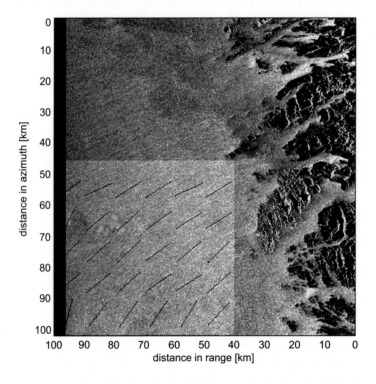

FIGURE 9.32

ERS-1 synthetic aperture radar image of the west cost of the Shetland Islands (Scotland, UK) taken on December 1, 1992 at 10:28 UTC. The *highlighted area* was used to investigate the wind variability. The superimposed *solid lines* represent the wind direction as computed from the wind streaks.

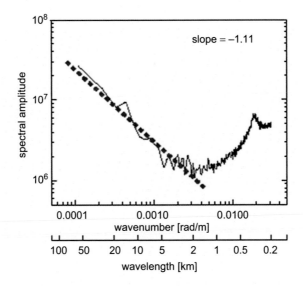

FIGURE 9.33

Spectral density versus wave number in log–log coordinates. The spectral density was computed from the wind-speed spectrum. Corresponding wavelengths are shown beneath the wave numbers in the x-axis.

different SAR images investigated. In these different cases the slop of the spectra between 100 and 2 km varied between −1.1 and −1.6, while the slope at wavelengths shorter than 2 km was approximately 1. The change in slope at approximately 300 m wavelength is due to the sea state.

9.3.2 DIRECTIONAL WAVE NUMBER SPECTRA FROM SYNTHETIC APERTURE RADAR IMAGERY

Ocean waves are ubiquitous features that cover the ocean's surface under most conditions. Surface waves range in wavelength from a few centimeters (capillary waves) to hundreds of meters (for wind waves or swell) with corresponding wave heights that range from millimeters to tens of meters. Surface waves are generated by the wind-generated turbulence above the sea at some location. Swell is generated at a remote location and propagates over the ocean. Ocean surface waves interact with the atmosphere, ocean currents, with the ocean bottom topography and with each other. Wave energy eventually dissipates through breaking in open water or upon shoaling.

SAR imagery is uniquely suited to an estimate of ocean surface wave conditions, since SAR is the only satellite sensor that can provide images from space with a high enough resolution independent of cloud cover and light conditions. First demonstrated by aircraft sampling in the early 1970s the two-dimensional surface wave field can be imaged by SAR. An ERS-1 SAR image of ocean surface waves refracting along a beach and around a headland is shown here in Fig. 9.34.

The earliest SAR-based mapping of surface waves was carried out using SeaSat SAR imagery (Beal et al., 1981). Later Beal et al. (1983) worked to measure the spatial evolution of SeaSat SAR-measured wave spectra from a SAR pass taken off the US East Coast. They found that the wave

(A) **(B)**

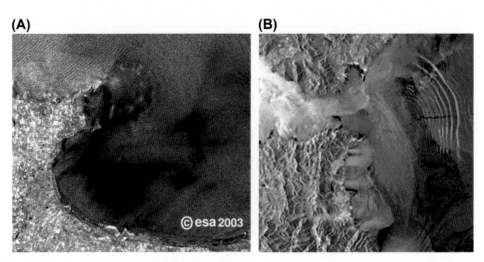

FIGURE 9.34

Sample ERS-1 synthetic aperture radar (SAR) image of ocean surface waves being (A) refracted around a headland, and (B) diffracted in the Strait of Gibraltar. Note the number of solitons in the wave packet and the *dark line* intersecting the internal wave packet originated from a discharge of oil from a ship enroute ship (https://earth.esa.int/web/guest/missions/esa-operational-eo-missions/ers/instruments/sar/applications/tropical/asset_publisher/tZ7pAG6SCnM8/content/oceanic-internal-waves-strait-of-gibraltar).

From ESA-AOES Medialab.

number of the dominant wave spectra changed in a way that was consistent with deepwater wave dispersion as the waves propagated away from a storm and interacted with the Gulf Stream.

A similar study was carried out by Sun and Kawamura (2009) using ERS-1 and ERS-2 data both of which carried the active microwave instrument, which has a SAR operation function in the C-band with VV polarization. These SARs have incidence angles of 20–26 degrees resulting in a SAR swath of about 100 km. The resulting SAR images are 6656 pixels in azimuth and 5344 in range. They used 5120 × 5120 pixels of this for their study. The SAR image was divided into 40 × 40 subimages, which have approximately 128 × 128 pixels (about 1.6 km × 1.6 km). Each subimage is first filtered using a Gaussian high-pass filter to remote the low wavenumber signal, which has no relation to surface waves. An example is presented here in Fig. 9.35 along with various version of the SAR wave number spectra.

Here the JONSWAP refers to a theoretical spectrum developed by Hasselmann et al. (1973, 1980) along with the directional spreading function proposed by Donelan et al. (1985). Note the symmetrical retrievals in the original SAR image spectrum and the filtered SAR image spectrum. At this point there is a 180 degrees ambiguity that is only resolved by invoking the JONSWAP model which uses the SAR-derived wind speed to give the spectrum a preferred direction as indicated by the spectra in Fig. 9.35D and E.

The SAR spectra are always affected by azimuth cutoff, the effect of SAR spectral roll-off in the azimuth direction. This azimuth cutoff is caused by the nonlinearity of the SAR surface-wave imaging mechanism in the azimuth direction and acts as a low-pass Gaussian filter to the SAR spectra. In the ERS SAR wave mode products (Johnsen et al., 1999), the cutoff wavelength is a parameter derived by fitting a Gaussian function to the range integral SAR spectrum (Vachon et al., 1994; Kerbaol et al., 1998; Schulz-Stellenfleth and Lehner, 2007).

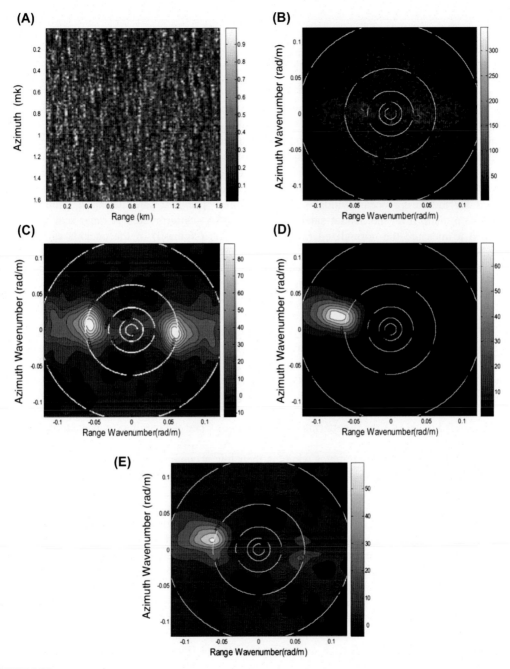

FIGURE 9.35

Example of the process for retrieving a wind-wave spectrum from a synthetic aperture radar (SAR) image. (A) SAR subimage of 128 × 128 pixels, (B) original SAR image spectrum, (C) filtered SAR image spectrum, (D) first guess spectrum constructed using the JONSWAP model using SAR wind speed, and (E) retrieved wave spectrum.

From Sun, J., Kawamura, H., 2008. Surface wave parameters retrieval in coastal seas from spaceborne SAR image mode data. PIERS Online 4 (4), 445–450. http://www.piers.org/piersonline/pdf/Vol4No4Page445to450.pdf.

9.4 OCEAN WIND SCATTEROMETRY
9.4.1 MAPPING THE OCEAN WIND VECTOR

One of the real observational challenges in physical oceanography is measuring the wind field over the oceans. Most of the ocean is unpopulated and wind measurements on ships are frequently compromised by ship superstructure influences on the wind sensor. As early as the 1940s, it was found that ocean waves caused large unwanted echoes on the radar screen when scanning at low elevation angles. These backscattered echoes from the sea surface became larger with increasing wind speed. This led to the development of ocean wind scatterometry. The most attractive method for measuring ocean winds is from satellites. The first satellite-based wind measurements were made from the SeaSat satellite with a NASA scatterometer. A radar scatterometer is designed to measure the normalized radar cross section (sigma-0) of the ocean's surface as an indication of the ocean surface wind. Scatterometers transmit a radar pulse toward the ocean and measure the reflected energy. A separate measurement of the noise-only power is made and subtracted from the reflected signal to determine the backscatter signal power. Sigma-0 is computed from the signal power measurement using the distributed target radar equation. Scatterometers must be very accurately calibrated to be able to make accurate backscatter measurements.

The backscattered energy measured by a scatterometer provides a measurement of near-surface ocean winds. By combining sigma-0 measurements from different azimuth angles, the near-surface wind vector over the ocean can be determined using a geophysical model function (GMF), which relates the wind magnitude and radar backscatter. From the ocean surface the backscatter results from wind-generated capillary–gravity waves, which are generated when the wind blows over the ocean. The capillary waves are generally in equilibrium with the surface wind over the ocean. The scattering mechanism is known as Bragg scattering, which happens when the capillary waves are in resonance with the radar wavelength.

One disadvantage of scatterometer winds is their low accuracy for low wind speeds, and at the inside of a swath, the increasing RMS error at high wind speeds due to representativeness errors (other geophysical parameters not included in the sigma-0 to wind transfer function become important at high wind speeds). In addition, the scatterometer operates on a near polar-orbiting weather satellite and hence measurements are not collected at strictly synoptic measurement times. This latter is less of a problem now that 4-D variational analysis is the primary method for data assimilation into weather forecast models. Other problems have been mentioned like the dependence of the transfer function on local static stability of the atmosphere (Brown, 1986).

The very first spaceborne scatterometer was the SeaSat-A scatterometer system (SASS) that flew for about 90 days in 1978. This was a "fan-beam" system with the two beam antennas varying from 25 to 55 degrees. SASS was a Ku-band radar with a wavelength of 2.1 cm at a frequency of 14.599 GHz. Thirteen years later ESA launched ERS-1, which carried a C-band scatterometer (wavelength of 5.7 cm). The ERS-1 scatterometer differed from SASS in that the radar antennas were mounted only on one side of the satellite illuminating a 500-km-wide swath to the right side of the subsatellite track. Instead of two antennas, it carried three: one pointed at 45 degrees (fore beam), one at 90 degrees (midbeam), and 135 degrees (aft beam) relative to the line of satellite propagation. Thus, the incidence angle of the radar beam varied from 18 to 47 degrees for the midbeam and from 25 to 569 degrees for the fore and aft beams. The 500-km-wide swath of the ERS-1 scatterometer was sampled every 25 km. The sampling distance along the swath is also 25 km. Adjacent measurements across the swath are not independent due to the effective spatial resolution of the instrument of 50 km (overlapping footprints).

The computation of the wind vector from the different scatterometer views is a nonlinear inversion process based on an accurate knowledge of the GMF (in an empirical or semiempirical form) that relates the radar backscatter to the vector wind retrieval. To retrieve the wind vector this requires that the scatterometer makes several backscatter measurements of the same spot from different azimuth angles. Many scatterometers use fan-beam stick antennas with complimentary orientations to enable the retrieval of the wind vector. These scatterometer designs require considerable power and mass and are often difficult to accommodate on spacecraft. An option is for a single-beam scanning scatterometer such as that flown on the QuikSCAT satellite (Fig. 9.36). This satellite carried NASA's Seawinds scatterometer. This satellite mission was conceived of as a "quick recovery" mission replacing the NASA scatterometer (NSCAT), which failed prematurely in June 1997 after just 9.5 months of operation. QuikSCAT exceeded all expectations and operated successfully for over a decade until a bearing failure on its antenna scan motor rendered the instrument useless on November 23, 2009 after a launch on June 19, 1999.

This instrument uses a rotating dish antenna with two spot beams that sweep in a circular pattern. The antenna consists of a 1 m diameter rotating dish that produces two spot beams sweeping in a circular pattern. It radiates 110 W microwave pulses at a pulse repetition frequency of 189 Hz. QuikSCAT operates at a frequency of 13.4 GHz, which is in the Ku-band of microwave frequencies. At this frequency the atmosphere is mostly transparent to nonprecipitating clouds and aerosols, although rain significant attenuates the signal.

The spacecraft is in a Sun-synchronous orbit, with equatorial crossing times of ascending swaths at about 06:00 LST ±30 min. Along the equator, consecutive swaths are separated by 2800 km. QuikSCAT orbits the Earth at an altitude of 802 km at a speed of about 7 km/s. During its operational lifetime, QuikSCAT measured winds in measurement swaths 1800 km wide centered on the

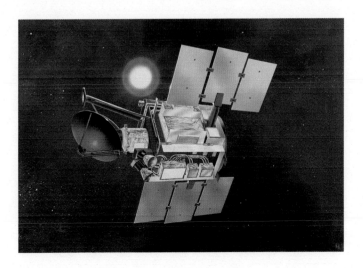

FIGURE 9.36

The QuikSCAT satellite with NASA's Seawinds instrument at the bottom.

satellite's ground track with no nadir gap such as occurs with fan-beam antennas. Because of its wide swath and lack of in swath gaps, QuikSCAT was able to collect at least one wind vector measurement over 93% of the World's Ocean each day. This improved significantly over the 77% coverage provided by NSCAT each day. Each day QuikSCAT recorded over 400,000 measurements of wind speed and direction. This is hundreds of times more surface wind measurements than are routinely collected from ships and buoys.

A limitation of QuikSAT is that wind vectors cannot be retrieved within 15—30 km of coastlines or in the presence of sea ice. Precipitation generally degrades the wind measurement accuracy although useful wind and rain information can still be obtained in midlatitudes and tropical cyclones (Fig. 9.37). In addition to retrieving the wind vector the scatterometer has proven to be useful in estimating the fractional coverage of sea ice, tracking large (>5 km long) icebergs, differentiating between different types of ice and snow, and detecting the freeze—thaw line in polar regions.

knots

| 0 | 5 | 10 | 15 | 20 | 25 | 30 | 35 | 40 | 45 | > 50 |

FIGURE 9.37

QuikSCAT image of Hurricane Katrina August 28, 2005.

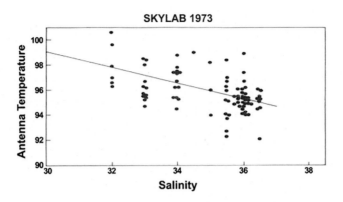

FIGURE 9.38

Antenna temperature and salinity correlation in Skylab data (http://aquarius.umaine.edu/images/ov_skylab_data.jpg).

9.4.2 SEA SURFACE SALINITY

SSS can be remotely measured using microwave radiometry at L-band (Swift and McIntosh, 1983). During the late 1940s and early 1950s, laboratory experiments were conducted to quantify the relationship between salinity and BT. Until ESA's SMOS, only two (experimental) spaceborne radiometers at L-band had been launched, in Cosmos 243 (1968) and in Skylab S-194 (1973), but the first evidence from space came only from the Skylab observations (Fig. 9.38).

The link between salinity and BT is through the dielectric constant of sea water. The "protected" 1.400–1.427 MHz band represents the optimum trade-off between sensitivity, atmospheric and ionospheric effects, and spectrum protection. Still, salinity measurements are very difficult to make because of the low sensitivity of BT to SSS, 0.2–0.8 K/psu (psu = practical salinity unit, roughly 1 g of salt per 1 kg of water). The dependence on the physical temperature (Fig. 9.39), surface roughness (induced by local wind or by far waves), incidence angle, and polarization, on one hand, and the technical difficulties to achieve a highly stable and accurately calibrated microwave radiometer, and the presence of RFI, make it a challenging observable.

The first SSS airborne map was obtained in 1996 using the scanning low-frequency microwave radiometer (SLFMR), a six-beam real aperture radiometer (Fig. 9.40A) (Miller et al., 1996). In 2000, the first SSS airborne map was obtained using the electronically steered thinned array radiometer (ESTAR) (LeVine et al., 2000) (Fig. 9.40B).

In 1995, ESA organized the Soil Moisture and Ocean Salinity Workshop to define the roadmap to remotely measure these two geophysical variables, and it was concluded that the most promising technique was aperture synthesis radiometry, which had been successfully demonstrated a few years earlier with the ESTAR instrument (Ruf et al., 1998). In 1998, an international team of scientists led by Dr. Y.H. Kerr (CESBIO, France) and Dr. J. Font (ICM/CSIC, Spain) proposed the SMOS mission to ESA, which was selected as the second Earth Explorer Opportunity Mission (Silvestrin et al., 2001) within the ESA Living Planet Programme. Finally, SMOS was successfully launched on November 2, 2009 from Plesetsk, Russia (Fig. 9.41).

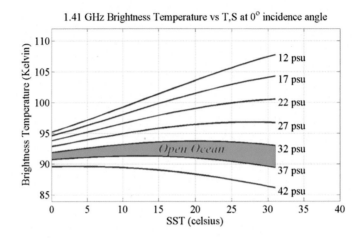

FIGURE 9.39

Brightness temperature at nadir dependence with sea surface temperature (SST) and sea surface salinity (Swift, 1980).

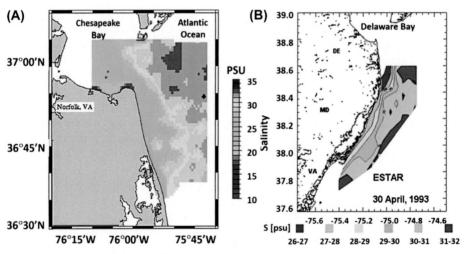

FIGURE 9.40

First sea surface salinity maps derived from L-band microwave radiometry using (A) a real aperture radiometer (Miller et al., 1996) and (B) a synthetic aperture radiometer in 2000 (LeVine et al., 2000).

However, when SMOS was selected in May 1999, there were still many uncertainties and many lessons to learn. In particular, for an accurate salinity retrieval, there were three key issues to be addressed: an accurate characterization of the impact of salinity and temperature on the dielectric constant, an accurate characterization of the impact of the sea state (wind speed, SWH, etc.), and foam on the BTs.

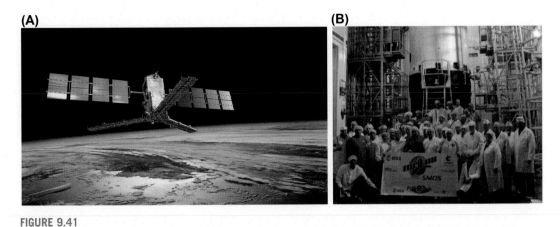

FIGURE 9.41

(A) SMOS artist's view in orbit, (B) SMOS launch team with ROCKOT launcher at Plesetsk, Russia.

(A) From ESA-Pierre Carril. http://www.esa.int/spaceimages/Images/2009/09/SMOS_in_orbit3

Two of the most widely used dielectric constant models are those of Klein and Swift (1977) and Ellison et al. (1998), and although they show similar results, their differences are too large to meet the salinity retrieval accuracy requirements, i.e., for an absolute accuracy of 0.2 psu, the BT must be on the order of 0.1K for cold waters (Fig. 9.43). Blanch and Aguasca (2004) and LeVine et al. (2011) developed new experimental setups to develop more accurate models. In general, there is a better agreement with the Klein and Swift model, while the Ellison model tends to overestimate the computed BT. However, it is difficult to warrant the absolute accuracy of this model because of the lack of absolute references.

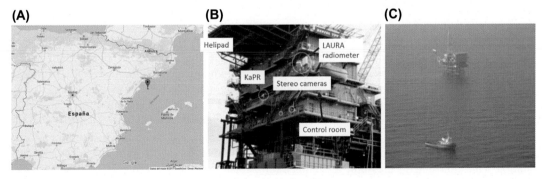

FIGURE 9.42

(A) Location of the REPSOL Casablanca oil rig in the Mediterranean coast of Catalonia. (B) Image of the north side showing the location of the instruments, (C) Heavy equipment was transported to Casablanca oil rig by ship, while delicate goods and personnel were flown by helicopter.

(A) **(B)** **(C)**

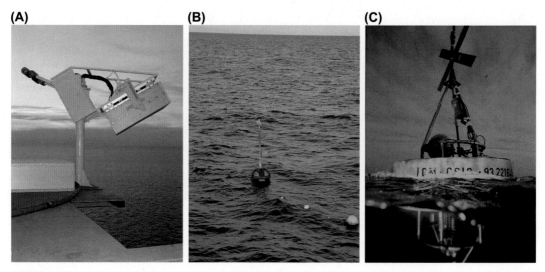

FIGURE 9.43

Some of the instrumentation deployed during WISE 2000 and 2001: (A) LAURA installed and acquiring data from the 32 m height deck (UPC), (B) Aanderaa CMB3280 (meteorological buoy, ICM/CSIC), (C) underwater view of the oceanographic buoy (ICM/CSIC).

Since 1971, no systematic measurements have been performed to assess the impact of the sea state, usually parameterized in terms of the wind speed, or the SWH. Additionally, only a few airborne experiments had been carried out, mostly with the SLFMR real aperture radiometer (Miller et al., 1996; Hollinger, 1971), and one with the ESTAR synthetic aperture radiometer (LeVine et al., 2000). Almost 30 years later, in 1999 ESA sponsored the WInd and Salinity Experiment (WISE) on the REPSOL Casablanca oil rig (40.72°N 1.36°E), ~50 km offshore of the coast of Tarragona (Spain), and ~115 km south of Barcelona (see Fig. 9.42).

Two WISE campaigns took place (November 2000 to January 2001, and October to November 2001). The instruments deployed were (Fig. 9.43) the LAURA (L-band AUtomatic RAdiometer) from the Universitat Politècnica de Catalunya (Spain), the first fully polarimetric (four Stokes parameters) L-band radiometer, a Ka-band polarimetric radiometer from the University of Massachussetts (USA), a stereocamera to determine surface topography and root mean square slopes of the sea surface from CETP (France), four oceanographic and climatological buoys for near-surface salinity and temperature measurements using the SeaBird MicroCAT system (model SBE37-SM), as well as wind speed and direction, wave height, and period, etc. from the Institute of Marine Sciences (ICM/CSIC, Spain) and LODYC (now LOCEAN, France), a portable meteorological station with atmospheric pressure, air temperature, relative humidity and rain rate, and a video camera mounted on the L-band radiometer pedestal to determine sea surface foam coverage from the Universitat Politècnica de Catalunya (Spain), and an IR radiometer to determine SST estimates from the Universitat de València (Spain). Additionally, during WISE 2000, temperature and salinity were recorded from the platform at 5 m below sea level, and simultaneously ocean color, wind vector, and SST were acquired by different satellites.

The main result of the WISE campaigns was the determination of the BT dependence on the 10 m height wind speed and the SWH required to perform the necessary corrections to retrieve SSS (Camps et al., 2002, 2004; Gabarró et al., 2004), including a weak azimuthal signature, which was more clearly seen during a very intense storm that occurred on November 2001. These results were used in one of the SMOS salinity retrieval algorithms, which was later refined, once satellite data became available (Guimbard et al., 2012).

One of the open questions from WISE 2000 and 2001 was the evident modulation of the BTs when a patch of foam appeared in the antenna beam (Fig. 9.44). This was confirmed by the IR images, in which the foam patches appeared as hot spots. To quantify this effect, as well as the impact of rain and oil spills, and to validate the existing numerical models, the FROG 2003 (foam, rain, oil spills, and GPS reflections) experiment was performed at the IRTA facilities at Poblenou del Delta (Tarragona, Spain) to study specifically the impact of foam, rain, and oil spills on sea surface emissivity (Camps et al., 2005). Results indicated that the foam impact is not negligible and should also be included in the emission model at L-band.

After the successful launch of SMOS in 2009 (Kerr et al., 2010; Font et al., 2010), and a long commissioning phase of 6 months, which was needed to learn many lessons from a new type of instrument, a new type of multiangular and polarimetric L-band measurements, SSS maps started to be produced (as well as soil moisture maps, see Chapter 11). Since ESA policy covers only up to the level 2 processing, individual ESA state members are responsible for the processing of the higher levels. In the case of SMOS, there are two of these centers: the CATDS (Centre d'Aval et Traitement des Données SMOS) in Toulouse, France, and the CP34 (Centro de Procesado de niveles 3 y 4 de SMOS) in the SMOS Barcelona Expert Center (BEC), Barcelona, Spain. The data produced at BEC are served in netCDF format and are freely distributed to the scientific community by means of OPENDAP, HTTPServer, and NetcdfSubset services through a THREDDS server. Maps produced by BEC can also be visualized online by means of a Web Map Service (http://cp34-bec.cmima.csic.es/data/available-products). Sample products from BEC are provided in Fig. 9.45, which show salinity maps derived using novel image reconstruction algorithms (González-Gambau et al., 2016) and novel retrieval techniques more robust to RFI (Olmedo et al., 2016).

Initially approved as a NASA Earth System Science Pathfinder program in 2001, Aquarius was confirmed as a mission in September 2005 (Le Vine et al., 2010). The Aquarius/SAC-D mission was developed collaboratively between NASA and Argentina's space agency, Comisión Nacional de Actividades Espaciales (CONAE) to best meet the goals of each agency while giving priority to salinity measurements. In June 2010, it was launched from Vandenberg Air Force Base, California. The technology used in Aquarius was a three beam push-broom noise-injection radiometer, and a radar to perform the sea surface roughness correction. In June 2015, after a failure in the power system, the mission is over. At present, there are no follow-on missions to measure the ocean salinity.

9.4.3 BATHYMETRY AND BENTHIC HABITATS MAPPING IN SHALLOW WATERS

Coastal ecosystems are very important for life quality (most of the world's population lives in the coasts) and global climate. The benthic zone extends from the coastline along the surface of the continental shelf out to sea. Efficient management of the benthic zone for ecological or economic

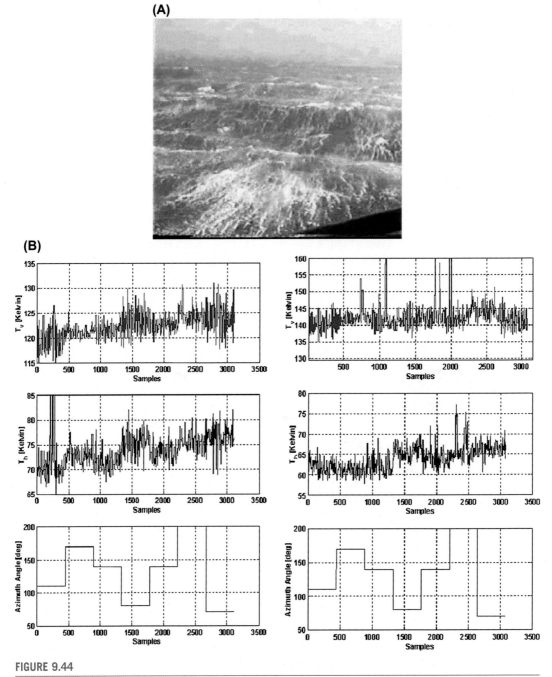

FIGURE 9.44

(A) Picture taken from 32-m-high deck of the Casablanca oil rig. Waves destroyed the 7 m deck and seriously harmed the 13 m deck. (B) Vertical polarization (upper row), horizontal polarization (central row), and azimuth angle (bottom row) for two azimuthal scans during the storm (Camps et al., 2004).

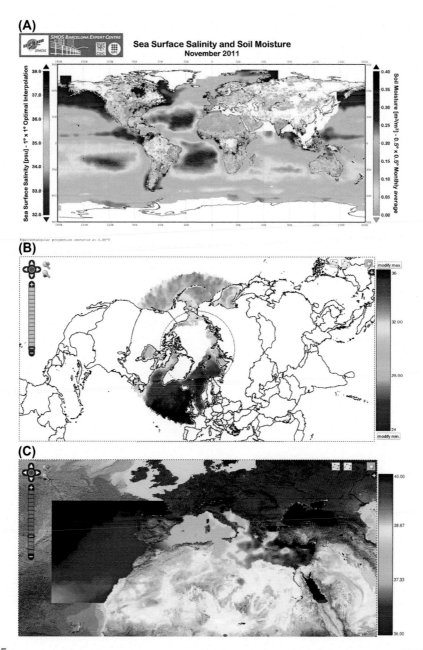

FIGURE 9.45

(A) Sample global soil moisture and ocean salinity maps, (B) new experimental SMOS sea surface salinity (SSS) maps at high latitudes, including Arctic Ocean open water regions computed at Barcelona Expert Center using a new methodological approach to substantially reduce land–sea and RFI contamination effects, as well as other instrumental biases, (C) new experimental SMOS SSS maps over the Mediterranean.

reasons requires adequate information on biogeochemicals, water turbidity, bathymetry, and distribution of habitats. Satellite-based imaging systems with spectral bands within the visible spectrum can provide reliable information at the spatial and temporal scales needed to estimate bathymetry and distribution of benthic habitats in the shallow water environments.

Coastal regions bathymetry can be estimated using a radiative transfer model of the radiance coming from the Sun, which is partly scattered in the sea surface producing the sunglint, and partly passes through the water column, gets reflected in the sea bottom, and passes through the water column again in the upwelling path (e.g., Zhongping et al., 2007; Lyons et al., 2011; Collin and Hench, 2012). In Eugenio et al. (2015), an improved algorithm is presented to perform the atmospheric correction and Sun deglinting to high resolution (0.5 m) WorldView-2 multispectral imagery prior to the development of bathymetric and benthic maps. The algorithm is based on the 6S accurate radiative transfer model, and it is applicable to turbid and optically shallow waters.

Bathymetric maps computed using the above method are able to reproduce depths up to approximately 25 m. The subsequent mapping of benthic habitats is a complex problem because only limited and noisy spectral information is available. Depending on the depth, only three to four WorldView-2 bands could be useful. The use of seafloor normalized indexes (see Table I of Eugenio et al., 2015), which include bathymetric information, provides an improved performance, with respect to the classical classification methods directly applied to the seafloor albedo or spectral bands. Fig. 9.46 shows an example of the Granadilla coast in Tenerife (Canary Islands, Spain) of the atmospheric and sunglint corrected WorldView-2 imagery, the seafloor albedo, and the bathymetry maps. After the bathymetry maps have been derived, the classification map of the benthic habitats for the same region is derived and satisfactorily compared to the CIMA 2008 map derived from in situ measurements. The obvious advantage of these new techniques as compared to traditional sampling techniques is the synoptic view of the whole area and the reduced revisit time (Fig. 9.47).

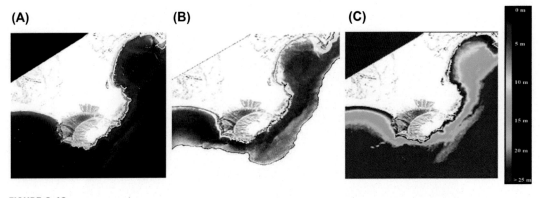

FIGURE 9.46

Selected areas on the WV2 images of Granadilla (Tenerife, Canary Islands) coastal area: (A) atmospheric and sunglint corrected imagery, (B) seafloor albedo (coastal blue-green-blue bands), and (C) maps of estimated depth (bathymetry).

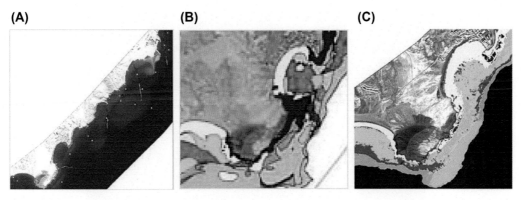

FIGURE 9.47

(A) Location of the test stations and transects. (B) CIMA 2008 map of benthic habitats for the area of Granadilla (Tenerife, Canary Islands): seagrass in green and light blue, sand in yellow and orange, maerl in pink, and algae in blue. (C) Classification map obtained by the combination of benthic indexes and Support Vector Machine supervised classification seagrass in green, sand in yellow and orange, maerl in pink, and algae in blue.

9.4.4 SARGASSUM SAGA: SPOTTING SEAWEED FROM SPACE

Each year a bizarre epidemic emerges upon thousands of miles of coastline, stretching from the Caribbean Islands to the Gulf States. This visitation has been considered an unpredictable act of nature and a detriment to beaches everywhere as long as it has existed. It seems now, however, that the tides of change are sweeping in. The visitor is called sargassum, a macroalgae related to kelp or seaweed. Despite sargassum being considered a nuisance for most of its existence, the Sargassum Early Advisory System (SEAS) of Texas A&M University at Galveston (TAMUG) is making huge leaps to learn more about the macroalgae. Thanks to their work it is now apparent that not only can sargassum be tracked and predicted through the USGS's Landsat imagery, but that providing this information to the public saves the taxpayers' money and takes away many of the inconveniences of the sargassums' arrival. SEAS uses many oceanographic techniques to deliver accurate predictions of the time, location, and severity of sargassum landings. The SEAS predictive model is 90% accurate, forecasting sargassums' arrival up to 2 weeks out. The primary means of tracking the large sargassum mats is through USGS's Landsat imagery. Weather buoys and beach cameras are also used for accurate monitoring of current weather conditions and confirmation of sargassum landings. The SEAS team has developed effective yet simple methods for spotting and correctly identifying sargassum through the use of USGS Landsat imagery (Fig. 9.48). Researchers look for irregularities in the water, caused by the sargassum mats, which are referred to as "slicks." These are areas of irregular disrupted water adjacent to a mat, which may be too small to be detected itself; the slicks appear as dark dendritic veins, through the water.

The SEAS team has focused a large part of its energy and resources on getting USGS's Landsat information out to the public whether that would be a school or beach managers. High schools were specifically targeted as a way to reach out to the public. Students were invited to TAMUG's computer laboratories and trained in the processes of analyzing Landsat images. In every instance the SEAS team was successful in training the students, so much so that they continued to analyze the images at

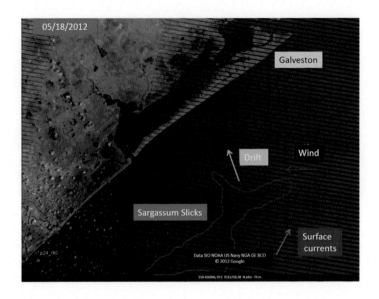

FIGURE 9.48

This Landsat image (Path 25 Row 40, acquired May 18, 2012) displays a mat of sargassum approaching the Texas shoreline.

school and sent in their products to the SEAS team as the year went on. The public is beginning to see that if they understand sargassum and its natural cycle that it can be planned around and no longer disrupt their beach plans. Beach managers were a target audience for the useful information that can be gleaned from analyzing the USGS Landsat data. It now allows them to know the amount of sargassum to expect as well as when and where to expect it. This means that the sargassum is cleaned up and dealt with more efficiently. This saves the taxpayers money and reduces the inconvenience to the tourists coming to enjoy a pristine beach. SEAS utilizes Landsat imagery to aid the local beach managers and communities in building a base of citizens who strive to solve problems in the most environmentally and economically savvy ways possible (Brandon N. Hill, SEAS/TAMUG).

9.5 CONCLUSIONS

Satellite remote sensing of the ocean has grown from early applications of weather satellite imagery for sensing and mapping SST to dedicated ocean measurement systems such as satellite altimeters. A major event in this development was the launch of the short-lived SeaSat satellite in 1979. While this satellite unfortunately only operated for 90 days, it demonstrated the potential of studying the ocean with a series of microwave instruments such as the altimeter, the scatterometer, and the SAR. It took a period of time for the community to finally organize satellites to make these measurements individually but SeaSat really showed the way.

At the same time the weather satellites have improved, but unfortunately their numbers have gone down. A big change, however, has been the change to the free distribution of all of the weather and other ocean satellite data. This online access to all of these data has revolutionized their study and opened avenues to research that would not have been possible without this ready and free-of-charge availability to these satellite data.

Today we have a relative long series of satellite altimeter data that continues. We also have moved optical measurements from just IR measurements of SST to instruments designed to measure ocean color as an indication of biological productivity in the ocean. Scatterometers have demonstrated their ability to provide high-resolution maps of winds over the ocean. Altimeters have proved useful not only in mapping mesoscale and longer-scale ocean currents but also for mapping SWH. SARs have provided high-resolution maps of wind fields, frequency-wave number spectra of wave fields and the mapping of ships on the sea surface.

9.6 STUDY QUESTIONS

1. What are the fundamental differences between the IR and microwave measurements of SST? What correction factors do you need to consider for each? How would you propose to merge these different SSTs together to take best advantage of each of their advantages? What measurements would you use to validate SST products?

2. How would you combine satellite altimetry, SST and ocean color measurements to map mesoscale ocean eddies? What benefits do you have by using multiple data sources to map these eddies? What ocean data sets might you use to validate these eddy maps?

3. List all of the corrections that need to be applied to satellite altimetry in the routine processing of these data. Give their general magnitudes and comment on how these corrections are routinely carried out. What method does one use to map ocean current variability with altimetry?

4. How are ocean color measurements routinely corrected for atmospheric aerosol contamination? How are the ocean instruments generally calibrated when they are on orbit? What is generally done to avoid the problem of sunglint in these reflected measurements? Name two biological quantities that are routinely measured with ocean color instruments from space.

5. How is ocean surface salinity measured from space? What accuracy in terms of magnitude and space–time resolutions are typical of these surface salinity sensors? What are the basic differences between the European SMOS mission and the American-Argentinian Aquarius mission? How are data from these missions validated with in situ measurements?

6. Describe how a satellite scatterometer measures wind speed and direction. Contrast this with the microwave radiometer in Windsat, which also measured wind speed and direction. Which is more accurate and which had a low starting sensitivity threshold. Discuss the differences between a "fan-beam" scatterometer and one that rotates like QuikSCAT.

7. What are the important components of an IR radiometer to make accurate measurements of SST? How can the accuracy of this onboard calibration system be validated when the satellite is flying? What about before launch, how is the calibration tested and evaluated?

8. How can SAR be used for ocean studies. How is SAR different from optical images of the same scene? What is the biggest problem with SAR and how can it be reduced?

9. How can satellite altimetry be used to map long-term sea level rise? What data are needed to combine with this altimetry to come up with the most accurate absolute measures of sea level rise? Are there significant regional differences in sea level rise?

10. Discuss how a satellite altimeter measures the SWH. How is this SWH used in practice and as an economic product?

LAND APPLICATIONS

10

10.1 HISTORICAL DEVELOPMENT

The US remote sensing of land parameters has an interesting history. Initially people started to use weather satellite imagery to estimate some important land characteristics like the health of vegetation. Soon it was clear that more focused remote sensing was needed for the land, and the early Earth Resources Technology Satellite (ERTS) was born out of the NASA NIMBUS program. This then gave rise to the US Landsat program, which was a collaboration between NASA and the United States Geological survey (USGS). This program went through a phase where it was unsuccessfully commercialized, and it is now back in the hands of NASA and the USGS.

During their evolution, the Landsat satellites have seen dramatic changes in the types of sensors that they flew to sense the Earth's surface. Vegetation was not the driving factor, but sensing geological resources were more important. Thus, the Landsat satellites did not fly in a diurnal repeating orbit, but rather in a 15-day repeat orbit. In addition the Landsat sensor swath widths were fairly narrow and did not overlap at the equator. It took a number of days for a single Landsat instrument to image the entire globe.

The first ERTS satellite was launched on July 23, 1972. This satellite carried two instruments: a camera built by the Radio Corporation of America (RCA) called the Return Beam Vidicon (RBV), and the multispectral scanner (MSS) system built by the Hughes Aircraft Company (El Segundo, California) under contract to NASA. The MSS recorded data in four spectral bands, green, red, and two near-infrared bands. These data were the source of over 300 scientific investigations funded by NASA. The RBV was intended to be the primary instrument, but the MSS data were quickly found to be superior to the RBV observations. In addition the RBV instrument was the source of an electrical transient that caused the satellite to briefly lose attitude control therefore compromising image data collection.

This first of the soon to become Landsat satellite series was based on NASA's successful NIMBUS series and used the same basic bus structure to support the mission (Fig. 10.1). In 1976 the then director of the USGS Dr. V.E. McKelvey reported "The ERTS spacecraft represents the first step in merging space and remote sensing technologies into a system for inventorying and managing the Earth's resources." Landsat 1 operated until January 1978, outliving its design life by 5 years. It demonstrated the value of the MSS collecting over 300,000 images of the Earth's land surfaces.

This satellite was placed into a 99 degrees polar orbit with a 10 a.m. local time equatorial crossing time. Its orbital period was approximately 100 min, which resulted in about 14–15 orbits per day. The satellite altitude was nominally 900 km.

Introduction to Satellite Remote Sensing. http://dx.doi.org/10.1016/B978-0-12-809254-5.00010-5

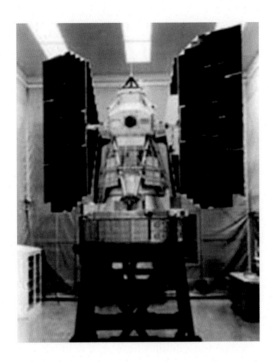

FIGURE 10.1

Earth resources technology satellite or Landsat 1 (https://landsat.gsfc.nasa.gov/landsat-data-continuity-mission/).

Landsat 2 also used the NIMBUS bus as the satellite platform and was again operated by NASA. It was launched on January 22, 1975. As this satellite was again considered experimental it carried the same instruments as Landsat 1. On February 25, 1982, after seven years of successful service Landsat 2 was taken out of operation due to serious yaw control problems that made it impossible to accurately point the satellite sensors.

Landsat 3 was again based on the NIMBUS satellite bus and was launched on March 5, 1978 and immediately put in standby mode. It began operation later in 1978 and operated until September 7, 1983. It carried the same sensors as Landsats 1 and 2. The MSS now had five spectral bands with two visible (0.5−0.6 µm, 0.6−0.7 µm), two near infrared (0.7−0.8 µm, 0.8−1.1 µm), and one thermal infrared (10.4−12.6 µm). The pixel size was roughly 57 × 79 m, and the scene size was 170 km × 185 km. As before, the swaths did not overlap at the equator, and it took a number of days for a single satellite to image the entire globe.

A major change came with Landsat 4 and 5. The spacecraft was completely redesigned and optimized for this Earth application. The RBV instrument was dropped and the MSS continued to fly. In addition, a new imager called the thematic mapper (TM) was added that had improved spectral and spatial resolutions. It could see a wider portion of the electromagnetic spectrum and had a better spatial resolution to see greater detail on the ground. The spectral channels, included the blue, green, red and near infrared, the mid-infrared (2 bands), and the thermal infrared portions of the spectrum. The spatial resolution improved from 80 m to about 30 m.

Landsat 4 lost two of its solar panels and both of its direct downlink antennas less than a year after its launch July 16, 1982. Thus, data downlink was not possible until NASA launched their Tracking and Data Relay Satellite (TDRSS). After that Landsat 4 could transmit data via its Ku-band transmitter from which TDRSS could then relay these data to the ground. This downlink via TDRSS continued until 1993 when this last remaining data downlink failed on Landsat 4. The satellite was kept on orbit for housekeeping telemetry command and tracking data until it was decommissioned in 2001. While NASA built and launched Landsat 4, it was initially operated by National Oceanic and Atmospheric Administration (NOAA) until the operations were contracted out to the Earth Observation Satellite Company (EOSAT) in 1984 as part of the Landsat commercialization effort.

The year Landsat 5 was launched (1984) the US Congress decided that land remote sensing satellites could be privatized in the 1984 Land Remote Sensing Commercialization Act. NOAA, the agency in charge of Landsat operations, was instructed to find a commercial vendor for Landsat data. NOAA selected the Earth Observation Satellite Company (EOSAT), which was a partnership between Hughes Aircraft and RCA. The contract gave EOSAT the responsibility for archiving, collecting, and distributing current Landsat data as well as the responsibility for building, launching, and operating the following two Landsat satellites (with government subsidies).

Commercialization was very difficult: EOSAT had limited commercial freedom under this congressional act and it therefore increased the price of Landsat imagery from $650 to $3700, and eventually $4400. This priced the data out of the reach of most of the researchers who wanted to work with it and as a result further reduced the income of EOSAT. Only large commercial users such as oil and mining companies could afford these data and since a lot of research was still needed to make these data useful this research was stopped as a consequence of this price hike. As a result, many users migrated to the free low-resolution land data being provided by meteorological satellites. In 1986, France launched SPOT a Landsat-like satellite, which broke the US monopoly further reducing the EOSAT market.

During this period of EOSAT commercialization even the data collection standards were reduced. Many Landsat image data collects were missed because there were no obvious and immediate buyers of these data. In a truly commercial enterprise, data are only collected for a customer with a scientific purpose. For a scientific program you collect all the data that you can for present and future study. In addition the Landsat 4 and 5 system calibration activities were not kept up during this same period of time again because the users of these data were not particularly interested in calibration.

In 1989, with two aging satellites and no operational budget, NOAA directed EOSAT to turn the satellite off. The program was only saved by a strong protest from Congress and foreign and domestic data users (such as oil companies) and a direct intervention by the US Vice President. In response, the Congress passed the Land Remote Sensing Policy Act of 1992, which instructed Landsat program management to build a government-owned Landsat 7, which followed the failure to achieve orbit of Landsat 6, which was built by EOSAT.

Two years after the launch of Landsat 7 EOSAT (later Space Imaging) returned operational responsibility for Landsat 4 and 5 back to the US government. On July 1, 2001, when this transfer occurred space imaging also relinquished their commercial rights to Landsat data, which were now transferred to the USGS to sell Landsat data according to USGS pricing policy. While this pricing was initially also too high, USGS has now realized that the best price for science users of Landsat data is free and all of the data are available on the web for no cost.

Landsat 8 (Fig. 10.2), called the "Landsat Data Continuity Mission", was developed by NASA. While the orbital characteristics were to be maintained, it was decided to have all new instruments. The

FIGURE 10.2

Landsat 8 on orbit configuration.

first would be a visible only sensor, which became known as the Operational Land Imager (OLI), built by Ball Aerospace. The thermal infrared was now assigned to a multichannel infrared instrument known as the Thermal Infrared Sensor (TIRS).

Landsat 8 was launched on February 11, 2013, from Vandenberg Air Force Base, California, on an Atlas-V 401 rocket, with the extended payload fairing (EPF) from United Launch Alliance, LLC. The operational land imager (OLI) and the TIRS provide seasonal coverage of the global landmass at a spatial resolution of 30 m (visible, NIR, SWIR); 100 m (thermal), and 15 m (panchromatic).

Landsat 8 was developed as collaboration between NASA and the USGS. NASA led the design, construction, launch, and on-orbit calibration phases, during which time the satellite was called the Landsat Data Continuity Mission (LDCM). On May 30, 2013, the USGS took over routine operations and the satellite became Landsat 8. USGS leads postlaunch calibration activities, satellite operations, data product generation, and data archiving at the Earth Resources Observation and Science (EROS) center.

Landsat 8 instruments represent an evolutionary advance in technology. OLI improves on past Landsat sensors using a technical approach demonstrated by a sensor flown on NASA's experimental EO-1 satellite. OLI is a push-broom sensor with a four-mirror telescope and 12-bit quantization. OLI collects data for visible, near infrared, and short-wave infrared spectral bands as well as a panchromatic band. It has a 5-year design life. The graphic below compares the OLI spectral bands to Landsat 7's ETM+ bands. OLI provides two new spectral bands, one tailored especially for detecting cirrus clouds, and the other for coastal zone observations.

The OLI collects data for two new bands, a coastal band (band 1) and a cirrus band (band 9), as well as the heritage Landsat multispectral bands. Additionally, the bandwidth has been refined for six of the heritage bands. The TIRS carries two additional thermal infrared bands. A comparison between the channels available with the extended thematic mapper (ETM) on Landsat 7 and the combination of the

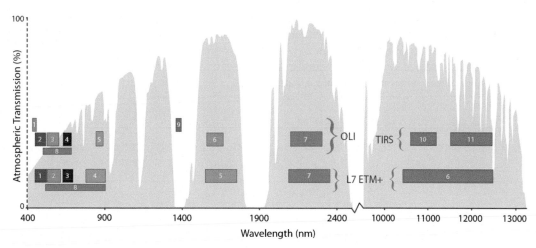

FIGURE 10.3

Comparison of Landsat 7 ETM bands with those of the OLI and TIS on Landsat 8. The background gray shades indicate the atmospheric windows in those wavelengths.

From https://deforestationwatch.wordpress.com/tag/landsat-8/.

OLI and TIS on Landsat 8 is shown here in Fig. 10.3. The background shows the atmospheric windows in these different bands.

From this figure it is very clear that the new combination of the OLI and TIS does an excellent job of covering the land reflected and emission spectrum.

TIRS collects data for two narrower spectral bands in the thermal region formerly covered by one wide spectral band on Landsats 4–7. The 100 m TIRS data are registered to the OLI data to create radiometrically, geometrically, and terrain-corrected 12-bit data products. Landsat 8 is required to return 400 scenes per day to the USGS data archive (150 more than Landsat 7 is required to capture). Landsat 8 has been regularly acquiring 550 scenes per day, while Landsat 7 is acquiring 438 scenes per day. This increases the probability of capturing cloud-free scenes for the global landmass. The Landsat 8 scene size is 185-km-cross-track by 180-km-along-track. The nominal spacecraft altitude is 705 km. Cartographic accuracy of 12 m or better, including compensation for terrain effects, is required for Landsat 8 data.

A good summary of the Landsat satellite series is provided here in this timeline (Fig. 10.4) beginning with Landsat 1 in 1972 and extending into the future past Landsat 8, which is now operating.

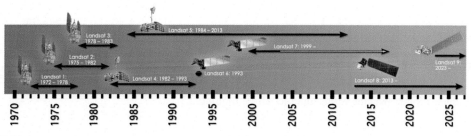

FIGURE 10.4

Landsat timeline.

It shows the durations and overlaps of each mission including our expectations for Landsat 8 and the present plans for Landsat 9, which will again be a collaboration between NASA and USGS. This timeline shows that at present both Landsat 7 and 8 are operating although Landsat 7 is somewhat restricted in its scan due to a problem with the scan correction, but still it provides useful imagery for comparison and combination with Landsat 8 imagery.

10.2 LANDSAT APPLICATIONS

The Landsat home page (http://landsat.gsfc.nasa.gov) offers an interesting matrix of applications that is reproduced here as Table 10.1. Some of these applications will be discussed in more detail in the following sections.

10.2.1 MONITORING DEFORESTATION

Brazil contains about 30% (3,562,800 km^2) the world's tropical forest. The estimated average deforestation rate from 1978 to 1988 was 15,000 km^2 per year. Systematic cutting of the forest vegetation starts along roads and then fans out to create "feather" or "fishbone" patterns as seen in the 1986 image (Fig. 10.5). Here we present a Landsat image of the state of Rondonia, Brazil. The 1975 and 1986 images are MSS data while the 1992 is TM. The deforested land and urban areas appear in light blue, while healthy vegetation appears red. Note that the area clear-cut in 1986 has partially recovered in terms of vegetation, which is no longer tropical forest, but agricultural crops in production and grasslands for cattle grazing.

An important concern for deforestation in Brazil is the global effect it produces on climate change. It has long been recognized that rainforests are of vital importance in sequestering carbon dioxide and reducing the amount of greenhouse gas being released to the atmosphere. These forests are second only to the oceans as a sink of carbon dioxide. Recent estimates suggest that deforestation in the Brazilian Amazon is responsible for as much as 10% of current greenhouse gas emissions. This problem is aggravated by the method generally used to clear the land where the forest is burned to the ground releasing large amount of carbon dioxide to the atmosphere These fires frequently burn more than the area intended, and in 1987 between July and October about 19,300 square miles (50,000 km^2) of rainforest was burned in the states of Para, Mato Grosso, Rondonia, and Acre, releasing more than 500 million tons of carbon, 44 million tons of carbon monoxide, and millions of tons of nitrogen oxides, and other poisonous chemicals into the atmosphere.

Deforestation rates in the Brazilian Amazon have slowed dramatically since peaking in 2004 at 27,423 km^2 per year. By 2009 deforestation had fallen to around 7000 km^2 per year, a decline of 74% from 2004, according to Brazil's National Institute for Space Research (INPE), which produces deforestation figures annually. Their deforestation estimates are derived from 100 to 220 images taken during the dry season in the Amazon by the China–Brazil Earth Resources Satellite program (CBERS), which may only consider the loss of the Amazon rainforest biome not the loss of natural fields of savannah within the rainforest. According to INPE, the original Amazon rainforest biome of 4,100,000 km^2 was reduced to 3,403,000 km^2 by 2005 representing a loss of 17.1%. A deforestation estimate chart is given here in Fig. 10.6. Note that the double increase for 1994 and 1995 was attributed to accidental forest burning rather than to active logging.

Table 10.1 Landsat Applications

Agriculture, Forestry, and Range Resources	Land Use Mapping	Geology	Hydrology	Coastal Resources	Environmental Monitoring
Discriminating vegetative, crop and timber types	Land uses classification	Mapping major geologic features	Determining water boundaries and surface water areas	Determining patterns and extent of turbidity	Monitoring deforestation
Measuring crop and timber acreage	Cartographic mapping and map updating	Revising geologic maps	Mapping floods and floodplain characteristics	Mapping shoreline changes	Monitoring volcanic flow activity
Precision farming land management	Categorizing and capabilities	Recognizing and classifying certain rock types	Determining area extent of snow and ice coverage	Mapping shoals, reefs, and shallow areas	Mapping and monitoring water pollution
Monitoring crop and forest harvests	Monitoring urban growth	Delineating unconsolidated rocks and soils	Measuring changes and extent of glacial features	Mapping and monitoring sea ice in shipping lanes	Determining effects of natural disasters
Determining range readiness, biomass and health	Aiding regional planning	Mapping volcanic surface deposits	Measuring turbidity and sediment patterns	Tracking beach erosion and flooding	Assessing drought impact
Determining soil conditions and associations	Mapping transportation networks	Mapping geologic landforms	Delineating irrigated fields	Monitoring coral reef health	Tracking oil spills
Monitoring desert blooms	Mapping land-water boundaries	Mineral and petroleum resources	Monitoring lake inventories and health	Coastal circulation patterns	Assessing and monitoring grass and forest fires
Assessing wildlife habitat	Citing transportation and power transmission routes	Determining regional geologic structures	Estimating snowmelt and runoff	Measuring sea surface temperature	Mapping and monitoring eutrophication
Characterizing forest range vegetation	Planning solid waste disposal sites, power plants, and other industries	Producing geomorphic maps	Characterizing tropical rainfall	Monitoring and tracking "red" tides	Monitoring mine waste pollution
Monitoring and mapping insect infestations	Mapping and managing floodplains	Mapping impact craters	Mapping watersheds	Coral reef health assessment	Monitoring volcanic ash plumes

Continued

Table 10.1 Landsat Applications—cont'd

Agriculture, Forestry, and Range Resources	Land Use Mapping	Geology	Hydrology	Coastal Resources	Environmental Monitoring
Monitoring irrigation practices	Tracking socioeconomic impacts on land use	Petrochemical exploration	Mapping closed-basin ponds	Global coral reef mapping	Assessing carbon stocks
Bison management	Online mapping	Mega-lake mapping	Monitoring wetlands	Coastal restoration	Cancer research
Crop production estimates	Cartographic discoveries	Soil carbon flux	Water management	Monitoring coastal erosion	Atmospheric modeling
Quantifying burn severity	Mapping		Wetland restoration	Coastal studies	Mapping rift valley fever risk areas
Fighting crop insurance fraud	Fighting hunger		Monitoring dam construction	Chesapeake Bay management	Assessing clear-cutting impacts
Forest trends in Madagascar	Urban sprawl and climate change		Groundwater discharge		Assessing impacts of industrial logging
Forest protection in Peru	Caribbean island mapping		Bushfire impact on water yields		Mapping urban heat islands
Better estimates of boreal forest loss	Tropical forest clearing for development				Landsat, potholes, and climate change
Crop water stress	Ancient site exploration				Disaster aftermath
Crop water demand	Landsat image mosaic of Antarctica				African environmental change
Rice production monitoring	Kansas mapping				Cyclone Nargis' impact on Burma
Demise of Papua New Guinea forests					Fire prevention
Monitoring conservation tillage					Surveying mangroves
North American forest disturbance					Greek fires
Sumatran deforestation					Algae monitoring
Forest damage caused by hurricane Katrina					Glacier monitoring
Insect outbreaks					

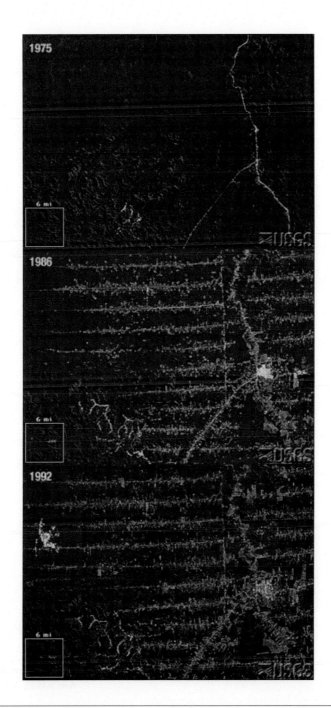

FIGURE 10.5

Landsat false-color infrared images of Rondônia, Brazil. Notice how the fishbone pattern of deforestation grows with time.

From https://landsat.gsfc.nasa.gov/monitoring-deforestation/.

■ Deforestation of the Brazilian rainforest km² (per year)

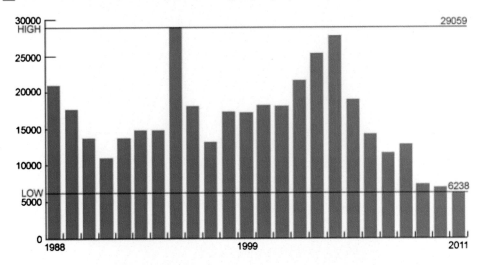

Deforestation in Rondônia, 1988-2011 (ha)

──Deforestation in Rondônia as % of total Brazilian Amazon deforestation

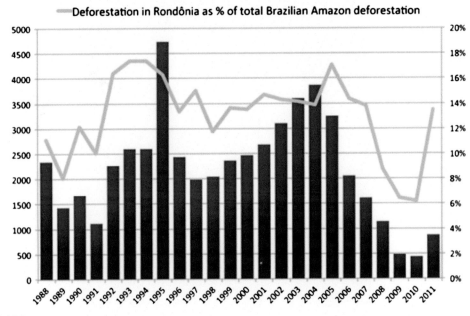

FIGURE 10.6

Deforestation chart for Brazil (http://rainforests.mongabay.com/amazon/charts.html).

From NASA's Scientific Visualization Studio.

10.2.2 MAPPING FLOODS AND FLOODPLAINS

National Oceanic and Atmospheric Administration's advanced very high resolution radiometer (NOAA's AVHRR) data along with Landsat TM imagery were used to map flood conditions, such as in the Rocheport, Missouri area in September 1993. Looking at conditions in the previous September (1992; Fig. 10.7, left) normal path of the river can be seen. The image in September 2013, clearly shows the flooded areas of basically the same image. While locating floods from these images is easier than mapping them from the ground, the process is a bit trickier than one would imagine. The water in the satellite images is frequently sitting under a forest canopy or is mixed in with the soil. Often the flood cannot be clearly seen until you compare the flood image with an image of normal conditions.

In the flood images the water changes the brightness values since water appears very dark and the surrounding land is relatively bright. Thus, one looks for changes between earlier nonflood conditions versus the image showing the flood. This is clearly seen in Fig. 10.7, (right) where the river has exceeded its normal bounds and flooded the surrounding areas. From the satellite image alone, it is an easy matter to calculate the amount of flooded area. Hence, in the second, or flooded, image the area of change is selected as the area flooded.

These floodplain maps have been used by many scientists for their applications. The USGS has used similar maps to study 1998 floods of the Missouri River on publicly owned conservation lands where levee breaks have occurred, trying to better understand how to restore the environment of large rivers and their floodplains.

Similar studies have been carried out in other parts of the world such as Pakistan (Hussain et al., 2011). However, the use of optical imagery for flood monitoring is limited by severe weather conditions, in particular the presence of clouds. In turn, SAR (synthetic aperture radar) measurements from space are independent of daylight and weather conditions and can provide valuable information to monitoring flood events. This is mainly due to the fact that a smooth water surface provides no return to the antenna in the microwave part of the spectrum and it therefore appears black in the SAR

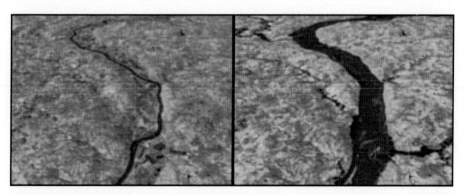

FIGURE 10.7

This pair of Landsat images, combined with elevation data of Rocheport, Missouri, were taken in September 1992 (left) and September 1993 (right). Note the brown and pink regions—agricultural fields—along the floodplain that can be seen in 1992 but are submerged in 1993. Floodplains are often fertile because of the sediment deposited there by previous floods.

Images by Jesse Allen, GSFC Visualization Analysis Lab,(http://earthobservatory.nasa.gov/Features/Floods/floods_3.php).

imagery. Flood monitoring using SAR imagery is very accurate ($\sim 95\%$), independent of cloud cover, provides a high revisit time, and easily detects smooth water. However, it also produces false alarms from shadows, smooth objects such as roads and sand, and it is more difficult to use in urban areas and vegetated areas, where usually at least two multitemporal images are required (Kussul et al., 2011). Fig. 10.8 shows an example of flood mapping using Sentinel-1 data from the European Union Copernicus program. Since December 2014, heavy rains affected Malawi (East of Nairobi, Kenia)

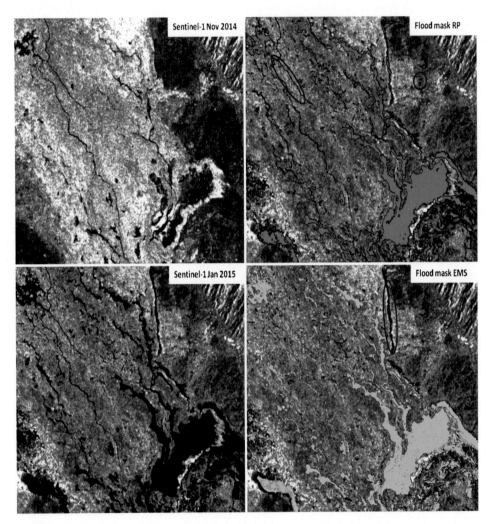

FIGURE 10.8

Flood mapping using Sentinel-1 data in Malawi, east of Nairobi, Kenia. SAR images show the changes in reflectivity since December 2014 (top left), when heavy rains affected Malawi causing rivers to overflow later in January 2015 (bottom left). Results for the UN best practices (top right) and Copernicus Emergency Services (bottom right) are shown, indicating the differences between both approaches (http://www.un-spider.org/advisory-support/recommended-practices/recommended-practice-flood-mapping).

causing rivers to overflow in January 2015. Results using two types of processing are presented: the United Nations recommended practice (top right) and the Copernius Emergency Service (bottom right). SAR data from Sentinel-1 is free of charge. Other SAR imagery may be used, but is not freely available. These include Radarsat-2, TerraSAR-X, and Cosmo-SkyMed.

10.2.3 CARBON STORAGE

USGS scientists have used Landsat data to determine that forests, wetlands, and farms in the eastern United States naturally store 300 million tons of carbon a year, which is nearly 15% of the greenhouse gas emissions. EPA estimates the country emits each year or an amount that exceeds and offsets yearly US car emissions. In conjunction with the national assessment, USGS released a new web tool that allows users to see the land and water carbon storage and change in their ecosystems between 2005 and 2050 in the lower 48 states. Biological carbon storage—also known as carbon sequestration—is the process by which carbon dioxide (CO_2) is removed from the atmosphere and stored as carbon in vegetation, soils, and sediment. The USGS estimates the ability of different ecosystems to store carbon now and in the future, providing vital information for land-use and land-management decisions. Management of carbon stored in our ecosystems and agricultural areas is relevant both for mitigation of climate change and for adaptation to such changes.

When disasters strike, pre- and post-event imagery can provide critical information regarding the extent, severity, and evolution of the event. The USGS Emergency Response supports the coordination of remotely sensed data acquisitions and the distribution of images and geospatial information products to aid in disaster response operations. This activity is based at the USGS EROS Center in Sioux Falls, South Dakota.

Over the past decade, the USGS has supported the response to hundreds of domestic and international disaster events. The USGS provides access to remotely sensed imagery and geospatial data sets in response to requests from agencies engaged in disaster response. One agency may submit the initial request for support, but the imagery that becomes available on HDDS can be shared across the response community.

10.2.4 DROUGHT MONITORING AND ITS IMPACT IN FOREST DECLINE AND FIRES OCCURRENCE

Drought strikes many regions in the world every year, turning green landscapes brown as precipitation falls below normal levels and water supplies dwindle. Drought is typically a temporary climatic aberration, but it is also an insidious natural hazard. It might last for weeks, months, or years and may have many negative impacts. Drought can threaten crops, livestock, and livelihoods, stress wildlife and habitats, and increase wildfire risks and threats to human health. Drought conditions can vary tremendously from place to place and week to week. Accurate drought monitoring is essential to understand a drought's progression and potential impacts, and to provide information necessary to support drought mitigation decisions. It is also crucial in light of climate change where droughts could become more frequent, severe, and persistent.

A team of researchers from the USGS's EROS Center, the National Drought Mitigation Center, and the High Plains Regional Climate Center are developing methods utilizing remote sensing for regional-scale mapping and monitoring of drought conditions for the conterminous United States. The ultimate goal of the project is to deliver timely georeferenced information (in the form of maps and data) about areas where the vegetation is impacted by drought. Research and methods for drought monitoring are developed in tandem with Remote Sensing Phenology.

Landsat 8 is demonstrating promising new capabilities for water quality assessment. Satellite-based instruments allow for more frequent observations over broader areas than physical water sampling. Four federal agencies—NASA, NOAA, EPA, and the USGS—are joining forces to develop an early warning system for toxic and nuisance algal blooms. Through this project, satellite data on harmful algal blooms will be converted to a format that stakeholders can use through mobile devices and web portals. This will improve detection of these blooms and help researchers better understand the conditions under which they occur. The 40-year archive of Landsat data is a valuable resource, supporting many different areas of focus for all users.

More recently, with the launch in 2009 of the ESA's Soil Moisture and Ocean Salinity (SMOS) mission, and in 2015 of NASA's SMAP mission carrying onboard L-band radiometers, the mapping of surface's soil moisture has been feasible with improved accuracy, despite the poorer spatial resolution as compared to other sensors (e.g., optical or SAR). The poorer spatial resolution prevents investigators from using microwave radiometry data for regional or local applications. To overcome this limitation, achieving a better spatial resolution, while keeping the good radiometric precision additional high quality and high resolution multispectral data can be merged with the radiometry data (Piles et al., 2011). Since June 2012, the SMOS Barcelona Expert Center (SMOS-BEC http://cp34-bec.cmima.csic.es/) is routinely producing near real time (morning and afternoon passes, if available) soil moisture maps of the Iberian peninsula[1] at 1 km, from SMOS and MODIS/AQUA and/TERRA data (Fig. 10.9), that are used by the regional Catalan authorities as a proxy for forest fires prevention during the summer seasons (Chaparro et al., 2016). The data set from 2009 to present has also been used to monitor the role of climatic anomalies and soil moisture in the decline of drought-prone forests (Chaparro et al., 2017).

The applicability of SMOS-derived soil moisture data to the study of climate change impacts is also promising. Seven years of data and the availability of high-resolution products (Piles et al., 2014) allow detecting abnormal drought periods involving natural hazards. In that regard, the complementary application of SMOS data and surface temperature information permits to estimate the potential propagation of forest fires (Chaparro et al., 2016). Also, the computation of moisture and temperature anomalies is crucial in the prediction of regional fire episodes. For instance, more than 200 fires burned in the northwestern Iberian Peninsula between end-September and mid-October 2011. This high fire activity period matched with a severe drought in the region, which was detected by negative SMOS soil moisture anomalies (very dry soils), and above normal surface temperatures (Fig. 10.10; Chaparro et al., 2017).

Additionally, the increasing number and duration of droughts rise the number of forest decline episodes in the world. The probability of forest decline was modeled in Catalonia as a function of climate and meteorological variables. In particular, the decline probability of oaks (deciduous Quercus) and beeches (*Fagus sylvatica*) in the region was high under dry soil conditions occurring in 2012 (Fig. 10.11; Chaparro et al., 2017). This shows the high potential of soil moisture data to study forest conditions, which are crucial in the context of climate change as forests act as carbon sinks.

---

[1]SMOS Barcelona Expert Center on Radiometric Calibration and Ocean Salinity (SMOS-BEC). Website: http://www.smos-bec.icm.csic.es/smos_bec.

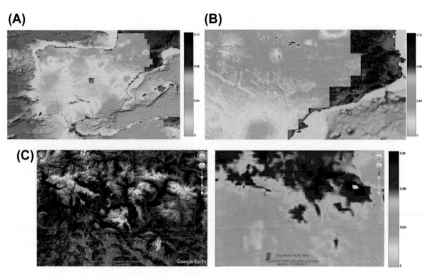

FIGURE 10.9

(A) 1 km down-scaled soil moisture map of the Iberian Peninsula generated from the combination of soil moisture and ocean salinity and MODIS/AQUA and/TERRA data, (B) zoom over the Ebro river valley, and (C) zoom over the Pyrenees. In (B) the cultivated regions near the river and its tributaries are clearly wetter than the rest. In (C) the north side of the mountains is clearly wetter than the south side, which is drier.

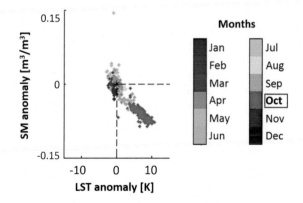

FIGURE 10.10

Anomalies of soil moisture and land surface temperature prior to fire occurrences in 2011. Both variables were calculated at a 30-day time scale. Fires are represented per month, and October is highlighted (high number of fires).

10.2.5 ANALYZING LANDSAT TO MITIGATE BIRD/AIRCRAFT COLLISIONS

The presence of birds near an aircraft runway is a constant concern. Although deadly crashes are rare, a bird strike to the windshield can cause visibility issues for pilots, and strikes to jet engines can cause

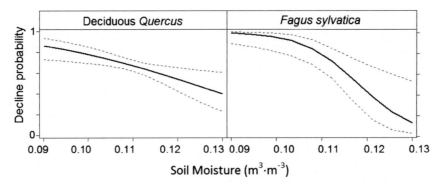

FIGURE 10.11

Estimated effect of soil moisture (SMOS, in m^3/m^3) on the decline of probability for oaks (deciduous Quercus) and beeches (*Fagus sylvatica*). Dashed lines show 95% confidence intervals. In both species, the effects were highly significant ($P < .01$). The effect in *Fagus sylvatica* was extrapolated for values of soil moisture below 0.11 m^3/m^3.

engine failure. Flocks of birds are particularly dangerous, with the threat of multiple strikes at the same time to the same aircraft. Flocks of formerly endangered Aleutian cackling geese enjoy the summertime in the Near Islands at the far western end of Alaska's Aleutian Islands, which is also home to Eareckson Air Station. The air station is located on Shemya Island, the only island in the group where Arctic foxes have been allowed to remain. The foxes ensure that the geese are not present on Shemya during the nesting and molting seasons of mid-summer. However, flocks of geese arrive in the spring as grasses grow early in the season abundant with nutrients. As they prepare to migrate south in the fall, the geese move to shrubby areas loaded with berries and to vegetation that greened up late in the summer, where they can eat well before their journey.

The increasing goose population in recent years has elevated the risk for bird strikes at Eareckson Air Station. The Bird Aircraft Strike Hazard (BASH) prevention program was implemented by the Department of Defense to provide the safest flying conditions possible. While you cannot see the birds from space, Matthew Macander and Christopher Swingley of ABR, Inc. Environmental Research & Services mapped the presence or absence of snow and spatial–temporal dynamics of the grassy and shrubby habitats in the Near Islands with Landsat satellite images. Based on the resulting maps, an ABR team visited the islands in spring 2008 and fall 2009 and conducted a habitat-use analysis by counting current-season scat. Back in the office, they mapped where the geese spent time in the spring and fall, and compared the scat counts with the Landsat images. They determined that there was a key 10-day period needed for BASH mitigation in spring. After their arrival, the geese rapidly moved on as snowmelt proceeded on other islands in the Aleutian chain. The ABR team additionally recommended mitigation strategies that include habitat modification such as vegetation removal or the types of seeds to avoid when revegetation is necessary due to disturbance.

The mapping of habitat with Landsat is hardly a unique application of the data; however, the information gained from the mapping of preferred geese habitat in this very remote region, as well as field surveys, enables new and valuable applied information on BASH mitigation. This knowledge can reduce costs of aircraft repair from multiple strikes and could save the lives of military personnel.

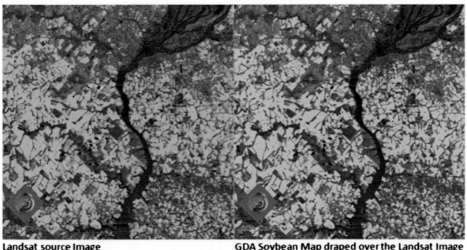

Landsat source Image   GDA Soybean Map draped over the Landsat Image

soybeans

FIGURE 10.12

Landsat 7 ETM+ image displaying Shemya Island October 8, 2003.

A Landsat 7 ETM+ image displaying Shemya Island with Eareckson Air Station, acquired October 8, 2003, is presented in Fig. 10.12.

10.2.6 LANDSAT ADDS TREMENDOUS VALUE TO DECISION MAKING

Dr. Stephanie Hulina, President of Geospatial Data Analysis Corporation (GDA Corp), discusses how access to free Landsat imagery from USGS enables her business to provide value-added products to her company's clients. The long-term continuity of the Landsat mission is essential to her company's ability to maintain a competitive edge in today's global economy. Access to readily available Landsat imagery has helped GDA to expand rapidly and serve agricultural, environmental, and resource management clients, some of whom were previously unaware of the tremendous value that Landsat data and their analysis can bring to their decision-making and ultimately their bottom line. It turns out that GDA is the second largest commercial downloader of Landsat imagery from the USGS world-wide. Not bad for a small business that formed in 2004 from a NASA Small Business Innovation Research (SBIR) project on developing a fully automated procedure for identifying clouds and their shadows in Landsat-class imagery (http://www.gdacorp.com/casa/).

From that one project, GDA has branched out to develop a handful of image-processing techniques for a range of commercial and government satellites and serves a multitude of clients across the globe. Landsat, however, is the sensor that GDA turns to repeatedly to solve client problems, either at a client's request or by suggestion from GDA. Dr. Hulina notes, "Landsat imagery is the workhorse sensor at GDA. Its 30 m resolution, large number of multispectral bands that include NIR and MIR data, its long historical record, along with excellent sensor calibration provides the absolute best bang for the buck compared to all other sensors currently out there for fine level monitoring of local and global land cover and land cover change." GDA client interests range from detection of data gaps in the imagery, image calibration to surface reflectance, and image mosaicking to full-blown image analysis to answer questions like when, where, and how much of a particular crop is planted in a particular region for any given year.

Landsat imagery is exceptionally well suited for agricultural analysis up to the field level. GDA is especially proud of their long-standing working relationship with the Foreign Agricultural Service (FAS) of the USDA. Working with the Office of Global Analysis/International Production Assessment Division within FAS, GDA has used its technology alongside the expertise of the USDA/FAS to provide timely and accurate global crop intelligence while consistently covering all major agricultural regions annually. GDA analysis of Landsat imagery and the final products generated from it factor into the global production forecasts that FAS releases every month.

The Food and Agriculture Organization (FAO) of the United Nations uses a combination of satellite remote sensing and a geographic information system (GIS) to aid in planning for the management of renewable resources in agriculture, forestry, and fisheries. One important application is concerned with land cover mapping and land management. The resulting land cover maps are necessary tools for management of any region. One important thing to remember is that land cover uses change with variations in human activity and natural disasters so it is important that these conditions can be easily updated. This is where the combination of satellite remote sensing and GIS come in as together they can provide a very current estimate of the present land cover uses.

10.3 LAND COVER MAPPING

Land cover maps need to be large enough to provide the information needed in the region of interest. At the same time they need to have a good enough resolution to be able to see the features of interest. It has been found that a map scale of 1:50,000 scale is ideal for this application. For optimal use these maps should be digital, which makes it easy to update and redisplay the land use maps. Combined with updated GIS information for the same area the results are a digital database, which describes the area and the relevant land surface uses.

The methodological approach is presented here in Fig. 10.13, which is a flow chart of the different elements involved in the mapping of land use. Inputs are the original satellite or aerial image, which has to be processed to calibrate the radiance reaching the sensor and to geometrically correct the image to fit a selected geographic map projection. Then, the ground truth information from the GIS is added to this mapped image. The result is an image of the land cover from the Land Cover and Classification System (LCCS). At this point new features are added to the land cover database and the GIS database.

Satellite imagery from Meteosat, NOAA AVHRR, Landsat, SPOT, ERS-SAR, WorldView 1-3, Ikonos, and from aerial photographs are used as inputs for this system. The GIS system combines all the ground measurement data from the area of interest to produce accurate overlays of ground truth information for the image in question. Information such as topography, water availability, soil type, forests and grasslands, climate, geology, population, landownership, highways, railway, electricity systems, and communication systems. There are many advantages of using satellite data for this application: they provide synoptic coverage, they can be acquired over an area at a high rate of repetition, usually provides information in a variety of wavelengths, and they can be collected for any part of the world.

FIGURE 10.13

FAO land cover mapping approach.

From http://www.fao.org/docrep/004/Y3642E/Y3642E00.HTM.

As an example FAO carried out a study of vineyards in Bulgaria in August of 2000. They used Landsat TM and Ikonos satellite imagery for this study. Image classification was carried out using the FAO LCCS, which is a comprehensive, standardized system created for mapping images of this type. The classification uses a set of independent diagnostics to find correlations with features in the new images. This system could serve as an internationally agreed reference database for land cover. The methodology could be used anywhere in the world.

This analysis produced 14 land cover maps at a 1:50,000 scale covering a total area of 5600 km^2. The resulting map includes the 49 land cover classes, which were identified for the study area. This analysis was tested by a similar analysis of part of the study region using a much higher spatial resolution optical satellite. The 1 m resolution (pan-sharpened imager) of the IKONOS satellite makes it possible to map land features with a scale of 1:5000, a marked improvement over the earlier analysis. This study only looked at vineyards in the Sandanski region. The results are summarized here in Fig. 10.14. From this image it was possible to identify existing, but active vineyard, abandoned vineyard, and regions where the vineyard had been destroyed.

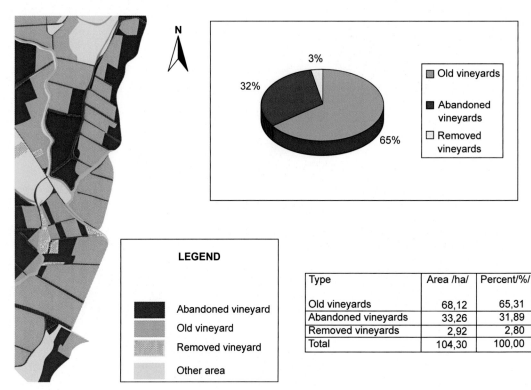

FIGURE 10.14

An example of a land cover map at a 1:5000 scale.

From http://www.fao.org/docrep/004/Y3642E/Y3642E00.HTM.

The conclusion of this study was that satellite remote sensing coupled with the LCCS analysis system provides an accurate, flexible, and cost-effective method to map land cover. This information is best merged with ground data from a GIS system and the new information is used to further update and improve the GIS system.

10.4 COMMERCIAL HIGH-RESOLUTION OPTICAL IMAGERY
10.4.1 SATELLITE POUR L'OBSERVATION DE LA TERRE

The Satellite Pour l'Observation de la Terre (SPOT) was one of the first commercial high-resolution optical Earth observing satellites. It is operated by SPOT Image, based in Toulouse, France. It was initiated by the French Space Agency (Center National d'Études Spatiales, CNES) in the 1970s and was developed in conjunction with SSTC (Belgian scientific, technical, and cultural services) and the Swedish National Space Board (SNSB). It was designed to improve the knowledge and management of the Earth's resources, detecting and forecasting phenomena in the areas of climatology, oceanography, and monitoring human activities and natural phenomena.

The SPOT series of satellites are given as follows:

1. SPOT 1 was launched on February 22, 1986, with a 10 m panchromatic and a 20 m multispectral image resolution capability. Its mission was terminated on December 31, 1990.
2. SPOT 2 was launched on January 22, 1990, with similar capabilities. It was deorbited on July 1, 2009.
3. SPOT 3 was launched on September 26, 1993, with similar capabilities. It ceased to operate on November 14, 1997.
4. SPOT 4 was launched on March 24, 1998, with similar capabilities and stopped functioning on January 11, 2013.
5. SPOT 5 (Fig. 10.15) was launched on May 4, 2002, with a nominally 5 m panchromatic resolution that could be operated in "supermode" with a 2.5 m resolution. The multispectral resolution improved to 10 m while the short wave infrared channel had a 20 m spatial resolution (Fig. 10.16). It stopped functioning on March 27, 2015
6. SPOT 6 (Fig. 10.17) was launched on September 9, 2012, and is still operating.
7. SPOT 7 was launched on June 30, 2014, and is still operating.

The SPOT orbit is polar, circular, sun-synchronous, and phased. The inclination of the orbital plane combined with the Earth's rotation makes it possible for a single SPOT satellite to fly over any point on the surface of the Earth every 26 days. The orbital altitude is 832 km and the inclination is 98.7 degrees resulting in the satellite completing 14 + 5/26 orbital revolutions per day. In its present configuration the SPOT constellation of two satellites makes it possible to acquire imagery from anywhere in the world every day. These two satellites are SPOT 4 and 5 and have identical high-resolution optical imaging

FIGURE 10.15

Illustration of the deployed SPOT-5 spacecraft.

Image credit: CNES. https://directory.eoportal.org/web/eoportal/satellite-missions/s/spot-5

FIGURE 10.16

Athens as seen by SPOT 5 in 2002.

FIGURE 10.17

A comparison between the 2.5 m resolution of SPOT 5 (left) and the 1.5 m resolution of SPOT 6 (right).

instruments, which can operate in the panchromatic band (P mode) or in a multispectral mode (xs). The orientation of each instrument's strip-selection mirror can be remotely steered by the ground station, offering oblique views with angles of ±27 degrees from the satellite's vertical axis. Using this capability the sampling interval can be shortened from 26 to 4–5 days in the temperature zones. SPOT 5 also carries the high-resolution stereoscopic (HRS) imaging instrument for the acquisition of optical stereo image pairs to compute digital elevation maps of the Earth's surface.

The continuity of the SPOT program is planned with the development of the Pléiades system, as well as SPOT 6 and 7. The latter offers 1.5 m spatial resolution images in a 60 × 60 km swath. A comparison between SPOT 5 and SPOT 6 is presented here in Fig. 10.17.

The advantage of this higher spatial resolution is readily obvious in Fig. 10.17 where one can now see expanded views of areas that are quite restricted in the lower resolution image. Features appear on the right that are impossible to see in the lower resolution image on the left.

One of the big changes with SPOT 5 (Fig. 10.15) was the addition of the HRS instrument, which was designed to take simultaneous stereo-pairs of a swath of 120 km across and 600 km long with ground resolutions of 10 × 10 m in the panchromatic and multispectral modes. This satellite has two high-resolution geometrical instruments that were derived from the HRVIR instrument that flew on SPOT 4. They have a spatial resolution of 2.5–5 m in panchromatic modes and 10 m in the multispectral mode. In addition the instrument has a 20 m horizontal resolution in the near infrared (1.58–1.75 μm). The HRS operates in the panchromatic mode and points both forward and backward of the satellite in creating the parallax views for stereo viewing and digital elevation computations. This rapid pointing capability made it possible to produce nearly coincident images of the same piece of ground very closely spaced in time.

For the high resolution geometric sensor the panchromatic bands had a native spatial resolution of 5 × 5 m or a supermode of 2.5 × 2.5 m. SPOT 6 (Fig. 10.18) satellite was built by AIRBUS Defense

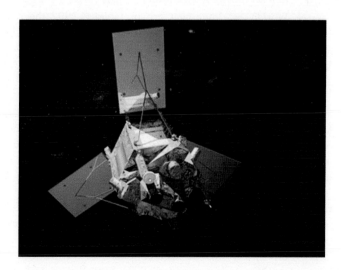

FIGURE 10.18

SPOT 6.

Image courtesy of AIRBUS Defense and Space.

and Space and was successfully launched on September 9, 2012 from the Satish Dhawan Space Center in India. On June 30, 2014, it was joined by SPOT 7.

The main changes in the SPOT 6, 7 instruments was an improvement to a multispectral resolution of 6.0 m and a panchromatic resolution of 1.5 m over a swath of 60 km at nadir. This instrument has the capability to collect imagery in a 120 km × 120 km bistrip or a 60 km × 180 km tristrip mapping in a single pass or the delivery of mosaic stereo and tristereo acquisitions of 60 × 60 km scenes for the production of digital elevation models. There are six different tasking plans and each takes 24 h. Up to 750 scenes are collected per day per satellite.

At the time of launch both SPOT 6 and SPOT 7 belonged to SPOT Image, a subsidiary of Astrium who funded the satellites and the launch. In December 2014 SPOT 7 was sold to Azerbaijan's space agency Azercosmos, who renamed it Azersky. The architecture is similar to that of other Pléiades satellites as depicted in Fig. 10.18. SPOT 6 and SPOT 7 are in the same orbit as Pléiades 1A and Pleiades 1B at an altitude of 694 km forming a constellation 90 degrees apart from each other.

The image product resolutions are as follows: 1.5 m in the panchromatic, 1.5 m in the color-merged (pansharpened), and 6 m in the multispectral. The spectral bands are as follows: panchromatic (450–745 nm), blue (450–5235 nm), green (530–590 nm), red (625–695 nm), and the near-infrared (760–890 nm). The image size on the ground is 60 km × 60 km, and the satellite has the capacity to acquire up to 3 million km^2 daily.

10.4.2 DIGITALGLOBE INC.

In the early 1990, a new era of commercial remote sensing was born with the object of flying high-resolution optical sensors to provide satellite imagery to the world for domestic and military applications. While initially there were a number of these companies today DigitalGlobe Inc. remains as the largest and most active while the others either folded or were merged with DigitalGlobe. Originally formed as the WorldView Imaging Corp in January 1992 the intention was to take advantage of the 1992 Land Remote Sensing Policy Act (enacted in October 1992) to establish a commercial company to provide high-resolution land satellite imagery. Its founder was Dr. Walter Scott and the company received the first high-resolution commercial remote sensing license issued under the 1992 Act. At the time Dr. Scott was the head of the Lawrence Livermore Laboratories "Brilliant Pebbles" and "Brilliant Eyes" projects, which were part of the then Strategic Defense Initiative. He was joined by Doug Gerull who was then the executive in charge of the Mapping Science division of Intergraph Corp. The company's first remote sensing license from the United States Department of Commerce allowed it to build a commercial satellite capable of collecting images with a 3 m spatial resolution.

In 1995, the company became EarthWatch Inc. merging WorldView with Ball Aerospace and Technologies Corp's commercial remote sensing operations. The first satellite launched for Earth-Watch was EarlyBird 1 on December 24, 1997. It was launched from the Svobodny Cosmodrome by a Start-1 rocket. The satellite carried a panchromatic camera with a 3 m resolution and a multispectral camera with a 15 m resolution (Fig. 10.19). This satellite was built by CTA Space Systems (now part of Orbital Sciences Corp.). EarlyBird 1 was the first commercial satellite to be launched from the Svobodny Cosmodrome. EarthWatch announced that all contact was lost with EarlyBird 1 on December 28, 1997.

FIGURE 10.19

EarlyBird 1 satellite.

Following the failure of EarlyBird 1 the new EarthWatch merge with Ball Aerospace became DigitalGlobe, and a new satellite QuickBird was built by Ball with a sensor built by ITT in Fort Wayne, Indiana. QuickBird was launched by a Delta-II launch vehicle on October 18, 2001, into a −98 degrees Sun-synchronous orbit. This followed an earlier launch effort that resulted in the loss of QuickBird-1. On orbit, the new QuickBird satellite provided a panchromatic band with a 60 cm spatial resolution and multispectral scanning with a 2.4 m resolution. QuickBird operated successfully until January 27, 2015, when it was deorbited, exceeding its initial design life by nearly 300%.

While QuickBird was still flying DigitalGlobe embarked on a new series of satellites that used the old WorldView name. WorldView-1 (WV-1) was again built by Ball Aerospace with a sensor from ITT. It was launched again on a Delta-II rocket from Vandenberg Air Force Base on September 18, 2007. Unlike QuickBird WV-1 was panchromatic only with a 50 cm spatial resolution. It was quickly followed by WorldView-2 launched on October 8, 2009, but this time the sensor had both the 50 cm panchromatic band from 450 to 800 nm, and an eight channel multispectral capability providing data in red, blue, green, yellow, red-edge, two near IR bands, and a coastal band.

The latest in the WorldView series is WV-3 (Fig. 10.20) that improves the panchromatic sensing to a 31 cm resolution, extends the multispectral sensing to bands that cover both the visible and near infrared with 12 bands having a 1.24 m resolution. There are also eight short-wave infrared bands (~3.7 μm) that can be used to sense and map wildfires. A new instrument called CAVIS is added to provide improved atmospheric correction for the imagery. All of the WorldView satellites are very agile meaning that they can be rapidly repointed to cover a region of interest. This increases DigitalGlobe's ability to cover one region from repeated passes.

A big change came in 2010 when it was announced that DigitalGlobe would merge with its competition the GeoEye Corp. This merger was completed on January 31, 2013, which joined the United States's only commercial satellite providers. The new company continues under the name DigitalGlobe, but now the satellite assets are operated by DigitalGlobe. Thus, the new DigitalGlobe satellite constellation consists of six operating satellites. The presently operating satellites are WorldView-1, WorldView-2, GeoEye-1, IKONOS, WorldView-3, and WorldView-4 launched on November 11, 2016. Both IKONOS and GeoEye-1 are assets taken over by DigitalGlobe in the

FIGURE 10.20

WorldView-3 satellite (http://www.satimagingcorp.com/satellite-sensors/worldview-3/).

merger with GeoEye. Built by General Dynamics GeoEye-1 provides panchromatic imagery with a 41 cm resolution and multispectral with a 1.65 m resolution. A product of Lockheed Martin Corp the IKONOS satellite was operated by the Space Imaging Corp. which later became GeoEye. IKONOS had an 80 cm panchromatic band, and four multispectral bands with a 4 m resolution.

10.4.2.1 Applications of High-Resolution Satellite Imagery
10.4.2.1.1 DigitalGlobe Imagery Use for Changchun Urban Planning Initiative

Changchun is at the center of China's automobile manufacturing industry and has also become an important center for biopharmaceutical, and other high technology activities. As such it is one of the fastest growing metropolises in northeast China. The Changchun Surveying and Mapping Institute (CSMI) was charged with keeping the municipality's database current, and they rely on high-resolution imagery from DigitalGlobe for this process. This municipality covers a total area or more than 20,000 km^2 including a wide variety of landscapes ranging from its urban core to rural mountainous areas with more than 200 rivers and lakes. Changchun has also been designated as one of four National Garden Cities in China an award made by the Chinese Ministry of Construction to reflect the city's high rate of "greening" of the city.

10.4.2.1.2 DigitalGlobe Satellite Imagery Helps Agricultural Development in the Philippines

Agriculture, livestock, and fishing employs more than 40% of the Philippines' 98 million inhabitants and represents 20% of the country's gross domestic product (GDP). The Department of Agriculture wishes to empower the farming and fishing communities to produce accessible and affordable food for every habitant and to provide a sufficient level of income for everyone involved in the industry. As a consequence this department is constantly seeking out new methods to improve its goods and services to its customers.

This department recognized that precise spatial information would provide a wealth of beneficial information to both identify vulnerabilities and to improve productivity says Josephine Minerva, executive vice president and chief operating officer for DigitalGlobe's information partner Geo-Surveys & Mapping (GSMI) Inc. They had been using SPOT imagery that was incomplete, out of date, and not of a very high spatial resolution. Established in 1994, GSMI is a pioneer in providing top-of-the-line geomatic solutions for surveying, mapping, remote sensing, GIS, and systems integration in the Philippines. GSMI houses a full line of sophisticated equipment for data acquisition, image processing, and mapping of satellite and other in situ geographic data.

In the fall of 2012 GSMI began delivery of GeoEye 1 and WV-2 imagery for both coastal and terrestrial areas. The project will eventually cover the entire Archipelago spanning an area of more than 300,000 square miles consisting of 7107 islands with more than 22,000 miles of coastline. "We are working very closely with the Department to help us prioritize and determine the applications of this technology," states DigitalGlobe. Using DigitalGlobe imagery they were able to offer the most competitive bid and satisfy their technical criteria. DigitalGlobe wants to ensure that they maximize their resources and develop applications of the data that will provide measureable benefits.

The Department considers the satellite imagery as the basis for improving the delivery of services with the establishment of a database for existing agriculture and fishery resources. With an extensive library of historic imagery the Department will have the ability to measure changes in the environment over time. The Department will have a true benchmark of change to make better informed decisions moving forward. At the same time the Department wishes for GSMI to optimize their investment in DigitalGlobe imagery.

10.4.2.1.3 City of Solvang, California

Every spring across the US potholes and other road damage result from drastic changes in temperature and regular road wear. Decades of spending on new road construction, instead of regular road repair, has left many states' roads in poor condition and in dire need of repair. To best deploy resources to repair the roads a method to quickly and easily survey road conditions is needed. At present most of the repair work is reactive in nature and waits for complaints to come in before repair crews are sent to repair road damage. In the city of Solvang, CA, the Department of Public Works uses DigitalGlobe imagery (Fig. 10.21) to quickly locate road damage and mobilize repair crews to rapidly repair and keep all roads travel ready.

FIGURE 10.21

Aerial photograph of damaged roads in Solvang, California.

The city of Solvang is a Danish replica village in California that is a major tourist attraction with windmills, horse ranches, and wineries. As such there are lots of automobile traffic mixed in with horse-and-buggy traffic in the town. In such a setting road damage can be a major distraction. The Public Works Department decided to use ortho-rectified aerial imagery provided by DigitalGlobe to assess the road damage in need of repair. These images were used in the deployment of road repaving crews to the areas in greatest need of repair as seen in the aerial images. Thus, the remote sensing imagery greatly reduced the cost of gaining information about the condition of the city's roads and made it possible to optimize the deployment of costly resources to repave the roads. The success in Solvang has now spread to other cities, which are also using high-resolution imagery to map road conditions in their areas.

Other related uses of high-resolution imagery for cities area:

1. Fire protection: Locate hydrants where they have the greatest access to buildings before a fire occurs.
2. Accidents: Identify stationary reference points, measure intersections, and evaluate preimpact and postimpacts conditions.
3. Code enforcement: Match permits, permit applications, and complaints to imagery to verify or determine violations.
4. Inventory: Quickly survey infrastructure objects from road markings and building to manholes and fire hydrants.
5. Purchasing: Determine quantities of material needed in a given area (concrete, equipment, manholes, hydrants, parking meters, etc.), and calculate the associated cost.
6. Parades and other events: Establish traffic control and station route marshals, sheriffs, and portable toilets.

10.4.2.1.4 Using DigitalGlobe Imagery for Planning Moscow's Green Space

As Europe's largest city and the world's sixth largest metropolis with 12 million people, Moscow is also one of the world's greenest cities. In addition to the landmark Gorky Park, Moscow is the home to nearly 100 additionally significant parks, numerous botanical gardens, 450 km^2 of green zones, and 100 km^2 of forest. Green space planning and monitoring has become a priority for city government as Moscow continues to grow.

With the arrival of a free market economy in the early 1990s, Moscow became one of the world's fastest growing cities. Western structures such as new office towers, roadways, and retail centers along with thousands of housing units quickly sprung up. Realizing the need to preserve its heritage, the city set out to create a vegetation geodatabase as a reference for future urban planning. "The city had no clear methodology to track the health and locations of its vegetation," says Olga Kolesnikova, head of the complex project department for DigitalGlobe's partner Sovzond JSC. "With all the building project and new roadways, the city sought a solution to help protect existing vegetation as well as identify opportunities to add new environmentally protected spaces."

Using DigitalGlobe's multispectral 8-band high-resolution satellite imagery Sovzond vectorized nearly 400,000 polygons throughout the city for the years 2009−11 (Fig. 10.22), about half related to grass and vegetation, the remainder to trees and shrubs. They developed methods to discriminate between grassy vegetation and trees/shrubs based on this high-resolution satellite data. They began using a small test area in the central and northeastern administrative districts of the city and later replicated this approach citywide.

FIGURE 10.22

8-Band satellite image of part of Moscow.

Using indices such as the Normalized Difference Vegetation Index (NDVI) to detect live green vegetation, Sovzond was able to create a rating system to measure vegetative health. This vector layer of vegetation cover ranged from perfect to unsatisfactory making it possible to rank the various parts of the city. This greatly aided the urban planners and other city officials in their decisions about preservation and planning.

The completion of the city's first vegetative database is only the beginning for Moscow city planners. The availability of DigitalGlobe's archival imagery combined with the new imagery collected for Moscow gives city officials a "living" vegetative map from which to compare how the city is managing its green space, while having an accurate and current database to use for future planning decisions.

10.4.2.1.5 Satellite Imagery Vital to Proactive Forestry Management

Forestry industry in Sweden accounts for 3% of the GDP and 12% of its exports. The forestry industry plays an important role in the economy of Sweden. Recently bark beetles, which had been in Central Europe for centuries, have started to make inroads into Sweden where normally the long cold winters kept them out. Now with warm winds and increasing global temperatures the bark beetles are starting to appear in the forests of Sweden. It is important to monitor the changes in Sweden's forests to be able to assess the damage caused by these beetles. High-resolution satellite imagery has proven to be an effective tool in monitoring the health of Sweden's vast forests (Fig. 10.23).

Sweden has a lot at stake in protecting their forests. It is the world's third largest exporter of paper and the second largest exporter of sawn timber. It is important to realize that bark beetles do not attack healthy trees, but prefer sick and dying trees. Thus, an indicator of tree health is critical to managing forest resources. To minimize the spread of bark beetles, it is necessary to identify the sick trees and remove them from the forest before they can be infested with bark beetles. Satellite imagery provides a tremendous advantage, both technically and economically, in helping to locate those trees most vulnerable to attack by bark beetles. These are the conclusions of Nils Erik Jorgensen, the chief technical officer and owner of TerraNor DigitalGlobe's alliance partner in Sweden. Founded in 1989 TerraNor is a software and consulting company working with GIS and remote sensing for applications in forestry, oil, agriculture, education, and research.

In 2011 TerraNor carried out a project to identify trees that had been attacked by bark beetles in the Öterlund and Sundsvall regions of northern Sweden. The DigitalGlobe satellite GeoEye-1 collected data over a 4-week period in July and August. With a 50 cm spatial resolution they were able to detect numerous species such as pine and spruce and determine the health of these trees. Once the poor trees are identified, on the basis of the satellite imagery, loggers go into the forest and

FIGURE 10.23

Near-infrared satellite image from WV-2 showing healthy (*green*) and sickly (*purple*) trees.

specifically remove the dead and dying trees to create borders around healthy trees to the hinder the spread of the bark beetles.

In addition to the bark beetle infestation study, DigitalGlobe satellite data were used in the middle of Norway and Sweden to identify areas with high concentrations of downed trees due to high wind speed. In January 2014, WorldView-2 panchromatic images (50 cm resolution) were used to survey a 1600 km² area to identify regions with high concentrations of downed trees. Quickly removing these downed trees is again critical to hindering the spread of the bark beetles. It is important to recognize that the weather conditions during these high wind events were such that ground operations were limited making the satellite information even more important.

10.5 FOREST FIRE DETECTION AND MAPPING

Satellite remote sensing provides a very valuable tool for the detection and monitoring of wildfires and the management of forests to prevent these wildfires. It is also an effective tool to map the damage caused by such a fire event. The primary advantages of satellite-based sensing are as follows: (1) its large area and synoptic coverage, (2) its frequent and repetitive coverage of an area of interest, (3) its ability to make quantitative measurements of ground features using radiometrically calibrated sensors, (4) the ability to do semiautomated computerized processing and analysis, and (5) the relatively low-cost per unit area of coverage. Ground and airborne surveys are much more manpower intensive and hence considerably more expensive (https://crisp.nus.edu.sg/~research/forest_fire/forest_fire.htm).

At present a number of operational satellite systems are being used in forest fire detection, monitoring, and assessment. Each of these satellite platforms carries sensors with slightly different characteristics in terms of wavelengths, resolutions, and radiometric fidelity. One of the oldest is the NOAA AVHRR sensor carried on the polar-orbiting weather satellites since the early 1980s. With a wide band visible channel centered in the red, a near infrared channel, a mid-range infrared channel, and two-thermal channels, the AVHRR is effective at mapping the smoke plumes from a wildfire and when the wind blows off the hot spot it can detect the thermal emissions from the first itself. The two thermal channels at 11 and 12 μm are too sensitive to Earth-emitted radiation and will saturate in the presence of a hot spot. The mid-range infrared channel, however, will respond to emissions from much higher temperatures. One must be careful, however, since this mid-range infrared channel also contains reflected radiation, so it is best to use it to sense the fire hot spots during the night when there is no reflected radiation. Depending on the satellite configuration, this instrument covers a spot on the Earth 2—4 times per day.

The DMSP-OLS instrument has a low-light capability that can be used at night to sense visible light emissions from fires at night. This unique capability is unfortunately compromised by a relatively low spatial resolution of 2.7 km. Again, depending on the number of satellites operating the coverage is approximately daily.

10.5.1 MODIS FIRE PRODUCTS

The MODIS instrument flying on NASA Terra (morning) and AQUA (afternoon) satellites has 36 bands ranging from the visible through the thermal infrared. It has two 250 m bands and five 500 m bands in the visible that can be used to more precisely locate the smoke plumes from the wildfires. The near-infrared, the mid-range infrared, and the thermal infrared channels all have the usual 1 km spatial resolution. This instrument has proven to be a very useful means of mapping wildfires.

There is a series of MODIS fire products as clearly discussed and described by Justice et al. (2002). At the time of this book, MODIS was producing the only current global daily active fire product. At that time data were also available from ASTER a high-resolution multispectrometer flying on the TERRA satellite. ASTER provided high spatial resolution thermal infrared imagery which was very useful in viewing fire hot spots and supplied a MODIS fire product validation source. In the paper by Justice et al. (2002) they discuss a specific fire validation study conducted in Africa. They also introduce a prototype MODIS burned area product, again with an example from Africa that was used to model pyrogenic emissions. They also discuss a MODIS Fire Rapid Response System and a web-based online image distribution system. As reported by Justice et al. (2002) the MODIS Fire Team developed and tested two types of fire products: active fire products to give location of active fires and a burned area product which gave the extent of burn scars over a specific time period. The active fire algorithm uses multiple MODIS channels to detect thermal anomalies on a per-pixel basis, which in addition to identifying fires includes high-temperature point sources such as gas flares and power plants. The burned area algorithm uses a multidate analysis and its development concentrated on southern Africa in the context of the SAFARI 2000 campaign (Swap et al., 2002).

10.5.2 MODIS ACTIVE FIRE DETECTION

The MODIS active fire detection algorithm was based on earlier similar algorithms developed for NOAA/AVHRR (Giglio et al., 1999), and the Tropical Rainfall Mapping Mission Visible Infrared Radiation Spectrometer (Kaufan et al., 1990). The 4 and 11 μm channels (T_4 and T_{11}) are used again. Unlike its predecessors MODIS has two 4 μm channels numbered 21 and 22 both of which are used in the detection algorithm. Channel 21 saturates at about 500 K while channel 22 saturates at 331 K. The low-saturation channel is less noisy and has a smaller quantization error, it is used the derive T_4 whenever possible. When it saturates, however, the calculation shifts to channel 21. The T_{11} value is computed from the single 11 μm channel number 31 which saturates at approximately 400 K. The 250 m spatial resolution near-infrared channel (0.86 μm) is averaged up to a 1 km resolution to be consistent with the other channels and is used to identify high reflective surfaces that maybe falsely identified as fires.

This fire detection algorithm is based on the absolute detection of the fire if the fire is strong enough, and the detection relative to the thermal emission of the surrounding pixels detect weaker fires. This latter part of the algorithm is designed to account for variability of the surface temperature and reflection by sunlight. For absolute fire detection the algorithm requires that at least one of two conditions be satisfied. These are as follows:

1. $T_4 > 360$ K (330 K at night)
2. $T_4 > 330$ K (315 K at night) and $T_4 - T_{11} > 25$ K (10 K at night)

If either of these two conditions is not met the algorithm then pursues a relative fire detection in which the fire is distinguished from the mean background by three standard deviation in T_4 and in $T_4 - T_{11}$. The mean, median, and standard deviations are computed for pixels within an expanding grid centered on the candidate fire pixel until a sufficient number of cloud-water and fire-free pixels are identified. A sufficient number is identified as 25% of all background pixels with a minimum of six. Water pixels are identified with an external water mask and cloud pixels are identified by the MODIS cloud mask. Fire-free pixels are those for which $T_4 < 325$ K (315 K at night) and $T_4 = T_{11} < 20$ K (10 K at night). If either standard deviation is less than 2 K, a value of 2 K is used instead. For daytime fire detection when sunglint may cause false alarms, a fire pixel is rejected if the MODIS 250 m red and near-infrared channels have a reflectance above 30% and lies within 40 degrees of the specular reflection position.

Each fire product is associated with a detection confidence estimate to help the users gauge the quality of the individual fire pixels. This confidence estimate assigns one of three fire classes (low-confidence, nominal-confidence, or high-confidence) to all fire pixels with the fire mask area. In addition total emitted power is assigned to each fire pixel (Kaufman et al., 1998). The MODIS fire products and their release dates are presented here in Table 10.2.

These standard MODIS fire products are distributed from the Earth Resources Observation Systems Data Center (EDC) Distributed Active Archive Center (DAAC). A description of these products can be found in Masuoka et al. (1998).

One unique product is the global browse images of the MODIS active fire product that are generated daily at 5 and 20 km spatial resolutions (Fig. 10.24).

These products show the global distribution of fires for each day and have been useful for preliminary QA, allowing easy identification of potential problems such as excessive false alarms. This browse type of product allows for a quick assessment of areas with distinct changes in fire activity.

The MODIS Rapid Response System was specifically developed in response to the need in the fire community for rapid detection and monitoring of a fire. The standard MODIS fire products run approximately 7 days behind the current image acquisition date. For fire management and response agencies this is too long an interval to be useful. The US Forest Service (USFS) generally wants information within a few hours of image acquisition to support fire management. In April, 2001 the Rapid Response System was initiated (Justice et al., 2000), and it provides MODIS fire data and imagery to the USFS Remote Sensing Applications Center in Salt Lake City and the National

Table 10.2 MODIS Fire Products

On Orbit Name	Before Launch	Launch Date	AVHRR Version	End of Mission
TIROS-N	TIROS-N	October 1978	1	
NOAA-6	NOAA-A	June 1979	1	November 11, 1986
NOAA-7	NOAA-C	June 1981	2	June 7, 1986
NOAA-8	NOAA-E	March 1983	1	October 31, 1985
NOAA-9	NOAA-F	December 1984	2	November 5, 1994
NOAA-10	NOAA-G	September 1986	1	October 2000
NOAA-11	NOAA-H	September 1988	2	September 13, 1994
NOAA-12	NOAA-D	May 1991	1	August 10, 2007
NOAA-13	NOAA-I	August 1993		Failed after launch
NOAA-14	NOAA-J	December 1994	2	October 3, 2001
NOAA-15	NOAA-K	May 1998	3	AM operational, secondary
NOAA-16	NOAA-L	September 2000	3	PM operational, secondary
NOAA-17	NOAA-M	June 2002	3	AM operational, backup
NOAA-18	NOAA-N	May 2005	3	PM operational, secondary
NOAA-19	NOAA-N′	February 2009	3	PM operational, primary
METOP-A	METOP-A	October 2006	3	AM operational, primary

Terra MODIS daily Surface Reflectance product
with daily daytime Active Fire Detection product overlay
ESDT: MOD14onMOD09
Start Sensor Acquisition Date: 2017158
Collection 006
Browse Image created on Jun 9 16:00:02 2017 UTC, MODLAND

FIGURE 10.24

Daily daytime active fire detection product over daily land surface reflectance (MOD14 over MOD09) for June 7, 2017.

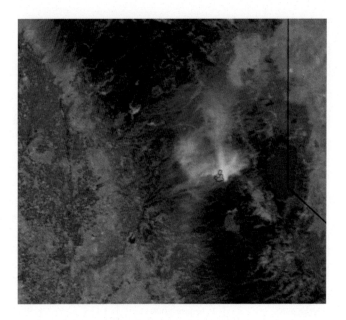

FIGURE 10.25

Star fire near Lake Tahoe, California, on August 27, 2001, 19—22 UTC. Pixels containing active fires, outlined in red, are superimposed on corrected true-color reflectances.

Courtesy of AIRBUS Defense and Space.

Interagency Fire Center in Boise, Idaho. The Rapid Response System takes data directly from the EDOS data, feed to NOAA at the NASA Goddard Space Flight Center, and generates Level 1B data products using the International MODIS/AIRS Processing Package (IMAPP) provided by the Univesity of Wisconsin. The MODIS fire algorithm is applied to this image data and images are then generated by superimposing the 1 km active fires on the 250 m corrected reflectance image for the same acquisition. An example is presented here in Fig. 10.25.

An interesting capability resulting from the MODIS fire products is the ability to make a global map of cumulative fire detection. An example is presented here in Fig. 10.26 that clearly shows those regions of persistent fire activity. You can see the biomass burning that takes place in eastern South America along with the fires in central and southern Africa. Less well known are the fires of India and east Asia along with the band in northern Australia. It is surprising to see that fires are more prevalent over this period in the eastern United States rather than in the west where fires are thought to frequently occur. There is a wide swath of fire activity from central Europe across central Russia.

10.5.3 MODIS FIRE VALIDATION

Validation requires an independent source of information on the fires to which the MODIS derived fire products can be compared. This validation source should have a high level of accuracy in that it represents an independent source of fire information. Sources of this type of validation information can

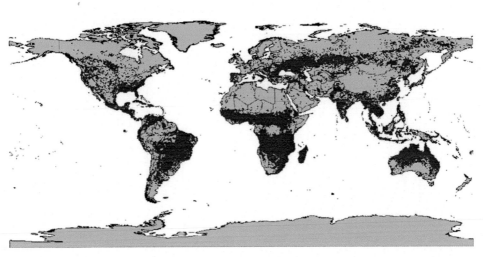

FIGURE 10.26

Global cumulative active fire detections (April 23, 2001 to February 25, 2002) from MODIS Land Rapid Response System. Each *red dot* represents a single 1-km MODIS active fire pixel detected during the time period.

From Justice, C.O., Giglio, L., Korontzi, S., Owens, J., Morisette, J.T., Roy, D., Descloitres, J., Alleaume, S., Petitcolin, F., Kaufman, Y., 2002. The MODIS fire products. Remote Sens. Environ. 83, 244–262.

be collected with aircraft, on the ground and from other satellite data. There are often great difficulties in aligning these independent fire observations with MODIS fire products due to the very variable nature of fires in both time and space.

As mentioned earlier the ASTER instrument operated onboard the TERRA spacecraft offering high resolution infrared data. This instrument has 15 spectral bands from 0.5 to 10 μm at resolutions from 15 to 90 m (Yamaguchi et al., 1998). Using ASTER data Justice et al. (2002) were able to obtain a global example of high-resolution fire detection. These high resolution fire retrievals were then mapped on to coincident MODIS pixels. Justice et al. (2002) describe a validation exercise using four ASTER scenes that fell within two MODIS scenes over southern Africa (Table 10.2). The ASTER data were first converted to a high-resolution fire map with a simple threshold. Fires less than 1000 m^2 will saturate the ASTER band 9 (2.4 μm) and therefore a simple saturation threshold for cloud-free scenes can be used to identify fires in ASTER data. On these high-resolution ASTER fire maps, fire pixels within 150 m of each other were grouped together to form a single "fire cluster." The ASTER fire clusters and MODIS pixels were then associated by mapping the 1 km^2 of each MODIS pixel onto the ASTER data. A registration accuracy of better than 300 m can be estimated according to the registration accuracy of each instrument, which is less than 100 m for ASTER and less than 200 m for MODIS (Wolfe et al., 2002).

As a validation Justice et al. (2002) looked at: (1) the number of ASTER clusters with a MODIS pixel, (2) the average size of these clusters, and (3) the maximum size of the ASTER fire clusters. The distributions of these variables were then sorted into two groups based on whether or not their corresponding MODIS pixel was classified as no-fire or fire. These distributions are presented here as box plots in Fig. 10.27.

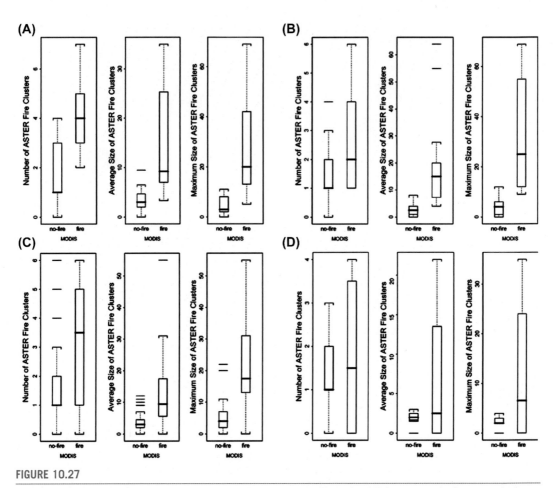

FIGURE 10.27

(A–D) box plot of ASTER-derived variables for MODIS14-no-fire (0) or fire (1) for scenes A–D, respectively.

From Justice, C.O., Giglio, L., Korontzi, S., Owens, J., Morisette, J.T., Roy, D., Descloitres, J., Alleaume, S., Petitcolin, F.,
Kaufman, Y., 2002. The MODIS fire products. Remote Sens. Environ. 83, 244–262.

These plots show that the three ASTER-derived variables are higher for those pixels classified by MOD14 as fire. This indicates that the MOD14 product behaves appropriately relative to the observed ASTER data; more and larger fire clusters are associated with those MOD14 pixels classified as fire. Scatter plots of these same data also confirmed the agreement between MODIS fire pixels and those identified in the ASTER data.

10.5.4 THE HAYMAN WILDFIRE IN COLORADO

The largest wildfire in Colorado was the Hayman Fire that was actually set by a volunteer forest worker to demonstrate her prowess in detecting and eliminating wildfires. The fire was first reported in the late afternoon of June 8, 2002. Low humidity, warm temperatures, and high wind speeds facilitated the fire's hot temperature and rapid spread along the South Platte River corridor. For 20 days the fire consumed public

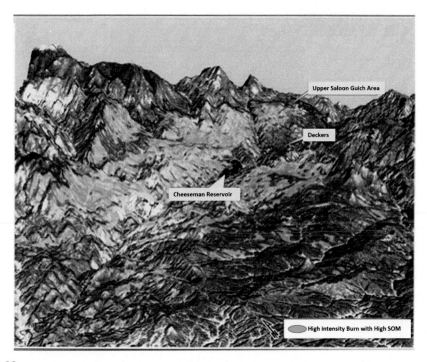

Upper Saloon Guich Area

Deckers

Cheeseman Reservoir

High Intensity Burn with High SOM

FIGURE 10.28

The Hayman Fire. Red outline is the total fire area, yellow = intensive burn area.

and private lands southwest of Denver in portions of Park, Teller, Jefferson, and Douglas Counties. The fire was very well studied by the USFS, the Colorado Forest Service, and the US Department of Agriculture. The fire was located in steep terrain and persisted over a fairly long period of time. Fanned by strong winds the fire grew explosively and was located in a fairly inaccessible area (Fig. 10.28).

Another map in Fig. 10.29 shows the location of the Hayman Fire relative to the South Platte River and the surrounding communities. The communities of Deckers and Lake George were threatened by the fire. The greatest burn severity was in the northeast sector of the fire where the fire raced through the South Platte River valley. Unburned patches represent higher elevations where the topmost trees were left unburned.

An AVHRR image from June 9 just after the start of the fire (Fig. 10.30) shows both the large smoke plume associated with the fire and the intense burn location on the ground. The strong winds blowing northwest up the canyon are clearly indicated by the smoke plume trending in this direction.

ASTER was again used to map the fire hot spot as can be seen on this mid-infrared (3.7 μm) ASTER channel image (Fig. 10.31). The maximum, minimum, and mean brightness temperatures are indicated in the upper left of the image. The distinct shape of the fire that was dictated by the local topography can be clearly seen in this thermal emission image.

A coincident long-wave thermal infrared (10.8 μm) ASTER image (Fig. 10.32) exhibits a similar pattern, but weaker emissions and much lower temperatures. This reflects the fact that in this wavelength

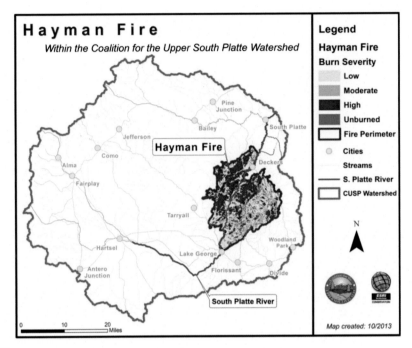

FIGURE 10.29

The Hayman Fire location and indicated burn severity.

FIGURE 10.30

Satellite image of Hayman Fire on June 9 shows the convection column and smoke plume extending across Denver into Wyoming carrying carbon monoxide and fine particulates.

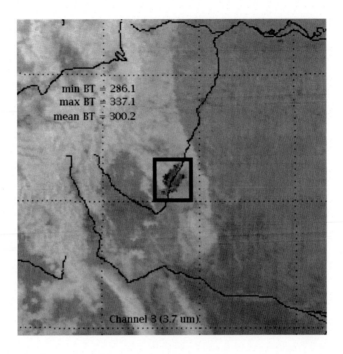

FIGURE 10.31

ASTER 3.7 μm channel image showing the Hayman Fire hot spot.

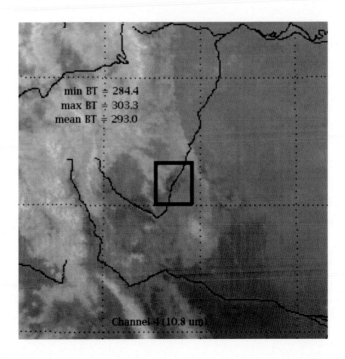

FIGURE 10.32

A 10.8 μm thermal infrared ASTER image of the Hayman Fire.

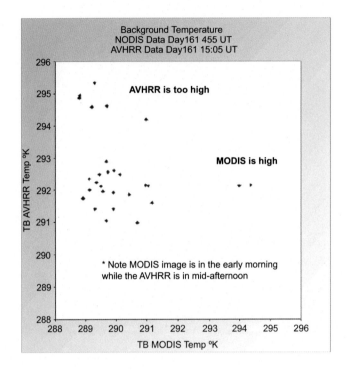

FIGURE 10.33

Comparison between an advanced very high resolution radiometer—derived fire product and a MODIS fire product. Note the difference in satellite overpass time.

the ASTER image saturates at these lower temperatures. This was used as a fire indicator algorithm in the Justice et al. (2002) study.

A comparison between an AVHRR fire product and the MODIS fire product for the Hayman Fire (Fig. 10.33) demonstrates that the AVHRR is in general too high while most of the MODIS fire pixels are appropriate. There are a couple of MODIS pixels that also read too high in this case.

10.6 MEASURING AND MONITORING VEGETATION FROM SPACE

The measurement of vegetative activity on the Earth's surface has been very actively studied with both optical and microwave satellite data. One of the most widely used indices of vegetative health is the NDVI. Defined as:

$$\text{NDVI} = \frac{R_{\text{NIR}} - R_{\text{VIS}}}{R_{\text{NIR}} + R_{\text{VIS}}}, \tag{10.1}$$

where R_{NIR} and R_{VIS} represent reflectances averaged over ranges of wavelengths in the visible ($\sim 0.6\ \mu\text{m}$ centered in the red) and near infrared ($\sim 0.8\ \mu\text{m}$) portions of the electromagnetic spectrum. This NDVI is not an intrinsic physical quantity, but rather it is correlated with certain physical properties of the vegetation canopy that it views from above such as: LAI, fractional vegetation cover, vegetation condition (health), and biomass. In spite of this limitation the NDVI is very useful in that it can be sensed by a wide variety of satellite sensors.

The NDVI has been criticized for the following limitations:

1. Differences between the "true" NDVI measured at the Earth's surface and that measured in space are created by the intervening atmosphere which can attenuate the NDVI signal due to the presence of aerosols in the atmosphere.
2. The sensitivity of NDVI to LAI becomes weaker as the LAI increases up to a threshold value somewhere between 2 and 3.
3. Variations in soil brightness may produce large variations in NDVI from one image to the next (Liu and Huete, 1995).

As a result many investigators have suggested alternate indices that are designed to correct for some of these limitations (Jasinski, 1996; Leprieur et al., 1996; Liu and Huete, 1995; Pinty and Verstraete, 1992).

The NDVI increases almost linearly with increasing LAI up to an asymptotic point where NDVI increases very slowly with increasing LAI. This is a region where the canopy is completely filled with leaves (Curran, 1983). This generally occurs for NDVI values of 0.5−0.8 for dense vegetation. The upper limit depends on vegetation type, age, and leaf water content (Paltridge and Barber, 1988).

In their study Carlson and Riply (1997) use a simple radiative transfer calculation to show: (1) the variation of NDVI with local and global LAI, (2) that the customary variation of NDVI with LAI can be largely explained by variations in fractional vegetation cover, (3) that NDVI decreases much less rapidly with increasing global LAI when the fractional vegetation cover reaches 100%, (4) that scaling the NDVI between values for bare soil and 100% vegetation cover factors out most of the needed atmospheric correction, and (5) provides further evidence to confirm the existence of a simple relation between scaled NDVI and fractional vegetation cover.

The radiative components of this model are represented here in Fig. 10.34. The NDVI was calculated from the reflectance values in two wavelength bands one in the visible (0.5−0.7 μm) and near infrared (0.7−0.9 μm) regions of the spectrum. It should be noted that this is an arbitrary selection and when observing the NDVI from space borne sensors the definition of the VIS and NIR channels is set by the particular instrument onboard the satellite.

In their radiative transfer simulations, they varied parameters such as LAI to determine how the NDVI would change. Fig. 10.35 shows NDVI as functions of LAI for both an NDVI uncorrected for atmospheric attenuation and an NDVI corrected using MODTRAN atmospheric simulations.

We can conclude from these simulations that the apparent NDVI at the asymptotic threshold is between 0.05 and 0.10 below the values corresponding to an infinitely large LAI (Carlson and Riply, 1997). From this, we can further conclude that for areas where the vegetation is just reaching 100% the NDVI will be adequately represented by values slightly less than the maximum found in the image over dense vegetation.

One of the big advantages of using NDVI as an index is that it is a normalized quantity and it really just indicates the relative health of the vegetation being viewed. It can not be used as an accurate measurement of LAI or biomass, but it can suggest where both are high or low. NDVI is sensitive to changes in the fractional vegetation cover until a full canopy is reached beyond which further increase in LAI results in small and asymptotic increases in NDVI. The identification of an NDVI threshold

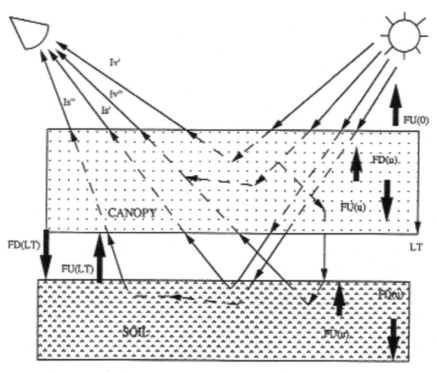

FIGURE 10.34

Schematic representation of the direct and diffuse radiance components interacting with the plant canopy and the soil. Single or direct scattering from the canopy and/or from the soil is indicated by a prime and double scattering by a double prime (Schluessel et al., 1994).

between full and partial vegetation cover allows one to scale NDVI between bare soil and 100% vegetation cover. In addition, the model radiative transfer simulations suggest that this relation holds equally well for NDVI whether or not it is corrected for atmospheric attenuation.

10.6.1 THE AVHRR NDVI 8-KM DATASET

One of the most widely used sensors for the computation of NDVI was the AVHRR. This group described this dataset in Tucker et al. (2005), which discusses the AVHRR NDVI dataset in the context of similar products based on MODIS and SPOT data. This product consisted of daily daytime AVHRR Global Area Coverage data reduced to an 8-km equal-area dataset from July 1981, through December 2004, for all continents except Antarctica. Also unique to this dataset were bimonthly composites and an effort to correct and compensate for volcanic aerosols in the atmosphere. In addition, for Africa-only there was a 10-day composite NDVI product. Considerable postprocessing was needed to correct for dropped scan lines, image navigation errors, other data dropouts, and edge-of-orbit composite discontinuities.

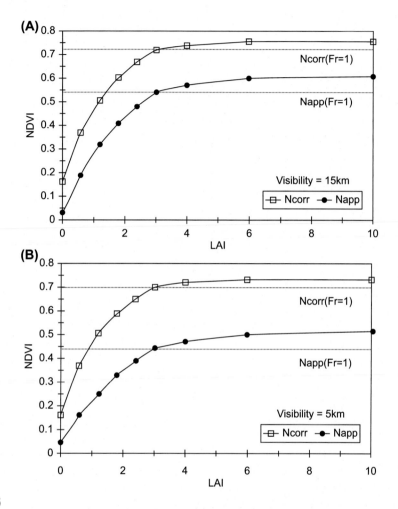

FIGURE 10.35

Uncorrected Normalized Difference Vegetation Index (NDVI) (Napp) and corrected NDVI (Ncorr) to surface values versus leaf area index (LAI); global LAI is equal to the functional vegetation cover times a fixed value of local LAI of 3 up to 100% vegetation cover above which local and global LAI are equal. Shown here are (A) for 15-km visibility and (B) for 5-km visibility. Horizontal dotted line denotes the value of NDVI at the threshold to the asymptotic regime in which fractional vegetation cover is just reaching 100%.

According to Tucker et al. (2005) the NDVI has become the most used vegetation product derived from NOAA AVHRR data (Cracknell, 2001), largely in the form of maximum value compositing (Holben, 1986). By compositing on maximum NDVI value any residual clouds are suppressed since clouds will only lower the NDVI value. Thus, an initial cloud mask is applied and then composites are formulated using the maximum NDVI procedure. Tucker et al. (2005) also provide an interesting

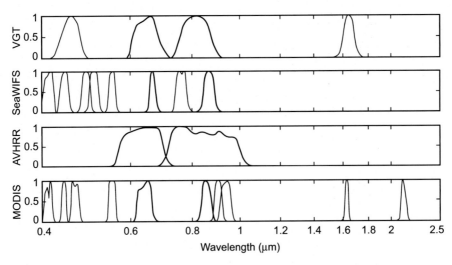

FIGURE 10.36

Spectral acceptance of the advanced very high resolution radiometer instruments in comparison to SPOT Vegetation, SeaWIFS, and the MODIS visible and near-infrared bands. The darker spectral acceptance bands are those used to calculate Normalized Difference Vegetation Indices (NDVIs). Differences in spectral bandwidth and spectral acceptance functions will result in identical surface reflectances producing different NDVIs.

comparison between the AVHRR sensor response and those of SPOT, SeaWIFS, and MODIS as reproduced in Fig. 10.36.

Here one can clearly see the broad bands of the AVHRR that do indeed separate the visible from the near infrared and have the advantage that they integrate a substantial portion of the spectrum. SPOT Vegetation (VGT) has similar bands that are a bit narrower than AVHRR while MODIS bands are very much narrower. Likewise, SeaWIFS (an ocean color instrument) has relatively narrow bands in both the visible and near infrared.

In their NDVI product (Tucker et al., 2005) every composite image was manually checked for image navigation accuracy by comparing the mapped data to a reference coastline for every continent. Images with >−1 pixel navigation error were individually investigated and the data reprocessed to maintain the 1 pixel navigation accuracy over the entire product dataset. This group decided to use maximum compositing as the method to correct for any and all atmospheric effects.

10.6.2 USING NDVI TO IDENTIFY AND MONITOR CORN GROWTH IN WESTERN MEXICO

A study in 2006 (Chen et al.) examined the use of a MODIS derived NDVI, a MODIS derived Enhanced Vegetation Index (EVI) and a SPOT VGT index to map and monitor the corn crop in western Mexico. Overall the NDVI derived from the SPOT-VGT sensor, based on a bidirectional compositing method, produced vegetation information most closely resembling actual crop conditions. By comparison the MODIS NDVI values were found to be saturated 30 days after planting but were well correlated with green leaf senescence in April. The temporal maximum value NDVI composites for the VGT sensor reached its maximum plateau at about 80 days, which then masked the important crop transformation from the vegetative stage to the reproductive stage. The EVI computed from MODIS reached a maximum plateau 40 days earlier than the maximum LAI and the maximum intercepted fraction of photosynthetic active radiation as derived from in situ measurements. This study demonstrated that the 250 m spatial resolution MODIS bands did not provide any more accurate vegetation information for corn growth description than the larger resolution 500 m and 1 km bands.

An example of the image data examined in this study is provided here in Fig. 10.37. The location of the study area is shown in a small map in the upper left of this figure.

10.6.3 MICROWAVE REMOTE SENSING OF VEGETATION AND SOIL MOISTURE

Passive and active microwave data have also been used to detect and map vegetation as well as the soil moisture beneath the plant canopy, as discussed in Section 10.2.4 for ESA's SMOS mission. A comparison between optical and microwave remote sensing of vegetation density over Mongolia was carried out by Liu et al. (2006). They find that optical and microwave data provide complementary sources of information on vegetation. They compared the NDVI derived from AVHRR imagery with a vegetation optical depth (VOD) computed from Special Sensor Microwave/Imager data collected over Mongolia from 1998 to 2006. Both of these data sources exhibited similar spatial distributions in terms of their annual averages and both are found to capture well the interannual variability of the vegetation. Longer term changes, however, are found to diverge in the two different data types. The microwave VOD shows long-term decreasing trends over grassland, whereas the optical NDVI exhibits weaker declines and in some cases even increases.

The similarities in long-term means is shown here in Fig. 10.38, which shows both the VOD and NDVI centered on Mongolia from 1988 to 2006.

Here, clearly the spatial patterns are very similar with the AVHRR-derived NDVI showing more detail due to the higher spatial resolution of the optical data. The passive microwave VOD has much smoother features, but still the basic patterns and relative values are essentially the same between the two fields. Low values are seen over the southern Gobi desert regions, and high values are found in the forest in northern Mongolia. Moderate values are found in the pastureland and cropland regions between these two different areas.

To demonstrate the similarities in the year-to-year variations both the VOD and NDVI are presented here in Fig. 10.39 for August 2003 (a wet year) and 2004 (a dry year).

The interannual variation between 2003 and 2004 can be easily seen in both the NDVI and the VOD by the increased area covered by red and yellow values in 2004 as compared with 2003. Thus, the dry character evidences itself more in terms of greater areas of low vegetation index values than in the severity of the decrease in the index value itself.

To test for long-term changes in these vegetation index data sets the Mann-Kendall index (Gilbert, 1987) was calculated for each (Fig. 10.40). In this comparison you can clearly see differences between the two estimates of vegetation. The VOD shows a very much larger decrease in vegetation index in northern central Mongolia than is seen in the corresponding NDVI. Again the smoother nature of the VOD is clearly apparent as is the higher spatial resolution of the AVHRR-derived NDVI. The NDVI actually exhibits a slight increase in northern Mongolia at about 100°E where the dark red values in

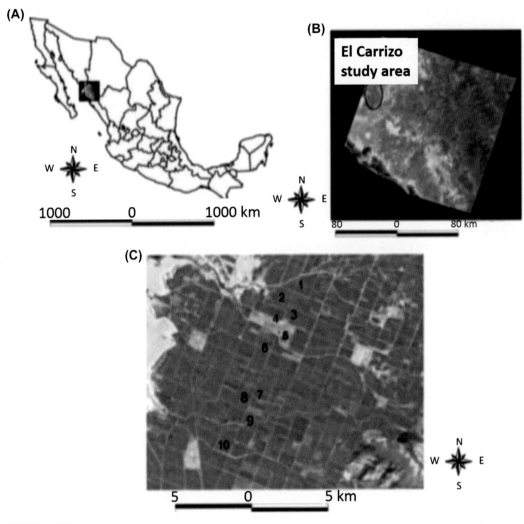

FIGURE 10.37

Irrigated corn fields in El Carrizo study area in Sinaloa state, western Mexico: (A) on Landsat-7 ETM+ image date on February 8, 2002, as a false color composite of the green, red, and near-infrared bands, (B) and 10 sampled locations, (C) pixels in red color were agricultural fields planted with corn, and pixels in other colors include fallow, lands, urban areas, water bodies, and agricultural field for other crops.

VOD display a very dramatic drop in vegetation. Thus, we must conclude that NDVI and VOD are not equivalent in all respects and that consideration must be given to the source of information in any and all analyses.

In a similar study Jones et al., (2010) studied the phenology of global vegetation using these same data sources (satellite optical and passive microwave). The VOD was computed by the Advanced

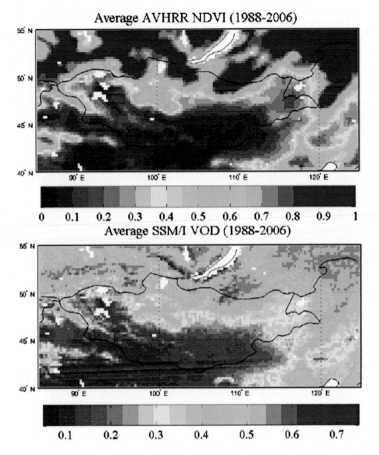

FIGURE 10.38

Annual average Normalized Difference Vegetation Index and vegetation optical depth (VOD) during 1988–2006.

Microwave Scanning Radiometer-Earth (AMSR-E) flying on NASA's Aqua afternoon satellite. The advantage of this VOD for monitoring the phenology of global vegetation is its lack of sensitivity to cloud cover and its independence of solar illumination. The VOD in this study was found to respond to changes in plant canopy biomass and water content. For comparison the EVI was calculated from the MODIS instrument also flying on the Aqua satellite, thus providing vegetation estimates coincident in space and time. This EVI was combined with the NDVI also computed from MODIS in what has been called the enhanced TIMESAT algorithm for retrieving vegetation phenology metrics from 250 to 500 m spatial resolution MODIS data. In addition to the satellite data, a bioclimatic phenology model was used to compare global vegetative phenology from 2003 to 2008. Jones et al. (2010) found good correspondence between the VOD and the optical vegetation indices (VIs) across most global biomes. This correspondence degraded over regions with persistent cloud cover due primarily to optical VI degradation and to phase-shifts or lags between the VI and VOD phenology signals. While the VOD

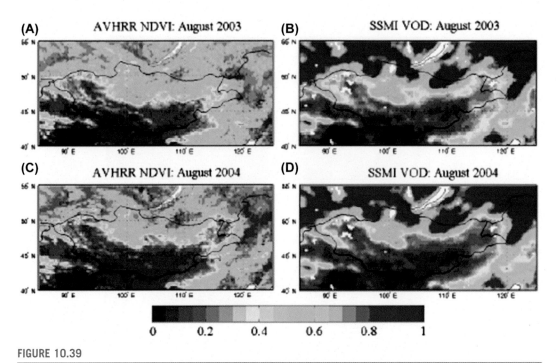

FIGURE 10.39

Monthly Normalized Difference Vegetation Index and vegetation optical depth for August 2003 and 2004.

and VIs covary, they do not detect identical changes in seasonal plant dynamics and are sensitive to independent, yet complementary, vegetation surface effects.

The mean global VOD is presented here in Fig. 10.41.

The algorithm to compute this VOD uses AMSR-E descending (AM, morning pass) Aqua satellite overpass horizontally (H) and vertically (V) polarized brightness temperatures for the 18.7 and 23.8 GHz frequencies to derive a slope parameter that is sensitive to vegetation and surface roughness. This slope parameter shows minimum sensitivity to changes in soil moisture, a parameter that is sensed using the lower frequency microwave signals. It is also largely independent of open water effects with a good sensitivity to canopy density. The smoothed slope parameter is then inserted into a tau-omega radiative transfer model and inverted to solve for VOD at 6.9, 10.7, and 18.7 GHz.

Jones et al. (2010) used the VOD data together with the TIMESAT 3.0 algorithm to compute the phenology metrics of the start, length, and end of growing season for the entire globe (Fig. 10.42) over the years 2003–08.

The areas with the latest start to their vegetation are in the Amazon and tropical South Africa There is some later start in parts of South China. The early starts take place in the polar regions, in central South America, and western North America, as well as in northern Australia. The maps at the end of the season are very different with the earlier values in tropical west Africa, the northern Midwest United States and in western Russia. There are also regions of early end date in central and southern India and in western China. The Korean peninsula also exhibits some early terminal dates. The late end

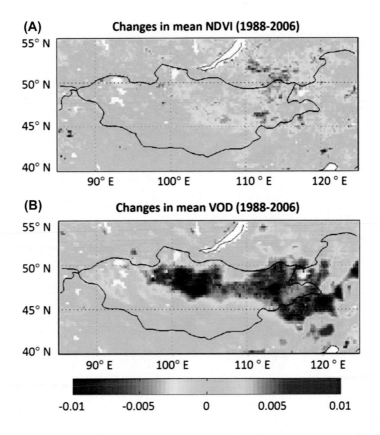

FIGURE 10.40

Trends in annual mean Normalized Difference Vegetation Index and vegetation optical depth between 1988 and 2006 using the Mann-Kendall analysis.

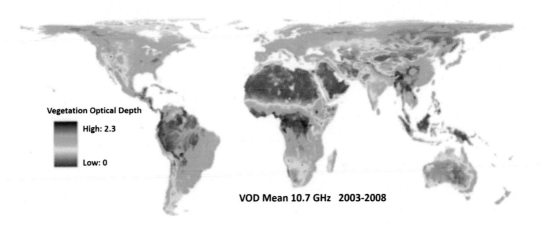

FIGURE 10.41

The vegetation optical depth computed from AMSR-E data for 2003—08.

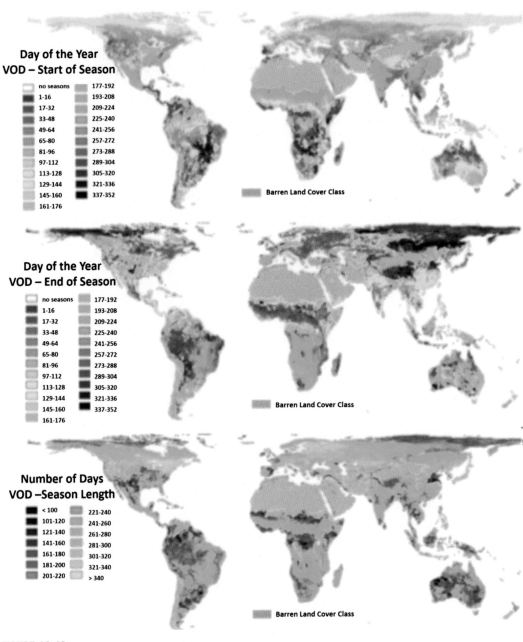

FIGURE 10.42

Phenology metrics of start (top), length (bottom), and end (middle) of the growing season for 2003—08.

of season regions are in central South America, all of southern Africa, high-latitude North America, and Russia along with pockets in western Australia.

The length of the growing season has its maximum in the mid-latitude northern hemisphere where crops are mainly grown. The shortest growing seasons (dark blues) are in tropical west Africa, central Africa, southern Colombia and Venezuela, southern United States (Texas), northern China, and western Australia.

Their overall conclusion was that the AMSR-E VOD time series provides an independent measure of the phenology signal that can be considered complementary to the optical-IR measured vegetation index information that has been routinely used in the past. The microwave nature of the VOD makes it possible to map global phenology even in areas that are persistently cloud covered. The complementary nature of these two types of independent systems makes it possible to take advantage of the best characteristics of each to get useful global maps of vegetative phenology.

There have been a number of studies of the measurement of soil moisture with passive microwave data. One of the earliest studies is by Jackson and Schmugge in 1989, which used data from a sensor operating at a 21 cm wavelength to sense the amount of water present in soils or on its surface. This is due to the large dielectric contrast between water and most solids at this wavelength. They report that this contrast causes variations in soil's emissivity from about 0.95 when dry to less than 0.7 when wet. Microwave remote sensing has the usual advantage of being weather independent, and it is also possible to measure soil moisture condition under moderate level of vegetation cover.

Previous studies of the interactions between the microwave observations and the target parameters have established that for passive microwave systems the optimal wavelength for observing soil moisture falls in the L-band range (20−30 cm) operating at near-nadir look angles and horizontal polarization. This past research has also established that the principle target parameter affecting the microwave emissivity is the volumetric soil moisture in the surface 5 cm of the soil. The conclusion is that if a passive microwave L-band sensor was placed in orbit it would be possible to map and monitor surface soil moisture on a regular basis. While the fundamental principles governing microwave emissivity from a smooth bare homogeneous soil with uniform vertical dielectric properties are well understood the problem is when you add in the complexity introduced by disruption of the soil, the presence of varied vegetation cover and alterations in soil moisture complicate the scene sufficiently that it is difficult to model the scene much less to later verify the conditions.

Jackson and Schmugge (1989) introduce a pictorial of an algorithm for estimating surface soil moisture with a passive microwave radiometer (Fig. 10.43).

The principle target variables in this model algorithm are soil properties, surface roughness, and vegetation characteristics. A soil target is made up of air, water, and soil. At microwave wavelengths, the real part of the complex numbers representing the dielectric constants are as follows: air $k = 1$, soil $k = 4$, and water $k = 80$. It is this very large contrast that makes the measurement of moisture in the soil possible. As more water is added to the soil the dielectric increases. Bound water has a lower dielectric constant because these water molecules are adsorbed by the surfaces of the ground particles, which immobilizes the dipoles. In soil this surface area is variable and depends primarily on the amount, size, and shape of the clay particles in the soil. Two physical parameters best describe this surface area and they are the specific surface area and the bulk density of the soil. The surface area will increase when either of these two variables increases. Typical soils range between 10 and 150 m^2/gm for the specific surface and bulk densities range between 1 and 1.5 gm/cm^3. The accepted approach to estimate these parameters is to assume a thickness for the bound water layer

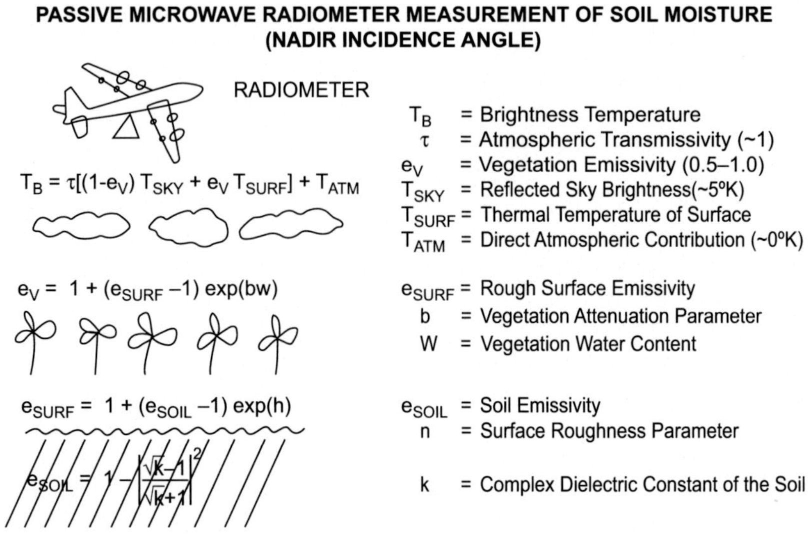

PASSIVE MICROWAVE RADIOMETER MEASUREMENT OF SOIL MOISTURE
(NADIR INCIDENCE ANGLE)

RADIOMETER

$$T_B = \tau[(1-e_V) T_{SKY} + e_V T_{SURF}] + T_{ATM}$$

T_B = Brightness Temperature
τ = Atmospheric Transmissivity (~1)
e_V = Vegetation Emissivity (0.5–1.0)
T_{SKY} = Reflected Sky Brightness(~5°K)
T_{SURF} = Thermal Temperature of Surface
T_{ATM} = Direct Atmospheric Contribution (~0°K)

$$e_V = 1 + (e_{SURF} - 1) \exp(bw)$$

e_{SURF} = Rough Surface Emissivity
b = Vegetation Attenuation Parameter
W = Vegetation Water Content

$$e_{SURF} = 1 + (e_{SOIL} - 1) \exp(h)$$

e_{SOIL} = Soil Emissivity
n = Surface Roughness Parameter

$$e_{SOIL} = 1 - \left| \frac{\sqrt{k}-1}{\sqrt{k}+1} \right|^2$$

k = Complex Dielectric Constant of the Soil

FIGURE 10.43

Schematic of an algorithm for estimating surface soil moisture using a passive microwave radiometer.

and a single average value of its dielectric constant. For bulk density it has been shown that regional estimates are adequate descriptions.

Jackson and Schmugge (1989) report that while there are many different ways to model the emission from a smooth bare soil, they all lead to the same conclusion, that microwave emission from soil is the result of the integration of the emission from all depths and the importance of each depth decreases with depth. The fraction of the upwelling radiation that is transmitted to the air will be called the emissivity, and it is determined by the dielectric properties of a transition layer at the surface. Studies have shown this layer to be a few tenths of a wavelength, thus, a long wavelength sensor provides more information on a thicker soil layer than does a shorter wavelength.

Reutov and Shutko (1986) performed a variety of simulations that clearly illustrate the contributing depth phenomena. Here Fig. 10.44 shows both the soil system that was modeled and the effects of wavelength penetration for a single soil moisture value. At the top (Fig. 10.44A) the soil moisture conditions are set showing a transition layer from an air dry at the surface laying over a wet subsurface layer with a uniform soil moisture (W_d) starting at a depth h_d. A set of simulations is represented in Fig. 10.44B, which clearly demonstrates that at shorter wavelength the sensor responds to a very shallow soil layer (<1 cm at a wavelength of 2.25 cm). As this wavelength increases the depth that the sensor sees through also increases. This layer is about 5 cm at L-band (wavelength of 21 cm). One other interesting result from Reutov and Shutko (1986) was that from their experimental program they found that the actual transition depths were <3 cm. This means that an L-band sensor would be responsive to most soil moisture conditions.

With a single-frequency sensor it is not possible to retrieve the profile of surface moisture. It is necessary to relate the microwave observation to the conditions in a representative soil layer that in most cases determines the emissivity. Experimental results have demonstrated that at L-band, a 5 cm

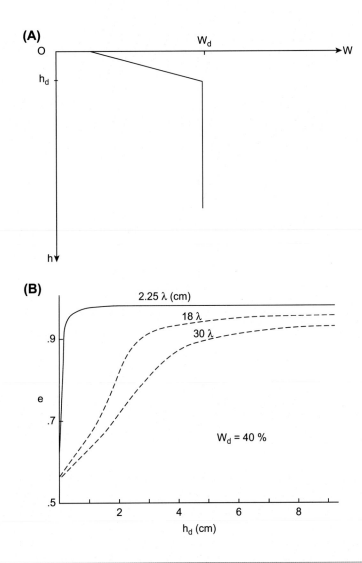

FIGURE 10.44

Effects of soil moisture transition depth on emissivity after Reutov and Shutko (1986). (A) Schematic of simulated conditions, volumetric soil moisture (W_d) as a function of depth (h), (B) Simulated emissivity at three wavelengths as a function of the thickness of the transition layer (h_d).

depth is representative of all except very sandy soils in arid environments. It is likely that a separate depth could be used for these conditions.

The curves in Fig. 10.45 show how well a combination of dielectric mixing models can be linked with a simple emission model. For this case of bare smooth soil, the models reproduce the patterns of the observations very well. As a result, the simplest approach to measuring soil moisture would then

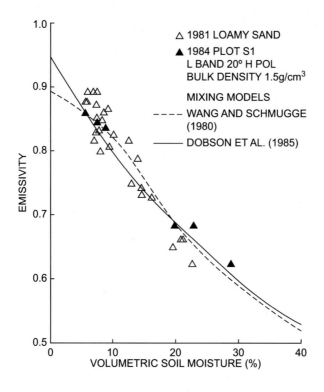

FIGURE 10.45

Observed and predicted relationships between emissivity and volumetric soil moisture (0–2 cm) for a bare smooth loamy sand soil at 1.4 GHz (Jackson and O'Neill, 1987).

involve the use of a single-frequency sensor to estimate the moisture in a 5 cm depth. These values could then be used on their own to determine infiltration and/or evaporation, and to extrapolate to deeper layers. Additional soil moisture information is available if two different wavelengths are used. Both simulations (Blanchard and Bausch, 1979) and experimental measurements together with simulations (Reutov and Shutko, 1986) were used to study the use of two wavelengths. The latter found that only one combination of transition depth and W_d can produce a particular pair of emissivity values at two wavelengths. This means that given two emissivity values one can determine the depth of the transition layer and the moisture below it. These authors also carried out airborne measurements as presented here in Fig. 10.46B where W_d was estimated from the aircraft measurements. Using the two-frequency approach, very good estimates of W_d were obtained from these measurements. A single-frequency radiometer adds very little information on W_d, but it can be used to estimate the average moisture in a surface layer, and not W_d.

As pointed out earlier microwave emissivity from a soil surface is related to the reflectivity of that surface which is why all of the modeling presented earlier addressed a bare, smooth soil surface where the variations in the surface structure are a fraction of the wavelength being used. Surface roughness

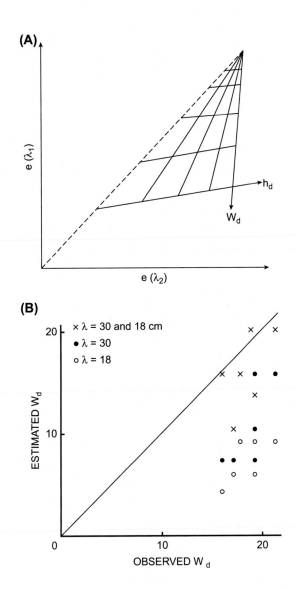

FIGURE 10.46

Reutov and Shutko (1986) two-frequency procedure. (A) Estimation of depth of transition layer (h_d), and (B) soil moisture below it (W_d) using emissivity measurements (e) at two wavelengths (λ_1, λ_2).

results in an increase in emissivity over that of equivalent smooth surfaces. With a rough surface one must also consider the scattering of the emitted waves. Several investigators have developed physically based models to account for these effects. These models generally result in formulations that cannot be verified through simple measurements or parameterizations that cannot be directly measured. Neither the geometry of the soil surfaces nor the physics of the radiative scattering from these surfaces is sufficiently well understood to calculate the roughness parameter (h) from first principles. Thus, the use of a very simple model is justified as presented by Choudhury et al. (1979) that relates the emissivity of the smooth surface (e_{soil}) and the roughness parameter to the emissivity of the surface (e_{surf}) as given here in Eq. (10.2)

$$e_{surf} = 1 + (e_{soil} - 1) \cdot e^h. \tag{10.2}$$

Jackson and Schmugge (1989) report that it does not appear that these effects are significantly different over geographic regions with similar land management practices and therefore representative values of h can be used for regional studies.

Vegetation cover must be considered for most application purposes. The complexity of the vegetation itself in terms of modeling limits our understanding. Deterministic approaches have, however, been successful and have shown that soil moisture can be determined under a wide variety of plant canopy conditions. In principle a plant canopy over the soil attenuates the emissivity of the soil and adds to the total radiative flux with its own emission. This assumes that scattering is negligible, which should be true for longer wavelengths. A theoretical model of this process that treats the problem using the electromagnetic relationships of a two-layer incoherent nonscattering medium was developed by Basharinov and Shutko (1975). In their model the vegetation is treated simply as an absorbing layer. The nature of the attenuation of this layer was modeled by Kirdiashev et al. (1979) as an optical depth that is dependent upon the plant shape, wavelength, look angle, and vegetation biomass. The plant shape parameter is empirical. Jackson et al. (1982) proposed that the variance in the plant shape parameter was small for agricultural crops and that the attenuation varied linearly as a function of vegetation water content. Beside the change in sensitivity with vegetation water content, attenuation is also a function of wavelength. Kirdiashev et al. (1979) evaluated this effect using a parameter called the slope reduction factor. This is the ratio of the slope of the emissivity-soil moisture function with a vegetation cover to the bare soil slope (Fig. 10.47).

A value of 1 indicates no effect while a small value means that most of the instrument sensitivity to soil moisture has been lost.

There are more sophisticated models available to account for vegetation effects all of which recognize the importance of the vegetation water content in determining the canopy effect. These models all require optimization or the establishment of an extensive database to develop parameter values for various plant canopies. The model proposed by Jackson et al. (1982) requires only an estimate of the vegetation wet biomass or water content. Biomass and water content can be estimated using visible and near infrared remote sensing or multifrequency microwave data and would have other uses in agricultural and hydrologic applications. Passive microwave data can also be used to estimate biomass when the soil moisture and background emissivity are known. These relationships require atmospheric corrections. If the biomass was known for several data cells, a calibration could be performed. One of the most likely targets for such a procedure would be flooded rice fields (Jackson and O'Neill, 1986), since the emissivity of the background water is predictable and the microwave sensor data could then be used to predict biomass. This biomass estimate could then be compared with that from the optical vegetation index value.

In a more recent paper (Jackson and Schmugge, 1991) explored the vegetation effects on microwave emissions of soils. They considered the vegetation present as requiring correction for the soil

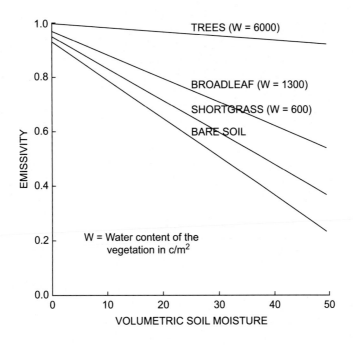

FIGURE 10.47

Predicted relationships between volumetric soil moisture and emissivity for bare soil and selected vegetation canopies.

microwave emission. Improving upon the model discussed earlier they found that it was critical to be able to estimate the vegetation parameter b that characterizes the canopy. This paper is an exploration of the computation of this parameter for various observed vegetation characteristics. They concluded that vegetation is semitransparent at longer microwave wavelengths. The degree of its influence on the measured data is dependent on the amount of vegetation, its type, and the parameters of the sensor system, that is, its wavelength.

Looking at soybeans they found that the net effect of vegetation is a reduction in sensitivity that makes it much more difficult to estimate soil moisture accurately (Fig. 10.48). An increase in the vegetation water content further reduces the sensitivity.

Returning to the 1982, Jackson and Schmugge model, these authors argue that this relatively simple model can be used to account for particular vegetation. This approach requires, in addition to the microwave brightness temperature, an estimate of the physical temperature at the surface, the vegetation water content, and a vegetation parameter (b). Physical temperature and vegetation water content can be determined using existing techniques, but the dependence of the vegetation parameter (b) on various physical and sensor variables has not been established. The purpose of the Jackson and Schmugge (1991) paper was to study this vegetation parameter for various plant canopies. For specifics of this model the reader is directed to the 1982 and 1991 Jackson and Schmugge papers.

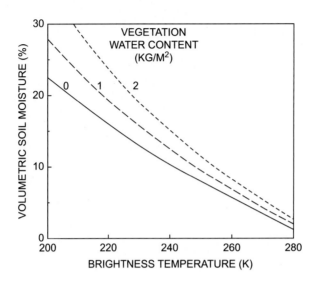

FIGURE 10.48

Predicted relationships between emissivity and soil moisture for a field with bare soil and soybeans. Curves are based on a wavelength of 21 cm, and at nadir.

Shutko (1986) suggested that in passive microwave remote sensing there are at least six significant categories of vegetation:

1. Small leaved culture such as cereal (small grains) crops in early stages, alfalfa (quasi-chaotic orientation).
2. Cereal crops near maturity (preferential orientation of stems).
3. Broad leaf cultures (scattering effects important).
4. Fruit and vegetables (fruit shadows the soil).
5. Bush and forest vegetation (high biomass/area).
6. Dense vegetation covers such as harvested grass and lodged vegetation.

This classification makes sense in terms of the physical parameters of the vegetation that can contribute to attenuation of the passive microwave signal through scattering. In their study (Jackson and Schmugge, 1991) only consider the first three categories on this list. They claim that there is very little information available for vegetable crops. The report that Chukhlantsev et al. (1989) present the results of a rather extensive data set collected over tomatoes. Their results show that a relationship between emissivity and soil moisture can be established throughout the growing season. Unfortunately, this Russian study only reported on wet biomass as a function of emissivity. This lack prevents the analysis of vegetation water content effects.

There is little point in attempting to estimate soil moisture under a forest canopy. Under these conditions, the vegetation water content is so large that it is unlikely that any useful information could be extracted. Jackson et al. (1984) reported on observations made using aircraft radiometer at wavelengths of 6 and 21 cm that clearly showed no soil moisture effect on emissivity for a tree canopy.

To address category 6 aforementioned one needs to look at grasses that have a buildup of thatch. It appears that under these conditions there is very little sensitivity to soil moisture and the signal is all driven by the plant canopy moisture. This behavior was observed in aircraft data collected over the Konza Praire Natural Area by Schmugge et al. (1988). In this particular tall grass prairie there were some watershed areas in which there was no grazing or burning and other areas that were burned yearly or every other year. The latter watersheds exhibited predictable variations in brightness temperature with moisture changes while the unburned areas showed little sensitivity to moisture falling from rainfall. It was assumed that this was due to the buildup of thatch or detritus matter of the soil surface over a number of years. This thatch layers acts as a highly emissive surface and this condition will only occur for lightly used or unburned grasslands and not in typical pastures.

In their study Jackson and Schmugge (1991) neglected atmospheric effects since all of the data they analyzed were taken from field towers or low flying aircraft where even at the shorter wavelengths the atmospheric effects are negligible. In addition, at the longer wavelengths that are most suitable for soil moisture observations, atmospheric effects will be minimal even from satellite altitudes. For their study they examined a wide range of existing studies for which they had the requirements that:

1. Wavelength between 0.8 and 30 cm.
2. Horizontal polarization.
3. Look angle correctable to 0 degree.
4. Data was available to determine optical depth τ.
5. Vegetation water content.
6. Cover description.

Table 10.3 briefly summarizes all of the studies that were used in the Jackson and Schmugge (1991) paper.

Using these data from historical studies, Jackson and Schmugge (1991) were able to construct a summary plot reproduced here as Fig. 10.49.

This plot includes all of the cases presented in Table 10.3, and it was used to calculate the vegetation parameter b versus the wavelength λ. All cases follow the trend of b decreasing with wavelength, which is expected from the analyses presented in Kirdiashev et al. (1979), Chukhlantsev (1988) and Pampaloni and Paloscia (1986). Kirdiashev et al. (1979) suggested a relationship that can be described as

$$b = b'/\sqrt{\lambda}, \tag{10.3}$$

for fine leaved vegetation which is Pampaloni and Paloscia (1986) is

$$b = b'/\lambda. \tag{10.4}$$

In each case b' is the wavelength-independent parameter. Using Fig. 10.49 as a guide, it appears that a linear relationship such as (10.3) might adequately describe the variation for a particular plant cover at wavelengths greater than 5 cm. At shorter wavelengths the b values change rapidly suggesting a form more like (10.4). It is also very possible that the entire wavelength regime cannot be described by a single function. As noted earlier, the effects of scattering are more important at shorter wavelengths.

Table 10.3 Vegetation Data Sets Used in Study

Source	Cover Type	Wavelength (cm)	Optical Depth	Vegetation Water Content (kg/m^2)	b
Shutko (1986)	Broad leaf	3	0.950	2.0	0.475
		20	0.150	2.0	0.075
Jackson and O'Neill (1990)	Corn	6	—	—	0.150
		21	—	—	0.115
	Soybeans	6	—	—	0.288
		21	—	—	0.086
Kirdiashev et al. (1979)	Winter rye	2.25	0.310	0.7	0.442
		10	0.160	0.7	0.229
		20	0.080	0.7	0.114
		30	0.030	0.7	0.043
Ulaby and Wilson (1985)	Wheat	2.8	1.980	5.2	0.380
		6.5	0.780	5.2	0.150
		18.2	0.270	5.2	0.050
	Soybeans	2.8	2.410	1.8	1.340
		6.5	0.800	1.8	0.440
		18.2	0.180	1.8	0.100
Pampaloni and Paloscia (1986)	Alfalfa	0.8	—	—	1.850
		3.1	—	—	0.930
	Corn	0.8	—	—	0.600
		3.1	—	—	0.340
Ulaby et al. (1983)	Corn	6	0.744	4.0	0.186
		21	0.452	4.0	0.113
Brunfeldt and Ulaby (1984)	Soybeans	5.9	1.025	2.8	0.366
		11.1	0.749	2.8	0.264
Chukhlantsev and Shutko (1988)	Cereals	18	0.284	2.0	0.142
	Alfalfa	18	0.364	2.0	0.182
Jackson et al. (1982)	Corn	6	0.199	1.2	0.162
		21	0.163	1.2	0.133
	Soybeans	6	0.240	1.0	0.240
		21	0.087	1.0	0.087
O'Neill et al. (1988)	Corn	6	0.785	6.0	0.131
		21	0.611	6.0	0.102
	Sweet sorghum	6	0.750	5.4	0.138
		21	0.613	5.4	0.105
Wang et al. (1980, 1982)	Short grass	6	0.550	0.3	1.770
		21	0.093	0.3	0.300
	Tall grass	6	0.797	0.4	1.990
		21	0.288	0.4	0.720
Wang et al. (1990)	Tall grass	21	—	0.5	0.150
Matzler (1990)	Oats	1.4	—	—	0.010
		2.9	—	—	0.621
		6.1	—	—	0.120
Vyas et al. (1990)	Broad leaf	19.3	—	—	0.092

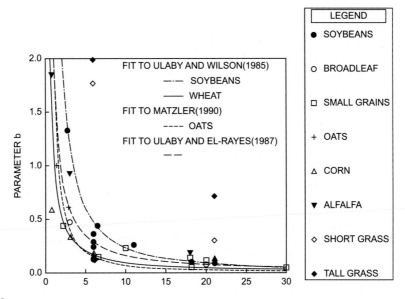

FIGURE 10.49

Summary of experimental data of the vegetation parameter b as a function of wavelength [cm] for a variety of vegetation types (Pampaloni and Paloscia, 1986).

As reported by Jackson and Schmugge (1991), a study by Ulaby and Wilson (1985) analyzed a data set that covered a wide range of wavelengths for the same plant canopy. They introduced an equation of the form

$$b = b' \cdot \lambda^x \tag{10.5}$$

where the value of x for wheat was -1.08 and for soybeans -1.38. Using the exponent for wheat we have an equation very similar to Eq. (10.4). Eq. (10.5) was also fit a power function to the oat observations reported by Matzler (1990). The resulting x exponent was -1.47 and the resulting curve is included here in Fig. 10.49. It appears very similar to the relationship developed by Ulaby and Wilson (1985) for wheat. Both crops have similar structures, and these results support the general results for a stem-dominated cereal crop.

The summary results in Fig. 10.49 reveal that except for the grasses most vegetation can be categorized as either leaf or stem dominated. All the corn and cereal crop data follow one pattern, and soybean and alfalfa another. There is variability in this generalization that becomes more obvious as the wavelengths get shorter. Part of the problem may be variations in the various cultures of the soybean plants where some are tall with fewer leaves while others are short and bushy. In addition, these crops have different growth stages when there will be variations between the relative numbers of leaves and stems.

One important observation from Fig. 10.49 is that if we exclude the grass observations, there is only a very small variation of b in the L-band range (18−21 cm). Thus, it appears that a single value of b might be used regardless of ground cover type, without introducing significant errors. At shorter wavelengths, the effects of crop type become more important and the uncertainty in estimating b for leafy vegetation increases.

The results for grass are very different than those for crops. The parameter b is much larger than that of other types of vegetation even at L-band. Fortunately for interpretation purposes, the vegetation water contents are relatively low, resulting in relatively low optical depths, at least at L-band. The primary concern here, however, is that the attenuation caused by grass is not just a function of changes in the nature of the canopy that result from growth (the bending of leaves and a thatch buildup).

In their paper Jackson and Schmugge (1991) concluded that soil moisture information can be retrieved using passive microwave remote sensing particularly at the lower L-band frequency. This is true even in the presence of a vegetation canopy, and they studied the dependence of the plant canopy parameter b on plant and sensor system characteristics using previous published results. They also concluded that the relationship between b and wavelength can be divided into three categories: leaf dominated (soybeans, cotton, alfalfa), stem dominated (corn, wheat), and grasses.

10.7 THE EUROPEAN *COPERNICUS* PROGRAM

Copernicus is the European system for Earth monitoring. It consists of a complex set of systems that collect data from multiple sources: Earth observation satellites (Fig. 10.50) and in situ sensors such as ground stations, airborne, and sea-borne sensors. It processes these data and provides free-of-charge to the users with reliable and up-to-date information through a set of services related to environmental and security issues (extracted from http://www.copernicus.eu/main/overview and https://sentinel.esa.int/web/sentinel/missions).

The services address six thematic areas: land, marine, atmosphere, climate change, emergency management, and security. They support a wide range of applications, including environment protection, management of urban areas, regional and local planning, agriculture, forestry, fisheries, health, transport, climate change, sustainable development, and civil protection, and they are coordinated and managed by the European Commission. The development of the observation infrastructure is performed by the European Space Agency (ESA), and by the European Environment Agency and the Member States for the in situ component of the Copernicus program, as illustrated in Fig. 10.51.

Through the different thematic areas it addresses (land, marine, atmosphere, climate change, emergency management, and security), Copernicus supports applications in a wide variety of domains. These include agriculture, forestry and fisheries, biodiversity and environmental protection, climate and energy, civil protection and humanitarian aid, public health, tourism, transport and safety, as well as urban and regional planning.

The Earth observation satellites that provide the data exploited by the Copernicus services are split into two groups of missions:

- The Sentinels (Fig. 10.52) that are currently being developed for the specific needs of the Copernicus program. Sentinel-1,-2, -3, -5P, and -6 are dedicated satellites, while Sentinel-4 and -5 are instruments onboard EUMETSAT's weather satellites.
  - Sentinel-1 is a two satellite constellation with the prime objectives of land and ocean monitoring. The goal of the mission is to provide C-Band SAR data continuity following the retirement of ERS-2 and the end of the Envisat mission.

(A)

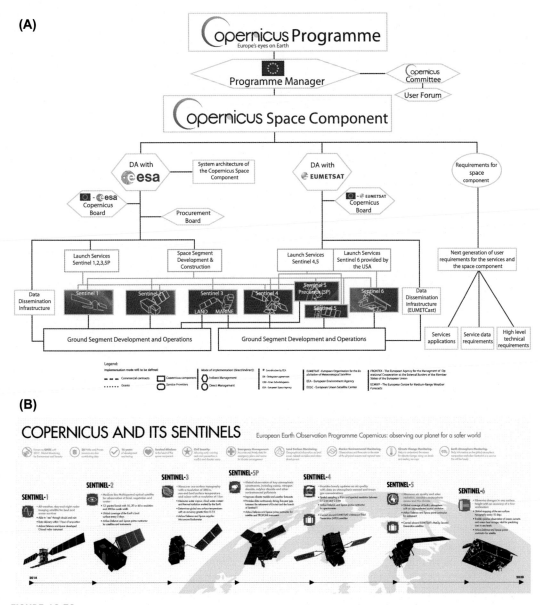

FIGURE 10.50

(A) Copernicus Space Segment: the fleet of Sentinel satellites, (B) overview of the Sentinel's: the Copernicus satellite fleet (http://www.space-airbusds.com/media/image/copernicus-poster-840x297_eng_1_1.jpg).

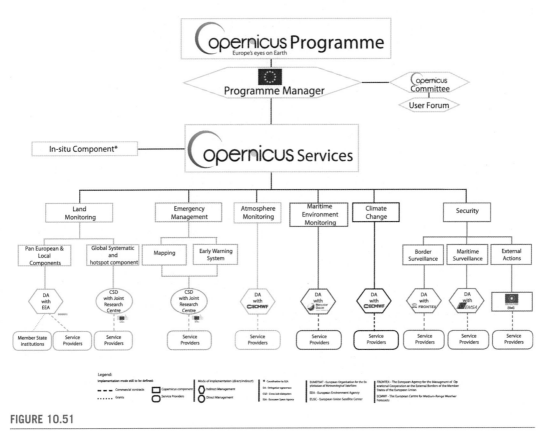

FIGURE 10.51

List of Copernicus Services and agencies working in their development.

- Sentinel-2 comprises twin polar-orbiting satellites in the same orbit, phased at 180 degrees to each other. The mission benefits from the heritage of SPOT and LANDSAT missions and will monitor variability in land surface conditions, and its wide swath width and high revisit time (10 days at the equator with one satellite, and 5 days with two satellites under cloud-free conditions, which results in 2–3 days at mid-latitudes) will support monitoring of changes in vegetation within the growing season. The coverage limits are between latitudes 56° south and 84° north. Its main payload is the multiSpectral instrument.
- Sentinel-3 mission's goal is to measure sea surface topography, sea and land surface temperature, and ocean and land surface color with high accuracy and reliability to support ocean forecasting systems, environmental monitoring, and climate monitoring. Its main payloads are Ocean and Land Color Instrument (OLCI), Sea and Land Surface Temperature Radiometer (SLSTR), SAR Radar Altimeter (SRAL), MicroWave Radiometer (MWR).

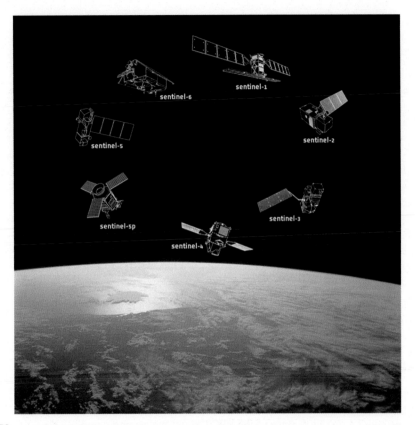

FIGURE 10.52

Sentinel family of satellites.

- Sentinel-4 mission's goal is to monitor key air quality trace gases and aerosols over Europe in support of the Copernicus Atmosphere Monitoring Service (CAMS) at high spatial resolution and with a fast revisit time. It consists of an ultraviolet−visible-near-infrared (UVN) light imaging spectrometer instrument embarked on the Meteosat Third Generation Sounder (MTG-S) satellite.
- Sentinel-5 mission consists of a high resolution spectrometer system operating in the ultraviolet to short wave infrared range with seven different spectral bands to support the Copernicus Atmospheric Monitoring Service.
- The Contributing Missions, which are operated by national, European or international organizations and already provide a wealth of data for Copernicus services.

ESA is responsible for the development of the space segment component of the Copernicus program and operates the Sentinel-1 and Sentinel-2 satellites. ESA will deliver the land mission from Sentinel-3 and will operate Sentinel-5P. EUMETSAT is responsible for operating the Sentinel-3 satellites and delivering the marine mission and will also operate and deliver products from the Sentinel-4, and -5 instruments, and the Sentinel-6 satellites.

As described earlier, Sentinels 1, 2, and partly 3, contribute to the Copernicus Land Monitoring Service in a number of ways:

- Producing data across a wide range of biophysical variables at a global scale (i.e., worldwide), which describe the state of vegetation [e.g., leaf area index (LAI), fraction of green vegetation cover, vegetation condition index], the energy budget (e.g., albedo, land surface temperature, top of canopy reflectance) and the water cycle (e.g., soil water index, water bodies).
- Producing five high-resolution data sets describing the main land cover types: artificial surfaces (e.g., roads and paved areas), forest areas, agricultural areas (grasslands), wetlands, and small water bodies. The pan-European component is also updating the Corine Land Cover dataset to the reference year 2012.
- Providing specific and more detailed information that is complementary to the information obtained through the Pan-European component. It focuses on "hotspots" that are prone to specific environmental challenges. The local component provides detailed land cover and land used information (over major European cities, which are the first type of "hotspots," the so-called Urban Atlas.

In addition to the aforementioned components, the service also supports the generation of a Pan-European digital elevation model (DEM).

10.8 STUDY QUESTIONS

1. Please compare the OLI's spectral bands in Landsat 8 to ETM+ bands in Landsat 7? What are they intended for?
2. Please comment a unique feature of the SPOT satellites, as compared to other optical imagers.
3. The EU Copernicus program consists of five constellations of satellites for different purposes. Which is the constellation that inherits the experience from LANDSAT and SPOT? Which is its main payload?
4. How do floods appear in LANDSAT imagery, and how this can be used to map floods? Which are the limitations of flood mapping using optical imagery?
5. How do floods appear in SAR imagery (e.g., Copernicus Emergency Service using Sentinel-1)? How this can be used to map floods? Which are the limitations of flood mapping using SAR imagery?
6. Surface soil moisture can be measured from space using L-band microwave radiometry, but the spatial resolution is modest. How can it be improved (i.e., the pixel size downscaled)?
7. Surface soil moisture observations using L-band radiometry are hampered by vegetation, which are the perturbing effects, and how can they be corrected?
8. Multispectral optical imagery is used for vegetation monitoring. Which are the bands used to compute the NDVI? How this "observable" is computed and which range of values does it take?
9. The VOD is related to the vegetation water content. Is this parameter somehow related to the NDVI?
10. MODIS sensor on NASA TERRA and AQUA is used—among other applications—for forest fire mapping. How active forest fires are detected?

CRYOSPHERE APPLICATIONS

11.1 INTRODUCTION

Satellite remote sensing of the cryosphere has dramatically changed polar sciences in general. Satellite sensing of these remote and inhospitable regions has opened a whole new world of information on these geographic areas. In the Arctic there was some information from the Vikings back to 870 AD with a more consistent record starting in 1600. Air temperature records extend back to 1880 and can be used as a proxy for sea ice, but even then, these temperatures were only measured at 11 coastal stations around the Arctic with no information on the ocean's interior. In the Antarctic the situation is even bleaker with little or no sea ice data prior to the era of satellite observations. There is some evidence that the catch of whales in the southern hemisphere can be related to sea ice extent. Because whales tend to congregate near the sea ice edge to feed, their locations could be used as a proxy for ice extent.

Initially, satellite remote sensing of polar regions was limited to the use of polar-orbiting weather satellite imagery and to the use of Landsat imagery. These optical data are not optimal for studying the persistently cloudy polar regions. Later, satellites were launched specifically to study the polar regions concentrating on both passive and active microwave sensing (radiometers and radars) that could penetrate the cloud cover. In addition LIDAR sensing optimized for retrieving polar sea ice height, ice sheet topography and snow cover was also later deployed. The long wavelengths of radar and passive microwave radiometer systems penetrate the persistent cloud cover that obscures the polar surfaces during a large part of the year. Together with the continued analysis of operational weather satellite data and Landsat the analysis of satellite data has revolutionized research in the polar regions.

This combination of optical and microwave data is very important; while the optical data are blocked by the persistent cloud cover they do have much greater spatial resolution and can often pick up details that are missed in the analysis of the microwave data with a much poorer spatial resolution. This is only true for the passive microwave systems, because the spatial resolution is determined by the antenna dimensions. Thus, passive microwave sensors generally have spatial resolutions in the tens of kilometers. Active microwave systems can be designed to overcome this spatial resolution problem, and in fact, synthetic aperture radar can produce very high spatial resolutions at the expense of increased data processing.

11.2 POLAR OBSERVATIONS

Before discussing the satellite techniques for observing the cryosphere, we wish to discuss some of the primary observables for this region that set it apart from the rest of the globe. The most distinctive feature of this region of the world is the presence of ice whether as sea ice or on the land as ice sheets or glaciers. Since the Arctic has no land in the center of the ocean it has been covered by sea ice throughout recorded history. Before the satellite era was ushered in about 1970, the nature of this sea ice was known only from expeditions that had crossed it or by submarine measurements that started as far back as 1958. Submarines collected upward-looking sonar profiles that could be converted to sea ice thickness. As these measurements were made for navigation and defense purposes the information was not released until many years had passed. Thus, much of this submarine-based ice thickness measurement data has only come to light since 1985. These ice thickness data were restricted to the tracks of the submarines under the Arctic ice pack.

As a result, one of the very important polar ice parameters became ice thickness, which was related to ice age with older sea ice being much thicker and more resilient than younger ice. In addition, sea ice rejects salt over time and becomes less salty resulting in a higher melting point. A simple two-category ice classification was introduced that discriminated between "first-year" (FY) ice and "multiyear" (MY) ice. Within the designation MY ice the ice is classified as second, third, fourth year ice, etc. An interesting study of Arctic ice age progression over time using satellite data can be found in Fowler et al. (2004). Here a merged set of microwave satellite data, infrared/visible satellite data, and in situ ice buoy data was used to compute ice motion over a 25-year period. For this period, weekly individual ice areas were followed over this duration and ice that exited the Arctic or melted were considered to be FY ice while ice that remained in the Arctic were then classified by the number of years that they persisted.

As an example, we present the Arctic sea ice age map for March 2012 in Fig. 11.1 along with a time series of Arctic sea ice age computed from this same data set. Notice the now well-known decline of older Arctic sea ice over time with a particular dip in 2007 when the Arctic ice cover hit a minimum (Maslanik et al., 2007). This minimum is most clearly marked by an increase in the FY ice cover. There is a modest recovery in ice cover after this minimum, but the ice cover never really reaches its earlier extent. More importantly as the map shows, the central ice pack is no longer made up primarily of older ice, and the older ice have retreated along the northern boundary of the Canadian archipelago. Also clear from this map is that a lot of the older sea ice is exiting the Arctic through Fram Strait along the east coast of Greenland.

The cycle of freeze up and melt can be clearly seen in these data as shown here in Fig. 11.2. In September 2014, there is a large amount of open water particularly along the Siberian and Alaskan coasts. As in earlier years, the thicker ice hugs in the northern limits of the Canadian archipelago and northern Greenland. Moving to March 2015, most of the open water has been replaced by FY ice, but at the same time the band of thicker ice has shrunk and continues to evidence the flow of ice out the Fram Strait. FY ice is mixed in with this outflow of MY ice. In both periods there is a tongue of third-year ice that extends across the central Arctic terminating in a counterclockwise twist.

Perhaps the most important ice parameter has to do with its cover. This is easily observed by passive microwave satellite instruments with the only limiting factor being the poor spatial resolution. This only becomes a problem near the ice boundaries or in discriminating the types of sea ice. See Section 4.3.4.5 for more details.

There are two basic ways to express the total polar sea ice cover: total ice area and ice extent. To estimate the area covered, one estimates the percentage of ice in each pixel, multiply by the pixel area, and total them up. To estimate ice extent, one sets a threshold ice cover percentage and counts every pixel meeting or exceeding that threshold as "ice covered." The common threshold level is 15% sea ice.

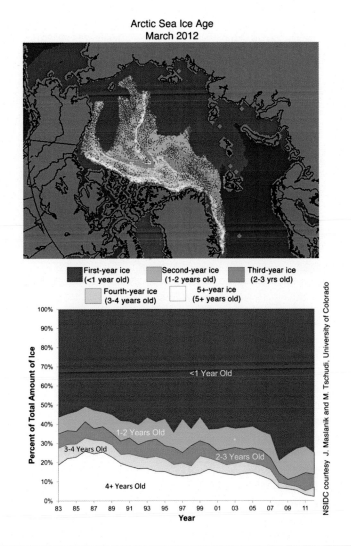

FIGURE 11.1

Ice age data show that the first-year ice made up 75% of the Arctic sea ice cover in March 2012. Thicker multiyear ice used to make up about a quarter of the Arctic sea ice cover. Note it constitutes only 2%.

NSIDC courtesy J. Maslanik and M. Tschudi, University of Colorado.

This threshold method appears to be less accurate, but it has the advantage of being more consistent. In general it is easier to say whether or not a pixel contains 15% ice or more as compared to saying if an ice cover is 70% or 75%. A reduction in the uncertainty of this estimation is necessary to be able to say if and how the ice cover is changing over time.

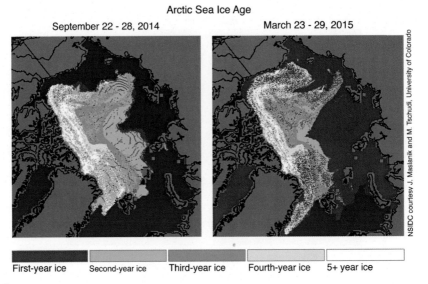

FIGURE 11.2

Arctic sea ice age for September 22–28, 2014, and March 23–29, 2015.

11.2.1 SATELLITE LASER ALTIMETRY

The first satellite dedicated to altimetric measurements of the polar regions was known as ICESat (Ice, Cloud and land Elevation Satellite), which carried as a primary instrument a laser altimeter for accurate measurement of ice sheet elevation over Greenland and Antarctica. These elevation measurements were essential for a proper ice sheet mass balance calculation. In addition, ICESat observed stratospheric cloud property information in polar regions. Over other latitude bands, ICESat was able to measure topography and vegetation type around the globe.

After significant delays ICESat was finally launched on January 12, 2003, into a high-inclination polar orbit. The sole instrument on ICESat was the Geoscience Laser Altimeter System (GLAS), which was initially commanded to turn-on February 20, 2003, using the GLAS-1 laser. GLAS combined a precision surface LIDAR with a sensitive dual-wavelength cloud and aerosol LIDAR. GLAS lasers emit infrared and visible laser pulses at 1064 and 532 nm wavelengths. Over an ICESat orbit, GLAS produces a series of approximately 70-m diameter laser spots that are separated by about 170 m along the spacecraft's ground track. During different phases of the mission, ICESat was placed in different repeat orbits. During the final decommissioning phase the satellite was placed in an 8-day repeat orbit while during its operational life in 2004 the satellite was placed in a 91-day repeat orbit.

GLAS was the first laser-ranging (LIDAR) instrument for continuous global observations of the Earth. GLAS was built by NASA's Goddard Space Flight Center in Greenbelt, MD. It is a facility instrument designed to measure ice sheet topography and associated temporal changes, cloud and atmospheric properties, and gives us information on the height and thickness of radiatively important cloud layers, which are needed to improve short-term weather prediction. GLAS comprises a laser system to measure distance, a global positioning system (GPS) receiver, and a star-tracker attitude determination system. The laser transmits short pulses (4 ns) of infrared light (1064 nm) and visible green light (532 nm). Photons are reflected back to the spacecraft from the surface of the Earth and from clouds. These signals are collected in a 1-m diameter telescope (Fig. 11.3). Laser pulses at 40 times per second illuminate spots (footprints) 70 m in diameter, spaced 170 m apart.

ICESat was designed to operate for 3–5 years. Ground testing indicated that each GLAS laser would last for 2 years, which meant that ICESat would have to carry three lasers to complete the mission. During the initial on-orbit test operation, a pump diode module on the first GLAS laser failed prematurely on March 29, 2003. An investigation indicated that corrosive degradation of the pump diodes, due to an unexpected, but known reaction between indium solder and gold bonding wires had resulted in this reduced life of the lasers. As a result the overall operational life of ICESat was reduced to just over a year. After 2 months of full operation in the fall of 2003, it was decided to operate the GLAS for 1 month out of every 3–6 months in an effort to extend the life of the satellite, which would be particularly important for changes in ice sheet elevation.

The last of the three lasers failed on October 11, 2009, and since following attempts to restart also failed the satellite was retired from service in February 2010. Between June 23, 2010, and July 14, 2010, the spacecraft was maneuvered into a lower orbit to speed up orbital decay. On August 14, 2010, the satellite was decommissioned and on 9:00 UTC on August 30, 2010, the satellite reentered the atmosphere and burned up. So by a clever operational schedule, NASA was able to squeeze out 7 years of data collection from the ICESat satellite.

FIGURE 11.3

Geoscience Laser Altimeter System Instrument on ICESat.

FIGURE 11.4

Artist view of CryoSat.

From ESA. http://www.esa.int/spaceinimages/Images/2009/09/CryoSat_seen_from_underneath.

11.2.2 SATELLITE RADAR ALTIMETRY

More recently satellite radar altimetry has been utilized to map precise changes in the thickness of the polar ice sheets and floating sea ice. Launched April 08, 2010, the European Space Agency's CryoSat-2 (Fig. 11.4) satellite carries a synthetic aperture radar (SAR)/interferometric radar altimeter (SIRAL), which has capabilities to measure ice sheet elevation and sea ice freeboard. To make this possible the satellite also carries three star trackers for measuring the precise orientation of the satellite. This orientation capability adds to orbital positioning accuracy provided by the French Doppler Orbit and Radiopositioning Integrated by Satellite (DORIS) system and laser ranging from an onboard reflector.

The satellite operates in a 717 km orbit that reaches to 88°N allowing it to map both sea ice and land covering ice sheets. Unlike most satellites, CryoSat does not have any deployable solar panels. Instead the solar panels are rigidly fixed to the satellite's body forming a "roof" with a carefully optimized angle designed to capture adequate power under all orbital conditions. In fact the satellite has no moving parts except for some valves in the propulsion system. This design has resulted in considerable cost savings in the construction of the satellite. Designed with a nominal 3.5-year lifetime CryoSat continues to operate properly well past this time.

11.3 SEA ICE

One of the most dramatic signs of global climate change has been seen in the extent and thickness changes in Arctic sea ice. In the past, typically 15 million square kilometer of the Arctic Ocean was covered by sea ice in winter. Satellite records have shown a constant downward trend in Arctic sea ice during all seasons. There was a distinct minimum in 2007, but that was exceeded by the Arctic ice cover minimum in 2012 when CryoSat-2 measured 6000 cubic kilometer in October 2012. It is interesting that CryoSat-2 observed an increase in Arctic sea ice cover up to 9000 cubic kilometer in October 2013. The autumn Arctic sea ice thickness from CryoSat-2 is presented here in Fig. 11.5.

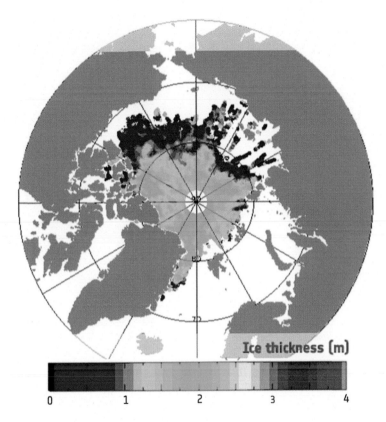

FIGURE 11.5

Autumn sea ice thickness from CryoSat 2010–13.

    Seasonal changes of sea ice have a significant influence on the global ocean circulation pattern in terms of the thermohaline circulation. As ice melts it introduces freshwater into the surrounding ocean reducing salinity and the density of the seawater. Conversely as sea ice forms under cooling conditions the salinity and density of the surface water (now being changed to ice) increase. This density increase causes the seawater to sink driving deep ocean currents toward the equator and away from the polar regions. Continuity requires a return flow of warmers, less dense surface northward from the lower latitudes. In the North Atlantic this current is the Gulf Stream, which carries warm surface water northward from the Gulf of Mexico to the subpolar regions east of Greenland. This flow of warm water is very important in moderating the climate in Western Europe. Fig. 11.6 shows the thermohaline circulation where the blue lines are the cold waters, and the red lines the warm waters.

11.4 **ICE SHEETS**

One of the biggest contributions to global sea level rise is the melting of land-fast ice sheets on Antarctica and Greenland in addition to the rapidly shrinking glaciers on the continental landmasses. It is widely accepted that after the year 2000, ice sheets are melting at their base. Between 1992 and 2012, the melting of the Antarctic and Greenland ice sheets alone added 11.1 mm to global sea levels,

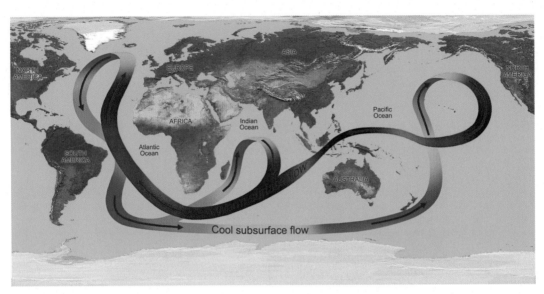

FIGURE 11.6

Schematic global oceanic thermohaline circulation: deep cold water flow (blue) and warm surface flow (red).

From Hogan, C., 2012. Thermohaline Circulation. Retrieved from: http://editors.eol.org/eoearth/wiki/thermohaline_circulation.

which is 20% of the overall sea level rise for that same period. CryoSat observations show that between 2010 and 2013, West Antarctica, East Antarctica, and the Antarctic Peninsula lost 134, 3, and 23 billion tons of ice each year, respectively. These recently measured losses from Antarctica alone (Fig. 11.7) are enough to raise global sea level by 0.45 mm each year. The improvement in resolution

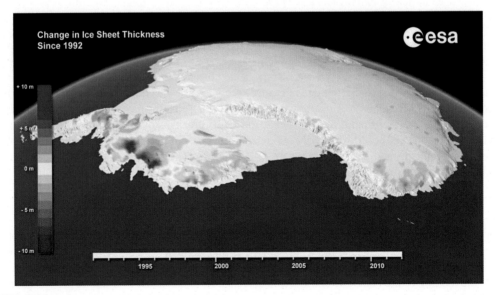

FIGURE 11.7

Antarctic ice-sheet loss.

of the CryoSat radar over that of its pulse-limited predecessors, coupled with its interferometric capability, has made spatially and temporally continuous measurements of the ice sheet margins possible for the first time.

11.5 CRYOSAT INSTRUMENTS

CryoSat's primary instrument is the SIRAL. It was designed to satisfy the requirements for both ice sheet and sea ice freeboard (height of the sea ice protruding from the water) measurements. Conventional radar altimeters send pulses at intervals long enough that the echoes are uncorrelated so that many echoes can be averaged to reduce noise. At a typical satellite orbital speed of 7 km/s, the interval between pulses is about 500 µs. By comparison, the SIRAL sends a burst of pulses at an interval of only 50 µs. The returning echoes are correlated and by treating the whole burst together, the data processor can separate the echoes into strips arranged across the track by exploiting the frequency shifts, caused by the Doppler effect, in the fore and aft looking parts of the radar beam. Each strip is about 250 m wide and the interval between bursts is arranged so that the satellite moves forward by 250 m each time. The strips laid down by successive bursts can therefore be superimposed on each other and averaged to reduce the instrument noise. This mode of operation is known as SAR mode (Wingham et al., 2006).

To measure the arrival angle, a second antenna receives the radar echo simultaneously. When the echo comes from a point not directly beneath the satellite, there is a difference in the path length of the radar signal, which is then measured. Geometry then provides the angle between the baseline joining the two antennas and the echo direction. In addition to the altimeter signal, a precise knowledge of the orientation of the baseline is required for the SAR height retrieval. CryoSat measures this baseline orientation using three star trackers mounted on the antenna support structure, which collects five pictures per second for comparison with the onboard star catalog.

This allows an accurate SIRAL antenna orientation, but there is also a need for accurate orbital height information for the accuracy of the altimeter height measurement. To achieve this level of orbital accuracy, CryoSat carries two systems: the DORIS and a laser retroreflector. DORIS detects and measures the Doppler shift in radio signals broadcast from a network of more than 50 radio beacons around the world. While the highest positional accuracy is obtained only after ground processing, it also provides a real-time, onboard estimate. A French system developed for the operation of radar altimeters, DORIS has been successfully deployed on a number of altimetric satellites. The laser retroreflector has several corner tetrahedrals, which reflect light in exactly the direction it came from. A global network of laser tracking stations fire short laser pulses at the satellite and times the interval before the pulse arrives back at the ground station. These two systems provide independent measurements of orbital height accurate to about 2.6 cm.

The design minimizes the effect of thermal distortion when the antennas are used as an interferometer. The bench and waveguide materials provide a very low thermal expansion coefficient and the radio frequency unit (RFU) location minimizes the waveguide length. Heat from the RFU and the digital processing unit (DPU) is dissipated by the radiator that forms the very front of the spacecraft. In flight, the entire system is wrapped in multilayer insulation, while thin sheets of germanium-coated kapton cover the antenna reflectors to reduce the effect of asymmetric solar heating.

The use of the antennas on reception, the timing of the transmitted chirps, and the transmitted bandwidth depend on the operating "mode." There are three operating modes: "low resolution" mode (LRM), "synthetic aperture" mode (SARM), and "synthetic aperture interferometric" mode (SARInM). LRM provides for conventional, pulse-limited altimetry using a single antenna. SARM provides for along-track aperture synthesis using a single antenna. SARInM allows for along-track aperture synthesis using two antennas and for phase comparison (interferometry) between the echoes received on each antenna. The SARM and SARInM are collectively referred to as high-bit-rate modes.

In LRM, the instrument operates with a pulse-repetition frequency (PRF) of 1.971 kHz. In LRM, 128, 8-bit samples are formed at a sampling rate of 0.35 μs (only 44.8 μs of the echo is recorded, which reduces the measurement bandwidth to 320 MHz). A complex spectrum sampled at 22.3 kHz is formed from these samples via a fast Fourier transform (FFT), and the spectrum is power detected. Ninety-one consecutive echoes are accumulated (to reduce speckle noise) and passed to the logic (the tracker) that provides closed-loop control of the timing of the deramp chirp and the receiver gain. These same, averaged "spectral domain" echo powers also form the measurement (i.e., the telemetered) data of the LRM mode. With these parameters, the LRM measurement and tracking spectrum spans a "range window" of 60 m with a range resolution (sampling interval) of 0.46875 m.

These LRM measurements are those of a conventional pulse-limited altimeter, except that (for reasons of launcher accommodation) the antennas are slightly narrowed in the along-track direction, resulting in a slightly asymmetric antenna pattern. At the nominal altitude the LRM signal-to-noise ratio (SNR) is 8 dB when the system is operated over a uniform spherical surface of an Earth radius, and backscattering coefficient of −10 dB. The antennas are linearly polarized with an orientation parallel to the interferometer baseline.

In SARM, the same receive chain is used, but the mode differs in the timing of its transmissions and the formation of the measurement data. The pulses are transmitted in groups "bursts," of 64 pulses with a PRF of 18.182 kHz. During a burst, the carrier phase is locked to the transmission timing so that the transmissions within the burst are coherent. Each burst is transmitted with a burst-repetition frequency (BRF) of 87.5 Hz. The length of the burst, 3.6 ms, and the interval between the bursts, 11.7 ms, are sufficient for the echoes from a transmitted burst to be sent via the duplexer to the receive data before the next burst is transmitted (a "closed" burst arrangement).

During the reception of a burst, the timing of the deramp chirp is held constant (to avoid introducing differential phase shifts between the transmissions of the burst). As in the LRM mode, echoes are power detected and accumulated (over 46.7 ms) to provide closed-loop control. However, in contrast to the LRM, the measurement data comprise the 128 "time domain" samples of each individual echo, directly from the A/D converters. As with the LRM, the SARM tracking spectrum spans a range window of 60 m with a range resolution of 0.469 m, as will the measurement spectrum, once the data are processed on the ground.

For the SARInM mode, a 350 MHz, 64 pulses, 18.182 kHz PRF burst is again transmitted, but with a lower BRF of 21.4 Hz. The echoes are directed by the duplexer to the two receive chains. The timing of the deramp chirp is identical for each chain (so as not to introduce differential phases between the two receive chains). In contrast to SARM, the sampling interval at the A/D converters is reduced to 0.0875 μs and 512, 8-bit samples are generated for each receive channel. The SARInM measurement data comprise these time-domain samples of each individual echo from each receive channel, directly from the A/D converters. The decrease of sampling interval means that the SARInM spectrum spans a range window of 240 m with a range resolution of 0.469 m, once the data are processed on the ground.

In contrast to the LRM and SARM modes, the SARInM measurement data are not used to inform the closed-loop control of the instrument. Instead, the longer interval (46.7 ms) between the measurement bursts is employed to transmit 40 MHz bandwidth pulses. The echoes from these pulses are received on a single channel, sampled at 0.35 μs, passed through the FFT, power detected and accumulated over the

46.7 s interval. The resulting average, spectral-domain power provides for the closed-loop control of the instrument. This spectrum spans a range window of 480 m with a resolution of 3.75 m. This arrangement provides the SARInM with a tracking range window larger than that of the measurement range window (480 vs. 240 m) to provide for robust closed-loop control over various regions of the Earth, such as significant topographic changes.

SIRAL also has an acquisition mode that is used to initialize the closed-loop control and two calibration modes. One mode calibrates the DPU while the other calibrates the RFU for the deramp chirp timing and the receiver gain, intraburst phase rotations (which provide phase calibration to the aperture synthesis) and the SARInM phase differences as functions of frequency and automatic gain control (AGC) setting. A 64-pulse burst is directed through an attenuating connection between the transmission and receive chains to achieve this. The deramp chirp frequency is offset to allow the calibration to be made at 11 frequencies lying within the measurement spectrum. The other calibration mode provides detailed corrections for the variations of receiver gain across the measurement spectrum. It is implemented by averaging repeated measurement of the noise power in the absence of a transmission. Finally a second calibration path is included to provide at a single frequency a SARInM phase difference calibration that includes the duplexer. This calibration is not a separate mode, but is performed with a repetition frequency of 1 Hz from within the SARInM. It provides a correction for phase difference as a function of time; the total phase difference correction (a function of frequency, AGC setting and tie) is obtained by combing all of these calibration procedures into a single value.

11.5.1 CRYOSAT ORBIT

CryoSat operates from a near circular, near polar orbit with an average altitude of 717.2 km, and an eccentricity of 0.0014. The orbit inclination is 92 degrees, which is a compromise between the desire to achieve a high density of orbit crossovers at high latitudes (for land/ice altimetry), while having more-or-less complete coverage of the Arctic Ocean and the Antarctic continent. The repeat period is 369 days with a 7.5 km intertrack spacing at the Equator, which provides a high orbit crossover density (10 crossovers/km^2 per year at 87 degrees). The orbit also has a 30-day subcycle, which provides uniform coverage of the Arctic sea ice every 30 days.

The attitude of the platform and the SIRAL instrument is measured with three independent star trackers. The trackers are mounted on the zenith side of the SIRAL optical bench to provide as accurately as possible the attitude of the bench and hence the SIRAL instrument. Two other attitude sensors are mounted on the platform: a coarse Earth–Sun sensor and two three-axis magnetometers (which can provide attitude and attitude rate change relative to the geomagnetic field). Satellite attitude control is provided by three magnetotorquers, which provide torques about axes orthogonal to the direction of the magnetic field. In addition, there are 16, 10 mN attitude control thrusters, fed by the cold gas system. The cold gas system is also responsible for the initial acquisition of the orbit following launch, and for other orbit maneuvers such as collision avoidance. Because CryoSat is limited to a cold gas system, the usage of the fuel is an important factor in the total lifetime of the mission.

11.5.2 CRYOSAT ERROR BUDGET

Errors that arise from CryoSat measurements may be classified into one of three kinds. First there are those that arise from the measurement system itself (e_{1b}). Fluctuations arising from radar speckle give us an example. Second, errors arise because they are being used to transform the point measured radar

Table 11.1 Instrument System Contributions to the Elevation Errors

	SARInM	SARM	LRM
Orbit error (cm)		6	
Range error (cm)	14	10	7
Angle error (cm)	1.5		
RMS total	15	11.6	9.2

LRM, *low resolution mode*; RMS, *root mean square*; SARInM, *synthetic aperture interferometric mode*; SARM, *synthetic aperture mode*.

data to point geophysical quantities, such as elevation (e_2). Finally, errors (e_{h1}) arise in higher-level products when forming local averages due to the sampling. Errors e_{1b} and e_2 describe errors of commission and e_{h1} errors of omission. A summary of the overall errors is given here in Table 11.1.

These errors are the "single-shot" contributions associated with the ~ 20 Hz measurements. It is of interest to compare the performance of Table 11.1 with conventional ocean, pulse-limited altimetry, for which a 2-cm range precision of a 1-s average is typical. Were the SARM used for this purpose, the equivalent figure for the range precision is 2.6 cm, which is about the same level of performance.

It must be remembered that CryoSat has difficulty measuring sea ice concentration in very thin sea ice and this represents a fundamental limitation to this measurement system. As a result, it is best to combine as many different satellite systems as possible to estimate the MY and FY sea ice concentrations and extents.

11.6 USING SCATTEROMETRY TO COMPUTE SEA ICE CONCENTRATION AND DRIFT

Cryospheric applications of scatterometry include mapping of global snow cover, sea ice extent, sea ice motion, classification of sea ice types, estimate of snow accumulation, and deriving the direction of the wind patterns. Most of these applications have to do with sea ice and sea ice—derived properties. Application to geographic regions range from Greenland to Antarctica.

Dr. David Long and his students at Brigham Young University (BYU) have used data from the operational satellite scatterometers to estimate sea ice concentration and to classify both FY and MY sea ice. As compared to the regional ice charts produced by the Canadian Ice Service, the difference is found to be only 6% for the Arctic winters of 2006—07 and 2007—08. For these periods, it was found that microwave backscatter produces more temporally stable results than passive microwave brightness temperature introduced earlier in this chapter. In other studies, Kwok (2004) used Ku-band backscatter from the SeaWinds instrument on QuikSCAT to classify Arctic FY and MY sea ice. In a separate study, Nghiem et al. (2006) classified Arctic sea ice as seasonal (FY), perennial (MY), and mixed.

The BYU method uses a seasonally varying threshold to discriminate between FY and MY sea ice during winter. Similar to the previous studies, backscatter measurements from the SeaWinds instrument aboard the QuikSCAT satellite are used to classify Arctic sea ice as either FY or MY. This classification is derived from observed trends in SeaWinds measurements over a period of 7 years. The resulting classification indicates that the coverage of MY ice has reduced from year to year relative to the FY ice between 2003 and 2009 as it has been reported by other studies.

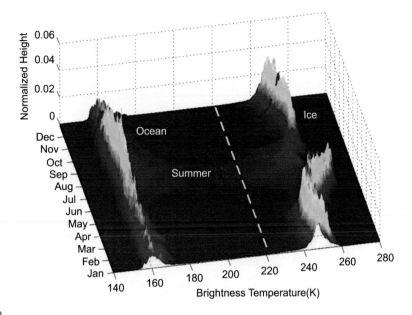

FIGURE 11.8

Daily time series of histograms of AMSR-E 6-GHz V brightness temperatures over the Arctic during 2004. Histograms are normalized and exclude measurements over land. The *dashed line* represents a threshold separating modes representative of ocean and ice. The summer melt period is indicated by the solid *red line*.

QuikSCAT measurements were processed by the BYU group using the scatterometer image reconstruction (SIR) algorithm. The resulting images have a nominal pixel resolution of 4.45 km with an estimated effective resolution of ~8–10 km. The daily SIR images are produced in a polar stereographic map projection. To derive ice classifications a mask is applied to remove scatterometer measurements over ocean and low sea ice concentrations where water dominates. Data from the advanced microwave scanning radiometer of NASA's Earth Observing System (AMSR-E) were selected for this purpose, which provided an independent compliment to the QuikSCAT data. AMSR-E provides measurements over six frequencies ranging from 6.9 to 89.0 GHz. Data are collected for both horizontal and vertical polarization resulting in a total of 12 channels of passive microwave radiometer data. The AMSR-E overlay allowed them to also exclude data from the marginal ice zone (MIZ) where a mixture of sea ice and open water result in confused backscatter signals. Also in the MIZ high winds influence the backscatter signal making it difficult to isolate high sea ice concentrations.

Daily time series of histograms of AMSR-E 6-GHz V brightness temperatures over the Arctic during 2004 (Fig. 11.8) clearly demonstrates how ice trends can be separated from open-ocean throughout the year.

To visualize the more detailed information derived from the QuikSCAT backscatter data, they plotted a series of histograms for the winter of 2006–07 where sea ice was selected out for analysis (Fig. 11.9).

These winter histograms have one large peak at about −20 dB, which represents the FY ice and another peak at about −10 dB, which corresponds to the presence of MY sea ice. Clearly the fraction of MY ice is far smaller than that for FY ice.

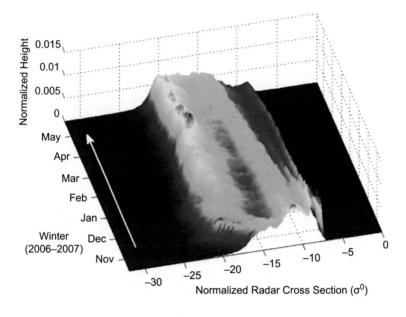

FIGURE 11.9

Temporal series of histograms of QuikSCAT VV backscatter over Arctic sea ice during the 2006—07 winter. The *white arrow* indicates the temporal progression.

Using the backscatter data, trends of FY and MY Arctic sea ice were calculated as presented here in Fig. 11.10. Here it can be seen that the backscatter signature of FY ice in combination with its snow cover is seasonally dependent. This pattern approximately describes the behavior of FY ice backscatter for every year between 2003 and 2009. These observations suggest that FY ice backscatter quickly moves to lower values between September and mid-November. It decreases slightly during December and January and then gradually moves to higher values until mid-March. After March, the backscatter moves to lower values until June. At that point it is difficult to distinguish the ice types by backscatter until late September due to ice melting. After September, new ice forms changing the amount of FY ice present. The backscatter of FY ice in Fig. 11.10A seems to stabilize by mid-November. The high backscatter of FY ice just prior to November may be partially explained by the presence of frost flowers, which are formed by the deposition of ice directly from the vapor phase. FY ice covered with frost flowers can have backscatter brightness temperatures that are similar to those seen for MY ice.

Trends in MY ice show a move to lower backscatter in the spring into summer with a rapid drop in the summer melt period. Swan and Long (2012) suggest that this drop in backscatter may indicate the presence of snow cover during the fall and spring periods.

The trends observed in sea ice backscatter for FY and MY ice appear to be interannually consistent. Each year, there is a clear separation of FY ice and MY ice. These differences suggest a method to discriminate the FY/MY classification using an average of the yearly distributions. Looking at the winter backscatter, there is a minimum region that separates the MY and FY ice maxima.

A limitation of this classification method is that MY ice can sometimes look like FY ice. Comiso (1990) suggested that ice floes near the MIZ that survive the summer melt, often have passive

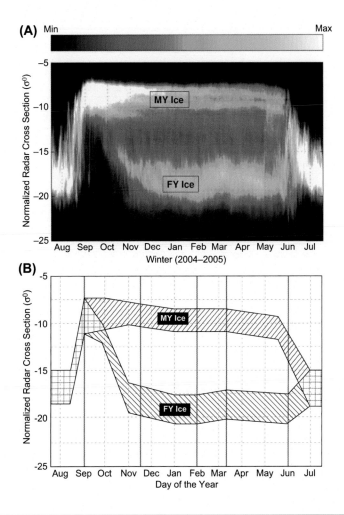

FIGURE 11.10

Seasonal trend in backscatter over first year (FY) and multiyear (MY) ice. (A) Time series of histograms of backscatter over Arctic sea ice for the winter of 2004—05. (B) Heuristic diagram showing approximate (nonstatistical) ranges of backscatter for FY and MY ice for each year between 2003 and 2009. The range of backscatter for each ice type appears to be seasonally dependent.

microwave signatures similar to those of FY ice. He suggests that this is due to the intrusion of seawater into the snow—ice interface during the summer.

11.7 THIN ICE THICKNESS ESTIMATION

Until present, the estimation of sea ice thickness has been limited by the capability of the electromagnetic waves used to penetrate into the ice. Recently, the new radiometers operating at the

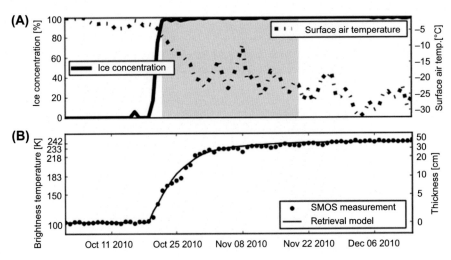

FIGURE 11.11

(A) Time series of AMSR-E ice concentration and National Centers for Environmental Prediction surface air temperature, and (B) Soil Moisture and Ocean Salinity (SMOS) observed and modeled brightness temperature with the corresponding ice thickness at 77.5°N, 137.5°E. The particular ice growth period discussed in the text is indicated in gray. The modeled brightness temperature is based on an exponential model and the ice thickness from Lebedev's growth parameterization.

From Kaleschke, L., Tian-Kunze, X., Maaß, N., Mäkynen, M., Drusch M., 2012. Sea ice thickness retrieval from SMOS brightness temperatures during the Arctic freeze-up period. Geophys. Res. Lett. 39 (L05501). http://dx.doi.org/10.1029/2012GL050916.

lowest frequency bands (e.g., ESA's Soil Moisture and Ocean Salinity mission at L-band) are capable of "seeing through" the ice layer, provided it is not too thick (<50 cm), and therefore are sensitivity to the sea surface emission below the ice layer as shown in Fig. 11.11B. At this frequency, for "thin" ice, the brightness temperature is sensitive to the seawater emission, and to the ice layer emission as well, following approximately the Lambert—Beer law:

$$I(x) = 234.1 \text{ K} + 133.9 \text{ K} \cdot e^{-x/12.7 \text{ cm}}. \tag{11.1}$$

where x is the ice thickness. As can be clearly seen in Fig. 11.12, the intensity $I = (T_v + T_h)/2$ (Fig. 11.12, bottom) increases with increasing ice thickness and saturates around ~50—60 cm.

Using an evolution of the Kaleschke et al. (2010) algorithm, Huntemann et al. (2014) used this variation of the intensity and the polarization difference $(T_v - T_h)$ (Fig. 11.12, top) to estimate maps of sea ice thickness, as shown in Fig. 11.13. Note that these maps do not have to match perfectly the sea ice cover maps derived by other sensors, or by microwave radiometers at 19 and 37 GHz, because of the different spatial resolution of the instruments, and due to the fact that they are sensing different aspects of the same phenomena (i.e., at 19 and 37 GHz, even a thin ice is classified as fully covered by ice, while this is not the case at lower frequencies).

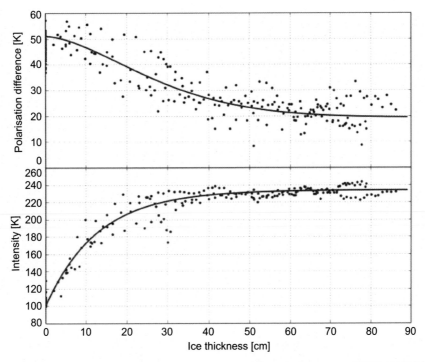

FIGURE 11.12

(Top) Dependence of polarization difference ($T_v - T_h$) and (bottom) intensity (($T_v + T_h$)/2) to the ice thicknesses obtained from the daily average temperature below the freezing point of seawater ($-1.8°C$) from the National Centers for Environmental Prediction and National Center for Atmospheric Research surface air temperature data. *Red line* shows a fit of an empirical function of polarization difference and intensity to ice thickness.

11.8 MULTIYEAR ARCTIC SEA ICE CLASSIFICATION USING OSCAT AND QUIKSCAT

Lindell and Long (2016) used these same methods with multiple scatterometers from the Oceansat-2 (OSCAT) together with NSCAT data from QuikSCAT to classify Arctic sea ice during the years 1999–2014. The idea is to combine the coverage by QuikSCAT for the period 1999–2009 and then to extend that with OSCAT on to 2014. The result is a 15-year record of scatterometer-based sea MY and FY sea ice classification. They used the NASA team sea ice concentration algorithm applied to passive microwave brightness temperatures to discriminate between sea ice and open water since these targets have similar backscatter properties. Over the 15-year period the MY ice fraction shows a systematic decrease, while the total area of sea ice cover remains constant throughout the winter seasons. Thus, FY ice is making up the area lost by the MY ice.

This progression is best seen in the winter maps of FY and MY Arctic sea ice in Fig. 11.14 where the MY ice is shown in white and the FY ice is shown as light blue.

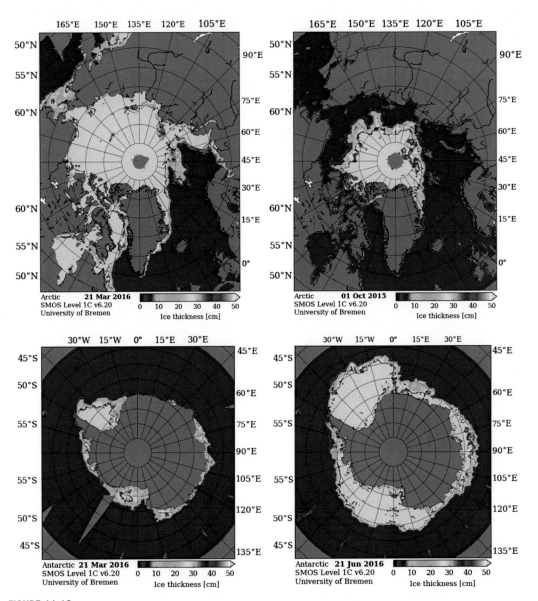

FIGURE 11.13

Sea ice thickness computed from Soil Moisture and Ocean Salinity (SMOS) vertical and horizontal brightness temperatures, and incidence angles from 40 to 50 degrees. (Top) Arctic, (bottom) Antarctic (http://www.iup. uni-bremen.de:8084/smos/).

From Huntemann, Heygster, Kaleschke, Krumpen, Makynen, Drusch, 2014. Empirical sea ice thickness retrieval during the freeze-up period from SMOS high incident angle observations. The Cryosphere 8 (2), 439–451.

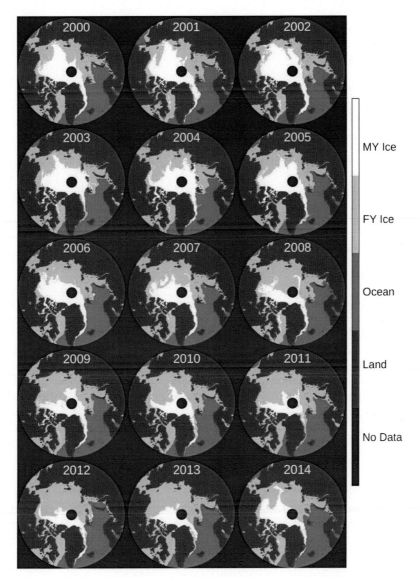

FIGURE 11.14

Winter Arctic sea ice cover for multiyear (MY) (white) and first-year (FY) (light blue) sea ice.

Here the decrease in MY ice appears to be due to the advection of MY ice into the Atlantic Ocean occurring primarily through Fram Strait. In 2014 MY ice recovers slightly and has almost reached the coverage levels of 2005 and 2006.

Unfortunately, the subsequent winter saw an additional decrease in MY sea ice cover as reported by the US National Snow and Ice Data Center in Fig. 11.15.

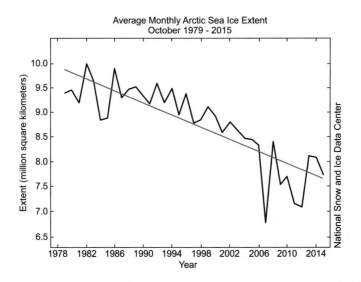

FIGURE 11.15

Average Arctic monthly sea ice extent estimated by the US National Snow and Ice Data Center, Boulder, CO.

As illustrated in Section 6.4.3.5, GNSS-R techniques also offer the potential to map sea ice extent, or to monitor subsurface ice layers down to ~200 m, especially in very dry conditions, where the penetration depth is larger. The recent launch of the UK TDS-1 satellite with an experimental GNSS-R payload onboard has demonstrated the capabilities of this technique, and it has also shown that the products at different frequencies "see" different things. Fig. 11.16 shows the AMSR2 sea ice concentration (SIC) maps (left column, data grid 6.25 km) derived using the ASI algorithm applied over AMSR2 data, based on the polarization difference of the 89 GHz channel, and the OSI-SAF SIC maps (right column, data grid 12.5 × 12.5 km) generated using the 19, 37, and 91 GHz of the SSMI/S sensor. Stripes correspond to the ground tracks of the specular reflection points of the GPS signals collected by TDS-1 (size of specular reflection zone blurred over the 1 s integration time: ~300−400 m × 6.5 km). Fig. 11.16A corresponds to November 15, 2014, and Fig. 11.16B to January 27, 2015. Note that, although there is a general agreement between the GNSS-R data and the well-accepted ASI AMSR2 and OSI-SAF algorithms, in the transition zones there are serious disagreements. This is especially true in Antarctica on January 27, 2015, where there are large tracks in pink detected as 100% ice, while the passive microwave products classify them as 0% ice. This may be due to the different spatial resolutions of both techniques and the different electromagnetic frequencies.

11.8.1 GREENLAND ICE SHEET

A primary concern in present day Earth science research is the question of sea level rise. It is presently not well known if the Greenland and Antarctic ice sheets are shrinking or growing overall. Greenland alone contains enough land ice to raise the global sea level by 5 m if it all melted. The larger ice sheets of Antarctica contain enough land ice to raise global sea level by 70 m if melted. The harsh climate of these regions makes in situ measurement campaigns difficult and this adverse weather also influences satellite measurements of mass balance in these areas difficult.

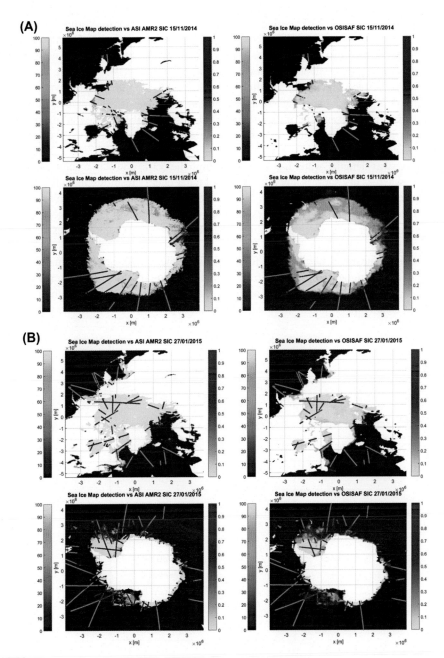

FIGURE 11.16

Sea ice concentration (SIC) maps computed from AMSR2 (left column) and OSI-SAF (right column) for (A) November 15, 2014, and (B) January 27, 2015. Stripes correspond to TDS-1 GNSS-R ground tracks (Alonso-Arroyo et al., 2016).

Scatterometry has a potential of helping to answer these questions. Monitoring snowmelt is critical to assessing sea level rise both for its direct contribution and also because it indirectly increases glacial discharge into the sea that also contributes to sea level rise (Zwally et al., 2002). Snowmelt percolates down to the base of the Greenland ice sheet, providing a lubricant between the ice sheet and the underlying ground surface that initiates sliding and accelerates the flow of ice toward the margins of the ice sheet into the surrounding ocean. Scatterometry is useful for measuring the extent of snowmelt due to its extreme sensitivity to the presence of liquid water. As it can be seen in Fig. 11.17 σ_0 decreases rapidly with a very small increase in snow wetness, dropping as much as 10 dB with only a 1% increase in wetness.

At microwave wavelengths, water has a relatively high dielectric constant resulting in the absorption of radar pulses, whereas dry snow is a good scatterer of radar energy resulting in higher σ_0 values. Thus, sudden decreases of backscatter have been used to estimate the amount of snowmelt (Wismann, 2000; Wismann and Boehnke, 1997). These studies have shown that detection of snowmelt with spaceborne scatterometry correlates well with the timing of in situ near surface temperatures rising above 0°C (Wismann, 2000).

The other important snow characteristic for ice mass balance is the snow accumulation. It is not yet certain whether or not higher temperatures from global warming might actually increase precipitation over Greenland, leading to a higher snow accumulation and hence a growth in the ice sheet. Snow accumulation is usually quantified as snow water equivalent (SWE), which is the height of a centimeter

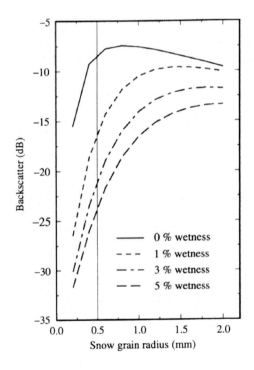

FIGURE 11.17

Ku-band backscatter (VV) at 54 degrees for different snow wetness (Nghiem et al., 2001).

of liquid water contained in a vertical snow column of 1 cm^2. Any relationship between backscatter and SWE must be empirically modeled. This is because the density and grain size vary with time and are not routinely measured over the entire ice sheet. It is expected that SWE and backscatter are inversely related: as the snow layer grows deeper, less of the incident radar energy penetrates down to lower layers penetrating to the high radar reflectivity of the firm resulting in lower σ_0. In addition, an increase in the snow layer contributes more to the volume scattering nature of the snow masking the higher radar reflectivity of the underlying firn resulting in lower σ_0.

Finally, scatterometer data have been used to map the different facies of the Greenland ice sheet. Greenland is conceptually divided into four different facies: the dry zone or accumulation zone which does not experience any melt throughout the year; the wet snow or percolation zone, which seasonally experience scattered snowmelt that eventually percolates into the underlying firn to be later refrozen into subsurface ice lenses; the saturated snow zone, which seasonally experiences enough melt to saturate the entire snow surface with melt water; and the ablation zone, which completely melts the snow cover each summer, creating numerous surface ponds and lakes and exposes the underlying ice elsewhere. These four facies are illustrated here in Fig. 11.18.

The backscatter of these facies is distinct enough to allow them to be mapped well using scatterometry. As mentioned earlier, snow tends to decrease σ_0 as it masks the more intense backscatter from the underlying firn. Thus, dry snow typically has a lower mean σ_0 as compared to the nearby percolation zone, which has less snow cover and relatively high backscatter from buried ice lenses and the sub-surface firn layer. This is particularly true during the late fall and early winter seasons when water is still present in this zone. The saturated and ablation zones look similar to the percolation zone in winter, but they decrease significantly in σ_0 in summer as the snowmelt begins. Hence the separation between the dry snow zone and the percolation zone is most visible in scatterometer images from late fall and early winter as shown here in Fig. 11.19. Some studies have used this mapping technique to map Greenland facies that show the retreat of some parts of the dry snow line toward the higher latitudes and increases in the summer melt extent (Drinkwater et al., 2001; Drinkwater and Long, 1998).

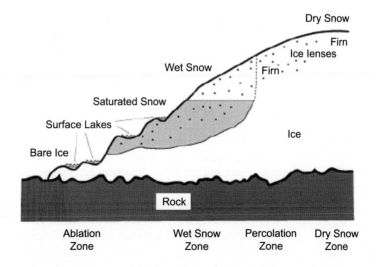

FIGURE 11.18

The four facies of the Greenland ice sheet (Drinkwater et al., 2001).

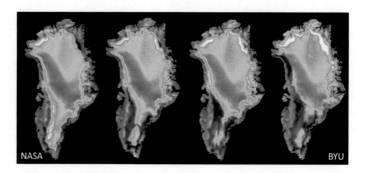

FIGURE 11.19

QuikSCAT backscatter imager over Greenland for days 203, 208, 213, and 218 in 1999 (from left to right), blue and white colors indicate dry surfaces while red and black indicate wet snow surfaces undergoing melt (NASA, 2017).

11.8.2 SEA ICE CONCENTRATION AND ICE MOTION

Sea ice is important to climate change due to its high albedo, which deflect solar radiation and prevents a significant amount of heat from the ocean warming the polar atmosphere. Thus, as the Arctic becomes ice-free, this albedo effect is sharply reduced. The thickness of sea ice also determines the amount of insulation it provides. Once again scatterometry responds to the surface roughness and changes in surface dielectric, which influences the backscatter amount.

The basic detection of sea ice is based on the distinct difference between the backscatter from open water and sea ice. The high backscatter from sea ice is so different that it is possible to set a limit in backscatter beyond which one can clearly say the target is sea ice. Classifying sea ice based on its thickness is far more difficult. Sea ice is usually classified by its age and can be lumped into the broad categories of MY ice (has persisted beyond a single year and you can have second year, third year, etc.) and FY ice, which will form and melt each year.

MY ice is usually very rough at its surface due to wind and wave deterioration over time while FY ice is generally smooth with some rough areas due to ridging and rafting when ice floes collide. As surface roughness increases the backscatter, response becomes more isotropic and the B value in the scatterometer geophysical model becomes greater. MY ice has the highest B values, while ridged FY ice has intermediate B values and smooth FY ice has the lowest B values. MY and FY ice differ greatly in their salinity. Ice formation forms brine pockets and this brine gradually leaks out of the ice thus making the MY ice fresher than the newly frozen FY ice. Since salinity increases the surface dielectric the ice increases it absorption of the incident microwave energy resulting in lower σ_0 values.

Sea ice concentration can also be estimated using the emissivities at 19 and 37 GHz at vertical polarization (e_{19V}, e_{37V}). The method to quantify the iced sea surface percentage consists of measuring the distance a given (e_{19V}, e_{37V}) point to the point ($e_{19V,ice}$, $e_{37V,ice}$) for an all ice pixel, normalized to the distance between the all ice ($e_{19V,ice}$, $e_{37V,ice}$) and all water ($e_{19V,water}$, $e_{37V,water}$) points. There is a difference between the Arctic and Antarctica. In the Antarctic the first-year ice melts every year, to freeze again the next year, while in the Arctic the ice layers can be maintained over the year to form multiyear ice. This translates in an emissivity of approximately 0.92 for the Arctic ice and 0.84 for the Antarctic at vertical polarization and 53 degrees incidence angle.

Comparisons of scatterometer sea ice classifications with those from passive microwave instruments show that the passive microwave instruments (such as the special sensor microwave imagery or SSM/I) overestimate the areal extent of MY ice versus FY ice by as much as 12% due to atmospheric interference at the higher frequencies used for these classifications (Voss et al., 2003). The overall ice cover is best characterized using a combination of both active and passive microwave sensors.

Scatterometry has also proven useful in sensing the date of melt onset and the duration of the melt season (Drinkwater and Liu, 2000; Winebrenner et al., 1998). These studies validated their scatterometer derived melt detection by comparing their onset periods with near-surface air temperatures recorded at nearby coastal meteorological stations which in both cases showed that the air temperatures increase above 0°C when the sea ice melt begins to be detected in the scatterometry. In a separate study, melt ponding was detected from a helicopter during the period for which the scatterometer had determined the melt onset (Drinkwater et al., 1998). The melt season can be estimated from the date of melt onset to the start of freeze up which is correlated with a sudden increase in σ_0.

Scatterometer data have also been used to compute sea ice motion. This is done with a sequence of ice images and a wavelet analysis has been employed in some of these studies. Comparisons with in situ sea ice motion from ice buoys have shown that scatterometer sea ice motion has an accuracy of about ±2−3 cm/s in speed and 30 degrees in direction (Zhao et al., 2002; Liu and Zhao, 1999). This accuracy is similar to SSM/I derived sea ice speeds and is slightly better that the SSM/I derived sea ice directions (±35−38 degrees). Because passive microwave radiometer and scatterometers respond differently to ice features (emissivity vs. backscatter), the spatial resolution of each will be different. Thus, it is best to merge sea ice motions from passive microwave with scatterometer and ice buoy motions to achieve the best possible coverage. In merging these different data sources, a weighted average is used proportional to their respective accuracies. Zhao et al. (2002) used a weighting scheme of 1/3, 1/2, and 1/6 to QuikSCAT, buoy, and SSM/I derived sea ice motion vectors, respectively and produced Arctic sea ice motion maps like that in Fig. 11.20.

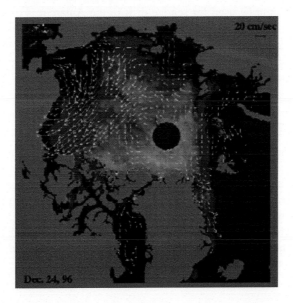

FIGURE 11.20

Sea ice motion (*white arrows*) from merged NSCAT, SSM/I, and ice buoy data (Liu and Zhao, 1999).

11.9 ARCTIC SEA ICE DRIFT ESTIMATION BY MERGING RADIOMETER AND SCATTEROMETER DATA

In a separate study, Girard-Ardhuin and Ezraty (2012) used a combination of passive microwave radiometry from the AMSR-E and the SSM/I along with SeaWinds/QuikSCAT and ASCAT scatterometry to estimate sea ice motion in the Arctic during the summer melt and transition season "freeze up" periods. Independent validation of their motion results was provided by the International Arctic Buoy Project buoys which give excellent ice movement information which is limited in spatial coverage by the relatively low number of buoy deployed across the Arctic. In their method the individual sensors were used separately to compute ice motion for the entire Arctic before any attempt at merging the resulting vectors was carried out. Thus, the goal of this study was to merge the resulting ice drift maps resulting from the use of the radiometer and scatterometer instruments individually.

Passive microwave radiometers, such as the SSM/I, have been used with substantial success to compute sea ice motion for many years. Able to see through the clouds the microwave data provide much better overall coverage along with a reduction in spatial resolution. Passive microwave imagery also operates both day and night with the same level of responsiveness. To minimize all atmospheric effects the ice motion is best calculated in the winter when that atmosphere is very dry. Microwave channels can be selected to minimize the influence of water vapor such as the Ka- and Ku-bands (19 and 37 GHz).

While several different methods have been used to compute sea ice motion in sequential imagery, the method selected in this research was the familiar maximum cross correlation (MCC; Emery et al., 1991). This method has been widely used and well validated in the literature. In this application a correlation threshold of 0.6 was used to filter out spurious vectors computed from the pair of images. The final spatial resolution was 12.5 km and an 11×11 pixel correlation array was used.

A time lag of 2−3 days was used for the MCC calculations, which is much longer than the less-than-one-day lag that should be used for sea ice drift in the Antarctic. For this 2−3 day lag period and the 12.5 km spatial resolution the minimum displacement for the MCC method is 6.25 km, which corresponds to a velocity of 7 cm/s. This is the same order of magnitude of the mean velocity for the buoy displacements measured over a winter. ASCAT and QuikSCAT data are also used for sea ice drift estimation using the same grid that we developed for SSM/I and AMSR-E.

11.10 MERGING THE SEA ICE DRIFT PRODUCTS

The main steps of merging the individual sea ice drift products are initialization phase, selection of a drift vector, and quality control. At any one-grid node, there are 14 possible cases ranging from a triplet of identical vectors to a single vector. Weights have to be given to all of these vectors at the grid point so that the weighted average can be computed for that grid point. The weight depends on the selection/quality control procedure and on the selected sensor or channel. Minimum weights are given to the scatterometer vectors. The cases where several drift vectors exist at a grid point are treated differently than at a grid point where only a single vector is present. Both cases use information from the surrounding grid points to produce a smooth resultant grid function.

An example of a merged sea ice motion field is presented here in Fig. 11.21 for April 24 and 27, 2007. Colors of the arrows correspond to the merged processing for each pixel: in red, the three estimations are identical (scatterometer and radiometer H and V polarizations), in green, two of the three drifts are identical, in blue, one channel has been chosen. The sea ice drift patterns are moving from day-to-day, and Fig. 11.21 presents a pattern, which is considered a "classical" drift pattern in the Beaufort gyres, a southward drift in Fram Strait, and the ice export out of the Kara Sea.

The validation of the merged sea ice drifts is based on data from five winters. The standard deviation of the difference to buoy displacements is 7.5 km. When converted to speed the mean velocity is 2.91 cm/s, which is similar to the buoy mean velocity of 2.6 cm/s. These results include small ice drifts where the MCC method is known to break down. Previous studies have made similar comparisons while excluding speeds below a certain threshold.

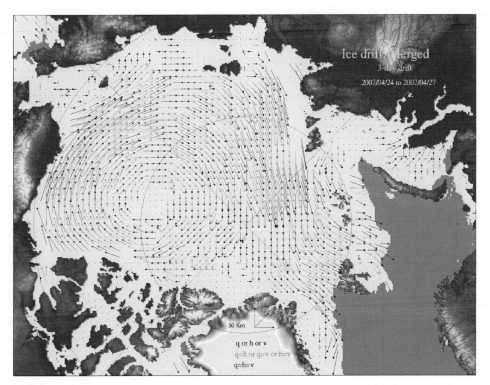

FIGURE 11.21

Arctic sea ice drift map from independent maps of drift vector from SSM/I H and SSM/I V polarization channels, and SeaWinds/QuikSCAT. The time lag is 3 days: April 24–27, 2007. Drift vectors less than one pixel are marked by a cross. Identical drift vectors for the three estimates is colored in red; identical drift vectors for any two of the estimates is colored in green; selection or validation of any single drift vector is colored in blue. Note that the *arrows* are not at the scale of the map.

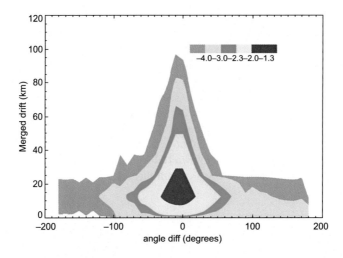

FIGURE 11.22

Angle difference between satellite merged drift vectors and buoys drift vectors as a function of the merged drift magnitude. The time lag is 3 days. The comparison spans over five winters. *Colors* represent the percentage of occurrence expressed in logarithm scale.

As expected, the angles of the drift vectors computed with the MCC method, exhibit a strong uncertainty for small drifts relative to the buoy drift directions. For comparisons including all drift vectors the standard deviation of the angle difference is 39.2 degrees; if drifts less than one pixel are excluded this reduces to 29.6 degrees. Thus, excluding small drifts significantly improves the angular accuracy. Fig. 11.22 presents the angle difference between the merged product and buoys as a function of merged drift magnitudes. The angle difference decreases sharply for drift magnitudes higher than 40 km (about three pixels).

Monitoring ice over land is also very important. Glaciers store 75% of Earth's freshwater and are sensitive indicators of long-term climate changes. Glacier extension and motion can also be precisely monitored using three different techniques:

- Repeated optical or SAR intensity satellite imagery by feature tracking (e.g., Willis et al., 2011; Herman et al., 2001), which requires accurate coregistration of the images, visible structures to be tracked, and enough displacement between data acquisitions in terms of the spatial resolution, so that they can be tracked. This is illustrated in Fig. 11.23 using a pair of TerraSAR-X images.
- Coherence (a measure of the similitude) tracking, which is based on the same principle and SAR intensity feature tracking, can work in smaller image regions with low texture, but may fail in areas of high speed, where coherence is lost.
- Differential SAR interferometry, in which two pairs of images in ascending and descending passes (to have two nonparallel displacement vectors), are used to calculate the interferograms (phase differences) from which the velocity fields can be calculated.

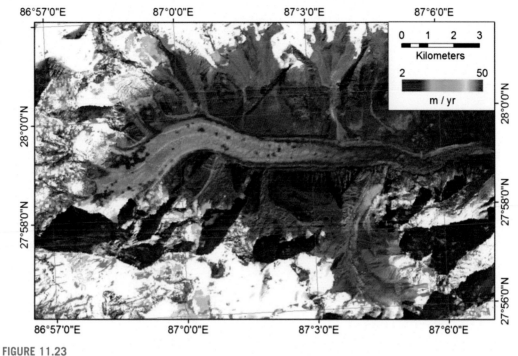

FIGURE 11.23

Velocity field derived by feature tracking of two TerraSAR-X images (January 7 to March 29, 2008) using a
128 × 128 tracking window background image: Landsat ETM+, © USGS, 2005.

From USGS.

11.11 STUDY QUESTIONS

1. How is sea ice created and how is it classified? Is there salt in sea ice and if so how much and how would you detect it from space?
2. What satellite sensors would you used to measure and monitor sea ice extent and its changes over time? What algorithms would you use for this measurement?
3. How does melting of sea ice affect coastal sea level rise? Is the Arctic sea ice cover increasing/decreasing? How does this change influence the global climate. What other impacts are consequences of sea ice melting?
4. What is happening to ice sheets in Greenland and Antarctica? How do you sense these changes from space? How does the future of these land ice sheets and glacier influence global sea level rise? What is the best method to measure these changes and how accurate can these measurements be?
5. Considering the change in Arctic sea ice cover in the figure do you believe that the Arctic will be completely ice-free by 2020?

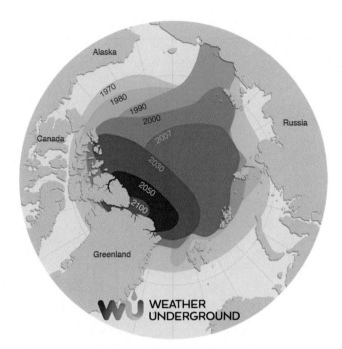

Sea ice extent observations (1970–2007) and forecast (2030–2100) reproduced using data from the NOAA GFDL model. Yearly extent represents an average 80% sea ice concentration, approximately. Click on the image for a larger view.

6. Do all sea ice cover products predict the same ice extent? Why?
7. How can the speed of the glaciers be estimated? Please list the different techniques and the advantages and disadvantages of them.
8. Which is the novelty in CryoSat radar altimeter? What does it achieve?
9. Please explain the difference in the intensity of the backscattered signal for multiyear and first-year ice, and why it can be due?
10. Please explain how the sea ice thickness can be inferred from L-band microwave radiometry?

REMOTE SENSING WITH SMALL SATELLITES

12.1 INTRODUCTION

At the beginning of the space age all satellites were "small." Sputnik-1, the first artificial satellite was launched on October 4, 1957, from the Baikonur Cosmodrome in the former USSR. It was an aluminum sphere of just 58 cm diameter and 83 kg of mass (Fig. 12.1A). Following Sputnik-1 and -2, Explorer-1 was the first US satellite (Fig. 12.1B). It was launched on January 31, 1958, from Cape Canaveral. Explorer-1 was about 205 cm long and 15 cm in diameter and weighted 13.4 kg. Vanguard-1 was the fourth artificial satellite in orbit (Fig. 12.1C). It was launched on March 17, 1958, from Cape Canaveral. Vanguard-1 was a sphere of just 16 cm, with a mass of 1.6 kg, and it was the first solar-powered satellite. The first radio amateur satellites also date from the early 1960s. The first "Orbiting Satellite Carrying Amateur Radio" OSCAR-1 (Fig. 12.1D) was designed and developed by a California-based group of radio amateurs, and it was launched on December 12, 1961, from Vandenberg Air Force Base. It was a parallelepiped of approximately 30 cm × 25 cm × 12 cm, battery-powered, and with a mass of 4.5 kg.

During the first two decades of the space age, each satellite had its own design in terms of shape, mass, size, stabilization methods, power, instrument mounting, onboard data handling, communications, etc. Standard spacecraft buses were practically unknown until the end of the 1970s. In the early 1980s microsatellites[1] emerged and, as compared to traditional spacecrafts, adopted a radically different design approach to reduce costs, focusing on available and existing technologies, and using "commercial off-the-shelf" (COTS) components. The definition of the "small satellite mission philosophy" now implies a design-to-cost approach, with strict cost and schedule constraints, mostly combined with a single mission objective so as to reduce the mission complexity and the associated management costs. To a large extent, this has been possible in the past decades thanks to the advances in electronic miniaturization and its associated performance. The appearance of new small launchers, for example, through the use of modified military missiles, and new companies such as Space X have also facilitated the access to space to many countries or even research institutes and universities. Table 12.1 from nanosats.eu (2016) summarizes the main launch providers as of October 2016.

Small satellite missions offer today a number of advantages such as more frequent mission opportunities, larger variety of missions and potential users, faster expansion of the technical and scientific knowledge base, and greater involvement of local and small industry (Kramer and Cracknell, 2008).

---

[1]Microsatellite is a satellite with mass between 10 and 100 kg.

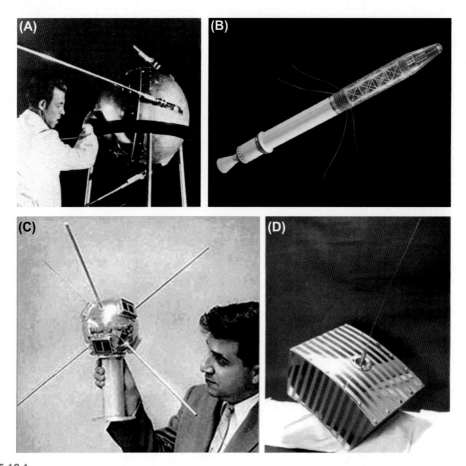

FIGURE 12.1

Images of the first USSR and US satellites (A) Sputnik-1 (Sputnik, 2012) and, (B) Explorer-1 (Howell, 2012), image of (C) Vanguard-1, the first solar-powered satellite (Pics about space, 2016), and (D) OSCAR-1 (OSCAR 1, 2, 2016).

In 1999, the so-called CubeSat "standard" was proposed by professors Jordi Puig-Suari of California Polytechnic State University and Bob Twiggs of Stanford University. The goal was to allow graduate students to conceive, design, implement, test, and operate in space a complete spacecraft, often using COTS components. Because of the simplicity of the CubeSat "standard" (CubeSat Developer Resources, 2016), which only defines the outer envelope, it ended up becoming a "de facto" standard. The first CubeSats were launched in June 2003 on a Russian Eurockot. While originally, CubeSats (now called one unit or 1U CubeSats) were nanosatellites[2] of roughly a 10 cm × 10 cm × 10 cm cube, with a mass smaller than 1.33 kg; soon other form factors were developed to accommodate larger payloads and

[2]Nanosatellite is a satellite with mass between 1 and 10 kg.

Table 12.1 Main Nanosatellite Launch Providers, Programs, and Prices

Provider	Number Launched	First Launch	Cost	Additional Information
Terran Orbital/ Tyvak	121+ (40+ planned)	2003		CubeSat integrator for most US Gov. Missions. Launches include organized by Cal Poly.
ISIS (Innovative Solutions In Space)	75+	2009	$210,000−270,000 for 3U LEO	
NASA CSLI and ELaNa	46+ (120 selected)	2011	Free	For US educational, NASA and nonprofit CubeSats. Different rockets used.
ESA Fly your Satellite!	10	2012	Free	European educational CubeSats. Used to be with Vega. ISS orbit from now on.
JAMSS/ JAXA	10+	2012		
Nanoracks	80+	2012	$85,000 for 1U	Only deployed from ISS. Now up to $1 \times 6U$ and soon up to 12U.
Spaceflight	77+	2013	$295,000 for 3U LEO $545,000 for 6U LEO $995,000 for 12U LEO GTO and Lunar also listed	
G.A.U.S.S.	12+	2013		Only available launches for PocketQubes from UniSat microsatellites.
ULA (United Launch Alliance)	0	2017	Free	6 units planned for now. Only for US colleges and universities.
Rocket Lab	0 (722 booked until middle 2019)	2017	$70,000−80,000 for 1U LEO $200,000−250,000 for 3U LEO	
Virgin Galactic LauncherOne	0	2017	Assuming $10 million and 40 CubeSats gives $250,000 per 3U.	Sky and Space has contract for 4 launches.
KiboCUBE (UNOOSA, JAXA)	0	2018	Free	Provide ISS launches for educational or research institutions from developing countries.
Vector Space	0	2018		Integrated CubeSat dispensers. ICEYE has booked 21 launches from them.

Extracted from Nanosatellite and CubeSat database; www.nanosats.eu.

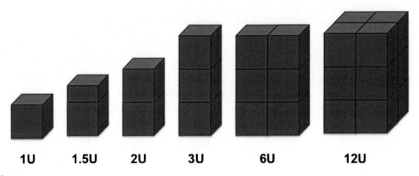

FIGURE 12.2

CubeSat size comparison.

From https://www.nasa.gov/content/what-are-smallsats-and-cubesats.

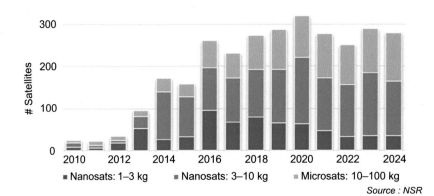

■ Nanosats: 1–3 kg   ■ Nanosats: 3–10 kg   ■ Microsats: 10–100 kg

Source : NSR

FIGURE 12.3

Global satellite launches by mass (Belle, 2015).

increase the power generation capability. The current CubeSat Design Specification (CDS rev. 13 [CubeSat Developer Resources, 2016]) defines the envelopes for 1U, 1.5U, 2U, 3U, and 3U + form factors, and a provisional CDS for 6U CubeSats has also been issued. Fig. 12.2 shows the CubeSat structures corresponding to 1U, 2U, 3U, 6U and the planned 12U (Radius Space, 2016).

Most CubeSats are nowadays either used as technology demonstrators, without the cost of a larger satellite, or to perform "risky" small scientific experiments. Biological research payloads have also been flown on several missions, and there are even several missions planned to the Moon and Mars (Gibney, 2016).

Fig. 12.3 shows the predicted number of nano- and microsatellite launches by mass. It is expected that the 1–3 kg nanosatellite segment (mostly 1U CubeSats) will grow at a 3% compound annual growth rate (CAGR) over the next decade, with over 500 1U CubeSats to be launched by 2024. Despite this growth, the single most active small satellite segment is expected to be 3–10 kg nanosatellite

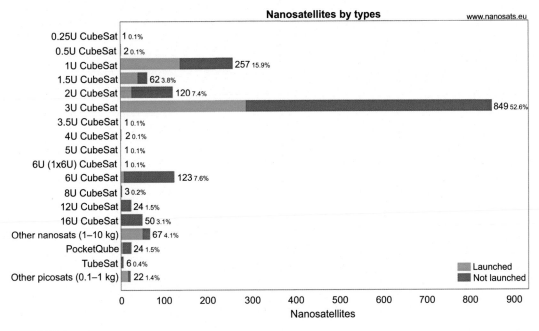

FIGURE 12.4

Number of nanosatellites by types as of November 2016 (nanosats.eu, 2016).

From Erik Kulu, www.nanosats.eu, 2016.

platforms up to 6U, while larger 10−100 kg microsatellite platforms are the most rapidly growing, at a 14% CAGR (Belle, 2015).

By 2014, the 3U CubeSat had become the most popular form factor in the market, due to the deployment of the Planet (www.planet.com) and Spire (https://spire.com/) constellations. This form factor is still dominating the scene (Fig. 12.4), and constellations of 3U CubeSats will sustain activity through the next decade. However, it is expected that the next wave of growth will be based on 6U and 12U CubeSats (Fig. 12.2), which achieve a balance between very capable payloads, and limited manufacturing and launch costs, leveraging the benefits of standardization (Belle, 2015). While in the early space age small satellites were simply ignored or overlooked by industry, space agencies, and the media, the international radio amateur community and associated universities must be regarded as the true pioneers of small satellite technology.

12.2 EARTH OBSERVATION USING CONSTELLATIONS OF SMALL SATELLITES

Since the review by Selva and Krejci (2012) the capability to perform many real science mission using small satellites has become a reality. Table 12.2 from Freeman et al. (2016) summarizes the main areas in which technology has matured enough so as to enable new types of Earth observation missions that were not feasible in 2012. Table 12.3 adapted from Freeman et al. (2016) summarizes the main CubeSat instruments as of November 2016.

Table 12.2 Technological Feasibility of Different Earth Observation Missions Using Small Satellites

Technology	Selva and Krejci (2012)	Freeman et al. (2016)	Justification
Atmospheric chemistry instruments	Problematic	Feasible	PICASSO, IR sounders
Atmospheric temperature and humidity sounders	Feasible	Feasible	–
Cloud profile and rain radars	Infeasible	Feasible	JPL RainCube demo
Earth radiation budget radiometers	Feasible	Feasible	SERB, RAVAN
Gravity instruments	Feasible	Feasible	No demo mission
High-resolution optical imagers	Infeasible	Feasible	Planet
Imaging microwave radars	Infeasible	Problematic	Ka-Band 12U design
Imaging multispectral radiometers (Vis/IR)	Problematic	Feasible	AstroDigital
Imaging multispectral radiometers (μWave)	Problematic	Feasible	TEMPEST
Lidars	Infeasible	Problematic	DIAL laser occultation
Lightning imagers	Feasible	Feasible	–
Magnetic fields	Feasible	Feasible	INSPIRE
Multiple direction/polarization radiometers	Problematic	Feasible	HARP Polarimeter
Ocean color instruments	Feasible	Feasible	SeaHawk

Adapted from Freeman, A., Waliser, D., Hyon, J., 2016. The Cube-Train Constellation for Earth Observation, 13th Annual CubeSat Developers Workshop, April 20–22, 2016. Cal Poly State University, San Luis Obispo, CA. Available online at: http://mstl.atl. calpoly.edu/~bklofas/Presentations/DevelopersWorkshop2016/4_AnthonyFreeman.pdf.
RAVAN, radiometer assessment using vertically aligned nanotubes; SERB, solar irradiance and earth radiation budget.

Despite this explosion of new capable instruments today, most of the applications are still focused on panchromatic or multispectral and near-infrared imaging, and achieving a very high spatial resolution, which ultimately cannot overcome the laws of Physics. Fig. 12.5 graphically summarizes the temporal and spatial requirements for some applications (Sandau and Brieβ, 2008).

While achieving submetric resolution will be a big challenge for CubeSat-like small satellites, achieving a metric or decametric resolution is nowadays feasible. For example, Planet (www.planet. com) is achieving 3.7 m resolution with a constellation of 3U CubeSats named "Dove" (see Fig. 12.6). The advantage of this much faster turnaround time of Earth imagery for commercial and humanitarian applications is the most important application for most constellations of small satellites (Foust, 2013, 2014).

Table 12.3　Overview of CubeSat Instruments

Technology	Some Applications	Organization or Instrument	Description	Status and Additional Information
Visible and near-IR cameras	Determine asteroid's shape, rotational properties, spectral class, local dust and debris field, regional morphology, and regolith properties.	**Planet Scope PS2**	29 MP detector capable of taking images with 3.7 m ground resolution and swath of 24.6 × 16.4 km from 475 km altitude.	4 Band imager with Two-Stripe NIR filter. Can be a single RGB or a split-frame (RGB half and an NIR half).
		Hera Systems	1-m resolution imaging satellite is built on a 12U CubeSat, 22-kg form factor.	First launch of 9 12U CubeSats in late 2016 or early 2017.
		Astro Digital (Aquila)	6U has 22 m resolution in RGB and NIR. 16U has 2.5 m resolution in RGB, red edge, and NIR using 70 MP sensor and butcher block filter.	First satellites now planned to launch in late 2016.
		Malin Space Systems	ECAM C-50 imager uses the Aptina MT9P031 sensor certified for deep space. 5 MP (2592 × 1944) CMOS.	NEA Scout 6U planned to have monochrome with narrow field of view (FOV) optics. Malin cameras on Curiosity, Juno, etc.
		JPL IntelliCam	20 MP, 15° FOV. 10 cm/pix at ~800 m. Asteroid (~5–12 m) detection from ~50K km. Science and optical (autonomous) navigation.	Flight on NEA Scout. Based on Mars 2020 rover EECAM and OCO-3.
Microwave radars	Precipitation profiling	**NASA KaPDA Ka-band antenna**	KaPDA parabolic deployable Ka-band antenna with 0.5 m diameter, 1.5U stowed size, 1.2 kg mass and 42.5 dB gain.	Scheduled to launch in 2017 onboard 6U RainCube. First 35.75 GHz Ka-band radar payload on a CubeSat.
Infrared imagers	Characterize volatiles and minerals.	**NASA BIRCHES**	1.5U, 2.5 kg, 5 W. Spectral resolution (5 nm) to characterize volatiles (water etc.) and minerals (oxides etc.). Microcryocooler to keep <140K.	Will fly on Lunar IceCube in 2018. Compact version of the volatile-seeking spectrometer on new horizons.
	Measure temperature and water vapor in atmosphere. Night-imaging, temperature mapping.	**Thoth Argus 1000**	Infrared Range: 1000 nm–1700 nm. Spectral resolution: 6 nm. 15 mm aperture, 0.15° FoV. Envelope: 45 × 50 × 80 mm. Mass <230 g.	First flown in space in 2008.

Continued

Table 12.3 Overview of CubeSat Instruments—cont'd

Technology	Some Applications	Organization or Instrument	Description	Status and Additional Information
		MWIR Grating Spectrometer (CIRAS)	Spatial: 13.5 × 0.32 km. Spectral: 4.8–5.1 µm. 625 Channels. HOT-BIRD detectors comparable to HgCdTe at much reduced cost.	Will fly on 6U CubeSat also called CIRAS in 2018. Cryocooler by Lockheed Martin is smallest available.
		Planetary Resources MWIR	Mid wave infrared imager (MWIR) in 3–5 µm with 15 m ground resolution.	Will launch on Arkyd-100.
Hyperspectral imagers/spectrometers	Material detection, crop identification, soil moisture, oil spill concentrations, monitoring pollutants, hazardous gases.	Harris Fourier Spectrometer (CubeSat-FTS)	4 cm aperture, MWIR band only (5.7–8.3 µm). Cooled to ~ 120K using an AVHRR-based passive cooler. Hundreds of hyperspectral bands.	Flight in 2019. More affordable and efficient 3D wind data sets. Utah State University provides spacecraft.
		Planetary Resources VNIR	Visible-NIR 40 channel hyperspectral imager with 10 m resolution.	Will launch on Arkyd-100.
		Planetary Resources CHAP	Visible, NIR, IR and far-UV (90–140 nm) possible, 400 channels. Coregistration of spatial and spectral information.	NASA SBIR Phase II contract.
		VTT VISION	Tunable spectral imager operating in the visible and near-infrared spectra (430–800 nm) for ozone vertical profile measurement.	Based on VTT Fabry-Perot Interferometer technology and Aalto-1 spectral imager.
		ESA Mini-TMA	90 m ground resolution, 565 km swath, 400–800 nm range, 300 channels, 15 nm spectral resolution, 150 g, 60 × 50 × 30 mm.	Miniaturization of Proba V design. Prototype engineering model was built by Dutch company VDL ETG.
		Snow and Water Imaging Spectrometer (SWIS)	4U, 1.9 kg, 350–1700 nm region with 5.7 nm sampling, 10° FoV, 160 m resolution from 500 km.	Mission not yet planned, but development well underway with testing in early 2017.
Neutron spectrometers	Map hydrogen (and water) abundances	NASA Mini-NS (Neutron Spectrometer)	2.5U detector using Cs2YLiCl6:Ce (CLYC) scintillator material to detect epithermal neutrons at spatial scales below 10 km.	Flight in 2018 on LunaH-Map to understand the relationship between hydrogen and permanently shadowed regions.

X-ray	Chemical composition	**Amptek X-123SDD**	7 × 10 × 2.5 cm, 180 g, 2.5 W, solar SXR spectral measurements in the 0.5–30 keV range (0.04–2.5 nm) with 0.15 keV energy resolution.	Flew on MinXSS 3U CubeSat. Planned for 6U CublXSS plus X123-CdTe. XR-100SDD flies on OSIRIS-REx.
		REDLEN M1770	Precise measurement of the cosmic (diffuse) X-ray background in the 20–50 KeV range.	Will launch on CXBN-2.
		Lobster Eye X-ray telescope	Wide-field optical system for X-ray monitoring in range 3–40 keV, based on Lobster Eye optics and Timepix detector	Will launch on 2U VZLUSAT-1 with QB50.
Mass spectrometer	Elemental composition	**Quadrupole Ion Trap Mass Spectrometer (QIT-MS)**	2.5 kg, 2U, isotopic accuracy <1%.	Leverages foldable edge-connected electronics.
Gamma ray spectrometer	Subsurface composition	**Miniaturized Sr12 Gamma Ray Spectrometer**	1U, 0.5 kg, 3 W. Europium-doped strontium iodide (Srl2) crystal is 15 × 15 × 10 mm.	Elemental composition with CubeSats and small landers. Prototype development in early phase.

From Erik Kulu, Nanosatellite and CubeSat database; www.nanosats.eu.

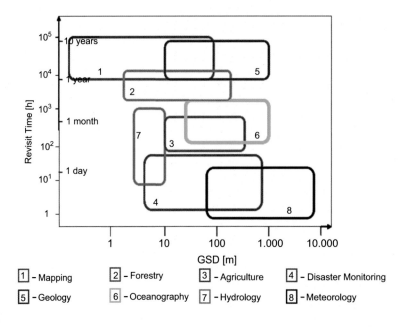

FIGURE 12.5

Spatio-temporal requirements for some Earth observation missions.

From Sandau, R., Brieß, K., 2008. Potential for Advancements in Remote Sensing Using Small Satellites, vol. XXXVII (part B1). The International Archives of the Photogrammetry, Remote Sensing and Spatial Information Sciences, Beijing, 919–924.

FIGURE 12.6

Two of the 28 Planet Labs Dove satellites that make up the Flock 1 constellation being launched into orbit from the International Space Station on February 11, 2013.

From NASA.

FIGURE 12.7

LEMUR-2 satellites carry two payloads: STRATOS GPS radio occultation payload and the SENSE AIS payload.

From Gunter's Space Page, "LEMUR-2"; http://space.skyrocket.de/doc_sdat/lemur-2.htm.

Another example is the contract awarded by the US National Oceanic and Atmospheric Administration (NOAA) on September 2016 to Spire as part of the government's Commercial Weather Data Pilot. Spire provides temperature, pressure, and water vapor content through GPS Radio Occultation (GPS-RO) data. These data are relied upon for weather forecasts. These data are gathered using a constellation of 3U CubeSats and are under evaluation for future commercial procurement to be used operationally in NOAA's weather forecast. The first four of Spire's "LEMUR-2" class satellites were launched piggy-back on a PSLV-XL rocket in 2015. The second batch was launched with Cygnus CRS-6 to the ISS. Four of the satellites to be deployed from the ISS airlock, five of them directly from the NanoRacks NRCSD-E deployer on the Cygnus vehicle. Deployment of the ISS satellites began on May 18, 2016, and the satellites on Cygnus were deployed on June 21, 2016. During the following OA-5 mission, which departed the ISS on November 21, 2016, four additional LEMUR-2 satellites were released from another Cygnus vehicle after it raised its orbit to an altitude of 500 km prior to a planned reentry November 27, 2016 (Foust, 2016). At the end of May 2017, Spire has launched a total of 36 satellites, from which 21 are still in orbit. Spire aims to launch new and upgraded satellites nearly every month, which will also include ADS-B payloads to track airplanes by the end of 2017 (LEMUR-2, 2016) (see Fig. 12.7).

A third example is the constellation of Arkyd-100 satellites, 12U CubeSat-compliant, from Planetary Resources (PlanetaryResources_1, 2016) offering visible-NIR 40 channel hyperspectral imagery with 10 m resolution, and mid wave infrared imager (MWIR) in 3−5 μm with 15 m resolution. It is claimed that these imagery cannot only be used for Earth observation, but for asteroid prospecting, in view of future space mining (PlanetaryResources_2, 2016).

Table 12.4, extracted from nanosats.eu (2016) shows the main Earth observation constellations. It is worth noting that among the top 18 constellation projects, 7 correspond to Earth observation, and the first 5 positions are occupied by the three projects listed above. Still, a major challenge for high-resolution imaging missions or for hyperspectral missions on small spacecraft is the thermal stability of the imaging instrument and the attitude stability of the bus.

Table 12.4 Main CubeSat Commercial Earth Observation Constellations

Organization	Launched/ Planned Network	First Launch	Form Factor	Field	Funding	Technical and Comments
Planet	179/150+	2013	3U	Earth observation	$183 million	29 MP sensor taking images with 3.7 m ground resolution and swath of 24.6 × 16.4 km from 475 km altitude.
Spire	17/100+	2013	3U	Weather/ AIS	$69.5 million	Measure change in GPS signal after passing through atmosphere to calculate precise profiles for temperature, pressure, and humidity.
Planetary resources	2/10	2014	12U	Earth observation	$21.1+ million	Visible-NIR 40 channel hyperspectral imager with 10 m resolution. Mid wave infrared imager (MWIR) in 3–5 μm with 15 m resolution.
Astro Digital (Aquila)	2/10 + 20	2014	6U & 16U	Earth observation	unknown	6U has 22 m resolution in RGB and NIR. 16U has 2.5 m resolution in RGB, red edge, and NIR using one 70 MP sensor.
Hera Systems	0/9–48	2017	12U	Earth observation	$4.2+ million	Capable of 1-m resolution with 22-kg form factor.
Harris	0/12	2019	6U	Weather	unknown	Immediate access to 3D wind data sets from Harris-owned HyperCubes. Utah State University provides the spacecraft.
Karten Space	0/?	?	6U	Earth observation/ AIS	unknown	Optical resolution is 3–5 m.

From Erik Kulu, Nanosatellite and CubeSat database; www.nanosats.eu.

Other examples of Earth observation missions include the NASA CYGNSS mission (Fig. 12.8) with eight enhanced 6U Cubesats to perform GPS-Reflectometry (GPS-R) and the QB-50 project, which plans to fly 50 university-built CubeSats on a single rocket in 2017 for technology demonstration and studies of the Earth's thermosphere. The Dutch company Innovative Solutions In Space (ISIS) has developed the payload dispenser to accommodate all 50 CubeSats in a rocket whose total payload capacity to LEO is only 220–230 kg.

(A)

(B)

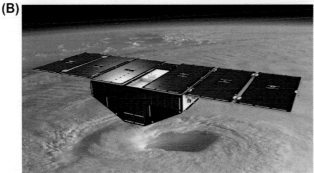

FIGURE 12.8

(A) Technicians with Orbital ATK have installed the first half of the Pegasus XL fairing around NASA's Cyclone Global Navigation Satellite System (CYGNSS) at Vandenberg Air Force Base in California. The second half is being installed. CYGNSS was launched on the Pegasus XL rocket from Cape Canaveral on December 15, 2016, (B) Artist's concept of one of the eight Cyclone Global Navigation Satellite System satellites deployed in space above a hurricane.

(A) Credits NASA/Mark Mackley, from Pegasus XL Payload Fairing Installed on CYGNSS Satellites, November 14, 2016; https:// www.nasa.gov/image-feature/pegasus-xl-mated-to-l-1011-stargazer-carrier-aircraft; (B) Credits NASA, from CYGNSS inorbit artist concept; https://www.nasa.gov/sites/default/files/thumbnails/image/cygnss-inorbit-artconcept_002.jpeg.

12.3 FUTURE TRENDS IN SMALL SATELLITES

While the CubeSat concept has been around for more than a decade, and it has originally been considered by many in the industry and space agencies little more than a novelty, nice engineering student projects, and an orbital debris nuisance as well, CubeSats are finding their ways in Earth observation, communications, and in planetary science and astrophysics/heliophysics also, typically first as demonstrators, then as constellations to exploit their full potential in terms of improved revisit time. The interested readers are encouraged to read the numerous web pages on these topics, and in particular the "Nano-satellite and CubeSat Database" (nanosats.eu) and the "CubeSat web page" at NASA/JPL (CubeSat Era, 2016).

Today, the technology development trends are characterized by: (1) higher performance of micro-/nanosatellites buses thanks to developments at component and subsystem level such as onboard computers, onboard data handling systems to preprocess data onboard and reduce the amount of data to be downloaded, higher speed transmitters, solar arrays, batteries, navigation receivers, etc., (2) higher performance optical payloads suitable for disaster and emergency monitoring, which are with high geometric and radiometric resolutions, and more spectral bands, (3) new passive (bistatic) radars using micro-/nanosatellites in formation-flying with an active radar transmitter or using signals of opportunity, (4) an explosion of satellite constellations to reduce the revisit time thanks to a decrease of the unitary satellite cost, and novel international partnerships to share costs, (5) an increased flexibility of the mission operations, (6) the creation of networks of ground stations to increase the responsivity according to user requirements and reduce latency, and (7) standardization of processing algorithms and delocalization of processing centers for near real-time data processing closer and to the end-user and better tailored to his needs. In summary, small satellites and their associated networks of ground-stations will provide faster geospatial data, and with a better match to user needs, improving the response time and the information extraction process for decision makers.

However, small satellites will not be the only elements of this change of paradigm in Earth observation. New airborne platforms such as drones and Unmanned Air Vehicles, tethered balloons, or dirigible airships will augment the space segment regionally or locally providing extremely high revisit time. At the same time, the fusion of incredible amounts of not-so-remotely sensed data (i.e., "parasensing" as opposed to "remote sensing") crowdsourced from all available sources, either from interconnected (the IoT or Internet of Things) fixed or mobile geolocated sensors will provide "ground truth" of different formats and levels of quality that will pose an incredible engineering challenge, but with the reward of the first truly global, synoptic view of our planet Earth.

References

Ackerman, S.A., 1997. Remote sensing aerosols using satellite infrared observations. JGR Atmos. http://dx.doi.org/10.1029/96JD03066.

Ackerman, S.A., Strabala, K.I., Menzel, W.P., Frey, R.A., Moeller, C.C., Gumley, L.E., 1998. Discriminating clear sky from clouds with MODIS. J. Geophys. Res. 103 (D24), 32141–32157. http://dx.doi.org/10.1029/1998JD200032.

Aguttes, J.P., July 21–25, 2003. The SAR train concept: required antenna area distributed over N satellites, increase of performance by N. In: Proc. IGARSS, pp. 4068–4070 (Toulouse, France).

Allan, D.W., 1987. Time and frequency (Time-Domain) characterization, estimation, and prediction of precision clocks and oscillators. IEEE Trans. Ultrason. Ferroelectr 6, 647–654.

Alonso-Arroyo, A., Camps, A., Aguasca, A., Forte, G., Monerris, A., Rudiger, C., Walker, J.P., Park, H., Pascual, D., Onrubia, R., December 2014a. Improving the accuracy of soil moisture retrievals using the phase difference of the dual-polarization GNSS-R interference patterns. IEEE Geosci. Remote Sens. Lett. 11 (12), 2090–2094.

Alonso-Arroyo, A., Camps, A., Monerris, A., Rüdiger, C., Walker, J.P., Forte, G., Pascual, D., Park, H., Onrubia, R., July 2014b. The light airborne reflectometer for GNSS-R observations (LARGO) instrument: initial results from airborne and Rover field campaigns. In: 2014 IEEE Geoscience and Remote Sensing Symposium, pp. 4054–4057.

Alonso-Arroyo, A., Zavorotny, V.U., Camps, A., 2016. Sea ice detection using GNSS-R data from UK TDS-1 mission. In: Proceedings of the IGARSS.

Anding, D., Kauth, R., 1970. Estimation of sea surface temperature from space. Remote Sens. http://dx.doi.org/10.1016/S0034-4257(70)80002-5.

Anguelova, M.D., Gaiser, P.W., 2013. Microwave emissivity of sea-foam layers with vertical profile of dielectric properties. Rem. Sens. Environ. 139, 81–96. http://dx.doi.org/10.1016/j.rse.2013.07.017.

Anterrieu, E., A, 2004. Resolving matrix approach for synthetic aperture imaging radiometers. IEEE Trans. Geosci. Remote Sens. 42 (8), 1649–2165.

Arroyo, A.A., Camps, A., Aguasca, A., Forte, G.F., Monerris, A., Rüdiger, C., Walker, J.P., Park, H., Pascual, D., Onrubia, R., 2014. Dual polarization GNSS-R interference pattern technique for soil moisture mapping. IEEE J. Sel. Tophic App. Remote Sens. http://dx.doi.org/10.1109/JSTARS.2014.2320792.

Arsac, J., 1955. Nouveau réseau pour l'observation radio astronomique de la brillance sur le soleil à 9359 MC. Compt. Rend. Acad. Sci. 240, 942–945.

Attema, E.P.W., 1991. The active microwave instrument on-board the ERS-1 satellite. Proc. IEEE 79 (6), 791–799.

Auber, J.C., Bibaut, A., Rigal, J.M., 1994. Characterization of multipath on land and sea at GPS frequencies. In: Proceedings of 7th International Technical Meeting of the Satellite Division of the Institute of Navigation, Part 2, ION GPS94, pp. 1155–1171.

Auterman, J.L., March 1984. Phase stability requirements for bistatic SAR. In: Proc. IEEE Nat. Radar Conf, pp. 48–52 (Atlanta, USA).

Bachmann, M., Borg, E., Fichtelmann, B., Günther, K., Krauß, T., Müller, A., Müller, R., Richter, R., Wurm, M. ESA contract 4000101779/10/I-LG, Technical Report D3.6.1-Algorithm Theoretical Basis Document, Volume I, Pre-processing ATBD I, Pre-Processing, Version 2, ESA CCI ECV Fire Disturbance (Fire_cci). Available at: https://www.esa-fire-cci.org/webfm_send/734.

Bamler, R., Hartl, P., 1998. Synthetic aperture radar interferometry. Inverse Probl. 14 (4), R1.

Bamler, R., July 1992. A comparison of range-Doppler and wavenumber domain SAR focusing algorithms. IEEE Trans. Geosci. Remote Sens. 30 (4), 706–713.

Bará, J., Camps, A., Torres, F., Corbella, I., 1998. Angular resolution of two-dimensional hexagonally sampled interferometric radiometers. Radio Sci. 33 (5), 1459–1473.

Bará, J., Camps, A., Torres, F., Corbella, I., 2000a. The correlation of visibility errors and its impact on the radiometric resolution of an aperture synthesis radiometer. IEEE Trans. Geosci. Remote Sens. GRS-38 (5), 2423–2426.

Bara, M., Scheiber, R., Broquetas, A., Moreira, A., 2000b. Interferometric SAR signal analysis in the presence of squint. IEEE Trans. Geosci. Remote Sens. 38 (5), 2164–2178.

Barnes, J.A., Chi, A.R., Cutler, L.S., Healey, D.J., Leeson, D.B., Mc-Gunigal, T.E., Mullen Jr., J.A., Smith, W.L., Sydnor, R.L., Vessot, R.F.C., Winkler, G.M.R., 1971. Characterization of frequency stability. IEEE Trans. Instrum. Meas. IM-20 (2), 105–120.

Barton, I.J., Zavody, M., O'Briwn, D.M., Cutten, D.R., Saunders, R.W., Llewellyn-Jones, D.T., 1989. Theoritical algorithms for satellite-derived sea surface temperatures. J. Geophys. Res. http://dx.doi.org/10.1029/JD094iD03p03365.

Barton, I.J., 1995. Satellite-derived sea surface temperatures: current status. J. Geophys. Res. http://dx.doi.org/10.1029/95JC00365.

Basharinov, A.E., Shutko, A.M., 1975. Simulation studies of the SHF radiation characteristics of soils under moist conditions. NASA Tech. Translation TT-F-16.

Bass, F.G., Fuks, I.M., 1979. Wave scattering from statistically rough surfaces. Int. Ser. Nat. Phil 93.

Basu, S., Bresler, Y., 2000. O(N2 log2 N) filtered backprojection reconstruction algorithm for tomography. IEEE Trans. Geosci. Remote Sens. 9 (10), 1760–1773.

Bauck, J.L., Jenkins, W.K., 1989. Convolution-backprojection image reconstruction for bistatic synthetic aperture radar. In: Proc. ISCAS, pp. 631–634 (Portland, USA).

Beal, R.C., DeLeonibus, P.S., Katz, I., 1981. Spaceborne Synthetic Aperture Radar for Oceanography. The Johns Hopkins University Press, 215 pp.

Beal, R.C., Tilley, D.G., Monaldo, F.M., 1983. Large and small-scale spatial evolution of digitally processed ocean wave spectra from Seasat synthetic aperture radar. J. Geophys. Res. 88 (C3), 1761–1778.

Behari, J., 2004. Microwave Dielectric Behaviour of Wet Soils, Remote Sensing and Digital Image Processing. Springer, The Netherlands isbn:10:1402032714.

Betz, J.W., 2001. Binary offset carrier modulations for radionavigation. Navigation 48. http://dx.doi.org/10.1002/j.2161-4296.2001.tb00247.

BIOMASS: ESA'S Forest Mission, 2017. http://esamultimedia.esa.int/docs/EarthObservation/BIOMASS_sheet_130611.pdf.

Birchak, J.R., Gardner, C.G., Hipp, J.E., Victor, J.M., 1974. High dielectric constant microwave probes for sensing soil moisture. Proc. IEEE 62 (1), 93–98.

Blake, L.V., 1962. A Guide to Basic Pulse-radar Maximum-Range Calculation. Naval Research Laboratory Report 5868.

Blanch, S., Aguasca, A., 2004. Seawater dielectric permittivity model from measurements at L band. Proc. IEEE Geosci. Remote Sens. Symp. 2, 1362–1365. http://dx.doi.org/10.1109/igarss.2004.1368671.

Blanchard, B.J., Bausch, W., 1979. Algorithms to Estimate Soil Moisture Storage from Microwave Measurements. Final Rep. RSC 3843. Remote Sensing Center, Texas A&M Univ. College Station, TX.

Bleistein, N., Handelsman, R.A., 1975. Asymptotic Expansions of Integrals. Holt, Rinehart, and Winston, New York, USA.

Bonnedal, M., Lindgren, T., Carlström, A., Christensen, J., December 8–10 2010b. MetOp GRAS: signal tracking performance results. In: 2010 5th ESA Workshop on Satellite Navigation Technologies and European Workshop on GNSS Signals and Signal Processing (NAVITEC), pp. 1,4.

Borsa, A.A., Moholdt, G., Fricker, H.A., Brunt, K.M., 2014. A Range Correction for Icesat and its Potential Impact on Ice-Sheet Mass Balance Studies. NASA Technical Reports Server (NTRS). http://dx.doi.org/10.5194/tc-8-345-2014.

Bosch-Lluis, X., Ramos-Perez, I., Camps, A., Rodriguez-Alvarez, N., Valencia, E., Park, H., 2011. A general analysis of the impact of digitization in microwave correlation radiometers. Sensors 11, 6066−6087. http://dx.doi.org/10.3390/s110606066.

Bounjakowsky, V., 1859. Sur quelques inegalités concernant les intégrales aux différences finis. Mem. Acad. Sci. St. Petersbourg 1 (7), 9.

Bracewell, R.N., 1966. Optimum spacings for radio telescopes with unfilled apertures. Natl. Acad. Sci. Natl. Res. Council. Publ. 1408, 243−244.

Braubach, H., Voelker, M., April 24, 2007. Method for Drift Compensation with Radar Measurements with the Aid of Reference Radar Signals. U.S. Patent 7209072 B. https://docs.google.com/viewer?url=patentimages.storage.googleapis.com/pdfs/US7209072.pdf.

Brown, O.B., Brown, J.W., Evans, R.H., 1985. Calibration of advanced very high resolution radiometer infrared observation. J. Geophys. Res. 90, 11667−11677.

Brown, R.A., 1986. On satellite scatterometer capabilities in air-sea interaction. J. Geophys. Res. 91 (C2), 2221−2232. http://dx.doi.org/10.1029/JC091iC02p02221.

Brown, S., Ruf, C., Keihm, S., Kitiyakara, K., 2004. Jason-1 microwave radiometer performance and on-orbit calibration. Mar. Geodesy. 27, 199−220.

Brown, S., Desai, S., Keihm, S., Lu, W., 2009. Microwave radiometer calibration on decadal time scales using on-earth brightness temperature references: application to the TOPEX Microwave Radiometer. J. Atmos. Oceanic Tech. http://dx.doi.org/10.1175/2009JTECHA1305.1.

Brown, W.M., March 1967. Synthetic aperture radar. IEEE Trans. Aerosp. Electron. Syst. AES-3 (2), 217−229.

Brown, G., 1977. The average impulse response of a rough surface and its applications. IEEE Trans. Ant & Prop 25.

Brunfeldt, D.R., Ulaby, F.T., 1984. Effect of Row Direction on the Microwave Emission from Vegetation Canopies, vol. 1. European Space Agency, (Special Publication) ESA SP, pp. 125−130.

Brunner, F.K., Welsh, W.M., 1993. Effect of the troposphere on GPS measurements. Smithsonian/NASA Astrophys. Data Sys 4, 42−46.

Buchner, R., Barthel, J., Stauber, J., 1999. The dielectric relaxation of water between 0°C and 35°C. Chem. Phys. Lett. 306, 57−63.

Cafforio, C., Prati, C., Rocca, F., 1991. SAR data focusing using seismic migration techniques. IEEE Trans. Aerosp. Electron. Syst. 27 (2), 194−207.

Camps, A.J., Swift, C.T., 2001. A two-dimensional Doppler-Radiometer for Earth observation. IEEE Trans. Geosci. Electron. 39 (7), 1566−1572.

Camps, A., Swift, C.T., 2002. New techniques in microwave radiometry for earth remote sensing: principles and applications. In: Stone, W.R. (Ed.), Review of Radio Science 1999−2002. J. Wiley & Sons Inc, Sommerset, NJ, USA, pp. 499−518.

Camps, A., Tarongi, J.M., 2009. RFI mitigation in microwave radiometry using Wavelets. Algorithms 2, 1248−1262.

Camps, A., Tarongí, J.M., 2010. Microwave radiometer resolution optimization using variable observation times. Remote Sens. 2, 1826−1843. http://dx.doi.org/10.3390/rs2071826.

Camps, A., Bara, J., Corbella, I., Torres, F., January 1997. The processing of hexagonally sampled signals with standard rectangular techniques: application to 2−D large aperture synthesis interferometric radiometers. IEEE Trans. Geosci. Remote Sens 35 (1), 183−190.

Camps, A., Corbella, I., Bará, J., Torres, F., 1998a. Radiometric sensitivity computation in aperture synthesis interferometric radiometry. IEEE Trans. Geosci. Remote Sens. GRS-35 (2), 680−685 (errata appeared in GRS-35, 5).

Camps, A., Bará, J., Torres, F., Corbella, I., Monzón, F., 1998b. Experimental validation of radiometric sensitivity in correlation radiometers. Electronics Lett. 34 (25), 2377−2379.

Camps, A., Torres, F., Bará, J., Corbella, I., Monzón, F., 1999. Automatic calibration of channels frequency response in interferometric radiometers. Electronics Lett. 35 (2), 115−116.

Camps, A., Font, J., Etcheto, J., et al., 2002. L-band sea surface emissivity radiometric observations under high winds: preliminary results of the Wind and Salinity Experiment WISE-2001. Proceed. IEEE Geosci. Remote Sens. Symp. 3, 1367–1369. http://dx.doi.org/10.1109/igarss.2002.1026118.

Camps, A., Torres, F., Lopez-Dekker, P., Frasier, S.J., December 20, 2003. Redundant space calibration of hexagonal and Y-shaped beamforming radars and interferometric radiometers. Int. J. Remote Sens. 24, 5183–5196.

Camps, A., Font, J., Vall-llossera, M., et al., 2004. The WISE 2000 and 2001 field experiments in support of the SMOS mission: sea surface L-band brightness temperature observations and their application to sea surface salinity retrieval. IEEE Trans. Geosci. Remote Sens. 42 (4), 804–823. http://dx.doi.org/10.1109/tgrs.2003.819444.

Camps, A., Vall-llossera, M., Villarino, R., et al., 2005. The emissivity of foam-covered water surface at L-band: theoretical modeling and experimental results from the FROG 2003 field experiment. IEEE Trans. Geosci. Remote Sens. 43 (5), 925–937. http://dx.doi.org/10.1109/TGRS.2004.839651.

Camps, A., Bosch-Luis, X., Ramos-Perez, I., Merchan-Hernandez, J.F., Izquiero, B., Rodriguez-Alvarez, N., 2007. New instrument concepts for ocean sensing: analysis of the PAU-radiometer. T. Geosci. Remote Sens. http://dx.doi.org/10.1109/TGRS.2007.894925.

Camps, A., Vall-llossera, M., Corbella, I., Duffo, N., Torres, F., January 2008. Improved image reconstruction algorithms for aperture synthesis radiometers. IEEE Trans. Geosci. Remote Sens. 46 (1), 146–157.

Camps, A., Gourrion, J., Tarongí, J.M., Vall·llossera, M., Gutiérrez, A., Barbosa, J., Castro, R., 2011a. Radio-frequency interference detection and mitigation algorithms for synthetic aperture radiometers. Algorithms 4, 155–182. http://dx.doi.org/10.3390/a40x000x.

Camps, A., Marchan, J.F., Valencia, E., Ramos, I., Bosch-Lluis, X., Rodriguez, N., Park, H., Alcayde, A., Mollfulleda, A., Galindo, J., Martinez, P., Chavero, S., Angulo, M., Rius, A., July 24–29, 2011b. PAU instrument aboard INTA MicroSat-1: a GNSS-R demonstration mission for sea state correction in L-band radiometry. In: 2011 IEEE International Geoscience and Remote Sensing Symposium, pp. 4126–4129.

Camps, A., Forte, G., Ramos, I., Alonso, A., Martinez, P., Crespo, L., Alcayde, A., 2012a. Recent advances in land monitoring using GNSS-R techniques. In: Reflectometry Using GNSS and Other Signals of Opportunity, Workshop, West Lafayette, IN. http://dx.doi.org/10.1109/GNSSR.2012.6408255.

Camps, A., Pascual, D., Park, H., Martin, F., 2012b. PARIS IOD: ID-16A Contribution to Performance and Error Budget's Report, PARIS-pha-ieec-upc-TN-008, Rev 3.0.

Camps, A., Martin, F., Park, H., Valencia, E., Rius, A., D'Addio, S., July 22–27, 2012c. Interferometric GNSS-R achievable altimetric performance and compression/denoising using the wavelet transform: an experimental study. In: IEEE International Geoscience and Remote Sensing Symposium (IGARSS), pp. 7512–7515.

Camps, A., Park, H., Alonso-Arroyo, A., July 2013a. Wind speed maping from the ISS using GNSS-R? A simulation study. In: 2013 IEEE International Geoscience and Remote Sensing Symposium, pp. 382–385.

Camps, A., Park, H., Valencia, E., Pascual, D., Martin, F., Rius, A., Ribó, S., Benito, J., Andres-Beivide, A., Saameno, P., Staton, G., Martín, M., d'Addio, S., Willemsen, P., 2014. Optimization and performance analysis of interferometric GNSS-R altimeters: application to the PARIS IoD mission. IEEE J. Selected Top. Appl. Earth Observations Remote Sens. 7 (5), 1436–1451.

Camps, A., Park, H., Pablos, M., Foti, G., Gommenginger, C., Liu, P.-W., Judge, J., October 2016. Sensitivity of GNSS-R spaceborne observations to soil moisture and vegetation. IEEE J. Selected Top. Appl. Earth Observations Remote Sens. 9 (10), 4730–4742. http://dx.doi.org/10.1109/JSTARS.2016.2588467.

Camps, A., 2003. A radar course at undergraduate level: an approach to systems engineering. IEEE Trans. Ed. 46 http://dx.doi.org/10.1109/TE.2003.816065.

Camps, A., 2010. Noise wave analysis of Dicke and noise injection radiometers: complete S-Paramater analysis and effect of temperature gradients. Radio Sci. http://dx.doi.org/10.1029/2009RS004198.

Cantalloube, H., Koeniguer, E.C., June 2008. Assessment of physical limitations of high resolution on targets at X-band from circular SAR experiments. In: European Conference on Synthetic Aperture Radar.

Caponi, E.A., Crawford, D.R., Yuen, H.C., Saffman, P.G., 1988. Modulation of radar backscatter from the ocean by a variable surface current. J. Geophys. Res. 93, 12249–12263.

Cardama, A., Jofre, L., Rius, J.M., Romeu, J., Blanch, S., 1998. Antenas. Ediciones UPC, Barcelona.

Cardellach, E., Fabra, F., Rius, A., Pettinato, S., D'Addio, S., 2012. Characterization of dry-snow sub-structure using GNSS reflected signals. Remote Sens. Environ. 124 http://dx.doi.org/10.1016/j.rse.2012.05.012.

Cardellach, E., Tomás, S., Oliveras, S., Padullés, R., Rius, A., de la Torre-Juárez, M., 2015. Sensitivity of PAZ LEO polarimetric GNSS radiom-occultation experiment to precipitation events. IEEE Trans. Geosci. Remote Sens. http://dx.doi.org/10.1109/TGRS.2014.2320309.

Cardellach, E., May 30, 2014. E-GEM Deliverable D4.1 State of the Art Description Document ref. FP7-E-GEM-D4.1_v1.

Carlson, T.N., Ripley, D.A., 1997. On the relation between NDVI, fractional vegetation cover, and leaf area index. Remote Sens. Environ. 62, 241–252.

Carlson, A.B., 1986. Communication Systems. McGraw-Hill International, Singapore.

Carlstrom, A., Bonnedal, M., Lindgren, T., Christensen, J., 2012. Improved GNSS radio occultation with the next generation GRAS instrument. In: 2012 6th ESA Workshop on Satellite Navigation Technologies and European Workshop on GNSS Signals and Signal Processing, (NAVITEC), vol. 1(4), pp. 5–7.

Carrara, W.G., Goodman, R.S., Majewski, R.M., 1995. Spotlight Synthetic Aperture Radar: Signal Processing Algorithms. Artech House, Norwood, Massachusetts.

Carreño,-Luengo, H., Camps, A., Ramos-Pérez, I., Rius, A., 2014. Experimental evaluation of GNSS-reflectometry altimetric precision using the P(Y) and C/A signals. IEEE J. Selec. Top. App. Earth Obs. Remote Sens. 7. http://dx.doi.org/10.1109/JSTARS.2014.2320298.

Carreño-Luengo, H., Camps, A., Perez-Ramos, I., Rius, A., 2012. Pycaro's instrument proof of concept. In: Reflectometry Using GNSS and Other Signals of Opportunity (GNSS+R), 2012 Workshop. http://dx.doi.org/10.1109/GNSSR.2012.6408251.

Carreno-Luengo, H., Park, H., Camps, A., Fabra, F., Rius, A., 2013. GNSS-R derived centimetric sea topography: an airborne experiment demonstration. IEEE J. Selec. Top. App. Earth Obs. Remote Sens. http://dx.doi.org/10.1109/JSTARS.2013.2257990.

Carreno-Luengo, H., Camps, A., Via, P., Munoz, J.F., Cortiella, A., Vidal, D., Jane, J., Catarino, N., Hagenfeldt, M., Palomo, P., Cornara, S., 2016. 3Cat-2—an experimental nanosatellite for GNSS-R earth observation: mission concept and analysis. IEEE J. Selected Top. Appl. Earth Observations Remote Sens.

Castro, S.L., Emery, W.J., Wick, G.A., 2004. Skin and bulk sea surface temperature estimates from passive microwave and thermal infrared satellite imagery and their relationships to atmospheric forcing. Gayana 68 (2), 96–101 supl. T.I.Proc. ISSN:0717652X.

Chaigneau, A., Pizarro, O., 2005. Eddy characteristics in the eastern South Pacific. J. Geophys. Res. http://dx.doi.org/10.1029/2004JC002815.

Chanzy, A., Raju, S., Wigneron, J.P., 1997. Estimation of soil microwave effective temperature at L and C bands. IEEE Trans. Geosci. Remote Sens. 35 (3), 570–580.

Chaparro, D., Vall-llossera, M., Piles, M., Camps, A., Rüdiger, C., Riera-Tatché, R., 2016. Predicting the extent of Wildfires using remotely sensed soil moisture and temperature trends. IEEE J. Selected Top. Appl. Earth Observations Remote Sens. 9 (6), 2818–2829. http://dx.doi.org/10.1109/JSTARS.2016.2571838.

Chaparro, D., Vayreda, J., Vall-llossera, M., Banqué, M., Piles, M., Camps, A., Martínez-Vilalta, J., February 2017. The role of climatic anomalies and soil moisture in the decline of drought-prone forests. IEEE J. Selected Top. Appl. Earth Observations Remote Sens. 10 (2), 503–514. http://dx.doi.org/10.1109/JSTARS.2016.2585505.

Chelton, D.B., Schlax, M.G., Samelson, R., 2011. Global observations of nonlinear mesoscale eddies. Prog. Oceanogr. http://dx.doi.org/10.1016/j.pocean.2011.01.002.

Chesters, D.C., Uccellini, L.W., Robinson, W.D., 1983. Low-level water vapor fields from the VISSR Atmospheric Sounder (VAS) "split-window" channels. J. Clim. Appl. Meteor. 22, 725–743.

Chevalier, L., 2002. Principles of Radar and Signal Processing. Artech House.

Chew, C., Shah, R., Zuffada, C., Hajj, G., Masters, D., Mannucci, A.J., 2016. Demonstrating soil moisture remote sensing with observations from the UK TechDemoSat-1 satellite mission. Geophys. Res. Lett. 43, 3317–3324.

Choudhury, B.J., Schmugge, T.J., Chang, A.T.C., Newton, R.W., 1979. Effect of surface roughness on the microwave emission of soils. J. Geophys. Res. 84, 5699–5705.

Choudhury, B.J., Schmugge, T.J., Mo, T., 1982. A parameterization of effective soil temperature for microwave emission. J. Geophys. Res. 87 (C2), 1301–1304. http://dx.doi.org/10.1029/JC087iC02p01301.

Christiansen, W.N., Högbom, J.A., 1985. Radiotelescopes, second ed. Cambridge University Press, pp. 163–167.

Christensen, J., Carlström, A., Ekstrom, H., Emrich, A., Embretsen, J., de Maagt, P., Colliander, A., July 23–28, 2007. GAS: the geostationary atmospheric sounder. In: 2007 IEEE International Geoscience and Remote Sensing Symposium, IGARSS 2007, pp. 223–226.

Chukhlantsev, A.A., Golovachev, S.P., Shutko, A.M., 1989. Experimental study of vegetable canopy microwave emissions. Adv. Space Res. 9, 317–321.

Chukhlantsev, A.A., 1988. The effective dielectric constant of vegetation SHF ban. Radiotekhnika 33, 2310–2319 (Eng Translation).

Chukhlantsev, A.A., Shutko, A.M., 1988. An account of the effect of vegetation during remote superhigh frequency radiometric sounding of terrestrial deposits. Issledovaniye Zemli iz Kosmosa 2, 67–72. English translation.

Chukhlantsev, A.A., 2006. Microwave Radiometry of Vegetation Canopies. Springer, The Netherlands. ISBN: 10-1-4020-4681-2.

Claassen, J.P., Fung, A.K., 1974. The recovery of polarized apparent temperature distributions of flat scenes from antenna temperature measurements. IEEE Trans. Antennas Propagation AP-22 (3), 433–442.

Clark, R.N., Gallagher, A.J., Swayze, G.A., 1990a. Material absorption band depth mapping of imaging spectrometer data using a complete band shape least-squares fit with library reference spectra. In: Proceedings of the Second Airborne Visible/Infrared Imaging Spectrometer (AVIRIS) Workshop. JPL Publication 90-54, pp. 176–186.

Clark, R.N., King, T.V.V., Klejwa, M., Swayze, G., Vergo, N., 1990b. High spectral resolution reflectance spectroscopy of minerals. J. Geophys. Res. 95, 12653–12680.

Clark, R.N., 1995. Reflectance spectra. In: Ahrens, T.J. (Ed.), Rock Physics & Phase Relations: A Handbook of Physical Constants. American Geophysical Union, Washington, DC. http://dx.doi.org/10.1029/RF003p0178.

Clarke, A.C., March 1968. Geostationary satellite communications. Spaceflight 10 (3), 85–86.

Cloude, S.R., Pottier, E., January 1997. An entropy based classification scheme for land applications of polarimetric SAR. IEEE Trans. Geosci. Remote Sens. 35 (1), 68–78.

Colliander, A., Ruokokoski, L., Suomela, J., Veijola, K., Kettunen, J., Kangas, V., Aalto, A., Levander, M., Greus, H., Hallikainen, M.T., Lahtinen, J., 2007. Development and calibration of SMOS reference radiometer. IEEE Trans. Geosci. Remote Sens. 45 (7), 1967–1977.

Collin, A., Hench, J.L., 2012. Towards deeper measurements of tropical reefscape structure using the WorldView-2 spaceborne sensor. Remote Sens. 4 (5), 1425–1447.

Comiso, J.C., 1986. Characteristics of Arctic winter sea ice from satellite multispectral microwave observations. J. Geophys. Res. 91 (C1), 975–994. http://dx.doi.org/10.1029/JC091iC01p00975.

Comiso, J., 1990. Arctic multiyear ice classification and summer ice cover using passive microwave satellite data. J. Geophys. Res. Oceans 95, 13411–13422.

Corbella, I., Duffo, N., Vall-llossera, M., Camps, A., Torres, F., 2004. The visibility function in interferometric aperture synthesis radiometry. Geosci. Remote Sens. IEEE Trans. 42 (8), 1677–1682.

Corbella, I., Torres, F., Camps, A., Duffo, N., Vall-llossera, M., Rautiainen, K., Martín-Neira, M., Colliander, A., 2005. Analysis of correlation and total power radiometer front-ends using noise waves. IEEE Trans. Geosci. Remote Sens. 43 (11), 2452–2459.

Corbella, I., Torres, F., Torres, F., Camps, A., Camps, A., Duffo, N., Duffo, N., Vall-llossera, M., Vall-llossera, M., 2009. Brightness-temperature retrieval methods in synthetic aperture radiometers. IEEE Trans. Geosci. Remote Sens. 47 (1), 285–294.

Cornillon, P., Gilman, C., Stramma, L., Brown, O., Evans, R., Brown, J., 1987. Processing and analysis of large volumes of satellite-derived thermal infra-red data. J. Geophys. Res. 92, 12993–13002.

Cosgriff, R.L., Peake, W.H., Taylor, R.C., 1960. Terrain scattering properties for sensor system design (Terrain Handbook II). In: Engineering Experiment Stations Bull No. 181. Ohio State Univ.

Cox, C., Munk, W., 1954. Measurement of the roughness of the sea surface from photographs of the sun's glitter. J. Opt. Soc. Am. 44, 838–850.

Cracknell, A.P., 2001. The exciting and totally unanticipated success of the AVHRR in applications for which it was never intended. Adv. Space Res. 28, 233–240.

Cressman, G.P., 1959. An operational objective analysis system. Month. Wea. Rev. http://dx.doi.org/10.1175/1520 0493(1959)087<0367:AOOAS>2.0.CO;2.

Crispin, J.W., Siegel, K.M. (Eds.), January 1, 1968. Methods of Radar Cross-section Analysis, first ed. Academic Press. ISBN-10: 0121977501.

Cumming, I.G., Bennett, J.R., April 1979. Digital processing of Seasat SAR data. In: Proc. ICASSP, pp. 710–718 (Washington, DC, USA).

Cumming, I., Wong, F., 1979. Digital Processing of SAR Data. Artech House, Norwood, MA.

Cumming, I.G., Wong, F.H., 2005. Digital Processing of Synthetic Aperture Radar Data. Algorithms and Implementation. Artech House, Boston, London.

Curlander, J.C., McDonough, R.N., 1991. Synthetic Aperture Radar: Systems and Signal Processing. John Wiley & Sons, New York, USA.

Curran, L.J., 1983. Multispectral remote sensing for the estimation of green leaf area index. Pidos. Trms. R. Land. Sr. A 309, 257–270.

Cutrona, J., Leigth, E.N., Palermo, C.J., Porcello, L.J., 1960. Optical data processing and filtering systems. IRE Trans. Inf. Theory IT-6, 386–400.

Cutrona, L.J., Leigth, E.N., Palermo, C.J., Porcello, L.J., 1961. A high resolution radar combat-surveillance system. IRE Trans. Milit. Electron. 127–131.

CYGNSS Handbook: Cyclone Global Navigation Satellite System — Deriving Surface Wind Speeds in Tropical Cyclones, April 1, 2016. Michigan Publishing. http://clasp-research.engin.umich.edu/missions/cygnss/reference/cygnss-mission/CYGNSS_Handbook_April2016.pdf.

D'Addio, S., Martin-Neira, M., 2013. Comparison of processing techniques for remote sensing of earth-exploiting reflected radio-navigation signals. Electronics Lett. 49 (4), 292–293.

D'Aria, D.A., Guarnieri, M., Rocca, F., 2004. Focusing bistatic synthetic aperture radar using dip move out. IEEE Trans. Geosci. Remote Sens. 42 (6), 1362–1810.

D'Addio, S., Martin-Neira, M., February 14, 2013. Comparison of processing techniques for remote sensing of earth-exploiting reflected radio-navigation signals. Electronics Lett. 49 (4), 292–293.

De Roo, R.D., Misra, S., 2008. Effectiveness of the sixth moment to eliminate a kurtosis blind spot in the detection of interference in a radiometer. In: Proceedings of the 2008 IEEE International Geoscience and Remote Sensing Symposium, IGARSS 2008, vol. 2, pp. 331–334 (Boston, MA, USA).

Derber, J.C., Wu, W.S., 1997. The use of TOVS cloud-cleared radiances in the NCEP SSI analysis system. Mon. Wea. Rev. 126, 2287–2299.

Desnos, Y.-L., Buck, C., Guijarro, J., Suchail, J.L., Torres, R., Attema, E., July 2000. The ENVISAT advanced synthetic aperture radar system. In: Proc. IGARSS, pp. 1171–1173 (Honolulu, USA).

Dibarboure, G., Purjol, M.I., Briol, F., LeTraon, P.Y., Larnicol, G., Picot, N., Mertz, F., Ablain, M., 2011. Jason-2 in DUACS: updated system description, first tandem results and impact on processing and products. Mar. Geod 34, 214–241.

Dicke, R.H., 1946. The measurement of thermal radiation at microwave frequencies. Rev. Sci. Instr. 17, 268–279.

Dobler, et al., 2002. Doppler lidar using aerosol backscatter and a frequency agile laser transmitter for profiling atmospheric winds in the planetary boundary layer. Geosci. Rem. Sens. Sym. http://dx.doi.org/10.1109/IGARSS.2002.1026742.

Dobson, M.C., Ulaby, F.T., Hallikainen, M., El-Rayes, M., 1985. Microwave dielectric behavior of we soil, Part II: dielectric mixing models. IEEETrans. Geosci. Remote Sens. 23, 35–46.

Donelan, Hamilton, J., Hui, W.H., 1985. Directional spectra of wind-generated surface gravity waves. Atmos. Ocean 30, 457—478.

Drinkwater, M.R., Long, D.G., 1998. Seasat, ERS-1/2 and NSCAT Scatterometer-observed Changes on the Large Ice Sheets. Eur. Space Agency Spec. Publ, pp. 91—96. ESA SP-424.

Drinkwater, M.R., Liu, X., Low, D., Wadhams, P., 1998. Interannual variability in Weddell sea ice from ERS wind scatterometer. Eur. Space Agency Spec. Publ. ESA SP-424, 119—123.

Drinkwater, M.R., Liu, X., 2000. Seasonal to interannual variability in Antarctic sea-ice surface melt. IEEE Trans. Geosci. Remote Sens. 38 (4), 1827—1842.

Drinkwater, M.R., Long, D.G., Bingham, A.W., 2001. Greenland snow accumulation estimates from satellite radar scatterometer data. J. Geophys. Res. 106, 33,935—33950. http://dx.doi.org/10.1029/2001JD900107.

Dubois-Fernandez, P., Dupuis, X., Garestier, F., 2005. POLINSAR calibration of a single-pass deramp-on—receive system: example of RAMSES airborne radar. Can. J. Remote Sens. J. canadien de télédétection 31 (1). http://dx.doi.org/10.5589/m04—055.

Dubovik, O., King, M.D., 2000. A flexible inversion algorithm for retrieval of aerosol optical properties from Sun and sky radiance measurements. J. Geophys. Res. 105 (D16), 20673—20696. http://dx.doi.org/10.1029/2000JD900282.

Dubovik, O., Sinyuk, A., Lapyonok, T., Holben, B.N., Mishenchenko, M., Yang, P., Eck, T.F., Voten, H., Muñoz, O., Veihelmann, B., van der Zande, W.J., Leon, J.F., Sorokin, M., Slutsker, I., 2006. Application of spheroid models to account for aerosol particle nonsphericity in remote sensing of desert dust. J. Geophys. Res. 111. http://dx.doi.org/10.1029/2005JD006619.

Duffo, N., Vall llossera, M., Camps, A., Corbella, I., Torres, F., 2009. Polarimetric emission of rain events: simulation and experimental results at X-band. Remote Sens. 1, 107—121.

Edelstein, W., Madsen, S.N., Moussessian, A., Chen, C., January 18, 2005. Concepts and technologies for synthetic aperture radar from MEO and geosynchronous orbits. Proc. SPIE 5659. http://dx.doi.org/10.1117/12.57898911111.

Edelsohn, C., Gurley, J., McCord, H., Donnelly, R., Virga, P., Butler, W., Jain, A., July 6—10, 1998. RADSDAR (RADiometric SAR) experimental results. In: Proceedings of the 1998 IEEE International Geoscience and Remote Sensing Symposium, pp. 372—374 (Seattle, WA, USA).

Edelsohn, C.R., August 8—12, 1994. Applications of synthetic aperture radiometry. In: Proceedings of the 1994 IEEE International Geoscience and Remote Sensing Symposium, pp. 1326—1328. Pasadena, California.

Egido, A.E., May 2013. GNSS Reflectometry for Land Remote Sensing Applications (Ph.D. thesis Dissertation). Starlab, Barcelona, Spain, 234 pp.

Elachi, C., Brown, W.E., Cimino, J.B., Dixon, T., Evans, D.L., Plant, J., 1982. Shuttle imaging radar experiment. Science 218, 996—1004.

Elfouhaily, T., Thompson, D., Vandermark, D., Chapron, B., 1999. Weakly nonlinear theory and sea state bias estimations. J. Geophys. Res. 104. http://dx.doi.org/10.1029/1998JC900128. ISSN: 0148-0227.

Ellingson, S.W., Johnson, J.T., 2006. A polarimetric survey of radio-frequency interference in C- and X-Bands in the continental United States using Windsat radiometry. IEEE Trans. Geosci. Remote Sens. 44, 540—548.

Ellison, W., Balana, A., Delbos, G., et al., 1998. New permittivity measurements of sea water. Radio Sci. 33 (3), 639—648. http://dx.doi.org/10.1029/97RS02223.

Emery, W.J., Thomas, A.C., Collins, M.J., Crawford, W.R., Mackas, D.L., 1986. An objective method for computing advective surface velocities from sequential infrared satellite images. J. Geophys. Res. 91, 12865—12878.

Emery, W.J., Fowler, C.W., Hawkins, J., Preller, R.H., 1991. Fram Strait satellite image—derived ice motions. J. Geophys. Res. Oceans 96, 4751—4768.

Emery, W.J., Yu, Y., Wick, G.A., Schlüssel, P., Reynolds, R.W., 1994. Correcting infrared satellite estimates of sea surface temperature for atmospheric water vapor attenuation. J. Geophys. Res. 99.

Emery, W.J., Crocker, R.I., Baldwin, D., 2010. Automated AVHRR image navigation. In: Le Moigne, J., Netanyahu, N.S., Eastman, R.D. (Eds.), Image Registration for Remote Sensing. Cambridge University Press, Cambridge, UK, pp. 383−399.

Ender, J.G.H., Walterschied, I., Brenner, A., 2004. New aspects of bistatic SAR: processing and experiments. In: Proc. IGARSS, pp. 1758−1762 (Anchorage, USA).

Ender, J.G.H., Walterschied, I., Brenner, A., June 2006. Bistatic SAR translational invariant processing and experimental results. IEE Proc. Radar Sonar Navig. 153, 177−183.

Entekhabi, D., Njoku, E.G., O'Neill, P.E., Kellogg, K.H., Crow, W.T., Edelstein, W.N., Entin, J.K., Goodman, S.D., Jackson, T.J., Johnson, J., Kimball, J., Piepmeier, J.R., Koster, R.D., Martin, N., McDonald, K.C., Moghaddam, M., Moran, S., Reichle, R., Shi, J.C., Spencer, M.W., Thurman, S.W., Tsang, L., Van Zyl, J., May.2010. The soil moisture active passive (SMAP) mission. Proc. IEEE 98 (5), 704−716.

Esbensen, S.K., Kushnir, V., 1981. The Heat Budget of the Global Ocean: An Atlas Based on Estimates from Surface Marine Observations. Climate Research Institute, Rep. No. 29. Oregon State University, 27 p.

Escorihuela, M.J., Kerr, Y.H., de Rosnay, P., Wigneron, J.-P., Calvet, J.-C., Lemaitre, F., 2007. A simple model of the bare soil microwave emission at L-band. IEEE Trans. Geosci. Remote Sens. 45, 1978−1987.

Evans, D.L., Farr, T.G., Zyl, J.J.V., Zebker, H.A., Nov.1988. Radar polarimetry: analysis tools and applications. IEEE Trans. Geosci. Remote Sens. 26 (6), 774−789.

Evans, D.L., Alpers, W., Cazenave, A., Elachi, C., Farr, T., Glackin, D., Holt, B., Jones, L., Liu, W.T., McCandless, W., Menard, Y., Moore, R., Njoku, E., 2005. Seasat-A 25-year legacy of success. Remote Sens. Environ. 94, 384−404.

Eyre, J.R., Kelly, G.A., McNally, A.P., Anderson, E., Persson, A., 1993. Assimilation of TOVS radiance information through one-dimensional variational analysis. Q. J. R. Meteorol. Soc. 119, 1427−1463.

Fabra-Cervella, F., Cardellach, E., Rius, A., Ribo, S., Oliveras, S., Nogues-Correig, O., Rivas, M.B., Semmling, M., D'Addio, S., 2012. Phase altimetry with dual polarization GNSS-R over sea ice. IEEE Trans. Geosci. Remote Sens. 50, 2112−2121. http://dx.doi.org/10.1109/TGRS.2011.2172797.

Fabra-Cervellera, F., 2013. GNSS-r as a Source of Opportunity for Remote Sensing of the Cryosphere (Ph.D. dissertation). UPC, Barcelona, Spain.

Fairall, C.W., Bradley, E.F., Rogers, D.P., Edson, J.B., Young, G.S., 1996. Bulk parameterization of air-sea fluxes for tropical ocean global atmosphere coupled-ocean atmosphere response experiment. J. Geophys. Res. http://dx.doi.org/10.1029/95JC03205.

Fang, F., Morrow, R., 2003. Evoluation, movement and decay of warm-core Leéuwin Current eddies. Deep Sea Res. Part II 50, 2,245−2,261.

Farr, T.G., Rosen, P.A., Caro, E., Crippen, R., Duren, R., Hensley, S., Kobrick, M., Paller, M., Rodriguez, E., Roth, L., Seal, D., Shaffer, S., Shimada, J., Umland, J., Werner, M., Oskin, M., Burbank, D., Alsdorf, D., 2007. The shuttle radar topography mission. Rev. Geophys. 45, 19617−19634.

Ferrazzoli, P., Guerriero, L., Wigneron, J.-P., 2002. Simulating L-band emission of forests in view of future satellite applications. IEEE Trans. Geosci. Remote Sens. 40, 2700−2708.

Ferrazzoli, P., Guerriero, L., Perdicca, N., Rahmoune, R., 2011. Forest biomass monitoring with GNSS-R: theoretical simulations. Adv. Space Res. 47, 1823−1832.

Fjeldbo, G., Eshleman, V.R., 1965. J. Geophys. Res. 70 (13), 3217.

Fong, C.J., Yen, N.L., Chu, C.H., Yang, S.K., Shiau, W.T., Huang, C.Y., Chi, S., Chen, S.S., Liou, Y.A., Kuo, Y.H., 2009. FORMOSAT-3/COSMIC spacecraft constellation system, mission results, and prospect for follow-on mission. Terr. Atmos. Ocean Sci. 20, 1−19. http://www.ocean-sci.net/20/1/2009/.

Font, J., Camps, A., Borges, A., Martin-Neira, M., Boutin, J., Reul, N., Kerr, Y.H., Hahne, A., Mecklenburg, S., 2010. SMOS: the challenging sea surface salinity measurement from space. Proc. IEEE 98 (5), 649−665.

Fontana, R.D., Wai, C., Novak, P.M., September 2001. The new L2 civil signal. In: Proc. ION GPS 2001, Salt Lake City, UT, pp. 617−631.

Forte, G.F., Tarongí Bauza, J.M., dePau, V., Vall·llossera, M., Camps, A., October 2013. Experimental study on the performance of RFI detection algorithms in microwave radiometry: toward an optimum combined test. In: IEEE Transactions on Geoscience and Remote Sensing, vol. 51, No. 10, pp. 4936–4944. http://dx.doi.org/10.1109/TGRS.2013.2273081.

Fortescue, S., 2003. In: Swiner (Ed.), Spacecraft Systems Engineering. John Wiley.

Fowler, C., Emery, W.J., Maslanik, J.A., 2004. Satellite derived evolution of Arctic sea ice age: Oct. 1978 to Mar. 2003. IEEE Geosci. Remote Sens. Lett. 1, 71–74.

Franceschetti, G., Lanari, R., 1999. Synthetic Aperture Radar Processing. CRC Press, Boca Raton, USA.

Frasier, S.J., Camps, A.J., February 2001. Dual-beam interferometry for ocean surface current vector mapping. IEEE Trans. Geosci. Remote. Sens. 39 (2), 401–414.

Frater, R.H., Williams, D.R., 1981. An active 'Cold' noise. IEEE Trans. Microwave Theory 29, 344–347.

Freeman, et al., 2016. The cube train. In: 2016 CubeSat Developpers Workshop. CalPoly.

Frankignoul, C., Müller, P., 1979. On the generation of geostrophic eddies by surface buoyancy flux anomalies. J. Phys. Oceanogr. 9, 1207–1213.

Frigo, M., Johnson, S.G., 1998. FFTW: an Adaptive software architecture for the FFT. In: Proceedings of the IEEE International Conference on Acoustics, Speech, and Signal Processing.

Fu, L.-L., Alsdorf, D., Rodríguez, E., Morrow, R., Mognard, N., Lambin, J., Vaze, P., Lafon, T., 2009. The SWOT (Surface Water and Ocean Topography) mission: spaceborne radar interferometry for oceanographic and hydrological applications. In: Hall, J., Harrison, D.E., Stammer, D. (Eds.), Proceedings of OceanObs'09: Sustained Ocean Observations and Information for Society (vol. 2), Venice, Italy, 21–25 September 2009. ESA Publication WPP-306. http://dx.doi.org/10.5270/OceanObs09.cwp.33.

Fu, L.-L., Chelton Dudley, B., Le Traon, P.-Y., Morrow, R., 2010. Eddy dynamics from satellite altimetry. Oceanography 23 (4), 14–25. Open Access version: http://archimer.ifremer.fr/doc/00026/13744/.

Fukui, H., 1966. The noise performance of microwave transistors. IEEE Trans. Electron Devices 13 (3), 329–341.

Fung, A.K., Pan, G.W., 1986. An integral equation model for rough surface scattering. In: Proceedings of the International Symposium on Multiple Scattering of Waves in Random Media and Random Rough Surface, 29 July–2 August 1985. Pennsylvania State University Press, University Park, PA, pp. 701–714.

Fung, A.K., Pan, G.W., 1987. A scattering model for perfectly conducting random surface: I Model development, II Range of validity. Int. J. Remote Sens. 8, 1579–1605.

Fung, A.K., 1994. Microwave Scattering and Emission. Models and Their Applications. Artech House, Norwood, MA. ISBN:10-0890065233.

Gabarró, C., Font, J., Camps, A., et al., 2004. A new empirical model of sea surface microwave emissivity for salinity remote sensing. Geophys. Res. Lett. 31, L01309. http://dx.doi.org/10.1029/2003GL018964.

Gabriel, A.K., Goldstein, R.M., Feb.1988. Crossed orbit interferometry: theory and experimental results from SIR-B. Int. J. Remote Sens. 9 (5), 857–872.

Gabriel, A.K., Goldstein, R.M., Zebker, H.A., 1989. Mapping small elevation changes over large areas: differential radar interferometry. J. Geophys. Res. 94, 9183–9191.

Galileo, I.C.D., September 2010. European GNSS (Galileo) Open Service Signal in Space Interface Control Document (OS SIS ICD), Issue 1.1, 2010. European Union.

Garrison, J.L., Katzberg, S.J., 2000. The application of reflected GPS signals to ocean remote sensing. Remote Sens. Environ. 73, 175–187.

Garrison, J.L., Katzberg, S.J., April 12–14, 1997. Detection of ocean reflected GPS signals: theory and experiment. In: Southeastcon'97, 'Engineering New New Century', Proc. IEEE, pp. 290–294.

Garrison, J.L., Katzberg, S.J., Howell, C.T., 1997. Detection of ocean reflected GPS signals: theory and experiment. In: Proc. IEEE Southeastern. IEEE, Blacksburg, VA, pp. 290–294.

Garrison, J.L., Komjathy, A., Zavorotny, V.U., Katzberg, S.J., 2002. Wind speed measurement using forward scattered GPS signals. IEEE Trans. Geosci. Remote Sens. 40 (1), 50–65.

Garrison, J.L., 2012. Modeling and simulation of bin-bin correlations in GNSS-R waveforms. IEEE Intl. Geosci. Remote Sens. Soc. Symp. (Munich, Germany).

Gary, B.L., Hereford, A.Z., 2013. Tutorial on Airborne Microwave Temperature Profilers. http://brucegary.net/MTP_tutorial/x.htm.

Gatebe, C.K., Butler, J.J., Cooper, J.W., Kowalewski, M., King, M.D., 2007. Characterization of errors in the use of integrating-sphere systems in the calibration of scanning radiometers. App. Opt. 46. http://dx.doi.org/10.1364/AO.46.007640.

Gatelli, F., Guarnieri, A.M., Parizzi, F., Pasquali, P., Prati, C., Rocca, F., 1994. TheWavenumber Shift in SAR interferometry. IEEE Trans. Geosci. Remote Sens. 32 (4), 855−863.

Gentemann, C.L., Meissner, T., Wentz, F.J., 2010. Acuracy of satellite sea surface temperatures at 7 and 11 GHz, 48. http://dx.doi.org/10.1109/TGRS.2009.2030322.

Germain, O., Ruffini, G., Soulat, F., Caparrini, M., Chapron, B., Silvestrin, P., July 2003. The GNSS-R Eddy Experiment II: L-band and optical speculometry for directional sea-roughness retrieval from low altitude aircraft. In: Paper Presented at 2003 Workshop on Oceanography with GNSS Reflections. Starlab, Barcelona, Spain. Available at: http://arxiv.org/abs/physics/0310093.

Germain, O., Ruffini, G., Soulat, F., Caparrini, M., Chapron, B., Silvestin, P., 2004. The eddy experiment: GNSS-R speculometry for directional sea-roughness retrieval from low altitude aircraft. Geophys. Res. Lett. http://dx.doi.org/10.1029/2004GL020991.

Ghavidel, A., Schiavulli, D., Camps, A., 2014. A numerical simulator to evaluate the electronmagnetic bias in GNSS-R altimetry. Geosci. Remote Sci. http://dx.doi.org/10.1109/IGARSS.2014.6947379.

Ghavidel, A., Schiavulli, D., Camps, A., 2016. Numerical computation of the electromagnetic bias in GNSS-R altimetry. IEEE Trans. Geosci. Remote Sens. 54 (1), 489−498.

Giglio, L., Kendall, J.D., Justice, C.O., 1999. Evaluation of global fire detection algorithms using simulated AVHRR infrared data. Int. J. Remote Sens. 20, 1947−1985.

Gilbert, R.O., 1987. Statistical Methods for Environmental Pollution Monitoring. Wiley, NY.

Girard-Ardhuin, F., Ezraty, R., 2012. Enhanced Arctic sea ice drift estimation merging radiometer and scatterometer data. Trans. Geosci. Remote Sens. 50, 2639−2648.

Gleason, S., Gebre-Egziabher, D. (Eds.), 2009. GNSS Applications and Methods. Artech House, Nordwood, MA, USA.

Gleason, S., Hodgart, S., Sun, Y., Gommenginger, C., Mackin, S., Adjrad, M., Unwin, M., 2005. Detection and processing of bistatically reflected GPS signals from low Earth orbit for the purpose of ocean remote sensing. IEEE Trans. Geosci. Remote Sens. http://dx.doi.org/10.1109/TGRS.2005.845643.

Goggins, W.B., 1967. A microwave feedback radiometer. IEEE Trans. Aerosp. Electron. Syst. 3, 83−90.

Goldstein, R.M., Engelhardt, H., Kamb, B., Frolich, R.M., 1993. Satellite radar interferometry for monitoring ice sheet motion: application to an Antarctic ice stream. Science 262, 1525−1530.

González-Gambau, V., Turiel, A., Olmedo, E., Martínez, J., Corbella, I., Camps, A., April 2016. Nodal sampling: a new image reconstruction algorithm for SMOS. IEEE Trans. Geosci. Remote Sens. 54 (4), 2314−2328. http://dx.doi.org/10.1109/TGRS.2015.2499324.

González-Tello, P., Camacho, F., Vicaria, J.M., González, P.A., 2008. A modified Nukiyama−Tanasawa distribution function and a Rosin−Rammler model for the particle-size-distribution analysis. Powder Technol. 186, 278−281.

Goodman, J.W., 1968. In: Introduction to Fourier Optics. Mc Graw-Hill.

Goodman, J.W., 1985. In: Statistical Optics. Wiley Interscience.

Gough, S.R., 1972. A low temperature dielectric cell and the permittivity of hexagonal ice to 2K. Can. J. Chem. 50, 3046−3051.

Graham, L.C., 1974. Synthetic interferometric radar for TopographicMapping. Proc. IEEE 62, 763−768.

Gray, A.L., Mattar, K.E., Sofko, G., 2000. Influence of ionospheric electron density fluctuations on satellite radar interferometry. Geophys. Res. Lett. 27, 1451−1454.

Guan, L., Kawamura, H., 2004. Merging satellite infrared and microwave SSTs: methodology and evaluation of the new SST. J. Oceanogr. 60, 905—912.

Guimbard, S., Gourrion, J., Portabella, M., et al., 2012. SMOS semi-empirical ocean forward model adjustment. IEEE Trans. Geosci. Remote Sens. 50, 1676—1687. http://dx.doi.org/10.1109/TGRS.2012.2188410.

Güner, B., Johnson, J.T., Niamswaun, N., 2007. Time and frequency blanking for radio frequency interference mitigation in microwave radiometry. IEEE Trans. Geosci. Remote Sens. 45, 3672—3679.

Guner, B., Frankford, M.T., Johnson, J.T., 2008. On the Shapiro—Wilk test for the detection of pulsed sinusoidal radio frequency interference. In: Proceedings of the 2008 IEEE International Geoscience and Remote Sensing Symposium, IGARSS 2008, vol. 2, pp. 157—160 (Boston, MA, USA).

Hach, J.P., 1968. A very sensitive airborne microwave radiometer using two reference temperatures. IEEE Trans. Microwave Theory 16, 629—636.

Hagen, J., Farley, D., 1973. Digital correlation techniques in radio science. Radio Sci. 8 (8—9), 775—784.

Haines, B., Bertiger, W., Desai, S., Kuang, D., Munson, T., Young, L., Willis, P., 2003a. Initial orbit determination results for Jason-1: towards a 1 cm orbit. Navigation. http://dx.doi.org/10.1002/j.2161-4296.2003.tb00327.x.

Haines, B.J., Dong, D., Born, G.H., Gill, S.K., 2003b. The harvest experiment: monitoring Jason-1 and TOPEX/POSEIDON from a California offshore platform. Special issue: Jason-1 calibration/validation. Marine Geodesy 26 (3—4).

Hajj, G.A., Zuffada, C., 2003. Theoretical description of a bistatic system for ocean altimetry using the GPS signal, 38. http://dx.doi.org/10.1029/2002RS002787.

Hall, C.D., Cordey, R.A., 1988. Multisatic scatterometry. IGARSS 88. http://dx.doi.org/10.1109/IGARSS.1988.570200.

Hallikainen, M.T., Ulaby, F.T., Dobson, M.C., El-Rayes, M.A., Wu, L., 1985. Microwave dielectric behaviour of wet soil — Part 1: empirical models and experimental observations. IEEE Trans. Geosci. Remote Sens. GE-23, 25—34.

Han, Y., Westwater, E.R., 2000. Analysis and improvement of tipping calibration for ground-based microwave radiometers. IEEE Trans. Geosci. Remote Sens. 38 (3), 1260—1276.

Hansen, J.E., Travis, 1974. Light scattering in planetary atmospheres. Space Sci. Rev. 16, 527—610.

Hardy, W.N., 1973. Precision temperature reference for microwave radiometry. IEEE Trans. Microwave Theory 21, 149—150.

Hasselmann, K., Barnett, T.P., Bouws, E., Carlson, H., Cartwright, D.E., Enke, K., Ewing, J.A., Gienapp, H., Hasselmann, D.E., Kruseman, P., Meerburg, A., Mller, P., Olbers, D.J., Richter, K., Sell, W., Walden, H., 1973. Measurements of Wind-Wave Growth and Swell Decay during the Joint North Sea Wave Project (JONSWAP). Ergnzungsheft zur Deutschen Hydrographischen Zeitschrift Reihe, A(8) (Nr. 12), p. 95.

Hecht, E., 1997. Optics, third ed. Addison-Wesley.

Heise, S., Wickert, J., Beyerle, G., Schmidt, T., Smit, H., Cammas, J.-P., Rothacher, M., 2008. Comparison of water vapor and temperature results from GPS radio occultation aboard CHAMP with MOZAIC aircraft measurements. IEEE Trans. Geosci. Remote Sens. 46 (11), 3406—3411.

Hersman, J.M.S., Poe, G.A., 1981. Sensitivity of the total power radiometer with periodic absolute calibration. IEEE Trans. Microwave Theory 29, 32—40.

Hilburn, K.A., Wentz, F.J., 2008. Intercalibrated passive microwave rain products from the unified microwave ocean retrieval algorithm (UMORA). J. Appl. Meteorol. Climatol. 47, 778—794.

Hipp, J.E., 1974. Soil electromagnetic parameters as a function of frequency, soil density and soil moisture. Proc. IEEE 98—103.

Holben, B.N., Eck, T.F., Slutsker, I., Tanre, D., Buis, J.P., Setzer, A., Vermote, E., Reagan, J.A., Kaufman, Y.J., Nakajima, T., Lavenu, F., Jankowiak, I., Smirmov, A., 1998. AERONET-A federated instrument network and data archive for aerosol characterization. Remote Sens. Environ. 66 (1), 1—16. http://dx.doi.org/10.1016/S0034-4257(98)00031-5.

Holben, B.N., 1986. Characteristics of maximum-value composite images from temporal AVHRR data. Int. J. Remote Sens. 7, 1417—1434.

Hollinger, J.P., 1971. Passive microwave measurements of sea surface roughness. IEEE Trans. Geosci. Electronics GE-9 (3), 165—169. http://dx.doi.org/10.1109/TGE.1971.271489.

Hollinger, J.P., Lo, J.R., Poe, G., Savage, R., Peirce, J., 1987. Special Sensor Microwave/Imager User's Guide. Naval Research Laboratory.

Holmes, T., de Rosnay, P., de Jeu, R., Wigneron, J.P., Kerr, Y., Calvet, J.C., Escorihuela, M.J., Saleh, K., Lemaitre, F., 2006. A new parameterization of the effective temperature for L- band radiometry. Geophys. Res. Lett. 33, 7405.

Horn, R., Moreira, J., Meier, E., 1992. A refined procedure to generate calibrated imagery from airborne synthetic aperture radar data. In: Proceedings of ZGARSS'92, Houston (USA).

Horn, R., May 27—31, 1996. The DLR airborne SAR project E-SAR. In: Proc. IGARSS, pp. 1624—1628. Lincoln, USA.

Horstmann, J., Koch, W., Lehner, S., Rosenthal, W., January 1998. Ocean wind fields and their variability derived from SAR. Earth Observation Q.

Hovis, W.A., Knoll, J.S., 1983. Characteristics of an internally illuminated calibration sphere. Appl. Opt. 22. http://dx.doi.org/10.1364/AO.22.004004.

Hülsmeyer, C., 1904. Verfahren um entfernte metallische Gegenstaede mittels elektrscher Wellen einem Beobachter zu melden. German Patent 165.

Huntemann, M., Heygster, G., Kaleschke, L., Krumpen, T., Mäkynen, M., Drusch, M., 2014. Empirical sea ice thickness retrieval during the freeze-up period from SMOS high incident angle observations. The Cryosphere 8, 439—451. http://dx.doi.org/10.5194/tc-8-439-2014.

Husar, R.B., Prospero, J.M., Stowe, L.L., 1997. Chracterization of tropospheric aerosols over the oceans with the NOAA advanced very high resolution radiometer optical thickness operational product. J. Geophys. Res. 102 http://dx.doi.org/10.1029/96JD04009.

Hussain, E., Ural, S., Malik, A., Shan, J., May 1—5, 2011. Mapping Pakistan 2010 floods using remote sensing data. In: ASPRS 2011 Annual Conference, Milwaukee, Wisconsin.

Isern-Fontanet, J., Garcia-Ladona, E., Font, J., 2003. Identification of marine eddies from altimetric maps. AMS. http://dx.doi.org/10.1175/1520-0426(2003)20<772:IOMEFA>2.0.CO;2.

Ishiguro, M., 1980. Minimum redundancy linear arrays for a large number of antennas. Radio Sci. 15 (6), 1163—1170.

Jackson, T.J., O'Neill, P.E., 1986. Microwave dielectric model for aggregated soils. IEEE Trans. Geosci. Remote Sens. 24, 920—929.

Jackson, T.J., O'Neill, P.E., 1987. Salinity effects on the microwave emission of soil. IEEE Trans. Geosci. Remote Sens GE-25, 214—220.

Jackson, T.J., O'Neill, P.E., 1990. Attenuation of soil microwave emission by corn and soy beans at 1.4 and 5 GHz. IEEE Trans. Geosci. Remote Sens. 28, 978—980.

Jackson, T.J., Schmugge, T.J., 1989. Passive microwave remote sensing system for soil moisture: some supporting research. IEEE Trans. Geosci. Remote Sens. GE-27, 225—235.

Jackson, T.J., Schmugge, T.J., 1991. Vegetation effects on the microwave emission of soils. Remote Sens. Environ. 36, 203—212.

Jackson, T.J., Schmugge, T.J., Wang, J.R., 1982. Passive microwave remote sensing of soil moisture under vegetation canopies. Water Resour. Res. 18, 1137—1142.

Jackson, T.J., Schmugge, T.J., O'Neil, P., 1984. Passive microwave remote sensing of soil moisture from an aircraft platform. Remote Sens. Environ. 14, 135—151.

Jacquemoud, S., Bacour, C., Poilve, H., Frangi, J.P., 2000. Comparison of four radiative transfer models to simulate plant canopies reflectance: direct and inverse mode. Remote Sens. Environ. 74, 471—481. http://dx.doi.org/10.1016/S0034-4257(00)00139-5.

Jasinski, M.F., 1996. Estimation of subpixel vegetation density of natural regions using satellite multispectral imagery. IEEE Trans. Geosci. Remote Sens. 34, 804–813.

Jayne, S.R., Marotzke, J., 2002. The oceanic eddy heat transport. J. Phys. Oceanogr. 32, 3328–3345.

Jin, S.G., Cardellach, E., Xie, F., 2014. GNSS Remote Sensing: Theory, Methods and Applications. Springer, Dordrecht, Netherlands, ISBN 978-94-007-7481-0.

Johnson, R.C., Jasik, H., 1993. Antenna Engineering Handbook. McGraw-Hill Professional.

Johnson, J., 1928. Thermal agitation of electricity in conductors. Phys. Rev. 32, 97. http://dx.doi.org/10.1103/PhysRev.32.97.

Johnsen, H., Engen, G., Hogda, K., Chapron, B., Desnos, Y., 1999. Validation of ENVISAT wave mode level 1b and level 2 product using ERS SAR data. In: Proc. CEOS SAR Workshop, Toulouse, 26–29 October 1999, ESA SP-450, pp. 59–64.

Jones, M.O., Kimball, J.S., McDonald, K.C., Jones, L.A., 2010. Enhanced Phenology Monitoring Using Microwave and Optical-infrared Satellite Remote Sensing (unpublished manuscript).

Justice, C.O., Wolfe, R.E., El Saleous, N., Descloitres, J., Vermote, E., Roy, D., Owens, J., Masuoka, E., 2000. The availability and status of MODIS land products. The Earth Observer 12 (6), 10–18.

Justice, C.O., Giglio, L., Korontzi, S., Owens, J., Morisette, J.T., Roy, D., Descloitres, J., Alleaume, S., Pertitcolin, F., Kaufman, Y., 2002. The MODIS fire products. Remote Sens. Environ. 83, 244–262.

Kachi, M., Murakami, H., Imaoka, K., Shibata, A., 2001. Sea surface temperature retrieved from TRMM microwave imager and visible infrared scanner. J. Meteorol. Soc. Jpn.

Kaleschke, L., Maaß, N., Haas, C., Hendricks, S., Heygster, G., Tonboe, R.T., 2010. A sea-ice thickness retrieval model for 1.4 GHz radiometry and application to airborne measurements over low salinity sea-ice. The Cryosphere 4, 583–592. http://dx.doi.org/10.5194/tc-4-583-2010.

Kaleschke, L., Tian-Kunze, X., Maaß, N., Mäkynen, M., Drusch, M., 2012. Sea ice thickness retrieval from SMOS brightness temperatures during the Arctic freeze-up period. Geophys. Res. Lett. 39, L05501. http://dx.doi.org/10.1029/2012GL050916.

Karsten, R.H., Marshall, J., 2002. Constructing the residual circulation of the ACC from observations. J. Phys. Oceanogr. 32, 3315–3327.

Katzberg, S.J., Garrison, J.L., December 1996. Utilizing GPS to determine ionospheric delay over the ocean. NASA Tech. Memorandum 4750.

Katzberg, S.J., Torres, O., Ganoe, G., 2006. Calibration of reflected GPS for tropical storm wind speed retrievals. Geophys. Res. Lett. 33, L18602. http://dx.doi.org/10.1029/2006GL026825.

Kaufman, Y.J., Setzer, A.W., Justice, C.O., Tucker, C.J., Pereira, M.C., Fung, I., 1990. Remote sensing of biomass burning in the tropics. In: Goldammer, J.G. (Ed.), Fire and the tropical biota: ecosystem pro- cesses and global challenges. Springer-Verlag, Berlin, pp. 371–399.

Kaufman, Y.J., Kleidman, R.G., King, M.D., 1998. SCAR-B fires in the tropics: properties and remote sensing from EOS-MODIS. J. Geophys. Res. 103, 31955–31968.

Kavak, A., Vogel, A.J., Xu, G., 1998. Using GPS to measure ground complex permittivity. Electronics Lett. 34 http://dx.doi.org/10.1049/el:19980180.

Kay, S.M., 1993. Fundamentals of Statistical Signal Processing: Estimation Theory. Prentice Hall, Upper Saddle River, NJ 07458.

Kell, R.E., 1965. On the derivation of bistatic RCS from monostatic measurements. Proc. IEEE 53. http://dx.doi.org/10.1109/PROC.1965.4077.

Kerbaol, V., Chapron, B., Vachon, P.W., 1998. Analysis of ERS-1/2 synthetic aperture radar wave mode imagettes. J. Geophys. Res. 103 http://dx.doi.org/10.1029/97JC01579.

Kerr, Y.H., Waldteufel, P., Wigneron, J.-P., Delwart, S., Cabot, F., Boutin, J., Escorihuela, M.J., Font, J., Reul, N., Gruhier, C., Juglea, S.E., Drinkwater, M.R., Hahne, A., Martin-Neira, M., Mecklenburg, S., 2010. The SMOS mission: new tool for monitoring key elements of the global water cycle. Proc. IEEE 98 (5), 666–687. http://dx.doi.org/10.1109/JPROC.2010.2043032.

Kidwell, K.B., 1997. Global Vegetation Index User's Guide (Camp Springs MD: US Department of Commerce. NOAA, National Environmental Satellite Data and Information Service, National Climatic Data Center, Satellite Data Services Division).

Kim, J.H., Younis, M., Prats-Iraola, P., Gabele, M., Krieger, G., January 2013. First spaceborne demonstration of digital beamforming for azimuth ambiguity suppression. IEEE Transactions Geoscience Remote Sens. 51 (1), 579.

King, M.D., Strange, M.G., Leone, P., Blaine, L.R., 1986. Multiwavelength scanning radiometer for airborne measurements of scattered radiation within clouds. J. Atmos. Oceanic Technol. 3, 513−522.

King, M.D., Radke, L.F., Hobbs, P.V., 1990. Determination of the spectral absorption of solar radiation by marine stratocumulus clouds from airborne measurements within clouds. J. Atmos. Sci. 47, 894−907.

King, M.D., Tsay, S.C., Platnick, S.E., Wang, M., Liou, K.N., 1997. Cloud Retrieval Algorithms for MODIS: Optical Thickness, Effective Particle Radius, and Thermodynamic Phase. Algorithm Theoretical Basis Document ATBD-mod-05. Goddard Space Flight Center, 79 pp.

Kirchhoff, G., 1860. On the relation between the radiating and absorbing powers of different bodies for light and heat. Phylosophical Magazine Ser. 4 20, 1−21.

Kirdiashev, K.P., Chukhlantsev, A.A., Shutko, A.M., 1979. Microwave radiation of the Earth's surface in the presence of vegetation cover. Radio Eng. Electron. Phys. 2, 37−56.

Kirk Jr., J.C., 1975a. A discussion of digital processing in synthetic aperture radar. IEEE Trans. Aerosp. Electron. Syst. AES-11 (3), 326−337.

Kirk Jr., J.C., 1975b. Motion compensation for synthetic aperture radar. IEEE Trans. Aerosp. Electron. Syst. 11 (3), 220−230.

Klein, L.A., Swift, C.T., 1977. An improved model for the dielectric constant of sea water at microwave frequencies. IEEE Trans. Anten. Propag. AP 25, 104−111. http://dx.doi.org/10.1109/TAP.1977.1141539.

Klobuchar, J.A., April 1991. Ionospheric effects on GPS. GPS World 2 (4).

Kogut, A., Fixsen, D., Fixsen, S., Levin, S., Limon, M., Lowe, L., Mirel, P., Seiffert, M., Singal, J., Lubin, P., Wollack, E., 2006. ARCADE: absolute radiometer for cosmology, astrophysics, and diffuse emission. New Astron. Rev. 50, 925−931.

Komiyama, K., Kato, Y., July 10−14, 2016. Two-dimensional supersynthesis radiometer for field experiment. In: Proceedings of the 1995 IEEE International Geoscience and Remote Sensing Symposium, pp. 2264−2266. Firenze, Italy.

Komiyama, K., Kato, Y., Iwasak, T., August 8−12, 1994. Indoor experiment of two-dimensional supersynthesis radiometer. In: Proceedings of the 1994 IEEE International Geoscience and Remote Sensing Symposium, pp. 1329−1331 (Pasadena, California).

Komiyama, K., Kato, Y., Furuya, K., August 3−8, 1997. Interpretation of the brightness temperature retrieved by supersynthesis radiometers. In: Proceedings of the International Geoscience and Remote Sensing Symposium, pp. 481−483 (Singapore).

Komiyama, K., July 6−10, 1998. Preliminary experiment of a one-dimensional imaging by microwave super-synthesis radiometer. In: Proceedings of the 1998 IEEE International Geoscience and Remote Sensing Symposium, pp. 1708−1710 (Seattle, WA, USA).

Komjathy, A., J. Maslanik, V.U. Zavorotny, P. Axelrad, and S.J. Katzberg, Sea ice remote sensing using surface reflected GPS signals, IEEE 2000 International Geoscience and Remote Sensing Symposium, vol. 7, pp. 2855−2857.

Komjathy, A., Armatys, M., Masters, D., Axelrad, P., Zavorotny, V., Katzberg, S., March 2004. Retrieval of ocean surface wind speed and wind direction using reflected GPS signals. J. Atmos. Oceanic Technol. 21, 515−526.

Kovaly, J.J., Newell, G.S., Prothe, W.C., Sherwin, C.W., November 1952. The Observations of Snorkels and Sea Clutter Using Coherent Airborne Radar.

Kraak, M.J., Ormeling, F.J., 1996. Cartography Visualization of Spatial Data. Adison Wesley, Essex.

Krieger, G., Gebert, N., Moreira, A., 2008. Multidimensional waveform encoding: a new digital beamforming technique for synthetic aperture radar remote sensing. IEEE Trans. Geosci. Remote Sens. 46 (1), 31–46.

Kristensen, S.S., Balling, J., Skou, N., Søbjærg, S.S., 2012. RFI in SMOS data detected by polarimetry. In: Proceedings of the 2012 IEEE International Geoscience and Remote Sensing Symposium, IGARSS 2012, pp. 3320–3322.

Kussul, N., Shelestov, A., Skakun, S., 2011. Flood monitoring on the basis of SAR data (2011). In: Kogan, F., Powell, A., Fedorov, O. (Eds.), Use of Satellite and In-situ Data to Improve Sustainability, NATO Science for Peace and Security Series C: Environmental Security, pp. 19–29. http://dx.doi.org/10.1007/978-90-481-9618-0_3.

Kwak, A.K., Rosenfeld, D.E., Chung, J.K., Fayer, M.D., 2008. J. Phys. Chem. B 112, 13906–13915.

Kwok, R., 2004. Annual cycles of multiyear sea ice coverage of the Arctic Ocean: 1999–2003. J. Geophys. Res. 109. http://dx.doi.org/10.1029/2003JC002238.

Lahtinen, J., Gasiewski, A.J., Klein, M., Corbella, I., 2003. A calibration method for fully polarimetric microwave radiometers. IEEE Trans. Geosci. Remote Sens. 41 (3), 558–602.

Larson, K.M., Nievinski, F.G., 2013. GPS snow sensing: results from the EarthScope plate boundary observatory. GPS Solutions 17. http://dx.doi.org/10.1007/s10291-012-0259-7.

Larson, K.M., Small, E.E., Gutmann, E., Bilich, A., Axelrad, P., Braun, J., 2008a. Using GPS multipath to measure soil moisture fluctuations: initial results. GPS Solutions 12, 173–177.

Larson, K.M., Small, E.E., Gutmann, E.D., Bilich, A.L., Braun, J.J., Zavorotny, V., 2008b. Use pf GPS receivers as a soil moisture network for water cycle studies. Geophys. Res. Lett. 35. http://dx.doi.org/10.1029/2008GL036013.

Laur, H., Bally, P., Measdows, P., Sanchez, J., Schaettler, B., Lopinto, E., 1997. Derivation of Backscattering Coefficient $\Sigma°$ in ESA ERS PRI Products ERS SAR Calibration Doc. No. ES-TN-rs-pm-hl09, 2(4). European Space Agency, Frascati, Italy.

Laws, J.O., Parsons, D.A., 1943. The relationship of raindrops size to intensity. Trans. AGU 452–460.

Le Traon, P.Y., Dibarboure, G., 2002. Velocity mapping capabilities of present and future altimeter missions: the role of high frequency signals. J. Atm. Ocean Tech. 19, 2077–2088.

Leech, J., 1956. On the representation of 1; 2;…; n by differences. J. London Math. Soc. 31, 160–169.

Le Chevalier, F., 1989. Principes de traitement des signaux radar et sonar. éditions Masson. ISBN: 2-2258-1423-6. (2002. Revised and corrected English edition: Principles of Radar and Sonar signal processing. Artech House, Boston, London, ISBN: 1-5805-3338-8).

Le Traon, P.Y., Dibarboure, G., 2004. An illustration of the contribution of the TOPEX/Poseidon-Jason-1tandem mission to mesoscale variability studies, 27. http://dx.doi.org/10.1080/01490410490489313.

Le Traon, P.Y., Morrow, R., 2001a. Chapter 3 Ocean currents and eddies. Intern. Geophys. 69, 171–215.

Le Traon, P.-Y., Morrow, R.A., 2001b. Ocean currents and mesoscale eddies. In: Fu, L.-L., Cazenave, A. (Eds.), Satellite Altimetry and Earth Sciences. A Handbook of Techniques and Applications. Academic Press, 171–215.

Le Traon, P.Y., Ogor, F., 1998. ERS-1/2 orbit improvement using TOPEX/Poseidon: the 2 cm challenge. J. Geophys. Res. 103. http://dx.doi.org/10.1029/97JC01917.

Le Traon, P.Y., 1991. Time scales of mesoscale variability and their relationship with space scales in the North Atlantic. J. Mar. Res. 49, 467–492.

Le Traon, P.Y., October 15–20, 2007. Satellite Altimetry, ESA-most Dragon Programme, 2nd Advanced Training Course in Ocean Remote Sensing, Hangzou, P.R. China slide #5. https://www.yumpu.com/en/document/view/21972852/principles-of-satellite-radar-altimetry/5.

Le Vine, D.M., Abraham, S., 2002. The effect of the ionosphere on remote sensing of sea surface salinity from space: absorption and emission at L band. Geosci. Remote Sens. IEEE Trans. 40 (4), 771–782.

Le Vine, D.M., Lagerloef, G.S.E., Torrusio, S.E., May 2010. Aquarius and remote sensing of sea surface salinity from space. Proc. IEEE 98 (5), 688–703. http://dx.doi.org/10.1109/JPROC.2010.2040550.

Leprieur, C., Kerr, Y.H., Pichon, J.M., 1996. Critical assessment of vegetation indices from AVHRR in a semi-arid environment. Int. J. Remote Sens. 17, 2549–2563.

LeVine, D.M., Good, J.C., 1983. Aperture synthesis for microwave radiometry in space. In: NASA Technical Memorandum 85033.

LeVine, D.M., Griffis, A.J., Swift, C.T., Jackson, T.J., 1992. ESTAR: a synthetic aperture microwave radiometer for remote sensing applications. Proc. IEEE 82 (2).

LeVine, D.M., Zaitzeff, J.B., D'Sa, E.J., et al., 2000. Sea surface salinity: toward an operational remote-sensing system. Satellites Oceanogr. Soc. Elsevier Oceanogr. Ser. 63, 321−335.

Li, W., Yang, D., D'Addio, S., Martín-Neira, M., 2014. Partial interferometric processing of reflected GNSS signals for ocean altimetry. IEEE Geosci. Remote Sens. Lett. 11 (9), 1509−1513.

Liu, H.Q., Huete, A.R., 1995. A feedback based modification of the NDV I to minimize canopy background and atmospheric noise. IEEE Trans.Geosci. Remote Sens. 33, 457−465.

Liu, A.K., Zhao, Y., 1999. Arctic sea ice drift from wavelet analysis of NSCAT and special sensor microwave imager data. J. Geophys. Res. 104 (C5), 11529−11538.

Liu, H., Crawford, J.H., Pierce, R.B., Norris, P., Platnick, S.E., Chen, G., Logan, J.A., Yantosca, R.M., Evans, M.J., Kittaka, C., Feng, Y., Tie, X., 2006. Radiative effect of clouds on tropospheric chemistry in a global three-dimensional chemical transport model. J. Geophys. Res. 111, D20303. http://dx.doi.org/10.1029/2005JD006403.

Liu, W.T., Tang, W., Xie, X., 2008. Wind power distribution over the ocean. Geophys. Res. Lett. 35, L13808. http://dx.doi.org/10.1029/2008GL034172.

Lindell, D.B., Long, D.G., 2016. Multiyear Arctic sea ice classification using OSCAT and QuikSCAT. Trans. Geosci. Remote Sens. 54, 167−175.

Llewellyn-Jones, D.T., Mutlow, C.T., Zavody, A.M., Murray, M.J., Allen, M.R., Saunders, R.W., 1993. SST measurements from ATSR on ESA's ERS-1 satellite - early results. In: Proc. of the IGARSS, Tokyo 1993, pp. 155−156.

Lopez-Dekker, P., Rodriguez-Cassola, M., Prats, P., De Zan, F., Kraus, T., Sauer, S., Mittermayer, J., 2014. Experimental Bidirectional SAR ATI acquisitions of the ocean surface with TanDEM-X," EUSAR 2014. In: 10th European Conference on Synthetic Aperture Radar, Berlin, Germany, 2014, pp. 1−4.

Lowe, S.T., Zuffada, C., LaBrecque, J.L., Lough, M., Lerma, J., Young, L.E., 2000. An ocean-altimetry measurement using reflected GPS signals observed from a low-altitude aircraft. In: 2000 IEEE International Geoscience and Remote Sensing Symposium, vol. 5, pp. 2185−2187.

Lowe, S.T., Zuffada, C., Chao, Y., Kroger, P., Young, L.E., LaBrrecque, J.L., 2002a. 5-cm precision aircraft ocean altimetry using GPS reflections. Geophys. Res. Lett. 29. http://dx.doi.org/10.1029/2002GL014759.

Lowe, S.T., Kroger, P., Franklin, G., LaBrecque, J.L., Lerma, J., Lough, M., Marcin, M.R., Muellerschoen, R.J., Spitzmesser, D., Young, L.E., May 2002b. A delay/Doppler-mapping receiver system for GPS-reflection remote sensing. IEEE Trans. Geosci. Remote Sens. 40 (5), 1150−1163.

Lowe, S.T., Meehan, T., Young, L., May.2014. Direct signal enhanced semicodeless, processing of GNSS surface-reflected signals. IEEE J. Selected Top. Appl. Earth Observations Remote Sens. 7 (5), 1469−1472.

Lozier, M.S., 1997. Evidence for large-scale eddy-driven gyres in the North Atlantic. Science 18. http://dx.doi.org/10.1126/science.277.5324.361.

Ludwig, A.C., 1973. The definition of cross-polarization. IEEE Trans. Antennas Propagation AP-21 (1), 116−119.

Lyons, M., Phinn, S., Roelfsema, C., 2011. Integrating Quickbird multispectral satellite and field data: mapping bathymetry, seagrass cover, seagrass species and change in Moreton Bay, Australia in 2004 and 2007. Remote Sens. 3 (1), 42−64.

Machin, K.E., Ryle, M., Vonberg, D.D., 1952. The design of an equipment for measuring small radio frequency noise powers. Proc. IEEE (London) 99, 127−134.

Maitre, H., 2008. Processing of Synthetic Aperture Radar Images. Wiley, London.

Marchan-Hernandez, J.F., Ramos-Perez, I., Bosch-Lluis, X., Camps, A., Rodriguez-Alvarez, N., Albiol, D., 2007. PAU-GNSS/R, a real-time GPS-reflectometer for earth observation applications: architecture insights and preliminary results. In: 2007. IEEE International Geoscience and Remote Sensing Symposium, pp. 5113−5116.

Marchan-Hernandez, J.F., Camps, A., Rodriguez-Alvarez, N., Valencia, E., Bosch-Lluis, X., Ramos-Perez, I., 2009. An efficient algorithm to the simulation of delay—Doppler maps of reflected global navigation satellite system signals. IEEE Trans. Geosci. Remote Sens. 47 (8), 2733—2740.

Marshall, J.S., Palmer, W.M., 1948. The distribution of raindrops with size. J. Meteorol. 5, 165—166.

Martin, F., Camps, A., Park, H., d'Addio, S., Martín, M., Pascual, D., 2014. Cross-correlation waveform analysis for conventional and interferometric GNSS-R approaches. IEEE J. Selected Top. Appl. Earth Observations Remote Sens. 7 (5), 1560—1572.

Martin, F., Camps, A., Fabra, F., Rius, A., Martin-Neira, M., D'Addio, S., 2015. A. Alonso, "mitigation of direct signal cross-talk and study of the coherent component in GNSS-R". IEEE Geosci. Remote Sens. Lett. 12 (2), 279—283.

Martín-Asín, F., 1990. In: Geodesia Y Cartografía Matemática, 3a ed. Paraninfo, Madrid.

Martinez-Vazquez, A., Camps, A., Lopez-Sanchez, J.M., Vall-llossera, M., Monerris, A., 2009. Numerical simulation of the full-polarimetric emissivity of vines and comparison with experimental data. Remote Sens. 1, 300—317.

Martin-Neira, M., Font-Rosello, J., 1997. Mechanically Scanned Interferometric Radiometer Using Sub-Arraying. ESTEC Internal Report, XR1/108.97/MMN.

Martin-Neira, M., Suess, M., Kainulainen, J., Martin-porqueras, F., 2008. The flat target transformation. IEEE Trans. Geosci. Remote Sens. 46 (3), 613—620.

Martín-Neira, M., D'Addio, S., Buck, C., Floury, N., Prieto-Cerdeira, R., June 2011. The Paris ocean altimeter in-orbit demonstrator. IEEE Tran. Geosci. Remote Sens, 49 (6), 2209. http://dx.doi.org/10.1109/TGRS.2010.2092431.

Martin-Neira, M., 1993. A passive reflectometry and interferometry System (PARIS) — application to ocean altimetry. ESA J. 17 (4), 331—355.

Martín-Neira, M., 2001. Introduction to Two-dimensional Aperture Synthesis Microwave Radiometry for Earth Observation: Polarimetric Formulation of the Visibility Function. In: ESTEC Working Paper, No 2130, ESA/ESTEC, Noordwijk, The Netherlands.

Maslanik, J., Fowler, C., Stroeve, J., Drobot, S., Zwally, J., Yi, D., Emery, W., 2007. A younger, thinner Arctic ice cover: increased potential for rapid, extensive sea-ice loss. Geophys. Res. Lett. 34, L24501. http://dx.doi.org/10.1029/2007GL032043.

Masters, D.S., 2004. Surface Remote Sensing Applications of GNSS Bistatic Radar: Soil Moisture and Aircraft Altimetry (Ph.D. thesis). Univ. Colorado, 189 pp.

Masuoka, E., Fleig, A., Wolfe, R.E., Patt, F., 1998. Key characteristics of MODIS data products. IEEE Trans. Geosci. Remote Sens. 36, 1313—1323.

Mätlzer, C., Standley, A., 2000. Relief effects for passive microwave remote sensing. J. Geophys. Res. 21, 2403—2412.

Mätzler, C., Wegmüller, U., 1987. Dielectric properties of fresh-water ice at microwave frequencies. J. Phys. D 20, 1623—1630.

Mätzler, C., 1990. Seasonable evolution of microwave radiation from an oat field. Remote Sens. Environ. 31, 161—173.

Mätzler, C. (Ed.), 2006. Thermal Microwave Radiation: Applications for Remote Sensing. IET.

May, D.A., Parmeter, M.M., Olszewski, D.S., McKenzie, B.D., 1998. Operational processing of satellite sea surface temperature retrievals at the Naval Oceanographic Office. Bull. Amer. Metero. Soc. 79 (3), 397—407.

Mayerhöfer, T.G., Mutschke, H., Popp, J., 2016-04-01. Employing theories far beyond their limits—the case of the (Boguer-) Beer—Lambert Law. ChemPhysChem. ISSN: 1439-7641. http://dx.doi.org/10.1002/cphc.201600114.

McClain, E.P., 1989. Global sea surface temperatures and cloud clearing for aerosol optical depth estimates. Int. J. Remote Sens. 10, 763—769.

McClain, E.P., Pichel, W.G., Walton, D.D., 1985. Comparative performance of AVHRR-based multichannel sea surface temperatures. J. Geophys. Res. 90 (6), 11587—11601. http://dx.doi.org/10.1029/JC090iC06p11587.

Meissner, T., Wentz, F.J., 2004. The complex relaxation of water between 9°C and 35°C. Chem. Phys. Lett. 306, 57–63.

Meissner, T., Wentz, F.J., 2012. The emissivity of the ocean surface between 6 and 90 GHz over a large range of wind speeds and Earth incidence angles. IEEE Trans. Geosci. Remote Sens. 50. http://dx.doi.org/10.1109/TGRS.2011.2179662.

Mel'nik, Y.A., 1972a. Potential applications of coherent processing of random signals. Radio Eng. Electron. Phys. 4, 575–578.

Mel'nik, Y.A., 1972b. Space-time handling of radiothermal signals from radiators that move in the near zone of an interferometer, Izvestiya Vysshikh Uchebnykh Zavedenii. Radiofiz 15, 1376–1380.

Mennella, A., Bersanelli, M., Butler, R.C., Maino, D., Mandolesi, N., Morgante, G., Valenziano, L., Villa, F., Gaier, T., Seiffert, M., Levin, S., Lawrence, C., Meinhold, P., Lubin, P., Tuovinen, J., Varis, J., Karttaavi, T., Hughes, N., Jukkala, P., Sjöman, P., Kangaslahti, P., Roddis, N., Kettle, D., Winder, F., Blackhurst, E., Davis, R., Wilkinson, A., Castelli, C., Aja, B., Artal, E., de la Fuente, L., Mediavilla, A., Pascual, J.P., Gallegos, J., Martinez-Gonzalez, E., de Paco, P., Pradell, L., 2003. Advanced pseudo-correlation radiometers for the Planck-LFI instrument. In: Proc 3rd ESA Workshop on Millimetre Wave Technology and Applications (ESPOO, 21–23 May 2003), p. 69. arXiv:astro-ph/0307116.

Menzel, W.P., Strabala, K.I., 1997. Cloud Top Properties and Cloud Phase: Algorithm Theoretical Basis Document. ATBD-MOD-04. NASA Goddard Space Flight Center, 55 p.

Menzel, W.P., Smith, W.L., Herman, L.D., 1981. Visible infrared spin scan radiometer atmospheric sounder radiometric calibration: an inflight evaluation from intercomparison with HIRS and radio-sonde measurements. Appl. Opt. 20, 3641–3644.

Miller, K.S., Rochwarger, M.M., 1972. A covariance approach to spectral moment estimation. IEEE Trans. Inform. Theory IT-18 (5), 588–596.

Miller, J., Goodberlet, M.A., Zaitzeff, J., 1996. Airborne salinity mapper makes debut in coastal zone. EOS Trans. AGU 79, 173–177. http://dx.doi.org/10.1029/98EO00126.

Milman, 1993. SAR imaging by ω–k migration. Int. J. Remote Sens. 14 (10), 1965–1979.

Minnett, P.J., 1990. The regional optimization of infrared measurements of sea surface temperature from space. J. Geophys. Res. 95. http://dx.doi.org/10.1029/JC095iC08p13497.

Mironov, V.L., Dobson, M.C., Kaupp, V.H., Komarov, S.A., Kleshchenko, V.N., 2004. Generalized refractive mixing dielectric model for moist soils. IEEE Trans. Geosci. Remote Sens. 42 (4), 773–785.

Misra, S., Mohammed, P.N., Güner, B., Ruf, C.S., Piepmeier, J.R., Johnson, J.T., 2009. Microwave radiometer radio-frequency interference detection algorithms: a comparative study. IEEE Trans. Geosci. Remote Sens. 47 (11), 3742–3754.

Mitchum, G.T., 1998. Monitoring the stability of satellite altimeters with tide gauges. J. Atm. Oceanic Tech. http://dx.doi.org/10.1175/1520-0426(1998)015<0721:MTSOSA>2.0.CO;2.

Mittermayer, J., Wollstadt, S., Prats-Iraola, P., López-Dekker, P., Krieger, G., Moreira, A., January 2013. Bidirectional SAR imaging mode. IEEE Trans. Geosci. Remote Sens. 51 (1), 601–614.

Mo, T., Schmugge, T., 1987. A parameterization ot the effect of surface roughness on microwave emission. IEEE Trans. Geosci. Remote Sens. 25, 47–54.

Moccia, A., Renga, A., October 2011. Spatial resolution of bistatic synthetic aperture radar: impact of acquisition geometry on imaging performance. IEEE Trans. Geosci. Remote Sens. 49 (10), 3487–3503.

Moffet, A., Mar 1968. Minimum-redundancy linear arrays. IEEE Trans. Antennas Propagation 16 (2), 172–175.

Monerris, A., Schmugge, T., October 1, 2009. Advances in geoscience and remote sensing. In: Jedlovec, G. (Ed.), Chapter 21: Soil Moisture Estimation Using L-band Radiometry, ISBN 978-953-307-005-6.

Monerris, A., Benedicto, P., Vall-llossera, M., Santanach, E., Piles, M., Prehn, R., 2008. Assessment of the topography impact on microwave radiometry at L-band. J. Geophys. Res. 113, B12202.

Montgomery, H.E., Uccellini, L.W. (Eds.), 1985. VAS Demonstration (VISSR Atmospheric Sounder): Description and Final Report, vol. 1151. National Aeronautics and Space Administration, Scientific and Technical Information Branch.

Moreira, A., Huang, Y., 1994. Chirp scaling algorithm for processing SAR data with high squint angle and motion error. SAR Data Process. Remote Sens. 268/SPIE 2316.

Moreira, A., Mittermayer, J., Scheiber, R., 1996. Extended chirp scaling algorithm for air- and spaceborne SAR data processing in stripmap and Scan SAR imaging modes. Trans. Geosci. Remote Sens. 34 (5).

Moreira, A., Pau, P.I., Younis, M., Krieger, G., Hajnsek, I., Papathanassiou, K., 2013. A tutorial on synthetic aperture radar. IEEE Geosci. Remote Sens. Magazine (GRSM) 1 (1), 6–43.

Morel, A., Prieur, L., 1977. Analysis of variations in ocean color. Limnol. Oceanogr. 22, 709–722.

Moreno, J.F., Meliá, J., 1993. A method for accurate geometric correction of NOAA AVHRR HRPT data. IEEE Trans. Geosci. Remote Sens. 31. http://dx.doi.org/10.1109/36.210461.

Morrow, R., Le Traon, P.-Y., 2012. Recent advances in observing meoscale ocean dynamics with satellite altimetry. Adv. Space Res. 50. http://dx.doi.org/10.1016/j.asr.2011.09.033.

Morrow, R.A., Coleman, R., Church, J.A., Chelton, D.B., 1994. Surface eddy momentum flux and velocity variances in the Southern Ocean from Geosat altimetry. J. Phys. Oceanogr. 24, 2050–2071.

Nakajima, T., King, M.D., 1990. Determination of the optical thickness and effective particle radius of cloud from reflected solar radiation measurement. Part 1: Theory. J. Atm. Sco. http://dx.doi.org/10.1175/1520-0469(1990)047<1878:DOTOTA>2.0.CO;2.

NASA Scatterometer Climate Record Pathfinder (SCP). http://www.scp.byu.edu/.

Navstar, September 2011. GPS space segment – user segment L5 interfaces, interface specification (IS-GPS-705), revision B. In: Global Positioning System Wing (GPSW) Systems Engineering & Integration. Available at: http://www.gps.gov/technical/icwg/IS-GPS-705B.pdf.

Nerem, R.S., Chambers, D.P., Choe, C., Mitchum, G.T., 2010. Estimating mean sea level change from the TOPEX and Jason altimeter missions. Mar. Geodes 33. http://dx.doi.org/10.1080/01490419.2010.491031.

Nghiem, S.V., Steffen, K., Kwok, R., Tsai, W.Y., 2001. Detection of snowmelt regions on the Greenland ice sheet using diurnal backscatter change. J. Glaciology 47. http://dx.doi.org/10.3189/172756501781831738.

Nghiem, S.V., Chao, Y., Neumann, G., Li, P., Perovich, D.K., Street, T., Clemente-Color, P., 2006. Depletion of perennial sea ice in the East Arctic Ocean. Geophys. Res. Lett. 33 http://dx.doi.org/10.1029/2006GL027198.

Niiler, P.P., Maximenko, N.A., McWilliams, J.C., 2003. Dynamically balanced absolute sea level of the global ocean derived from near-surface velocity observations. Geophys. Res. Lett. 30. http://dx.doi.org/10.1029/2003GL018628.

Njoku, E.G., 1999. Retrieval of land surface parameters using passive microwave measurements at 6–18 GHz. IEEE Trans. Geosci. Rem. Sens. 37, 79–93.

Njoku, E.G., Ashcroft, P., Chan, T.K., Li, L., 2005. Global survey and statistics of radio-frequency interference in AMSR-E land observations. IEEE Trans. Geosci. Remote Sens. 43, 938–947.

Nogues, O., Sumpsi, A., Camps, A., Rius, A., 21-25 July 2003. A 3 GPS-channels Doppler-delay receiver for remote sensing applications. In: 2003 IEEE International Geoscience and Remote Sensing Symposium, vol. 7, pp. 4483–4485.

Nogués-Correig, O., Cardellach-Galí, E., Sanz-Campderrós, J., Rius, A., January 2007. A GPS-reflections receiver that computes Doppler/delay maps in real time. IEEE Trans. Geosci. Remote Sens. 45 (1), 156–174.

Nogues-Correig, O., Ribo, S., Arco, J.C., Cardellach, E., Rius, A., Valencia, E., Tarongi, J.M., Camps, A., Van Der Marel, H., Martin-Neira, M., 2010. The proof of concept for 3-cm altimetry using the Paris interferometric technique. In: 2010 IEEE International Geoscience and Remote Sensing Symposium, pp. 3620–3623.

North, D.O., 1963. An analysis of the factors which determine signal/noise discrimination in pulsed carrier systems. Proc. IEEE 51, 1016–1027.

Nyquist, H., 1928. Thermal agitation of electric charge in conductors. Phys. Rev. 32, 110. http://dx.doi.org/10.1103/PhysRev.32.110.

Offringa, A.R., de Bruyn, A.G., Biehl, M., Zaroubi, S., Bernardi, G., Pandey, V.N., 2010. Post-correlation radio frequency interference classification methods. Monthly Notices Royal Astronomical Soc. 405 (1), 155–167.

Olive, R., Amezaga, A., Carreno-Luengo, H., Park, H., Camps, A., 2016. Implementation of a GNSS-R payload based on software-defined radio for the 3CAT-2 mission. IEEE J. Selected Top. Appl. Earth Obser. Remote Sens.

Olmedo, E., Martínez, J., Umbert, M., Hoareau, N., Portabella, M., Ballabrera-Poy, J., Turiel, A., 2016. Improving time and space resolution of smos salinity maps using multifractal fusion. Remote Sens. Environ. 180, 246–263.

O'Neill, R.V., Krummel, J.R., Gardner, R.H., Sugihara, G., Jackson, B., DeAngelis, D.L., Milne, B.T., Turner, M.G., Zygmunt, B., Christensen, S.W., Dale, V.H., Graham, R.L., 1988. Indices of landscape pattern. Landscape Ecol. 1 (3), 153–162.

Oppenheim, A., Willsky, A.S., Young, I.T., 1983. Signals and Systems. Prentice Hall, Englewood Cliffs, New Jersey.

Orhaug, T., Waltman, W., 1962. A switched load radiometer. Publ. Nat. Radio Astron. Obs. 1, 179–204.

Paltridge, G., Barber, J., 1988. Monitoring grassland dryness and fire potential in Australia with NOAA/AVHRR data. Remote Sens. Environ. 28, 384–393.

Pampaloni, P., Paloscia, S., 1986. Microwave emission and plant water content: a comparison between field measurements and theory. IEEE Trans. Geosci. Remote Sens. GE-24, 900–905.

Papoulis, A., 1962. The Fourier Integral and its Applications. McGraw-Hill, Inc., New York, USA.

Papoulis, A., 1965. Probability, Random Variables, and Stochastic Processes, International student edition. McGraw-Hill Kogakusha, Ltd, Tokyo, Japan.

Papoulis, A., 1968. System and Transforms with Applications in Optics. McGraw-Hill, Inc., New York, USA.

Park, H., Kim, Y.-H., 2009a. Microwave motion induced synthetic aperture radiometer using sparse array. Radio Sci. 44, RS3012. http://dx.doi.org/10.1029/2008RS003998.

Park, H., Kim, Y.-H., 2009b. Improvement of a Doppler-radiometer using a sparse array. IEEE Geosci. Remote Sens. Lett. 6 (2), 229–233.

Park, H., Valencia, E., Rodriguez-Alvarez, N., Bosch-Lluis, X., Ramos-Perez, I., Camps, A., 2011. New approach to sea surface wind retrieval from GNSS-R measurements. In: 2011 IEEE International Geoscience and Remote Sensing Symposium, Vancouver, BC, pp. 1469–1472.

Park, H., Camps, A., Valencia, E., Rodriguez, N., Bosch, X., Ramos, I., Carrero-Luengo, H., 2012a. Retracking considerations in spaceborne GNSS-R altimetry. GPS Solutions 16 (4), 507–518.

Park, H., Camps, A., Valencia, E., 2012b. Impact of Doppler frequency compensation errors on spaceborne GNSS-R altimetry. In: Proc. IEEE Int. Geosci.Remote Sens. Symp. (IGARSS 12), July 22–27, pp. 2661–2664.

Park, H., Camps, A., Valencia, E., Rodriguez-Alvarez, N., Bosch-Lluis, X., Ramos-Perez, I., Carreno-Luengo, H., 2013. Delay tracking in spaceborne GNSS-R ocean altimetry. IEEE Geosci. Remote Sens. Lett. 10 (1), 57–61.

Park, H., Pascual, D., Camps, A., Martin, F., Alonso-Arroyo, A., Carreno-Luengo, H., 2014a. Analysis of spaceborne GNSS-R delay-Doppler tracking. IEEE J. Selected Top. Appl. Earth Observation Remote Sens. 7 (5), 1481–1492.

Park, H., Camps, A., Pascual, D., Alonso-Arroyo, A., Martin, F., Carreno-Luengo, H., Onrubia, R., 2014b. Simulation study on tropical cyclone tracking from the ISS using GNSS-R measurements. In: 2014 IEEE Geoscience and Remote Sensing Symposium, Quebec City, QC, pp. 4062–4065.

Pascual, A., Pujol, M-I, Larnicol, G., Le Traon, P-Y, Rio, M-H, 2007. Mesoscale mapping capabilities of multisatellite altimeter missions: First results with real data in the Mediterranean Sea. J. Mar. Sys. 65, pp 190-211.

Parkinson, C.L., 2003. Aqua: an earth-observing satellite mission to examine water and other climate variables. IEEE Trans. Geosci. Remote Sens. 41 (2), 173–183.

Pascual, D., Camps, A., Martin, F., Par, H., Arroyo, A.A., Onrubia, R., 2014. Precision bounds in GNSS-R ocean altimetry. J. Sel. Top. Appl. Earth Obs. Remote Sens. 7. http://dx.doi.org/10.1109/JSTARS.2014.2303251.

Pavelyev, A.G., Zhang, K., Matyugov, S.S., Liou, Y.A., Wang, C.S., Yakovlev, O.I., Kucherjavenkov, I.A., Kuleshov, Y., 2011. Analytical model of bistatic reflections and radio occultation signals. Radio Sci. 46. http://dx.doi.org/10.1029/2010RS004434.

Peake, W.H., 1959. Interaction of electromagnetic waves with some natural surfaces. IRE Trans. Antennas Propag. AP-7, S324—S329 (Special Supplement).

Picardi, G., Seu, R., Sorge, S.G., Neira, M.M., 1998. Bistatic model of ocean scattering. Trans. Antennas Propag. 46. http://dx.doi.org/10.1109/8.725286.

Piles, M., Camps, A., Vall-llossera, M., Corbella, I., Panciera, R., Rudiger, C., Kerr, Y.H., Walker, J., September 2011. Downscaling smos-derived soil moisture using MODIS visible/infrared data. IEEE Trans. Geosci. Remote Sens. 49 (9), 3156—3166.

Piles, M., Sánchez, N., Vall-llossera, M., Camps, A., Martínez-Fernández, J., Martínez, J., González, V., 2014. A downscaling approach for SMOS land observations: evaluation of high-resolution soil moisture maps over the Iberian Peninsula. IEEE J. Selected Top. Appl. Earth Observations Remote Sens. 7 (9), 3845—3857. http://dx.doi.org/10.1109/JSTARS.2014.2325398.

Pinty, B., Verstraete, H.M., 1992. GEMI: a non-linear index to monitor global vegetation from satellites. Vegetation 101, 15—20.

Planck, M., 1901. 1928: on the law of distribution of energy in the normal spectrum. Annalen der Physik 4, 553 (Potocnik).

2016 Plank, V.G., Implications of the Khrgian-Mazin, 1991. Distribution Function for Water Clouds and Distribution Consistencies with Aerosols and Rain, Technical rept. 1970-1991, ADA273810. PHILLIPS LAB HANSCOM AFB MA. Available online: http://www.dtic.mil/cgi-bin/GetTRDoc?Location=U2&doc=GetTRDoc.pdf&AD=ADA273810.

Pozar, D., 2000. Microwave and RF Design of Wireless Systems. John Wiley & Sons, New York.

Prabhakara, C., Dalu, G., Kunde, V.G., 1974. Estimation of Sea Surface Temperature from Remote Sensing in the 11-13 Micron Window Region. NASA-TM-X-70649, X-911-74-60, 30 p.

Prats-Iraola, P., Scheiber, R., Marotti, L., Wollstadt, S., Reigber, A., 2012. TOPS interferometry with TerraSAR-X. Geosci. Remote Sens. IEEE Trans. 50 (8), 3179—3188.

Prats-Iraola, R., Scheiber, M., Rodriguez-Cassola, J., Mittermayer, S., Wollstadt, F., De Zan, B., Bräutigam, Schwerdt, M., Reigber, A., Moreira, A., October 2014. On the processing of very high resolution spaceborne SAR data. IEEE Trans. Geosci. Remote Sens. 52 (10), 6003—6016.

Price, R., 1958. A useful theorem for nonlinear devices having Gaussian inputs. Inf. Theory, IRE Trans. 4 (2), 69—72.

Querol Borràs, J., Camps Carmona, A.J., March 14, 2017. System and Method for Detecting and Eliminating Radio Frequency Interferences in Real Time. United States Patent, US 9,596,610 B1.

Querol, X., Alasfuey, A., Pandolfi, M., Reche, C., Pérez, N., Minguillón, M.C., Moreno, T., Viana, M., Escudero, M., Orio, A., Pallarés, M., Reina, F., 2014. 2001—2012 trens on air quality in Spain. Sci. Total Environ. 15. http://dx.doi.org/10.1016/j.scitotenv.2014.05.074.

Querol, J., Alonso-Arroyo, A., Onrubia, R., Pascual, D., Camps, A., 2015. Assessment of back-end RFI mitigation techniques in passive remote sensing. In: 2015 IEEE International Geoscience and Remote Sensing Symposium (IGARSS), Milan, 2015, pp. 4746—4749.

Querol, J., Onrubia, R., Pascual, D., Alonso-Arroyo, A., Park, H., Camps, A., 2016. Comparison of real-time time-frequency RFI mitigation techniques in microwave radiometry. In: 2016 14th Specialist Meeting on Microwave Radiometry and Remote Sensing of the Environment (MicroRad), Espoo, Finland, pp. 68—70.

Quilfen, Y., Bentamy, A., 1994. Calibration/validation of ERS-1 scatterometer precision products. Geosci. Remote Sens. Symp. http://dx.doi.org/10.1109/IGARSS.1994.399308.

Rädel, G., Stubenrauch, C.J., Holz, R., Mitchelll, D.L., 2003. Retrieval of effective ice crystal size in the infrared: sensitivity study and global measurements from TIROS-N operational vertical sounder, 108. http://dx.doi.org/10.1029/2002JD002801.

Ramo, S., Whinnery, J.R., Van Duzer, T., 1994. Fields and Waves in Communication Electronics. Wiley.

Ramos-Pérez, I., Bosch-Lluis, X., Camps, A., Rodriguez-Alvarez, N., Marchán-Hernandez, J.F., Valencia-Domènech, E., Vernich, C., De la Rosa, S., Pantoja, S., Correction: Ramos-Pérez, 2009. Calibration of correlation radiometers using pseudo-random noise signals. Sensors 9, 6131—6149. Sensors 2009, 9, 7430.

Raney, R.K., Runge, H., Bamler, R., Cumming, I.G., Wong, F.H., 1994. Precision SAR processing using chirp scaling. Trans. Geosci. Remote Sens. 32 (4).

Raney, R.K., 1986. Doppler properties of radar in circular orbits. Int. J. Remote Sens. 7 (9).

Raney, R.K., 1998. The delay/Doppler radar altimeter. Trans. Geosci. Remote Sens. 36. http://dx.doi.org/10.1109/36.718861.

Reutov, E.A., Shutko, A.M., 1986. Opredeleniye vlazhnosti neodnorodono uvlazhnennykh porchvogruntov s poverkhnostnym perekhodnym sloyem po dannym spektral'nykh svch-radiometricheskikh izmereniy. Issledovaniye Zemiliz Komosa 1, 71–78 (in Russian).

Reynolds, R.W., Smith, T.M., 1994. Improved global sea surface temperature analyses using optimum interpolation. AMS. http://dx.doi.org/10.1175/1520-0442(1994)007<0929:IGSSTA>2.0.CO;2.

Ricciardulli, L., Wentz, F., 2004. Uncertainties in sea surface temperature retrievals from space:O comparison of microwave and infrared observations from TRMM. J. Geophys. Res. 109. http://dx.doi.org/10.1029/2003JC002247.

Rihaczek, A.W., 1969. Principles of High-resolution. Artech House Radar Library Radar, p. 498.

Rius, A., Fabra, F., Ribo, S., Arco, J.C., Oliveras, S., Cardellach, E., Camps, A., Nogues-Correig, O., Kainulainen, J., Rohue, E., Martin-Neira, M., July 22–27, 2012. PARIS Interferometric Technique proof of concept: sea surface altimetry measurements. In: IEEE Geoscience and Remote Sensing Symposium (IGARSS), pp. 7067–7070.

Rivas, M.B., Maslanik, J.A., Axelrad, P., March 2010. Bistatic scattering of GPS signals off Arctic sea ice. IEEE Trans. Geosci. Remote Sens. 48 (3), 1548–1553.

Wolfe, R.E., Nishihama, M., Fleig, A.J., Kuyper, J.A., Roy, D.P., Storey, J.C., Patt, F.S., November 2002. Achieving sub-pixel geolocation accuracy in support of MODIS land science. Remote Sens. Environ. ISSN: 0034-4257 83 (1–2), 31–49. http://dx.doi.org/10.1016/S0034-4257(02)00085-8.

Rodriguez Alvarez, N., 2011. Contributions to Earth Observation Using GNSS-r Opportunity Signals (Ph.D. Dissertation). Universitat Politecnica de Catalunya. Available online: http://www.tdx.cat/handle/10803/53636.

Rodriguez-Alvarez, N., Garrison, J.L., 2016. Generalized linear observables for ocean wind retrieval from calibrated GNSS-R delay–Doppler maps. IEEE Trans. Geosci. Remote Sens. 54 (2), 1142–1155.

Rodriguez-Alvarez, N., Bosch-Lluis, X., Camps, A., Vall-llossera, M., Valencia, E., Marchan-Hernandez, J.F., Ramos-Perez, I., November 2009. Soil moisture retrieval using GNSS-R techniques: experimental results over a bare soil field. IEEE Trans. Geosci. Remote Sens. 47 (11), 3616–3624. http://dx.doi.org/10.1109/TGRS.2009.2030672.

Rodriguez-Alvarez, N., Bosch-Lluis, X., Acevo, R., Aguasca, A., Camps, A., Vall-llossera, M., Ramos-Perez, I., Valencia, E., July 2010. Study of maize plants effects in the retrieval of soil moisture using the interference pattern GNSS-R technique. In: 2010 IEEE International Geoscience and Remote Sensing Symposium, pp. 3813–3816.

Rodriguez-Alvarez, N., Camps, A., Vall-llossera, M., Bosch-Lluis, X., Monerris, A., Ramos-Perez, I., Valencia, E., Marchan-Hernandez, J.F., Martinez-Fernandez, J., Baronini-Turricchia, G., Perez-Gutierrez, C., Sanchez, N., N, January 2011. Land geophysical parameters retrieval using the interference pattern GNSS-R technique. IEEE Trans. Geosci. Remote Sens. 49 (1), 71–84.

Rodriguez-Alvarez, N., Aguasca, A., Valencia, E., Bosch-Luis, X., Camps, A., Ramos-Perez, I., Vall-Ilossera, M., 2012. Snow thickness monitoring using GNSS measurements. IEEE Geosci. Remote Sens. Lett. http://dx.doi.org/10.1109/LGRS.2012.2190379.

Rodriguez-Alvarez, N., Akos, D., Zavorotny, V.U., Smith, J.A., 2013. Airborne GMSS-R wind retrievals using delay-Doppler maps. IEEE Trans. Geosci. Remote Sens. 51. http://dx.doi.org/10.1109/TGRS.2012.2196437.

Rodriguez-Cassola, M., Prats, P., Schulze, D., Tous-Ramón, N., Steinbrecher, U., Marotti, L., Nanninni, M., Younis, M., Zink, M., Reigber, A., López-Dekker, P., Krieger, G., Moreira, A., 2012. First bistatic spaceborne SAR experiments with TanDEM-X. IEEE Geosci. Remote Sens. Lett. 7 (1), 108–112.

Rosbourgh, G.W., Baldwin, D.G., Emery, W.J., 1994. Precise AVHRR image navigation, 32. http://dx.doi.org/10.1109/36.297982.

Rosin, P., Rammler, E., 1993. The laws governing the fineness of powdered coal. J. Inst. Fuel 7, 29–36.

Rossouw, M.J., Joubert, J., McNamara, D.A., 1997. Thinned arrays using a modified minimum redundancy synthesis technique. Electron. Lett. 33 (10), 826–827.

Roth, C.H., Malicki, M.A., Plagge, R., 1992. Empirical evaluation of the relationship between soil dielectric constant and volumetric water content as the basis for calibrating soil moisture measurements by TDR. J. Soil Sci. 43, 1–3.

Ruf, C.S., Li, J., 2003. A correlated noise calibration standard for interferometric, polarimetric, and autocorrelation microwave radiometers. IEEE Trans. Geosci. Remote Sens. 41 (10), 2187–2196.

Ruf, C.S., Swift, C.T., Tanner, A.B., LeVine, D.M., 1988a. Interferometric synthetic aperture radiometry for the remote sensing of the earth. IEEE Trans. Geosci. Remote Sens. GRS-26 (5), 597–611.

Ruf, C.S., Swift, C.T., Tanner, A.B., et al., 1998b. Interferometric synthetic aperture microwave radiometry for the remote sensing of the Earth. IEEE Trans. Geosci. Remote Sens. 26, 597–611. http://dx.doi.org/10.1109/36.7685.

Ruf, C.S., Gross, S.M., Misra, S., March 2006. RFI detection and mitigation for microwave radiometry with an agile digital detector. IEEE Trans. Geosci. Remote Sens. 44 (3), 694–706. http://dx.doi.org/10.1109/TGRS.2005.861411.

Ruf, C., Chang, P., Paola-Clarizia, M., Jelenak, Z., Ridley, A., Rose, R., 2014. CYGNSS: NASA Earth ventur tropical cyclone mission. In: Proc. SPIE 9241, Sensors, Systems, and Next-Generation Satellites XVIII, 924109. http://dx.doi.org/10.1117/12.2071404.

Ruf, C.S., January 1993. Numerical annealing of low-redundancy linear arrays. IEEE Trans. Antennas Propagation 41 (1), 85–90.

Ruf, C.S., 2000. Detection of calibration drifts in spaceborne microwave radiometers using a Vicarious cold reference. IEEE Trans. Geosci. Remote Sens. 38 (1), 44–52.

Ruf, C., April 2016. CYGNSS Handbook: Cyclone Global Navigation Satellite System. Michigan Publishing. Available online: http://clasp-research.engin.umich.edu/missions/cygnss/reference/cygnss-mission/CYGNSS_Handbook_April2016.pdf.

Ruffini, G., Flores, A., Rius, A., 1998. GPS tomography of the ionospheric electron content with a correlations functional. Trans. Geosci. Remote Sens. 36. http://dx.doi.org/10.1109/36.655324.

Sakaida, F., Kawamura, H., 1992a. Accuracies of NOAA/NESDIS sea surface temperature estimation technique in the oceans around Japan. J. Oceanogr. 48, 345–351.

Sakaida, F., Kawamura, H., 1992b. Estimation of sea surface temperature around Japan using the advanced very high resolution radiometer (AVHRR)/NOAA-11. J. Oceanogr. 48, 179–192.

Sakaida, F., Kudoh, J.I., Kawamura, H., 2000. A_HIGHERS-The system to produce the high spatial resolution sea surface temperature maps of the western North Pacific using the AVHRR/NOAA. J. Oceanogr. 56. http://dx.doi.org/10.1023/A:1011181918048.

Salzmann, N., Kaab, A., Huggel, C., Allgower, B., Haeberli, W., 2004. Assessment f the hazard potential of ice avalances using remote sensing and GIS-modelling. Norsk Geografisk Tidsskrift-Nowegian J. Geogr. 58, 74–84.

Sandwell, D.T., Smith, W.H.F., 1995. Ocean basin tectonics revealed with declassified Geosat altimeter data. EOS Trans. Am. Geophys. Union 76, 149.

Sansosti, E., Berardino, P., Manunta, M., Serafino, F., Fornaro, G., October 2007. Geometrical SAR image registration. IEEE Trans. Geosci. Remote Sens. 44 (10), 2861–2870.

Schiavulli, D., Ghavidel, A., Camps, A., Migliaccio, M., December 2015. GNSS-R wind-dependent polarimetric signature over the ocean. IEEE Geosci. Remote Sens. Lett. 12 (12), 2374–2378. http://dx.doi.org/10.1109/LGRS.2015.2477685.

Schmugge, T.J., Wang, K.R., Asrar, G., 1988. Results from the pushbroom microwave radiometer flights over the Konza Prairie in 1985. IEEE Trans. Geosci. Remote Sens. GE-26, 590–596.

Schneeberger, K., Schwank, M., Stamm, Ch, de Rosnay, P., Mätzler, Ch, Flühler, H., 2004. Topsoil structure influencing soil water retrieval by microwave radiometry. Vadose Zone J. 3, 1169–1179.

Schott, J., 2007a. In: Remote Sensing: The Image Chain Approach. Oxford University Press.

Schott, J.R., 2007b. Remote Sensing: The Image Chain Approach. Oxford University Press, 666 pp.

Schreiner, W., Sokolovskiy, S., Hunt, D., Rocken, C., Kuo, Y.H., 2011. Analysis of GPS radio occultation data from the FORMOSAT-3/COSMIC and Metop/GRAS missions at CDAAC. Atmos. Meas. Tech. 4. http://dx.doi.org/10.5194/amt-4-2255-2011.

Schulz-Stellenfleth, J., Lehner, S., 2007. Meansurement of 2D sea surface elevation fields using complex synthetic aperture radar data. Trans. Geosci. Remote Sens. 42, 1149–1160.

Schwarz, H.A., 1888. Über ein Flächen kleinsten Flächeninhalts betreffendes Problem der Variationsrechnung. Acta Soc. Sci. Fenn. XV, 318.

Schwank, M., Stähli, M., Wydler, H., Leuenberger, J., Mätzler, C., Flühler, H., 2004. Microwave L-band emission of freezing soil. IEEE Trans. Geosci. Remote Sens. 42, 1252–1261.

Sherwin, C.W., Ruina, J.P., Rawcliffe, R.D., 1962. Some early developments in synthetic aperture radar systems. IRE Trans. Milit. Electronics MIL-6, 111–115.

Shi, Z., Fung, K.B., August 8–12, 1994. A comparison of digital speckle filters. In: Proceedings of IGRASS 94, pp. 2129–2133.

Shi, J., Chen, K.S., Li, Q., Jackson, T., O'Neill, P.E., Tsang, L., 2002. A parameterized surface reflectivity model and estimation of bare surface soil moisture with l-band radiometer. IEEE Trans. Geosci. Remote Sens. 40 (12), 2674–2686.

Shibata, A., Imaoka, K., Kachi, M., Murakami, H., 1999. SST observation by TRMM microwave imager aboard tropical rainfall measuring mission. Umi no Kenkyu 8, 135–139.

Shutko, A.M., 1986. Microwave Radiometer of Water Surface and Grounds. Nauka Moscow, 192 pp. (English translation).

Silvestrin, P., Berger, M., Kerr, Y., et al., 2001. ESA's second earth explorer opportunity mission: the soil moisture and ocean salinity mission— SMOS. IEEE Geosci. Remote Sens. News Lett. 118, 11–14.

Skolnik, M.I., 1980a. Introduction to Radar Systems, second ed. McGraw-Hill, Sinapore.

Skolnik, M.I., 1980b. Radar Handbook, second ed. McGraw-Hill, Sinapore.

Skou, N., LeVine, D., 2006. Microwave Radiometer Systems, Design and Analysis. Artech House.

Skou, N., 1989. In: Microwave Radiometer Systems: Design and Analysis. Artech House, pp. 81–96.

Slater, P.N., 1980. Remote Sensing; Optics and Optical Systems. Addison-Wesley, Reading MA.

Smith Jr., W.L., Minnis, P., Young, D.F., Chen, Y., June 28, 1999a. Satellite- derived surface emissivity for ARM and CERES. In: Proc. AMS 10th Conf. Atmos. Rad. Madison, WI, pp. 410–413.

Smith, D.L., Delderfield, J., Drummond, D., Edwards, T., Godfrey, J., Mutlow, C.T., Read, P.D., Toplis, G.M., December 28, 1999b. Prelaunch calibration of the Advanced Along-Track Scanning Radiometer (AATSR). In: Proc. SPIE 3870, Sensors, Systems, and Next-generation Satellites III. http://dx.doi.org/10.1117/12.373206.

Smith Jr., W.O., Marra, J., Hiscock, M.R., Barber, R.T., 2000. The seasonal cycle of phytoplankton biomass and primary productivity in the Ross Sea, Antarctica. Deep-Sea Res. Part II 47. http://dx.doi.org/10.1016/S0967-0645(00)00061-8.

Smith, J., 1985. Modern Communication Circuits. McGraw-Hill International, Singapore.

Smullin, L.D., Fiocco, G., 1962. Optical echoes from moon. Nature. ISSN: 0028-0836 194 (4835), 1267. http://dx.doi.org/10.1038/1941267a0.

Strong, A.E., McClain, E.P., 1984. Improved ocean surface temperatures from space-comparisons with drifting buoys. Bull. Am. Meteorol. Soc. 65, 138–142.

Soulat, F., Caparrini, M., Germain, O., Lopez-Dekker, P., Taani, M., Ruffini, G., 2004. Sea state monitoring using coastal GNSS-R. Geophys. Res. Lett. 31, L21303. http://dx.doi.org/10.1029/2004GL020680.

Souza, J.M.A.C., de Boyer Montégut, C., CAbanes, C., Klein, P., 2011. Estimation of the Agulhas ring impacts on meridional heat fluxes and transport using ARGO floats and satellite data. Geophys. Res. Lett. 38. http://dx.doi.org/10.1029/2011GL049359.

Spilker, J.J., 1996. Satellite constellation and geometric dilution of precision. In: Global Positioning System: Theory and Applications, vol. 1. AIAA Press.

Stammer, D., Theiss, J., 2004. Velocity statistics inferred from the TOPEX/POSEIDON-JASON tandem mission data. Marine Geodesy 27 (3–4), 551–575. http://dx.doi.org/10.1080/01490410490902052.

Stammer, D., Wunsch, C., 1999. Temporal changes in eddy energy of the oceans. Deep Sea Res. 46, 77–108.

Stoffelen, A.C.M., Anderson, D.L.T., 1997. Ambiguity removal and assimilation of scatterometer data. Q. J. Roy. Meteo. Soc. 123, 491–518.

Stubenrauch, C.J., Eddounia, F., Sauvage, L., 2005. Cloud heights from TOVS path-B: evaluation using LITE observations and distributions of highest cloud layers. J. Geophys. Res. 110. http://dx.doi.org/10.1029/2004JD005447.

Sun, J., Kawamura, H., 2009. Retrieval of surface wave parameters from SAR image and their validation in the coastal seas around Japan. J. Oceanogr. 65. http://dx.doi.org/10.1007/s10872-009-0048-2.

Svendsen, E., Kloster, K., Farrelly, B., Johannessen, O.M., Johannessen, J.A., Campbell, W.J., Gloersen, P., Cavalieri, D., Matzler, C., 1983. Norwegian remote sensing experiment: evaluation of the Nimbus-7 scanning multichannel microwave radiometer for sea ice research. J. Geophys. Res. 88 (C5), 2781–2791.

Svensson, J., 1985. Remote Sensing of Atmospheric Temperature Profiles by TIROS Operational Vertical Sounder. Norrkoping: Swedish Meteorological and Hydrological Institute, Rep. RNK 45.

Swan, A.M., Long, D.G., 2012. Mulltiyear Arctic sea ice classification using QuikSCAT. Trans. Geosci. Remote Sens. 50, 3317–3326.

Swap, R.J., et al., 2002. The Southern African Regional Science Initiative (SA-FARI-2000): dry-season field campaign, an overview. S. Afr. J. Sci. 98, 125–130.

Swift, C.T., McIntosh, R.E., 1983. Considerations for microwave remote sensing of ocean-surface salinity. IEEE Trans. Geosci. Elec 21, 480–491. http://dx.doi.org/10.1109/tgrs.1983.350511.

Swift, C.T., 1980. Passive microwave remote sensing of the ocean – a review. Boundary – Layer Meteorol. 18, 25–54. http://dx.doi.org/10.1007/BF00117909.

Talone, M., Camps, A., Monerris, A., Vall-llossera, M., Ferrazzoli, P., Piles, M., 2007. Surface topography and mixed pixel effects on the simulated L-band brightness temperatures. IEEE Trans. Geosci. Remote Sens. 42, 786–794.

Tanahashi, M., Fujimura, M., Miyauchi, T., 2000. Coherent fine-scale eddies in turbulent premixed flames. Proc. Combust. Inst. 28.

Tanner, A.B., Swift, C.T., January 1993. Calibration of a synthetic aperture radiometer. IEEE Transaction Geosci. Remote Sens. 31 (1), 257–267.

2013 Tanner, A.B., January 1, 1990. Aperture Synthesis for Passive Microwave Remote Sensing: The Electronically Scanned Thinned Array Radiometer, Electronic Doctoral Dissertations for UMass Amherst. http://scholarworks.umass.edu/dissertations/AAI9022749.

Tarongi, J.M., March 2013. Radio Frequency Interference in Microwave Radiometry: Statistical Analysis and Study of Techniques for Detection and Mitigation (Ph.D. thesis dissertation). Universitat Politècnica de Catalunya. http://www.tdx.cat/handle/10803/117023.

Tarongi, J.M., Camps, A., 2010. Normality analysis for RFI detection in microwave radiometry. Remote Sens. 2, 191–210. http://dx.doi.org/10.3390/rs2010191.

Tarongi, J.M., Camps, A., 2011. Radio frequency interference detection and mitigation algorithms based on spectrogram analysis. Algorithms 4, 239–261. http://dx.doi.org/10.3390/a40x000x.

Tenerelli, J., Reul, N., 2010. Analysis of L1PP Calibration Approach Impacts in SMOS Tbs and 3-Days SSS Retrievals over the Pacic Using an Alternative Ocean Target Transformation Applied to L1OP Data. Technical report. IFREMER/CLS.

Thompson, D.R., Beal, R.C., 2000. Mapping High-resolution Wind Fields Using Synthetic Aperture Radar, vol. 21. Johns Hopkins APL Technical Digest, pp. 58–67.

Thompson, A.R., Moran, J.M., Swenson, G.W., 1986. Interferometry and Synthesis in Radio Astronomy. John Wiley and Sons, pp. 41–77.

Thompson, A.R., Emerson, D.T., Schwab, F.R., 2007. Convenient formulas for quantization efficiency. Radio Sci. 42, RS3022. http://dx.doi.org/10.1029/2006RS003585.

Thomsen, F., 1984. On the resolution of Dicke type radiometers. IEEE Trans. Microwave Theory 32, 145–150.

Thourel, L., 1971. Les Antennes. Editions Dunod, Paris.

Ticconi, F., Pulvirenti, L., Pierdicca, N., 2011. Models for scattering from rough surfaces. In: Zhurbenko, V. (Ed.), Electromagnetic Waves. In Tech, ISBN 978-953-307-304-0, 2011 under CC BY-NC-SA 3.0 license (Chapter 10). http://cdn.intechopen.com/pdfs-wm/16082.pdf.

Tiuri, M.E., 1964. Radiometer astronomy receivers. IEEE Trans. Antennas Propag. 12, 930–938.

Tomiyasu, K., Pacelli, J.L., 1983. Synthetic aperture radar imaging from an inclined geosynchronous orbit. IEEE Trans. Geosci. Remote Sens. GE-21 (3), 324–329.

Topp, G.C., Davies, J.L., Annan, A.P., 1980. Electromagnetic determination of soil water content: measurements in coaxial transmission lines. Water Resour. Res. 16, 574–582.

Trucco, A., Omodei, E., Repetto, P., 1997. Synthesis of sparse planar arrays. Electron. Lett. 33 (22), 1834–1835.

Tsang, L., 1991. Polarimetric passive microwave remote sensing of random discrete scatterers and rough surfaces. J. Electr. Wav. Appl. 5, 41–57.

Tsao, T., Slamani, M., Varshney, P., Weiner, D., Schwarzlander, H., Borek, S., 1997. Ambiguity function for bistatic radar. IEEE Trans. Aerospace Electron. Syst. AES-33, 1041–1051.

Tsui, J.B.-Y., 2000. Fundamentals of Global Positioning System Receivers. Wiley-Interscience.

Tucker, C.J., Pinzon, J.E., Brown, M.E., Slayback, D.A., Pak, E.W., Mahoney, R., Vermote, E.F., ElSaleous, N., 2005. An extended AVHRR 8-km NDVI dataset compatible with MODIS and SPOT vegetation NDVI data. Int. J. Remote Sens. 26, 4485–4498.

Ulaby, F.T., Wilson, E.W., 1985. : microwave dielectric spectrum of vegetation. II. Dual dispersion model. IEEE Trans. Geosci. Remote Sens. GE-25, 550–557.

Ulaby, F., Moore, R.K., Fung, A.K., 1981. Microwave Remote Sensing. Active and Passive: Microwave Remote Sensing Fundamentals and Radiometry, vol. I. Addison-Wesley.

Ulaby, F.T., Kouyate, F., Brisco, B., Lee Williams, T.H., 1986. Textural information in SAR images. Trans. Geosci. Remote Sens. GE-24. http://dx.doi.org/10.1109/TGRS.1986.289643.

Urick, R.J., 1986. Principles of Underwater Sound. McGraw-Hill International, Singapore.

Vachon, P.W., Johannessen, O.M., Johannessen, J.A., 1994. An ERS-1 synthetic aperture radar image of atmospheric lee waves. J. Geophys. Res. 99. http://dx.doi.org/10.1029/94JC01392.

Valencia, E., Camps, A., Marchan-Hernandez, J.F., Bosch-Lluis, X., Rodriquez-Alvarez, N., Ramos-Perez, I., 2010. Advanced architetures for real-time delay-Doppler map GNSS-reflectometers: the GPS reflectometer instrument for PAU (gruPAU). Adv. Space Res. 46. http://dx.doi.org/10.1016/j.asr.2010.02.002.

Valencia, E., Camps, A., Vall llossera, M., 2013b. GNSS-R Delay-Doppler Maps over land: preliminary results of the GRAJO field experiment. In: 2013 IEEE International Geoscience and Remote Sensing Symposium.

Valencia, E., Camps, A., Bosch, X., Rodriguez, N., Ramos, I., Eugenio, E., Marcello, J., 2011a. On the use of GNSS-R data to correct L-band brightness temperatures for sea-state effects: results of the ALBATROSS field experiments. IEEE Trans. Geosci. Remote Sens. 49 (9), 3225–3235.

Valencia, E., Camps, A., Rodriguez-Alvarez, N., Ramos-Perez, I., Bosch-Lluis, X., Park, H., 2011b. Improving the accuracy of sea surface salinity retrieval using GNSS-R data to correct the sea state effect. Radio Sci. 46, RS0C02. http://dx.doi.org/10.1029/2011RS004688.

Valencia, E., Camps, A., Rodriguez-Alvarez, N., Park, H., Ramos-Perez, I., 2013a. Using GNSS-R imaging of the ocean surface for oil slick detection. IEEE J. Selected Top. Appl. Earth Observations Remote Sens. 6 (1), 217–223.

Valencia, E., Zavorotny, V., Akos, D.M., Camps, A., 2014. Using DDM asymmetry metrics for wind direction retrieval from GPS ocean-scattered signals in airborne experiments. IEEE Trans. Geosci. Remote Sens. 52 (7), 3924–3936.

Valenzuela, G.R., 1978. Theories for the interaction of electromagnetic and oceanic waves — a review. Boundary-Layer Meteorol. 13, 61. http://dx.doi.org/10.1007/BF00913863.

Van de Griend, A.A., Wigneron, J.P., 2004. The b-factor as a function of frequency and canopy type at H-polarization. IEEE Trans. Geosci. Remote Sens. 42 (4), 786–794.

Van Trees, H.L., 2002. Optimum Array Processing. Wiley, NY.

Van Vleck, J.H., Middleton, D., 1966. The spectrum of clipped noise. Proc. IEEE 54 (1), 2–19.

Voronovich, A.G., 1996. Non-local small-slope approximation for wave scattering from rough surfaces. Waves Random Media 6, 151–167.

Voss, S., Heygster, G., Ezraty, R., 2003. Improving sea ice type discrimination by the simultaneous use of SSM/I and scatterometer data. Polar Res. 22 (1), 35–42.

Vyas, A., Trivedi, A.J., Calla, O.P.N., Rana, S.S., Sharma, S.B., Vora, A.B., 1990. Experimental separation of vegetation and soil and estimation of soil moisture using passive microwaves. Int. J. Remote Sens 11 (8), 1421–1438.

Walton, C.C., McClain, E.P., Sapper, J.F., 1990. Recent changes in satellite-based multi-channel sea surface temperature algorithms. Mar. Tech. Soc. 90 (Washington, DC).

Walton, C.C., Pichel, W.G., Sapper, F., May, D.A., 1998. The development and operational application of nonlinear algorithms for the measurement of sea surface temperatures with the NOAA polar-orbiting environmental satellites. J. Geophys. Res. 103. http://dx.doi.org/10.1029/98JC02370.

Walton, C.C., 1988. Nonlinear multichannel algorithms for estimating sea surface temperature with AVHRR satellite data. J. Appl. Meteorol. 27, 115–124.

Wang, J.R., Choudhury, B.J., 1981. Remote sensing of soil moisture content over bare field at 1.4 GHz frequency. J. Geophys. Res. 86, 5277–5282.

Wang, J.R., Schmugge, T.J., Gould, W.I., Glazar, W.S., Fuchs, J.E., McMurtrey, J.E., 1982. A multi-frequency radiometric measurement of soil moisture content over bare and vegetated fields,. Geophys. Res. Lett. 19, 416–419.

Wang, J.R., Shiue, J.C., Schmugge, T.J., Engman, E.T., September 1990. The L-band PBMR measurements of surface soil moisture in FIFE. IEEE Trans. Geosci. Remote Sens. 28 (5), 906–914. http://dx.doi.org/10.1109/36.58980.

Wang, J.R., Schmugge, T.J., 2000. An empirical model for the complex dielectric permittivity of soils as a function of water content. IEEE Trans. Geosci. Remote Sens. 18, 288–295.

Wang, J.R., Shiue, J.C., McMurtrey, J.E., 1980. Microwave remote sensing of soil moisture content over bare and vegetated fields. Geophys. Res. Lett. 7, 801–804. http://dx.doi.org/10.1029/GL007i010p00801.

Wang, Y., Hess, L.L., Filoso, S., Melack, J.M., 1995. Understanding the radar backscattering from flooded and non-flooded Amazonian forests: results canopy backscatter modeling. Remote Sens. Environ. 54, 324–332.

Wang, J.J.H., July 2012. Antennas for global navigation satellite system (GNSS). Proc. IEEE 100 (7), 2349–2355.

Weatherspoon, M.H., Dunleavy, L.P., 2006. Experimental validation of generalized equations for FET cold noise source design. IEEE Trans. Microwave Theory Tech. 54 (2), 608–614.

Wentz, F.J., Meissner, T., 2000. AMSR Ocean Algorithm. Algorithm Theor. Basis Doc, 121599A 1.

Wentz, F.J., Gentemann, C., Smith, D., Chelton, D., 2000. Satellite measurements of sea surface temperature through clouds. Science 288, 847–850.

Wentz, F.J., 1992. Measurement of oceanic wind vector using satellite microwave radiometers. Trans. Geosci. Remote Sens. 30. http://dx.doi.org/10.1109/36.175331.

Wentz, F.J., 1997. A well-calibrated ocean algorithm for special sensor microwave/imager. J. Geophys. Res. 102. http://dx.doi.org/10.1029/96JC01751.

Wentz, F.J., 1998. Algorithm theoretical basis document: AMSR ocean algorithm. Remote Sens. Sys. Tech. Rep. 110398. Santa Rosa, CA, 28 pp.

Werner, M., May 23–25, 2000. Shuttle radar topography mission (SRTM)-Mission overview. In: Proc. EUSAR, pp. 247–253. Munich, Germany.

Werninghaus, R.W., Balzer, S., Buckreuss, J., Mittermayer, P., February 2010. Mühlbauer, the TerraSAR-X mission. IEEE Trans. Geosci. Remote Sens. 48 (2), 606–614.

Wick, G.A., Emery, W.J., Schluessel, P., 1992. A comprehensive comparison between satellite-measured skin and multichannel sea surface temperature. J. Geophys. Res. 97 (C4), 5569–5595.

Wick, G., Jackson, D.L., Castro, S.L., 2004. Production of an enhanced blended infrared and microwave sea surface temperature product. In: Geosci. Rem. Sens Symp. IGARSS 04. http://dx.doi.org/10.1109/IGARSS.2004.1368534.

Wickert, J., Beyerle, G., Cardellach, E., Förste, C., Gruber, T., Helm, A., Hess, M.P., Hoeg, P., Jakowski, N., Montenbruck, O., Rius, A., Rothacher, M., Shum, C.K., Zuffada, C., 2011. GNSS REflectometry, Radio Occultation and Scatterometry Onboard ISS for Long-term Monitoring of Climate Observations Using Innovative Space Geodetic Techniques On-board the International Space Station. Proposal in response to Call: ESA Research Announcement for ISS Experiments relevant to study of Global Climate Change. http://www.gfz-potsdam.de/en/research/organizational-units/departments/department-1/gpsgalileo-earth-observation/projects/geros-iss/.

Wickert, J., Cardellach, E., Bandeiras, J., Bertino, L., Andersen, O., Camps, A., Catarino, N., Chapron, B., Fabra, F., Floury, N., Foti, G., Gommenginger, C., Hatton, J., Høeg, P., Jaggi, A., Kern, M., Lee, T., Li, Z., Martin-Neira, M., Park, H., Pierdicca, N., Ressler, G., Rius, A., Rosello, J., Saynisch, J., Soulat, F., Shum, C.K., Semmling, M., Sousa, A., Xie, J., Zuffada, C., October 2016. GEROS-ISS: GNSS reflectometry, radio occultation, and scatterometry onboard the International Space Station. IEEE J. Selected Top. Appl. Earth Observations Remote Sens. 9 (10), 4552–4581. http://dx.doi.org/10.1109/JSTARS.2016.2614428.

Wigneron, J.-P., Chanzy, A., de Rosnay, P., Rüdiger, C., Calvet, J.C., 1997. Estimating the effective soil temperature at L-band as a function of soil properties. IEEE Trans. Geosci. Remote Sens. 46 (3), 797–807.

Wiley, C., 1985. Synthetic aperture radars. IEEE Trans. Aerosp. Electron. Syst. 21 (3), 440–443.

Wilheit, T.T., 1979. The effect of wind on the microwave emission from the ocean's surface at 37 GHz. J. Geophys. Res. 84. http://dx.doi.org/10.1029/JC084iC08p04921.

Willis, N., D'Aria, D., Guarnieri, A.M., Rocca, F., 2004. Focusing bistatic synthetic aperture radar using dip move out. IEEE Trans. Geosci. Remote Sens. 42 (6), 1362–1810.

Willis, M.J., Melkonian, A.K., Pritchard, M.E., Ramage, J.M., 2011. Ice loss rates at the Northern Patagonian Icefield derived using a decade of satellite remote sensing. Remote Sens. Environ. 117, 184–198.

Willis, N., 1991. Bistatic Radar. Artech House, Norwood, USA.

Wilson, W.J., Tanner, A., Pellerano, F., 2003. Development of a high stability l-band radiometer for ocean salinity measurements. In: Proceedings of 2003 IEEE International Geoscience and Remote Sensing Symposium, Pasadena, CA, vol. 2, pp. 1238–1240.

Wilson, W.J., Tanner, A., Pellerano, F., 2005. Ultra Stable Microwave Radiometers for Future Sea Surface Salinity Missions. JPL Report D-31794. California Institute of Technology, Pasadena, CA, USA.

Winebrenner, D.P., Long, D.G., Holt, B., 1998. Mapping the progression of melt onset and freeze-up on Arctic sea ice using SAR and scatterometry. In: Tsatsoulis, C., Kwok, R. (Eds.), Analysis of SAR Data of the Polar Oceans. Springer-Verlag, New York, pp. 129–144.

Wingham, D.J., Francis, C.R., Baker, S., Bouzinac, C., Brockley, D., Cullen, R., de Chateau-Thierry, P., Laxton, S.W., Mallow, U., Mavrocordators, C., Phalippou, L., Ratier, G., Rey, L., Rostan, F., Viau, P., Wallis, D.W., 2006. CryoSat: a mission to determine the fluctuation in Earth's land and marine ice fields. Adv. Space Res 37, 841–871.

Winkel, B., Kerp, J., Stanko, S., 2007. RFI detection by automated feature extraction and statistical analysis. Astronomische Nachrichten 328 (1), 68–79.

Wismann, V., Boehnke, K., 1997. Snow on Greenland Land Surface Observations Using the ERS Wind Scatterometers. Institute for Applied Remote Sensing, Wedel, Germany, pp. 7–10.

Wismann, V., 2000. Monitoring of seasonal thawing in Siberia with ERS scatterometer data. Trans. Geosci. Remote Sens. 38. http://dx.doi.org/10.1109/36.851764.

Woodward, P.M., 1980. Probability and Information Theory, with Applications to Radar. Pergamon Press, 1953; reprinted by Artech House.

Wu, C., He, C., October 21–23, 2011. Interference analysis among modernized GNSS, Computational Problem-Solving (ICCP). In: 2011 International Conference on, pp. 669–673.

Wu, L., Corbella, I., Torres, F., Duffo, N., Martin-Neira, M., July 24–29, 2011. Correction of spatial errors in SMOS brightness temperature images. In: 2011 IEEE International Geoscience and Remote Sensing Symposium (IGARSS), pp. 3752–3755 (Vancouver, Canada).

Yakovlev, O.I., 2002. Space Radio Science. CRC Press.

Yamaguchi, U., Kahle, A.B., Tsu, H., Kawakami, T., Pniel, M., 1998. Overview of Advanded spaceborne thermal emission and reflection radiometer (ASTER). IEEE Trans. Geosci. Remote Sens. 36, 1062–1071.

You, H., Garrison, J.L., Heckler, G., Zavorotny, V.U., 2004. Stochastic voltage model and experimental measurement of ocean-scattered GPS signal statistics. IEEE Trans. Geosci. Remote Sens. 42 (10), 2160–2169.

Younis, M., Maurer, J., Fortuny-Guasch, J., Schneider, R., Wiesbeck, W., Gasiewski, A.J., 2007. Interference from 24-GHz automotive radars to passive microwave Earth remote sensing satellites. IEEE Trans. Geosci. Remote Sens. 42, 1387–1398.

Zavody, A.M., Mutlow, C.T., Llewellyn-Jones, D.T., 1995. A radiative transfer model for sea surface temperature retrieval for the along-track scanning radiometer. J. Geophys. Res. http://dx.doi.org/10.1029/94JC02170.

Zavorotny, V.U., Voronovich, A.G., 2000. Scattering of GPS signals from the ocean with wind remote sensing application. IEEE Trans. Geosci. Remote Sens. 38 (2), 951–964.

Zebker, H.A., Villasenor, J., 1992. Decorrelation in interferometric radar echoes. Geosci. IEEE Trans. Geosci. Electron. 30.5, 950–959.

Zebker, H.A., Rosen, P., 1994. On the derivation of coseismic displacement fields using differential radar interferometry: the Launders earthquake. In: Int. Geoscience & Remote Sensing Symposium IGARSS'94, Pasadena, CA, USA, pp. 286–288.

Zhang, C., Wu, J., Liu, H., Yan, J., Sun, W. Sampling Patterns and Perspective applications of clock scanning synthetic aperture imaging radiometer, 2013 IEEE International Geoscience.

Zhao, Y., Liu, A.K., Long, D.G., 2002. Validation of sea ice motion from QuikSCAT with those from SSM/I buoy. IEEE Trans. Geosci. Remote Sens. 40 (6), 1241–1246.

Zhongping, L., Casey, B., Arnone, R., Weidemann, A., Parsons, R., Montes, M.J., Gao, B.-C., Goode, W., Davis, C., Dye, J., January 2007. Water and bottom properties of a coastal environment derived from Hyperion data measured from the EO-1 spacecraft platform. J. Appl. Remote Sens. 1 (1), 011502.

Zwally, H.J., Schutz, B., Abdalati, W., Abshire, J., Bentley, C., Brenner, A., Bufton, J., Dezio, J., Hancock, D., Harding, D., Herring, T., Minster, B., Quinn, K., Palm, S., Spinhirne, J., Thomas, R., 2002. ICEsat's laser measurements of polar ice, atmosphere, ocean and land. J. Geodynamics 34. http://dx.doi.org/10.1016/S0264-3707(02)00042-X.

FURTHER READING

Bonnedal, M., Christensen, J., Carlström, A., Berg, A., 2010a. Metop-GRAS in-orbit instrument performance. GPS Solut. 14, 109–120.

Camps, A., Marchan, J.F., Valencia, E., Ramos, I., Bosch-Luis, X., Rodriguez, N., Park, H., Alcayde, A., Mollfuleda, A., Galindo, J., Martinez, P., Chavero, S., Angulo, J., Ruis, A., 2011c. PAUL Instrument Aboard Into MicroSAT-1: A GNSS-R Demonstration Mission for Sea State Correction in L-band Radiometry.

Camps, A., Rodriguez-Alvarez, N., Valencia, E., Forte, G., Ramos, I., Alonso-Arroyo, A., Bosch-Lluis, X., July 2013b. Land monitoring using GNSS-R techniques: a review of recent advances. In: 2013 IEEE International Geoscience and Remote Sensing Symposium, pp. 4026−4029.

Lowe, S.T., LaBrecque, J.L., Zuffada, C., Romans, L.J., Young, L.E., Hajj, G.A., 2000b. First spaceborne observation of an earth-reflected GPS signal. Radio Sci. 37. http://dx.doi.org/10.1029/2000RS002539.

Index

'*Note*: Page numbers followed by "f" indicate figures and "t" indicate tables.'